MECHANICS OF CONTINUOUS MEDIA

MATHEMATICS & ITS APPLICATIONS

Series Editor: Professor G. M. Bell

Chelsea College, University of London

Mathematics and its applications are now awe-inspiring in their scope, variety and depth. Not only is there rapid growth in pure mathematics and its applications to the traditional fields of the physical sciences, engineering and statistics, but new fields of application are emerging in biology, ecology and social organisation. The user of mathematics must assimilate subtle new techniques and also learn to handle the great power of the computer efficiently and economically.

The need of clear, concise and authoritative texts is thus greater than ever and our series will endeavour to supply this need. It aims to be comprehensive and yet flexible. Works surveying recent research will introduce new areas and up-to-date mathematical methods. Undergraduate texts on established topics will stimulate student interest by including applications relevant at the present day. The series will also include selected volumes of lecture notes which will enable certain important topics to be presented earlier than would otherwise be possible.

In all these ways it is hoped to render a valuable service to those who learn, teach, develop and use mathematics.

The Foundation Programme includes:

MODERN INTRODUCTION TO CLASSICAL MECHANICS AND CONTROL
David Burghes, Cranfield Institute of Technology and Angela Downs, University of Sheffield.

VECTOR & TENSOR METHODS
Frank Chorlton, University of Aston, Birmingham.

LECTURE NOTES ON QUEUEING SYSTEMS
Brian Conolly, Chelsea College, London University.

MULTIPLE HYPERGEOMETRIC FUNCTIONS
Harold Exton, The Polytechnic, Preston.

APPLIED LINEAR ALGEBRA
Ray J. Goult, Cranfield Institute of Technology.

GENERALISED FUNCTIONS: Theory, Applications
Roy F. Hoskins, Cranfield Institute of Technology.

MECHANICS OF CONTINUOUS MEDIA
S.C. Hunter, University of Sheffield.

USING COMPUTERS
Brian Meek and Simon Fairthorne, Queen Elizabeth College, University of London.

ENVIRONMENTAL AERODYNAMICS
R.S. Scorer, Imperial College of Science and Technology, University of London.

PHYSICS OF THE LIQUID STATE: A Survey for Scientists and Technologists
H.N.V. Temperley, University College of Swansea, University of Wales and
H.D. Trevena, University of Wales, Aberystwyth.

MECHANICS OF CONTINUOUS MEDIA

S. C. HUNTER
Professor of Mathematics
Department of Mathematics and Computing Science
University of Sheffield

ELLIS HORWOOD LIMITED

Publisher Chichester

Halsted Press: a division of
JOHN WILEY & SONS INC.
New York · London · Sydney · Toronto

The publisher's colophon is reproduced from James Gillison's drawing of the Ancient Market Cross, Chichester.

First published in 1976 by

ELLIS HORWOOD LIMITED PUBLISHER
Coll House, Westergate, Chichester, Sussex, England

Distributors:

Australia, New Zealand, South-east Asia:
JOHN WILEY & SONS, AUSTRALASIA PTY LIMITED
1–7 Waterloo Road, North Ryde, N.S.W. Australia

Canada:
JOHN WILEY & SONS CANADA, LIMITED
22 Worcester Road, Rexdale, Ontario

Europe, Africa:
JOHN WILEY & SONS LIMITED
Baffins Lane, Chichester, Sussex, England

U.S.A., Mexico, S. America and the rest of the world:
HALSTEAD PRESS a division of
JOHN WILEY & SONS INC.,
605 Third Avenue, New York, N.Y. 10016, U.S.A.

Library of Congress Cataloging in Publication Data
Hunter, S. C.
Mechanics of continuous media.
1. Continuum mechanics. I. Title
QA808.2.H86 531 76-7923
ISBN 0-470-15092-0 (Halsted)
ISBN 85312-042-0 (Ellis Horwood, Publisher: Library Edn.)
ISBN 85312-043-9 (Ellis Horwood, Publisher: Student Edn.)

Set in 10 point Press Roman by Speedpress Repro Services,
14 St. John's Precinct, Hebburn, Tyne & Wear

Printed Offset Litho in Great Britain by Cox & Wyman Ltd.

Table of Contents

Preface

Acknowledgements

Chapter 1 Scope of the Subject . . . 13

Chapter 2 Description of Continua and Kinematics

2.1 Notion of a continuum . . . 20
2.2 Motion of deformable bodies (material description) . . . 22
2.3 Motion of deformable bodies (spatial description) . . . 26
2.4 Example . . . 35
Summary . . . 39
Appendix to Chapter 2 . . . 40
Problems . . . 40

Chapter 3 Forces in a Continuum: Stress and Equations of Motion

3.1 Forces in a continuum . . . 45
3.1.1 Internal Forces . . . 45
3.1.2 External Forces . . . 46
3.1.3 Surface Contact Forces . . . 46
3.2 Integral forms of the equations of motion . . . 47
3.2.1 Balance of Linear Momentum . . . 47
3.2.2 Balance of Angular Momentum . . . 49
3.3 Cartesian components of stress . . . 50
3.4 Equations of motion . . . 51
Summary . . . 57
Problems . . . 57

Chapter 4 Suffix Notation, Cartesian Tensors and Matrices

4.1 Suffix notation . . . 60
4.2 Definition of Cartesian Tensors . . . 65
4.3 Elementary tensor operations and the quotient theorem . . . 72
4.3a Addition of Tensors . . . 72
4.3b Multiplication of Tensors . . . 73
4.3c Contraction of Indices . . . 73
4.3d Differentiation with Respect to Time and Space Variables . . 73
4.3e The Quotient Theorem . . . 75
4.4 The stress tensor . . . 76
4.5 Properties of second order symmetric tensors; principal stress and principal axes . . . 82
4.6 The polar decomposition theorem . . . 89
Summary . . . 92
Problems . . . 93

Chapter 5 Continuum Deformation

5.1 Stretch and Rotation 101
5.2 Strain 106
5.3 The 'small displacement gradient' approximation 109
5.4 Rate of strain and spin tensor 111
Problems 113

Chapter 6 Geometrical Restrictions on the Form of Constitutive Equations

6.1 Nature of constitutive equations 116
6.2 Material frame indifference 118

Chapter 7 Constitutive Equations for Fluids

7.1 Ideal fluids 125
7.2 Reiner-Rivlin fluids 130
7.3 Energy theorems for barotropic fluids 138
Summary 142

Chapter 8 Elastic and Thermo-elastic Materials

8.1 Definition and introduction 144
8.2 Material frame indifference 145
8.3 Isotropic thermoelastic materials 146
8.4 Constitutive equation in linear (isothermal) isotropic elasticity . 149
8.5 Elastic strain energy function and isothermal Green elasticity . . 150
8.6 Strain energy and strain rate in the theory of linear elasticity . . 158
Summary 159
Problems 160

Chapter 9 Shear Flow Solutions for Reiner-Rivlin Fluids

9.1 Introduction 165
9.2 Rectilinear shear flows between parallel plates 167
9.3 Rectilinear shear flows with axial symmetry 169
9.4 Azimuthal (Couette) shear flows with axial symmetry 174

Chapter 10 Some Solutions of the Navier-Stokes Equations

10.1 Introduction 179
10.2 Boundary conditions 180
10.3 Rectilinear shear flow solutions 180
10.3.1 Steady Flow Between Parallel Fixed Plates 182
10.3.2 Shearing of Liquid Between Two Parallel Plates in Relative Motion 185
10.3.3 Rectilinear Shear Flow Along the Annulus Between Two Parallel Cylindrical Walls $C_1(x, y) = 0$, $C_2(x, y) = 0$ . . 186
10.3.4 Rectilinear Flow with Axial Symmetry 189
10.3.5 Rectilinear Flow Along a Tube with Elliptical Cross Section. 191
10.4 Couette flow 193
10.5 Unsteady rectilinear flow 196
10.6 Appendix 205
Problems 207

Chapter 11 General Theorems in Inviscid Hydrodynamics

11.1 Introduction 214
11.2 Kelvin's circulation theorem 215
11.3 The Bernoulli equations 218
11.4 Applications in hydraulic engineering of the Bernoulli equation . 221
11.5 The role played by the Bernoulli equation in irrotational inviscid incompressible flow 223
11.6 Cavitation in inviscid, incompressible and irrotational flow . . . 223
11.7 Some simple (unsteady) hydraulics examples solved by appeal to the Bernoulli equation 224
Problems 228

Chapter 12 Potential Theory of Classical Hydrodynamics

12.1 Basic equations 233
12.2 Spherically symmetric flows and the underwater explosion problem 234
12.3 Potential due to a point dipole 237
12.4 Boundary conditions in classical hydrodynamics 238
12.5 Point sources, dipoles and planes. Method of Images 240
12.6 Plane two dimensional flow solutions using plane polar coordinates, line sources and rectilinear vortices 242
12.6.1 General solution of Laplace equation for plane polar coordinates 242
12.6.2 The rectilinear vortex 246
12.6.3 Circulation around a cylinder 252
12.6.4 Translatory motion of rigid cylinder immersed in a fluid . . 254
12.7 Axisymmetric flow problems in spherical polar coordinates . . 259
12.7.1 General solution of axisymmetric Laplace equation . . . 259
12.7.2 Translatory motion of a rigid sphere immersed in a fluid . . 265
12.7.3 Flow due to a point source located outside a sphere . . . 267
12.8 Solution of plane two dimensional flow problems by complex variable theory 270
12.8.1 General results 270
12.8.2 Some simple examples of complex potentials 275
12.8.3 Superposition of Line Sources, Rectilinear Vortices, Streams in Absence of Boundaries 278
12.8.4 Images in plane boundaries 279
12.8.5 The Milne-Thomson circle theorem 281
12.8.6 The Blasius theorems 284
12.8.7 Conformal mapping technique (basic results) 287
12.8.8 Some simple conformal mappings 289
12.8.9 The Kutta-Joukowski transformation 291
12.8.10 Flows past plates and elliptic cylinders 292
12.8.11 Aerofoil theory 296
12.9 Problems with free streamlines 303
12.9.1 Plane jet incident normally on rigid wall 303
12.9.2 Emergence of a plane jet from a slit 309
12.9.3 Stream incident normally on a plate of finite width . . . 312
Problems 317

Chapter 13 Compressible Flow of Inviscid Fluids – The Acoustic Approximation
13.1 Introduction and basic equations . . . 332
13.2 General solutions of the wave equation . . . 334
13.3 Magnitudes of c and applications of the acoustic approximation . 337
13.4 Some simple wave propagation problems . . . 338
13.4.1 Piston Problem for Plane Waves . . . 338
13.4.2 Expanding Cylindrically Symmetric Waves – A Similarity Solution . . . 341
13.4.3 Spherically Symmetric Waves – Oscillating Sphere . . . 343
13.4.4 Reflection of a Plane Wave at Non-Normal Incidence by a Rigid Plane Barrier . . . 345
13.4.5 Room Acoustics, Musical and Wind Instruments . . . 346
Problems . . . 354

Chapter 14 The Theory of Classical Linear Elasticity for Isotropic Solids
14.1 Introduction and basic equations . . . 361
14.2 Boundary conditions and notation . . . 363
14.3 Young's modulus, Poisson's ratio: Engineering version of the theory 365
14.4 Static problems . . . 368
14.4.1 Prism of Uniform Cross Section Suspended in Earth's Gravitational Field . . . 368
14.4.2 Bending of Beams by Terminal Couples . . . 372
14.4.3 Spherical Pressure Vessel . . . 377
14.4.4 Cylindrical Pressure Vessel . . . 381
14.4.5 Stresses in Rotating Shafts . . . 384
14.4.6 Torsion of a Cylinder of Arbitrary Cross Section . . . 386
14.5 Approximate theories in classical elasticity; Bending of beams . . 393
14.6 Elastodynamics . . . 398
14.6.1 Basic Equations and General Solutions . . . 398
14.6.2 Torsional Waves . . . 402
14.6.3 Rayleigh Waves . . . 403
14.6.4 Approximate Elastodynamic Theories . . . 408
Problems . . . 418

Chapter 15 The Theory of Linear Viscoelasticity
15.1 Introduction . . . 444
15.2 Basic behaviour and one dimensional constitutive equations for linear viscoelastic materials . . . 444
15.3 Equation for three dimensional viscoelasticy–Quasi static problems 456
15.4 One dimensional viscoelastic waves . . . 460

Chapter 16 Plasticity Theory
16.1 Introduction: The experimental behaviour of metals . . . 465
16.2 Mathematical idealisations . . . 467
16.3 The initiation of yielding . . . 468
16.4 Constitutive equations of plasticity and the plastic potential . . 473
16.5 Elastic-Plastic torsion of a cylindrical bar . . . 483
16.6 Elastic-Plastic expansion of cylindrical shell . . . 484
Problems . . . 487

Answers to Problems . . . 491
Appendix – Results in Vector Analysis . . . 556

Preface

This text is based on an introductory course in Continuum Mechanics first given at the University of Strathclyde and subsequently at the University of Sheffield. Over a period of six years the course necessarily evolved, but its predominant aim, and that of the book, has remained a unified treatment of the topic.

Many separate University courses have been given on many of the topics in this book (classical hydrodynamics, classical elasticity, viscous flow). In those separate courses it has not always been clear to students that many of the principles of continuum mechanics are common to all such topics, e.g. the notion of stress, equations of motion, equations of mass conservation, etc.

One of the principal aims of this book is to make clear these common ideas and also the further necessity of constitutive equations to describe material behaviour.

Of course with different constitutive equations, together possibly with various kinematic and geometrical approximations, the various branches of the subject emerge. It is my hope that this book helps make clear the nature of constitutive equations and their limited validity, any approximations involved and the complete sets of equations used in the various branches of the subject.

Specific branches of continuum mechanics which are dealt with in some detail are those of classical hydrodynamics, viscous flow, and classical elasticity. There are further chapters on finite elasticity, Reiner-Rivlin fluids, linear viscoelasticity and plasticity. Where possible I have attempted to give physical motivation for the nature of any constitutive equations assumed and also to discuss the types of (real) materials envisaged, and the limitations of the equations used.

A secondary aim of the book is to introduce and use many of the mathematical techniques used in the solution of problems. Two of the techniques discussed at length are the method of separation of variables (and the resulting Eigenfunction concept), and the use of complex variable theory in the solution of the two dimensional Laplace equation. There are also a few isolated examples which introduce some of the other techniques such as the use of Fourier transforms and similarity solutions.

The theory of Cartesian tensors is discussed and used throughout the book. No appeal has been made to non-Cartesian tensors, nor in general have equations of motion or constitutive equations been formulated in terms of physical stress components, displacements and velocity in orthogonal curvilinear coordinates.

On the other hand where necessary I have presupposed complete familiarity with standard vector analysis material expressed in terms of both Cartesian and the standard orthogonal curvilinear coordinates, i.e. spherical and cylindrical polar coordinates. [A detailed treatment of this subject area can be studied in F. Chorlton's *Vector and Tensor Methods,* Ellis Horwood (1976).] In places I have introduced physical components of stress for spherical and cylindrical polar coordinate systems, primarily to simplify the description of solutions obtained, e.g. in the solution of the classical elastic problem of a spherical shell subjected to internal pressure.

Considerable thought was given to the question of whether or not to include an account of continuum thermodynamics. With some reluctance I decided the subject had to be omitted, principally because of the already considerable length of the book, but also because it seems questionable whether the topic is suitable material for a first course in continuum mechanics. However at various places in the text reference has been made to some very limited aspects of thermodynamics and also to the thermal properties of some materials. In the former context the text is not entirely self-contained (e.g. in reference to the adiabatic and isothermal behaviour of perfect gases) but the missing material would certainly be familiar to anyone who has undertaken an A-level or an equivalent physics course. For those who have not undertaken such a course the omitted material is, in the general context of the book, of minor significance.

Of course not all the material of this book could be given in 50 hours teaching. In particular I have never taught in an introductory course any of the material offered here on linear viscoelasticity or plasticity. These chapters are meant primarily as an introduction to the subjects for students who wish to pursue the topics further, and also to indicate that many real everyday materials behave quite differently from the familiar incompressible inviscid liquids and classical elastic solids that have dominated University courses for many years.

The book contains a number of examples scattered throughout the text and at the ends of most chapters. Some of the latter expand the material in the text. For all except the most straightforward problems, answers are provided either during the course of the question (by leading the reader step by step) or alternately at the end of the book.

S. C. Hunter
Sheffield University.

Acknowledgements

In the course of writing this book I had access to lecture notes prepared by Professor A. J. M. Spencer, University of Nottingham, and Professor P. Chadwick, University of East Anglia. Part of the proof offered here on the isotropic symmetric tensor function theorem of Chapter 7 is based on ideas in Professor Spencer's notes, while I am indebted to Professor Chadwick's notes for a number of details and in particular his explicit account of the nature of forces acting in solids. Needless to say the responsibility for the presentation of the material mentioned, and for the entire book, is mine.

After writing a first draft of Chapter 1, I invited comments from Professor Chadwick and Professor J. B. Helliwell, University of Bradford. Both responded generously with their time and gave me a long list of invaluable comments which not only caused me to re-write the chapter in question but also influenced considerably my attitude to the rest of the book.

I am indebted to Professor H. Kolsky, Brown University, Providence, Rhode Island, U.S.A., and the Editor and Proprietors of Philosophical Magazine for permission to reproduce Fig. 15.8 taken from H. Kolsky, Phil. Mag. Vol. 8 (1956), p. 693. Also I am indebted to Pergamon Press for permission to reproduce Fig. 15.9 appearing originally in Reference 16 given on p. 560.

The book is based predominantly on a 50 hours lecture course which originated at the University of Strathclyde. The course was given for 3 successive years at Strathclyde and for a further 3 years at the University of Sheffield. I am grateful to a number of students who passed through these courses and who made many useful comments on the material.

Finally, and not least, I must record my gratitude to Mrs. Diana Wood, who patiently, and over several years, coped with my hand-writing to produce an excellent typescript.

References, Further Reading and Numbering of Equations

The text is divided into sixteen Chapters, each further subdivided into Sections. The system of equation numbering adopted identifies both the Chapter and Section. Thus equation (7.4.9) refers to the ninth numbered equation in Section

4 of Chapter 7 while equation (5.3.10) refers to the tenth numbered equation of Section 3 of Chapter 5. Exceptionally in Chapter 1, which is not subdivided, the equations are numbered consecutively from (1.1) to (1.3).

For the convenience of the reader, bibliographic details of references are given in the text where they occur.

A list of texts for further study is given on p. 559-560.

Bold face notation

In this text vectors labelled by English letters are denoted by a bold roman face, e.g. **r**, **R**, **U**, while matrices are denoted by a bold Gill face, e.g. **R**, **Q**, **I**.

Exceptionally the same bold face, e.g. $\boldsymbol{\beta}$, $\boldsymbol{\mu}$, is used for both vectors and matrices denoted by Greek letters. There should be no confusion since only rarely are Greek letters used to denote vectors, and where this occurs it is clear from the context that a vector is implied. An example occurs on p. 109-110 where the symbol $\boldsymbol{\Omega}$ denotes both a vector and an (antisymmetric square) matrix.

Chapter 1
Scope of the Subject

The mathematical theory of continuum mechanics is concerned with (a) the formulation of equations which describe the motion and mechanical (and thermal) behaviour of materials, (b) the solution of these equations for specific problems. While the former of these objectives is a most important topic, which is discussed at some length in this book, the raison d'être of the subject lies in the solution of specific problems. Many problems in the theory stem from engineering design, and from laboratory tests set up to elucidate the basic mechanical properties of materials. Typical examples of practical problems, which are sufficiently simple to be covered in this book, are as follows

Problem 1 A weight W is suspended from the end of a horizontal cantilever beam of length l (Fig. 1.1). The beam is initially straight and horizontal, of uniform cross section, and fabricated from an elastic material whose Young's Modulus is E. What is the maximum deflexion of the beam (y) and what is the maximum safe load W?

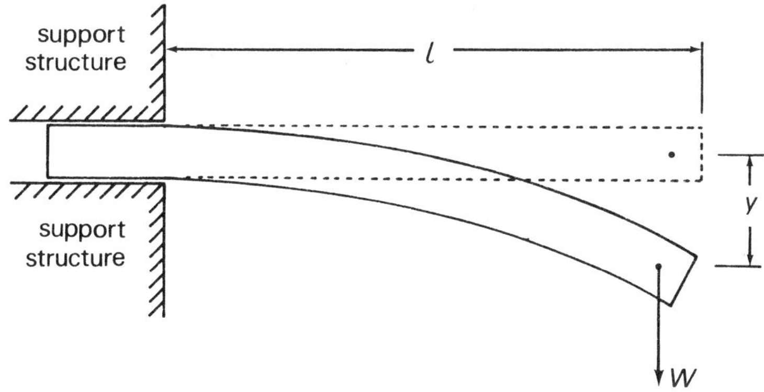

Fig. 1.1 Deflexion of a cantilever beam by a terminal load W

Problem 2 A cylindrical pressure vessel of internal radius a and thickness t contains gas at a pressure P (Fig. 1.2). How much does the vessel expand and what is the maximum safe pressure?

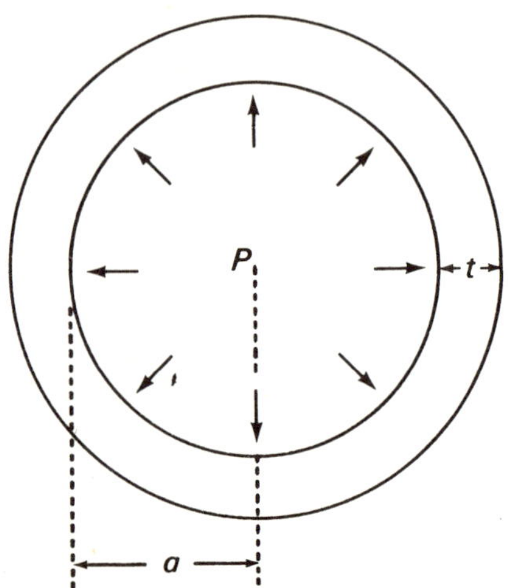

Fig. 1.2 Cylindrical pressure vessel subject to internal pressure P

Problem 3 An explosive charge is detonated under water. Describe the motion of the water due to the gas bubble formed by the explosive debris.

Problem 4 Oil of viscosity μ enters one end of a capillary tube of internal radius a and length l at pressure P_1. The oil emerges at the other end of the tube at pressure P_2 (Fig. 1.3). Show that the volume rate at which oil is discharged from the tube is given by the Poiseuille formula

$$Q = \pi a^4(P_1 - P_2)/8\mu l . \tag{1.1}$$

[This result is used commonly to provide a measurement of the quantity μ.]

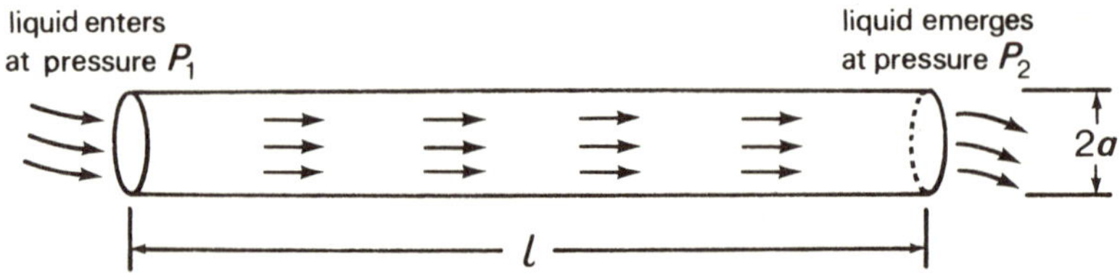

Fig. 1.3 Capillary tube measurement of viscosity (after Poiseuille)

Problem 5 Calculate the forces required to accelerate a rigid sphere immersed in water.

Problem 6 An aeroplane is travelling at uniform velocity through the atmosphere. Describe the motion of the air around the wings, and calculate the lift and drag forces exerted on the wings.†

Problems 1 and 2 lie in the field of 'solid mechanics' while the remainder of the problems are in the field of 'fluid mechanics'. Fluids are usually subdivided into liquids (as in Problems 3, 4 and 5) and gases (as in Problem 6). For present purposes it is sufficient to rely on the reader's intuitive concepts as to what constitutes a gas, liquid or solid. More precise definitions, in conformity with the intuitive ones, are given subsequently.

Essential mathematical ingredients of continuum mechanics which apply to all materials are the law of conservation of mass, the laws of linear and angular momentum balance, and the law of energy balance.

The law of conservation of mass, that matter is neither created nor destroyed, is a familiar law of physics. Exceptionally the principle requires an extended interpretation in the presence of nuclear reactions, where mass is converted into other energy forms, but such situations are rare in most applications of continuum mechanics.

The momentum balance laws may be regarded as generalisations of Newton's second and third laws of motion. If, for a system of particles, the internal forces between pairs of particles are equal in magnitude but opposite in sign (Newton's third law of action and reaction), then the second law of motion leads to the results that (a) the rate of change of total linear momentum of the system is equal to the sum of the external forces, (b) the rate of change of angular momentum of the system is equal to the sum of the external couples.

The general form of the energy balance law states that the total rate of increase of all forms of energy within the system is equal to the rate of working of the external forces less the rate at which energy is lost to the external surroundings.

The energy balance law in theoretical mechanics does not always play as crucial a role as the laws of motion. Consider for example the simplest situation in rigid body dynamics in which an isolated system is mechanically conservative. For such a system energy is essentially a derived rather than fundamental concept. The statement that energy is conserved is simply a statement that the equations of motion possess a certain first integral for which the integration constant is labelled 'energy'.

In physics generally however, the energy balance equation is as fundamental as the laws of motion, and yields information that cannot be derived otherwise. Even then the equation may not contribute to our understanding of the mechanical aspects of the situation. Suppose, for example, we have an isolated system in rigid body dynamics which is not mechanically conservative e.g. by the

† The drag problem is not discussed in the present text.

inclusion of frictional forces, but whose mechanical aspects are completely independent of temperature. The mechanical behaviour of the system is determined solely by appeal to the equations of motion and initial conditions. Of course mechanical energy is no longer conserved, but this is irrelevant to the fact that the motion is determinable. In these circumstances the energy balance law merely serves to determine the quantity of heat generated by the frictional forces – in fact if the heat capacity of the system is known and we assume instantaneous distribution of heat, the energy balance law determines the rise in temperature. The temperature rise is here the information which is not derivable otherwise. Circumstances where the energy balance equation plays a crucial role in analyzing mechanical behaviour are those where the mechanical aspects of the system are affected by temperature. For example if in the system discussed above the frictional forces themselves depend on temperature, then the equation of energy balance is as vital as the equations of motion in determining the motion of the system. The motion of the system and the temperature are inextricably coupled, and the determination of motion and temperature must be effected concomitantly, using the laws of motion and the energy balance equation.

The situation in continuum mechanics is similar to that outlined above. There are many circumstances where the energy balance equation does not play a crucial role and where the mechanical behaviour is determinable without appeal to energy considerations. This book is concerned solely with such situations, which are either mechanical conservative or where temperature changes are assumed sufficiently small not to influence mechanical properties. Perhaps we should indicate that situations where temperature effects are important complicate the mathematics, and require detailed application of the principles of thermodynamics.

The laws of mass conservation and momentum balance yield a set of basic equations which are valid for all continuous media. However, in sharp contrast to rigid body mechanics, it becomes apparent that these laws alone are insufficient to determine the motion of a medium. The additional information required comprises a set of equations which describe the mechanical properties of the material. Since evidently different materials exhibit quite different mechanical behaviour, e.g. compare air, oil, steel, glass and clay, it is clear that the set of equations will vary from one medium to another. Equations which describe mechanical properties are known as **constitutive equations** and are peculiar to the material in question. Of course broadly similar types of constitutive equations might be expected to apply to broadly similar materials, and indeed this is usually the case. In these circumstances what differentiates one material from another are just the numerical values of the parameters which enter the similar sets of equations. For instance, the difference between oils of various grades is described essentially by different numerical values of the coefficient of viscosity, and perhaps slightly different values of the density and compressibility.

Simple examples of constitutive equations which will be familiar from elementary physics are the 'perfect' gas law and Hooke's law for elastic solids. The perfect gas law, which is a combination of Boyle's law and Charle's law, states that the pressure P in a gas is proportional to the density ρ and the absolute temperature T. In symbols

$$P = R\rho T/m \tag{1.2}$$

where R is the gas constant and m the molecular weight. Hooke's law in its simplest form relates the tensile stress σ in a uniform wire to the tensile strain e through the equation

$$\sigma = Ee \tag{1.3}$$

where the constant E is the Young's modulus of the wire (Fig. 1.4).

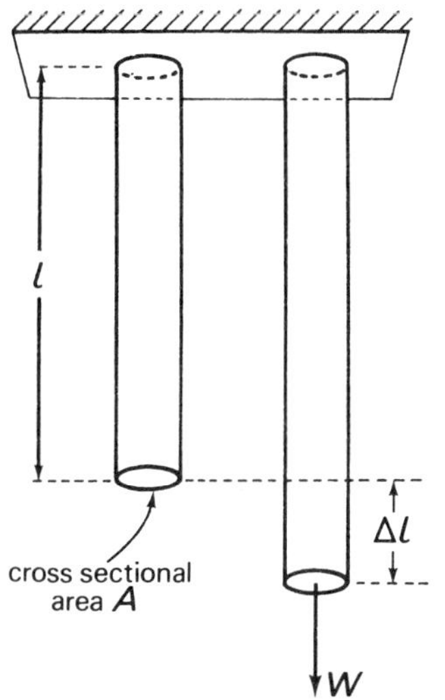

Fig. 1.4 Hooke's law experiment for an elastic solid

Equation (1.2) contains the parameter m which distinguishes different gases. Similarly in equation (1.3) the quantity E varies from one solid to another e.g. for steel $E = 2.10^{11}$ N/m^2, for aluminium $E = 6.10^{10}$ N/m^2.

Except for the realms of the very small (atomic dimensions), the very large (astrophysical dimensions) and the very fast (velocities approaching the velocity of light), the Newtonian laws of motion are well established; their validity is assumed throughout this book, which deals with problems on a scale encompassed by everyday experience. In contrast, the postulated constitutive equations for real materials are usually severely restricted in their validity. For example the assumption that steel behaves as an elastic solid, so that on removal of an applied force it recovers its original shape, is obviously false in a situation in which two automobiles collide. Nevertheless the assumption is used legitimately and successfully every day in civil and mechanical engineering design; essential knowledge for the design engineer are the precise circumstances for which the assumption is valid.†

This book is concerned predominantly with the three idealised materials of classical applied mathematics. These three model materials are (i) the **isotropic linear elastic solid** (ii) the **ideal fluid** (iii) the **Newtonian (or Stokes') viscous fluid**. The justification, both historically and in the present book, for the emphasis on these three materials, is that these mathematical models provide a good description of the behaviour of many real materials for limited ranges of stress, deformation and temperature.

The model of a linear isotropic elastic solid, discussed in Chapter 14, provides a good description of the behaviour of polycrystalline metals for limited ranges of stress, i.e. for stress levels below the initial yield stress. Thus linear elasticity is of paramount importance in the stress analysis of steel, which is the commonest engineering structural material. To a lesser extent linear elasticity describes the mechanical behaviour of other common solid materials, e.g. concrete, wood and coal (again with restrictions on the magnitude of the stresses). However, the theory does not apply to the behaviour of many of the new synthetic materials of the elastomer and polymer type, e.g. polymethylmethacrylate (perspex), polyethylene, polyvynil chloride. These materials display time-dependent mechanical properties and a more appropriate theory at low stress levels, is that of linear viscoelasticity.(see Chapter 15).

The ideal fluid model, discussed in Chapters 11 and 12 is a particular case of the Newtonian fluid (Chapter 10); both models are experimentally appropriate to the behaviour of gases except under extremely rarified conditions, when the continuum model itself is in question, and it becomes necessary to employ the methods of statistical mechanics. For many purposes viscosity effects are negligible and an adequate approximation to the behaviour of a gas is that of an ideal fluid; however there are circumstances where the viscosity of the gas contributes significantly to the flow pattern and it is then necessary to adopt the more sophisticated Newtonian model, e.g. in evaluating drag forces on aircraft wings. Both models provide in appropriate circumstances a description of some liquids notably air and water. However many liquids display properties requiring a more complex description, e.g. blood, printer's ink, paints and molten glass near the solidification temperature. In certain very limited circumstances the theory of linear viscoelasticity provides a description of these materials, but usually the behaviour is more complicated and requires a non-linear theory of viscoelastic liquids. Such theories are mathematically complicated and have not yet been integrated effectively with experimental observations on real materials. Theories of this type are too mathematically complex for the present volume.

The ultimate test of a set of constitutive equations for a given material demands appeal to experiment. If, for a given set of constitutive equations, the results of mathematical analysis lead to predictions which conform to experimental

† The relevant knowledge in respect of mechanical properties is the initial yield stress of the steel and, in circumstances involving vibration, the fatigue curve relating maximum stress amplitude to the number of stress cycles before fracture. Of course certain non-mechanical properties may be very important e.g. corrosion resistance.

measurements, then we are inclined to believe that the postulated constitutive equations provide a good model for the behaviour of the material. However the results of one type of experiment are insufficient to establish empirically a set of constitutive equations. For example it is often said in connection with Problem 4 above that experimental confirmation of the well known Poiseuille formula [equation (1.1)] in respect of the dependence of Q on a, $(P_1 - P_2)$ and l is sufficient to establish that a real liquid behaves as a Stokesian fluid with viscosity μ. However the argument is false, since there exist more complex constitutive equations which yield precisely the same dependence of Q on a, $(P_1 - P_2)$ and l. Of course for these materials the proportionality factor is no longer interpretable as $\pi/8\mu$.

Nevertheless if the results of experiment showed that the behaviour of a liquid conformed to both the Poiseuille formula and (say) Stokes' formula in a somewhat different geometrical situation [see Problem (8) Chapter 10], then we would be inclined to have more faith that the Stokesian fluid model provides a valid description of the mechanical behaviour. Further confirmatory experiments would strengthen this belief. Previous statements regarding the constitutive equations for metals, liquids and gases are made in the context of countless experimental observations in different geometrical situations. For example, there is no doubt about the validity of the theory of classical elasticity for metals subject to sufficiently small stress levels.

Studies in mathematical physics have contributed significantly to the simpler forms of constitutive equations. The perfect gas equation (1.2), the equations of classical linear elasticity and the equations of a Stokesian fluid and the simplest form of the equations of finite elasticity are examples of this. These 'fundamental' studies originate at the sub-macroscopic level with considerations of gas molecules, electron configurations in metals and molecular chains in polymers. In the sense that the constitutive equations and the values of the parameters appearing therein (e.g. elastic constants) are interpretable in terms of basic physical elements e.g. electron charge and mass, interatomic distance, Planck's constant etc.), the description 'fundamental' is legitimate. However it should be remembered that the theoretical physicist is replacing one set of hypotheses (the constitutive equation postulates) by another set or sets, such as the Schroedinger wave equation. Historically, fundamental studies have had little effect on continuum theories, primarily because the most commonly used constitutive equations had all been well established empirically at the time the fundamental studies were undertaken. It seems possible that theoretical physics will play an increasingly important role in predicting explicit constitutive equations for materials with more complex mechanical properties than those considered in the present book.

Chapter 2

Description of Continua and Kinematics

2.1 Notion of a continuum

Continuum mechanics capitalizes on the apparent continuity of structure displayed by materials like liquids, metals and polymers. Of course the continuity of structure apparent to the naked eye is an illusion, since we know that at sufficiently large magnifications, e.g. by means of an electron microscope or field ion microscope or more indirectly by X-ray diffraction patterns, all real materials display a molecular or atomic structure.

Most liquids consist of molecules loosely bound together by weak electrical forces; there are 'gaps' between the molecules allowing them considerable mobility. A similar picture emerges for gases except that the intermolecular forces are even weaker so that the molecules are further apart and possess greater mobility. In contrast the electrical bonding forces in metals are much stronger, so that a typical picture here is a regular array of charged metallic ions immersed in a sea of highly mobile electrons. Finally some naturally occurring materials like rubber and also the new synthetic organic solids like polyethylene comprise long flexible polymer chains built up from recurring molecular units. The chains are randomly oriented and interweave with one another; in some materials they intersect that is to say they are 'cross linked' either directly or via the presence of a foreign bonding atom e.g. sulphur in rubber. The greater the number of cross links the less deformable is the polymer material.

Continuum mechanics ignores all the fine detail of the molecular or atomic structure discussed in the previous paragraph and assumes that the highly discontinuous structures of real materials may be replaced by a smoothed out hypothetical 'continuum', every portion of which, however small, is assumed to exhibit macroscopic physical properties which are essentially properties of the bulk material. In particular, the physical characteristics of material contained within an infinitesimal volume element $d\Omega$ are assumed to be the same as those determined experimentally from measurement on samples with finite dimensions. For example experiments on the mechanical behaviour of many gases and liquids suggest that the pressure P, density ρ and temperature T are related through a constitutive equation of the form

$$P = f(\rho,T) \tag{2.1.1}$$

where f is an experimentally determined function.† The experiments determining the form of f for a given substance also show that the result (2.1.1) is independent of the sample size, at least for samples sufficiently large to be handled experimentally. However in continuum mechanics it is assumed that (2.1.1) remains valid even for samples of infinitesimal volume $d\Omega$.

In view of the molecular structure of matter, the latter assumption is false. Indeed it is known from the kinetic theory of gases (which deals in a statistical manner with the motion of individual gas molecules) that P and T are themselves valid concepts only for samples sufficiently large to contain many thousands of millions of molecules; in other words equation (2.1.1) provides a valid description of the behaviour of a finite volume element $\Delta\Omega$, whose linear dimensions are of order Δl where Δl is (say) about 10^4 times the mean separation distance between molecules. In what follows we take Δl to be the lowest bound of the linear dimensions of a volume element for which (2.1.1) provides a good description. For liquids Δl is of order 10^{-6}m while for gases at normal temperature and pressure Δl is of order 10^{-5}m; of course with decreasing density Δl increases e.g. for air at 20,000m above the earth $\Delta l \sim 10^{-4}$m. Finally for metals Δl may be taken as 10^4 times the interatomic distance and is also about 10^{-6}m.

Since the continuum hypothesis is 'notional' it is necessary to enquire under what circumstances the continuum model provides a valid description of the flow and deformation of real materials. A mathematical answer to this question is not possible within the confines of continuum mechanics alone, but requires considerations of atomic or molecular structure. A complete discussion of the problem is possible only within the framework of statistical mechanics; however from physical considerations we can obtain necessary restrictions in terms of the quantity Δl introduced above. Suppose we have found a solution to (say) a fluid dynamics problem using a continuum theory. Among other results the solution yields the pressure as a definite function of the space variables (x,y,z) and of time t.

$$P = P(x, y, z, t) .$$

Now we know from the previous discussion that P is a valid concept only if the variation of P throughout a region whose linear dimensions are of order Δl is negligible. If follows that a necessary restriction for the validity of the solution is that in the vicinity of every point (x,y,z), $P(x,y,z,t)$ should be essentially constant over linear distances of order Δl

$$\text{i.e.} \quad |\,\text{grad}\, P\,|_{\max} \ll P/\Delta l \tag{2.1.2}$$

where the notation implies the maximum value of $|\,\text{grad}\, P\,|$ with respect to direction. Usually the inequality (2.1.2) is readily satisfied in the solution of continuum mechanics problems. Exceptionally (2.1.2) may be violated at points where P is very small, or is rapidly varying or is discontinuous. If (2.1.2) is violated, then strictly speaking the continuum model is inapplicable. However

† A relation like (2.1.1) connecting pressure, density and temperature is sometimes known as an equation of state.

regions of violation are almost always of very limited extent, and usually the continuum model then provides a realistic description of the deformation (or flow) in the remaining regions.†

Inequalities similar to (2.1.2) may be derived for other macroscopically measured quantities such as density and temperature.

As stated earlier it is not possible to give a satisfactory mathematical discussion of the validity of the continuum theory; perhaps the ultimate justification of the model is empirical, i.e. that for a wide range of problems the theory provides answers which are well checked experimentally.

The reader may wonder whether the use of a continuum theory completely obliterates the details of atomic or molecular structure from the picture. The answer is no, because even within a continuum theory, the precise forms of the constitutive equations are determined by the fine details of the atomic or molecular structure. A simple qualitative illustration is provided by the contrast between the well-known physical characteristics of two solids, steel and rubber. We know that large forces are required to stretch steel while for a rubber band of the same initial cross sectional area much smaller forces suffice. In the former case the large forces reflect the very strong interatomic metallic bonding while in rubber elasticity the small forces reflect the ease with which the extremely flexible molecular chains can be uncoiled and straightened out. Both physical facts are quantified in the constitutive equations of the two materials; indeed for these two examples it is possible to derive constitutive equations from atomic and molecular considerations. In such an analysis (which is outside the framework of the present book), various parameters appearing in the constitutive equations become expressed in terms of atomic and molecular parameters.

2.2 Motion of deformable bodies (material description)

The mathematical description of the motion of a deformable body calls for the introduction of reference axes whose origin and orientation are most conveniently chosen to be fixed in space. More precisely the axes are to be chosen to yield an astronomical reference frame for which the fixed stars are at rest and for which the Newtonian laws of motion are well attested, (in practice for most earth-bound problems it is sufficient to choose a 'laboratory' frame, at rest with respect to the earth).

We choose a conventional right handed set of orthogonal Cartesian axes $Oxyz$. Unit vectors directed along the positive x, y and z axes are denoted respectively by $\mathbf{i}$, $\mathbf{j}$ and $\mathbf{k}$ so that the position vector $\mathbf{r}$ of a point with coordinates (x, y, z) is given by

$$\mathbf{r} = x\mathbf{i} + y\mathbf{j} + z\mathbf{k} .$$

† In fact the appearance of discontinuities in P in the solutions of problems is commonly due to the use of oversimplified constitutive equations. By employing more complicated (and more physically realistic) constitutive equations, the discontinuities in P are smoothed out into a continuous function which may however vary rapidly. Unless the variation is very large indeed the inequality (2.1.2) will be then satisfied. In the event (2.1.2) remains violated, the detailed structure of the region of rapidly varying pressure can only be discussed on the basis of a non-continuum model.

In order to define a body in mathematical terms we make use of a familiar property of physical bodies. A physical body occupies a region of space. Because in general a body is in motion and also is subject to distortion, different regions of space are occupied at different times. Each of these regions may be termed a configuration of the body.

Let us arbitrarily select any configuration C_0 which the body may be made to occupy. Suppose C_0 is bounded by the surface S_0, which encloses the open set of points $\mathbf{R} = (X,Y,Z)$. We refer to C_0 as the reference configuration. In terms of C_0 the body is defined to consist of the set of material points (or particles) which lie in one to one correspondence with the set of position vectors **R**. Perhaps it should be emphasised that from the viewpoint of atomic physics or molecular structure, the particles defined above are mythical mathematical entities which in no sense are to be confused with atoms or molecules.

By definition the correspondence 'particle $\rightleftarrows$ **R**' is bijective so that each particle is uniquely labelled by its coordinate vector **R**. Consider now the actual configuration C_t occupied by the body at time t (Fig. 2.1). The particle labelled **R** will have moved to a position **r**, which depends on **R** and t. Thus one possible description of the motion of the body is given by the vector equation

$$\mathbf{r} = \mathbf{r}(\mathbf{R}, t) . \tag{2.2.1}$$

In writing down (2.2.1), and subsequently, extensive use is made of the succinct notation whereby a symbol is used to denote both the dependent variable and the function.

Expressed in component form equation (2.2.1) may be written

$$x = x(X,Y,Z,t) ; \quad y = y(X,Y,Z,t) ; \quad z = z(X,Y,Z,t) \tag{2.2.2}$$

which are three parametric equations, with parameter t, for the particle paths. Equation (2.2.1), or equivalently equations (2.2.2), define the **material** or **referential** description of the motion of a continuum. Conventionally the terminology **Lagrangian** description is attached to (2.2.1) or (2.2.2). In the present text we use the term **material** description.

In the material description it is not imperative that the body should ever actually occupy the reference configuration C_0. However it is often convenient to choose C_0 to be one of the configurations C_t, say for $t = t_0$. Under these circumstances we have the identity

$$\mathbf{r}(\mathbf{R}, t_0) = \mathbf{R}$$

Again, without loss of generality, it is often convenient to choose $t_0 = 0$ so that

$$\mathbf{r}(\mathbf{R}, 0) = \mathbf{R}$$

and $\mathbf{r}(\mathbf{R}, t)$ is then the current position of the particle which was located at **R** at $t = 0$. For this case the configuration C_0 may be termed an initial reference configuration. In many applications the initial reference configuration is most conveniently chosen to be the state of the continuum prior to the imposition of external forces. For example for the cantilever beam problem of Chapter 1 (Fig. 1.1) the obvious reference configuration of the beam is that indicated by the broken lines for which $W = 0$.†

We assume that the functions x, y and z appearing on the right hand sides of equations (2.2.2) may be differentiated as many times as we require with respect to the arguments X, Y, Z and t. With this assumption the particle velocity and particle acceleration in the material description are given by

$$\mathbf{V}(\mathbf{R}, t) = \frac{\partial \mathbf{r}(\mathbf{R}, t)}{\partial t} \tag{2.2.3}$$

$$\mathbf{F}(\mathbf{R}, t) = \frac{\partial \mathbf{V}(\mathbf{R}, t)}{\partial t} \equiv \frac{\partial^2 \mathbf{r}(\mathbf{R}, t)}{\partial t^2} . \tag{2.2.4}$$

Capital letters are used to denote velocity and acceleration in the material description; in both cases the independent spatial variables are the reference coordinates X, Y and Z. It is most important to realise that these spatial arguments of $\mathbf{V}$ and $\mathbf{F}$ do not denote the current position of the particle (which is of course $\mathbf{r}$), but the initial position. For many purposes it is more convenient to express velocity and acceleration in terms of $\mathbf{r}$ rather than $\mathbf{R}$. This is accomplished in Section 2.3 using the spatial description of motion.

We assume that the particle located initially at $\mathbf{R}$ moves to one and only one point $\mathbf{r}$, and conversely that no two separate particles in the initial configuration arrive at the same point $\mathbf{r}$. If follows,that the mapping $\mathbf{r} \rightarrow \mathbf{R}$ is one to one (bijective), and that equation (2.2.1) can be solved to obtain $\mathbf{R}$ as a function of $\mathbf{r}$ and t

$$\mathbf{R} = \mathbf{R}(\mathbf{r}, t) . \tag{2.2.5}$$

For differentiable functions $\mathbf{r}(\mathbf{R}, t)$, a necessary condition for the existence of the unique solution (2.2.5) of equation (2.2.1) is the non-vanishing of the

† The reader may be curious about the absence of the time variable in this problem. Clearly the problem posed is one without time dependence so that $\mathbf{r}$ is a function of $\mathbf{R}$ rather than of both $\mathbf{R}$ and t. Let $\mathbf{r}_S(\mathbf{R})$ (s denoting static) denote this final configuration. The physical act of loading the beam must involve dynamic behaviour. For example if W is suddenly attached the beam will vibrate. (Even if the loading is carried out via small increments ΔW, there will be time dependent motion.) For an isolated elastic beam, the vibrations would persist indefinitely. However in reality the vibrations will be damped out ultimately either by external forces (air resistance), or through internal frictional damping in the material, or by the transmission of energy through the cantilever support. Thus the quantity $\mathbf{r}_S(\mathbf{R})$ is essentially the limit of $\mathbf{r}(\mathbf{R}, t)$ for large times. However usually it is unnecessary to calculate the complete dynamic cycle $\mathbf{r}(\mathbf{R},t)$ and proceed to the limit at large time to find $\mathbf{r}_S(\mathbf{R})$. Similar considerations apply to all 'static' problems in continuum mechanics.

Jacobian J, i.e. $J \neq 0$ where J is given by the determinant

$$J = \frac{\partial(x, y, z)}{\partial(X, Y, Z)} = \begin{vmatrix} \frac{\partial x}{\partial X} & \frac{\partial x}{\partial Y} & \frac{\partial x}{\partial Z} \\ \frac{\partial y}{\partial X} & \frac{\partial y}{\partial Y} & \frac{\partial y}{\partial Z} \\ \frac{\partial z}{\partial X} & \frac{\partial z}{\partial Y} & \frac{\partial z}{\partial Z} \end{vmatrix} . \tag{2.2.6}$$

For $J \neq 0$ an equivalent result is

$$J^{-1} = \frac{\partial(X, Y, Z)}{\partial(x, y, z)} = \begin{vmatrix} \frac{\partial X}{\partial x} & \frac{\partial X}{\partial y} & \frac{\partial X}{\partial z} \\ \frac{\partial Y}{\partial x} & \frac{\partial Y}{\partial y} & \frac{\partial Y}{\partial z} \\ \frac{\partial Z}{\partial x} & \frac{\partial Z}{\partial y} & \frac{\partial Z}{\partial z} \end{vmatrix} . \tag{2.2.7}$$

Equation (2.2.6) expresses J in terms of $\mathbf{R}$ while (2.2.7) gives J in terms of $\mathbf{r}$.

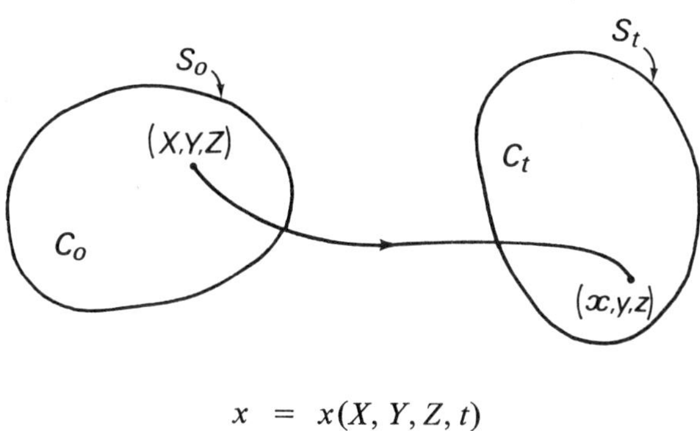

$$\begin{aligned} x &= x(X, Y, Z, t) \\ y &= y(X, Y, Z, t) \\ z &= z(X, Y, Z, t) \end{aligned}$$

Fig. 2.1 Material description of motion

The quantity J arises naturally from considerations of mass conservation. Consider an arbitrary portion of a body and suppose the portion occupies a reference configuration C_0 and current configuration C_t. We suppose for each configuration the existence of a continuous density function ρ defined as the limit

$$\rho = Lt_{\Delta\Omega\to 0}\left(\frac{\Delta m}{\Delta\Omega}\right)$$

where Δm is the mass contained within the element of volume $\Delta\Omega$. The quantities $\rho(\mathbf{r},t)$ and $\rho_0(\mathbf{r})$ are defined to be respectively the density in the current and reference configuration at the point **r**. The assumption of conservation of mass leads to the equation

$$\iiint_{\Omega_t} \rho(\mathbf{r}, t)dx\,dy\,dz = \iiint_{\Omega_0} \rho_0(\mathbf{r})dx\,dy\,dz \equiv \iiint_{\Omega_0} \rho_0(\mathbf{R})dX\,dY\,dZ \tag{2.2.8}$$

where Ω_t and Ω_0 denote the volumes associated respectively with C_t and C_0. Using the well known rule for changing the integration variables in the first of these integrals from x, y and z to X, Y and Z yields

$$\iiint_{\Omega_0} \rho\,[\mathbf{r}\,(\mathbf{R}, t), t]\;J\;dX\,dY\,dZ = \iiint_{\Omega_0} \rho_0(\mathbf{R})dX\,dY\,dZ \tag{2.2.9}$$

Or, since the integrands are assumed continuous while Ω_0 may be chosen arbitrarily, there follows from the mean value theorem for integrals

$$\rho\,[\mathbf{r}(\mathbf{R}, t), t]\,J = \rho_0(\mathbf{R}) \tag{2.2.10}$$

A useful result, equivalent to (2.2.10) is

$$\rho d\Omega = \rho_0 d\Omega_0 \tag{2.2.11}$$

where $d\Omega = dxdydz$, $d\Omega_0 = dXdYdZ$. Equation (2.2.10) expresses the density in the current configuration in terms of the density in the initial configuration. Note that ρ_0 is measured at the particle labelled **R** while ρ is measured at the same particle (which is now located at **r** rather than **R**).

Equation (2.2.10) is one form of the equation of conservation of mass. Since physically ρ and ρ_0 are both positive then evidently J is essentially a positive quantity.

In general the result (2.2.10) is fairly complicated, since both **r** and **R** appear as independent variables. For the particular case where the density of the reference state is constant, equations (2.2.7) and (2.2.10) combine to express $\rho(\mathbf{r}, \mathrm{t})$ in terms of **r** alone. An alternative form of the conservation of mass equation is given in the section following immediately.

2.3 Motion of deformable bodies (spatial description)

The results (2.2.3) and (2.2.4) express particle velocity and acceleration in terms of the reference coordinates **R**. For some purposes it is more convenient to work with a velocity vector $\mathbf{v}(\mathbf{r}, t)$ which depends on the current position $\mathbf{r}$ rather than $\mathbf{R}$. Formally we define

$$\mathbf{v}(\mathbf{r}, t) = \mathbf{V}[\mathbf{R}(\mathbf{r}, t), t] \tag{2.3.1}$$

where $\mathbf{v}$ is obtained from $\mathbf{V}$ by substituting for $\mathbf{R}$ from (2.2.5) into (2.2.3). It is important to recognise that $\mathbf{v}$ as a function of $\mathbf{r}$ and t is a totally different function from $\mathbf{V}$ as a function of $\mathbf{R}$ and t. This is readily illustrated by the one dimensional motion

$$x = X/(1 + tX)\,; \quad y = Y\,; \quad z = Z \tag{2.3.2}$$

for which
$$V_x = -X^2/(1 + tX)^2\,; \quad V_y = V_z = 0 \tag{2.3.3}$$

On the other hand solving (2.3.2) for X, Y and Z in terms of x, y, z and substituting into (2.3.3) leads to

$$v_x = -x^2\,; \quad v_y = v_z = 0 \tag{2.3.4}$$

comparing the two functions

$$v_x(x, t) = -x^2\,; \quad V_x(X, t) = -X^2/(1 + tX)^2 \tag{2.3.5}$$

illustrates the important point concerning the difference in functional dependence. Here the dependence of v_x on x is quite different from that of V_x on X; as it happens here v_x is also independent of t. A further example (in this case of two dimensional motion) appears at the end of the chapter.

The quantity $\mathbf{v}(\mathbf{r}, t)$ defined by (2.3.1) is the particle velocity referred to **spatial** coordinates $\mathbf{r}$. Lower case letters are utilised for such functions. Commonly $\mathbf{v}(\mathbf{r}, t)$ is designated the **Eulerian** velocity field but in the present text, and in conformity with much current practice, $\mathbf{v}(\mathbf{r}, t)$ is called the **spatial** velocity field. One way of interpreting $\mathbf{v}(\mathbf{r}, t)$ is as follows. Consider some fixed point $\mathbf{r}$ in space. At this location suppose that an apparatus is installed to record the velocity of the particles passing through $\mathbf{r}$ as a function of time. The measurements yield $\mathbf{v}(\mathbf{r}, t)$ where in the first instance $\mathbf{r}$ is a fixed triplet of parameters (x, y, z) identifying the recording station. However, by locating recording apparatus at all points $\mathbf{r}$ and assembling the resultant data, we obtain the quantity $\mathbf{v}$ as a function of both $\mathbf{r}$ and t.

An important result is the formula expressing particle acceleration $\mathbf{f}(\mathbf{r}, t)$ $(\equiv \mathbf{F}[\mathbf{R}(\mathbf{r}, t), t])$ in the spatial description in terms of $\mathbf{v}(\mathbf{r}, t)$. It is evident that $\mathbf{f}$ is not given simply by $\partial\mathbf{v}(\mathbf{r}, t)/\partial t$ since the latter derivative is the rate of change of the velocity field with time at the fixed point $\mathbf{r}$. To obtain the correct formula for $\mathbf{f}$ we make use of equation (2.3.1) in the inverse form

$$\mathbf{V}(\mathbf{R}, t) = \mathbf{v}[\mathbf{r}(\mathbf{R}, t), t] \tag{2.3.6}$$

i.e. temporarily we suppose $\mathbf{r}$ in $\mathbf{v}(\mathbf{r}, t)$ to be replaced by the function $\mathbf{r}(\mathbf{R}, t)$.

On differentiating (2.3.6) with respect to time and using (2.2.3), (2.2.4) and (2.3.1) we find

$$\mathbf{F}(\mathbf{R},t) = \frac{\partial \mathbf{v}}{\partial t} + \frac{\partial x}{\partial t}\frac{\partial \mathbf{v}}{\partial x} + \frac{\partial y}{\partial t}\frac{\partial \mathbf{v}}{\partial y} + \frac{\partial z}{\partial t}\frac{\partial \mathbf{v}}{\partial z} = \frac{\partial \mathbf{v}}{\partial t} + \left(\frac{\partial \mathbf{r}}{\partial t}\cdot \text{grad}\right)\mathbf{v}(\mathbf{r},t)$$

$$= \frac{\partial \mathbf{v}}{\partial t} + (\mathbf{V}\cdot\text{grad})\mathbf{v} = \frac{\partial \mathbf{v}}{\partial t} + (\mathbf{v}\cdot\text{grad})\mathbf{v} \quad (2.3.7)$$

By definition $$\mathbf{f}(\mathbf{r},t) = \mathbf{F}[\mathbf{R}(\mathbf{r},t),t] \quad (2.3.8)$$

so that from (2.3.7)

$$\mathbf{f}(\mathbf{r},t) = \frac{\partial \mathbf{v}(\mathbf{r},t)}{\partial t} + (\mathbf{v}\cdot\text{grad})\mathbf{v}(\mathbf{r},t) \quad (2.3.9)$$

expressing the spatial form of the particle acceleration (of the particle currently passing through **r**) in terms of $\mathbf{v}(\mathbf{r},t)$.

Equation (2.3.9) is a very important result in the spatial description. The result is often written in the form

$$\mathbf{f}(\mathbf{r},t) = \frac{D\mathbf{v}}{Dt} \quad (2.3.10)$$

where the scalar operator D/Dt defined by

$$\frac{D}{Dt} = \frac{\partial}{\partial t} + \mathbf{v}\cdot\text{grad} \quad (2.3.11)$$

is known as a **convected** or **mobile** derivative, i.e. a derivative with respect to time following the particle.

The argument leading to (2.3.10) is concerned with the rate of change of velocity for a particle when the velocity field is expressed in spatial coordinates. A precisely similar calculation carried out for any vector or scalar quantity $\xi(\mathbf{r},t)$ leads to the result that the rate of change of ξ following the particle is given by

$$D\xi/Dt\,.$$

For example if $\rho(\mathbf{r},t)$ denotes the density, the rate of change of density measured by an observer moving with the particle passing through **r** at time t is given by

$$D\rho/Dt \equiv \frac{\partial \rho}{\partial t} + (\mathbf{v}\cdot\text{grad})\rho\,. \quad (2.3.12)$$

We note that although D/Dt involves a vector (**v**) and a vector operator (grad), the operator is nevertheless a scalar operator which may be applied to either a vector quantity [as in (2.3.10)] to yield a vector, or to a scalar quantity [as in (2.3.12)] to yield a scalar.

The spatial form of the equation of conservation of mass is expressed in terms of $\rho(\mathbf{r}, t)$ and $\mathbf{v}(\mathbf{r}, t)$. The equation may be obtained as follows; consider a closed surface S, fixed in space, enclosing a fixed volume Ω. Through each surface element $\mathbf{dS}$ of S the mass rate of flow of material out of Ω is

$$\rho v_n dS \equiv \rho \mathbf{v} \cdot \mathbf{dS} \tag{2.3.13}$$

where v_n is the outward normal component of $\mathbf{v}$ in the direction of $\mathbf{dS}$. Integrating the right hand side of (2.3.13) over the entire surface leads to a total mass efflux

$$\frac{dm}{dt} = \iint_S \rho \mathbf{v} \cdot \mathbf{dS}\,. \tag{2.3.14}$$

However if material is neither created nor destroyed, the mass efflux dm/dt out of Ω is achieved only at the expense of a decrease in density within Ω. The rate of decrease of mass within Ω due to change in density is given by

$$\frac{dm}{dt} = -\iiint_\Omega \frac{\partial \rho}{\partial t}\, d\Omega\,. \tag{2.3.15}$$

Equating the two values of dm/dt given by (2.3.14) and (2.3.15) leads to

$$-\iiint_\Omega \frac{\partial \rho}{\partial t}\, d\Omega = \iint_S \rho \mathbf{v} \cdot \mathbf{dS}\,. \tag{2.3.16}$$

From Green's integral theorem

$$\iint_S \rho \mathbf{v} \cdot \mathbf{dS} \equiv \iiint_\Omega \operatorname{div}(\rho \mathbf{v})\, d\Omega$$

so that (2.3.16) may be re-written

$$\iiint_\Omega \left[\frac{\partial \rho}{\partial t} + \operatorname{div}(\rho \mathbf{v})\right] d\Omega = 0\,. \tag{2.3.17}$$

Finally since Ω may be chosen arbitrarily we derive (assuming the integrand is continuous)

$$\frac{\partial \rho}{\partial t} + \operatorname{div}(\rho \mathbf{v}) = 0 \tag{2.3.18}$$

which is the equation of conservation of mass expressed in differential form. Equation (2.3.18) [and equation (2.2.10)] is sometimes known as the **equation of mass balance**, although commonly in texts on fluid mechanics, equation (2.3.18) is known as the 'equation of continuity'. The equation expresses entirely in spatial coordinates the same physical facts as equation (2.2.10). Indeed equation (2.3.18) may be derived from equation (2.2.10) as follows.

Since $\rho_0(\mathbf{R})$ is independent of t, equation (2.2.10) implies

$$\frac{\partial}{\partial t}\left[\rho\,[\mathbf{r}(\mathbf{R}, t), t]\; J\right] = 0\,. \tag{2.3.19}$$

where temporarily we regard **R** as the independent space variable. From previous arguments (2.3.19) may be written

$$J\,\frac{D\rho(\mathbf{r}, t)}{Dt} + \rho(\mathbf{r}, t)\,\frac{\partial J}{\partial t} = 0 \tag{2.3.20}$$

where we have reverted to the spatial description [independent variable **r**] everywhere except for the term $\partial J/\partial t$. It is shown below that

$$\frac{\partial J}{\partial t} = J\,\mathrm{div}\,[\mathbf{v}(\mathbf{r}, t)]\;. \tag{2.3.21}$$

On substituting this result into (2.3.20) and cancelling the non-zero factor J, there results

$$\frac{D\rho}{Dt} + \rho\,\mathrm{div}\,\mathbf{v} = 0 \tag{2.3.22}$$

or on using (2.3.11)

$$\frac{\partial \rho}{\partial t} + \rho\,\mathrm{div}\,\mathbf{v} + (\mathbf{v}.\mathrm{grad})\rho = 0\,. \tag{2.3.23}$$

With the help of the vector identity

$$\mathrm{div}(\rho\mathbf{v}) = \rho\,\mathrm{div}\,\mathbf{v} + (\mathbf{v}\cdot\mathrm{grad})\,\rho \tag{2.3.24}$$

it is clear that (2.3.18) and (2.3.23) are identical.

The proof of (2.3.21) is cumbersome. In evaluating $\partial J/\partial t$ in (2.3.20) it is implied that the independent space variable is taken to be **R**. Accordingly we choose J to be given by (2.2.6). From the usual rules for differentiating determinants

$$\frac{\partial J}{\partial t} = \begin{vmatrix} \dfrac{\partial^2 x}{\partial X \partial t} & \dfrac{\partial^2 x}{\partial Y \partial t} & \dfrac{\partial^2 x}{\partial Z \partial t} \\ \dfrac{\partial y}{\partial X} & \dfrac{\partial y}{\partial Y} & \dfrac{\partial y}{\partial Z} \\ \dfrac{\partial z}{\partial X} & \dfrac{\partial z}{\partial Y} & \dfrac{\partial z}{\partial Z} \end{vmatrix} + \begin{vmatrix} \dfrac{\partial x}{\partial X} & \dfrac{\partial x}{\partial Y} & \dfrac{\partial x}{\partial Z} \\ \dfrac{\partial^2 y}{\partial X \partial t} & \dfrac{\partial^2 y}{\partial Y \partial t} & \dfrac{\partial^2 y}{\partial Z \partial t} \\ \dfrac{\partial z}{\partial X} & \dfrac{\partial z}{\partial Y} & \dfrac{\partial z}{\partial Z} \end{vmatrix} + \begin{vmatrix} \dfrac{\partial x}{\partial X} & \dfrac{\partial x}{\partial Y} & \dfrac{\partial x}{\partial Z} \\ \dfrac{\partial y}{\partial X} & \dfrac{\partial y}{\partial Y} & \dfrac{\partial y}{\partial Z} \\ \dfrac{\partial^2 z}{\partial X \partial t} & \dfrac{\partial^2 z}{\partial Y \partial t} & \dfrac{\partial^2 z}{\partial Z \partial t} \end{vmatrix}$$

We consider in detail the first of these determinants and make use of the result

$$\frac{\partial x}{\partial t} = V_x(\mathbf{R}, t) \equiv v_x[\mathbf{r}(\mathbf{R}, t), t] \tag{2.3.25}$$

where V_x and v_x are respectively x components of V and v. From (2.3.25)

$$\frac{\partial^2 x}{\partial X \partial t} = \frac{\partial v_x}{\partial x}\frac{\partial x}{\partial X} + \frac{\partial v_x}{\partial y}\frac{\partial y}{\partial X} + \frac{\partial v_x}{\partial z}\frac{\partial z}{\partial X}$$

$$\frac{\partial^2 x}{\partial Y \partial t} = \frac{\partial v_x}{\partial x}\frac{\partial x}{\partial Y} + \frac{\partial v_x}{\partial y}\frac{\partial y}{\partial Y} + \frac{\partial v_x}{\partial z}\frac{\partial z}{\partial Y}$$

$$\frac{\partial^2 x}{\partial Z \partial t} = \frac{\partial v_x}{\partial x}\frac{\partial x}{\partial Z} + \frac{\partial v_x}{\partial y}\frac{\partial y}{\partial Z} + \frac{\partial v_x}{\partial z}\frac{\partial z}{\partial Z} .$$

On substituting these results into the first of the three determinants in (2.3.24) and using the rules for the manipulation of determinants we find

$$\begin{vmatrix} \frac{\partial^2 x}{\partial X \partial t} & \frac{\partial^2 x}{\partial Y \partial t} & \frac{\partial^2 x}{\partial Z \partial t} \\ \frac{\partial y}{\partial X} & \frac{\partial y}{\partial Y} & \frac{\partial y}{\partial Z} \\ \frac{\partial z}{\partial X} & \frac{\partial z}{\partial Y} & \frac{\partial z}{\partial Z} \end{vmatrix} = \frac{\partial v_x}{\partial x} \begin{vmatrix} \frac{\partial x}{\partial X} & \frac{\partial x}{\partial Y} & \frac{\partial x}{\partial Z} \\ \frac{\partial y}{\partial X} & \frac{\partial y}{\partial Y} & \frac{\partial y}{\partial Z} \\ \frac{\partial z}{\partial X} & \frac{\partial z}{\partial Y} & \frac{\partial z}{\partial Z} \end{vmatrix}$$

$$+ \frac{\partial v_x}{\partial y} \begin{vmatrix} \frac{\partial y}{\partial X} & \frac{\partial y}{\partial Y} & \frac{\partial y}{\partial Z} \\ \frac{\partial y}{\partial X} & \frac{\partial y}{\partial Y} & \frac{\partial y}{\partial Z} \\ \frac{\partial z}{\partial X} & \frac{\partial z}{\partial Y} & \frac{\partial z}{\partial Z} \end{vmatrix} + \frac{\partial v_x}{\partial z} \begin{vmatrix} \frac{\partial z}{\partial X} & \frac{\partial z}{\partial Y} & \frac{\partial z}{\partial Z} \\ \frac{\partial y}{\partial X} & \frac{\partial y}{\partial Y} & \frac{\partial y}{\partial Z} \\ \frac{\partial z}{\partial X} & \frac{\partial z}{\partial Y} & \frac{\partial z}{\partial Z} \end{vmatrix} . \tag{2.3.26}$$

The first determinant on the right hand side of (2.3.26) is simply J, while the other two determinants vanish (because in each case two rows are identical). It follows that the first determinant on the right hand side of (2.3.24) is $J(\partial v_x/\partial x)$. Similarly the second and third determinants are respectively $J(\partial v_y/\partial y)$ and $J(\partial v_z/\partial z)$ and the result (2.3.21) follows.

The most immediate proof of the equation of continuity makes use of the following standard theorem of vector analysis and kinematics.

Let S_t denote a time-dependent closed surface which always encloses the same particles of a continuum. The time-dependent volume of space enclosed by S_t is denoted by Ω_t. Also let $\xi(\mathbf{r}, t)$ denote any differentiable scalar quantity. The theorem states that if $I(t)$ denotes the integral

$$I(t) = \iiint_{\Omega_t} \xi(\mathbf{r}, t) d\Omega \tag{2.3.27}$$

then the time derivative of $I(t)$ is given by the formula†

$$\frac{dI}{dt} = \iiint_{\Omega_t} \left[\frac{\partial \xi}{\partial t} + \operatorname{div}(\mathbf{v}\xi) \right] d\Omega\,. \tag{2.3.28}$$

On identifying ξ with $\rho(\mathbf{r}, t)$, $I(t)$ becomes the mass m contained within S_t. Since S_t always bounds the same particles, $dm/dt = 0$ and therefore

$$\iiint_{\Omega_t} \left[\frac{\partial \rho}{\partial t} + \operatorname{div}(\rho \mathbf{v}) \right] d\Omega = 0$$

If this is to be true for arbitrary choice of Ω_t and the integrand is continuous, the integrand vanishes and (2.3.18) follows.

Both the material description $[\mathbf{r} = \mathbf{r}(\mathbf{R}, t)]$ and the spatial description $[\mathbf{v} = \mathbf{v}(\mathbf{r}, t)]$ are used in continuum mechanics. The material description contains more information (namely the quantity **R** which is equivalent to knowing the initial position of the particle)†† and is imperative for most problems in solids, where (usually) one of the quantities of interest is the displacement $\mathbf{U}(\mathbf{R}, t)$ of points from their initial positions; for the case where C_0 is the initial configuration the displacement **U** is

$$\mathbf{U} = \mathbf{r}(\mathbf{R}, t) - \mathbf{R}\,. \tag{2.3.29}$$

(An example is the cantilever beam problem of Chapter 1. Here one of the quantities of practical interest is the vertical deflexion of the beam caused by the weight W.) More important, stresses generated in solids depend on the space derivatives of the quantity $\mathbf{r}(\mathbf{R}, t)$. If, as in the spatial description, $\mathbf{v}(\mathbf{r}, t)$ is taken as the dependent variable where

$$\mathbf{v}(\mathbf{r}, t) \equiv \mathbf{V}(\mathbf{R}, t) = \frac{\partial \mathbf{r}(\mathbf{R}, t)}{\partial t}$$

it is not possible to express quantities like $\partial x/\partial Y$, $\partial z/\partial Z$ etc. simply in terms of $\mathbf{v}(\mathbf{r}, t)$.

† A proof of this result is given at the end of this Chapter.

†† If C_0 is an initial reference configuration, **R** is already the initial particle position.

On the other hand most problems in fluid mechanics are formulated in terms of the spatial description of motion. A few problems in fluid mechanics are more easily worked in the material description, but for the simpler fluids, i.e. the ideal fluid and the Stokesian fluid, the spatial description usually proves easier to handle. Commonly in problems in fluid mechanics particle displacements are of no direct interest. For example in the problem of the flow of air around an aircraft wing, the quantities of interest are the spatial velocity field $\mathbf{v}(\mathbf{r}, t)$ and the resultant pressure distribution on the wing. More important, for fluids the stresses generated by the deformation are expressible directly in terms of the space derivatives of $\mathbf{v}(\mathbf{r}, t)$ and in terms of the density $\rho(\mathbf{r}, t)$.

If required, the material description of motion is recoverable in principle from the spatial description i.e. from $\mathbf{v}(\mathbf{r}, t)$, by quadrature. For if $\mathbf{r}(\mathbf{R}, t)$ denotes the particle paths, then $\mathbf{r}$ is the solution of the ordinary differential equation

$$\frac{d\mathbf{r}}{dt} = \mathbf{v}(\mathbf{r}, t) . \tag{2.3.30}$$

The solution of (2.3.30) necessarily entails constants of integration. Since the equation is a vector equation of first order, one vector constant of integration is required and essentially the vector in question is $\mathbf{R}$ (or a quantity equivalent to $\mathbf{R}$).

Particle paths are often unimportant in problems of fluid mechanics. Of more direct interest are the streamlines, which are lines drawn everywhere rallel to the velocity field. The streamline pattern in general changes from one instant of time to another although there are important exceptions when the streamline pattern is stationary. At any given instant of time the streamlines are defined by the (partial) equations

$$\mathbf{dr} = \mathbf{v}(\mathbf{r}, t)\, d\lambda \tag{2.3.31}$$

where $d\lambda$ is an arbitrary infinitesimal quantity of dimension time. Equation (2.3.31) merely expresses the condition that if $\mathbf{dr}$ is the vector element of a streamline at the point $\mathbf{r}$, then $\mathbf{dr}$ is parallel to $\mathbf{v}$.

Provided $\mathbf{v} \neq 0$ equation (2.3.31) predicts a unique direction for $\mathbf{dr}$. It follows that unless $\mathbf{v} = 0$, only one streamline passes through a given point. Thus an important property of streamlines is that, except possibly at a 'stagnation' point ($\mathbf{v} = 0$), streamlines do not intersect. Exceptionally, at a stagnation point, a number of streamlines may meet.

Mathematically one way of looking at equation (2.3.31) is to eliminate $\mathbf{d}\lambda$ and write the equation in the 'Pfaffian' form

$$\frac{dx}{v_x(\mathbf{r}, t)} = \frac{dy}{v_y(\mathbf{r}, t)} = \frac{dz}{v_z(\mathbf{r}, t)} . \tag{2.3.32}$$

Formally the integration of equations of this type leads to lines formed by the intersection of two surfaces

$$q_1(x, y, z, t) = c_1 , \quad q_2(x, y, z, t) = c_2 \tag{2.3.33}$$

where c_1 and c_2 are constants of integration. Different choices of c_1 and c_2 lead to different streamlines. Note that in general q_1 and q_2 are time dependent, although of course t is treated as a constant in the integration of (2.3.32). Equations (2.3.33) define the streamline pattern at each instant of time and in general the pattern changes with time.

Exceptionally if $\mathbf{v}(\mathbf{r}, t)$ is independent of t, the streamline pattern is stationary. Moreover under these circumstances the streamlines are identical with the particle paths. This latter result is most easily derived by looking at equations (2.3.31) from a different viewpoint. Instead of writing (2.3.31) in the Pfaffian form (2.3.32), we determine the streamlines parametrically from the ordinary differential equations

$$\frac{dx}{d\lambda} = v_x(x, y, z, t) ; \quad \frac{dy}{d\lambda} = v_y(x, y, z, t) ; \quad \frac{dz}{d\lambda} = v_z(x, y, z, t) \tag{2.3.34}$$

where (i) as before t is taken as constant in the integration process, (ii) x, y and z are now regarded as functions of the parameter λ. Equations (2.3.34) yield parametric representations of the streamline pattern in the form

$$x = x(\lambda, t) \quad y = y(\lambda, t) \quad z = z(\lambda, t) .$$

However for the particular case where $\mathbf{v}(\mathbf{r}, t)$ is independent of t equations (2.3.34) assume the form

$$\frac{dx}{d\lambda} = v_x(x, y, z) ; \quad \frac{dy}{d\lambda} = v_y(x, y, z) ; \quad \frac{dz}{d\lambda} = v_z(x, y, z) . \tag{2.3.35}$$

and since t is no longer present there is no objection to setting $\lambda \equiv t$ in which case equations (2.3.35) are identical with the particle path equations

$$\frac{dx}{dt} = v_x(x, y, z) ; \quad \frac{dy}{dt} = v_y(x, y, z) ; \quad \frac{dz}{dt} = v_z(x, y, z) .$$

Thus for steady (or stationary) flow conditions where $\mathbf{v}(\mathbf{r}, t)$ is independent of t, the streamlines and particle paths coincide.

The converse theorem is not true. Consider for example the accelerated rectilinear motion of a rigid body.

2.4 Example

The following example of a possible motion in an inviscid fluid is chosen to illustrate many of the points discussed in Section 2.2 and 2.3; the motion in question is that associated with an isolated rectilinear vortex whose axis coincides with the z axis. (This problem is discussed using the spatial description in Section 12.6.2). Here we begin with the form of the solution in material coordinates given by

$$x = (X^2 + Y^2)^{1/2} \cos\left[\frac{\omega t}{(X^2 + Y^2)} + \tan^{-1}\left(\frac{Y}{X}\right)\right] \tag{2.4.1}$$

$$y = (X^2 + Y^2)^{1/2} \sin\left[\frac{\omega t}{(X^2 + Y^2)} + \tan^{-1}\left(\frac{Y}{X}\right)\right] \tag{2.4.2}$$

$$z = Z \tag{2.4.3}$$

where ω is a constant. It is readily verified that at $t = 0$

$$\mathbf{r}(\mathbf{R}, 0) = \mathbf{R}$$

so that the reference configuration is here the initial configuration.

In view of (2.4.3) the motion is two dimensional so that the same flow pattern occurs on each plane $z = \text{const}$. Clearly since

$$x^2 + y^2 = X^2 + Y^2 \tag{2.4.4}$$

the particle paths are circles centred on the z axis.

The material velocity and acceleration fields are given by

$$\frac{\partial x}{\partial t} = V_x = -\frac{\omega}{(X^2 + Y^2)^{1/2}} \sin\left[\frac{\omega t}{(X^2 + Y^2)} + \tan^{-1}\left(\frac{Y}{X}\right)\right] \tag{2.4.5}$$

$$\frac{\partial y}{\partial t} = V_y = \frac{\omega}{(X^2 + Y^2)^{1/2}} \cos\left[\frac{\omega t}{(X^2 + Y^2)} + \tan^{-1}\left(\frac{Y}{X}\right)\right] \tag{2.4.6}$$

$$\frac{\partial z}{\partial t} = V_z = 0 \tag{2.4.7}$$

$$F_x = \frac{\partial^2 x}{\partial t^2} = -\frac{\omega^2}{(X^2 + Y^2)^{3/2}} \cos\left[\frac{\omega t}{(X^2 + Y^2)} + \tan^{-1} \frac{Y}{X}\right] \tag{2.4.8}$$

$$F_y = \frac{\partial^2 y}{\partial t^2} = -\frac{\omega^2}{(X^2+Y^2)^{3/2}} \sin\left[\frac{\omega t}{(X^2+Y^2)} + \tan^{-1}\frac{Y}{X}\right] \tag{2.4.9}$$

$$F_z = \frac{\partial^2 z}{\partial t^2} = 0\,. \tag{2.4.10}$$

The spatial velocity field may be obtained by substituting for X, Y, Z from (2.4.1)–(2.4.3) in terms of x, y and z into (2.4.5)–(2.4.7). In view of (2.4.4) we have immediately

$$v_x = -\frac{\omega y}{x^2+y^2}; \quad v_y = \frac{\omega x}{x^2+y^2}; \quad v_z = 0\,; \tag{2.4.11}$$

so that here the motion is steady. Note that there is no resemblence between the functional forms of **V** and **v**.

Similarly we find for the particle acceleration in the spatial description

$$f_x = -\frac{\omega^2 x}{(x^2+y^2)^2}; \quad f_y = -\frac{\omega^2 y}{(x^2+y^2)^2}; \quad f_z = 0\,; \tag{2.4.12}$$

which results also follow from

$$\mathbf{f} = \frac{D\mathbf{v}}{Dt} \equiv \frac{\partial \mathbf{v}}{\partial t} + (\mathbf{v}\cdot\text{grad})\mathbf{v}\,.$$

Here $(\partial \mathbf{v}/\partial t) = 0$ so that f_x is given by

$$\begin{aligned} f_x &= -\frac{\omega}{x^2+y^2}\left(y\frac{\partial}{\partial x} - x\frac{\partial}{\partial y}\right)\left(-\frac{\omega y}{x^2+y^2}\right) \\ &= -\frac{\omega^2 x}{(x^2+y^2)^2} \end{aligned}$$

in agreement with the first of (2.4.12). Similarly calculations of f_y and f_z are in agreement with the remaining equations in (2.4.12).

It is left to the reader to undertake the algebraically cumbersome calculation of J from (2.2.6). The ultimate result is simply $J = 1$ so that equation (2.2.10) reads

$$\rho\,[\mathbf{r}(\mathbf{R}, t), t] = \rho_0(\mathbf{R})\,.$$

In other words the density at the particle labelled **R** is unchanged by the motion.† From the spatial viewpoint this means

$$\frac{D\rho}{Dt} = 0$$

or equivalently [from (2.3.22)]

$$\operatorname{div} \mathbf{v}(\mathbf{r}, t) = 0 .$$

It is readily verified that the velocity components (2.4.11) satisfy this condition.

To recover the material description (2.4.1), (2.4.2) and (2.4.3) from the spatial velocity field (2.4.11), we have the ordinary differential equations

$$\frac{dx}{dt} = -\frac{\omega y}{x^2 + y^2}, \quad \frac{dy}{dt} = \frac{\omega x}{x^2 + y^2}, \quad \frac{dz}{dt} = 0 . \tag{2.4.13}$$

The third of these equations leads immediately to $z =$ const.

$$\text{i.e.} \qquad z = Z . \tag{2.4.14}$$

A first integral of the coupled equations for x and y is found by noting that

$$x\frac{dx}{dt} + y\frac{dy}{dt} = 0$$

i.e.

$$x^2 + y^2 = X^2 + Y^2 . \tag{2.4.15}$$

Solving (2.4.15) for y and substituting the result into the first of (2.4.13) leads to

$$(X^2 + Y^2)\frac{dx}{dt} = -\omega[X^2 + Y^2 - x^2]^{1/2}$$

with solution $$\cos^{-1}\left[\frac{x}{(X^2 + Y^2)^{1/2}}\right] = \frac{\omega t}{X^2 + Y^2} + \text{const.}.$$

Finally choosing the constant so that $x = X$ at $t = 0$ leads to (2.4.1). A similar analysis for y leads to (2.4.2).

† In fact the vortex motion under discussion is almost always considered in the context of an incompressible fluid for which $\rho = \rho_0 =$ constant, independent of **R**, for example as in Chapter 12.

Because the motion is steady the particle paths and streamline coincide. From (2.3.32) and (2.4.11) the Pfaffian equations for the streamlines are

$$-\frac{dx}{y} = \frac{dy}{x} = \frac{dz}{0}$$

which on integration show that the streamlines are formed by the intersection of the surfaces

$$z = c_1, \qquad x^2 + y^2 = c_2 \tag{2.4.16}$$

where c_1 and c_2 are constants of integration. Different choices of c_1 and c_2 lead to different streamlines. As previously indicated the streamlines are here identical with the particle paths, e.g. compare equations (2.4.16) and (2.4.14) and (2.4.15).

2.5 Summary

The important results of the present chapter are summarised below.

The motion of a continuum may be specified either in the material description (basic independent variables **R** and t) or the spatial description (basic independent variables **r** and t). In each case **r** is the current particle position. Usually, in the present text, **R** will be chosen to be the initial position of the particle.

The material description is complete in the sense that the detailed history $\mathbf{r}(\mathbf{R}, t)$ of the position vector of each particle is chosen as the dependent variable.

In the spatial description the motion is described by $\mathbf{v}(\mathbf{r}, t)$. However it is common to choose both **v** and ρ as dependent variables and incorporate the equation of mass balance (2.3.18) into the basic equations. The reason is clear; (2.3.18) is a partial differential equation which [unlike (2.2.10)] does not yield an explicit value of ρ immediately.

Usually the material description is used in problems of solid mechanics, while commonly problems in fluid mechanics are solved using the spatial description. The following table lists the important results in both descriptions.

	MATERIAL	SPATIAL
Independent variables	$\mathbf{R}, t$	$\mathbf{r}, t$
Dependent variables	$\mathbf{r}(\mathbf{R}, t)$ (or $\mathbf{U}$)	$\mathbf{v}(\mathbf{r}, t)$; $\rho(\mathbf{r}, t)$
Displacement	$\mathbf{U}(\mathbf{R}, t) = \mathbf{r}(\mathbf{R}, t) - \mathbf{R}$ (if $\mathbf{R}$ is initial position)	No Description in terms of $(\mathbf{v}, \rho)$ but formally $\mathbf{u} = \mathbf{r} - \mathbf{R}(\mathbf{r}, t)$.
Particle velocity	$\mathbf{V}(\mathbf{R}, t) = \dfrac{\partial \mathbf{r}}{\partial t} \left(\equiv \dfrac{\partial \mathbf{U}}{\partial t}\right)$	$\mathbf{v}(\mathbf{r}, t)$
Particle acceleration	$\mathbf{F}(\mathbf{R}, t) = \dfrac{\partial^2 \mathbf{r}}{\partial t^2} \left(\equiv \dfrac{\partial^2 \mathbf{U}}{\partial t^2}\right)$	$\mathbf{f}(\mathbf{r}, t) = \dfrac{\partial \mathbf{v}(\mathbf{r}, t)}{\partial t} + [\mathbf{v}(\mathbf{r}, t) . \text{grad}]\, \mathbf{v}(\mathbf{r}, t)$
Equation of mass balance	$\rho_0(\mathbf{R}) = \rho\,[\mathbf{r}(\mathbf{R}, t), t]\, J$	$\dfrac{\partial \rho(\mathbf{r}, t)}{\partial t} + \text{div}\,[\rho(\mathbf{r},t)\mathbf{v}(\mathbf{r},t)] = 0$
Particle paths	$\mathbf{r} = \mathbf{r}(\mathbf{R}, t)$	No description but spatial form obtained from integration of ordinary differential equation $\dfrac{d\mathbf{r}}{dt} = \mathbf{v}(\mathbf{r}, t)$
Streamlines	No description	From integration of Pfaffian equations $\dfrac{dx}{v_x} = \dfrac{dy}{v_y} = \dfrac{dz}{v_z}$.

In the above J is defined by

$$J = \frac{\partial(x, y, z)}{\partial(X, Y, Z)}$$

Appendix to Chapter 2: Proof of (2.3.28)

In view of previously derived results the simplest proof of (2.3.28) is as follows.

To avoid the difficulty of differentiating with respect to time a volume integral evaluated over a time dependent region of space Ω_t, we transform the integration coordinates from spatial (x, y, z) to referential (X, Y, Z). Thus

$$I(t) = \iiint_{\Omega_0} \xi\,[\mathbf{r}(\mathbf{R}, t), t]\, J\, d\Omega_0$$

where Ω_0 is the reference configuration, independent of t, and $d\Omega_0 = dX dY dZ$. The differentiation yields now

$$\frac{dI}{dt} = \iiint_{\Omega_0} \left(\frac{D\xi}{Dt} J + \xi \frac{\partial J}{\partial t}\right) d\Omega_0$$

or, on using (2.3.21) and reverting to spatial coordinates

$$\frac{dI}{dt} = \iiint_{\Omega_t} \left(\frac{D\xi}{Dt} + \xi \operatorname{div} \mathbf{v}\right) d\Omega$$

$$= \iiint_{\Omega_t} \left(\frac{\partial \xi}{\partial t} + \operatorname{div}(\mathbf{v}\xi)\right) d\Omega$$

Problems

1. The material description of a motion is

$$x = \alpha(t)X, \quad y = \beta(t)Y, \quad z = \gamma(t)Z$$

where α, β and γ are (as indicated) functions of time.

Show that the material velocity and acceleration are given by

$$\mathbf{V} = (\dot{\alpha}X, \dot{\beta}Y, \dot{\gamma}Z)\,; \qquad \mathbf{F} = (\ddot{\alpha}X, \ddot{\beta}Y, \ddot{\gamma}Z)\,.$$

Deduce that the spatial velocity and acceleration fields are given by

$$\mathbf{v} = (\dot{\alpha}x/\alpha, \dot{\beta}y/\beta, \dot{\gamma}z/\gamma)\,; \qquad \mathbf{f} = (\ddot{\alpha}x/\alpha, \ddot{\beta}y/\beta, \ddot{\gamma}z/\gamma)$$

and verify the relation

$$\mathbf{f} = \frac{\partial \mathbf{v}}{\partial t} + (\mathbf{v}\cdot\operatorname{grad})\mathbf{v}\,.$$

Calculate ρ/ρ_0 from $\rho_0 = \rho J$ and verify the relation

$$\frac{\partial \rho}{\partial t} + \operatorname{div}(\rho\mathbf{v}) = 0\,.$$

2. The quantities r and R are defined by

$$r = (x^2 + y^2 + z^2)^{1/2}, \qquad R = (X^2 + Y^2 + Z^2)^{1/2}$$

and a spherically symmetric motion is specified in the material description by the equation

$$r = r(R, t).$$

Show that in Cartesian coordinates the motion is given by

$$x = Xr/R, \qquad y = Yr/R, \qquad z = Zr/R, \qquad [r = r(R,t)]$$

and evaluate $J \equiv \partial(x, y, z)/\partial(X, Y, Z)$.

Deduce that for an incompressible material

$$r = [R^3 + h(t)]^{1/3}$$

where $h(t)$ is an arbitrary function of time. If $a(t)$ denotes the current radius of material which at $t = 0$ was located at a_0 show that

$$r^3 - a^3(t) = R^3 - a_0^3$$

and provide a simple geometrical interpretation of this last result for a material of initially uniform density.

Show also that in the spatial description the velocity and acceleration are directed radially and given by

$$\mathbf{v} = (a^2\dot{a}/r^2)\mathbf{i}_r$$

$$\mathbf{f} = [(a^2\ddot{a} + 2a\dot{a}^2)r^{-2} - 2a^4\dot{a}^2 r^{-5}]\,\mathbf{i}_r$$

where $\mathbf{i}_r$ is a unit vector in the r direction.

3. A motion is specified by

$$x = \lambda(t)\{X\cos[\tau(t)Z] - Y\sin[\tau(t)Z]\}$$

$$y = \lambda(t)\{Y\cos[\tau(t)Z] + X\sin[\tau(t)Z]\}$$

$$z = \Lambda(t)Z$$

where λ, Λ and τ are (as indicated) functions of time. Also

$$\lambda(0) = \Lambda(0) = 1, \qquad \tau(0) = 0.$$

Show that at $t = 0$, x, y, z take the values X, Y, Z and that circles of initial radius a (given by $X^2 + Y^2 = a^2$) become circles of radius λa.

Show also that if X, Y and x, y are expressed in polar form

$$X = R\cos\Theta\,, \qquad Y = R\sin\Theta\,; \qquad x = r\cos\theta\,, \qquad y = r\sin\theta$$

then

$$\tan\theta = \tan[\Theta + \tau Z]\,.$$

Infer that the motion describes the simultaneous extension (in the z direction), expansion (in the radial direction) and twisting (about the Z axis) of the material. Show that the spatial velocity field is given by

$$v_x = (\dot{\lambda}/\lambda)x - (\dot{\tau}/\Lambda)yz\,, \qquad v_y = (\dot{\lambda}/\lambda)y + (\dot{\tau}/\Lambda)xz\,, \qquad v_z = (\dot{\Lambda}/\Lambda)z$$

and deduce that if the motion occurs in an incompressible material then

$$2(\dot{\lambda}/\lambda) + (\dot{\Lambda}/\Lambda) = 0$$

with solution

$$\lambda^2\Lambda = \text{const}\,.$$

[However since $\lambda(0) = \Lambda(0) = 1$, the constant is unity and $\lambda = \Lambda^{-1/2}$.]

4. If x, y and z are written

$$x = X + U_x(X,Y,Z;t), \quad y = Y + U_y(X,Y,Z;t), \quad z = Z + U_z(X,Y,Z;t)$$

defining the displacement vector $\mathbf{U}$, show that

$$J = \begin{vmatrix} 1 + \partial U_x/\partial X\,, & \partial U_x/\partial Y\,, & \partial U_x/\partial Z \\ \partial U_y/\partial X\,, & 1 + \partial U_y/\partial Y\,, & \partial U_y/\partial Z \\ \partial U_z/\partial X\,, & \partial U_z/\partial Y\,, & 1 + \partial U_z/\partial Z \end{vmatrix}$$

and deduce that if all the dimensionless quantities $\partial U_x/\partial X$, $\partial U_x/\partial Y$ *etc.* are small compared with unity

$$\text{e.g.} \qquad |\partial U_x/\partial Y| = O(\epsilon) \text{ where } \epsilon \ll 1$$

then to first order in ϵ (i.e. neglecting terms in ϵ^2)

$$\rho = \rho_0(1 - \Delta)$$

where

$$\Delta = \operatorname{div}\mathbf{U} \equiv (\partial U_x/\partial X + \partial U_y/\partial Y + \partial U_z/\partial Z)\,.$$

5. Show that the streamlines associated with the velocity field

$$v_x = kx, \quad v_y = -ky, \quad v_z = 0, \quad (k \text{ constant})$$

are formed by the intersections of the planes $z = C$ and the rectangular hyperbolae sheets $xy = A$ where A and C are constants.

Show directly that the only points where streamlines intersect are points for which $\mathbf{v} = 0$.

6. An initially rectangular parallelopiped of material $0 \leqslant X \leqslant L$, $0 \leqslant Y \leqslant M$, $0 \leqslant Z \leqslant N$ is subject to the deformation

$$x = X + aY, \quad y = Y + bX, \quad z = Z$$

where a and b are constants (or possibly functions of time).

Show that the deformation is possible in a real material only if

$$ab < 1.$$

Show that the material assumes the form of a parallelopiped bounded by the planes

$$x - ay = 0, \quad x - ay = (1 - ab)L; \quad y - bx = 0,$$

$$y - bx = (1 - ab)M; \quad z = 0, \quad z = N.$$

Comment on the geometrical significance of the restriction $ab < 1$.

7. A two dimensional flow is one for which

$$v_x = v_x(x, y; t), \quad v_y = v_y(x, y; t), \quad v_z = 0.$$

Show that for incompressible two dimensional flows the equation $\operatorname{div} \mathbf{v} = 0$ is satisfied automatically by choosing

$$v_x = -\partial\psi/\partial y \qquad v_y = \partial\psi/\partial x$$

where $\psi = \psi(x, y, t)$ is a function of the indicated variables.

The quantity ψ is known as the 'stream function'. Show that the streamlines are given by the intersection of the planes $z =$ const. with the surfaces $\psi(x, y, t) =$ const. (thereby justifying the terminology). [Usually in two dimensional incompressible flows it is usual to state simply that the streamlines are given by $\psi(x, y, t) =$ const. – omitting all reference to intersections of surfaces.]

Show that for two dimensional compressible steady flows there exists a stream function $\psi(x, y)$ for which

$$\rho v_x = -\partial\psi/\partial y, \qquad \rho v_y = \partial\psi/\partial x.$$

8. In the kinematics of the motion of a rigid body it is shown that the velocity of the particle located currently at **r** (where **r** is referred to a fixed system of coordinates) is given by

$$\mathbf{v} \equiv \frac{d\mathbf{r}}{dt} = \dot{\mathbf{a}} + \boldsymbol{\omega} \wedge (\mathbf{r} - \mathbf{a}).$$

Here $\boldsymbol{\omega}(t)$ is the angular velocity and $\mathbf{a}(t)$ is the motion of any arbitrary particle in the body. (The result is independent of the choice of particle.)

Calculate the acceleration of the particle at **r**, (i) directly from $\mathbf{f} = (d\mathbf{v}/dt)$ and also (ii) from the convective derivative formula $\mathbf{f} = (\partial\mathbf{v}/\partial t) + (\mathbf{v}\cdot\text{grad})\mathbf{v}$.

Find the particle paths for the particular case where $\mathbf{a} = 0$ and $\boldsymbol{\omega} = \omega(t)\mathbf{k}$ is fixed in direction but variable in magnitude. Deduce that in this particular case the motion is indeed that of a rigid body [rotating at angular velocity $\omega(t)$ about the z axis].

Chapter 3

Forces in a Continuum: Stress and Equations of Motion

3.1 Forces in a continuum

To obtain a description of the forces acting on the material of a continuum, we consider an arbitrary portion of the body which always contains the same particles. The portion is supposed bounded by a closed surface S_t which encloses a volume Ω_t. The outward unit normal vector to S_t is denoted by $\mathbf{n}$. Of course $\mathbf{n}$ points in different directions at different points of S_t.†

We distinguish three types of force acting on the material contained within S_t.

(1) Internal forces between pairs of particles belonging to Ω_t, e.g., electrical forces and mutual gravitational forces.

(2) External forces acting on the material within S_t, e.g., the force of the earth's gravitational field, gravitational forces due to the material outside S_t, forces of electromagnetic origin.

(3) Surface contact forces transmitted across S_t from the region outside S_t.

3.1.1. Internal Forces

If it is assumed that internal forces between pairs of particles are equal in magnitude but opposite in direction then the internal forces do not contribute to the balance laws governing the changes of the total linear and angular momentum of the material contained within S_t, [e.g., Chapters V and XII, *Principles of Mechanics*, Synge and Griffiths, McGraw-Hill (1942)], and a detailed knowledge of the internal forces is not required. [There are some difficulties here in that the mathematical particles of the continuum are not the point masses of classical particle mechanics. A conceivable way out of the difficulty is to remember that the continuum picture is itself a model of a non-continuum situation. Then if the internal forces between pairs of particles in real materials are equal in magnitude but opposite in sign, it is only reasonable to incorporate into the continuum model the assumption that these forces are not relevant to the laws governing total linear and angular momentum. The difficulties are not entirely eliminated because the 'real' particles of most materials are governed by the laws of quantum rather than classical mechanics and the forces between pairs of particles are not necessarily central. The only entirely satisfactory way out of the difficulty mathematically is to lay down as postulates the momentum balance laws of Section 3.2.]

† The subscript t signifies the time dependence of S_t and Ω_t.

3.1.2 EXTERNAL FORCES

Most external forces which act on particles in classical mechanics are proportional to the particle mass. For example, if g denotes the vector representing the acceleration due to the earth's gravitational field, then the associated force on a particle of mass m is mg. We assume here that the external forces which can act on the material of a continuum are likewise proportional to mass. Explicitly for the external force $d\mathbf{F}_e$ acting on the elementary volume $d\Omega$ we write

$$d\mathbf{F}_e = \rho\mathbf{b}d\Omega \tag{3.1.1}$$

where **b** is a vector which in general may be a function of both **r** and t. Usually **b** is a known quantity. Commonly the only significant external force in most applications of continuum mechanics is the local earth's gravitational field; for this case

$$\mathbf{b} = \mathbf{g}$$

and in almost all applications it is adequate to assume g is constant.
Commonly **b** is known as the 'body force' vector.
From (3.1.1) the sum of the external forces acting on the material contained within S_t is

$$\mathbf{F}_e = \iiint_{\Omega_t} \rho\mathbf{b}d\Omega \tag{3.1.2}$$

3.1.3 SURFACE CONTACT FORCES

It remains to discuss the contact forces acting on the surface S_t of the material contained within S_t. We assume that the contact forces are governed by principles originally enunciated by Euler and Cauchy. The Euler-Cauchy stress principle asserts that the effect of the material outside S_t on the material inside S_t is equipollent to the existence of a force $d\mathbf{F}_s$ acting on each element of surface dS; here $d\mathbf{F}_s$ is given by

$$d\mathbf{F}_s = \mathbf{T}_{(\mathbf{n})}(\mathbf{r},t)dS \tag{3.1.3}$$

where **n** is the outward normal unit vector to dS and where the vector field **T** is a continuous function of **n** as well as being a function of both **r** and t. It will be seen in Chapter 4 that $\mathbf{T}_{(\mathbf{n})}$ is a rather special function of **n**, and this is the reason for the subscript notation.

The quantity $\mathbf{T}_{(\mathbf{n})}$ is known as the **stress vector** and, as indicated immediately above, depends on the orientation of the element of surface **dS** ($\equiv \mathbf{n}dS$) as well as the position of **dS**. It should be said immediately that in general the direction of $\mathbf{T}_{(\mathbf{n})}$ is not that of **n**.

The stress vector may always be resolved into **stress components** normal and parallel to **dS** (i.e., in the direction **n** and in some direction perpendicular to **n** lying in the plane of dS). Commonly the normal stress component is termed tensile or compressive, according to whether it is respectively of positive or negative sign. Irrespective of sign, the component in the plane of dS is termed a shear component of stress.

The assumption that contact forces are described by (3.1.3) is of course a hypothesis. In mathematical terms (3.1.3) asserts that the force $\Delta \mathbf{F}_s$ on an element of surface ΔS necessarily tends to zero as $\Delta S \to 0$, but that the ratio $\Delta \mathbf{F}_s/\Delta S$ tends to a definite limit $\mathbf{T}_{(\mathbf{n})}$. Some physical motivation for the introduction of the quantity $\mathbf{T}_{(\mathbf{n})}$ is provided in the elementary physics of the Hooke's law wire experiment mentioned in Chapter 1 and illustrated by Fig. 1.4. For sufficiently small extensions, experimental results obtained for most metals with the apparatus of Fig. 1.4 are well described by the equation

$$\sigma = Ee \tag{3.1.4}$$

where σ is the ratio (force/area) while e is the ratio (extension/length). Finally E is Young's modulus for the metal in question. For a single wire (3.1.4) is equivalent to the statement

$$\text{Force} \propto \text{extension} \tag{3.1.5}$$

and indeed this was the result given by Hooke. However (3.1.4) is much more informative since, for a given metal (i.e., fixed value of E), it correlates experiments on wires of different lengths and cross sections. Thus from elementary physics it is known that the concept (force/area) is a profitable one. In fact for the experiment of Fig. 1.4 the quantity σ is the (here uniform) magnitude of the stress vector which acts on all cross sections perpendicular to the wire axis. (In this case and for these cross sections, $\mathbf{T}_{(\mathbf{n})}$ is parallel to $\mathbf{n}$.)

As indicated above the stress vector is not an arbitrary function of $\mathbf{n}$. In fact as will be seen subsequently (Chapter 4) $\mathbf{T}_{(\mathbf{n})}$ is completely specified at all points in the body, and for all orientations $\mathbf{n}$ at each point by six functions of $\mathbf{r}$ and t. For the moment we note the result (due to Cauchy)

$$\mathbf{T}_{(\mathbf{n})} = -\mathbf{T}_{(-\mathbf{n})} \tag{3.1.6}$$

which may be regarded as a slight generalisation of Newton's third law. Alternatively (3.1.6) follows from the laws of motion discussed below (Section 3.2).

From (3.1.3) the sum of the contact forces acting on the material within S_t is given by

$$\mathbf{F}_s = \iint_{S_t} \mathbf{T}_{(\mathbf{n})}(\mathbf{r},t)dS \ . \tag{3.1.7}$$

3.2 Integral forms of the equations of motion

3.2.1 Balance of Linear Momentum

The balance law of linear momentum states that the sum of the body forces (3.1.2) together with the sum of the contact forces (3.1.7) is equal to the rate of change of the linear momentum of the material contained within S_t. The resulting vector equation is

$$\iint_{S_t} \mathbf{T}_{(\mathbf{n})}(\mathbf{r},t)dS + \iiint_{\Omega_t} \rho\mathbf{b}(\mathbf{r},t)d\Omega = \frac{d}{dt}\iiint_{\Omega_t} \rho\mathbf{v}(\mathbf{r},t)d\Omega \tag{3.2.1}$$

It is convenient to transform the right hand side of (3.2.1) into the form

$$\iiint_{\Omega_t} \rho\mathbf{f}(\mathbf{r},t)d\Omega$$

where **f** is the (spatial) acceleration vector. The result is not immediately obvious since in general Ω_t is time dependent and ρ is not constant. The result is most simply obtained by re-writing the momentum integral in the form

$$\iiint_{\Omega_t} \rho\mathbf{v}d\Omega = \iiint_{\Omega_0} \rho_0(\mathbf{R})\mathbf{V}(\mathbf{R},t)d\Omega_0 \tag{3.2.2}$$

where Ω_0 is the volume of the material contained in S_t in the reference configuration. Equation (3.2.2) follows from changing the integration variables from x, y and z to X, Y and Z, and recalling (2.2.11). On differentiating (3.2.2) with respect to time we find immediately (since ρ_0 and Ω_0 are independent of time)

$$\frac{d}{dt}\iiint_{\Omega_t} \rho\mathbf{v}d\Omega = \iiint_{\Omega_0} \rho_0(\mathbf{R})\mathbf{F}(\mathbf{R},t)d\Omega_0 \tag{3.2.3}$$

where $\mathbf{F}(\equiv \partial\mathbf{V}/\partial t)$ is the material acceleration. Finally on transforming back to spatial variables we derive.

$$\frac{d}{dt}\iiint_{\Omega_t} \rho\mathbf{v}d\Omega = \iiint_{\Omega_t} \rho\mathbf{f}(\mathbf{r},t)d\Omega\ . \tag{3.2.4}$$

From (3.2.1) and (3.2.4)

$$\iint_{S_t} \mathbf{T}_{(\mathbf{n})}(\mathbf{r},t)dS + \iiint_{\Omega_t} \rho\mathbf{b}d\Omega = \iiint_{\Omega_t} \rho\mathbf{f}(\mathbf{r},t)d\Omega \tag{3.2.5}$$

which is an alternate integral form of the linear momentum balance law.

Equation (3.2.5) is used to derive the basic partial differential equations for the motion of a continuum (Sections 3.4 and 4.4).

Cauchy's relation (3.1.6) may be derived from (3.2.5) by considering Ω_t to be a small disc shaped element of plane area ΔS and uniform thickness Δh [Fig. 3.1]. We denote the normals to the two plane surfaces by **n** and −**n**. On evaluating the integrals in (3.2.5) with the help of the mean value theorem, there results

$$(\mathbf{T}_{(\mathbf{n})} + \mathbf{T}_{(-\mathbf{n})})\Delta S + O(\Delta S^{1/2}\Delta h) + \rho\mathbf{b}\Delta S\Delta h = \rho\mathbf{f}\Delta S\Delta h \quad (3.2.6)$$

where the quantities $\mathbf{T}_{(\mathbf{n})}$, $\mathbf{T}_{(-\mathbf{n})}$, $\rho\mathbf{b}$ and $\rho\mathbf{f}$, assumed continuous, are evaluated at (usually different) points all lying within or on the disc. The terms not identified explicitly in (3.2.6) derive from integrals of the stress vector over the peripheral surfaces of thickness Δh. On dividing through (3.2.6) by ΔS and proceeding to the limit $\Delta h \to 0$ and $\Delta S \to 0$, equation (3.1.6) follows provided

$$\underset{\substack{\Delta h \to 0 \\ \Delta S \to 0}}{Lt} \; [\Delta h/(\Delta S)^{1/2}] \to 0 \quad .$$

This latter requirement is easily arranged, for example, by choosing at all stages of the limiting process

$$\Delta h = k(\Delta S)^{(1/2 + \epsilon)}$$

where k and ϵ are positive constants.

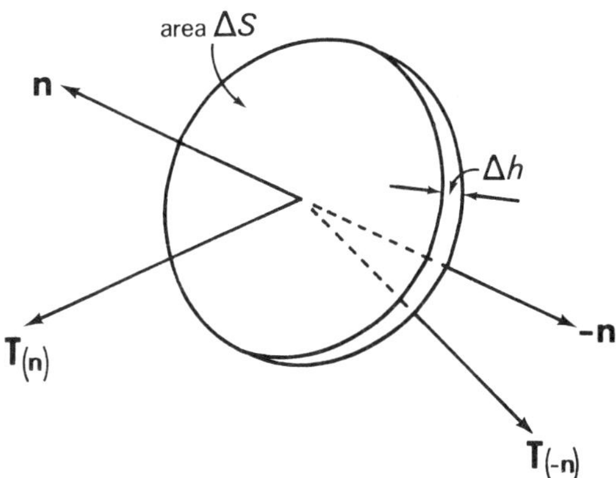

Fig. 3.1 Stress vectors $\mathbf{T}_{(\mathbf{n})}$ and $\mathbf{T}_{(-\mathbf{n})}$ acting on flat surfaces of laminar disc shaped element

3.2.2 Balance of Angular Momentum

The balance law of angular momentum states that the total couple exerted by the body forces and contact forces is equal to the rate of change of the angular momentum contained within S_t. The couple exerted by an element of force $d\mathbf{F}$ is $\mathbf{r} \wedge d\mathbf{F}$ while the angular momentum of the volume element $d\Omega$ is $\rho(\mathbf{r} \wedge \mathbf{v})d\Omega$. From these results and equations (3.1.1) and (3.1.3), the mathematical formulation of the angular momentum balance law is

$$\iint_{S_t} [\mathbf{r} \wedge \mathbf{T}_{(\mathbf{n})}]\, dS + \iiint_{\Omega_t} \rho(\mathbf{r} \wedge \mathbf{b})d\Omega = \frac{d}{dt}\iiint_{\Omega_t} \rho(\mathbf{r} \wedge \mathbf{v})d\Omega \; . \quad (3.2.7)$$

As in the case of (3.2.1), the derivative d/dt on the right hand side of (3.2.7) may be transferred into the integrand. Using exactly the same transformation of variables as before [so that here $\rho(\mathbf{r} \wedge \mathbf{v})d\Omega$ becomes $\rho_0(\mathbf{r} \wedge \mathbf{V})d\Omega_0$ – in which only $\mathbf{V}$ and $\mathbf{r}$ depend on time) yields

$$\iint_{S_t} (\mathbf{r} \wedge \mathbf{T}_{(\mathbf{n})})dS + \iiint_{\Omega_t} \rho(\mathbf{r} \wedge \mathbf{b})d\Omega = \iiint_{\Omega_t} \rho(\mathbf{r} \wedge \mathbf{f})d\Omega \tag{3.2.8}$$

for an alternate integral form of the angular momentum balance equation. In deriving (3.2.8) use is made of $d\mathbf{r}/dt = \mathbf{V}$ and $\mathbf{V} \wedge \mathbf{V} = 0$.

The consequences of (3.2.8) are exploited in Section 3.4 and Section 4.4.

3.3 Cartesian components of stress

As mentioned in Section 3.1 the stress vector $\mathbf{T}_{(\mathbf{n})}$ is not an arbitrary function of $\mathbf{n}$. In fact if the components of $\mathbf{T}_{(\mathbf{n})}$ are known for three distinct non-coplaner values of $\mathbf{n}$, (say $\mathbf{n}_1, \mathbf{n}_2, \mathbf{n}_3$) then $\mathbf{T}_{(\mathbf{n})}$ may be determined for all $\mathbf{n}$. Thus $\mathbf{T}_{(\mathbf{n})}$ is expressible in terms of (at most) nine components of stress. In fact as we shall see below, there are only six independent components of stress.

The choice $\mathbf{n}_1 = \mathbf{i}$, $\mathbf{n}_2 = \mathbf{j}$, $\mathbf{n}_3 = \mathbf{k}$, where $\mathbf{i}$, $\mathbf{j}$ and $\mathbf{k}$ are the unit vectors directed along the Cartesian axes, lead to nine stress components defined by the equations

$$\begin{aligned} \mathbf{T}_{(\mathbf{i})} &= \sigma_{xx}\mathbf{i} + \sigma_{xy}\mathbf{j} + \sigma_{xz}\mathbf{k} \\ \mathbf{T}_{(\mathbf{j})} &= \sigma_{yx}\mathbf{i} + \sigma_{yy}\mathbf{j} + \sigma_{yz}\mathbf{k} \\ \mathbf{T}_{(\mathbf{k})} &= \sigma_{zx}\mathbf{i} + \sigma_{zy}\mathbf{j} + \sigma_{zz}\mathbf{k} \end{aligned} \tag{3.3.1}$$

which express the three stress vectors in question in terms of their Cartesian components.

In equations (3.3.1) the nine quantities σ_{xx}, σ_{xy} etc., are known as Cartesian components of stress. The rules governing the determination of the algebraic suffixes is as follows. The first suffix refers to the direction of the normal to the surface on which the stress vector acts; the second suffix refers to the particular component of the stress vector. Thus for example σ_{zy} is the component in the y direction of the stress vector acting on the surface whose outward normal is in the positive z direction. Fig. 3.2 shows the sense in which the various stress components are measured by reference to a rectangular parallelopiped, whose sides lie parallel to the coordinate axes.

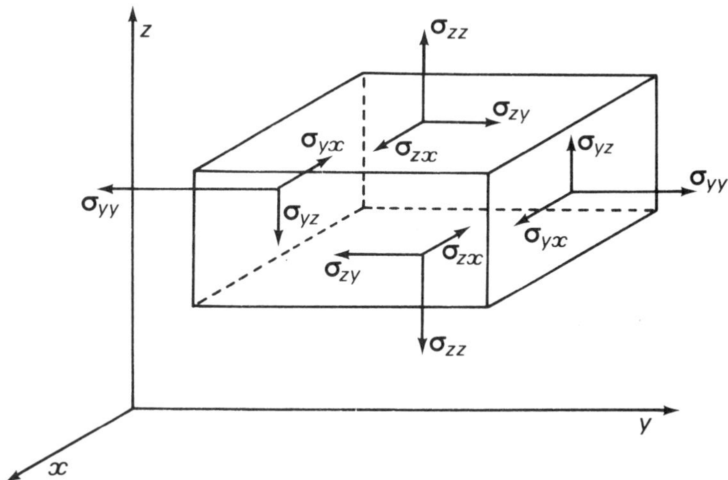

Fig. 3.2 Stress components acting on faces of a rectangular parallelopiped whose sides are parallel to the coordinate axes. For clarity the stress components acting on the surfaces with normals in the ± x direction are omitted.

Pairs of opposite faces of the rectangular parallelopiped have outward normals in opposite directions. Consider the two faces whose outward normals lie in the $\pm y$ directions. The stress vectors on the two surfaces are $\mathbf{T}_{(\mathbf{j})}$ and $\mathbf{T}_{(-\mathbf{j})}$ and these are of opposite sign; this result is reflected in Fig. 3.2.

The quantities σ_{xx}, σ_{yy}, σ_{zz} are normal components of stress while the remaining quantities σ_{xy}, σ_{yz} etc., are shear components of stress.

On occasions it is useful to think of the nine quantities σ_{xx}, σ_{xy} etc., as the components of a second order square matrix $\boldsymbol{\sigma}$ given by

$$\boldsymbol{\sigma} = \begin{bmatrix} \sigma_{xx} & \sigma_{xy} & \sigma_{xz} \\ \sigma_{yx} & \sigma_{yy} & \sigma_{yz} \\ \sigma_{zx} & \sigma_{zy} & \sigma_{zz} \end{bmatrix} \tag{3.3.2}$$

and we shall make use of this representation in Chapter 4.

The question of expressing $\mathbf{T}_{(\mathbf{n})}$ for arbitrary $\mathbf{n}$ in terms of σ_{xx}, σ_{xy} etc., is deferred until the next Chapter.

3.4 Equations of motion

With the representation (3.3.1) of the stress vectors, an elementary derivation of the partial differential equations of motion of a continuum is obtained as follows. In the integral equation of motion (3.2.5), we suppose Ω_t to be an elementary rectangular parallelopiped with sides of length Δx, Δy, Δz. (Fig. 3.2.)

Consider now the x component of (3.2.5). From (3.3.1), contributions to the surface integral arise from integrals of σ_{xx} over the two surfaces of area $\Delta y\,\Delta z$, from integrals of σ_{yx} over the two surfaces of area $\Delta x\,\Delta z$ and from integrals of σ_{zx} over the two surfaces of area $\Delta y\,\Delta x$. Remembering the result $\mathbf{T}_{(\mathbf{n})} = -\mathbf{T}_{(-\mathbf{n})}$, so that contributions from opposite pairs of faces are of opposite sign (see also Fig. 3.3), we obtain for the x component of the first term in (3.2.5)

$$\iint_{\Delta y \Delta z} [\sigma_{xx}(x + \Delta x, y, z, t) - \sigma_{xx}(x, y, z, t)]\, dydz +$$

$$\iint_{\Delta x \Delta z} [\sigma_{yx}(x, y + \Delta y, z, t) - \sigma_{yx}(x, y, z, t)]\, dxdz +$$

$$\iint_{\Delta x \Delta y} [\sigma_{zx}(x, y, z + \Delta z, t) - \sigma_{zx}(x, y, z, t)]\, dxdy \; . \tag{3.4.1}$$

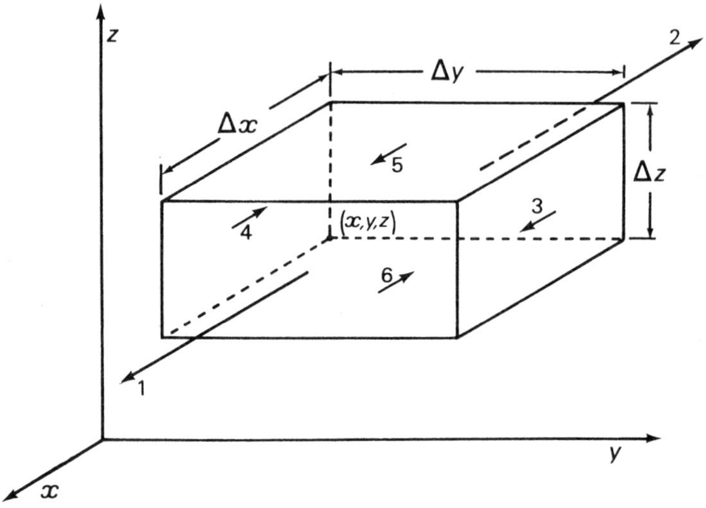

(1) $\displaystyle\iint_{\Delta y \Delta z} \sigma_{xx}(x + \Delta x, y, z)dydz$ (2) $\displaystyle\iint_{\Delta y \Delta z} \sigma_{xx}(x, y, z)dydz$

(3) $\displaystyle\iint_{\Delta x \Delta z} \sigma_{yx}(x, y + \Delta y, z)dxdz$ (4) $\displaystyle\iint_{\Delta x \Delta z} \sigma_{yx}(x, y, z)dxdz$

(5) $\displaystyle\iint_{\Delta x \Delta y} \sigma_{zx}(x, y, z + \Delta z)dxdy$ (6) $\displaystyle\iint_{\Delta x \Delta y} \sigma_{zx}(x, y, z)dxdy$

Fig. 3.3 Surface forces acting in x direction on faces of rectangular parallopiped

The first of these integrals may be evaluated if σ_{xx} is a differentiable function of x. From the mean value theorem for derivatives

$$\sigma_{xx}(x + \Delta x, y, z, t) - \sigma_{xx}(x, y, z, t) = \frac{\partial \sigma_{xx}}{\partial x} \Delta x \qquad (3.4.2)$$

where on the right side of (3.4.2), the derivative is evaluated at some point in the interval $x,\ x + \Delta x$. Integrating (3.4.2) over the surface of area $\Delta y \Delta z$ and applying the mean value theorem for integrals leads to

$$\iint_{\Delta y \Delta z} [\sigma_{xx}(x + \Delta x, y, z, t) - \sigma_{xx}(x, y, z, t)]\, dydz = \frac{\partial \sigma_{xx}}{\partial x} \Delta x \Delta y \Delta z$$

where $\partial \sigma_{xx}/\partial x$ is evaluated at some point within the rectangular parallelopiped.

Similar considerations apply to the remaining two integrals in (3.4.1) so that the x component of the surface integral in (3.2.5) becomes

$$\left(\frac{\partial \sigma_{xx}}{\partial x} + \frac{\partial \sigma_{yx}}{\partial y} + \frac{\partial \sigma_{zx}}{\partial z}\right) \Delta x \Delta y \Delta z\ . \qquad (3.4.3)$$

In (3.4.3) the differentials are evaluated in general at different points inside the parallelopiped. Ultimately in the limiting process in which Δx, Δy and Δz tend to zero, these points necessarily coalesce.

The x components of the volume integrals appearing in (3.2.5) are readily evaluated directly from the mean value theorem for integrals,

$$\iiint_{\Omega_t} \rho(b_x - f_x) d\Omega = \rho(b_x - f_x) \Delta x \Delta y \Delta z$$

where again $\rho(b_x - f_x)$ is evaluated at some point within the parallelopiped. Combining these results leads to

$$\left(\frac{\partial \sigma_{xx}}{\partial x} + \frac{\partial \sigma_{yx}}{\partial y} + \frac{\partial \sigma_{zx}}{\partial z}\right) \Delta x \Delta y \Delta z + \rho(b_x - f_x) \Delta x \Delta y \Delta z = 0\ .$$

Cancelling the factor $\Delta x \Delta y \Delta z$ and proceeding to the limit $\Delta x \to 0$, $\Delta y \to 0$, $\Delta z \to 0$ leads finally to the partial differential equation

$$\frac{\partial \sigma_{xx}}{\partial x} + \frac{\partial \sigma_{yx}}{\partial y} + \frac{\partial \sigma_{zx}}{\partial z} + \rho b_x = \rho f_x \qquad \left(\begin{array}{c}\text{equation of motion}\\ \text{in } x \text{ direction}\end{array}\right) \qquad (3.4.4)$$

for the equation of motion in the x direction.

A similar analysis for the y and z components of (3.2.5) yields two further equations

$$\frac{\partial \sigma_{xy}}{\partial x} + \frac{\partial \sigma_{yy}}{\partial y} + \frac{\partial \sigma_{zy}}{\partial z} + \rho b_y = \rho f_y \quad \begin{pmatrix}\text{equation of motion}\\ \text{in } y \text{ direction}\end{pmatrix} \tag{3.4.5}$$

$$\frac{\partial \sigma_{xz}}{\partial x} + \frac{\partial \sigma_{yz}}{\partial y} + \frac{\partial \sigma_{zz}}{\partial z} + \rho b_z = \rho f_z \quad \begin{pmatrix}\text{equation of motion}\\ \text{in } z \text{ direction}\end{pmatrix} \tag{3.4.6}$$

Equations (3.4.4), (3.4.5) and (3.4.6) are partial differential equations for the motion of any continuum. They have been derived on the assumption that the stress components are differentiable functions of position with continuous derivatives, and that $\rho(\mathbf{b} - \mathbf{f})$ is continuous.

Equations (3.4.4), (3.4.5) and (3.4.6) look complicated initially. However there is a clear structure to the equations. In each case the second suffix attached to the stress components is attached also to the body force and acceleration components. The remaining suffixes appearing in the stress components run through x, y and z, as do the associated partial derivatives. This structure is made clearer by the use of the succint notation developed in the next chapter for handling the three scalar equations of motion (and other equations).

It remains to discuss the consequences of the angular momentum equation (3.2.8). Again Ω_t is chosen to be a rectangular parallelopiped with sides $\Delta x, \Delta y, \Delta z$. However additionally we make use of the fact that the origin of coordinates may be chosen arbitrarily. Here it is convenient to choose the coordinates axes as shown in Fig. 3.4, with the origin lying at one corner of the parallelopiped and with axes directed along three adjacent edges.

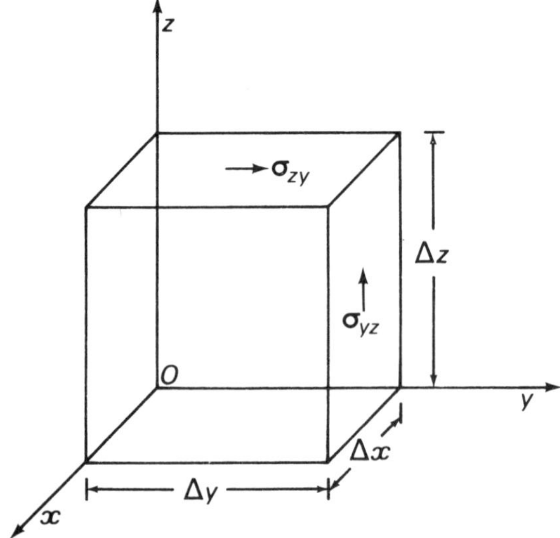

Fig. 3.4 Diagram for proof of $\sigma_{yz} = \sigma_{zy}$

The surface integral in (3.2.8) is the total couple exerted by the surface stresses on the material contained within S_t. Consider in Fig. 3.4 the component of the couple exerted by the surface forces about the x axis. Dominant contributions arise from the stress components σ_{zy} and σ_{yz} acting (as indicated in the figure) on the surfaces located respectively at $z = \Delta z$ and $y = \Delta y$. From Fig. 3.4 the joint contribution of these two stress components to the couple about the x axis is

$$[\sigma_{zy}\Delta x\Delta y]\Delta z - [\sigma_{yz}\Delta x\Delta z]\Delta y = (\sigma_{zy} - \sigma_{yz})\Delta x\Delta y\Delta z \ .$$

In this result σ_{zy} and σ_{yz} are evaluated at points within the respective surfaces $z = \Delta z$ and $y = \Delta y$.

There are further contributions to the x component of the couple about the x axis, e.g., the shear stress σ_{xy} acting on the surfaces $x = 0, x = \Delta x$ leads to a total contribution

$$\iint_{\Delta y\Delta z} [\sigma_{xy}(x + \Delta x, y, z, t) - \sigma_{xy}(x, y, z, t)]z\,dy\,dz = \frac{1}{2}\frac{\partial \sigma_{xy}}{\partial x}\Delta x\Delta y(\Delta z)^2$$

where the derivative is evaluated at some point within the parallelopiped. Clearly this contribution is of higher order than the dominant contribution. Similarly for all remaining contributions; thus the x component of the couple about the x axis may be written

$$\left(\iint_{S_t}[\mathbf{r} \wedge \mathbf{T}_{(\mathbf{n})}]\, dS\right)_x = (\sigma_{zy} - \sigma_{yz})\Delta x\Delta y\Delta z + O(\Delta x\Delta y\Delta z^2, \text{etc.}) \ . \qquad (3.4.7)$$

The volume integrals in (3.2.8) are also of higher order than $\Delta x\Delta y\Delta z$, since the integrands contain the vector $\mathbf{r}$ whose magnitude is necessarily restricted to be an infinitesimal. It follows from these results that

$$(\sigma_{zy} - \sigma_{yz})\Delta x\Delta y\Delta z = O(\Delta x\Delta y\Delta z^2, \text{etc.})$$

so that on dividing through by $\Delta x\Delta y\Delta z$, and proceeding to the limit $\Delta x \to 0$, $\Delta y \to 0$, $\Delta z \to 0$, we derive the simple result

$$\sigma_{zy} = \sigma_{yz} \ . \qquad (3.4.8)$$

Similar arguments lead to

$$\sigma_{xy} = \sigma_{yx} \ , \quad \sigma_{xz} = \sigma_{zx} \ . \qquad (3.4.9)$$

The important results (3.4.8) and (3.4.9) show that the matrix σ is symmetric, and that the number of independent components of Cartesian stress is six.

Because of the symmetry conditions (3.4.8) and (3.4.9) it is permissible to interchange the σ suffixes in the equations of motion (3.4.4), (3.4.5) and (3.4.6). Conventionally, the equations of motion are written in this alternate form as in the Summary at the end of this chapter.

The equations of motion (3.4.4) - (3.4.6) and the symmetry conditions (3.4.8) and (3.4.9) exhaust all the information obtainable from the general principles of mechanics. Since these principles are sufficient for the solution of problems in particle mechanics and rigid body dynamics, it is worth enquiring whether the continuum equations obtained so far are adequate to solve problems in deformable media. The problem is discussed in the material description.

To specify the motion in the material description necessitates knowledge of three functions $x(\mathbf{R},t)$, $y(\mathbf{R},t)$ and $z(\mathbf{R},t)$ which may be taken as the basic dependent variables. In terms of x, y and z it is possible in principle to derive the quantities ρ and $\mathbf{f}$ appearing in the equations of motion. Since in the latter $\mathbf{b}$ is to be regarded as known, the three equations of motion contain nine unknowns, namely six independent components of stress (σ_{xx}, σ_{yy}, σ_{zz}, σ_{xy}, σ_{yz}, σ_{zx}) and the three quantities $x(\mathbf{R},t)$, $y(\mathbf{R},t)$, $z(\mathbf{R},t)$.

There is a deficiency of six equations which must be sought from a source other than general mechanical principles. The six equations in question depend on the mechanical properties of the continuum and relate the stress components to the deformation of the material. Such equations are known as constitutive equations and are different for different materials. Unlike the equations of motion which are valid for all materials, the constitutive equations are of very limited validity. Even for one material it may be necessary to postulate different types of constitutive equation in different circumstances (as for example is apparent from comparing the mechanical properties of H_2O at $-20°C$, $50°C$ and $150°C$).

Constitutive equations for various materials are discussed in many chapters of the present book. A general discussion is necessarily postponed until we have considered geometrical aspects of the deformation. However it is worth writing down one set of constitutive equations (as an illustration of the type of relation involved). The equations are those for an 'ideal fluid' and are

$$\sigma_{xx} = \sigma_{yy} = \sigma_{zz} = -P(\rho,T); \ \sigma_{xy} = \sigma_{yz} = \sigma_{zx} = 0 \qquad (3.4.10)$$

where $P(\rho,T)$, the pressure, is a known function of the density ρ and temperature T. Many liquids and gases are well described by equations (3.4.10). For most liquids P is temperature insensitive and is adequately approximated by

$$P = K(\rho - \rho_0)/\rho_0$$

where ρ_0 is the density in the stress free state, and K is the bulk modulus. For the simplest gases

$$P = R\rho T/m \qquad (3.4.11)$$

where R is the gas constant, m is the molecular weight and T is the (absolute) temperature. [As indicated in Chapter 1 if T is not constant, the equations of motion and constitutive equations require further supplementation. The additional equation, required to account for the introduction of the new variable T, is the energy balance equation. Exceptionally under so called 'adiabatic conditions', when density changes occur sufficiently rapidly to permit neglect of heat transfer within the gas, T may be determined in terms of ρ (and thereby eliminated from the problem). Under these circumstances the pressure-density relation for a gas is of the form

$$P = k\rho^{\gamma} \quad (k, \gamma \text{ constants}) \tag{3.4.12}$$

in place of (3.4.11)].

Summary

The important results of this chapter are the equations of motion expressed in terms of the Cartesian stress components, and the symmetry of the stress components in respect of the suffixes.

Equations of Motion

$$\frac{\partial \sigma_{xx}}{\partial x} + \frac{\partial \sigma_{xy}}{\partial y} + \frac{\partial \sigma_{xz}}{\partial z} + \rho b_x = \rho f_x$$

$$\frac{\partial \sigma_{yx}}{\partial x} + \frac{\partial \sigma_{yy}}{\partial y} + \frac{\partial \sigma_{yz}}{\partial z} + \rho b_y = \rho f_y$$

$$\frac{\partial \sigma_{zx}}{\partial x} + \frac{\partial \sigma_{zy}}{\partial y} + \frac{\partial \sigma_{zz}}{\partial z} + \rho b_z = \rho f_z \ .$$

Symmetry of Stress Components

$$\sigma_{xy} = \sigma_{yx}, \ \sigma_{xz} = \sigma_{zx}, \ \sigma_{yz} = \sigma_{zy} \ .$$

In writing down the partial differential equations of motion immediately above, which differ slightly from equations (3.4.4) - (3.4.6), use has been made of the symmetry of the stress components.

Problems, Chapter 3

1. Verify that the following stress fields satisfy the equilibrium equations in the absence of body forces

(a) $\sigma_{xx} = \sigma_{yy} = \sigma_{zz} = \sigma_{xy} = 0, \ \sigma_{xz} = -Ay(x^2 + y^2)^{-1},$

$\sigma_{yz} = Ax(x^2 + y^2)^{-1}$

(b) $\sigma_{xx} = -By(3x^2 + y^2)(x^2 + y^2)^{-2}, \ \sigma_{yy} = By(x^2 - y^2)(x^2 + y^2)^{-2},$

$\sigma_{xy} = Bx(x^2 - y^2)(x^2 + y^2)^{-2}, \ \sigma_{zz} = \sigma_{zx} = \sigma_{zy} = 0$

where A and B are constants.

2. 'Two dimensional' stress situations are specified by $\sigma_{xz} = \sigma_{yz} = 0$, while $\sigma_{xx}, \sigma_{yy}, \sigma_{zz}$ and σ_{xy} are functions only of x and y, e.g. as in Problem 1b above. Show that in the absence of body forces the equations of equilibrium are satisfied by choosing

$$\sigma_{xx} = \partial^2\chi/\partial y^2 \ , \quad \sigma_{yy} = \partial^2\chi/\partial x^2 \ , \quad \sigma_{xy} = -\partial^2\chi/\partial x\partial y$$

where $\chi = \chi(x, y)$ is usually known as the 'Airy' stress function.

Verify that the stress field of Problem 1b above is given by

$$\chi = -\tfrac{1}{2}By \log(x^2 + y^2) \ .$$

[It is to be noted that while an arbitrary differentiable function χ satisfies the equilibrium equations, the quantity χ is not arbitrary but satisfies an equation which is determined by the mechanical properties of the continuum. For example, for a classical elastic solid, χ satisfies the biharmonic equation

$$\frac{\partial^4\chi}{\partial x^4} + \frac{\partial^4\chi}{\partial y^4} + 2\frac{\partial^4\chi}{\partial x^2\partial y^2} = 0$$

(see Problem 17 of Chapter 14).]

3. A rigid body of uniform density ρ_0 and in the form of a rectangular parallelopiped

$$0 \leqslant X \leqslant a, \ 0 \leqslant Y \leqslant b, \ 0 \leqslant Z \leqslant c$$

moves in translation in the z direction so that

$$x = X, \quad y = Y, \quad z = Z + \alpha(t) \ .$$

The body force vector due to the earth's gravitational field is $\mathbf{b} = (0,0,g)$ where g is a constant. There are no external forces acting on the faces $X = 0, \ X = a$, $Y = 0, \ Y = b$ while forces normal to the faces $Z = 0, \ Z = c$, given respectively by $[0,0,F_0(t)]$ and $[0,0,F_c(t)]$, are distributed uniformly over the faces.

Show that all the conditions on $X = 0, \ X = a, \ Y = 0, \ Y = b$ are met by the assumption $\sigma_{xx} = \sigma_{yy} = \sigma_{xy} = \sigma_{xz} = \sigma_{yz} = 0$ (everywhere in the body) and that the equations of motion reduce to

$$\frac{\partial\sigma_{zz}}{\partial Z} = \rho_0(\ddot{\alpha} - g)$$

with solution $$\sigma_{zz} = \rho_0(\ddot{\alpha} - g)Z + A(X, Y, t)$$

where A is an arbitrary function of the indicated variables.

Show that the conditions on $Z = 0$ are met by choosing A independent of X and Y,

i.e. $$A = -F_0/ab$$

and deduce that $$\ddot{\alpha} = g + (F_0 + F_c)/\rho_0 abc \ .$$

Is this result in concurrence with the predictions of rigid body mechanics?

4. A perfect fluid is characterised by the assumptions

$$\sigma_{xy} = \sigma_{yz} = \sigma_{zx} = 0 \, , \quad \sigma_{xx} = \sigma_{yy} = \sigma_{zz} = -P(\rho, T)$$

where P (pressure) is a function of density ρ and temperature T.

Show that the equations of motion may be expressed in the vector form

$$-\frac{\partial P}{\partial \rho} \operatorname{grad} \rho - \frac{\partial P}{\partial T} \operatorname{grad} T + \rho \mathbf{b} = \rho \mathbf{f} \, .$$

The behaviour of water at a fixed temperature is well modelled by the assumption

$$P = K(\rho - \rho_0)/\rho_0$$

where ρ_0 is the density in the absence of pressure and K (bulk modulus) is a constant. Show that in these circumstances if $\mathbf{b} = (0,0,g)$ is the body force due to gravity and there is no motion, then ρ depends only on z and satisfies the equation

$$K\frac{d\rho}{dz} = \rho_0 \rho g \, .$$

If $\rho = \rho_0$ at $z = 0$, deduce that

$$\rho = \rho_0 \exp\{\rho_0 g z/K\} \, , \quad P = K[\exp\{\rho_0 g z/K\} - 1]$$

and derive the corresponding results for an incompressible liquid ($K \to \infty$).

A lake of depth 1000 metres is subjected to (negligible) atmospheric pressure at the surface $z = 0$. If the bulk modulus of water is 2×10^9 N/m^2, $g \simeq 10$ m/s^2 and $\rho_0 = 10^3$ kg/m^3 show that the maximum error perpetrated in estimating the pressure by the assumption of incompressibility is about $\frac{1}{4}$%.

5. The gaseous atmosphere surrounding the earth is well modelled by the assumption of a perfect gas,

$$\sigma_{xx} = \sigma_{yy} = \sigma_{zz} = -P(\rho, T) \, ; \quad \sigma_{xy} = \sigma_{yz} = \sigma_{zx} = 0$$

where the pressure is related to density and (absolute) temperature through the equation

$$P = R\rho T/m \, .$$

Here R is the gas constant and m the mean molecular weight.

Show that in the absence of motion the pressure at height z is given by

$$P = P_0 \exp\left\{ -mgR^{-1} \int_0^z [T(z')]^{-1} dz' \right\}$$

where g is the (constant) acceleration due to gravity, P_0 is the air pressure at ground level and the temperature $T(z)$ varies with height.

Show that if T varies with x, y and z, no hydrostatic solution is possible, i.e. that there must occur motion of the air.

Chapter **4**

Suffix Notation, Cartesian Tensors and Matrices

4.1 Suffix notation

The concept of Cartesian tensors plays an important role in the equations of continuum mechanics. As a necessary preliminary to defining a Cartesian tensor, we introduce some minor changes in notation.

The notational changes involved associate the numbers 1, 2 and 3 respectively with the x, y and z directions. In place of x, y and z themselves we write

$$x \equiv x_1 , \quad y \equiv x_2 , \quad z \equiv x_3$$

and similarly for the reference coordinates

$$X \equiv X_1 , \quad Y \equiv X_2 , \quad Z \equiv X_3 .$$

Similar changes of notation are employed for other variables. For example the component of the vector **v** in the y direction, i.e. the x_2 direction, is denoted by v_2, while the normal component of stress acting on the surface whose outward normal lies in the positive z (or x_3) direction is denoted by σ_{33}. Similarly σ_{21} denotes the shear component of stress acting in the x (or x_1) direction on a surface whose outward normal points in the positive y (or x_2) direction. No notational changes are necessary for scalar quantities like ρ.

We also adopt the notation

$$\mathbf{i} = \mathbf{i}_1 , \quad \mathbf{j} = \mathbf{i}_2 , \quad \mathbf{k} = \mathbf{i}_3$$

for the unit vectors directed along the positive x_1, x_2 and x_3 axes.

In terms of the new (suffix) notation the matrix of the stress components $\boldsymbol{\sigma}$ becomes

$$\boldsymbol{\sigma} = \begin{bmatrix} \sigma_{11} & \sigma_{12} & \sigma_{13} \\ \sigma_{21} & \sigma_{22} & \sigma_{23} \\ \sigma_{31} & \sigma_{32} & \sigma_{33} \end{bmatrix} . \qquad (4.1.1)$$

An abbreviated notation is

$$\boldsymbol{\sigma} = (\sigma_{ij})$$

in which the matrix brackets enclose a typical element σ_{ij} where i and j take any of the values 1, 2 or 3.

Frequently it is convenient to denote the numerical suffixes by a symbol (as above). Conventionally, and in the present text only lower case English letters are used for suffixes (*i, j, k . . . p, q, r . . .*). Perhaps the simplest general statement that can be made in continuum mechanics using symbols for suffixes is

$$\sigma_{ij} = \sigma_{ji} \tag{4.1.2}$$

which asserts the symmetry of the matrix (4.1.1). In (4.1.2) i and j may assume independently any one of the values 1, 2 or 3. For identical choices of i and j, (4.1.2) is a trivial identity. For different choices of i and j, (4.1.2) implies the three equations (3.4.8 and 3.4.9) derived in the last chapter.

The results (4.1.2) are equally expressed with different choices of suffixes

e.g. $\sigma_{pq} = \sigma_{qp}\,, \quad \sigma_{mn} = \sigma_{nm}$ etc.

where in all cases it is understood that the suffixes p, q, m, n etc., may be assigned any one of the values 1, 2 or 3.

Frequently we shall encounter algebraic expressions involving sums over indices. For example the expression of a vector **v** in component form is given by

$$\mathbf{v} = v_1\mathbf{i}_1 + v_2\mathbf{i}_2 + v_3\mathbf{i}_3 = \sum_{s=1}^{3} v_s\mathbf{i}_s\,. \tag{4.1.3}$$

Similarly the quantity div **v** entails a sum

$$\operatorname{div}\mathbf{v} = \sum_{i=1}^{3} \frac{\partial v_i}{\partial x_i}\,. \tag{4.1.4}$$

Of course when a sum over a suffix is involved, it is no longer possible to assign arbitrarily a value to the suffix in question. This situation arises only when a suffix is repeated in an expression. For example in (4.1.3) the suffix s appears twice while in (4.1.4) the suffix i is duplicated.

In summations like (4.1.3) and (4.1.4), the summation sign, together with the associated symbols indicating which index is involved and also the range of the index, is redundant. The reason is simply because the summation sign, together with the associated symbols, appears in conjunction with a repeated suffix. In these circumstances the mere repetition of the suffix is alone sufficient to imply a sum.

The summations appearing in the general theory of continuum mechanics almost always involve expressions with repeated suffixes as in (4.1.3) and (4.1.4). Throughout the remainder of this book we adopt the universally accepted summation convention whereby in these circumstances (a) we dispense with summation signs and their associated symbols, (b) repetition of a suffix automatically implies a sum over that suffix. [However the repeated suffix associated with a sum is always denoted by an English lower case letter and not by an integer, so that, for example, $\sigma_{ii} \equiv \sigma_{11} + \sigma_{22} + \sigma_{33}$ whereas σ_{33} simply denotes one particular component (namely σ_{zz}) of stress.]

With the adoption of the summation convention, equations (4.1.3) and (4.1.4) assume the forms

$$\mathbf{v} = v_s \mathbf{i}_s, \quad \operatorname{div} \mathbf{v} = \frac{\partial v_i}{\partial x_i} .$$

Some minor penalties, exacted by the use of the convention, are discussed immediately.

Usually summations appear in expressions containing suffixes additional to the repeated ones. For example we shall meet equations like

$$x'_i = a_{ij} x_j \tag{4.1.5}$$

which relate components of the vector **r** in two different systems of Cartesian coordinates. In the same context there appear more complicated equations involving more than one summation as, for example, in the equations

$$\sigma'_{ij} = a_{ir} a_{js} \sigma_{rs} \tag{4.1.6}$$

In (4.1.5) the suffix j is repeated once; in (4.1.6) each of r and s is repeated once.

The repetition of a suffix more than once is prohibited. No exceptions to this rule are allowed so that expressions like

$$a_{mm} x_m x_m \,, \quad a_{mm} b_m \,, \quad a_{ij} a_{ij} x_j x_j \tag{4.1.7}$$

are absolutely prohibited. Quantities like (4.1.7) rarely arise (nor are such quantities easily handled) in tensor calculus, so there are no difficulties in adhering to the rule. It is sometimes stated that the reason for the rule is to avoid ambiguities of interpretation. For example the third expression of (4.1.7) is supposed open to either of the (different) interpretations

$$a_{ij} a_{ij} x_j x_j \equiv \sum_{i=1}^{3} \sum_{j=1}^{3} a_{ij} a_{ij} \, x_j x_j \tag{i}$$

$$a_{ij} a_{ij} x_j x_j \equiv \sum_{i=1}^{3} x'_i x'_i \text{ where } x'_i = a_{ij} x_j \tag{ii}$$

In fact strict adherence to a rule whereby the appearance of multiple suffixes implies summation (no matter how many times the suffix appears) only allows the first of these interpretations, so there is no real ambiguity. The essential point is that the elementary operations of multiplication of quantities involving indices, [as for example in forming the sum $x'_i x'_i$ in (ii) above where x'_i is given by (4.1.5)], do not lead to expressions like the right hand side of (i). It will be seen below that the expression of product sums like $x'_i x'_i$ in terms of the x_i naturally conform to the rule whereby a suffix is only allowed to be repeated once.

A relation like (4.1.5) summarises three equations since the index i, which appears only once on each side of the equation, is free to take any of the values 1, 2 or 3. On the other hand the index j on the right hand side of (4.1.5) is repeated and hence implies a sum over the values $j = 1, 2$ and 3. In full the three equations represented by (4.1.5) are

$$\begin{aligned} x_1' &= a_{11}x_1 + a_{12}x_2 + a_{13}x_3 \quad (\text{Choice } i = 1) \\ x_2' &= a_{21}x_1 + a_{22}x_2 + a_{23}x_3 \quad (\text{Choice } i = 2) \\ x_3' &= a_{31}x_1 + a_{32}x_2 + a_{33}x_3 \quad (\text{Choice } i = 3) \end{aligned} \tag{4.1.8}$$

so that the use of suffix notation together with the summation convention leads to a compact formalism.

Non-repeated suffixes are sometimes known as free suffixes since they are free to assume any of the values 1, 2 or 3, while repeated suffixes are known as dummy suffixes. The origin of the term 'dummy' derives from the possibility of replacing the dummy suffix symbol by any other lower case English letter not already in use. For example equations (4.1.8) are equally represented by

$$x_i' = a_{is}x_s \text{ or } x_i' = a_{il}x_l \text{ etc. .}$$

In fact we are also free to choose different symbols for the free suffixes, again with the restriction that the chosen symbol is not already in use. For example, equations (4.1.8) are also represented by

$$x_p' = a_{pq}x_q \text{ or } x_s' = a_{sr}x_r \quad \text{etc. .}$$

The possibility of changing suffix symbols is much exploited in tensor calculus.

In equations (4.1.6) there are two free suffixes (i and j) and two dummy suffixes. Thus (4.1.6) summarises 9 equations. To write out even one of these in full is cumbersome. For example the choice $i = 1, j = 2$ leads to

$$\begin{aligned} \sigma_{12}' = a_{1r}a_{2s}\sigma_{rs} &= \sum_{r=1}^{3}\sum_{s=1}^{3} a_{1r}a_{2s}\sigma_{rs} \qquad \text{(old notation)} \\ &= a_{11}a_{21}\sigma_{11} + a_{11}a_{22}\sigma_{12} + a_{11}a_{23}\sigma_{13} + \\ &\quad a_{12}a_{21}\sigma_{21} + a_{12}a_{22}\sigma_{22} + a_{12}a_{23}\sigma_{23} + \\ &\quad a_{13}a_{21}\sigma_{31} + a_{13}a_{22}\sigma_{32} + a_{13}a_{23}\sigma_{33} \end{aligned}$$

which demonstrates again the compactness of the notation. Alternative representations of equations (4.1.6) are

$$\sigma_{mn}' = a_{mp}a_{nq}\sigma_{pq}, \quad \sigma_{jk}' = a_{ju}a_{kv}\sigma_{uv} \quad \text{etc. .}$$

In learning to handle equations in suffix notation it is worthwhile checking that each product group of terms in the equation contains the same free suffixes, and that dummy suffixes only occur in pairs. For example the equation

$$\sigma_{ij} = \lambda e_{mm} \delta_{ij} + 2\mu e_{ij}$$

is acceptable, while $\sigma_{ij}\sigma_{js} = \lambda q_{js} + 2\mu e_{mm}$ is nonsense.

Frequently in the course of analysis it is necessary to utilise the possibility of changing the notation of both free and dummy suffixes. For example suppose we are given the two sets of equations

$$p_i = a_{ij} q_j , \quad q_i = b_{ij} s_j$$

and it is required to eliminate the q_i, expressing the p_i in terms of the s_i. It is tempting to write

$$p_i = a_{ij} q_j = a_{ij} b_{jj} s_j \text{ (wrong)} \tag{4.1.9}$$

but the last step is wrong because of the quadruple appearance of j on the right hand side of (4.1.9).† However we may write

$$q_j = b_{jk} s_k$$

so that $p_i = a_{ij} b_{jk} s_k$

which is correct. Note the same free suffix i appears on each side of the equation, while the remaining suffixes on the right side are grouped into two pairs of dummy suffixes.

Computation of the quantity $x_i' x_i'$ (mentioned previously) with x_i' given by (4.1.5) follows a similar pattern

$$x_i' x_i' = a_{ij} x_j a_{ik} x_k \equiv a_{ij} a_{ik} x_j x_k \quad .$$

Here it has been necessary to choose a symbol other than j for the dummy suffix in one of the quantities x_i'. In this example there are no free suffixes.

An important example in suffix notation is provided by the equations of motion of a continuum. In suffix notation these equations, given in the Summary of Chapter 3, assume the form

$$\frac{\partial \sigma_{ij}}{\partial x_j} + \rho b_i = \rho f_i \quad . \tag{4.1.10††}$$

So far it might appear that the virtue of suffix notation, together with the summation convention, is principally one of compactness. Certainly the

† The result (4.1.9) is not wrong simply because the suffix rule has been flaunted. It is mathematically wrong because if the right side of (4.1.9) has any interpretation, it is

$\sum_{j=1}^{3} a_{ij} b_{jj} s_j$ and this is not the correct result!

†† The original form of these equations [(3.4.4) - (3.4.6)] is $\partial \sigma_{ji} / \partial x_j + \rho b_i = \rho f_i$. The form (4.1.10) follows after using the symmetry condition $\sigma_{ij} = \sigma_{ji}$.

succinctness of the notation has been amply demonstrated. However the real value of the notation becomes evident only when the notation is exploited in the context of tensor calculus. For example knowledge that the quantities $\partial\sigma_{ij}/\partial x_j$, ρb_i and ρf_i in (4.1.10) are each components of a tensor of order one conveys immediately the information that although (4.1.10) necessarily makes overt reference to a specific coordinate system, nevertheless the equation is valid for all coordinate systems. In other words in a quite real sense equation (4.1.10) is independent of the choice of coordinate system.

4.2 Definition of Cartesian Tensors

Many of the equations of mechanics and physics are often expressed in vector notation. Familiar examples are (i) the equation of motion of a particle of mass m

$$\frac{d}{dt}\left(m\,\frac{d\mathbf{r}}{dt}\right) = \mathbf{F}(t)$$

where $\mathbf{r}(t)$ is the position vector of the particle and $\mathbf{F}(t)$ the external force, (ii) the force on a charged particle in an electromagnetic field

$$\mathbf{F} = q[\mathbf{E} + \mathbf{v} \wedge \mathbf{B}]$$

where $\mathbf{v}$ is the particle velocity, q the charge, $\mathbf{E}$ the electric field and $\mathbf{B}$ magnetic induction. Numerous other examples are easily found.

The principal reason for writing the equations of physics, or rather some of the equations, in vector form is because it is then evident immediately that such equations are independent of the choice of coordinate system. Indeed the equations in question make no reference to any specific coordinate system. Of course in the solution of specific problems, it is usually necessary to choose a specific coordinate system and deal with the vector equation(s) in component form.

The underlying reason for writing equations in vector form is the assumption that the laws of the material world do not depend in any real sense on the choice of coordinate system. However, unfortunately not all the entities of physics are classifiable as either vectors or scalars. In fact we have already met the quantity σ_{ij} which falls into neither category.

The tensor concept is one which enables us to extend the notion that equations containing quantities like σ_{ij}, while not expressible in vector form, are nevetheless also independent of the choice of coordinate system as in equations like (4.1.10). It is no longer possible to suppress overt reference to the coordinates, but it is possible to write the equations in an 'invariant' form with respect to the coordinates. By 'invariant', we mean that the equation assumes the same form for any choice of coordinate system.

Cartesian tensors are associated with different Cartesian frames of reference in Euclidean space.† As a (final) necessary preliminary to defining Cartesian tensors we consider the relations between the coordinates of two such frames of reference.

Figure 4.1 shows two right handed Cartesian reference frames $0'x_1'x_2'x_3'$ (the primed frame) and $0x_1x_2x_3$ (the unprimed frame). The immediate problem is to derive relations connecting the x_i' with the x_i.

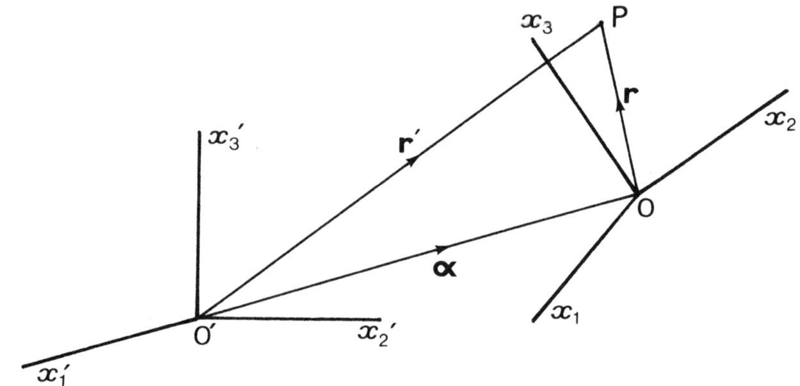

Fig. 4.1 Relation between primed and unprimed coordinate axes

Let the vector joining O′ to O be denoted by $\boldsymbol{\alpha}$ and let $\mathbf{r}$ and $\mathbf{r}'$ denote respectively the position vectors $\overrightarrow{\mathrm{OP}}$, $\overrightarrow{\mathrm{O'P}}$ for some arbitrary point P; evidently

$$\mathbf{r}' = \mathbf{r} + \boldsymbol{\alpha} . \tag{4.2.1}$$

Unit vectors directed along the Cartesian axes of the two systems are denoted by $\mathbf{i}_1, \mathbf{i}_2, \mathbf{i}_3$ (unprimed system) and $\mathbf{i}_1', \mathbf{i}_2', \mathbf{i}_3'$ (primed system). In terms of these unit vectors we write

$$\mathbf{r} = x_p\mathbf{i}_p \ \ (4.2.2)\,, \quad \mathbf{r}' = x_p'\mathbf{i}_p' \ \ (4.2.3)\,, \quad \boldsymbol{\alpha} = \alpha_p'\mathbf{i}_p' = \alpha_p\mathbf{i}_p \ \ (4.2.4)$$

Since the $\mathbf{i}_p$ and the $\mathbf{i}_p'$ form two complete orthogonal sets of unit vectors in Euclidean space, it is possible to express either set in terms of the other. We find

$$\mathbf{i}_p = a_{qp}\mathbf{i}_q' \tag{4.2.5}$$

where

$$a_{qp} = \mathbf{i}_p \, . \, \mathbf{i}_q' \tag{4.2.6}$$

is the cosine of the angle between the $0x_p$ and $0'x'_q$ axes. Similarly

$$\mathbf{i}_p' = a_{pq}\mathbf{i}_q \; . \tag{4.2.7}$$

† General tensor theory is concerned with non-Cartesian frames of reference in both Euclidean and non-Euclidean spaces. The present text is concerned solely with Cartesian tensors.

From equations (4.2.1) - (4.2.5)

$$x'_p \mathbf{i}'_p = \alpha'_p \mathbf{i}'_p + x_p \mathbf{i}_p = \alpha'_p \mathbf{i}'_p + a_{qp} x_p \mathbf{i}'_q = (\alpha'_p + a_{pq} x_q)\mathbf{i}'_p \quad (4.2.8)$$

where in the last step we have interchanged the notation for the indices in the term $a_{qp} x_p \mathbf{i}'_q$. Since the $\mathbf{i}'_p$ are independent unit vectors equation (4.2.8) implies

$$x'_p = \alpha'_p + a_{pq} x_q$$

or equivalently

$$x'_i = \alpha'_i + a_{ij} x_j \; . \quad (4.2.9)$$

If $\boldsymbol{\alpha}$ is expressed in terms of its unprimed components (4.2.9) becomes

$$x'_i = a_{ij}(\alpha_j + x_j) \quad (4.2.10)$$

since from (4.2.4) and (4.2.5) components of $\boldsymbol{\alpha}$ in the two frames are related by

$$\alpha'_i = a_{ij} \alpha_j \; . \quad (4.2.11)$$

Relations inverse to (4.2.9), (4.2.10) and (4.2.11) are easily derived from similar arguments [using (4.2.7) in place of (4.2.5)]. The results are

$$x_i = -\alpha_i + a_{ji} x'_j \quad (4.2.12)$$

$$= a_{ji}[-\alpha'_j + x'_j] \quad (4.2.13)$$

$$\alpha_i = a_{ji} \alpha'_j \; . \quad (4.2.14)$$

Important results concerning the coefficients a_{ij} follow from noting that the square of the length of the vector $\boldsymbol{\alpha}$ is given by either of the formulae

$$|\boldsymbol{\alpha}|^2 (\equiv \boldsymbol{\alpha}.\boldsymbol{\alpha}) = \alpha_i \alpha_i = \alpha'_i \alpha'_i \; .$$

Forming $\alpha'_i \alpha'_i$ from (4.2.11) results in

$$\alpha'_i \alpha'_i = a_{ij} a_{ik} \alpha_j \alpha_k \quad (4.2.15)$$

Since the right side of (4.2.15) is also $\alpha_j \alpha_j$, while $\boldsymbol{\alpha}$ may be an arbitrary vector, it follows that

$$a_{ij} a_{ik} = \begin{cases} 1 & (j = k) \\ 0 & (j \neq k) \end{cases} \quad (4.2.16)$$

Similarly it may be proved that

$$a_{ji} a_{ki} = \begin{cases} 1 & (j = k) \\ 0 & (j \neq k) \end{cases} \quad (4.2.17)$$

It is convenient to define the 'Kronecker delta' given by

$$\delta_{ij} = \begin{cases} 1 & (i = j) \\ 0 & (i \neq j) \end{cases} . \quad (4.2.18)$$

In terms of δ_{ij}, equations (4.2.16) and (4.2.17) may be written

$$a_{ij} a_{ik} = \delta_{jk} \quad (4.2.19a)$$

$$a_{ji} a_{ki} = \delta_{jk} \; . \quad (4.2.19b)$$

Equations (4.2.19) are often known as the orthogonality relations for the quantities a_{ij}.

Equations (4.2.19) play a crucial role in the theory of Cartesian Tensors. For many purposes it is useful to regard the a_{ij} as components of a three by three matrix $\mathbf{A}$

$$\mathbf{A} = (a_{ij}) \equiv \begin{bmatrix} a_{11} & a_{12} & a_{13} \\ a_{21} & a_{22} & a_{23} \\ a_{31} & a_{32} & a_{33} \end{bmatrix}$$

with transpose

$$\mathbf{A}^T = (a_{ji}) = \begin{bmatrix} a_{11} & a_{21} & a_{31} \\ a_{12} & a_{22} & a_{32} \\ a_{13} & a_{23} & a_{33} \end{bmatrix}$$

In terms of $\mathbf{A}$ and $\mathbf{A}^T$, the relations (4.2.19) are summarised by

$$\mathbf{A}\mathbf{A}^T = \mathbf{A}^T\mathbf{A} = \mathbf{I} \tag{4.2.20}$$

where

$$\mathbf{I} = (\delta_{ij}) \equiv \begin{bmatrix} 1 & 0 & 0 \\ 0 & 1 & 0 \\ 0 & 0 & 1 \end{bmatrix}$$

is the unit square matrix of dimensions three.

Since the determinants of $\mathbf{A}$ and $\mathbf{A}^T$ are equal, (4.2.20) shows that $\mathbf{A}$ is a non-singular square matrix with determinant $|\mathbf{A}|$ which satisfies the equation $|\mathbf{A}|^2 = 1$ so that $|\mathbf{A}| = \pm 1$. The following argument determines that the correct choice of sign is positive so that

$$|\mathbf{A}| = 1 \ . \tag{4.2.21}$$

It is possible to conceive a smooth motion of the primed coordinate system axes, carrying the frame from a position coincident initially with the unprimed frame to the current position. For motions sufficiently smooth for the a_{ij} to be continuous functions (of time), $|\mathbf{A}|$ is also continuous. However $|\mathbf{A}| = \pm 1$ and since a trivial calculation shows that initially $|\mathbf{A}| = +1$, we conclude that (4.2.21) is true for all possible choices of the a_{ij}. [The negative choice of sign is associated with the cases where either the primed system is left handed and the unprimed system right handed, or vice versa.]

Since $|\mathbf{A}|$ is non-zero the matrix $\mathbf{A}$ possesses a unique inverse $\mathbf{A}^{-1}$ such that

$$\mathbf{A}\mathbf{A}^{-1} = \mathbf{A}^{-1}\mathbf{A} = \mathbf{I} \ .$$

Clearly from (4.2.20)

$$\mathbf{A}^{-1} = \mathbf{A}^T \ . \tag{4.2.22}$$

A square matrix $\mathbf{A}$ with real elements which possesses the properties (4.2.21) and (4.2.22) is termed a 'proper orthogonal' matrix.

The product of two proper orthogonal matrices $\mathbf{A}_1$ and $\mathbf{A}_2$ is also proper orthogonal for

(i) $|\mathbf{A}_1||\mathbf{A}_2| = |\mathbf{A}_1\mathbf{A}_2| = 1$

(ii) $(\mathbf{A}_1\mathbf{A}_2)^T\mathbf{A}_1\mathbf{A}_2 = \mathbf{A}_2^T\mathbf{A}_1^T\mathbf{A}_1\mathbf{A}_2 = \mathbf{A}_2^T\mathbf{I}\mathbf{A}_2 = \mathbf{A}_2^T\mathbf{A}_2 = \mathbf{I}$.

We return now to equations (4.2.9), which may also be written in matrix form. Column matrices $\mathbf{x}$, $\mathbf{x}'$, $\boldsymbol{\alpha}$, $\boldsymbol{\alpha}'$ may be defined as follows

$$\mathbf{x} = \begin{pmatrix} x_1 \\ x_2 \\ x_3 \end{pmatrix}; \quad \mathbf{x}' = \begin{pmatrix} x'_1 \\ x'_2 \\ x'_3 \end{pmatrix}; \quad \boldsymbol{\alpha} = \begin{pmatrix} \alpha_1 \\ \alpha_2 \\ \alpha_3 \end{pmatrix}; \quad \boldsymbol{\alpha}' = \begin{pmatrix} \alpha'_1 \\ \alpha'_2 \\ \alpha'_3 \end{pmatrix}$$

when equations (4.2.9) may be written

$$\mathbf{x}' = \boldsymbol{\alpha}' + \mathbf{A}\mathbf{x} \tag{4.2.23}$$

with 'solution' [equations (4.2.12)]

$$\mathbf{x} = -\boldsymbol{\alpha} + \mathbf{A}^T\mathbf{x}' . \tag{4.2.24}$$

It is evident from Fig. 4.1 that the primed axes are derivable from the unprimed axes by means of a translation of origin ($\boldsymbol{\alpha}$) together with a rotation of axes about some fixed line passing through O$'$. Clearly in (4.2.23) the rotation is described by the proper orthogonal matrix $\mathbf{A}$ and conversely every proper orthogonal matrix may be associated with a rotation of axes. Determination of the fixed line and the amount of rotation in terms of the elements a_{ij} of $\mathbf{A}$ are not important here [for details see Mirsky's *An Introduction to Linear Algebra,* p. 236-243, (Oxford University Press), 1961]. The order in which the translation and rotation operations are carried out is immaterial since (4.2.23) may also be written

$$\mathbf{x}' = \mathbf{A}(\boldsymbol{\alpha} + \mathbf{x}) .$$

Clearly in (4.2.24) the inverse matrix $\mathbf{A}^T(\equiv \mathbf{A}^{-1})$ is associated with rotation of axes by the same amount about the same line but in the opposite sense to the rotation associated with $\mathbf{A}$.

In Chapter 6 we shall regard $\mathbf{A}$ and $\boldsymbol{\alpha}$ as functions of time. For the remainder of this chapter, $\mathbf{A}$ and $\boldsymbol{\alpha}$ are assumed constant. If the unprimed coordinate system is astronomical then any other astronomical system is derivable with an appropriate choice of constant $\mathbf{A}$ and $\boldsymbol{\alpha}$.

We are now in a position to define Cartesian tensors. A Cartesian tensor T of order n is defined in terms of 3^n components

$$\underbrace{T_{ijk\,\cdots}}_{n \text{ suffixes}}$$

which may be chosen arbitrarily. Usually the $T_{ijk}\ldots$ are functions of position and time; in any event the $T_{ijk}\ldots$ are assumed to be referred to a given astronomical coordinate system. Here we choose the given system to be unprimed. The defining properties of T are that components

$$\underbrace{T'_{ijk}\ldots}_{n\text{ suffixes}}$$

referred to a primed coordinate system are given by the formulae

$$T'_{ijk}\ldots = a_{ir}a_{js}a_{kt}\ldots T_{rst}\ldots \quad (4.2.25)$$

where the a_{ij} are the elements of the matrix $\mathbf{A}$. Thus tensors of order n are concerned with the transformation properties of 3^n quantities from one coordinate system to a system with different orientation; the translation vector $\boldsymbol{\alpha}$ does not enter the definition of $T'_{ijk}\ldots$, and for most purposes it is sufficient to think of the $T'_{ijk}\ldots$ as referred to a primed system $Ox'_1x'_2x'_3$ where $\mathbf{x}' = \mathbf{Ax}$.

It is important to realise that in order for T to be a tensor, the primed coordinate system may be chosen arbitrarily and that (4.2.25) holds for all choices of primed system. (Of course each distinct orientation of the primed system is associated with a distinct matrix $\mathbf{A}$). If components of T in the primed and unprimed systems are related to each other through equations (4.2.25), then we say that $T_{ijk}\ldots$ (or $T'_{ijk}\ldots$) is a component of the tensor T. Sometimes it is convenient to speak (less precisely) of the tensor $T_{ijk}\ldots$.

A tensor of order zero contains only one component T (and is therefore not labelled with a suffix). From (4.2.25), T' is related to T by the equation

$$T' = T \quad (4.2.26)$$

which is an assertion that the numerical value of T is independent of the choice of coordinate system. Clear examples of zero order tensors are physical quantities like temperature, density, specific heat, length, and speed. Such quantities are often known as scalars in physics and a Cartesian tensor of order zero is commonly called a scalar.

A statement like (4.2.26) does not assert that the functional dependence of T on position in the unprimed system is the same as that of T' on position in the primed system. Indeed unless T is independent of position the functional dependences are different, for example suppose T is given to be

$$T(\mathbf{r},t) = x_1x_2e^{-t}$$

then from (4.2.26) and (4.2.13)

$$T'(\mathbf{r}',t) = a_{j1}a_{k2}(-\alpha'_j + x'_j)(-\alpha'_k + x'_k)e^{-t}\ .$$

However the numerical values of T and T' at the point labelled (x_1, x_2, x_3) in the unprimed system and (x'_1, x'_2, x'_3) in the primed system, are the same. Similar remarks apply to the definitions of Cartesian tensors of higher order.

Cartesian tensors of order one comprise three components T_i which satisfy the transformation laws

$$T'_i = a_{ij}T_j\ . \quad (4.2.27)$$

These are precisely the transformation laws for the components of a vector – compare (4.2.27) with (4.2.11). For this reason Cartesian tensors of order one are commonly known as vectors.† Clear physical examples occurring in mechanics are displacement, velocity, force, acceleration, angular velocity. Many other examples occur in physics (e.g., electric and magnetic fields, current density).

Cartesian tensors of order two comprise nine components T_{ij} which satisfy the transformation laws

$$T'_{ij} = a_{ir}a_{js}T_{rs} \; . \tag{4.2.28}$$

To date we have not identified any tensor quantity of second order. In fact the components of Cartesian stress σ_{ij} form a second order (symmetric) tensor, but a proof of this result depends on the laws of mechanics (Section 4.4). The simplest second order tensor is the Kronecker delta. For suppose δ_{ij} to be defined by (4.2.18) for the unprimed system. The statement that δ_{ij} is a tensor means that δ'_{ij} defined by the transformation laws

$$\delta'_{ij} = a_{ir}a_{js}\delta_{rs} \tag{4.2.29}$$

is the same as δ_{ij}.

Products involving the Kronecker delta δ_{rs}, as on the right hand side of (4.2.29) are much simplified if either r or s (or both) appear elsewhere in the product. Under these circumstances the defining properties (4.2.18) lead to the result that the effect of the Kronecker delta is to delete the Kronecker delta from the product and set the index r equal to s (or vice versa) in what remains. In the case of (4.2.29) we find

$$\delta'_{ij} = a_{ir}a_{js}\delta_{rs} = a_{ir}a_{jr} = \delta_{ij}$$

where the last step follows directly from the orthogonality relations (4.2.19b).

Cartesian tensors of order two play an important role in continuum mechanics, as might be expected from the knowledge that the stress components form a tensor. The transformation laws (4.2.28) for second order tensors are elegantly expressed in terms of matrices. For if $\mathbf{T}$ and $\mathbf{T}'$ denote the matrices

$$\mathbf{T} = (T_{ij}), \quad \mathbf{T}' = (T'_{ij})$$

equations (4.2.28) are summarised by the matrix equation

$$\mathbf{T}' = \mathbf{A}\mathbf{T}\mathbf{A}^T \tag{4.2.30}$$

with 'solution'

$$\mathbf{T} = \mathbf{A}^T\mathbf{T}'\mathbf{A} \; . \tag{4.2.31}$$

The matrix representations of the transformation laws is useful subsequently. (Section 4.5 and Chapters 7 and 8.)

† Exceptionally the vector $\mathbf{r}$ itself does not transform according to the law (4.2.27) unless $\boldsymbol{\alpha} = 0$

Many second order tensors arising in continuum mechanics are classifiable as symmetric ($T_{ij} = T_{ji}$, i.e. $\mathbf{T}^T = \mathbf{T}$) or anti-symmetric ($T_{ij} = -T_{ji}$, i.e. $\mathbf{T}^T = -\mathbf{T}$). Symmetry and anti-symmetry properties are unchanged by the transformation laws. Suppose in (4.2.30) $\mathbf{T}$ is symmetric so that $\mathbf{T}^T = \mathbf{T}$. We have

$$(\mathbf{T}')^T = (\mathbf{ATA}^T)^T = (\mathbf{A}^T)^T\mathbf{T}^T\mathbf{A}^T = \mathbf{ATA}^T = \mathbf{T}'$$

and similarly for the anti-symmetric case.

Symmetric tensors contain six independent components while anti-symmetric tensors contain three independent components. The nine components of an arbitrary second order tensor may always be decomposed into symmetric and anti-symmetric parts:

$$T_{ij} = \tfrac{1}{2}(T_{ij} + T_{ji}) + \tfrac{1}{2}(T_{ij} - T_{ji}) \; . \qquad (4.2.32)$$

Tensors of order three and higher are defined by (4.2.25). Little use is made in the present text of tensors of order higher than two. The importance of tensors in continuum mechanics derives from the fact that the physical variables of the subject-density, displacement, velocity, acceleration, stress, strain (yet to be defined), etc., are all classifiable as tensors of order zero, one or two.

Tensor components are usually functions of position and time. In respect of dependence on position the independent variables may be either the x_i or the X_i. Since under coordinate transformations from one frame to another both the x_i and X_i transform similarly, the dependence on material or referential coordinates is irrelevant to the question of whether a set of quantities form a tensor. Thus both quantities v_i and V_i (or f_i and F_i) are components of a first order tensor.

In conclusion it should be emphasised that the quantities a_{ij} do not form a tensor. In fact the question does not arise, for the a_{ij} themselves enter the defining properties of tensors.

4.3 Elementary tensor operations and the quotient theorem

Elementary operations which are performed with tensors are addition, multiplication, contraction of indices and differentiation with respect to time and space variables.

4.3a ADDITION OF TENSORS

Only tensors of the same order may be added. The resulting quantity is a tensor. For example if A_{ij} and B_{ij} are components of second order tensors, the quantities C_{ij} given by

$$C_{ij} = A_{ij} + B_{ij}$$

also form the components of a second order tensor. The proof that the components C_{ij} form a tensor is straightforward and left to the reader.

4.3b MULTIPLICATION OF TENSORS

Tensors of different orders (n and m) may be multiplied to yield a tensor of order ($n + m$). We consider in detail the case $n = 1,\ m = 2$. Let A_i and B_{ij} form the components of tensors respectively of order one and two. Then

$$A'_i = a_{ip}A_p\,, \quad B'_{jk} = a_{jq}\,a_{kr}B_{qr}\,. \tag{4.3.1}$$

If now we define
$$C_{pqr} = A_pB_{qr}$$

then from equations (4.3.1)

$$C'_{ijk} = A'_iB'_{jk} = a_{ip}a_{jq}a_{kr}A_pB_{qr} = a_{ip}a_{jq}a_{kr}C_{pqr}$$

so that the C_{ijk} form the components of a tensor of order three. Clearly the proof generalises to any tensor product. Also products of three or more tensors also lead to a tensor (by an obvious extension of the above proof) so that for example

$$C_{ijklm} = A_iB_{jk}D_{lm}$$

is a tensor provided A_i, B_{jk} and D_{lm} are tensors.

4.3c CONTRACTION OF INDICES

If in the tensor $A_{ijkl}\ldots$ of order n any two of the free indices are put equal (thereby effecting a sum), the resultant quantity is a tensor of order ($n - 2$). For suppose the indices in question are j and k. Then the quantity

$$C_{il}\ldots = A_{ijjl}\cdots$$

is a tensor, for

$$\begin{aligned} C'_{il}\ldots &= A'_{ijjl}\ldots = a_{ir}a_{js}a_{jp}a_{lq}\ldots A_{rspq}\cdots \\ &= a_{ir}a_{lq}\delta_{sp}\ldots A_{rspq} \\ &= a_{ir}a_{lq}\ldots A_{rssq} \\ &= a_{ir}a_{lq}\ldots C_{rq}\ldots \end{aligned}$$

The process is known as contraction of indices.

4.3d DIFFERENTIATION WITH RESPECT TO TIME AND SPACE VARIABLES

Here we are concerned with tensors $A_{ijk\ldots}(x_1, x_2, x_3; t)$ which are differentiable functions of x_1, x_2, x_3 and t. Differentiation of a tensor of order n with respect to time results in a tensor also of order n. The proof is left to the reader.

Differentiation of a tensor of order n with respect to position leads to a tensor of order ($n + 1$). We consider the case $n = 2$. Let A_{ij} form the components of a tensor so that

$$A'_{ij}(x'_k, t) = a_{ip}a_{jq}A_{pq}(x_r, t)\,. \tag{4.3.2}$$

In this equation we are regarding the A'_{ij} as functions of the x'_k, and the A_{ij} as functions of the x_k. Also the notation $A'_{ij}(x'_k, t)$ is a shorthand notation implying that each of the A'_{ij} depend on all the x'_k and t. Define

$$C_{ijk} = \partial A_{ij}/\partial x_k .$$

Differentiating (4.3.2) with respect to x'_k leads to

$$C'_{ijk} = \frac{\partial A'_{ij}}{\partial x'_k} = a_{ip}a_{jq}\frac{\partial A_{pq}}{\partial x_r}\frac{\partial x_r}{\partial x'_k} \equiv a_{ip}a_{jq}\frac{\partial x_r}{\partial x'_k}C_{pqr} \qquad (4.3.3)$$

However from (4.2.13) $$\frac{\partial x_r}{\partial x'_k} = a_{kr}$$

so that (4.3.3) leads immediately to

$$C'_{ijk} = a_{ip}a_{jq}a_{kr}C_{pqr}$$

i.e. the C_{ijk} form the components of a tensor of order three. Clearly the proof generalises to tensors of any order.

In view of the penultimate paragraph of Section (4.2), if A_{ij} is regarded as a function of the X_k and t, the derivatives

$$\partial A_{ij}/\partial X_k$$

are also the components of a third order tensor.

Two important second order tensors are formed by the two sets of displacement gradients $\partial x_i/\partial X_k$ and $\partial X_i/\partial x_k$ where

$$x_i = x_i(X_k, t) \qquad (4.3.4)$$

are the current positions of particles expressed as functions of the reference coordinates while the $X_i(x_k, t)$ are the inverse relations [equations (2.2.2) and (2.2.5) of Chapter 2].†

In view of the hypothesis that

$$J = \left|\frac{\partial x_i}{\partial X_k}\right| \neq 0$$

the quantities $\partial x_i/\partial X_k$ also form the nine components of a non-singular square matrix $\mathbf{F}$

$$\mathbf{F} = \left(\frac{\partial x_i}{\partial X_k}\right) . \qquad (4.3.5)$$

† There is a very minor difficulty here. By the above $\partial x_i/\partial X_j$ and $\partial X_i/\partial x_j$ are tensors if respectively x_i and X_i are tensors. In fact unless $\boldsymbol{\alpha} = 0$, x_i and X_i are not tensors. However the reader will show easily that the presence of a non-vanishing constant vector $\boldsymbol{\alpha}$ does not affect the tensor character of the derivatives $\partial x_i/\partial X_j$ and $\partial X_i/\partial x_j$.

Since **F** is non-singular it possesses a unique inverse. Further the inverse is easily found for on differentiating (4.3.4) with respect to x_j we have

$$\delta_{ij} = \frac{\partial x_i}{\partial X_k}\frac{\partial X_k}{\partial x_j}$$

so that

$$\mathbf{F}^{-1} = \left(\frac{\partial X_i}{\partial x_k}\right) . \tag{4.3.6}$$

No use is made of $\mathbf{F}^{-1}$ in the present text.

Another important tensor of the second order is formed by the nine components of velocity gradient in the spatial description

$$l_{ij} = \partial v_i / \partial x_j . \tag{4.3.7}$$

Symmetric and antisymmetric tensors formed from l_{ij} are

$$d_{ij} = \tfrac{1}{2}(l_{ij} + l_{ji}) = \frac{1}{2}\left(\frac{\partial v_i}{\partial x_j} + \frac{\partial v_j}{\partial x_i}\right) \tag{4.3.8}$$

$$\omega_{ij} = \tfrac{1}{2}(l_{ij} - l_{ji}) = \frac{1}{2}\left(\frac{\partial v_i}{\partial x_j} - \frac{\partial v_j}{\partial x_i}\right) . \tag{4.3.9}$$

The symmetric tensor is often known as the 'rate of strain tensor' (the origin of the teminology is made clear in Chapter 5). The antisymmetric tensor ω_{ij} may be written (in matrix form)

$$(\omega_{ij}) = \begin{pmatrix} 0 & -\omega_3 & \omega_2 \\ \omega_3 & 0 & -\omega_1 \\ -\omega_2 & \omega_1 & 0 \end{pmatrix} \tag{4.3.10}$$

where the vector $\boldsymbol{\omega} \equiv (\omega_1, \omega_2, \omega_3)$ is also recognisable as

$$\boldsymbol{\omega} = \tfrac{1}{2} \operatorname{curl} \mathbf{v} .$$

4.3e THE QUOTIENT THEOREM

The quotient theorem discussed here provides a method for testing whether a set of quantities A_{ij} form the components of a second order Cartesian tensor. One method is to check whether the transformation laws are satisfied. However, in some circumstances, the quotient theorem provides a more convenient test.

Suppose the set of nine quantities A_{ij} are suspected of forming the components of a tensor. If B_{ij} is an arbitrary second order tensor and the product

$$A_{ij}B_{ij} = \phi$$

is such that ϕ is a scalar (i.e. a tensor of zero order) then the quantities A_{ij} form the components of a tensor. The proof is as follows. Since by hypothesis ϕ is a scalar, $\phi = \phi'$ and

$$A_{ij}B_{ij} = A'_{ij}B'_{ij} . \tag{4.3.11}$$

Further B_{ij} is a tensor so that

$$B'_{ij} = a_{ir}a_{js}B_{rs} .$$

Substituting this latter result into (4.3.11) leads to

$$A'_{ij}a_{ir}a_{js}B_{rs} - A_{ij}B_{ij} = 0 .$$

If in the first of these products the roles of i and r and also the roles of j and s are interchanged, there results

$$(A'_{rs}a_{ri}a_{sj} - A_{ij})B_{ij} - 0 . \qquad (4.3.12)$$

However since B_{ij} is an arbitrary tensor we may choose the B_{ij} so that in the unprimed coordinate system only one of the components does not vanish. Performing this operation for every component in turn leads to

$$A'_{rs}a_{ri}a_{sj} - A_{ij} = 0 \qquad (4.3.13)$$

or on multiplying by $a_{pi}a_{qj}$ and using the orthogonality relations (4.2.19b)

$$A'_{pq} = a_{pi}a_{qj}A_{ij}$$

so that the A_{ij} do form the components of a tensor.

A slightly more elaborate version of the theorem, required subsequently, is as follows. If B_{ij} is symmetric, but otherwise arbitrary, and also A_{ij} is symmetric, then again the A_{ij} form the components of a tensor. For if $B_{ij} = B_{ji}$ the conclusion from (4.3.12) is now

$$A'_{rs}a_{ri}a_{sj} + A'_{rs}a_{rj}a_{si} - (A_{ij} + A_{ji}) = 0 . \qquad (4.3.14)$$

Interchanging the roles of r and s in the second term of (4.3.14) yields

$$(A'_{rs} + A'_{sr})a_{ri}a_{sj} - (A_{ij} + A_{ji}) = 0 . \qquad (4.3.15)$$

However if $A_{ij} = A_{ji}$ (and hence, $A'_{rs} = A'_{sr}$), (4.3.15) reduces to (4.3.13). (However if B_{ij} is symmetric and it is not known that A_{ij} is symmetric, it is not possible to conclude that A_{ij} is a tensor.)

A more general version of the quotient theorem for tensors of any order is given in Spain's 'Tensor Calculus' p14-15, Oliver and Boyd (1953).

4.4 The stress tensor

An important result in continuum mechanics is that the components of Cartesian stress σ_{ij} form the components of a second order symmetric tensor, i.e.

$$\sigma'_{ij} = a_{ir}a_{js}\sigma_{rs} . \qquad (4.4.1)$$

Here σ'_{ij} denotes the component of stress acting in the $\mathbf{i}'_j$ direction on a surface whose positive normal lies in the $\mathbf{i}'_i$ direction.

The proof of (4.4.1) devolves on the laws of mechanics. Fig. 4.2 shows an infinitesimal tetrahedron formed by the three coordinate planes $x_1 = 0$,

$x_2 = 0$, $x_3 = 0$ and a fourth plane formed by a triangle of area ΔS, whose vertices lie on the coordinate axes. The areas of the triangular faces with outward normals in the negative x_i directions are denoted by ΔS_i.

Let $\mathbf{i}'_1$ denote a unit vector pointing in the outward normal direction of ΔS and choose unit vectors $\mathbf{i}'_2$, $\mathbf{i}'_3$ so that $\mathbf{i}'_1$, $\mathbf{i}'_2$, $\mathbf{i}'_3$ form an orthogonal right handed triad. [There is freedom in the choice of the orientation of one of $\mathbf{i}'_2$, $\mathbf{i}'_3$ within the plane ΔS.]

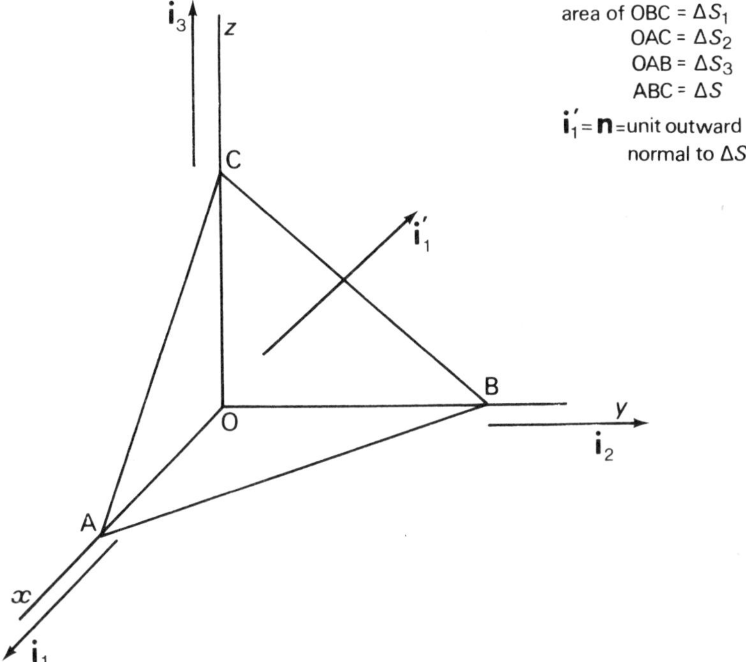

Fig. 4.2 Tetrahadron for deriving transformation laws for stress components

We assume for the moment that the areas ΔS_i and ΔS are sufficiently small to neglect variations of stress over these areas; with this assumption the contact force acting on ΔS is

$$\Delta S \sigma'_{1p} \mathbf{i}'_p$$

while the sum of the contact forces on the remaining sides of the tetrahedron is

$$-\Delta S_l \sigma_{lm} \mathbf{i}_m$$

Here the minus sign derives from the directions of the outward normals of the ΔS_i. From the linear momentum balance laws for the material within the tetrahedron.

$$\iiint\limits_{\Omega} \rho \mathbf{b} d\Omega + \sigma'_{1p}\mathbf{i}'_p \Delta S - \sigma_{lm} \Delta S_l \mathbf{i}_m = \iiint\limits_{\Omega} \rho \mathbf{f} d\Omega \qquad (4.4.2)$$

where the integrals are taken over the volume Ω of the tetrahedron. The triple integrals contribute terms of order $\Delta x_1 \Delta x_2 \Delta x_3$ in (4.4.2) where the Δx_i are the distances from the origin to the other vertices of the tetrahedron in Fig. 4.2. In contrast the terms in ΔS and ΔS_l are of order $\Delta x_1 \Delta x_2$, etc., so that (4.4.2) is adequately approximated by

$$\sigma'_{1p}\mathbf{i}'_p \Delta S = \sigma_{lm} \Delta S_l \mathbf{i}_m \ . \tag{4.4.3}$$

[For continuous functions σ_{ij}, a similar argument justifies the initial neglect of variations in stress over ΔS and the ΔS_i.] In (4.4.3)

$$\mathbf{i}_m = a_{pm}\mathbf{i}'_p \tag{4.4.4}$$

(from 4.2.5) while

$$\Delta S_l = \Delta S(\mathbf{i}'_1 . \mathbf{i}_l) \tag{4.4.5}$$

where $(\mathbf{i}'_1 . \mathbf{i}_l)$ is the cosine of the angle between the normals of ΔS and ΔS_l. From (4.2.6)

$$\mathbf{i}'_1 . \mathbf{i}_l = a_{1l} \ . \tag{4.4.6}$$

Substituting from (4.4.4), (4.4.5) and (4.4.6) into (4.4.3), and cancelling a common factor ΔS, and equating the coefficients of the (independent) unit vectors $\mathbf{i}'_p$ yields

$$\sigma'_{1p} - a_{1l}\, a_{pm}\, \sigma_{lm} = 0$$

or more generally

$$\sigma'_{qp} = a_{ql} a_{pm} \sigma_{lm} \tag{4.4.7}$$

since any choice of label is possible for the normal to the surface ΔS.

Equations (4.4.7), showing that the quantities σ_{ij} form the components of a Cartesian tensor, solves the problem of calculating stress vectors for surfaces of arbitrary orientation in terms of the stress components referred to a fixed set of axes. A particularly useful form of this result is provided by (4.4.3). From (4.4.3) the element of force $d\mathbf{F}$ acting on an element of surface dS with outward unit normal $\mathbf{n} = (n_1, n_2, n_3)$ is

$$\begin{aligned} d\mathbf{F}(\equiv \sigma'_{1p}\, \mathbf{i}'_p\, dS) &= \sigma_{lm}\, dS_l \mathbf{i}_m \\ &= a_{1l}\sigma_{lm}\, \mathbf{i}_m\, dS \end{aligned}$$

so that the component of $d\mathbf{F}$ in the direction m is

$$dF_m = a_{1l}\sigma_{lm}\, dS \ .$$

We write this result now in the form

$$dF_m = \sigma_{lm}\, n_l dS$$

i.e.

$$dF_i = \sigma_{ji} n_j dS \tag{4.4.8}$$

where $n_j (\equiv a_{1j})$ is the component in the j direction of the unit vector $\mathbf{n}(\equiv \mathbf{i}'_1)$ which points in the outward normal direction of dS. From (4.4.8) the stress vector $\mathbf{T}_{(\mathbf{n})}$ for the surface dS has components

$$T_i = \sigma_{ji} n_j \ . \tag{4.4.9}$$

which renders explicit the dependence of $\mathbf{T}_{(\mathbf{n})}$ on $\mathbf{n}$.

Equation (4.4.8), expressing the ith component of the force per unit area acting on a surface element dS of arbitrary orientation $\mathbf{n}$, leads to a more elegant derivation of the differential equations of motion for a continuum. In integral form the equations of motion are given by equations (3.2.5). Taking the ith component of this equation and using (4.4.9) for T_i yields

$$\iint_{S_t} \sigma_{ji} n_j dS + \iiint_{\Omega_t} \rho b_i = \iiint_{\Omega_t} \rho f_i d\Omega \ . \tag{4.4.10}$$

The surface integral may be transformed into a volume integral (for differentiable σ_{ij}) using Green's theorem. In vector analysis the latter theorem is commonly written

$$\iint_S \mathbf{A} \cdot \mathbf{dS} = \iiint_\Omega \operatorname{div} \mathbf{A}\, d\Omega \tag{4.4.11}$$

where $\mathbf{dS} \equiv \mathbf{n} dS$ and $\mathbf{A}$ is a vector. In the surface integral in (4.4.10), σ_{ji} may be regarded temporarily as the component in the jth direction of a vector $\mathbf{A}$ (remembering i is a free index assuming one of the values 1, 2 or 3). The integral in question is now precisely of the form of the surface integral on the left hand side of (4.4.11). Applying Green's theorem leads therefore to

$$\iiint_{\Omega_t} \left[\frac{\partial \sigma_{ji}}{\partial x_j} + \rho(b_i - f_i) \right] d\Omega = 0$$

or since Ω_t may be chosen arbitrarily

$$\frac{\partial \sigma_{ji}}{\partial x_j} + \rho(b_i - f_i) = 0 \ . \tag{4.4.12}$$

On recognition of the symmetry conditions $\sigma_{ij} = \sigma_{ji}$, equations (4.4.12) assume the more conventional form

$$\frac{\partial \sigma_{ij}}{\partial x_j} + \rho b_i = \rho f_i \ . \tag{4.4.13}$$

Until the last step none of the analysis above presupposes the symmetry of the stress tensor. In fact equations (4.4.12) are precisely the equations (3.4.4) - (3.4.6), which also were derived in the absence of knowledge of the symmetry condition. We know from Section 3.4 that the symmetry conditions are derived from the angular momentum equations. An analysis which parallels that leading to (4.4.12) is as follows. The angular momentum equations in integral form are [equations (3.2.8)]

$$\iint_{S_t} (\mathbf{r} \wedge \mathbf{T_{(n)}}) dS = \iiint_{\Omega_t} \rho \mathbf{r} \wedge (\mathbf{f} - \mathbf{b}) d\Omega \ . \tag{4.4.14}$$

On making use of (4.4.9) for the components of $\mathbf{T}_{(\mathbf{n})}$ and transforming the surface integral into a volume integral (again using Green's theorem), the x_1 component of the left side of (4.4.14) becomes

$$\left[\iint_{S_t}(\mathbf{r}\wedge\mathbf{T}_{(\mathbf{n})})dS\right]_1 = \iiint_{\Omega_t}\frac{\partial}{\partial x_j}(x_2\sigma_{j3} - x_3\sigma_{j2})d\Omega$$

$$= \iiint_{\Omega_t}\left(x_2\frac{\partial\sigma_{j3}}{\partial x_j} - x_3\frac{\partial\sigma_{j2}}{\partial x_j}\right)d\Omega + \iiint_{\Omega_t}(\sigma_{23} - \sigma_{32})d\Omega$$

$$= \iiint_{\Omega_t}\rho[\mathbf{r}\wedge(\mathbf{f} - \mathbf{b})]_1 d\Omega + \iiint_{\Omega_t}(\sigma_{23} - \sigma_{32})d\Omega \tag{4.4.15}$$

where the last step has made use of equations (4.4.12). Comparing (4.4.15) with the x_1 component of (4.4.14) leads immediately to

$$\iiint_{\Omega_t}(\sigma_{23} - \sigma_{32})d\Omega = 0$$

or since Ω_t is arbitrary $\qquad \sigma_{23} = \sigma_{32}$.

The results $\sigma_{13} = \sigma_{31}$ and $\sigma_{12} = \sigma_{21}$ follow similarly from the x_2 and x_3 components of (4.4.14).

In view of the result $\sigma_{ij} = \sigma_{ji}$ it is usual to write (4.4.8) in the form

$$dF_i = \sigma_{ij}n_j dS \ . \tag{4.4.16}$$

An energy integral theorem which makes use of (4.4.16) is

$$\iint_{S_t}\sigma_{ij}v_i n_j dS + \iiint_{\Omega_t}\rho b_i v_i d\Omega = \frac{d}{dt}\iiint_{\Omega_t}\tfrac{1}{2}\rho v_i v_i d\Omega + \iiint_{\Omega_t}\sigma_{ij}d_{ij}d\Omega \tag{4.4.17}$$

where d_{ij} is the rate of strain tensor defined in equations (4.3.8).

To prove (4.4.17) we transform the surface integral as follows

$$\iint_{S_t}\sigma_{ij}v_i n_j dS = \iiint_{\Omega_t}\left(v_i\frac{\partial\sigma_{ij}}{\partial x_j} + \sigma_{ij}\frac{\partial v_i}{\partial x_j}\right)d\Omega$$

$$= \iiint_{\Omega_t}\left(v_i\frac{\partial\sigma_{ij}}{\partial x_j} + \sigma_{ij}d_{ij}\right)d\Omega$$

where in deriving the last step use has been made of the symmetry condition

$\sigma_{ij} = \sigma_{ji}$. The left hand side of equation (4.4.17) now becomes

$$\iiint_{\Omega_t} v_i\left(\frac{\partial \sigma_{ij}}{\partial x_j} + \rho b_i\right) d\Omega + \iiint_{\Omega_t} \sigma_{ij} d_{ij} d\Omega$$

or, on using (4.4.13)

$$\iiint_{\Omega_t} \rho v_i f_i d\Omega + \iiint_{\Omega_t} \sigma_{ij} d_{ij} d\Omega \ .$$

Equation (4.4.17) follows provided

$$\iiint_{\Omega_t} \rho v_i f_i d\Omega = \frac{d}{dt} \iiint_{\Omega_t} \tfrac{1}{2} \rho v_i v_i d\Omega \ . \qquad (4.4.18)$$

This latter result is most easily obtained by changing the integration variables on the right hand side of (4.4.18) to the reference coordinates. We have

$$\begin{aligned} \frac{d}{dt} \iiint_{\Omega_t} \tfrac{1}{2} \rho v_i v_i d\Omega &= \frac{d}{dt} \iiint_{\Omega_0} \tfrac{1}{2} \rho_0 V_i V_i d\Omega_0 \\ &= \iiint_{\Omega_0} \rho_0 V_i F_i d\Omega_0 \\ &= \iiint_{\Omega_t} \rho v_i f_i d\Omega \end{aligned}$$

where in the final step we revert to spatial coordinates.

Equation (4.4.17) is important because it introduces for the first time the concepts of energy and work done by external forces. The left hand side of the equation is the rate of working of the external forces on the material within S_t while the first term on the right hand side is the rate of change of kinetic energy. Thus the integral

$$\iiint_{\Omega_t} \sigma_{ij} d_{ij} d\Omega$$

measures the rate of supply of energy, other than kinetic energy, by the external forces. With the geometrical interpretation of d_{ij} in Chapter 5, it is useful sometimes to think of $\sigma_{ij} d_{ij} d\Omega$ as the rate of supply of energy required to deform the material in the volume element $d\Omega$.

It is worth re-examining the equations of motion (4.4.13) in the light of the information that σ_{ij} is a tensor. Equation (4.4.13) is an acceptable tensor equation since each term in the equation is the component of a tensor of the first order.

This result is immediately evident for the terms ρf_i and ρb_i since ρ is a scalar while the f_i and b_i are components of vectors. The term $\partial\sigma_{ij}/\partial x_j$ is the component of a vector, since from Section 4.3d the quantities

$$\frac{\partial\sigma_{ij}}{\partial x_k}$$

form the components of a third order tensor, which on contraction of the indices j and k leads to components of a tensor of order one.

Since (4.4.13) is a correct tensor equation we know immediately that for any other coordinate system

$$\frac{\partial\sigma'_{ij}}{\partial x'_j} + \rho' b'_i = \rho' f'_i \;.$$

The sense in which a tensor equation, while making overt reference to a specified coordinate system, is nevetheless independent of the system chosen, is now clear. An obvious requirement of all equations purporting to describe the physical world is that they be expressible in tensor form.

4.5 Properties of second order symmetric tensors; principal stresses and principal axes

In the theory of square matrices the determinental equation

$$|b_{ij} - b\delta_{ij}| = 0 \tag{4.5.1}$$

i.e.

$$\begin{vmatrix} b_{11}-b & b_{12} & b_{13} & \cdots \\ b_{21} & b_{22}-b & b_{23} & \cdots \\ b_{31} & b_{32} & b_{33}-b & \cdots \\ \cdots & \cdots & \cdots & \cdots \\ \cdots & \cdots & \cdots & \cdots \\ b_{n1} & b_{n2} & \cdots & b_{nn}-b \end{vmatrix} = 0$$

is known as the characteristic equation of the matrix **b**. If **b** is an $n \times n$ matrix, (4.5.1) is a polynomial equation in b of degree n, with n roots

$$b(1), b(2), b(3) \ldots b(n) \;.$$

The roots $b(r)$ are known as the 'eigenvalues' or 'characteristic roots' or 'latent roots' of **b**. In general the eigenvalues may be real or complex although clearly if the b_{ij} are real and n is odd, at least one of the $b(r)$ is real.

A well known theorem in matrix algebra states that if **b** is 'Hermitian' (defined by $b_{ij} = \overline{b_{ji}}$ where the bar denotes complex conjugate) then the $b(r)$ are all real. In particular if b_{ij} is a real symmetric matrix, the eigenvalues are real. [e.g. see Mirsky's *An Introduction to Linear Algebra,* p.209, Oxford

University Press (1961).] Below we give a restricted proof of this last result for $n = 3$ in a context natural to continuum mechanics. As a preliminary we note the result that if $\mathbf{S}$ is a non-singular square matrix of order n, then the eigenvalues of

$$\mathbf{SbS}^{-1} \tag{4.5.2}$$

are also those of $\mathbf{b}$. This theorem is true for an arbitrary square matrix $\mathbf{b}$. The proof is trivial since the characteristic equation for $\mathbf{b}$ is

$$|\mathbf{b} - b\mathbf{I}| = 0 \tag{4.5.3}$$

where $\mathbf{I}$ is the $n \times n$ unit matrix. However since

$$|\mathbf{S}||\mathbf{S}^{-1}| = 1$$

equation (4.5.3) may be written

$$|\mathbf{S}||\mathbf{b} - b\mathbf{I}||\mathbf{S}^{-1}| = 0$$

or from the theory of determinants

$$|\mathbf{S}(\mathbf{b} - b\mathbf{I})\mathbf{S}^{-1}| = 0$$

i.e.

$$|\mathbf{SbS}^{-1} - b\mathbf{I}| = 0 \, .$$

In particular on identifying $\mathbf{b}$ with $\boldsymbol{\sigma}$ and $\mathbf{S}$ with the orthogonal matrix $\mathbf{A}$ (and remembering that the σ_{ij} form the components of a tensor of second order), it is seen that the eigenvalues of $\boldsymbol{\sigma}$ given by

$$|\boldsymbol{\sigma} - \sigma\mathbf{I}| = 0$$

are the same as the eigenvalues of $\boldsymbol{\sigma}'$. Here the prime denotes a new reference frame for which $\boldsymbol{\sigma}'$ is given by

$$\boldsymbol{\sigma}' = \mathbf{A}\boldsymbol{\sigma}\mathbf{A}^T \, .$$

It follows that the eigenvalues $\sigma(1)$, $\sigma(2)$ and $\sigma(3)$ of the matrix

$$|\sigma_{ij} - \sigma\delta_{ij}| = 0 \tag{4.5.4}$$

are independent of the choice of coordinate system in which the stress components are measured. We say that $\sigma(1)$, $\sigma(2)$ and $\sigma(3)$ are 'invariants' of the stress tensor $\boldsymbol{\sigma}$.

Any quantity which is solely a function of the $\sigma(r)$, $(r = 1, 2, 3)$, is also called an invariant of $\boldsymbol{\sigma}$. For example the cubic equation (4.5.4) for σ may be written

$$[\sigma(1) - \sigma]\,[\sigma(2) - \sigma]\,[\sigma(3) - \sigma] = 0$$

i.e.

$$-\sigma^3 + \mathrm{I}_\sigma\sigma^2 - \mathrm{II}_\sigma\sigma + \mathrm{III}_\sigma = 0 \tag{4.5.5}$$

where

$$\mathrm{I}_\sigma = \sigma(1) + \sigma(2) + \sigma(3) \tag{4.5.6}$$

$$\mathrm{II}_\sigma = \sigma(1)\sigma(2) + \sigma(2)\sigma(3) + \sigma(3)\sigma(1) \tag{4.5.7}$$

$$\mathrm{III}_\sigma = \sigma(1)\sigma(2)\sigma(3) \tag{4.5.8}$$

are invariants of the stress tensor.

On expanding (4.5.4) directly and comparing coefficients of powers of σ in (4.5.4) and (4.5.5) we also find†

$$\mathrm{I}_\sigma = \sigma_{ii} \equiv \text{trace}\, \boldsymbol{\sigma} \tag{4.5.9}$$

$$\mathrm{II}_\sigma = \tfrac{1}{2}[\sigma_{ii}\sigma_{jj} - \sigma_{ij}\sigma_{ji}] \equiv \tfrac{1}{2}[(\text{trace}\, \boldsymbol{\sigma})^2 - \text{trace}\, \boldsymbol{\sigma}^2] \tag{4.5.10}$$

$$\begin{aligned}\mathrm{III}_\sigma &= |\sigma_{ij}| = \tfrac{1}{6}[\sigma_{ii}\sigma_{jj}\sigma_{kk} - 3\sigma_{ii}\sigma_{jk}\sigma_{kj} + 2\sigma_{ij}\sigma_{jk}\sigma_{ki}] \\ &= \tfrac{1}{6}[(\text{trace}\, \boldsymbol{\sigma})^3 - 3(\text{trace}\, \boldsymbol{\sigma})(\text{trace}\, \boldsymbol{\sigma}^2) + 2\,\text{trace}\,(\boldsymbol{\sigma}^3)].\end{aligned} \tag{4.5.11}$$

It is not surprising that the quantities on the right hand sides of equations (4.5.9), (4.5.10) and (4.5.11) are invariants, since in each case they are scalar quantities formed in a correct tensor manner from tensor components σ_{ij}.

The eigenvalue equation (4.5.4) arises naturally with the question as to whether at any given point in a continuum it is possible to choose the orientation of an element of surface ΔS so that the stress vector $\mathbf{T}_{(\mathbf{n})}$ is normal to the plane of ΔS. Denoting the unit normal to ΔS by $\mathbf{n}$, the conditions for $\mathbf{T}_{(\mathbf{n})}$ to be in the direction $\mathbf{n}$ may be written

$$T_i = \sigma n_i$$

where σ is a proportionality constant. Substituting for T_i from (4.4.9) in terms of the stress components and using $\sigma_{ij} = \sigma_{ji}$ leads to the equations

$$\sigma_{ij}n_j = \sigma n_i \tag{4.5.12}$$

i.e.

$$(\sigma_{ij} - \sigma\delta_{ij})n_j = 0\ . \tag{4.5.13}$$

Equations (4.5.13) are three homogeneous equations for the components n_i. It is a well known result in the theory of linear equations that these equations possess non-trivial solutions, i.e. solutions for which at least one of the n_i is non-zero, only if σ satisfies the determinant equation

$$|\sigma_{ij} - \sigma\delta_{ij}| = 0$$

i.e. only if σ is an eigenvalue of the matrix $\boldsymbol{\sigma}$.

The eigenvalue equation is a cubic in σ and therefore at least one of the eigenvalues [say $\sigma(1)$] is real.

With the eigenvalue $\sigma(1)$ we associate a real unit vector $\mathbf{n}(1)$ determined by equations (4.5.13) together with the relation

$$n_i(1)n_i(1) = 1\ . \tag{4.5.14}$$

With $\sigma = \sigma(1)$ in equations (4.5.13), the equations are necessarily consistent and the rank of the matrix of coefficients $(\sigma_{ij} - \sigma(1)\delta_{ij})$ is less than three. Usually the rank is two so that two of the three equations (4.5.13) are

† See Problem 11, at the end of the chapter, for an easy derivation of the results (4.5.10) and (4.5.11).

independent. These two independent equations may be written in terms of ratios of the n_i and are of the form

$$a_1 h + b_1 k = c_1 , \quad a_2 h + b_2 k = c_2 \qquad (4.5.15)$$

where $a_1, b_1, c_1, a_2, b_2, c_2$ are given in terms of some of the σ_{ij} and $\sigma(1)$, and where h and k are (say)

$$h = n_1/n_3 , \quad k = n_2/n_3$$

[Conceivably the equations (4.5.13) yield $n_3 = 0$ in which case we choose h and k to have one of n_1, n_2 as denominator.] Equations (4.5.15) may now be solved for h and k and the results $n_1 = hn_3, n_2 = kn_3$ substituted into (4.5.14), thereby determining $n_3^2(1)$ uniquely and $n_3(1)$ to within a choice of sign. Evidently the process determines $\mathbf{n}(1)$ uniquely to within a choice of sign.

Exceptionally the rank of the matrix $(\sigma_{ij} - \sigma(1)\delta_{ij})$ may be one (or even zero). In the former case, which subsequently will be seen to be associated with two identical eignenvalues, equation (4.5.13) contains only one independent equation which will be of the form

$$a_1 n_1 + b_1 n_2 - c_1 n_3 = 0$$

where at least one of a_1, b_1, c_1 is non-zero. Clearly it is not possible to determine $\mathbf{n}(1)$ uniquely from this last equation and (4.5.14). On the other hand any choice of unit vector $\mathbf{n}(1)$ lying within the plane whose normal possesses direction cosine ratios $a_1 : b_1 : -c_1$ now satisfies equations (4.5.13) and (4.5.14). We select arbitrarily one of these unit vectors for $\mathbf{n}(1)$. [If the matrix of coefficients is of rank zero then, as will be seen subsequently, all three eigenvalues are equal and $\boldsymbol{\sigma}$ is already in the diagonal form $\boldsymbol{\sigma} = \sigma\mathbf{I}$.]

We now show that the remaining eigenvalues $\sigma(2)$ and $\sigma(3)$ are also real. Let a new (primed) coordinate system be chosen for which the direction of the positive $O'x_1'$ axis is chosen to be $\mathbf{n}(1)$. The choice of direction for the positive $O'x_2'$ axis is restricted to lie in a plane whose normal is $\mathbf{n}(1)$, but is otherwise arbitrary. With the $O'x_2'$ axis chosen, the $O'x_3'$ axis is determined so that the primed reference axes form a right handed orthogonal system. Referred to the primed system the stress component σ_{11}' is $\sigma(1)$ while $\sigma_{12}' = \sigma_{13}' = 0$. Also from the symmetry of the σ_{ij}, $\sigma_{21}' = \sigma_{31}' = 0$ so that the characteristic equation now takes the form

$$\begin{vmatrix} \sigma(1) - \sigma & 0 & 0 \\ 0 & \sigma_{22}' - \sigma & \sigma_{23}' \\ 0 & \sigma_{32}' & \sigma_{33}' - \sigma \end{vmatrix} = 0 . \qquad (4.5.16)$$

One root of the resultant cubic in $\sigma(1)$, as expected. The remaining roots are solutions of the quadratic equation

$$(\sigma_{22}' - \sigma)(\sigma_{33}' - \sigma) - \sigma_{23}'\sigma_{32}' = 0$$

i.e. $$\sigma^2 - (\sigma_{22}' + \sigma_{33}')\sigma + \sigma_{22}'\sigma_{33}' - \sigma_{23}'\sigma_{32}' = 0 .$$

The condition for real roots $\sigma(2)$, $\sigma(3)$ is

$$(\sigma'_{22} + \sigma'_{33})^2 - 4(\sigma'_{22}\sigma'_{33} - \sigma'_{23}\sigma'_{32}) \geqslant 0$$

i.e.

$$(\sigma'_{22} - \sigma'_{33})^2 + 4\sigma'_{23}\sigma'_{32} \geqslant 0 \ . \tag{4.5.17}$$

However since $\sigma'_{23} = \sigma'_{32}$ this condition is satisfied automatically and all the eigenvalues $\sigma(r)$, $(r = 1, 2, 3)$, are real.

In general $\sigma(1)$, $\sigma(2)$ and $\sigma(3)$ are distinct, in which case equations (4.5.13) with $\sigma = \sigma(r)$ $(r = 1, 2, 3)$ together with

$$n_i(r)n_i(r) = 1 \quad (r = 1, 2, 3; r \text{ not summed})$$

furnish sufficient information to determine (within a choice of sign) $\mathbf{n}(1)$, $\mathbf{n}(2)$ and $\mathbf{n}(3)$. We shall show that $\mathbf{n}(1)$, $\mathbf{n}(2)$ and $\mathbf{n}(3)$ are orthogonal.

For $\mathbf{n} = \mathbf{n}(1)$ and $\mathbf{n} = \mathbf{n}(2)$ equations (4.5.12) may be written

$$\sigma_{ij}n_j(1) = \sigma(1)n_i(1)$$

$$\sigma_{ij}n_j(2) = \sigma(2)n_i(2) \ .$$

Multiplying the first of these by $n_i(2)$ and the second by $n_i(1)$ and subtracting leads to

$$\sigma_{ij}n_j(1)n_i(2) - \sigma_{ij}n_j(2)n_i(1) = [\sigma(1) - \sigma(2)]\, n_i(1)n_i(2) \ . \tag{4.5.18}$$

However since $\sigma_{ij} = \sigma_{ji}$ the left hand side of (4.5.18) may be re-written

$$\sigma_{ij}n_j(1)n_i(2) - \sigma_{ji}n_j(2)n_i(1) \equiv \sigma_{ij}n_j(1)n_i(2) - \sigma_{ij}n_i(2)n_j(1) = 0 \ .$$

(from a suffix interchange)

It follows that if $\sigma(1) \neq \sigma(2)$ then

$$n_i(1)n_i(2) = 0\,, \quad \text{i.e. } \mathbf{n}(1)\,.\,\mathbf{n}(2) = 0 \ .$$

Clearly if $\sigma(1)$, $\sigma(2)$ and $\sigma(3)$ are all distinct then $\mathbf{n}(1)$, $\mathbf{n}(2)$ and $\mathbf{n}(3)$ form a mutually orthogonal set of unit vectors. Bearing in mind the possible choice of sign of the $\mathbf{n}(r)$, evidently the unit vectors may be chosen to form a right handed system so that $\mathbf{n}(1) \wedge \mathbf{n}(2) = \mathbf{n}(3)$. If we choose a system of coordinate axes $O''x''_1x''_2x''_3$ such that $O''x''_1$, $O''x''_2$, $O''x''_3$ lie in the directions $\mathbf{n}(1)$, $\mathbf{n}(2)$ and $\mathbf{n}(3)$, then with respect to these axes the stress matrix takes the diagonal form

$$\boldsymbol{\sigma}'' = \begin{pmatrix} \sigma(1) & 0 & 0 \\ 0 & \sigma(2) & 0 \\ 0 & 0 & \sigma(3) \end{pmatrix} . \tag{4.5.19}$$

The process of 'diagonalising' the original stress matrix $\boldsymbol{\sigma}$ (referred to axes $Ox_1x_2x_3$) is achieved by a suitable choice of $\mathbf{A}$ so that

$$\boldsymbol{\sigma}'' = \mathbf{A}\boldsymbol{\sigma}\mathbf{A}^T \ . \tag{4.5.20}$$

[In fact $\mathbf{A}$ is determined explicitly in terms of the components of the $\mathbf{n}(r)$, for from (4.2.7) $\mathbf{i}''_p \equiv \mathbf{n}(p) \equiv n_s(p)\mathbf{i}_s = a_{ps}\mathbf{i}_s$, i.e. $a_{ps} = n_s(p)$.]

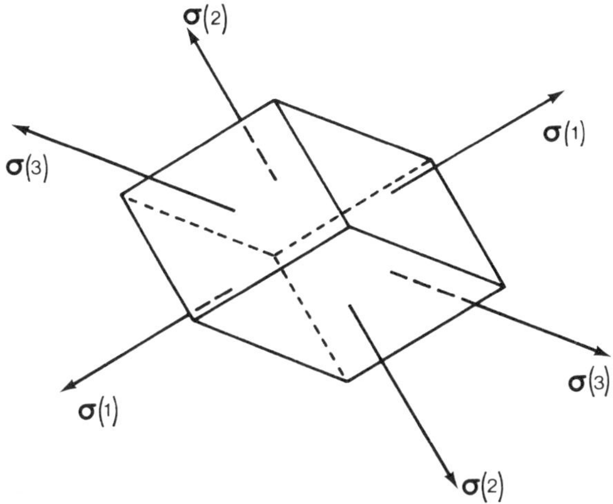

Fig. 4.3 Stress vectors acting on the faces of a parallelopiped whose edges lie in the directions **n**(1), **n**(2), **n**(3)

It should be emphasised that usually at different points in a continuum $\boldsymbol{\sigma}$ is different so that the $\mathbf{n}(r)$ [and the $\sigma(r)$] vary with position (and perhaps with time).

The normal stress components $\sigma(r)$ referred to the double primed system are known as principal stresses and the directions specified by the $\pm\mathbf{n}(r)$ are known as principal axes. If a small rectangular parallelopiped is located with sides parallel to the $\mathbf{n}(r)$, the faces of the parallelopiped are subject only to normal stress vectors of magnitude $\sigma(1)$, $\sigma(2)$, $\sigma(3)$ as in Fig. (4.3).

Principal stresses are important both to the engineer and the technologist since for many metals an adequate criterion for the occurrence of fracture is

$$\text{Max}\ [\sigma(1), \sigma(2), \sigma(3)] = \sigma_f$$

where $\sigma_f (> 0)$, a constant for the material, is termed the fracture stress. Principal stresses also enter the theory of the plastic deformation of metals; here it is commonly assumed that in order for metals to deform in an irrecoverable manner the principal stresses satisfy the equation

$$\text{Max}\ [\,|\sigma(1) - \sigma(2)|,\ |\sigma(2) - \sigma(3)|,\ |\sigma(3) - \sigma(1)|\,] = Y$$

where Y, again a positive constant characteristic of the material, is called the 'yield stress' [see Chapter 16].

The exceptional cases where two (or all three) of the eigenvalues are equal involves a modified discussion. We return to consideration of the case where $\mathbf{n}(1)$ is determined only to satisfy

$$a_1 n_1 + b_1 n_2 - c_1 n_3 = 0$$

Let **n**(1) be chosen arbitrarily to lie within the plane whose normal is in the direction $a_1 : b_1 : -c_1$ and choose the $O'x'_1$ axis to lie in the **n**(1) direction. With **n**(1) chosen, select **n**(2) and the direction of the $O'x'_2$ axis also to lie in the plane with **n**(2) perpendicular to **n**(1). Complete the choice of unit vectors and primed coordinate system with **n**(3) normal to the plane so that **n**(1), **n**(2), **n**(3) form a right handed orthogonal triad. With reference to the primed coordinate system the stress components σ'_{12}, σ'_{13} vanish since **n**(1) is a principal stress direction. Also the stress components σ'_{21}, σ'_{23} vanish since for *any* unit vector lying in the plane equations (4.5.12) are satisfied. Therefore the stress vector acting on surfaces with normal in both the **n**(1) and **n**(2) directions are normal and of equal magnitude $\sigma(1)$. Combining these results with the symmetry conditions for σ_{ij} shows finally that for the primed system the stress matrix is diagonal

$$\boldsymbol{\sigma}' = \begin{pmatrix} \sigma(1) & 0 & 0 \\ 0 & \sigma(1) & 0 \\ 0 & 0 & \sigma(3) \end{pmatrix} \tag{4.5.21}$$

so that circumstances in which **n**(1) is not determined uniquely (to within a choice of sign) are associated with the case of two equal eigenvalues $[\sigma(2) = \sigma(1)]$. Clearly *any* choice of axes $O'x'_1$, $O'x'_2$ lying within the plane with normal **n**(3) leads to the diagonal matrix (4.5.21).†

The case $\sigma(1) = \sigma(2) = \sigma(3) = \sigma$ (say) arises if the stress matrix is initially diagonal with three equal eigenvalues, i.e. if

$$\boldsymbol{\sigma} = \begin{pmatrix} \sigma & 0 & 0 \\ 0 & \sigma & 0 \\ 0 & 0 & \sigma \end{pmatrix}$$

or equivalently

$$\sigma_{ij} = \sigma\delta_{ij} \ . \tag{4.5.22}$$

Here any choice of mutually orthogonal unit vectors defines a set of principal axes, as is also clear if we recollect that δ_{ij} (and hence $\sigma\delta_{ij}$) is a second order symmetric tensor. In other words for all choices of the primed coordinate system, $\boldsymbol{\sigma}' = \sigma\mathbf{I} = \boldsymbol{\sigma}$.

For some materials, notably solids such as steel and concrete, which are readily able to support significant shear stresses, stress states of the form (4.5.22) usually arise (if at all) only at isolated points. Engineering books refer to such stress states as 'hydrostatic' or more specifically as 'hydrostatic tension' ($\sigma > 0$) or 'hydrostatic pressure' ($\sigma < 0$).

A more important example of (4.5.22) is the class of materials which are unable to sustain significant shear stresses. Under some flow conditions the shear stresses occurring in liquids and gases are small compared with the normal stresses. In these circumstances the behaviour of the material is described adequately by the equation

† Also, evidently $\pm$**n**(3) lies in the direction $a_1 : b_1 : -c_1$.

$$\sigma_{ij} = -P\delta_{ij} \tag{4.5.23}$$

where P, the pressure, is positive.

The statement that P is positive is a reflexion of the fact that the materials in question are unable to sustain states of hydrostatic tension. This is evident, for example, if we consider the possibility of exerting a tensile traction on the surface of a liquid. However it is known that in exceptional circumstances liquids may sustain appreciable tensile stresses for short time durations. This occurs during the reflexion of explosively generated pressure pulses at the free surface of a liquid. Probably under these conditions the liquid is able to sustain significant shear stresses; so that in any event the model $\sigma_{ij} = -P\delta_{ij}$ is quite inadequate. Under less drastic flow conditions, attempts to impose negative pressure result in the formation of voids or bubbles, i.e. the material breaks up or *cavitates*.

For materials described by (4.5.23), P is usually a function of density and temperature (known from experiment) and the resulting equations

$$\sigma_{ij} = -P(\rho, T)\delta_{ij} \tag{4.5.24}$$

are then constitutive equations for these materials.

Materials whose mechanical behaviour is described by (4.5.24) are known as 'ideal fluids'. In circumstances where also the variation of P with T is unimportant so that

$$\sigma_{ij} = -P(\rho)\delta_{ij}$$

the material is known as a 'barotropic' fluid.

In the absence of motion, virtually all liquids and gases conform to the ideal fluid model. For many purposes the model also serves to describe the mechanical properties of liquids and gases in motion.

Because an ideal fluid is unable to support shear stress, the force exerted by the fluid on a small element of surface of an immersed object is normal to the surface element. This is a familiar result in elementary hydrostatics. Sometimes this fact appears in elementary texts in the unfortunate form 'the pressure is the same in all directions'.

4.6 The polar decomposition theorem

This chapter concludes with an isolated result in matrix algebra which is of some subsequent importance.

The polar decomposition theorem of matrix algebra is concerned with the representation of a non-singular square matrix, possibly with complex elements, in terms of products of Hermitian and Unitary matrices. Here we shall be concerned with a restricted form of the theorem for real 3×3 matrices $\mathbf{F} \equiv (F_{ij})$ with positive determinant.

The theorem states that any such matrix **F** may be represented in either of the forms

$$\mathbf{F} = \mathbf{RU} \tag{4.6.1}$$

$$\mathbf{F} = \mathbf{VR} \tag{4.6.2}$$

where (i) **R** is a proper orthogonal matrix (a matrix associated with a rotation of coordinate axes), (ii) **U** and **V** are real symmetric matrices with positive eigenvalues which satisfy the matrix equations

$$\mathbf{U}^2 = \mathbf{F}^T\mathbf{F}, \quad \mathbf{V}^2 = \mathbf{F}\mathbf{F}^T . \tag{4.6.3}$$

Further the representations (4.6.1) and (4.6.2) are unique; also **U** and **V** possess the same set of (positive) eigenvalues.

The proof now given is 'constructive' in the sense that by following the steps through for a given **F** it is possible to construct the three matrices **U**, **V** and **R**.

In principle therefore it is possible to give 'formulae' for the elements of the three matrices; in practise this is a virtual impossibility, since one of the steps entails solving a cubic equation, and the resulting formulae would be extremely complicated. We can readily verify the degree of complexity by inspecting Cardan's solution of the cubic. Even if the complications of the latter are acceptable, it transpires that the elements of **R**, **U** and **V** are themselves complicated functions of the roots of the cubic, as well as of other quantities appearing in **F**. Fortunately for almost all purposes in continuum mechanics only the form of the polar decomposition theorem is important, and not the precise details of the matrices **R**, **U** and **V**.

We begin by noting that both $\mathbf{F}^T\mathbf{F}$ and $\mathbf{F}\mathbf{F}^T$ are symmetric matrices whose (i, j) components are respectively

$$F_{ki}F_{kj}, \quad F_{ik}F_{jk} .$$

Consider in detail the product $\mathbf{F}^T\mathbf{F}$ and the proof of (4.6.1). Since $\mathbf{F}^T\mathbf{F}$ is symmetric there exists a proper orthogonal matrix (say $\mathbf{A}^T$) which transforms $\mathbf{F}^T\mathbf{F}$ into diagonal form

i.e. $\qquad \mathbf{A}^T\mathbf{F}^T\mathbf{F}\mathbf{A} \qquad$ is diagonal.

Moreover the diagonal elements of $\mathbf{A}^T\mathbf{F}^T\mathbf{F}\mathbf{A}$ are positive. This last result is most easily seen by noting that since $\mathbf{A}\mathbf{A}^T = \mathbf{I}$ the diagonalised matrix also has the representation

$$\mathbf{A}^T\mathbf{F}^T\mathbf{F}\mathbf{A} = (\mathbf{F}')^T\mathbf{F}'$$

where $\qquad \mathbf{F}' = \mathbf{A}^T\mathbf{F}\mathbf{A} .$

The diagonal elements of $(\mathbf{F}')^T\mathbf{F}'$ are

$$F'_{k\alpha}F'_{k\alpha} , \; \alpha = 1, \; 2 \text{ or } 3 \quad (\alpha \text{ not summed})$$

and evidently are formed from sums of squares. Writing $\lambda^2(1)$, $\lambda^2(2)$ and $\lambda^2(3)$

respectively in place of these sums shows that the matrix $\mathbf{F}^T\mathbf{F}$ has the representation

$$\mathbf{F}^T\mathbf{F} = \mathbf{A}\,\mathrm{diag}\,[\lambda^2(1), \lambda^2(2), \lambda^2(3)]\,\mathbf{A}^T \tag{4.6.4}$$

Also for non-singular $\mathbf{F}$ it follows easily from taking the determinant of each side of (4.6.4) that $\lambda^2(1)\lambda^2(2)\lambda^2(3) > 0$ so that none of $\lambda^2(1), \lambda^2(2), \lambda^2(3)$ is zero. Without loss of generality we choose $\lambda(1)$, $\lambda(2)$ and $\lambda(3)$ to be positive.

The matrix $\mathbf{F}^T\mathbf{F}$ possesses $2^3 (= 8)$ 'square roots'

$$\mathbf{U} = \mathbf{A}\,\mathrm{diag}\,[\pm\lambda(1), \pm\lambda(2), \pm\lambda(3)]\,\mathbf{A}^T$$

as is readily verified. For present purposes $\mathbf{U}$ is defined uniquely by choosing all the eigenvalues of $\mathbf{U}$ to be positive, i.e. we define

$$\mathbf{U} = \mathbf{A}\,\mathrm{diag}\,[\lambda(1), \lambda(2), \lambda(3)]\,\mathbf{A}^T \tag{4.6.5}$$

It is perhaps worth noting that although $\mathbf{U}$ is unique, the representation (4.6.5) is not. For distinct $\lambda^2(1), \lambda^2(2), \lambda^2(3)$, a finite choice of matrices $\mathbf{A}$ exist. The various choices may be associated with labelling the eigenvalues in a different order or, equivalently, labelling the associated (primed) principal axes of $\mathbf{U}$ in a different order. Even with the order fixed there remains some choice in direction of principal axes, e.g. it is possible to interchange simultaneously the positive and negative directions of the Ox'_1 and Ox'_2 axes. The actual number of distinct proper diagonalising matrices is 18. However for any of these choices, $\mathbf{U}$ itself is invariant. For example the transformation $x'_1 \to -x'_2,\ x'_2 \to x'_1,\ x'_3 \to x'_3$ in (4.6.5) leads to the representation

$$\mathbf{U} = \hat{\mathbf{A}}\,\mathrm{diag}\,[\lambda(2), \lambda(1), \lambda(3)]\,\hat{\mathbf{A}}^T$$

where

$$\hat{\mathbf{A}} = \mathbf{A}\begin{bmatrix} 0 & -1 & 0 \\ 1 & 0 & 0 \\ 0 & 0 & 1 \end{bmatrix}$$

but it is readily verified that $\mathbf{U}$ is the same for both representations. For the case where two of $\lambda^2(1), \lambda^2(2), \lambda^2(3)$ are equal there are an infinite number of proper diagonalising matrices; again it is possible to show that $\mathbf{U}$ is invariant. In the case where all three eigenvalues are equal $\mathbf{U}$ is necessarily in the diagonal form $\lambda\mathbf{I}$.

$\mathbf{U}$ is a non-singular matrix since

$$|\mathbf{U}| = \lambda(1)\lambda(2)\lambda(3) > 0\ .$$

It follows that a matrix $\mathbf{R}$ is defined uniquely by

$$\mathbf{F} = \mathbf{R}\mathbf{U} \tag{4.6.6}$$

and further since by hypothesis $|\mathbf{F}| > 0$ while $|\mathbf{U}| = \lambda(1)\lambda(2)\lambda(3)$ is also positive it follows that $|\mathbf{R}| > 0$. Since also $|\mathbf{F}|^2 = |\mathbf{U}|^2$ then $|\mathbf{F}| = |\mathbf{U}|$ and

$$|\mathbf{R}| = 1$$

It remains to show that $\mathbf{R}^T\mathbf{R} = \mathbf{I}$. From (4.6.6)

$$\mathbf{R} = \mathbf{F}\mathbf{U}^{-1}, \quad \mathbf{R}^T = (\mathbf{U}^{-1})^T\mathbf{F}^T = \mathbf{U}^{-1}\mathbf{F}^T \tag{4.6.7}$$

since $\mathbf{U}$, and therefore $\mathbf{U}^{-1}$, is symmetric. From (4.6.7)

$$\mathbf{R}^T\mathbf{R} = \mathbf{U}^{-1}\mathbf{F}^T\mathbf{F}\mathbf{U}^{-1} = \mathbf{U}^{-1}\mathbf{U}^2\mathbf{U}^{-1} = \mathbf{I}$$

so that $\mathbf{R}$ is a proper orthogonal matrix. Also $\mathbf{R}\mathbf{R}^T = \mathbf{I}$ since $\mathbf{R}^T = \mathbf{R}^{-1}$ which commutes with $\mathbf{R}$.

The representation (4.6.6) is unique since $\mathbf{U}$ is unique.

A similar argument shows that $\mathbf{F}$ may also be written uniquely in the form

$$\mathbf{F} = \mathbf{V}\mathbf{R}_2 \tag{4.6.8}$$

(where for the moment $\mathbf{R}_2$ is not necessarily the same as $\mathbf{R}$). To prove $\mathbf{R}_2 = \mathbf{R}$ we write

$$\mathbf{F} = \mathbf{R}\mathbf{U} = \mathbf{R}\mathbf{U}\mathbf{R}^T\mathbf{R} = (\mathbf{R}\mathbf{U}\mathbf{R}^T)\mathbf{R}\ .$$

However $\mathbf{R}\mathbf{U}\mathbf{R}^T$ is a symmetric matrix and the representation (4.6.8) is unique. It follows that $\mathbf{R}_2 = \mathbf{R}$ and

$$\mathbf{V} = \mathbf{R}\mathbf{U}\mathbf{R}^T, \quad \mathbf{U} = \mathbf{R}^T\mathbf{V}\mathbf{R}$$

and hence that the eigenvalues of $\mathbf{U}$ and $\mathbf{V}$ are identical.

In general the calculation of $\mathbf{R}$, $\mathbf{U}$ and $\mathbf{V}$ from $\mathbf{F}$ is tedious.

Fortunately in subsequent applications of the theorem it is sufficient to know only the forms of the decompositions and not the details.

Summary

Important results of the present Chapter are the following:

(a) Physical laws are expressible in terms of tensor equations, as, for example, the equations of motion (4.4.13)

$$\frac{\partial \sigma_{ij}}{\partial x_j} + \rho b_i = \rho f_i$$

(b) The energy identity (4.4.17)

$$\iint\limits_{S_t} \sigma_{ij}v_i n_j dS + \iiint\limits_{\Omega_t} \rho b_i v_i d\Omega = \frac{d}{dt}\iiint\limits_{\Omega_t} \tfrac{1}{2}\rho v_i v_i d\Omega + \iiint\limits_{\Omega_t} \sigma_{ij} d_{ij} d\Omega$$

where

$$d_{ij} = \frac{1}{2}\left(\frac{\partial v_i}{\partial x_j} + \frac{\partial v_j}{\partial x_i}\right)$$

and where S_t encloses the same particles.

(c) In matrix notation the components of a second order tensor T satisfy the matrix equation

$$\mathbf{T}' = \mathbf{A}\mathbf{T}\mathbf{A}^T$$

where $\mathbf{A}$ is a proper orthogonal matrix relating primed and unprimed coordinate systems. If $T_{ij} = T_{ji}$ there is an $\mathbf{A}$ for which $\mathbf{T}'$ is diagonal and the diagonal elements are the eigenvalues of $\mathbf{T}$.

(d) The characteristic equation of the stress matrix determines three principal stresses. If these are distinct, three mutually orthogonal principal axes are determined uniquely. When two principal stresses are equal only one of the principal axes (associated with the third principal stress) is determined uniquely. For the case where all three principal stresses are equal, any set of orthogonal axes form a set of principal axes.

(e) A non-singular square matrix $\mathbf{F}$ with real elements F_{ij} and positive determinant may be decomposed uniquely into either of the forms

$$\mathbf{F} = \mathbf{R}\mathbf{U} = \mathbf{V}\mathbf{R}$$

where $\mathbf{U}$ and $\mathbf{V}$ are symmetric matrices with the same set of positive eigenvalues, and $\mathbf{R}$ is a proper orthogonal matrix.

Problems, Chapter 4

1. Solely from the viewpoint of suffix convention, state which of the following expressions or equations are well formed. Where the expressions (or equations) are well formed, assuming a_{ij}, b_i and c_{ij} are components of tensors which depend on x_i and t, state the (tensor) order of the expressions (or equations).

(a) $a_{ij}b_j$ (b) $a_{ij}c_{ij}$

(c) $a_{ij}c_{jj}$ (d) $a_{ii}c_{jj}$

(e) $a_{ij}c_{ji}$ (f) $a_{ij} = b_ib_j$

(g) $a_{ii} = c_{jj}$ (h) $\partial a_{ij}/\partial x_k = c_{ik}b_j$

(i) $b_i\partial b_i/\partial t = b_m b_m$ (j) $c_{ij} = (\partial b_i/\partial x_j + \partial b_j/\partial x_i)$

(k) $c_{ij} = (\partial b_i/\partial x_i) + (\partial b_j/\partial x_j)$.

2. Verify that the matrix

$$\mathbf{A} = \begin{pmatrix} \frac{1}{4}\sqrt{6} & \frac{1}{4}\sqrt{2} & \frac{1}{2}\sqrt{2} \\ \frac{1}{4}\sqrt{6} & \frac{1}{4}\sqrt{2} & -\frac{1}{2}\sqrt{2} \\ -\frac{1}{2} & \frac{1}{2}\sqrt{3} & 0 \end{pmatrix}$$

is proper orthogonal.

Two coordinate systems are related by

$$\begin{pmatrix} x_1' \\ x_2' \\ x_3' \end{pmatrix} = \mathbf{A}\begin{pmatrix} x_1 \\ x_2 \\ x_3 \end{pmatrix} .$$

Referred to the unprimed system, the stress matrix at a point is

$$\boldsymbol{\sigma} = \begin{pmatrix} \frac{1}{2} & -\frac{1}{2}\sqrt{3} & \frac{1}{2}\sqrt{3} \\ -\frac{1}{2}\sqrt{3} & \frac{3}{2} & \frac{1}{2} \\ \frac{1}{2}\sqrt{3} & \frac{1}{2} & 0 \end{pmatrix}$$

Find $\boldsymbol{\sigma}'$, the stress matrix referred to the primed coordinate system, and write down the principal stresses $\sigma(1)$, $\sigma(2)$, $\sigma(3)$. Verify directly that

$$|\sigma| = \sigma(1)\sigma(2)\sigma(3) .$$

3. In Problem 2 find the magnitude and direction of the stress vector acting on the surface whose normal is equally inclined to all three (positive) unprimed coordinate axes.

4. (i) Show that a rotation of coordinates about the x_3 (i.e. z) axis is specified by the proper orthogonal matrix

$$\mathbf{A}(\theta) = \begin{pmatrix} \cos\theta & \sin\theta & 0 \\ -\sin\theta & \cos\theta & 0 \\ 0 & 0 & 1 \end{pmatrix}$$

where θ is the angle of rotation.

(ii) Verify that

$$\mathbf{A}(\theta)\mathbf{A}(\phi) = \mathbf{A}(\theta + \phi)$$

and interpret this result geometrically.

(iii) If

$$\boldsymbol{\sigma} = \begin{pmatrix} \sigma_{xx} & \sigma_{xy} & 0 \\ \sigma_{yx} & \sigma_{yy} & 0 \\ 0 & 0 & \sigma_{zz} \end{pmatrix}$$

show that on referring $\boldsymbol{\sigma}$ to the primed coordinates (x_1', x_2', x_3') where $x_1' = x_1 \cos\theta + x_2 \sin\theta$, $x_2' = -x_1 \sin\theta + x_2 \cos\theta$, $x_3' = x_3$, the stress matrix $\boldsymbol{\sigma}'$ assumes the form

$$\begin{pmatrix} \sigma_{xx}\cos^2\theta + \sigma_{yy}\sin^2\theta + \sigma_{xy}\sin(2\theta) & \sigma_{xy}\cos 2\theta + \frac{1}{2}(\sigma_{yy} - \sigma_{xx})\sin(2\theta) & 0 \\ \sigma_{xy}\cos(2\theta) + \frac{1}{2}(\sigma_{yy} - \sigma_{xx})\sin(2\theta) & \sigma_{xx}\sin^2\theta + \sigma_{yy}\cos^2\theta - \sigma_{xy}\sin(2\theta) & 0 \\ 0 & 0 & \sigma_{zz} \end{pmatrix}$$

Deduce that if θ satisfies

$$\tan(2\theta) = 2\sigma_{xy}/(\sigma_{xx} - \sigma_{yy})$$

then $\boldsymbol{\sigma}'$ is diagonal with diagonal elements (principal stresses) given by

$$\tfrac{1}{2}[\sigma_{xx} + \sigma_{yy} \pm \{(\sigma_{xx} - \sigma_{yy})^2 + 4\sigma_{xy}^2\}^{\frac{1}{2}}], \quad \sigma_{zz} .$$

Show directly that the eigenvalues of $\boldsymbol{\sigma}$ are given by the above expressions.

5. Show that the normal component of the stress vector acting on a surface with normal **n** is

$$\mathbf{T}_{(\mathbf{n})}\cdot\mathbf{n} = \sigma_{ij}n_in_j$$

and that the shear component is of magnitude

$$S = [\mathbf{T}^2 - (\mathbf{T}\cdot\mathbf{n})^2]^{\frac{1}{2}} \equiv [\sigma_{ij}\sigma_{ik}n_jn_k - \sigma_{ij}\sigma_{kl}n_in_jn_kn_l]^{\frac{1}{2}} .$$

If $\boldsymbol{\sigma}$ is in principal axes form deduce that

$$S^2 = \sigma^2(1)n_1^2 + \sigma^2(2)n_2^2 + \sigma^2(3)n_3^2 - [\sigma(1)n_1^2 + \sigma(2)n_2^2 + \sigma(3)n_3^2]^2$$

where n_1, n_2, n_3 are restrained to satisfy

$$n_1^2 + n_2^2 + n_3^2 = 1 .$$

By eliminating n_3 from the formula for S^2 show that S^2 (and hence S) is a maximum with respect to n_1, n_2 for either

$$n_1 = 0,\ n_2^2 = n_3^2 = \tfrac{1}{2} \text{ with } S = \tfrac{1}{2}|\sigma(2) - \sigma(3)|$$

or

$$n_2 = 0,\ n_1^2 = n_3^2 = \tfrac{1}{2} \text{ with } S = \tfrac{1}{2}|\sigma(3) - \sigma(1)| .$$

[From symmetry arguments there is also a maximum for $n_3 = 0, n_1^2 = n_2^2 = \frac{1}{2}$ of value $S_{\max} = \frac{1}{2}|\sigma(1) - \sigma(2)|$. This may be found, for example, by eliminating originally n_1 or n_2 instead of n_3. Alternately all three local maxima may be derived simultaneously from the theory of Lagrange multipliers.]

Deduce that the absolute maximum shear stress (i.e. maximum with respect to orientation) is

$$S_{max} = \tfrac{1}{2}\,\mathrm{Max}\,\{|\sigma(1) - \sigma(2)|,\ |\sigma(2) - \sigma(3)|,\ |\sigma(3) - \sigma(1)|\} .$$

[This last result is important in the theory of yield of many solid materials. For some metals, for example, the behaviour is elastic until S_{max} attains a critical value, at which stage the material deforms plastically (see Chapter 16). The formula above shows how to calculate S_{max} in terms of principal stresses.]

6. In many theories of plane stress the stress matrix everywhere assumes the form

$$\boldsymbol{\sigma} = \begin{pmatrix} \sigma_{xx} & \sigma_{xy} & 0 \\ \sigma_{yx} & \sigma_{yy} & 0 \\ 0 & 0 & \sigma_{zz} \end{pmatrix}$$

where the non-vanishing components of stress are functions solely of x and y. For incompressible materials it is also true usually that

$$\sigma_{zz} = \tfrac{1}{2}(\sigma_{xx} + \sigma_{yy}) .$$

Use the results of Problems 4 and 5 to show that in the above circumstances, the maximum shear stress is given by

$$S_{max} = \tfrac{1}{2}[(\sigma_{xx} - \sigma_{yy})^2 + 4\sigma_{xy}{}^2]^{\frac{1}{2}} .$$

7. Show that if an arbitrary state of stress is augmented by a hydrostatic pressure P, so that the stress matrix assumes the form

$$\boldsymbol{\sigma} = \begin{pmatrix} \sigma_{xx} - P & \sigma_{xy} & \sigma_{xz} \\ \sigma_{yx} & \sigma_{yy} - P & \sigma_{yz} \\ \sigma_{zx} & \sigma_{zy} & \sigma_{zz} - P \end{pmatrix}$$

then the maximum shear stress (see Problem 5) is independent of P.

8. Show that for the stress matrix

$$\boldsymbol{\sigma} = \begin{pmatrix} \dfrac{41}{50} & -\dfrac{6}{25} & -\dfrac{9}{10} \\ -\dfrac{6}{25} & \dfrac{17}{25} & -\dfrac{6}{5} \\ -\dfrac{9}{10} & -\dfrac{6}{5} & \dfrac{1}{2} \end{pmatrix}$$

the principal stresses are 1, -1 and 2, and determine the directions of the principal axes. Verify that the principal axes are mutually orthogonal.

9. Show that for the stress matrix

$$\boldsymbol{\sigma} = \begin{pmatrix} 34 & 12 & 0 \\ 12 & 41 & 0 \\ 0 & 0 & 25 \end{pmatrix}$$

two of the principal stresses are equal. Show that $\mathbf{n} = \pm(\tfrac{3}{5}, \tfrac{4}{5}, 0)$ determines the direction of one principal stress axis and that the other axes are not uniquely determined, but specified merely to lie in the plane whose normal is in the direction $\mathbf{n}$.

10. For some problems it is more convenient to use stress components referred to coordinates other than Cartesian. An example is given by cylindrical polar coordinates (r, θ, z) where z is the usual Cartesian coordinate while r and θ are the usual plane polar coordinates defined by

$$x = r\cos\theta\,, \quad y = r\sin\theta\ .$$

The r and θ directions are as shown in the diagram [i.e. the directions of increasing r (constant θ and z) and increasing θ (constant r and z)]. The choice of the positive z direction (out of the plane of the paper) shows that the r, θ and z directions form a right handed orthogonal system.

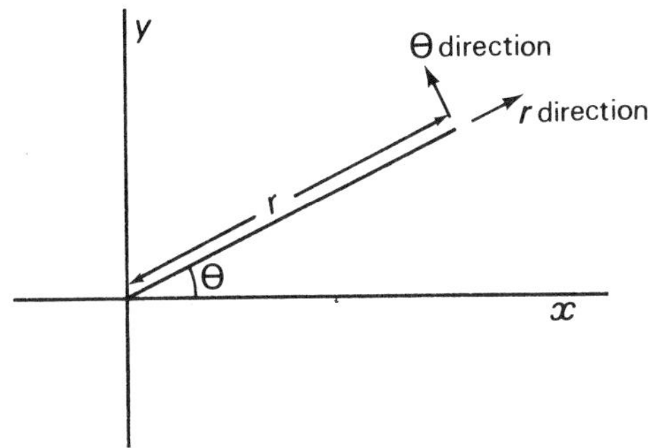

Referred to the orthogonal curvilinear coordinates (r, θ, z) the stress matrix may be written

$$\sigma = \begin{bmatrix} \sigma_{rr} & \sigma_{r\theta} & \sigma_{rz} \\ \sigma_{\theta r} & \sigma_{\theta\theta} & \sigma_{\theta z} \\ \sigma_{zr} & \sigma_{z\theta} & \sigma_{zz} \end{bmatrix}$$

where the notation is self evident; e.g. σ_{rr} denotes the component in the r direction of the stress vector acting on the surface whose normal is in the positive r direction, while $\sigma_{\theta z}$ denotes the component in the z direction of the stress vector acting on the surface with normal in the positive θ direction.

Show that

$$\begin{bmatrix} \sigma_{xx} & \sigma_{xy} & \sigma_{xz} \\ \sigma_{yx} & \sigma_{yy} & \sigma_{yz} \\ \sigma_{zx} & \sigma_{zy} & \sigma_{zz} \end{bmatrix} = \mathbf{A} \begin{bmatrix} \sigma_{rr} & \sigma_{r\theta} & \sigma_{rz} \\ \sigma_{\theta r} & \sigma_{\theta\theta} & \sigma_{\theta z} \\ \sigma_{zr} & \sigma_{z\theta} & \sigma_{zz} \end{bmatrix} \mathbf{A}^T$$

where

$$\mathbf{A} = \begin{bmatrix} \cos\theta & -\sin\theta & 0 \\ \sin\theta & \cos\theta & 0 \\ 0 & 0 & 1 \end{bmatrix}$$

and hence (in detail), that

$$\begin{aligned}
\sigma_{xx} &= \cos^2\theta\,\sigma_{rr} + \sin^2\theta\,\sigma_{\theta\theta} - \sin(2\theta)\sigma_{r\theta}\\
\sigma_{xy} &= \tfrac{1}{2}\sin(2\theta)\,[\sigma_{rr} - \sigma_{\theta\theta}] + \cos(2\theta)\sigma_{r\theta}\\
\sigma_{xz} &= \cos\theta\,\sigma_{rz} - \sin\theta\,\sigma_{\theta z}\\
\sigma_{yy} &= \cos^2\theta\,\sigma_{\theta\theta} + \sin^2\theta\,\sigma_{rr} + \sin(2\theta)\sigma_{r\theta}\\
\sigma_{yz} &= \sin\theta\,\sigma_{rz} + \cos\theta\,\sigma_{\theta z}\,.
\end{aligned}$$

Show also in changing from (x, y, z) to (r, θ, z) coordinates that the partial derivatives $\partial/\partial x$ and $\partial/\partial y$ become

$$\begin{aligned}
\frac{\partial}{\partial x} &= \cos\theta\,\frac{\partial}{\partial r} - r^{-1}\sin\theta\,\frac{\partial}{\partial\theta}\\
\frac{\partial}{\partial y} &= \sin\theta\,\frac{\partial}{\partial r} + r^{-1}\cos\theta\,\frac{\partial}{\partial\theta}\,.
\end{aligned}$$

Deduce that

$$\begin{aligned}
\frac{\partial\sigma_{xx}}{\partial x} + \frac{\partial\sigma_{xy}}{\partial y} + \frac{\partial\sigma_{xz}}{\partial z} &= L\cos\theta - M\sin\theta\\
\frac{\partial\sigma_{yx}}{\partial x} + \frac{\partial\sigma_{yy}}{\partial y} + \frac{\partial\sigma_{yz}}{\partial z} &= L\sin\theta + M\cos\theta\\
\frac{\partial\sigma_{zx}}{\partial x} + \frac{\partial\sigma_{zy}}{\partial y} + \frac{\partial\sigma_{zz}}{\partial z} &= N
\end{aligned}$$

where

$$\begin{aligned}
L &= \frac{\partial\sigma_{rr}}{\partial r} + \frac{\sigma_{rr} - \sigma_{\theta\theta}}{r} + \frac{1}{r}\frac{\partial\sigma_{r\theta}}{\partial\theta} + \frac{\partial\sigma_{rz}}{\partial z}\\
M &= \frac{1}{r}\frac{\partial\sigma_{\theta\theta}}{\partial\theta} + \frac{1}{r}\frac{\partial\sigma_{r\theta}}{\partial r} + \frac{2\sigma_{r\theta}}{r} + \frac{\partial\sigma_{\theta z}}{\partial z}\\
N &= \frac{\partial\sigma_{rz}}{\partial r} + \frac{\sigma_{rz}}{r} + \frac{1}{r}\frac{\partial\sigma_{\theta z}}{\partial\theta} + \frac{\partial\sigma_{zz}}{\partial z}\,.
\end{aligned}$$

Finally deduce the equations of motion in the r, θ and z directions

$$\begin{aligned}
L + \rho b_r &= \rho f_r\\
M + \rho b_\theta &= \rho f_\theta\\
N + \rho b_z &= \rho f_z
\end{aligned}$$

where (b_r, b_θ, b_z) and (f_r, f_θ, f_z) are the components of $\mathbf{b}$ and $\mathbf{f}$ referred to the (r, θ, z) coordinates.

Notes

(i) Equations of motion and stress components may be referred to other choices of curvilinear coordinates (notably spherical polar coordinates). With the exception of two examples in Chapter 16 which use the above results, little use is made in the present text of the basic equations of motion in other than Cartesian coordinates (although certain derived equations are sometimes referred to cylindrical and spherical polar coordinates).

(ii) The physical stress components σ_{rr}, $\sigma_{r\theta}$, etc., do *not* form the components of a *Cartesian* tensor, since they are referred to a coordinate system whose orientation varies with position. It is possible to define components of *non-Cartesian* tensors referred to (r, θ, z) coordinates (or indeed to any set of coordinates). However the quantities so defined in the case of the stress tensor (and not used in the present text) are not σ_{rr}, $\sigma_{r\theta}$, etc., but are obtained from the latter by multiplying by various geometrical factors defined by the curvilinear coordinate system.

11. The quantities I_σ, II_σ, III_σ defined by

$$\begin{aligned} \mathrm{I}_\sigma &= \sigma_{ii} \\ \mathrm{II}_\sigma &= \tfrac{1}{2}[\sigma_{ii}\sigma_{jj} - \sigma_{ij}\sigma_{ji}] \\ \mathrm{III}_\sigma &= \tfrac{1}{6}[2\sigma_{ij}\sigma_{jk}\sigma_{ki} - 3\sigma_{mm}\sigma_{ij}\sigma_{ji} + \sigma_{mm}\sigma_{pp}\sigma_{qq}] \end{aligned}$$

are evidently tensors of order zero (i.e. scalars), and therefore invariant with respect to rotation of coordinates.

By considering the particular case in which σ_{ij} assumes the principal axes form

$$\boldsymbol{\sigma} = \mathrm{diag}\,[\sigma(1), \sigma(2), \sigma(3)]$$

show that

$$\begin{aligned} \mathrm{I}_\sigma &= \sigma(1) + \sigma(2) + \sigma(3)\,, \\ \mathrm{II}_\sigma &= \sigma(1)\sigma(2) + \sigma(2)\sigma(3) + \sigma(3)\sigma(1)\,, \\ \mathrm{III}_\sigma &= \sigma(1)\sigma(2)\sigma(3)\,. \end{aligned}$$

Also show that

$$\mathrm{trace}\,\boldsymbol{\sigma} = \sigma_{ii}\,, \quad \mathrm{trace}\,\boldsymbol{\sigma}^2 = \sigma_{ij}\sigma_{ji}\,, \quad \mathrm{trace}\,\boldsymbol{\sigma}^3 = \sigma_{ij}\sigma_{jk}\sigma_{ki}$$

and hence deduce the results

$$\begin{aligned} \mathrm{I}_\sigma &= \mathrm{trace}\,\boldsymbol{\sigma} \\ \mathrm{II}_\sigma &= \tfrac{1}{2}[(\mathrm{trace}\,\boldsymbol{\sigma})^2 - \mathrm{trace}\,\boldsymbol{\sigma}^2] \\ \mathrm{III}_\sigma &= \tfrac{1}{6}[2\,\mathrm{trace}\,\boldsymbol{\sigma}^3 - 3(\mathrm{trace}\,\boldsymbol{\sigma})(\mathrm{trace}\,\boldsymbol{\sigma}^2) + (\mathrm{trace}\,\boldsymbol{\sigma})^3]\,. \end{aligned}$$

[In attempting to express I_σ, II_σ and III_σ in terms of σ_{ij} by using the original definitions of the text for I_σ, II_σ

i.e.

$$\begin{aligned} \mathrm{I}_\sigma &= \sigma(1) + \sigma(2) + \sigma(3) \\ \mathrm{II}_\sigma &= \sigma(1)\sigma(2) + \sigma(2)\sigma(3) + \sigma(3)\sigma(1) \\ &= \tfrac{1}{2}[\{\sigma(1) + \sigma(2) + \sigma(3)\}^2 - \{\sigma^2(1) + \sigma^2(2) + \sigma^3(3)\}] \end{aligned}$$

together with invariance concepts, it is not difficult to arrive at the results

$$\mathrm{I}_\sigma = \sigma_{ii}$$

$$\mathrm{II}_\sigma = \tfrac{1}{2}[\sigma_{ii}\sigma_{jj} - \sigma_{ij}\sigma_{ji}] \ .$$

On the other hand the expression for III_σ is not so obvious. Obtain the expression for III_σ by taking the trace of the Cayley-Hamilton result that $\boldsymbol{\sigma}$ satisfies its own characteristic equation

i.e. $$\boldsymbol{\sigma}^3 - \mathrm{I}_\sigma\boldsymbol{\sigma}^2 + \mathrm{II}_\sigma\boldsymbol{\sigma} - \mathrm{III}_\sigma\mathbf{I} = 0 \ .]$$

12. If $\mathbf{F}$ is the square matrix of order 2

$$\mathbf{F} = \begin{bmatrix} \frac{2}{5} & -1 \\ \frac{11}{5} & 2 \end{bmatrix}$$

find the matrices $\mathbf{R}$, $\mathbf{U}$ and $\mathbf{V}$ in the polar decomposition

$$\mathbf{F} = \mathbf{RU} = \mathbf{VR}$$

where $\mathbf{R}$ is proper orthogonal and $\mathbf{U}$ and $\mathbf{V}$ are symmetric matrices with the (same) positive eigenvalues.

Chapter 5

Continuum Deformation

5.1 Stretch and Rotation

This chapter is concerned with geometric aspects of the 'deformation' of a continuum. Primitive notions about deformable media include compression (or expansion) and distortion of volume elements of material. These ideas are here quantified in terms of the material description $\mathbf{r}(\mathbf{R}, t)$.

In general the amounts of compression (expansion) and distortion of material vary with position throughout the continuum. Accordingly a description of continuum deformation must be made on a local basis; in what follows we enquire into the geometrical aspects of the motion of material in the immediate vicinity of the particle labelled **R**.

The nature of the deformation of material in the immediate vicinity of the particle labelled **R** is made clear by examining the behaviour of all line vector elements connecting the material particle **R** with an arbitrary neighbouring material particle $\mathbf{R} + \mathbf{dR}$. In the reference configuration the line vector element is **dR**. At time t the line vector element connecting the same particles, now located at $\mathbf{r}(\mathbf{R}, t)$ and $\mathbf{r}(\mathbf{R} + \mathbf{dR}, t)$, is

$$\mathbf{dr} = \mathbf{r}(\mathbf{R} + \mathbf{dR}, t) - \mathbf{r}(\mathbf{R}, t) \; . \tag{5.1.1}$$

The effect of the motion is to carry the line vector element from **R** to **r** and concomitantly to transform the line vector element from **dR** to **dr**. It is the nature of this transformation, and not the translation of the vector line element, which determines the local deformation.

With the assumption that **r** is a differentiable function, equation (5.1.1) may be re-written (to first order in infinitesimal quantities)

$$\mathbf{dr} = (\mathbf{dR} \,.\, \mathrm{grad})\mathbf{r}(\mathbf{R}, t) \tag{5.1.2}$$

where the operator grad refers to the reference coordinates **R**.

Denoting components of **dr** and **dR** respectively by dx_i and dX_i and expressing (5.1.2) in component form gives

$$dx_i = \frac{\partial x_i}{\partial X_j}\, dX_j \; . \tag{5.1.3}$$

In writing down (5.1.3) use has been made of the summation convention.

It is readily verified that the 'deformation gradients' $\partial x_i/\partial X_j$ form the components of a second order Cartesian tensor, and use is made of this fact subsequently in formulating constitutive equations for elastic materials (Chapter 8). However for present purposes the nature of the deformation is made clearer by writing (5.1.3) in matrix form.

We define column matrices

$$\mathbf{dx} = \begin{pmatrix} dx_1 \\ dx_2 \\ dx_3 \end{pmatrix}, \qquad \mathbf{dX} = \begin{pmatrix} dX_1 \\ dX_2 \\ dX_3 \end{pmatrix} \tag{5.1.4}$$

and define $\mathbf{F}$ to be the *deformation gradient matrix*

$$\mathbf{F} = \left(\frac{\partial x_i}{\partial X_j}\right) \tag{5.1.5}$$

$\mathbf{F}$ is non-singular since from (2.2.6) and (2.2.10)

$$|\mathbf{F}| = J = \rho_0/\rho > 0 .$$

Accordingly $\mathbf{F}$ possesses an inverse $\mathbf{F}^{-1}$ such that

$$\mathbf{F}\mathbf{F}^{-1} = \mathbf{F}^{-1}\mathbf{F} = \mathbf{I} .$$

$\mathbf{F}^{-1}$ is expressed in terms of the reference coordinates from inversion of the right side of (5.1.5); alternately $\mathbf{F}^{-1}$ is given directly in terms of spatial coordinates by (4.3.6).

In terms of $\mathbf{dx}$ and $\mathbf{F}$, equations (5.1.3) assume the form

$$\mathbf{dx} = \mathbf{F}\mathbf{dX} \tag{5.1.6}$$

or equivalently

$$\mathbf{dX} = \mathbf{F}^{-1}\mathbf{dx} . \tag{5.1.7}$$

We consider in detail (5.1.6). Since $\mathbf{F}$ is non-singular with positive determinant we may use the polar decomposition theorem to express $\mathbf{F}$ in either of the forms

$$\mathbf{F} = \mathbf{RU}\,,\ \text{(a)}; \qquad \mathbf{F} = \mathbf{VR}\,,\ \text{(b)} \tag{5.1.8}$$

where $\mathbf{R}$ is a proper orthogonal matrix satisfying

$$\mathbf{R}\mathbf{R}^T = \mathbf{R}^T\mathbf{R} = \mathbf{I}\,, \qquad |\mathbf{R}| = 1 \tag{5.1.9}$$

while $\mathbf{U}$ and $\mathbf{V}$ are symmetric matrices with identical sets of positive eigenvalues $\lambda(1), \lambda(2), \lambda(3)$. The matrices $\mathbf{R}$, $\mathbf{U}$ and $\mathbf{V}$ are uniquely determined and satisfy the equation

$$\mathbf{V} = \mathbf{R}\mathbf{U}\mathbf{R}^T . \tag{5.1.10}$$

Both of the symmetric matrices $\mathbf{U}$ and $\mathbf{V}$ possess a set of principal axes. Referred to these (in general different) axes $\mathbf{U}$ and $\mathbf{V}$ assume the diagonal form

$$\text{diag}\,[\lambda(1), \lambda(2), \lambda(3)] .$$

As shown below the quantities $\lambda(r)$, $(r = 1, 2, 3)$, known as the principal stretches, provide direct measures of the deformation in the vicinity of the material particle **R**.

Let unit vectors directed along the principal axes of $\mathbf{U}$ be denoted by $\mathfrak{l}(1)$, $\mathfrak{l}(2)$ and $\mathfrak{l}(3)$ and define the associated column matrices

$$\mathfrak{l}(r) = \begin{pmatrix} l_1(r) \\ l_2(r) \\ l_3(r) \end{pmatrix} \quad (r = 1, 2, 3) .$$

The (r) satisfy the equations†

$$\mathbf{U}\mathfrak{l}(r) = \lambda(r)\mathfrak{l}(r) \quad (r = 1, 2, 3) \tag{5.1.11}$$

and the orthogonality conditions

$$\mathfrak{l}^T(r)\mathfrak{l}(s) = \delta_{rs} . \tag{5.1.12}$$

Column matrices satisfying (5.1.11) and (5.1.12) are known as 'orthonormal eigenvectors' of the (symmetric) matrix $\mathbf{U}$. Normalised eigenvectors of $\mathbf{V}$ are given by

$$\mathbf{R}\mathfrak{l}(r) \quad (r = 1, 2, 3) \tag{5.1.13}$$

since on pre-multiplying (5.1.11) by $\mathbf{R}$ and inserting $\mathbf{R}^T\mathbf{R}$ $(\equiv \mathbf{I})$ between $\mathbf{U}$ and $\mathbf{I}$ on the left side, yields

$$\mathbf{R}\mathbf{U}\mathbf{R}^T\mathbf{R}\mathfrak{l}(r) = \lambda(r)\mathbf{R}\mathfrak{l}(r)$$

i.e.

$$\mathbf{V}[\mathbf{R}\mathfrak{l}(r)] = \lambda(r)[\mathbf{R}\mathfrak{l}(r)] .$$

It is readily verified that the eigenvectors (5.1.13) are orthonormal. The geometrical interpretation of these results is that the principal axes of $\mathbf{V}$ are derived from those of $\mathbf{U}$ by the rotation specified by $\mathbf{R}$.

With these results equation (5.1.6) admits the following interpretation. We write

$$\mathbf{dy} = \mathbf{U}\mathbf{dX} , \text{ (a) }; \quad \mathbf{dx} = \mathbf{R}\mathbf{dy} , \text{ (b)} \tag{5.1.14}$$

which decomposes the transformation $\mathbf{dR} \to \mathbf{dr}$ into two transformations $\mathbf{dR} \to \mathbf{dy}$ and $\mathbf{dy} \to \mathbf{dx}$. The geometrical interpretation of the first operation is most easily seen if we refer equation (5.1.14a) to the principal axes of $\mathbf{U}$. We have

$$\begin{pmatrix} dy'_1 \\ dy'_2 \\ dy'_3 \end{pmatrix} = \begin{pmatrix} \lambda(1) & 0 & 0 \\ 0 & \lambda(2) & 0 \\ 0 & 0 & \lambda(3) \end{pmatrix} \begin{pmatrix} dX'_1 \\ dX'_2 \\ dX'_3 \end{pmatrix} \tag{5.1.15}$$

where the primes refer to components of $\mathbf{dy}$ and $\mathbf{dR}$ resolved along the principal axes of $\mathbf{U}$. From (5.1.15)

$$dy'_1 = \lambda(1)dX'_1, \quad dy'_2 = \lambda(2)dX'_2, \quad dy'_3 = \lambda(3)dX'_3$$

i.e. the deformation represented by (5.1.14a) is obtained by resolving the vector $\mathbf{dR}$ in the directions of the principal axes of $\mathbf{U}$, and multiplying the resulting

† In the equation that follows, there is no summation over r. There is no confusion if it is remembered that the summation convention is employed only in respect of suffixes.

components by the 'stretch ratios' $\lambda(1)$, $\lambda(2)$, and $\lambda(3)$. In particular an infinitesimal rectangular parallelopiped of material with sides dX'_1, dX'_2, dX'_3 is transformed by (5.1.14a) into a second rectangular parallelopiped with sides $\lambda(1)dX'_1$, $\lambda(2)dX'_2$, $\lambda(3)dX'_3$, lying respectively in the directions dX'_1, dX'_2, dX'_3. The origin of the term principal stretches for $\lambda(1)$, $\lambda(2)$ and $\lambda(3)$ is now clear.

In the second of the transformations (5.1.14b), $\mathbf{R}$ is a proper orthogonal matrix. Previously matrix equations of the form (5.1.14b) have been interpreted as relating the components of a vector in one coordinate system to the components of the same vector referred to a second coordinate system. The matrix $\mathbf{R}$ determines the axis and angle of rotation. An alternative interpretation, more appropriate in the present context, is that the vector **dr** is obtained from **dy** by a rotation specified by $\mathbf{R}$. Thus the combined operations (5.1.14a) and (5.1.14b) represent a stretching of the material by amounts $\lambda(1)$, $\lambda(2)$, $\lambda(3)$ in the direction of the principal axes of $\mathbf{U}$, followed by a rigid body rotation specified by $\mathbf{R}$. Fig. 5.1 depicts the effects of the combined operations on the (initial) rectangular parallelopiped with sides dX'_1, dX'_2, dX'_3

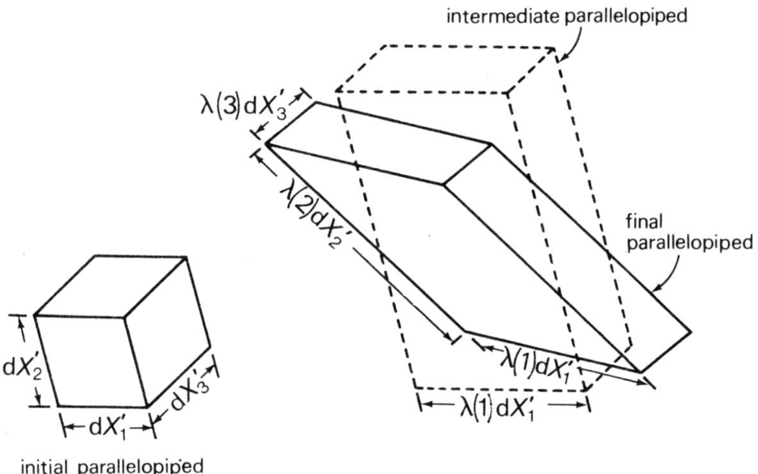

Fig. 5.1 Geometrical interpretation of $\mathbf{dx} = \mathbf{RUdX}$

Evidently material deformation, as distinct from rotation, is associated with $\mathbf{U}$ rather than $\mathbf{R}$. $\mathbf{U}$ is known as the **right stretch matrix**. In general the transformation

$$\mathbf{dy} = \mathbf{UdX}$$

alters both the dimensions and shape of small arbitrary shaped volume elements of material centered on **R**. Again, in general, only for the particular rectangular parallelopiped with sides lying in the directions of the principal axes of $\mathbf{U}$, is the rectangularity preserved. Other choices of initial rectangular parallelopiped lead to a rhomboid in the deformed configuration rather than a rectangular parallelopiped. [See the discussion in Section 5.2, for the initial parallelopiped with sides lying in the directions of the Cartesian axes.] Other initial choices of shape are also distorted. For example the sphere of radius dq, i.e.

$$(dX'_1)^2 + (dX'_2)^2 + (dX'_3)^2 = dq^2$$

becomes the ellipsoid

$$\frac{(dy'_1)^2}{\lambda^2(1)} + \frac{(dy'_2)^2}{\lambda^2(2)} + \frac{(dy'_3)^2}{\lambda^2(3)} = dq^2 .$$

The final state depicted in Fig. 5.1 is attained [using equations (5.1.14)] from a stretch followed by a rigid body rotation. Clearly from geometrical considerations the same final state may be obtained by performing the operations stretch and rotation in the reverse order, i.e. rotation followed by stretch. This is accomplished mathematically by the second polar decomposition of $\mathbf{F}$ (5.1.8b) when equation (5.1.6) may be decomposed into

$$\mathbf{dz} = \mathbf{RdX} \text{ (a)}, \qquad \mathbf{dx} = \mathbf{Vdz} \text{ (b)} .$$

The first of these equations (a) shows that the vector $\mathbf{dz}$ is obtained from $\mathbf{dR}$ by the rigid body rotation specified by $\mathbf{R}$; the second equation (b) describes the stretching process. The stretch ratios are $\lambda(1)$, $\lambda(2)$, $\lambda(3)$ as before, since $\mathbf{U}$ and $\mathbf{V}$ possess identical sets of eigenvalues. However the principal axes of stretching are now specified by the eigenvectors (5.1.13), reflecting merely the fact that the rotation has been effected prior to the stretching. The matrix $\mathbf{V}$, known as the **left stretch matrix**, describes the material deformation equally as well as $\mathbf{U}$. In fact equation (5.1.10) shows that the components of $\mathbf{U}$ and $\mathbf{V}$ may be regarded simply as components of the same tensor, referred to different axes.

The matrices $\mathbf{R}$, $\mathbf{U}$ and $\mathbf{V}$ are defined in terms of $\mathbf{F}$. However calculations of $\mathbf{R}$, $\mathbf{U}$ and $\mathbf{V}$ from $\mathbf{F}$ are algebraically impractical. For this reason it is simpler to express the deformation, as distinct from the rotation, in terms of quantities more readily calculated than $\mathbf{U}$ and $\mathbf{V}$. Quantities easily obtained from $\mathbf{F}$ are

$$\mathbf{C} \equiv \mathbf{U}^2 = \mathbf{F}^T\mathbf{F} \text{ (a)}, \qquad \mathbf{B} \equiv \mathbf{V}^2 = \mathbf{F}\mathbf{F}^T \text{ (b)} \qquad (5.1.16)$$

where $\mathbf{C}$ and $\mathbf{B}$ are known as the right and left Cauchy-Green matrices. Further $\mathbf{C}$ and $\mathbf{B}$ are clearly associated with deformation, rather than rotation, since, referred to (different) principal axes, both $\mathbf{C}$ and $\mathbf{B}$ take the form

$$\text{diag } [\lambda^2(1), \lambda^2(2), \lambda^2(3)]$$

and are associated only with the stretching.

The components of $\mathbf{C}$ admit a geometrical interpretation which determines the size and shape of the rhomboid obtained from the deformation of an initial rectangular parallelopiped, whose sides dX_1, dX_2, dX_3 lie in the Cartesian directions. This follows immediately.

5.2 Strain

To obtain an interpretation of the components of $\mathbf{C}$ compare the squares of the lengths of the vectors $\mathbf{dR}$ and $\mathbf{dr}$ respectively in the reference and deformed states. We define

$$dS^2 = dX_i dX_i\,, \quad ds^2 = dx_i dx_i$$

and consider the difference

$$ds^2 - dS^2\ .$$

This latter quantity is determined solely by the deformation (and not the rotation) since in circumstances where locally near the particle $\mathbf{R}$

$$ds^2 = dS^2$$

i.e.

$$|\mathbf{dr}| = |\mathbf{dR}|$$

for all local $\mathbf{dR}$, it is geometrically intuitive that the shape and magnitude of an arbitrary small volume element is preserved in the transformation, i.e. there is no deformation.

The quantity ds^2 may be expressed in terms of $\mathbf{C}$. From (5.1.6)

$$ds^2 = dx_i dx_i = \mathbf{dx}^T\mathbf{dx} = \mathbf{dX}^T\mathbf{F}^T\mathbf{F}\mathbf{dX}$$

while from the first of (5.1.8)

$$\mathbf{F}^T\mathbf{F} = \mathbf{U}\mathbf{R}^T\mathbf{R}\mathbf{U} = \mathbf{U}^2 = \mathbf{C}\ .$$

It follows that

$$\begin{aligned} ds^2 - dS^2 &= \mathbf{dX}^T(\mathbf{C} - \mathbf{I})\mathbf{dX} \\ &= 2\mathbf{dX}^T\mathbf{E}\mathbf{dX} \qquad (5.2.1) \end{aligned}$$

where $\mathbf{E}$, the matrix of the 'strain components' is defined by

$$\mathbf{E} = \tfrac{1}{2}(\mathbf{C} - \mathbf{I})\ .$$

In component form equation (5.2.1) becomes

$$ds^2 - dS^2 = 2E_{ij}dX_i dX_j \qquad (5.2.2)$$

where [from (5.1.5) and (5.1.16a)] the strain components E_{ij} are given by

$$E_{ij} = \tfrac{1}{2}\left[\frac{\partial x_m}{\partial X_i}\frac{\partial x_m}{\partial X_j} - \delta_{ij}\right]\ . \qquad (5.2.3)$$

The quantities E_{ij}, which from the quotient theorem form the components of a second order symmetric tensor, provide an easily obtained measure of the deformation. A geometrical interpretation of the E_{ij} follows from considering the fate of the rectangular parallelopiped with sides dX_1, dX_2, dX_3 lying in the directions of the coordinate axes. Consider first the vector

$$\mathbf{dR} = (dX_1, 0, 0)$$

of length

$$dS = |dX_1|\ .$$

After deformation the length of the transformed vector is, from (5.2.2),

$$\begin{aligned} ds^2 &= dS^2 + 2E_{11}dX_1dX_1 \\ &= (1 + 2E_{11})dS^2 \end{aligned}$$

so that the side dX_1 of the initial parallelopiped is stretched in the ratio

$$(1 + 2E_{11})^{\frac{1}{2}} .$$

Similar interpretations follow for the other diagonal components E_{22} and E_{33}.

After deformation the sides of the deformed parallelopiped are not perpendicular. To obtain an interpretation of the diagonal components of **E**, consider the angular deviation θ_{12} from perpendicularity of the two sides initially of length dX_1 and dX_2 [Fig. (5.2)].

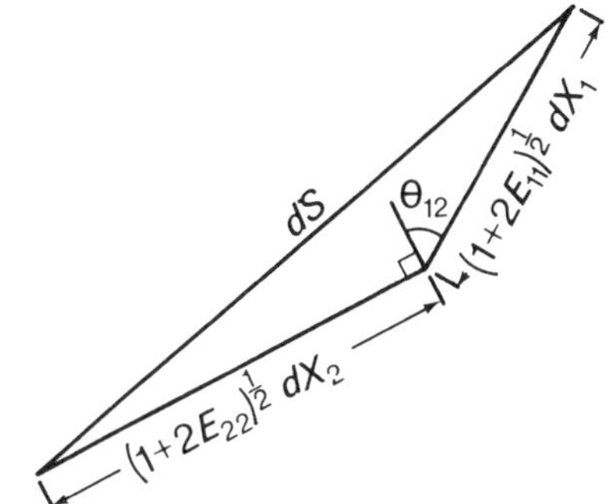

Fig. 5.2 Geometry for interpretation of E_{12}

From the figure

$$\begin{aligned} ds^2 &= (1 + 2E_{11})dX_1{}^2 + (1 + 2E_{22})\, dX_2{}^2 \\ &\quad + 2(1 + 2E_{11})^{\frac{1}{2}}(1 + 2E_{22})^{\frac{1}{2}} \sin(\theta_{12})\, dX_1 dX_2 \end{aligned} \qquad (5.2.4)$$

Also ds^2 is the square of the length of the vector derived initially from the vector

$$(dX_1, dX_2, 0)$$

so that from (5.2.2)

$$ds^2 - dX_1{}^2 - dX_2{}^2 = 2E_{11}dX_1{}^2 + 2E_{22}dX_2{}^2 + 2E_{12}dX_1dX_2 + 2E_{21}dX_2dX_1$$

i.e. $$ds^2 = (1 + 2E_{11})dX_1{}^2 + (1 + 2E_{22})dX_2{}^2 + 4E_{12}dX_1dX_2 \, . \qquad (5.2.5)$$

(In writing down (5.2.5) we recall that E_{ij} is symmetric so that $E_{12} = E_{21}$). Finally comparing (5.2.4) and (5.2.5) yields

$$\sin(\theta_{12}) = \frac{2E_{12}}{(1 + 2E_{11})^{\frac{1}{2}}(1 + 2E_{22})^{\frac{1}{2}}} \qquad (5.2.6)$$

which provides an interpretation for E_{12}. Similar interpretations follow for E_{13} and E_{23}. Fig. 5.3 shows that the effect of the deformation is to transform the initial rectangular parallelopiped into a rhomboid.

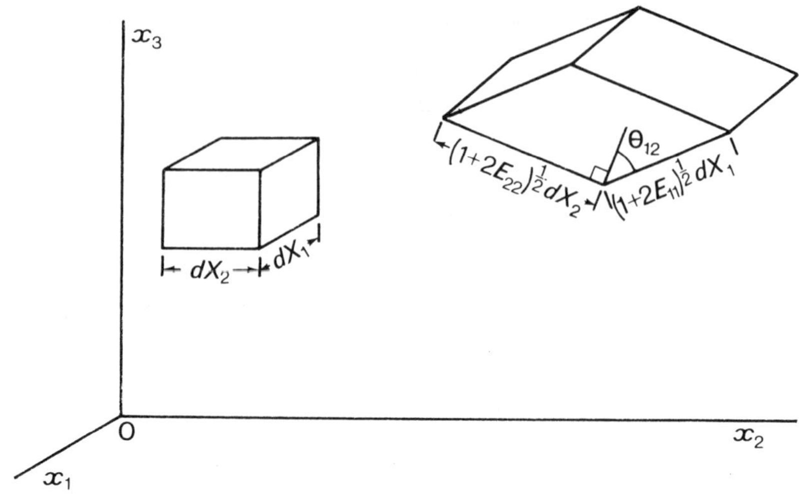

Fig. 5.3 Deformation of an initial rectangular parallelopiped with sides dX_1, dX_2 and dX_3 into a rhomboid.

The matrix $\mathbf{B}^{-1} = \mathbf{V}^{-2}$ is susceptible to a similar interpretation; we leave the reader to derive [using (5.1.7)]

$$dS^2 \equiv \mathbf{dX}^T\mathbf{dX} = \mathbf{dx}^T\mathbf{B}^{-1}\mathbf{dx}$$

so that the components of $(\mathbf{B}^{-1} - \mathbf{I})$ measure the changes in length and angular deviations from perpendicularity of the sides of an initial rhomboid whose final state is a rectangular parallelopiped of sides dx_1, dx_2, dx_3. The details are similar to those given above.

It is sometimes convenient to write

$$\mathbf{r} = \mathbf{R} + \mathbf{U}(\mathbf{R}, t) \tag{5.2.7}$$

and express E_{ij} in terms of the displacement gradients

$$\partial U_i / \partial x_j \ .$$

Writing (5.2.7) in component form

$$x_m = X_m + U_m(X_s, t)$$

and calculating E_{ij} from (5.2.3) leads to

$$\begin{aligned} E_{ij} &= \frac{1}{2}\left[\left(\delta_{mi} + \frac{\partial U_m}{\partial X_i}\right)\left(\delta_{mj} + \frac{\partial U_m}{\partial X_j}\right) - \delta_{ij}\right] \\ &= \frac{1}{2}\left[\frac{\partial U_i}{\partial X_j} + \frac{\partial U_j}{\partial X_i} + \frac{\partial U_m}{\partial X_i}\frac{\partial U_m}{\partial X_j}\right] \ . \end{aligned} \tag{5.2.8}$$

5.3 The 'small displacement gradient' approximation

Important simplifications occur in **F**, **U**, **V**, **R** and **E** if all nine dimensionless quantities $\partial U_i/\partial X_j$ may be assumed small compared with unity

i.e. $$\partial U_i/\partial X_j = O(e) \quad \text{(for all } i, j\text{)} \tag{5.3.1}$$

where $$e \ll 1 \ .$$

The approximation (5.3.1) is often realised for solid materials like steel which are able to withstand large external forces without appreciable deformation, e.g. as in most engineering applications.

For circumstances in which the approximation (5.3.1) prevails, E_{ij}, given by equation (5.2.8), may be approximated by

$$E_{ij} = \frac{1}{2}\left(\frac{\partial U_i}{\partial X_j} + \frac{\partial U_j}{\partial X_i}\right) \tag{5.3.2}$$

and also is of order e.

Again $$\mathbf{U}^2 = \mathbf{C} = \mathbf{I} + 2\mathbf{E} \ ,$$

or to first order in e $$\mathbf{U} = \mathbf{C}^{\frac{1}{2}} \simeq \mathbf{I} + \mathbf{E} \ . \tag{5.3.3}$$

In component form equation (5.3.3) becomes

$$U_{ij} = \delta_{ij} + \frac{1}{2}\left(\frac{\partial U_i}{\partial X_j} + \frac{\partial U_j}{\partial X_i}\right) \ .$$

It is possible to find also a simple approximate expression for **R**. Irrespective of the present approximation F_{ij} is given in terms of the U_i by

$$F_{ij} = \frac{\partial x_i}{\partial X_j} = \delta_{ij} + \frac{\partial U_i}{\partial X_j}$$

$$= \delta_{ij} + \frac{1}{2}\left(\frac{\partial U_i}{\partial X_j} + \frac{\partial U_j}{\partial X_i}\right) + \frac{1}{2}\left(\frac{\partial U_i}{\partial X_j} - \frac{\partial U_j}{\partial X_i}\right) \ . \tag{5.3.4}$$

However in the light of the assumption (5.3.1), (5.3.4) may now be approximated by

$$\mathbf{F} = (\mathbf{I} + \boldsymbol{\Omega})(\mathbf{I} + \mathbf{E}) \tag{5.3.5}$$

where $\boldsymbol{\Omega}$ is the anti-symmetric matrix with components

$$\Omega_{ij} = \frac{1}{2}\left(\frac{\partial U_i}{\partial X_j} - \frac{\partial U_j}{\partial X_i}\right) \ . \tag{5.3.6}$$

Comparing (5.3.3) and (5.3.5) with (5.1.8a) yields

$$\mathbf{R} \simeq \mathbf{I} + \boldsymbol{\Omega} \ . \tag{5.3.7}$$

Also since to order e, the factors on the right hand side of (5.3.5) may be commuted, we find on comparing with (5.1.8b),

$$\mathbf{V} \simeq \mathbf{I} + \mathbf{E} \simeq \mathbf{U}$$

so that in this approximation **V** and **U** are the same.

We note for future reference the associated approximations

$$\mathbf{B} \simeq \mathbf{C} = \mathbf{I} + 2\mathbf{E}\,, \quad \mathbf{B}^{-1} \simeq \mathbf{I} - 2\mathbf{E}\,. \tag{5.3.8}$$

It is illuminating to write $\boldsymbol{\Omega}$ in the form

$$\boldsymbol{\Omega} = \begin{pmatrix} 0 & -\Omega_3 & \Omega_2 \\ \Omega_3 & 0 & -\Omega_1 \\ -\Omega_2 & \Omega_1 & 0 \end{pmatrix}$$

where Ω_1, Ω_2 and Ω_3 are easily shown to be components of the vector

$$\boldsymbol{\Omega} = \tfrac{1}{2}\,\text{curl}\,\mathbf{U}(\mathbf{R}, t) \tag{5.3.9}$$

and 'curl' refers to the reference coordinates **R**. With the help of this result and (5.3.7) equation (5.1.14b), representing now small rotations of a rigid body, may be expressed in the vector form

$$\mathbf{dr} = \mathbf{dy} + \boldsymbol{\Omega} \wedge \mathbf{dy} \tag{5.3.10}$$

where $\boldsymbol{\Omega}$ is a dimensionless vector quantity of magnitude e. The axis of rotation is the direction of $\boldsymbol{\Omega}$, while the magnitude $|\boldsymbol{\Omega}|$ specifies the (small) angle of rotation (Fig. 5.4).

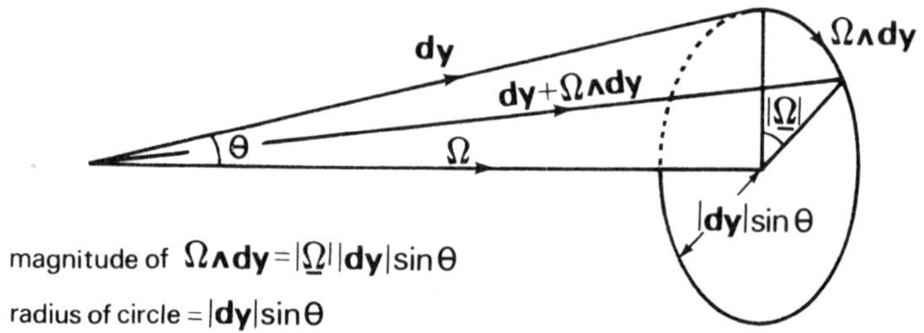

Fig. 5.4 Geometry of small rotations

To first order in e the magnitude of **dr** is the same as **dy** (as must be the case exactly for a rigid body rotation); the representation of angular rotation approximately by a vector equation like (5.3.10) is only possible for small rotations $|\boldsymbol{\Omega}| \ll 1$.

The equation (5.3.10) is familiar in rigid body mechanics as the starting point for the derivation of the velocity of a point in a rotating rigid body. Similar ideas are employed in Section 5.4 below in arriving at a geometrical interpretation of the spin tensor.

Other approximations consistent with those given above are that in the geometrical interpretations of the strain components (Fig. 5.2) the sides of the rhomboid are approximated by $(1 + E_{11})dX_1$, etc., while the angles θ_{12} etc. are approximated [from (5.2.6)] by

$$\theta_{12} = 2E_{12}\,, \text{ etc.}$$

Finally we note the result

$$J = |\mathbf{F}| \simeq 1 + E_{mm} = (1 + \operatorname{div}\mathbf{U}) \cong (1 - \operatorname{div}\mathbf{U})^{-1}$$

so that in this approximation the conservation of mass equation [equation (2.2.10)] reads

$$\rho = \rho_0(1 - \operatorname{div}\mathbf{U}) \ . \qquad (5.3.11)$$

The quantity div **U** is sometimes known as the dilatation.

The approximations discussed in this section are pertinent to the theory of classical linear elasticity (Section 8.4 and Chapter 14). Commonly the approximation is known as the 'small strain approximation'; better descriptions would be 'small strain and rotation approximation' or 'small displacement gradient approximation'.

5.4 Rate of strain tensor and spin tensor

Two important tensors defined in terms of spatial derivatives of the spatial velocity field are the (symmetric) rate of strain tensor (4.3.8) with components

$$d_{ij} = \frac{1}{2}\left(\frac{\partial v_i}{\partial x_j} + \frac{\partial v_j}{\partial x_i}\right) \qquad (5.4.1)$$

and the (anti-symmetric) spin tensor (4.3.9)

$$\omega_{ij} = \frac{1}{2}\left(\frac{\partial v_i}{\partial x_j} - \frac{\partial v_j}{\partial x_i}\right) \ . \qquad (5.4.2)$$

We have already encountered (5.4.1) in the energy theorem (4.4.17).

The origin of the terminology stems from the following interpretation of the components d_{ij} and ω_{ij}. Consider the configurations of the continuum at times t and $t + \Delta t$, and, for immediate purposes, suppose the configuration at time t to be the reference configuration. At time $t + \Delta t$ the new configuration is described by the displacement vector

$$\Delta\hat{\mathbf{U}} = \mathbf{v}(\mathbf{r}, t)\Delta t + O(\Delta t)^2$$

Here the circumflex notation is used to emphasise that $\Delta\hat{\mathbf{U}}$ is a measure of displacement from a reference state chosen to coincide with the current state. From the results of the previous paragraph the resulting (small) strains $\Delta\hat{E}_{ij}$ produced by $\Delta\hat{\mathbf{U}}$ are

$$\Delta\hat{E}_{ij} = \frac{1}{2}\left(\frac{\partial v_i}{\partial x_j} + \frac{\partial v_j}{\partial x_i}\right)\Delta t + O(\Delta t^2) \ . \qquad (5.4.3)$$

In deriving this result, spatial differentiations are taken with respect to spatial coordinates, reflecting the choice of reference configuration. From (5.4.3)

$$Lt_{\Delta t \to 0}\frac{\Delta\hat{E}_{ij}}{\Delta t} = \frac{1}{2}\left(\frac{\partial v_i}{\partial x_j} + \frac{\partial v_j}{\partial x_i}\right) \equiv d_{ij} \qquad (5.4.4)$$

i.e. the d_{ij} provide a measure of the rate at which continuum volume elements are attempting to change their shape and dimensions from the current configuration.

Because d_{ij} is defined by the limiting process on the left side of (5.4.4), the set of components d_{ij} are commonly known as the 'rate of deformation' or 'rate of strain' tensor. The latter terminology is slightly unfortunate since in general the d_{ij} are not to be confused with $\partial(E_{ij})/\partial t$ when the E_{ij} are given by (5.2.8), (for which the reference configuration is other than the current configuration). [Exceptionally d_{ij} is given approximately by $\partial(E_{ij})/\partial t$ in the small displacement gradient approximation, as in the theory of classical linear elasticity and the theory of linear viscoelasticity].

The analysis for the interpretation of the tensor components ω_{ij} is similar. We denote by $\boldsymbol{\omega}$ the vector

$$\boldsymbol{\omega} = \tfrac{1}{2} \operatorname{curl} \mathbf{v}$$

so that the matrix of the quantities ω_{ij} assumes the form

$$\begin{pmatrix} 0 & -\omega_3 & \omega_2 \\ \omega_3 & 0 & -\omega_1 \\ -\omega_2 & \omega_1 & 0 \end{pmatrix} .$$

Consider now the configurations of the continuum at times t and $t + \Delta t$, and as before choose the configuration at time t to be the reference configuration. Substituting $\Delta\hat{\mathbf{U}} = \mathbf{v}\Delta t$ into (5.3.9) shows that at the point $\mathbf{r}$, the (infinitesimal) rigid body rotation experienced by the continuum at time $t + \Delta t$ is given by

$$\Delta\boldsymbol{\Omega} = \tfrac{1}{2} \operatorname{curl} \mathbf{v}\, \Delta t \ .$$

Here $|\Delta\boldsymbol{\Omega}|$ measures the infinitesimal angle of rotation, while the axis of rotation is in the direction $\Delta\boldsymbol{\Omega}$. In other words

$$\boldsymbol{\omega} = Lt_{\Delta t \to 0}\left(\frac{\Delta\boldsymbol{\Omega}}{\Delta t}\right) \equiv \tfrac{1}{2} \operatorname{curl} \mathbf{v}$$

measures the instantaneous angular velocity of the rigid body rotation.

The result that $\frac{1}{2}$ curl $\mathbf{v}$ is the instantaneous angular velocity associated with the rigid body motion, is precise; this is in contrast to (5.3.7) which is only approximately correct for finite (albeit small) rotations. In the light of these comments, one interpretation of the vector $\boldsymbol{\omega}$ is as follows:

Suppose at time t a small elementary volume of material located at $\mathbf{r}$ is instantaneously frozen, i.e. inhibiting from shearing and stretching. Then the instantaneous motion of the element is that of a rigid body with angular velocity $\boldsymbol{\omega}$.

The origin of the term spin tensor is now clear. Sometimes the vector $\boldsymbol{\xi}$ defined by

$$\boldsymbol{\xi} = \operatorname{curl} \mathbf{v} \equiv 2\boldsymbol{\omega}$$

is known as the *vorticity*.

It might be expected that tensor quantities like B_{ij}, C_{ij}, E_{ij} and d_{ij} which provide measures of deformation and rates of deformation, figure prominently in the fomulation of constitutive equation. This is indeed true. In particular the tensors, B_{ij}, C_{ij} and E_{ij}, which are defined in terms of an initial reference configuration, are important in the constitutive equations of solids, while d_{ij} appears in the constitutive equations of fluids.

The reason for the importance of a reference state in the constitutive equations of solids is that the mechanical behaviour of solids depends on the deviation of the current configuration from some 'remembered' initial configuration†. This is clear for example in the familiar experience of stretching a rubber band. The greater its extension, i.e. the greater the deviation from the unstretched state, the greater is the force required; further on release of the force the rubber band recovers to it's unstretched (or 'remembered') state.

On the other hand the mechanical behaviour of many fluids, though by no means all, does not suggest the existence of a remembered configuration. For instance there is no tendency for water to return to its initial configuration after being stirred in a vessel. For this reason it is the tensor d_{ij} which features prominently in the constitutive equations of the simpler fluids. This tensor, which measures rates of deformation referred to the current configuration, contains no reference to an initial configuration.

Problems

1. Show that the equations $E_{ij} = 0$

where
$$E_{ij} = \frac{1}{2}\left[\frac{\partial x_m}{\partial X_i}\frac{\partial x_m}{\partial X_j} - \delta_{ij}\right]$$

and where the x_m are assumed twice differentiable functions of the X_i, imply in fact that the derivatives $\partial x_m/\partial X_i$ are independent of the X_i (but possibly dependent on time) and that

$$x_i = a_{ij}X_j + b_i$$

where also the b_i are at most dependent on time, and the matrix $\mathbf{A} \equiv (a_{ij})$ is orthogonal

i.e.
$$\mathbf{A}\mathbf{A}^T = \mathbf{A}^T\mathbf{A} = \mathbf{I}\ .$$

2. Show that the equations $E_{ij} = 0$

where the E_{ij} are given by the approximate formulae

$$E_{ij} = \frac{1}{2}\left[\frac{\partial U_i}{\partial X_j} + \frac{\partial U_j}{\partial X_i}\right]$$

† For materials with complex mechanical behaviour, the history of the deviation is also important.

imply that
$$\mathbf{U} = (\mathbf{\Omega} \wedge \mathbf{R}) + \mathbf{b}$$
where $\mathbf{\Omega}$ and $\mathbf{b}$ are vectors independent of the X_i but dependent possibly on t.

In what sense is the result obtained different from that of Question 1?

3. The symbols r, θ, z denote the current cylindrical polar coordinates of a particle located initially at R, Θ, Z. If the motion is specified by
$$r = r(R), \quad z = z(Z), \quad \theta = \Theta$$
show that in Cartesian coordinates
$$x = XR^{-1}r(R), \quad y = YR^{-1}r(R), \quad z = z(Z)$$
where
$$R = (X^2 + Y^2)^{\frac{1}{2}} .$$
Calculate the deformation gradient matrix
$$\mathbf{F} = (\partial x_i/\partial X_j)$$
and show that in the polar decomposition $\mathbf{F} = \mathbf{RU}$,
$$\mathbf{R} = \mathbf{I}, \qquad \mathbf{U} = \mathbf{F} .$$
By referring $\mathbf{F}$ to a set of orthogonal axes lying in the R, Θ and Z directions deduce that the principal stretches are
$$\lambda_R = |dr/dR|, \quad \lambda_\theta = r/R, \quad \lambda_z = |dz/dZ|$$
and interpret these results geometrically.

If the material is incompressible deduce that
$$z = aZ + b, \quad r^2 = a^{-1}(R^2 + c)$$
where a, b and c are quantities which are constant or perhaps depend on time.

4. **(i)** Show that there is no volume or density change associated with the deformation
$$x = \lambda X + kY, \quad y = \lambda^{-1}Y, \quad z = Z$$
where λ and k are functions only of time.

If λ and k are continuous functions with $\lambda(0) = 1$, $k(0) = 0$ (so that there is no deformation at $t = 0$), by considering the behaviour of an initially rectangular parallelopiped, show from physical arguments that λ remains necessarily positive.

Calculate also the matrix
$$\mathbf{F}^T\mathbf{F}$$

(ii) An incompressible material is subject to the deformation above. The material is fabricated from a fibre reinforced resin where the (densely packed) fibres lie initially parallel at $t = 0$ in the direction with unit vector
$$\mathbf{N} = (\cos\Phi, \sin\Phi, 0)$$

where Φ is a constant and where, without loss of generality, Φ may be assumed to lie in the range $0 \leqslant |\Phi| \leqslant \frac{1}{2}\pi$. The fibres are very strong and with a large elastic (Young's) modulus compared to the resin so that it is a good approximation to assume that the composite material is inextensible in the direction of the fibres (although of course the fibre orientation may change with deformation). Show that if the fibres are assumed inextensible, the quantities λ and k are related by

$$\lambda^2 \cos^2 \Phi + (\lambda^{-2} + k^2) \sin^2 \Phi + 2k\lambda \sin \Phi \cos \Phi = 1$$

and solve this equation for k. Infer that $\lambda > |\sin \Phi|$ and that the fibres remain parallel but pointing in the direction

$$\mathbf{n} = (\cos \phi, \sin \phi, 0)$$

where

$$\tan \phi = \sin \Phi / [\lambda^2 - \sin^2 \Phi]^{\frac{1}{2}} .$$

Chapter 6

Geometrical Restrictions on the Form of Constitutive Equations

6.1 Nature of constitutive equations

In general terms constitutive equations connect the forces acting at a material particle with the history of both the local geometry of deformation and temperature. The only relevant forces derive from the stresses, since, for a small volume element of linear dimensions Δl, the contact forces are of magnitude $|\sigma_{ij}|\,\Delta l^2$ while the body force integral is of order $|\sigma_{ij}|\Delta l^3$. In the limit $\Delta l \to 0$, the contact forces dominate entirely the mechanical behaviour, and the general form of the constitutive equations is taken to be of the general form

$$\sigma_{ij} \;=\; f_{ij}\,(\text{history of deformation, history of temperature}) \qquad (6.1.1)$$

where the quantities f_{ij} are components of a second order symmetric tensor.†

The constitutive equations discussed in the present text are particular cases of (6.1.1). We shall be concerned primarily with (a) elastic materials for which the σ_{ij} are functions of the current deformation and temperature and (b) Reiner-Rivlin fluids for which the σ_{ij} are functions of density, temperature and rate of deformation. Limiting cases of these materials lead to the important classical materials of continuum mechanics, linear elastic solids, and Newtonian fluids. These 'theoretical' materials are still important because they provide good models for the behaviour of many 'real' materials which are important technologically. However these models are not adequate to describe many other materials; in particular polymer and elastomer materials, e.g. nylon and polyvinyl chloride, require in general constitutive equations of the form (6.1.1) in which the history of deformation is taken into account.††

Historically constitutive equations were postulated directly from experimental observation, as for instance in the cases of Boyle's law, Charle's law and the 'one dimensional' linear elastic Hooke's law; or by judicious mathematical generalisations of experimental observations as in the case of the generalised Hooke's law.

Experimentation is of course still crucial in determining various quantities such as the constants and functions which enter constitutive equations. However in recent years the principles discussed in this chapter have played an important role in limiting the possible forms of constitutive equations.

† In view of (6.1.1), continuum mechanics is a branch of classical deterministic physics.

†† Exceptionally the 'static' behaviour of these materials is described adequately by the theory of finite elasticity in Chapter 8.

The current theoretical approach to determining the equations is as follows. The general form of the equation is decided in advance in the sense that the independent variables appearing in the f_{ij} in (6.1.1) are assumed. For example in the theory of isothermal finite elasticity it is assumed that the f_{ij} are functions of the components of the deformation gradient tensor $\mathbf{F} = (\partial x_i/\partial X_j)$. Once the choice of independent variables has been made, (commonly the choice is motivated by experimental results), the possible forms of the f_{ij} are restricted by appeal to a variety of physical principles.

Two of the principles are inviolate. They are

(a) 'Invariance' of the equations with respect to different stationary coordinate systems.

(b) 'Invariance' of the equations with respect to different coordinate systems moving in arbitrary relative motion.

The first of these is readily met by expressing the constitutive equations in tensor form. We have already discussed at length the necessity of expressing equations of motion etc. in tensor form; the requirement (a) above is simply an extension of the principle to constitutive equations.

Requirement (b) has not been discussed previously, and is connected with the absence of a unique 'stationary' reference frame in contemporary physics. Section 6.2 discussed the point in detail.

Two other physical assumptions are often imposed on constitutive equation. Commonly it is assumed that materials are homogeneous, so that equations (6.1.1) assume an identical form for each material particle, i.e. the same equations apply throughout the continuum. Finally materials are often assumed to be isotropic, i.e. their material properties in some configuration, usually either the reference or current configuration, look the same in all directions. In the present text we shall be concerned exclusively with homogeneous and isotropic materials.

Conformity to invariance requirements, together with the assumption of isotropy, impose severe restrictions on the forms of the quantities f_{ij}, and lead to almost explicit forms of constitutive equations for the simpler materials. In these nearly explicit forms of the equations, there appear various functions, or in limiting cases various constants. Determination of these functions or constants is not possible within the framework of continuum mechanics. The problem is one for either (or both) the experimentalist or theoretical physicist. With the determination of these functions (constants), the constitutive equations are rendered entirely explicit.

It should perhaps be added that the idealised marriage of theory and experiment implied immediately above has been realised only for a very limited class of real materials. This class includes primarily substances which conform to the classical models of continuum mechanics, for which the experimentalist's task is simplified to the determination of various **parameters** (elastic constants, coefficients of viscosity). More recently the **functions** entering the theories of linear viscoelasticity and finite elasticity have been measured for a few elastomer and polymer materials. On the other hand while vast quantities of experimental data appertaining to large scale and rapid deformation of metals and polymers has been amassed, there has been virtually no attempt to correlate the data with constitutive equations. The reasons are twofold. One is the extremely complex form assumed by constitutive equations for complicated materials whose behaviour depends on the history of deformation. Here the problem posed to the experimentalist is incredibly difficult. It is indeed impossible in the absence of theoretical solutions for idealised experimental situations†. The second reason is more prosaic; there exists, with rare exceptions, an almost complete lack of communication between experimentalists and theoreticians.

6.2 Material frame indifference

In the discussion on Cartesian tensors in Chapter 4 the various primed reference frames were assumed to be relatively at rest. The assertion that an equation is a tensor equation is a statement that the law expressed by the equation is independent of the choice of coordinate system.

However, adopting the viewpoint of twentieth century physics we now assert the much more extensive principle that the laws of physics must also hold for reference frames moving arbitrarily with respect to each other. This assertion of 'frame indifference' stems essentially from the failure of experimental physicists to detect any 'preferred' reference frame; an example is the failure of the Michelson-Morley experiment to detect the motion of the earth through the aether. That the Michelson-Morley experiment led to the theory of special relativity is irrelevant in the present context. The more universal physical truth deriving from the absence of a preferred coordinate system is that it is impossible to detect the relative motion of a chosen frame from experiments carried out with reference to that frame alone. In other words the assertion of a physical law implies that the law is true for all possible frames of reference, irrespective of their relative motion. [There is no 'absolute' motion in the absence of a preferred reference frame.] We utilise the principal in the context of Newtonian, i.e. non-relativistic, mechanics.

A simple example is given by Newton's second law of motion for point masses. Let $\mathbf{F}$, $\mathbf{f}$ and m denote respectively the force, acceleration and mass measured in one coordinate system C and let $\mathbf{F}'$, f' and m' denote the corresponding quantities measured in a second system C' moving relative to C. The statement that Newton's second law is a valid physical law implies both

$$\mathbf{F} = m\mathbf{f} \quad \text{and} \quad \mathbf{F}' = m'\mathbf{f}' \,. \tag{6.2.1}$$

It is a matter of experiment that outside the context of large velocities (i.e. within the framework of classical physics) these equations are satisfied for systems in relative motion and further the two inertial masses m and m' are the same.

† Ideally the 'solutions' desired should be explicit in terms of the unknown functions appearing in the constitutive equations. In other words it must be possible to obtain a solution of a problem in terms of the unknown functions and without the necessity of specifying in detail the nature of these functions.

The assertions (6.2.1) do not imply $\mathbf{F} = \mathbf{F}'$ and $\mathbf{f} = \mathbf{f}'$, and indeed this is not in general true. For suppose the instantaneous acceleration of the point $\mathbf{r}'$ in C' with respect to C is $\mathbf{a}$ then

$$\mathbf{f} = \mathbf{f}' + \mathbf{a} \tag{6.2.2}$$

and the two equations (6.2.1) are reconcilable only if the law of transformation of force is

$$\mathbf{F}' = \mathbf{F} - m\mathbf{a} \tag{6.2.3}$$

A useful illustration is provided by the example of a pilot who has bailed out of an aeroplane. Prior to opening his parachute, and neglecting air resistance, the man is in free fall, accelerating towards the earth. From the viewpoint of an observer stationed on the earth (unprimed system) the particle mechanics of the pilot's motion are summarised by Newton's law

$$\mathbf{F} = m\mathbf{f} \tag{6.2.4}$$

supplemented by the recognition that the only force acting on the pilot is the body force due to the earth's gravitational field

i.e.
$$\mathbf{F} = m\mathbf{g} \tag{6.2.5}$$

where $\mathbf{g}$ is the acceleration due to gravity. From (6.2.4) and (6.2.5) the earth-bound observer deduces $\mathbf{f} = \mathbf{g}$ in accordance with the experimental observations. On the other hand in the pilot's (primed) reference frame (moving with the pilot), the pilot experiences no force, consistent with the weightless experience of free fall. The result $\mathbf{F}' = 0$ also follows from (6.2.3) on identifying $\mathbf{a}$ with $\mathbf{g}$ and using (6.2.5).

Consider now the slightly modified example which occurs after the pilot has opened his parachute. We suppose, for simplicity, that the effect of the parachute is to provide a constant force $-\mathbf{T}$ where $\mathbf{T}$ is in the direction $\mathbf{g}$. We suppose further that $\mathbf{T}$ can be measured by a spring incorporated in the harness and seen by both the pilot and the earthbound observer. In fact since distances between points are preserved for different coordinate systems in relative motion, both pilot and observer measure the same spring extension and deduce the same value for $\mathbf{T}$. In addition the pilot experiences $-\mathbf{T}$ physiologically via the contact forces exerted by the shoulder straps of the harness. The Newtonian mechanics of the pilot for the earthbound observer are now $\mathbf{F} = m\mathbf{f}$ supplemented by $\mathbf{F} = m\mathbf{g} - \mathbf{T}$ so that $\mathbf{f} = \mathbf{g} - \mathbf{T}/m$ while for the pilot's frame of reference $\mathbf{F}' = 0$ as before. However the pilot is no longer weightless since for $\mathbf{F}' = 0$ the contact force $-\mathbf{T}$ is necessarily counterbalanced by a body force $\mathbf{T}$.

A crucial point that emerges from this trivial example is that for both the pilot and earth bound observer the contact force as measured by the extension of the spring is the same, while the body forces are different. A further example cited by Truesdall [in *Principles of Continuum Mechanics* (1961) published by Socony Mobil Oil Co. Inc.] is that of a spring mounted on a rotating table with one end attached to the table and the other fixed to a mass. As the table spins the spring extends. From the viewpoint of an external observer the force in the spring is merely that required to satisfy Newton's law for the accelerated motion of the mass. For an observer utilising a reference frame fixed in the rotating table the spring force would derive seemingly from a counter balancing body force exerted on the mass by an external field. Nevertheless for both observers the force in the spring is the same, and hence so are the contact forces exerted on both the table and mass by the spring. This example is somewhat more general than the earlier one since it incorporates a rotating coordinate system.

We generalise from these examples and assert as a fundamental proposition that contact forces and hence stresses are **frame indifferent** quantities, i.e. the same stress tensor is measured by all observers irrespective of their relative motions. The mathematical consequences of this are explored below.

Not all forces are frame indifferent, as has been seen in the example of body forces, nor are velocity and accelerations, as is evident physically. We shall see that the requirement for a tensor quantity to be frame indifferent is that the transformation law relating components in different frames in relative motion is exactly the same as the ordinary tensor transformation law.

The mathematical apparatus required for discussing whether or not quantities are frame indifferent differs from that for tensors only in that the relation between the coordinates of two orthogonal systems involves translation vectors $\boldsymbol{\beta}(t)$ and proper orthogonal matrices $\mathbf{Q}(t)$ which are both time dependent. Let x_i be the material coordinates of a particle for one reference system C and x_i' the corresponding coordinates for a second system C' in relative motion with respect to the first. The from the discussion of Section 4.2 the relations connecting the primed and unprimed coordinates are necessarily of the form

$$x_i' = \beta_i(t) + Q_{ij}(t)x_j \tag{6.2.6}$$

where the $Q_{ij}(t)$ are components of a proper orthogonal matrix so that

$$\mathbf{Q}\mathbf{Q}^T = \mathbf{Q}^T\mathbf{Q} = \mathbf{I} \tag{6.2.7}$$

and

$$|\mathbf{Q}| = 1 \ . \tag{6.2.8}$$

[The notation $\mathbf{Q}$ and Q_{ij} is in common use in the present context.] The quantities β_i are the components of the vector joining the origins of C and C'.

Suppose the motion of a continuum is specified by

$$x_i = x_i(X_1, X_2, X_3, t) \tag{6.2.9}$$

in C. Then in C' the motion is given by

$$x_i' = \beta_i(t) + Q_{ij}x_j(X_1, X_2, X_3, t) \tag{6.2.10a}$$

with inverse

$$x_i = Q_{ji}(x_j' - \beta_j) \ . \tag{6.2.10b}$$

Note the crucial point that no transformation of the quantities X_1, X_2, X_3 is involved. The referential coordinates are the same for both C and C'. The origin of this most important point (which differs from the parallel point in the discussion of tensors) is most easily visualised by imagining observers O′ and O) attached to the origins respectively of C and C'. At some instant of time suppose C and C' coincide and at that instant both observers O and O′ attach identical sets of labels X_1, X_2 and X_3 to the material particles. Thereafter each observer sees different motions. Observer O see the motion (6.2.9) while O′ observes the motion (6.2.10a); however for both O and O′ the labels X_1, X_2 and X_3 are the same.

Frame indifferent quantities may now be defined as tensor quantities whose transformation law is given by

$$T'_{ijk\,\ldots} = Q_{ir}Q_{js}Q_{kt}\ldots T_{rst\,\ldots} \tag{6.2.11}$$

for arbitrary **time dependent** Q_{ij} satisfying (6.2.7) and (6.2.8) and for arbitrary $\boldsymbol{\beta}(t)$. Of course for arbitrary time independent $\boldsymbol{\beta}$ and $\mathbf{Q}$, all tensor quantities satisfy (6.2.11). However a frame indifferent quantity (i.e. a quantity which is truly independent of the relative motion of reference frames) is required to satisfy (6.2.11) for arbitrary proper orthogonal matrices $\mathbf{Q}(t)$ and arbitrary $\boldsymbol{\beta}(t)$. The relation (6.2.11) then merely implies that even a frame indifferent quantity will possess different components in differently oriented coordinate systems. [In terms of the observers O and O′ it is as if at any arbitrary instant of time t, O suddenly took notice of the primed coordinate frame and posed himself the question as to what would be the components of $T_{ijk\,\ldots}$ in the primed system. For a frame indifferent quantity like σ_{ij} the answer is given by (6.2.11), and this merely reflects that at that instant of time the direction cosines of the primed frame axes relative to the unprimed frame are given by the Q_{ij}.]

By hypothesis σ_{ij} is frame indifferent so that

$$\sigma'_{ij} = Q_{ir}Q_{js}\sigma_{rs} \tag{6.2.12a}$$

or equivalently

$$\boldsymbol{\sigma}' = \mathbf{Q}\boldsymbol{\sigma}\mathbf{Q}^T \;. \tag{6.2.12b}$$

Also the vector $\partial\sigma_{ij}/\partial x_j$ is frame indifferent since from (6.2.12a)

$$\begin{aligned}
\frac{\partial\sigma'_{ij}}{\partial x'_j} &= Q_{ir}Q_{js}\frac{\partial\sigma_{rs}}{\partial x_q}\frac{\partial x_q}{\partial x'_j} \\
&= Q_{ir}Q_{js}Q_{jq}\frac{\partial\sigma_{rs}}{\partial x_q} && \text{(from 6.2.10b)} \\
&= Q_{ir}\delta_{sq}\frac{\partial\sigma_{rs}}{\partial x_q} && \text{(from 6.2.7)} \\
&= Q_{ir}\frac{\partial\sigma_{rs}}{\partial x_s} \;. && (6.2.13)
\end{aligned}$$

which is the transformation law for a first order tensor.

We also hypothesise that the temperature θ is a frame indifferent scalar.

A second indifferent quantity is the density ρ. From (2.2.10)

$$\rho_0/\rho = J \equiv |\partial x_i/\partial X_j| \;. \tag{6.2.14}$$

For the primed system

$$\rho_0/\rho' = |\partial x_i'/\partial X_j| = |Q_{ir}\partial x_r/\partial X_j| = |\mathbf{Q}|\,|\partial x_r/\partial X_j| = |\partial x_r/\partial X_j| = \rho_0/\rho$$

(from 6.2.10a)

i.e.
$$\rho' = \rho \tag{6.2.15}$$

as might have been anticipated on physical grounds.

On the other hand, as also might be anticipated from physical arguments, velocity is not a frame indifferent quantity since from (6.2.10a).

$$V_i' = \dot{\beta}_i + \dot{Q}_{ij}x_j + Q_{ij}V_j \tag{6.2.16}$$

i.e.
$$v_i' = \dot{\beta}_i + \dot{Q}_{ij}x_j + Q_{ij}v_j \,. \tag{6.2.17}$$

Neither is the tensor $\partial v_i/\partial x_j$ frame indifferent since from (6.2.17)

$$\begin{aligned}\frac{\partial v_i'}{\partial x_k'} &= \dot{Q}_{ij}\frac{\partial x_j}{\partial x_k'} + Q_{ij}\frac{\partial v_j}{\partial x_l}\frac{\partial x_l}{\partial x_k'} \\ &= \dot{Q}_{ij}Q_{kj} + Q_{ij}\frac{\partial v_j}{\partial x_l}Q_{kl} \,.\end{aligned} \tag{6.2.18}$$

On the other hand, forming from (6.2.18) the symmetric quantity

$$d_{ik}' = \frac{1}{2}\left(\frac{\partial v_i'}{\partial x_k'} + \frac{\partial v_k'}{\partial x_i'}\right) ,$$

there results (in matrix notation)

$$\mathbf{d}' = \tfrac{1}{2}(\dot{\mathbf{Q}}\mathbf{Q}^T + \mathbf{Q}\dot{\mathbf{Q}}^T) + \mathbf{Q}\mathbf{d}\mathbf{Q}^T \tag{6.2.19}$$

However on differentiating the first of (6.2.7) we have

$$\dot{\mathbf{Q}}\mathbf{Q}^T + \mathbf{Q}\dot{\mathbf{Q}}^T = \mathbf{0}$$

so that
$$\mathbf{d}' = \mathbf{Q}\mathbf{d}\mathbf{Q}^T \tag{6.2.20}$$

is a frame indifferent quantity.

The anti-symmetric spin tensor

$$\omega_{ij} = \tfrac{1}{2}(\partial v_i/\partial x_j - \partial v_j/\partial x_i)$$

(whose three independent components are also components of the vector $\frac{1}{2}$ curl $\mathbf{v}$) is not frame indifferent. From (6.2.18) we find, again in matrix notation

$$\boldsymbol{\omega}' = \tfrac{1}{2}(\dot{\mathbf{Q}}\mathbf{Q}^T - \mathbf{Q}\dot{\mathbf{Q}}^T) + \mathbf{Q}\boldsymbol{\omega}\mathbf{Q}^T = \dot{\mathbf{Q}}\mathbf{Q}^T + \mathbf{Q}\boldsymbol{\omega}\mathbf{Q}^T \tag{6.2.21}$$

This last equation may be written

$$\boldsymbol{\omega}' = \mathbf{Q}[\boldsymbol{\omega} + \mathbf{Q}^T\dot{\mathbf{Q}}]\mathbf{Q}^T \tag{6.2.22}$$

in which $\mathbf{Q}^T\dot{\mathbf{Q}}$ is an anti-symmetric matrix (since $\mathbf{Q}^T\dot{\mathbf{Q}} + \dot{\mathbf{Q}}^T\mathbf{Q} = 0$). Bearing in mind the physical result that the angular velocity $\boldsymbol{\omega}'$ in the C' system must be the vector sum of $\boldsymbol{\omega}$ together with the angular velocity of C' with respect to C, it is evident that the three independent components of $\mathbf{Q}^T\dot{\mathbf{Q}}$ determine the relative angular velocity of C' with respect to C.†

Acceleration is not a frame indifferent quantity, as might be expected physically, for on differentiating (6.2.16) with respect to time and expressing the result in spatial quantities, we find

$$f_i' = \ddot{\beta}_i + \ddot{Q}_{ij}x_j + 2\dot{Q}_{ij}v_j + Q_{ij}f_j \ . \tag{6.2.23}$$

On the other hand because of the result that $\partial\sigma_{ij}/\partial x_j$ and ρ are frame indifferent quantities, then in order for the equations of motion to be frame indifferent, it is necessary that the vector $b_i - f_i$ be frame indifferent

i.e.
$$b_i' - f_i' = Q_{ij}(b_j - f_j) \tag{6.2.24}$$

where $\mathbf{b}$ is the body force. In other words the body force must transform according to

$$b_i' = \ddot{\beta}_i + \ddot{Q}_{ij}x_j + 2\dot{Q}_{ij}v_j + Q_{ij}b_j \tag{6.2.25}$$

i.e. in exactly the same manner as f_i.

The results (6.2.24), (6.2.25) (which do not enter any subsequent arguments), may be regarded as reflecting a general law of physics which states that it is impossible to distinguish the effects of reference frame motion from the effects of an external force field by experiments carried out solely with respect to the reference frame.

The deformation gradient tensor $\partial x_i/\partial X_j$ is not frame indifferent since from (6.2.10a)

$$\frac{\partial x_i'}{\partial X_k} = Q_{ij}\frac{\partial x_j}{\partial X_k}$$

or in matrix notation
$$\mathbf{F}' = \mathbf{QF} \ . \tag{6.2.26}$$

There is a difference in behaviour in the present context for the right and left stretch tensors, for from (6.2.26)

$$(\mathbf{U}')^2 \equiv (\mathbf{F}')^T\mathbf{F}' = \mathbf{F}^T\mathbf{Q}^T\mathbf{QF} = \mathbf{U}^2 \tag{6.2.27}$$

while
$$(\mathbf{V}')^2 \equiv \mathbf{F}'(\mathbf{F}')^T = \mathbf{QFF}^T\mathbf{Q}^T = \mathbf{QV}^2\mathbf{Q}^T \ . \tag{6.2.28}$$

Thus the square of the left stretch tensor is frame indifferent while the square of the right stretch tensor is not. The result (6.2.27) is not surprising in view of the result

$$dx_idx_i = \mathbf{dX}^T\mathbf{U}^2\mathbf{dX}$$

since for all observers the quantities $\mathbf{dX}$

and $ds^2 = dx_i'dx_i' = dx_idx_i$ are the same.

† The matrices $\mathbf{Q}$ and $\mathbf{Q}^T$ outside the brackets in (6.2.22) merely reflect the current orientations of C' with respect to C.

On taking the (unique) square root of (6.2.28)

$$\mathbf{V}' = (\mathbf{Q}\mathbf{V}^2\mathbf{Q}^T)^{\frac{1}{2}} = (\mathbf{Q}\mathbf{V}\mathbf{Q}^T\mathbf{Q}\mathbf{V}\mathbf{Q}^T)^{\frac{1}{2}} = \mathbf{Q}\mathbf{V}\mathbf{Q}^T$$

so that $\mathbf{V}$ is also frame indifferent. Similarly $\mathbf{U}' = \mathbf{U}$. The results that $\mathbf{V}$ is frame indifferent while all observers see the same $\mathbf{U}$, is reflected in the form of the equations of finite elasticity given in Chapter 8.

We conclude by re-iterating the requirement that the equations of continuum mechanics must be frame indifferent. Since σ_{ij} is frame indifferent, this means that the f_{ij} on the right side of (6.1.1) are frame indifferent. The consequences of this result are exploited in Chapter 7 and 8.

Chapter 7

Constitutive Equations for Fluids

7.1 Ideal fluids

The simplest constitutive equations encountered in continuum mechanics are those for an ideal fluid. These equations are

$$\sigma_{ij} = -P(\rho, \theta)\delta_{ij} \tag{7.1.1a}$$

i.e.

$$\boldsymbol{\sigma} = -P(\rho, \theta)\mathbf{I} \tag{7.1.1b}$$

where the positive quantity P, a scalar function of density ρ and temperature θ is called the 'pressure'. This 'pressure' is, of course, exactly the same 'pressure' that is encountered in elementary physics.

That P depends on the variables ρ and θ is a well known result which has been verified many times over during several centuries of experimentation on liquids and gases. Certainly in the absence of motion equation (7.1.1) describes with entire accuracy the thermo-mechanical behaviour of real liquids and gases.

Equation (7.1.1) is a correctly formulated tensor equation (since σ_{ij} and δ_{ij} are second order tensors, while P is a scalar which depends on scalars). Also (7.1.1b) satisfies the more restrictive frame indifference requirement for if

$$\boldsymbol{\sigma}' = \mathbf{Q}\boldsymbol{\sigma}\mathbf{Q}^T$$

while also ρ and θ are frame indifferent, then

$$\boldsymbol{\sigma}' = P\mathbf{Q}^T\mathbf{I}\mathbf{Q} = -P\mathbf{I} \tag{7.1.2a}$$

i.e.

$$\sigma'_{ij} = -P(\rho, \theta)\delta_{ij} \ . \tag{7.1.2b}$$

Moreover in view of the result (7.1.2), valid for arbitrary proper orthogonal $\mathbf{Q}$ (irrespective of whether $\mathbf{Q}$ is time dependent), the thermo-mechanical properties of an ideal fluid look the same in all directions, i.e. the material is isotropic.†

† The reader may find it difficult to see why isotropy is not always a consequence of a properly tensor formulated constitutive equation, since it has been emphasised that such an equation is necessarily independent of the coordinate system. The following counter example may be helpful. For small strains the constitutive equations of anisotropic elastic crystals are approximated by

$$\sigma_{ij} = C_{ijkl}E_{kl}$$

where the C_{ijkl} are a set of constants (the elastic constants) which form the components of a fourth order tensor. In a primed coordinate system the transformed equations take the form

$$\sigma'_{ij} = C'_{ijkl}E'_{kl}$$

and the material is isotropic if and only if $C_{ijkl} = C'_{ijkl}$.

Also since (7.2.1) holds everywhere in the fluid, the material is also homogeneous.†

In view of (7.1.2b), the constitutive equations of an ideal fluid assert that for all surfaces within or bounding the fluid the shear stresses are zero. Further the normal stress ($-P$) acting at a point on any surface passing through the point is the same for all surface orientations, and depends on the deformation in a particularly simple manner, i.e. only through the density. Finally since $P > 0$ [see the discussion below (4.5.23) in Chapter 4], the normal stresses are compressive.

That liquids are unable to withstand shear forces under no flow (hydrostatic) conditions is evident from the familar experience in which a vessel containing liquid initially at rest is tilted. On tilting, the instantaneous effect of the earth's gravitational field is to induce shear stresses within the liquid. Immediately the liquid flows in response to the shear stresses; however flow ceases when the liquid has attained a new equilibrium configuration for which the free surface is again horizontal (in the tilted vessel) and for which the shear stresses again are everywhere zero.

All liquids and gases are able to sustain shear stresses during motion; nevertheless for many such materials the shear stresses occurring in motion are small compared with P. In some circumstances, but not all, it is a legitimate approximation to assume equations (7.1.1) to be valid for fluids in motion. It is not possible to state the precise circumstances for which neglect of shear stresses is a valid approximation. This decision has to be made in the context of the particular problem in hand. The solutions of some problems, particularly in aerodynamics, have been obtained using the approximation (7.1.1) in part of the field of flow, and using more sophisticated equations, which allow non-zero shear stresses, in the remaining regions. Constitutive equations of this type are discussed in Section 7.2.

In the event equations (7.1.1) are valid under flow conditions, substitution into the equations of motion

$$\frac{\partial \sigma_{ij}}{\partial x_j} + \rho b_i = \rho \left[\frac{\partial v_i}{\partial t} + (\mathbf{v} \,.\, \text{grad}) v_i\right]$$

leads to the Euler equations of motion for an ideal fluid

$$-\,\text{grad}\, P + \rho \mathbf{b} = \rho \left[\frac{\partial \mathbf{v}}{\partial t} + (\mathbf{v} \,.\, \text{grad})\mathbf{v}\right] \tag{7.1.3}$$

which are expressed in terms of spatial coordinates. If necessary the term grad P may be expressed in terms of ρ and θ

$$\text{grad}\, P = \frac{\partial P}{\partial \theta}\, \text{grad}\, \theta + \frac{\partial P}{\partial \rho}\, \text{grad}\, \rho \tag{7.1.4}$$

but conventionally Euler's equations are written in the form (7.1.3).

† For non-homogeneous material (not discussed in the present text) the coordinates X_i, necessarily appear in the constitutive equations, reflecting the variation of thermomechanical properties from one material particle to another.

Equations (7.1.3) together with the equation of mass conservation

$$\frac{\partial \rho}{\partial t} + \operatorname{div}(\rho \mathbf{v}) = 0 \tag{7.1.5}$$

provide four (scalar) equations for the density ρ, temperature θ and the three components of $\mathbf{v}$. A fifth equation is provided by energy conservation considerations (see discussion in Chapter 1). In circumstances where P is expressible solely as a function of ρ (see below) equations (7.1.3) and (7.1.5) provide four scalar equations for four dependent variables.

An explicit functional form of $P(\rho, \theta)$ valid for gases over wide ranges of temperature and density is the well known expression

$$P = R\rho\theta/m \tag{7.1.6}$$

where R is the universal gas constant, m is the mean molecular weight of the gas and θ is now the absolute temperature.

Equation (7.1.6), which incorporates the laws of both Charles and Boyle, is obtainable theoretically from the statistical mechanics of weakly interacting molecules. The equation is one of the few constitutive relations, accurate in detail, which is derivable from consideration of the real particles (in this case molecules) which constitute the material.

For liquids the function $P(\rho, \theta)$ is well approximated by the linear relation

$$P = K_1(\theta)[\rho - \rho_0(\theta)]/\rho_0(\theta) \tag{7.1.7}$$

where $K_1(\theta)$ is the bulk modulus and $\rho_0(\theta)$ is the density in the absence of pressure. The dependence of ρ_0 on θ arises because of thermal expansion. Usually K_1 and ρ_0 are insensitive functions of temperature for temperature variations of a few degrees.

Typically K_1 is of order 10^9 N/m^2 and equation (7.1.7) is valid for values of $(\rho - \rho_0)/\rho_0$ up to about 0·05.

For isothermal behaviour of liquids and gases θ is constant and $P(\rho, \theta)$ is essentially dependent only on ρ. Fluids characterised by

$$P = P(\rho) \tag{7.1.8}$$

are called 'barotropic'. Clearly the barotropic model is applicable under isothermal conditions. A second important application of the model occurs when the fluid is assumed to behave adiabatically. Under adiabatic conditions, i.e. conditions when no heat transfer occurs in the fluid, variations in temperature may not be negligible; however utilisation of the no heat transfer condition enables determination of the temperature in terms of the density, with the result that $P(p, \theta)$ assumes the barotropic form. In particular for a perfect gas there results

$$P = k\rho^\gamma \tag{7.1.9}$$

In this equation the exponent γ is a true constant, characteristic of the gas, given by

$$\gamma = c_p/c_v > 1$$

where c_p and c_v are respectively specific heats at constant pressure and volume. The 'constant' k may vary from one continuum particle to another. However for many problems k is uniform throughout the flow field and determined by an initial uniform ambient pressure P_0 and density ρ_0. Under these circumstances, everywhere

$$P = P_0(\rho/\rho_0)^\gamma \tag{7.1.10}$$

and the material is homogeneous.

Physical conditions justifying use of the adiabatic approximation arise when local fluctuations in temperature occur so rapidly that heat liberated locally by gas compression is reabsorbed by the gas during the expansion phase and before any appreciable fraction of the heat is able to escape into neighbouring regions. This situation occurs in the propagation of sound pulses in gases. Here it is known that the magnitude of the sound velocity is accurately predicted in the adiabatic approximation, in contrast to predictions based on the assumption of isothermal behaviour.

For small departures of ρ from ρ_0, equation (7.1.10) may be approximated by the first two terms of a Taylor series

$$P = P_0 + (\rho - \rho_0)(dP/d\rho)_{\rho_0} \equiv P_0 + c^2(\rho - \rho_0)$$

where $$c^2 := \gamma P_0/\rho_0 = \gamma R\theta_0/m \tag{7.1.11}$$

and where P_0, ρ_0 and θ_0 are ambient values of pressure, density and (absolute) temperature.

The quantity

$$c = (dP/d\rho)_{\rho_0}^{\frac{1}{2}} \tag{7.1.12}$$

is of dimensions $[LT^{-1}]$ and is shown subsequently to be the speed of sound (Chapter 13). From (7.1.12) and (7.1.6) we find for the corresponding isothermal behaviour

$$c = (R\theta_0/m)^{\frac{1}{2}} = (P_0/\rho_0)^{\frac{1}{2}}$$

which differs from the adiabatic result (7.1.11) by the factor $\gamma^{\frac{1}{2}}$.

Experimental evidence in gases (in particular in air) is convincingly in favour of (7.1.11), justifying the assumption of adiabatic behaviour.

For liquids behaving isothermally equation (7.1.7) becomes

$$P = K_1(\theta_0)[\rho - \rho_0(\theta_0)]/\rho_0(\theta_0) \tag{7.1.13}$$

where $K_1(\theta_0)$ is the 'isothermal' bulk modulus and where θ_0 is the (constant) ambient temperature. However commonly for liquids where θ_0 is not too near the extreme values (freezing and boiling point), it may be shown from a detailed thermodynamic analysis that for adiabatic behaviour near θ_0, equation (7.1.13) is replaced by a similar equation

$$P = K_2(\theta_0)[\rho - \rho_0(\theta_0)]/\rho_0(\theta_0) \qquad (7.1.14)$$

where $K_2(\theta_0)$, the 'adiabatic' bulk modulus, is nearly identical with $K_1(\theta_0)$. [This latter result derives from the relative insensitivity of K_1 and ρ_0 to θ_0]. For small temperature excursions around the value θ_0 the extremes of adiabatic and isothermal behaviour bracket the mechanical response of a material. For liquids initially with ambient uniform temperature θ_0, the mechanical response of the material to pressure under conditions which are neither adiabatic nor isothermal, is bounded by (7.1.13) and (7.1.14). Because of the result $K_1 \simeq K_2$, commonly the pressure of a liquid is adequately approximated by the barotropic equation

$$P = K[\rho - \rho_0(\theta_0)]/\rho_0(\theta_0) \qquad (7.1.15)$$

where K and $\rho_0(\theta_0)$ are temperature independent. In the approximation (7.1.15) the velocity of sound in a liquid is given by

$$c^2 = \left(\frac{dP}{d\rho}\right)_{\rho_0} = (K/\rho_0) \; . \qquad (7.1.16)$$

The simplest model of a fluid (either liquid or gas) is that of an incompressible material. In the incompressible model the equation

$$\rho = \rho_0 = \text{constant} \qquad (7.1.17)$$

replaces the equation

$$P = P(\rho, \theta) \qquad (7.1.18)$$

while the equation of continuity simplifies to

$$\text{div } \mathbf{v} = 0 \; . \qquad (7.1.19)$$

Because P is no longer determined by an equation of the form (7.1.18), the pressure enters the theory only through the Euler equation (7.1.3) For the incompressible model it is no longer possible (or necessary) to eliminate grad P from the Euler equation by means of (7.1.4); rather the quantity P is to be regarded as a dependent variable, which together with $\mathbf{v}$, is determined by the four (scalar) equations (7.1.3) and (7.1.19).

The constitutive equations of an ideal incompressible fluid may be stated formally in the form

$$\sigma_{ij} = -P\delta_{ij} \; ; \quad \rho = \rho_0$$

where the (now arbitrary) scalar P is determined by the motion.

Motivation for adoption of the incompressible model for liquids is given by the large magnitude of K in (7.1.15) compared with the pressures occurring in many flow problems. For example, the hydraulics engineer is rarely faced with pressures larger than $10^{-3}K$, so that for most purposes the assumption $\rho = \rho_0$ is adequate.

In circumstances where throughout the field of flow of an ideal fluid $|\rho - \rho_0| \ll \rho_0$ the incompressible model is often adequate to describe the motion.

However the criterion $|\rho - \rho_0| \ll \rho_0$ is not in itself adequate to justify the use of the incompressible model. For, in particular, incompressible materials transmit mechanical disturbances at infinite speeds (note the incompressible limits discussed above for liquids and solids derive essentially from the limit $c \to \infty$). Clearly such a model is inappropriate for discussing problems of sound propagation, independent of whether or not the inequality $|\rho - \rho_0| \ll \rho_0$ is satisfied.

7.2 Reiner-Rivlin fluids

We turn now to the question of deriving constitutive equations for fluids which sustain shear stresses during motion. That fluids in motion sustain shear stresses is evident from a number of experimental results. For example if a thin plane lamina is placed parallel to the flow of a stream (Fig. 7.1) the lamina experiences a drag force in the direction of flow. To maintain the lamina in position requires imposition of a force of equal magnitude exerted in the opposite direction.

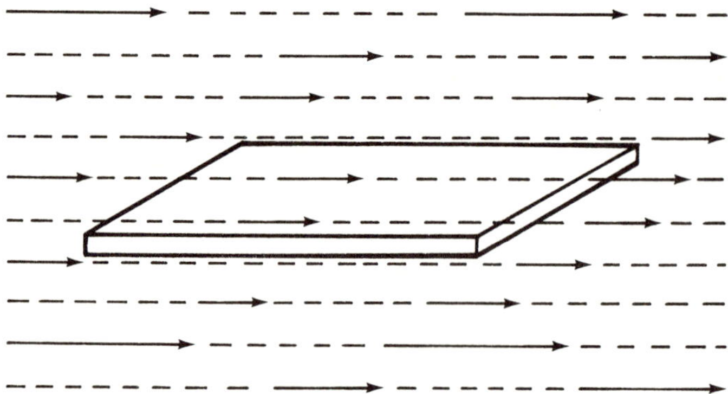

Fig. 7.1 Flow parallel to plane lamina

The magnitude of the force varies with the fluid, stream velocity and shape and size of the lamina.

Evidently such a drag force would never arise with constitutive equations of the type (7.1.1) since, irrespective of the fine details of the solution of the flow problem associated with Fig. 7.1, the lamina would experience forces only normal to the plane surfaces.

The drag force experienced by an aircraft wing provides a well known similar example of the ability of gases to sustain shear stresses.

We write the constitutive equations for a fluid in motion in the form

$$\sigma_{ij} = -P\delta_{ij} + \hat{\sigma}_{ij}$$

where $P(\rho, \theta)$ is the quantity discussed in Section 7.1 and where $\hat{\sigma}_{ij}$ vanishes in the absence of motion, i.e. $\hat{\sigma}_{ij}$ depends in some sense on the velocity field. The quantity $\hat{\sigma}_{ij}$ is commonly known as the extra stress.

We reject the hypothesis that $\hat{\sigma}_{ij}$ depends directly on the velocity components because this would lead to the result that $\hat{\sigma}_{ij}$ would be different for observers in different (moving) reference frames. A much more plausible assumption is that $\hat{\sigma}_{ij}$ depends on the spatial derivatives of the velocity field

i.e.
$$\hat{\sigma}_{ij} = \hat{f}_{ij}(\partial v_r/\partial x_s) \tag{7.2.1}$$

since the quantities $\partial v_r/\partial x_s$ provide a measure of the way in which the continuum elements are deforming (and rotating).

The assumption (7.2.1) is restrictive, i.e. we could write down more complicated equations in which $\hat{\sigma}_{ij}$ depends on the past motion. Such fluids exist, for instance certain polymer solutions, but a wide variety of liquids and gases conform to the description (7.2.1). Equation (7.2.1) contains no reference to any initial state, (i.e. no dependence on **R**) and asserts that the extra stress is determined entirely by the current motion. Thus fluids described by (7.2.1) are unable to 'remember' their preceding configurations. [Although it is sometimes said that a fluid satisfying (7.2.1) remembers its immediately preceding state. This interpretation follows from writing

$$\begin{aligned}\frac{\partial v_r}{\partial x_s} &= \frac{1}{2}\left(\frac{\partial v_r}{\partial x_s} + \frac{\partial v_s}{\partial x_r}\right) + \frac{1}{2}\left(\frac{\partial v_r}{\partial x_s} - \frac{\partial v_s}{\partial x_r}\right) \\ &\equiv d_{rs} + \omega_{rs}\end{aligned} \tag{7.2.2}$$

and recalling in particular the geometrical interpretation of the rate of deformation tensor (d_{rs}) discussed in Section 5.4.]

It is convenient to re-write (7.2.1) in the matrix form

$$\hat{\boldsymbol{\sigma}}' = \hat{\mathbf{f}}(\mathbf{d}, \boldsymbol{\omega}, \rho, \theta) \tag{7.2.3}$$

where we include in the dependence of $\hat{\mathbf{f}}$, the frame indifferent quantities ρ and θ. In (7.2.3) the dependence on $\mathbf{d}$ and $\boldsymbol{\omega}$ has been separated out for reasons which become immediately apparent.

We invoke now the principal of material frame indifference. In a rotating reference frame the quantity $\boldsymbol{\sigma}$ (and therefore $\hat{\boldsymbol{\sigma}}$) transforms according to the usual tensor law.

$$\hat{\boldsymbol{\sigma}}' = \mathbf{Q}\hat{\boldsymbol{\sigma}}\mathbf{Q}^T$$

and similarly [from (6.2.20)] for $\mathbf{d}$

$$\mathbf{d}' = \mathbf{Q}\mathbf{d}\mathbf{Q}^T$$

while the scalar quantities ρ and θ are unchanged. However from (6.2.22) $\boldsymbol{\omega}$ is not frame indifferent and transforms to

$$\boldsymbol{\omega}' = \mathbf{Q}(\boldsymbol{\omega} + \mathbf{Q}^T\dot{\mathbf{Q}})\mathbf{Q}^T .$$

Inserting these results into (7.2.3) now yields

$$\mathbf{Q}\hat{\boldsymbol{\sigma}}\mathbf{Q}^T = \hat{\mathbf{f}}[\mathbf{Q}\mathbf{d}\mathbf{Q}^T, \mathbf{Q}(\boldsymbol{\omega} + \mathbf{Q}^T\dot{\mathbf{Q}})\mathbf{Q}^T, \rho, \theta] \tag{7.2.4}$$

which is to be true for arbitrary proper orthogonal $\mathbf{Q}(t)$.

Consider in particular a choice of $\mathbf{Q}$ for which the orientation of the primed frame is instantaneously coincident with the unprimed frame

i.e. $$\mathbf{Q} = \mathbf{Q}^T = \mathbf{I}$$

but for which $\dot{\mathbf{Q}} \neq 0$ (so that the primed frame is rotating with arbitrary non-zero angular velocity). Equation (7.2.4) reduces to

$$\hat{\boldsymbol{\sigma}} = \hat{\mathbf{f}}[\mathbf{d}, \boldsymbol{\omega} + \dot{\mathbf{Q}}, \rho, \theta]$$

and this is reconcilable with (7.2.3) for arbitrary $\dot{\mathbf{Q}}$, if and only if $\hat{\mathbf{f}}$ is independent of $\boldsymbol{\omega}$. In other words

$$\hat{\boldsymbol{\sigma}} = \hat{\mathbf{f}}[\mathbf{d}, \rho, \theta] \; . \tag{7.2.5}$$

This result is readily inferred on physical grounds, namely that the stress must be independent of any rigid body motion.

Reverting now to the case of general $\mathbf{Q}(t)$ and omitting the dependence on $\boldsymbol{\omega}$, equation (7.2.4) becomes

$$\mathbf{Q}\hat{\boldsymbol{\sigma}}\mathbf{Q}^T = \hat{\mathbf{f}}[\mathbf{Q}\mathbf{d}\mathbf{Q}^T, \rho, \theta] \tag{7.2.6}$$

which is reconcilable with (7.2.3) if and only if

$$\mathbf{Q}\hat{\mathbf{f}}[\mathbf{d}, \rho, \theta]\mathbf{Q}^T = \hat{\mathbf{f}}[\mathbf{Q}\mathbf{d}\mathbf{Q}^T, \rho, \theta] \tag{7.2.7}$$

for all proper orthogonal $\mathbf{Q}$.

Equation (7.2.7), which is a functional equation for $\hat{\mathbf{f}}$, restricts the nature of the matrix $\hat{\mathbf{f}}$. One immediate consequence of (7.2.7) is that the fluid is isotropic. This follows from the fact that if $\boldsymbol{\sigma}$ and $\mathbf{d}$ are measured in two different reference frames, say primed and unprimed so that

$$\hat{\boldsymbol{\sigma}}' = \mathbf{Q}\hat{\boldsymbol{\sigma}}\mathbf{Q}^T, \qquad \mathbf{d}' = \mathbf{Q}\mathbf{d}\mathbf{Q}^T$$

then (7.2.6) may be re-written $\hat{\boldsymbol{\sigma}}' = \hat{\mathbf{f}}(\mathbf{d}', \rho, \theta)$

so that $\hat{\boldsymbol{\sigma}}'$ is related to $\mathbf{d}'$ in precisely the same manner as $\hat{\boldsymbol{\sigma}}$ is related to $\mathbf{d}$. In other words the properties of the material are the same in all directions.

Any second order symmetric tensor $\mathbf{f}$ which satisfies the relation (7.2.7) is said to be an isotropic symmetric tensor function of the second order symmetric tensor $\mathbf{d}$. We can prove immediately that the principal axes of $\mathbf{d}$ and $\hat{\boldsymbol{\sigma}}$ coincide. For suppose in (7.2.4) $\mathbf{d}$ is in diagonal form

$$\mathbf{d} = \operatorname{diag}[d(1), d(2), d(3)] \; .$$

(Since $\mathbf{d}$ is symmetric it is always possible to choose the coordinate system so that this is true.) Consider now the choice

$$\mathbf{Q} = \operatorname{diag}(1, -1, -1)$$

so that $$\mathbf{Q}\mathbf{d}\mathbf{Q}^T = \mathbf{d} \tag{7.2.8}$$

However on evaluating $\hat{\boldsymbol{\sigma}}' = \mathbf{Q}\hat{\boldsymbol{\sigma}}\mathbf{Q}^T$ there results

$$\mathbf{Q}\hat{\boldsymbol{\sigma}}\mathbf{Q}^T = \begin{pmatrix} \hat{\sigma}_{11} & -\hat{\sigma}_{12} & -\hat{\sigma}_{13} \\ -\hat{\sigma}_{21} & \hat{\sigma}_{22} & \hat{\sigma}_{23} \\ -\hat{\sigma}_{31} & \hat{\sigma}_{32} & \hat{\sigma}_{33} \end{pmatrix} . \tag{7.2.9}$$

Substituting (7.2.8) and (7.2.9) into (7.2.7) and comparing with (7.2.5) now leads to

$$\begin{pmatrix} \hat{\sigma}_{11} & \hat{\sigma}_{12} & \hat{\sigma}_{13} \\ \hat{\sigma}_{21} & \hat{\sigma}_{22} & \hat{\sigma}_{23} \\ \hat{\sigma}_{31} & \hat{\sigma}_{32} & \hat{\sigma}_{33} \end{pmatrix} = \begin{pmatrix} \hat{\sigma}_{11} & -\hat{\sigma}_{12} & -\hat{\sigma}_{13} \\ -\hat{\sigma}_{21} & \hat{\sigma}_{22} & \hat{\sigma}_{23} \\ -\hat{\sigma}_{31} & \hat{\sigma}_{32} & \hat{\sigma}_{33} \end{pmatrix}$$

so that $\hat{\sigma}_{12} = \hat{\sigma}_{13} = 0.$

Similarly the choice $\quad \mathbf{Q} = \text{diag}(-1, -1, 1)$

leads to $\hat{\sigma}_{13} = \hat{\sigma}_{23} = 0$ so that if $\mathbf{d}$ is in diagonal form then so is $\hat{\boldsymbol{\sigma}}$. Thus the principal axes of the extra stress coincide with those of the rate of deformation tensor. It follows that in a principal axes representation (7.2.5), is equivalent to three equations of the form

$$\begin{aligned} \hat{\sigma}(1) &= \phi[d(1), d(2), d(3), \rho, \theta] \\ \hat{\sigma}(2) &= \phi[d(2), d(3), d(1), \rho, \theta] \quad . \\ \hat{\sigma}(3) &= \phi[d(3), d(1), d(2), \rho, \theta] \end{aligned} \qquad (7.2.10)$$

Only one function ϕ is involved in the three equations because of the isotropy result [consider for example interchange of the labelling of the axes 1, 2, 3 to 2, 3, 1 and 3, 1, 2]. Also, from consideration of interchanging axes 2 and 3 (and the direction of the x_1 axis)

$$\phi[d(1), d(2), d(3), \rho, \theta] = \phi[d(1), d(3), d(2), \rho, \theta]$$

i.e.

$$\phi[x, y, z, \rho, \theta] = \phi[x, z, y, \rho, \theta] \qquad (7.2.11)$$

We now obtain the important result which renders (7.2.5) in the more explicit form

$$\hat{\boldsymbol{\sigma}} = \alpha\mathbf{I} + \beta\mathbf{d} + \gamma\mathbf{d}^2 \qquad (7.2.12)$$

where α, β and γ are functions of ρ, θ and the three scalar invarients of $\mathbf{d}$, namely

$$\mathrm{I}_d = d(1) + d(2) + d(3) \equiv d_{ii} \qquad (7.2.13\text{a})$$

$$\begin{aligned} \mathrm{II}_d &= d(1)d(2) + d(2)d(3) + d(3)d(1) \\ &\equiv \tfrac{1}{2}(d_{ii}d_{jj} - d_{ij}d_{ji}) \end{aligned} \qquad (7.2.13\text{b})$$

$$\mathrm{III}_d = d(1)d(2)d(3) \equiv |d_{ij}| \qquad (7.2.13\text{c})$$

[compare with equation (4.5.6)-(4.5.11)].

We write

$$\phi[d(1), d(2), d(3), \rho, \theta] = \alpha + \beta d(1) + \gamma d^2(1) \qquad (7.2.14\text{a})$$

$$\phi[d(2), d(3), d(1), \rho, \theta] = \alpha + \beta d(2) + \gamma d^2(2) \qquad (7.2.14\text{b})$$

$$\phi[d(3), d(1), d(2), \rho, \theta] = \alpha + \beta d(3) + \gamma d^2(3) \qquad (7.2.14\text{c})$$

where α, β, γ are functions of $d(1)$, $d(2)$, $d(3)$, ρ and θ.

This is always possible since on solving equations (7.2.14) for α, β, γ we have (from Cramer's rule)

$$\Delta\alpha = \begin{vmatrix} \phi[d(1), d(2), d(3), \rho, \theta] & d(1) & d^2(1) \\ \phi[d(2), d(3), d(1), \rho, \theta] & d(2) & d^2(2) \\ \phi[d(3), d(1), d(2), \rho, \theta] & d(3) & d^2(3) \end{vmatrix} \tag{7.2.15a}$$

$$\Delta\beta = \begin{vmatrix} 1 & \phi[d(1), d(2), d(3), \rho, \theta] & d^2(1) \\ 1 & \phi[d(2), d(3), d(1), \rho, \theta] & d^2(2) \\ 1 & \phi[d(3), d(1), d(2), \rho, \theta] & d^2(3) \end{vmatrix} \tag{7.2.15b}$$

$$\Delta\gamma = \begin{vmatrix} 1 & d(1) & \phi[d(1), d(2), d(3), \rho, \theta] \\ 1 & d(2) & \phi[d(2), d(3), d(1), \rho, \theta] \\ 1 & d(3) & \phi[d(3), d(1), d(2), \rho, \theta] \end{vmatrix} \tag{7.2.15c}$$

where Δ is the 'alternant' determinant

$$\Delta = \begin{vmatrix} 1 & d(1) & d^2(1) \\ 1 & d(2) & d^2(2) \\ 1 & d(3) & d^2(3) \end{vmatrix} = [d(1) - d(2)]\,[d(2) - d(3)]\,[d(3) - d(1)]\ . \tag{7.2.16}$$

For differentiable functions ϕ there are no difficulties associated with the singular case $\Delta = 0$. The latter occurs when two, (or possibly all three) of the $d(r)$ are equal. Under these conditions the determinants on the right sides of equations (7.2.15) also vanish. Consider for example the determinant on the right side of (7.2.15a) for the case $d(2) = d(3)$. The determinant becomes

$$\begin{vmatrix} \phi[d(1), d(2), d(2), \rho, \theta] & d(1) & d^2(1) \\ \phi[d(2), d(2), d(1), \rho, \theta] & d(2) & d^2(2) \\ \phi[d(2), d(1), d(2), \rho, \theta] & d(2) & d^2(2) \end{vmatrix}\ . \tag{7.2.17}$$

However because of (7.2.11) the last two rows of (7.2.17) are identical, and by a well known theorem, the determinant vanishes. Since also $[d(2) - d(3)]$ is a factor of Δ, α is determinable by a limiting process provided ϕ is a differentiable function of the $d(r)$.

More important, it is clear from equations (7.2.15) that α, β and γ are symmetric functions of $d(1)$, $d(2)$ and $d(3)$, i.e. interchange of pairs of the numbers 1, 2 and 3 leave α, β, γ unchanged.† There are three independent symmetric functions of three variables. These may be chosen in a variety of ways. For present purposes we choose the independent symmetric functions to be

† Consider for example interchange of 1 with 2 in (7.2.15a). Δ changes sign; however, by virtue of (7.2.11), the effect of the interchange on the determinant on the right hand side of (7.2.15a) is to interchange the first two rows which (by a well known theorem) changes the sign of the determinant in question. Similarly for other interchanges.

$$d(1) + d(2) + d(3) = \mathrm{I}_d; \quad d(1)d(2) + d(2)d(3) + d(3)d(1) = \mathrm{II}_d;$$
$$d(1)d(2)d(3) = \mathrm{III}_d$$

so that

$$\alpha = \alpha(\mathrm{I}_d, \mathrm{II}_d, \mathrm{III}_d\ \rho, \theta) \quad (7.2.18a); \quad \beta = \beta(\mathrm{I}_d, \mathrm{II}_d, \mathrm{III}_d, \rho, \theta) \quad (7.2.18b)$$
$$\gamma = \gamma(\mathrm{I}_d, \mathrm{II}_d, \mathrm{III}_d, \rho, \theta) \quad (7.2.18c)$$

where I_d, II_d and III_d are the invariants of d_{ij}, defined by equations (7.2.13).

With α, β, γ given as invariants of **d**, equations (7.2.10) and (7.2.14) may be written

$$\mathrm{diag}[\hat{\sigma}(1), \hat{\sigma}(2), \hat{\sigma}(3)] =$$
$$\alpha\mathbf{I} + \beta\,\mathrm{diag}[d(1), d(2), d(3)] + \gamma\,\mathrm{diag}[d^2(1), d^2(2), d^2(3)]$$

which establishes, in principal axes form, the more general matrix equation

$$\hat{\sigma} = \alpha\mathbf{I} + \beta\mathbf{d} + \gamma\mathbf{d}^2 \ . \qquad (7.2.19)$$

In suffix notation (7.2.19) may be written

$$\hat{\sigma}_{ij} = \alpha\delta_{ij} + \beta d_{ij} + \gamma d_{is}d_{sj} \qquad (7.2.20)$$

which is the most general form for the quantity $\hat{\sigma}_{ij}$ consistent with the hypothesis (7.2.1). Equations (7.2.20) were derived first by Reiner (*Am. J Math.* **67**, 350, 1945) and by Rivlin (*Nature* **160**, 611, 1947). Materials obeying (7.2.20) are usually known as Reiner-Rivlin fluids. For incompressible materials I_d is identically equal to zero. For this case α, β and γ are to be regarded as function of II_d, III_d and θ.

Since $\hat{\sigma}_{ij}$ vanishes in the absence of motion we must have [from (7.2.20)]

$$\alpha(0, 0, 0, \rho, \theta) = 0 \ .$$

Assuming α may be expanded in a Taylor series we have therefore

$$\alpha(\mathrm{I}_d, \mathrm{II}_d, \mathrm{III}_d, \rho, \theta) = \alpha_0(\rho, \theta)\mathrm{I}_d + \alpha_1(\rho, \theta)\mathrm{II}_d + \alpha_2(\rho, \theta)\mathrm{III}_d + \ldots$$

The classical Newtonian fluid is the particular case of (7.2.20) which is linear in the velocity gradients. This is derived by writing $\gamma = 0$, approximating α by $\alpha_0\mathrm{I}_d$ and assuming β is independent of the invariants:

$$\gamma = 0, \quad \alpha = \alpha_0(\rho, \theta)\mathrm{I}_d \equiv \alpha_0 d_{ii}, \quad \beta = 2\mu(\rho, \theta) \ .$$

The resulting equation may be written

$$\hat{\sigma}_{ij} = (\alpha_0 + \tfrac{2}{3}\mu)d_{mm}\delta_{ij} + 2\mu(d_{ij} - \tfrac{1}{3}d_{mm}\delta_{ij}) \qquad (7.2.21)$$

where $(\alpha_0 + \frac{2}{3}\mu)$ is known as the 'coefficient of bulk (or volume) viscosity' and μ is the 'coefficient of shear viscosity' (or more commonly the 'coefficient of viscosity'). The terminology derives from considering the two possible motions

(i) $\quad \mathbf{v} = d\mathbf{r} \quad (d = \text{constant})$

for which $\quad d_{ij} = d\delta_{ij}$

and (ii) $v_1 = dx_2, \quad v_2 = v_3 = 0$

for which $d_{12} = d_{21} = \frac{1}{2}d,\ d_{11} = d_{22} = d_{33} = d_{23} = d_{13} = 0$.

For the first of these motions, corresponding to uniform compression or expansion (see Fig. 7.2a)

$$\hat{\sigma}_{ij} = 3(\alpha_0 + \tfrac{2}{3}\mu)d\delta_{ij}$$

while for the second, corresponding to rectilinear shear flow (see Fig. 7.2b)

$$\hat{\sigma}_{12} = \hat{\sigma}_{21} = \mu d\ ,\ \hat{\sigma}_{11} = \hat{\sigma}_{22} = \hat{\sigma}_{33} = \hat{\sigma}_{23} = \hat{\sigma}_{13} = 0\ .$$

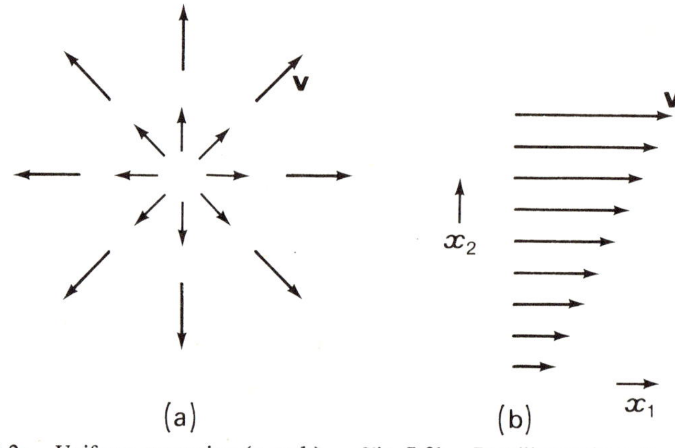

Fig. 7.2a Uniform expansion ($\mathbf{v} = d\mathbf{r}$) Fig. 7.2b Rectilinear shear flow $v_1 = dx_2$

Commonly it is assumed further that the coefficient of bulk viscosity vanishes. The resulting equation

$$\hat{\sigma}_{ij} = 2\mu(d_{ij} - \tfrac{1}{3}d_{mm}\delta_{ij}) \tag{7.2.22}$$

describes a fluid whose 'non ideal' properties are characterised by the single quantity $\mu(\rho, \theta)$. The material described by (7.2.22) is sometimes known as a Stokes' fluid and provides the most commonly used model for calculations in the flow of non ideal fluids. The term 'real fluid', also encountered, is less apt in view of the fact that some real fluids behave in a much more complex manner.

Equation (7.2.22) provides an adequate model for many liquids and gases, and in particular for the two most commonly encountered fluids, water and air.

For gases, equation (7.2.22) may be obtained from a molecular model based on the kinetic theory of gases. In this theory the existence of shear viscosity derives from the transfer of linear momentum caused by molecules moving between neighbouring layers of shearing flow†. The coefficient μ is obtained

† There seems to be no parallel mechanism which leads to a non-vanishing coefficient of bulk viscosity.

in terms of molecular parameters. For the simplest model of a gas of rigid spherical moledules of radius a, there results the formula

$$\mu = Aa^{-2}(mk\theta)^{\frac{1}{2}}$$

where m is the molecular weight, k is Boltzmann's constant, A is a numerical constant and θ is the absolute temperature. More complicated intermolecular force models lead to a similar dependence on μ on θ. The increase of μ with θ is broadly consistent with experimental observations on gases (Table 7.1).

In contrast, for liquids, μ decreases with θ (Table 7.1); here, molecular calculations are much less reliable.

The dimensions of μ are $[ML^{-1}T^{-1}]$ and the c.g.s. unit of μ is denoted by 'Poise' (after Poiseuille). The S.I. unit (Ns/m^2) is 10 Poise. Values of μ for different materials and temperatures vary by several orders of magnitude (see table 7.1 – taken from Handbook of Chemistry and Physics).

Table 7.1 Viscosity (μ) of some materials

Air at –100°C	113.10^{-7}	Ns/m^2
Air at 0°C	171.10^{-7}	,,
Air at 638°C	401.10^{-7}	,,
Water at 0°C	$1.79\ .10^{-3}$	,,
Water at 30°C	$0.801.10^{-3}$	,,
Water at 60°C	$0.469.10^{-3}$	,,
Pitch at 0°C	51.10^{9}	,,
Wax, shoe makers at 8°C	$4.7\ .10^{5}$	,,
Glucose at 22°C	$9.1\ .10^{12}$	,,
Glucose at 60°C	$9.3\ .10^{4}$	,,
Glycerin at –42°C	$6.7\ .10^{3}$	,,
Glycerin at 0°C	$1.21\ .10$	,,

Evidently viscosity is very sensitive to θ ; the dependence on density is usually of negligible significance.

On combining (7.2.22) with

$$\sigma_{ij} = -P\delta_{ij} + \hat{\sigma}_{ij}$$

and expressing d_{ij} in terms of the velocity field we derive

$$\sigma_{ij} = -P\delta_{ij} + \mu\left(\frac{\partial v_i}{\partial x_j} + \frac{\partial v_j}{\partial x_i}\right) - \frac{2}{3}\mu \operatorname{div} \mathbf{v}\delta_{ij} \qquad (7.2.23)$$

which is the usual form of the equation for a Stokes' fluid. On substituting (7.2.23) into the equations of motion

$$\frac{\partial \sigma_{ij}}{\partial x_j} + \rho b_i = \rho\left[\frac{\partial v_i}{\partial t} + (\mathbf{v}\,.\,\text{grad})v_i\right]$$

and assuming μ constant, independent of θ and ρ, there results the Navier-Stokes equation. In vector form the equation is

$$-\text{grad}\, P + \rho \mathbf{b} + \mu \nabla^2 \mathbf{v} + \tfrac{1}{3}\mu\, \text{grad}\,(\text{div}\, \mathbf{v}) = \rho \left[\frac{\partial \mathbf{v}}{\partial t} + (\mathbf{v} \,.\, \text{grad})\mathbf{v} \right] \tag{7.2.24}$$

which differs from the Euler equation by the appearance of the terms involving μ. If it is necessary to take account of the dependence of μ on θ, then extra terms appear in (7.2.24).

For the barotropic case where $P(\rho)$ depends only on ρ, equation (7.2.24) together with the equation of mass conservation provide four scalar equations for ρ and $\mathbf{v}$. In the particular case of an incomressible material with $\rho = \rho_0$, the basic equations for the flow of a Newtonian fluid are

$$\begin{aligned} -\text{grad}\, P + \rho_0 \mathbf{b} + \mu \nabla^2 \mathbf{v} &= \rho_0 [\partial \mathbf{v}/\partial t + (\mathbf{v} \,.\, \text{grad})\mathbf{v}] \\ \text{div}\, \mathbf{v} &= 0 \end{aligned} \tag{7.2.25}$$

which are four scalar equations for P and $\mathbf{v}$.

There arises the question of the relative merits of (7.2.20) versus (7.2.21) [or (7.2.22)] as models of real materials. For air and water there seems little doubt that (7.2.22) provides a perfectly good description and that the more general form (7.2.20) is unnecessary. This is probably generally the case for gases, where there is support for (7.2.22) from the kinetic theory, and also for some liquids besides water. However some liquids do *not* conform to (7.2.22), and even in cases where there are published values of μ in the literature, there is room for doubt. The reason is that much of the early data in the literature derives from a single type of experiment such as measurements based on Poiseuille flow (Chapters 1 and 10). It is possible that a value of μ can be assigned from such an experiment, even when (7.2.22) fails to describe correctly the material behaviour (see the discussion in Chapter 2).

Attempts have been made to correlate the behaviour of some liquids, known not to conform to the Newtonian model (7.2.22), with the Reiner-Rivlin model (7.2.20). These attempts, discussed in more detail in Section 9.4, are not entirely successful, and it appears that the Reiner-Rivlin model is too simple to describe the behaviour of real non-Newtonian fluids.

7.3 Energy theorems for barotropic fluids

We are now in a position to exploit the consequences of the energy equation (4.4.17) for Reiner-Rivlin fluids with constitutive equations

$$\sigma_{ij} = -P(\rho)\delta_{ij} + \alpha\delta_{ij} + \beta d_{ij} + \gamma d_{ik} d_{kj} \tag{7.3.1}$$

for the case where P, α, β and γ are independent of temperature.

The left hand side of (4.4.17) is the rate at which external forces are working on the material contained within S_t. Denoting this rate by dW/dt and writing

$$K = \iiint_{\Omega_t} \tfrac{1}{2}\rho \mathbf{v}^2 d\Omega$$

for the kinetic energy, equation (4.4.17) becomes

$$\frac{dW}{dt} = \frac{dK}{dt} + \iiint_{\Omega_t} \sigma_{ij} d_{ij}\, d\Omega \ . \tag{7.3.2}$$

We evaluate the integral appearing in (7.3.2) in the first instance for an 'ideal barotropic fluid' for which

$$\sigma_{ij} = -P(\rho)\delta_{ij} \ .$$

On using the result $\qquad d_{ij}\delta_{ij} = d_{ii} = \operatorname{div}\mathbf{v}$

together with the equation of mass conservation in the form (2.3.22)

$$\frac{D\rho}{Dt} + \rho \operatorname{div}\mathbf{v} = 0$$

we find

$$\iiint_{\Omega_t} \sigma_{ij} d_{ij} d\Omega = \iiint_{\Omega_t} \frac{P(\rho)}{\rho}\frac{D\rho}{Dt}\, d\Omega \ . \tag{7.3.3}$$

The right hand side of (7.3.3) may be expressed in the form

$$\iiint_{\Omega_t} \frac{P(\rho)}{\rho}\frac{D\rho}{Dt}\, d\Omega = \frac{d}{dt}\iiint_{\Omega_t} I(\rho)\, d\Omega \tag{7.3.4}$$

for on using the theorem (2.3.28)

$$\begin{aligned}\frac{d}{dt}\iiint_{\Omega_t} I(\rho) d\Omega &= \iiint_{\Omega_t} \left[\frac{\partial}{\partial t}[I(\rho)] + \operatorname{div}(\mathbf{v}I)\right] d\Omega \\ &= \iiint_{\Omega_t} \left\{ I'(\rho)\frac{\partial \rho}{\partial t} + I \operatorname{div}\mathbf{v} + I'(\rho)\mathbf{v} \, . \operatorname{grad}\rho \right\} d\Omega \\ &= \iiint_{\Omega_t} \left[I'(\rho) - I(\rho)/\rho \right] \frac{D\rho}{Dt}\, d\Omega \ . \end{aligned} \tag{7.3.5}$$

In obtaining the last line we have appealed again to (2.3.22). Comparing (7.3.4) and (7.3.5) now yields the ordinary differential equation

$$I'(\rho) - I(\rho)/\rho = P(\rho)/\rho$$

with solution $$I(\rho) = \rho \int_{\rho_1}^{\rho} [P(\rho')/(\rho')^2]\, d\rho' \qquad (7.3.6)$$

where ρ_1 is an arbitrary constant of integration.

With $I(\rho)$ defined by (7.3.6), equation (7.3.2) takes the form

$$\overset{\circ}{W} = \dot{K} + \overset{\circ}{V} \qquad (7.3.7)$$

where $$V = \iiint_{\Omega_t} I(\rho) d\Omega \; . \qquad (7.3.8)$$

For a motion where the external forces do no work, equation (7.3.7) yields the 'energy conservation' result

$$K + V = \text{const.} \qquad (7.3.9)$$

Since V depends on the state of the material (through ρ), and not on the velocity field, equation (7.3.9) identifies V as a 'potential energy' of compression. For the more general case where the external forces do not vanish

$$W = K + V + \text{const.} \qquad (7.3.10)$$

and the work done by the external forces contributes to the sum of the kinetic and potential energies. By analogy with particle mechanics, the ideal barotropic fluid which satisfies (7.3.10) could be termed 'conservative'. For example in a motion which carries the material from a state of rest to an identical state of rest and with the same value of V, the total work done by the external forces is zero.

With the above interpretation for V we identify $I(\rho)$ as the 'potential energy' per unit volume and

$$U(\rho) \equiv I(\rho)/\rho \equiv \int_{\rho_1}^{\rho} [P(\rho')/(\rho')^2]\, d\rho' \qquad (7.3.11)$$

as the potential energy per unit mass.

The quantity $I(\rho)$ may be termed the 'internal energy' per unit volume and $U(\rho)$ the 'internal energy' per unit mass.† For a barotropic fluid the internal energy is a stored energy of compression which is recoverable on release of the pressure.

In (7.3.6) the integration constant ρ_1 may be chosen arbitrarily. This merely reflects the indeterminateness of $U(\rho)$ to within an arbitrary constant. However a natural choice for ρ_1 is the density in the stress free state. In this case $U(\rho)$ is then the energy/unit mass required to compress (slowly) the material from the stress free state to the state in which the density is ρ; also if W is identified with the work done by the external forces in bringing the fluid from the stress free and motionless state to the current state then the constant in (7.3.10) is zero.

† Under isothermal conditions U is more appropriately called a 'free energy', and is not to be identified with what is termed internal energy in a more general thermodynamic context.

For the case of a liquid specified by (7.1.13), $U(\rho)$ is given by

$$U(\rho) = \frac{K_1}{\rho_0}\left[\log(\rho/\rho_0) - (1 - \rho_0/\rho)\right]$$

$$= \frac{K_1}{2\rho_0^{\,3}}(\rho - \rho_0)^2 \quad [\text{for } (\rho - \rho_0) \ll \rho_0] \,. \qquad (7.3.12)$$

For the gas specified by (7.1.9), with the choice $\rho_1 = 0$

$$U(\rho) = \frac{k\rho^{\gamma-1}}{\gamma - 1} \equiv \frac{P_0\rho^{\gamma-1}}{(\gamma - 1)\rho_0{}^{\gamma}} \,. \qquad (7.3.13)$$

For an incompressible fluid ρ is constant and P is indeterminate (except through the motion). However in (7.3.6), ρ is constant while P is certainly finite, so that for an incompressible fluid we must have $I = U = 0$.

We turn now to the more general problem of the Reiner-Rivlin material (7.3.1). With the same definition as in equation (7.3.8) for V, and using (7.3.2), equation (7.3.7) is replaced by

$$\overset{\circ}{W} = \overset{\circ}{K} + \dot{V} + Q \qquad (7.3.14)$$

where

$$Q = \iiint_{\Omega_t} D d\Omega \qquad (7.3.15)$$

and where

$$D = \alpha d_{ii} + \beta d_{ij} d_{ij} + \gamma d_{ik} d_{kj} d_{ji} \qquad (7.3.16)$$

is a scalar invariant of the tensor d_{ij}.

To interpret D consider a motion, such as the stirring of a liquid in a vessel, in which the material of a continuum is taken from an initial state of rest at $t = t_1$ $(K = 0,\ V = V_0)$ to a similar final state at $t = t_2$. From (7.3.14) the total work done by the external forces is

$$W \equiv \int_{t_1}^{t_2} \overset{\circ}{W} dt = \int_{t_1}^{t_2} Q dt \qquad (7.3.17)$$

In (7.3.17) the quantity W must be non-negative since otherwise the mechanical properties of the material could be used to provide a source of energy. If W is to be non-negative for all possible fluid motions then it is implied that $Q \geqslant 0$ and therefore $D \geqslant 0$.

The quantity Q is the rate of energy expenditure by the external forces over and above the rate of supply of kinetic and potential energy. Accordingly Q represents a rate of energy dissipation and D is the dissipation rate per unit volume. For a barotropic material the dissipated energy appears in the form of heat.

For the Newtonian fluid (7.2.21), D is given by

$$\begin{aligned} D &= (\alpha_0 + \tfrac{2}{3}\mu)d_{mm}d_{ii} + 2\mu\, d_{ij}(d_{ij} - \tfrac{1}{3}d_{mm}\delta_{ij}) \\ &= (\alpha_0 + \tfrac{2}{3}\mu)d_{mm}d_{ii} + 2\mu\, \hat{d}_{ij}\hat{d}_{ij} \end{aligned}$$

where

$$\hat{d}_{ij} = d_{ij} - \tfrac{1}{3}d_{mm}\delta_{ij}\,. \qquad (7.3.18)$$

The inequality $D \geqslant 0$, requires $(\alpha_0 + \frac{2}{3}\mu) \geqslant 0$, $\mu \geqslant 0$.

For the case of a Stokes' fluid $(\alpha_0 + \frac{2}{3}\mu) = 0$, and D is given by

$$D = 2\mu\, \hat{d}_{ij}\, \hat{d}_{ij} \quad (\mu \geqslant 0)\,. \qquad (7.3.19)$$

The quantity $\hat{d}_{ij}$, sometimes known as the 'deviatoric rate of deformation tensor', possesses the property

$$\hat{d}_{ii} = 0\,.$$

For an incompressible Stokes fluid $d_{ii} \equiv \operatorname{div} \mathbf{v} = 0$ and

$$\hat{d}_{ij} = d_{ij}\,.$$

In these circumstances

$$D = 2\mu\, \hat{d}_{ij}\, \hat{d}_{ij} \equiv 2\mu\, d_{ij}\, d_{ij}\,.$$

Summary

CONSTITUTIVE EQUATIONS OF REINER-RIVLIN FLUID

$$\sigma_{ij} = -P(\rho, \theta)\delta_{ij} + \alpha\,\delta_{ij} + \beta\, d_{ij} + \gamma\, d_{ik}\, d_{kj}$$

where α, β, γ are functions of

$$\mathrm{I}_d,\ \mathrm{II}_d,\ \mathrm{III}_d,\ \rho,\ \theta\,.$$

Energy dissipation rate/unit volume of Reiner-Rivlin fluid

$$D = \alpha\, d_{ii} + \beta\, d_{ij}\, d_{ij} + \gamma\, d_{ik}\, d_{kj}\, d_{ji}\,.$$

Internal energy/unit mass for $P = P(\rho)$ (barotropic case).

$$U = \int_{\rho_1}^{\rho} [P(\rho')/(\rho')^2]\, d\rho'\,.$$

Incompressible case. P indeterminate, $U = 0$. α, β, γ depend on II_d, III_d, and θ.

STOKES' FLUID

$$\begin{aligned} \sigma_{ij} &= -P(\rho, \theta)\,\delta_{ij} + 2\mu(d_{ij} - \tfrac{1}{3}d_{mm}\,\delta_{ij}) \\ &\equiv -P\,\delta_{ij} + 2\mu\, \hat{d}_{ij}\,. \end{aligned}$$

where

$$P = P(\rho, \theta)\,; \qquad \mu = \mu(\rho, \theta)$$

The rate of energy dissipation per unit volume is

$$D = 2\mu\, \hat{d}_{ij}\, \hat{d}_{ij}\,.$$

For the barotropic case $[P = P(\rho)]$, U is as above.

The Navier-Stokes equations, derived from equations of motion and constitutive equations (assuming $\mu =$ constant), are

$$-\operatorname{grad} P + \rho\,\mathbf{b} + \mu\nabla^2\mathbf{v} + \tfrac{1}{3}\mu \operatorname{grad}(\operatorname{div}\mathbf{v}) = \rho\left(\frac{\partial\mathbf{v}}{\partial t} + (\mathbf{v}\cdot\operatorname{grad})\mathbf{v}\right).$$

Complete equations for ρ and $\mathbf{v}$ for incompressible case ($\rho = \rho_0$, $\operatorname{div}\mathbf{v} = 0$ with μ independent of θ)

$$-\operatorname{grad} P + \rho_0\mathbf{b} + \mu\nabla^2\mathbf{v} = \rho_0\left(\frac{\partial\mathbf{v}}{\partial t} + (\mathbf{v}\cdot\operatorname{grad})\mathbf{v}\right)$$

$$\operatorname{div}\mathbf{v} = 0$$

$$(D = 2\mu\, d_{ij}\, d_{ij},\ U = 0)\,.$$

CONSTITUTIVE EQUATIONS FOR IDEAL FLUID

$$\sigma_{ij} = -P(\rho, \theta)\,\delta_{ij}\,.$$

The Euler equations of motion are

$$-\operatorname{grad} P + \rho\,\mathbf{b} = \rho\left(\frac{\partial\mathbf{v}}{\partial t} + (\mathbf{v}\cdot\operatorname{grad})\mathbf{v}\right).$$

Complete equations for barotropic case $[P = P(\rho)]$ given by above together with equation of mass conservation

$$\frac{\partial\rho}{\partial t} + \operatorname{div}(\rho\mathbf{v}) = 0\,.$$

Incompressible case,

$$-\operatorname{grad} P + \rho_0\mathbf{b} = \rho_0\left(\frac{\partial\mathbf{v}}{\partial t} + (\mathbf{v}\cdot\operatorname{grad})\mathbf{v}\right)$$

$$\operatorname{div}\mathbf{v} = 0$$

where ρ_0 is constant.

Chapter 8

Elastic and Thermo-elastic Materials

8.1 Definition and Introduction

Thermo-elastic materials are defined by constitutive equations of the general form

$$\sigma_{ij} = f_{ij}(\partial x_r/\partial X_s, \theta) \tag{8.1.1}$$

or, in matrix notation

$$\boldsymbol{\sigma} = \mathbf{f}(\mathbf{F}, \theta) . \tag{8.1.2}$$

Here $\mathbf{f}$ is a matrix function of the displacement gradient matrix $\mathbf{F}$, and also a function of temperature.

More generally it is possible to include the particle label $\mathbf{R}$ as one of the independent variables of $\mathbf{f}$; under these circumstances the dependence of $\boldsymbol{\sigma}$ on $\mathbf{F}$ and θ varies from particle to particle and the resulting thermomechanical properties are then inhomogeneous with respect to the reference configuration. Often, however, it is possible to assume the existence of a reference configuration, usually the stress free state, with respect to which the thermomechanical properties are homogeneous, as in (8.1.1) and (8.1.2). The present text is concerned only with homogeneous materials.

Equation (8.1.2) asserts that the stress components depend on the components of $\mathbf{F}$. In view of the analysis of Section 5.1, an equivalent statement is that the stress components are functions of the material stretches and the rotation. It will be seen below that the precise manner in which $\boldsymbol{\sigma}$ depends on the material rotation is made clear by the principle of material frame indifference.

The one characteristic common to all materials conventionally regarded as solid is the ability of the materials to sustain within limits tensile and shear stresses without flowing indefinitely. Equation (8.1.2) describes such a substance, and materials with constitutive equations of this form are called **thermoelastic solids.** In circumstances where the dependence of σ on θ is unimportant, as in isothermal problems, the equation simplifies further to the form

$$\boldsymbol{\sigma} = \mathbf{f}(\mathbf{F}) . \tag{8.1.3}$$

Materials defined by (8.1.3) are known as **Cauchy elastic solids.**

It is a familiar experience that provided the stresses are not too large, many materials conform to (8.1.3). For example in stretching an elastic band, the extent of stretch depends on the applied force, that is to say it depends on the axial stress induced in the band. Similarly the amount of flexure experience by a metal rule depends on the transverse forces.

Almost all solid materials conform to (8.1.2), or in appropriate circumstances (8.1.3), for limited ranges of magnitude of the stresses under static loading conditions.

8.2 Material frame indifference

Equation (8.1.2) is required to be invariant under the transformation

$$x_i' = \beta_i(t) + Q_{ij}(t)x_j$$

where $\mathbf{Q}$ is a proper orthogonal matrix. By hypothesis the stress matrix transforms in accordance with the usual tensor law

$$\sigma' = \mathbf{Q}\sigma\mathbf{Q}^T \qquad (8.2.1)$$

while from (6.2.26), the deformation gradient matrix in the primed frame is given by

$$\mathbf{F}' = \mathbf{QF} \; . \qquad (8.2.2)$$

Since the 'law' (8.1.2) is supposed to hold for all material reference frames we have

$$\sigma' = \mathbf{f}(\mathbf{F}', \theta) \; . \qquad (8.2.3)$$

In writing down this result, we recall that θ is hypothesised to be a frame indifferent quantity.

From equations (8.2.1), (8.2.2) and (8.2.3) we deduce

$$\mathbf{Q}\sigma\mathbf{Q}^T = \mathbf{f}(\mathbf{QF}, \theta)$$

i.e.

$$\sigma = \mathbf{Q}^T\mathbf{f}(\mathbf{QF}, \theta)\mathbf{Q} \qquad (8.2.4)$$

which result must hold for all choices of proper orthogonal $\mathbf{Q}$. In (8.2.4) we decompose $\mathbf{F}$ into the product.

$$\mathbf{F} = \mathbf{RU} \qquad (8.2.5)$$

where $\mathbf{R}$ is a proper orthogonal matrix and $\mathbf{U}$ is the right stretch tensor. On substituting for $\mathbf{F}$ from (8.2.5) into (8.2.4) and adopting the particular choice

$$\mathbf{Q} = \mathbf{R}^T$$

we derive the representation

$$\sigma = \mathbf{Rf}(\mathbf{U}, \theta)\mathbf{R}^T \qquad (8.2.6)$$

as a *necessary* consequence of material frame indifference.

The form (8.2.6) is also *sufficient* to guarantee that the requirements of material frame indifference are met. For under the transformation

$$x_i' = \beta_i + Q_{ij}x_j$$

we have from (6.2.27)

$$\mathbf{U}' = \mathbf{U} \qquad (8.2.7)$$

which together with (8.2.2) and (8.2.5) implies

$$\mathbf{R}' = \mathbf{QR} \qquad (8.2.8)$$

Accordingly pre-multiplying (8.2.6) by $\mathbf{Q}$, post-multiplying by $\mathbf{Q}^T$, and using (8.2.7), (8.2.8), and (8.2.1) leads to the equation

$$\boldsymbol{\sigma}' = \mathbf{R}'\mathbf{f}(\mathbf{U}', \theta)(\mathbf{R}')^T$$

which is of exactly the same form as the original equation (8.2.6).

The representation (8.2.6) for the general form of the constitutive equations of a thermoelastic solid shows the precise manner in which $\boldsymbol{\sigma}$ depends on the rotation matrix $\mathbf{R}$. In contrast the dependence of $\boldsymbol{\sigma}$ on the stretch through $\mathbf{f}(\mathbf{U}, \theta)$ is arbitrary.

In contrast to the parallel state of affairs in the theories for fluids [Chapter 7, equation (7.2.7) and following] the requirements of frame indifference have led to a general form of constitutive equation which in general is not isotropic. This is not a surprising result because there exist a number of real elastic materials whose mechanical behaviour, while conforming to the description (8.1.2) for sufficiently small stresses, are distinctly anisotropic; the best known examples are metal crystals. However many elastic and thermoelastic solids such as polycrystalline metals, natural rubber and synthetic elastomers are mechanically isotropic with respect to a 'natural' unstressed state. For these materials, the constitutive equations may be obtained in a much more 'explicit' form, as follows below.

8.3 Isotropic thermoelastic materials

A thermoelastic material is isotropic with respect to the reference state if the relation (8.2.6) is completly independent of the choice of orientation of the coordinate reference system.

From a slightly different, and more physical viewpoint, consider a deformation gradient matrix whose components are fixed with respect to the axes of a rectangular Cartesian coordinate system whose orientation may be chosen arbitrarily. If for all possible choices of orientation the stress response measured with respect to the coordinate axes is the same, then the material is isotropic.

It is well to remind the reader that we are concerned here with the family of different, relatively fixed, orthogonal coordinate systems, any one of which may be used for the measurement of both the X_i and the x_i. Thus in this section we are concerned with the tensor transformation properties of $\boldsymbol{\sigma}$, $\mathbf{F}$ etc.,

e.g. $$\boldsymbol{\sigma}' = \mathbf{A}\boldsymbol{\sigma}\mathbf{A}^T, \quad \mathbf{F}' = \mathbf{A}\mathbf{F}\mathbf{A}^T \tag{8.3.1}$$

where $\mathbf{A}$ is a time independent proper orthogonal matrix defined in Section 4.2. In particular the transformations of $\mathbf{F}$ to $\mathbf{F}'$ is the normal transformation law of a second order tensor and is different from the transformation law $\mathbf{F}' = \mathbf{Q}\mathbf{F}$ discussed in the last section in connection with the entirely different concept of material frame indifference.

In exploiting the mathematical consequences of isotropy it is profitable to consider (8.2.6) in the slightly different form

$$\boldsymbol{\sigma} = \mathbf{R}\mathbf{M}(\mathbf{C}, \theta)\mathbf{R}^T \tag{8.3.2}$$

where $$\mathbf{C} = \mathbf{F}^T\mathbf{F} \equiv \mathbf{U}^2 \tag{8.3.3}$$

and where $$\mathbf{M}(\mathbf{U}^2, \theta) \equiv \mathbf{f}(\mathbf{U}, \theta) \ .$$

The reason for working with $\mathbf{C}$ rather than $\mathbf{U}$ is that the constitutive equation obtained ultimately is expressed in terms of the quantity $\mathbf{B}(\equiv \mathbf{F}\mathbf{F}^T)$ which is easily expressed in terms of the partial derivatives $\partial x_i/\partial X_j$. If the form (8.3.2) is retained in the ensuing analysis, the constitutive equation is expressed in terms of

$$\mathbf{V}(\equiv \mathbf{B}^{\frac{1}{2}})$$

which is not readily expressed in terms of the deformation gradients.

Consider now two coordinate systems, primed and unprimed, whose orientations are related by the proper orthogonal matrix $\mathbf{A}$ so that

$$\mathbf{X}' = \mathbf{A}\mathbf{X}\,, \qquad \mathbf{x}' = \mathbf{A}\mathbf{x} \ .$$

In the unprimed system let $\mathbf{F}$ be the deformation gradient matrix and σ the associated stress matrix so that σ is given by (8.3.2) with $\mathbf{C}$ defined by (8.3.3). Also $\mathbf{R}$ and $\mathbf{U}$ are defined uniquely by the right polar decomposition

$$\mathbf{F} = \mathbf{R}\mathbf{U} \tag{8.3.4}$$

where $\mathbf{R}$ is proper orthogonal and $\mathbf{U}$ symmetric with positive eigenvalues.

Consider now the related deformation gradient matrix, expressed with respect to the unprimed system,

$$\mathbf{F}_1 = \mathbf{A}^T\mathbf{F}\mathbf{A} \ . \tag{8.3.5}$$

$\mathbf{F}_1$ referred to the primed axes becomes $\mathbf{F}$ and represents therefore the same deformation with respect to the primed axes as $\mathbf{F}$ represents with respect to the unprimed axes. The assumption of isotropy asserts that the stress with respect to the primed axes is $\boldsymbol{\sigma}$; transforming back to the unprimed axes we see that $\mathbf{F}_1$ is associated with a stress

$$\boldsymbol{\sigma}_1 = \mathbf{A}^T\sigma\mathbf{A} \ . \tag{8.3.6}$$

However the relationship connecting $\boldsymbol{\sigma}_1$ and $\mathbf{F}_1$ is the same as (8.3.2) since (8.3.2) is the 'law' for the unprimed reference frame. Accordingly

$$\sigma_1 = \mathbf{R}_1\mathbf{M}(\mathbf{C}_1, \theta)\mathbf{R}_1{}^T \tag{8.3.7}$$

where $$\mathbf{F}_1 = \mathbf{R}_1\mathbf{U}_1 \tag{8.3.8}$$

and $$\mathbf{C}_1 = \mathbf{U}_1{}^2 \ . \tag{8.3.9}$$

From (8.3.4) and (8.3.5) $$\mathbf{F}_1 = (\mathbf{A}^T\mathbf{R}\mathbf{A})(\mathbf{A}^T\mathbf{U}\mathbf{A})$$

and since (i) $\mathbf{A}^T\mathbf{R}\mathbf{A}$ is proper orthogonal, (ii) $\mathbf{A}^T\mathbf{U}\mathbf{A}$ is symmetric, (iii) the decomposition (8.3.8) is unique, we have

$$\mathbf{R}_1 = \mathbf{A}^T\mathbf{R}\mathbf{A}\,, \tag{8.3.10a}$$

$$\mathbf{U}_1 = \mathbf{A}^T\mathbf{U}\mathbf{A} \tag{8.3.10b}$$

from the second of which we also deduce

$$\mathbf{C}_1 \equiv \mathbf{U}_1{}^2 = \mathbf{A}^T\mathbf{C}\mathbf{A} \ . \tag{8.3.11}$$

Combining now (8.3.2), (8.3.6), (8.3.7) and substituting from (8.3.10a) for $\mathbf{R}_1$ and $\mathbf{R}_1{}^T$ and from (8.3.11) for $\mathbf{U}_1{}^2$ leads to the equation

$$\mathbf{A}^T\mathbf{R}\mathbf{M}(\mathbf{C},\theta)\mathbf{R}^T\mathbf{A} = \mathbf{A}^T\mathbf{R}\mathbf{A}\mathbf{M}(\mathbf{A}^T\mathbf{C}\mathbf{A},\theta)\mathbf{A}^T\mathbf{R}^T\mathbf{A}$$

or on cancelling non-singular factors $\mathbf{A}^T\mathbf{R}$ and $\mathbf{R}^T\mathbf{A}$

$$\mathbf{M}(\mathbf{C},\theta) = \mathbf{A}\mathbf{M}(\mathbf{A}^T\mathbf{C}\mathbf{A},\theta)\mathbf{A}^T \qquad (8.3.12)$$

which is a functional relation for $\mathbf{M}$ required to hold for any proper orthogonal $\mathbf{A}$. A more familiar version of the equation is obtained if we write $\mathbf{A} = \mathbf{Q}^T$ (where $\mathbf{Q}$ is proper orthogonal) and re-arrange in the form

$$\mathbf{M}(\mathbf{Q}\mathbf{C}\mathbf{Q}^T,\theta) = \mathbf{Q}\mathbf{M}(\mathbf{C},\theta)\mathbf{Q}^T\ . \qquad (8.3.13)$$

Equation (8.3.13) for $\mathbf{M}$ as a function of $\mathbf{C}$ is exactly the same as the functional relation (7.2.7) for $\hat{\mathbf{f}}$ as a function of $\mathbf{d}$. Using the subsequent results of Section 7.2 it is seen that $\mathbf{M}$ is expressible in the form

$$\mathbf{M} = \alpha\mathbf{I} + \beta\mathbf{C} + \gamma\mathbf{C}^2 \qquad (8.3.14)$$

where α, β and γ are functions of θ and of the invariants of $\mathbf{C}$. These invariants are those of $\mathbf{U}^2$ whose principal axes form is

$$\operatorname{diag}\,[\lambda^2(1), \lambda^2(2), \lambda^2(3)]$$

with invariants

$$\begin{aligned}
\mathrm{I}_C &= \mathrm{I}_B = \lambda^2(1) + \lambda^2(2) + \lambda^2(3)\\
\mathrm{II}_C &= \mathrm{II}_B = \lambda^2(1)\lambda^2(2) + \lambda^2(2)\lambda^2(3) + \lambda^2(3)\lambda^2(1)\\
\mathrm{III}_C &= \mathrm{III}_B = \lambda^2(1)\lambda^2(2)\lambda^2(3)
\end{aligned}$$

As indicated, these are also the invariants of $\mathbf{V}^2$ ($\equiv \mathbf{B}$).

From (8.3.2) and (8.3.14) we derive

$$\begin{aligned}
\boldsymbol{\sigma} &= \mathbf{R}(\alpha\mathbf{I} + \beta\mathbf{C} + \gamma\mathbf{C}^2)\mathbf{R}^T\\
&= \alpha\mathbf{I} + \beta\mathbf{B} + \gamma\mathbf{B}^2 \qquad (8.3.15)
\end{aligned}$$

In obtaining (8.3.15) use has been made of the results

$$\mathbf{R}\mathbf{C}\mathbf{R}^T = \mathbf{R}\mathbf{U}^2\mathbf{R}^T = (\mathbf{R}\mathbf{U})(\mathbf{U}\mathbf{R}^T) = \mathbf{F}\mathbf{F}^T = \mathbf{V}^2 = \mathbf{B}$$

and similarly $\qquad \mathbf{R}\mathbf{C}^2\mathbf{R}^T = (\mathbf{R}\mathbf{C}\mathbf{R}^T)(\mathbf{R}\mathbf{C}\mathbf{R}^T) = \mathbf{B}^2\ .$

Equation (8.3.15) is one form for the constitutive equation of an isotropic thermo-elastic solid. Another form in common use is

$$\boldsymbol{\sigma} = h_0\mathbf{I} + h_1\mathbf{B} + h_{-1}\mathbf{B}^{-1} \qquad (8.3.16)$$

where h_0, h_1 and h_{-1} are coefficients which, like α, β and γ are functions of the invariants of $\mathbf{B}$ and also functions of temperature

$$\begin{aligned}
h_0 &= h_0(\mathrm{I}_B, \mathrm{II}_B, \mathrm{III}_B, \theta), \quad h_1 = h_1(\mathrm{I}_B, \mathrm{II}_B, \mathrm{III}_B, \theta),\\
h_{-1} &= h_{-1}(\mathrm{I}_B, \mathrm{II}_B, \mathrm{III}_B, \theta) \qquad (8.3.17)
\end{aligned}$$

Equation (8.3.16) is obtained from (8.3.15) by making use of the Cayley-Hamilton theorem

$$\mathbf{B}^3 - \mathrm{I}_B\mathbf{B}^2 + \mathrm{II}_B\mathbf{B} - \mathrm{III}_B\mathbf{I} = 0 \qquad (8.3.18)$$

which, after multiplying by $\mathbf{B}^{-1}$, allows $\mathbf{B}^2$ in (8.3.15) to be expressed in terms of $\mathbf{B}$, $\mathbf{B}^{-1}$, the unit matrix and the invariants. With some restriction on the nature of the coefficients (Section 8.5), equations of the form (8.3.16) were first obtained by Finger in 1894.

In the reference state $\lambda(1) = \lambda(2) = \lambda(3) = 1$ and

$$\mathbf{B} = \mathbf{B}^{-1} = \mathbf{I}$$

$$\mathrm{I}_B = \mathrm{II}_B = 3, \quad \mathrm{III}_B = 1 .$$

It follows that if the reference state is chosen to be a stress free state at some given temperature, say $\theta = \theta_0$, then the coefficients satisfy the relation

$$h_0(3, 3, 1, \theta_0) + h_1(3, 3, 1, \theta_0) + h_{-1}(3, 3, 1, \theta_0) = 0 \qquad (8.3.19)$$

8.4 Constitutive equation in linear (isothermal) isotropic elasticity

The constitutive equations of linear elasticity theory are obtainable from (8.3.16) by introducing the small displacement gradient approximations of Section 5.3. In this approximation we have [equations (5.3.8)]

$$\mathbf{B} = \mathbf{I} + 2\mathbf{E}, \quad \mathbf{B}^{-1} = \mathbf{I} - 2\mathbf{E} \qquad (8.4.1)$$

where the E_{ij}, defined by

$$E_{ij} = \frac{1}{2}\left(\frac{\partial U_i}{\partial X_j} + \frac{\partial U_j}{\partial X_i}\right) \qquad (8.4.2)$$

are of order $e(\ll 1)$ and terms of order e^2 and higher are neglected. In this approximation also

$$\begin{aligned} \mathrm{I}_B &= \text{trace } \mathbf{B} = 3 + 2E_{mm} + O(e^2) \\ \mathrm{II}_B &= \tfrac{1}{2}[(\text{trace } \mathbf{B})^2 - \text{trace } \mathbf{B}^2] = 3 + 4E_{mm} + O(e^2) \\ \mathrm{III}_B &= |\mathbf{B}| = 1 + 2E_{mm} + O(e^2) \end{aligned}$$

so that to order e the invariants are functions solely of E_{mm}

$$E_{mm} = \operatorname{div} \mathbf{U} .$$

It follows that if (8.3.16) is expanded in a Taylor series as far as terms of order e, and use is made of (8.3.19),

$$\boldsymbol{\sigma} = [\lambda(\theta_0) \operatorname{div} \mathbf{U}]\mathbf{I} + 2\mu(\theta_0)\mathbf{E} \qquad (8.4.3)$$

where $\lambda(\theta_0) = \left(2\dfrac{\partial}{\partial \mathrm{I}_B} + 4\dfrac{\partial}{\partial \mathrm{II}_B} + 2\dfrac{\partial}{\partial \mathrm{III}_B}\right)(h_0 + h_1 + h_{-1})_0$

$$2\mu(\theta_0) = 2(h_1 - h_{-1})$$

and where the functions and derivatives are evaluated for the stress free state ($I_B = II_B = 3, III_B = 1, \theta = \theta_0$).

In suffix notation equations (8.4.3) become

$$\sigma_{ij} = \lambda\Delta\delta_{ij} + 2\mu E_{ij} \tag{8.4.4}$$

where

$$\Delta = \text{div}\,\mathbf{U} \tag{8.4.5}$$

is called the dilatation, † and λ and μ are known as the Lamé constants. The quantity μ is also called the shear modulus. More familiar elastic constants – Young's modulus and Poisson's ratio – are defined in terms of λ and μ in Chapter 14.

Equations (8.4.4) are valid for the isothermal case ($\theta = \theta_0$); if it is necessary to take account of variations in temperature, further terms, not considered in the present text, appear in the equations.

Equations of the form (8.4.4) were first introduced into elasticity theory by Cauchy in 1828. The equations are the most general linear isotropic relations connecting the stress components σ_{ij} and the strain components E_{ij}.

The equations (8.4.4), together with equations (8.4.2) and (approximated) equations of motion lead to the technically important subject of **linear elasticity theory**, also known as **classical elasticity** and **infinitesimal elasticity**. The basic equations of this theory are discussed in Chapter 14 together with some applications.

8.5 Elastic strain energy function and isothermal Green elasticity

If W denotes the work down by the external forces acting on material contained within a volume Ω_t, the energy equation (4.4.17) becomes

$$\frac{dW}{dt} = \frac{dK}{dt} + \iiint_{\Omega_t} \sigma_{ij} d_{ij} d\Omega \tag{8.5.1}$$

where K is the kinetic energy and where the integral

$$\iiint_{\Omega_t} \sigma_{ij} d_{ij} d\Omega \tag{8.5.2}$$

is the rate of expenditure of energy on material deformation, i.e. expenditure distinct from that required to increase the kinetic energy.

In elementary physics we are familiar with the idea that an 'elastic spring' stores the energy of deformation, which is recovered when the spring reverts to the unloaded state. The energy stored in the spring depends on the strain. Extrapolating

† Because in this approximation $\rho_0 = \rho|\mathbf{F}| = \rho[1 + \Delta + O(e^2)] = \rho(1 + \Delta)$, i.e. $\rho = \rho_0(1 - \Delta)$ so that Δ measures the relative volume expansion or dilatation of the material.

these ideas to the present context suggests that for an elastic material the integral (8.5.2) may be expressed in the form (at fixed temperature θ_0)

$$\iiint_{\Omega_t} \sigma_{ij} d_{ij} d\Omega = \frac{d}{dt} \iiint_{\Omega_t} \rho U(\mathbf{F}, \theta_0) d\Omega \tag{8.5.3}$$

where $U(\mathbf{F})$, the internal energy per unit mass is a function of the components of the deformation gradient matrix $\mathbf{F}$.† [We have already encountered a similar quantity in the simpler context of a barotropic fluid (7.3.11)]. In what follows the parameter θ_0 is suppressed and the analysis is pertinent to isothermal deformations. Thus U is a 'free' energy.

Materials described by a scalar function

$$U = U(\mathbf{F}) \tag{8.5.4}$$

which satisfy (8.5.3) were first investigated by Green in the mid-nineteenth century and are known as **Green elastic solids.** Green elastic materials form a sub-class of the more general Cauchy elastic materials as will be seen below for the isotropic case. There seems to be no experimental evidence pointing to the existence of Cauchy elastic materials which are not Green elastic.

Material frame indifference requires that for the scalar quantity U

$$U(\mathbf{F}') = U(\mathbf{F}) \tag{8.5.5}$$

where $\mathbf{F}' = \mathbf{QF}$ and where $\mathbf{Q}$ is a proper orthogonal matrix. Writing $\mathbf{F}$ in the form $\mathbf{F} = \mathbf{RU}$ and considering in particular the choice $\mathbf{Q} = \mathbf{R}^T$ shows that any U of the form (8.5.4) is necessarily of the more restricted form

$$U = U(\mathbf{U})$$

i.e. the stored energy depends essentially on the components of the right stretch tensor.†† We re-write the last result in the equivalent form

$$U = U(\mathbf{U}^2) \equiv U(\mathbf{C}) \tag{8.5.6}$$

which is the general form of the stored energy function for an elastic solid. The terminology in common use for U in the present context is 'strain energy function'.

Without loss of generality U may be assumed symmetric in C_{rs} and C_{sr}; for, if to the contrary $U(C_{rs}) \neq U(C_{sr})$, define a new strain energy function $\frac{1}{2}[U(C_{rs}) + U(C_{sr})]$ which is symmetric and which takes the same numerical values as $U(C_{sr})$ (since $C_{sr} = C_{rs}$).

With U of the form (8.5.6) the constitutive equations are derived by appeal to (8.5.3). On transforming the integration variables on the right hand side of (8.5.3) to material coordinates, recalling the conservation of mass result

† The scalar function U is not to be confused with the right stretch matrix $\mathbf{U}$.

†† This is also a sufficient condition for U to be frame indifferent since from (6.2.27), $\mathbf{U}$ is the same for all observers.

$\rho d\Omega = \rho_0 d\Omega_0$ (where the suffix refers to the reference state for which ρ_0 is time independent) and carrying out the differentiation, there obtains, after reverting to spatial coordinates,

$$\frac{d}{dt}\iiint_{\Omega_t} \rho U d\Omega = \iiint_{\Omega_t} \rho \frac{\partial U}{\partial C_{rs}} \frac{\partial C_{rs}}{\partial t} d\Omega \qquad (8.5.7)$$

where we are regarding U as a function of the X_i via the variables C_{rs}.

Since Ω_t may be chosen arbitrarily (8.5.3) and (8.5.7) lead to

$$\sigma_{ij} d_{ij} = \rho \frac{\partial U(\mathbf{C})}{\partial C_{rs}} \frac{\partial C_{rs}}{\partial t} \qquad (8.5.8)$$

in which the derivative $\partial C_{rs}/\partial t$ is given by

$$\frac{\partial C_{rs}}{\partial t} = \frac{\partial}{\partial t}(\mathbf{F}^T\mathbf{F})_{rs} = \frac{\partial}{\partial t}\left(\frac{\partial x_i}{\partial X_r}\frac{\partial x_i}{\partial X_s}\right)$$

$$= \frac{\partial V_i}{\partial X_r}\frac{\partial x_i}{\partial X_s} + \frac{\partial x_i}{\partial X_r}\frac{\partial V_i}{\partial X_s} . \qquad (8.5.9)$$

In order to obtain material derivatives of the spatial velocity field on the right hand side of (8.5.8), as appear on the left side, we replace V_i in (8.5.9) by $v_i(x_j, t)$ obtaining from the chain rule

$$\frac{\partial C_{rs}}{\partial t} = \frac{\partial v_i}{\partial x_j}\frac{\partial x_i}{\partial X_s}\frac{\partial x_j}{\partial X_r} + \frac{\partial v_i}{\partial x_j}\frac{\partial x_i}{\partial X_r}\frac{\partial x_j}{\partial X_s} . \qquad (8.5.10)$$

On interchanging the roles of the dummy indices i and j in the second of these expressions there results

$$\frac{\partial C_{rs}}{\partial t} = 2\frac{\partial x_i}{\partial X_s}\frac{\partial x_j}{\partial X_r} d_{ij} \qquad (8.5.11)$$

where

$$d_{ij} = \frac{1}{2}\left(\frac{\partial v_i}{\partial x_j} + \frac{\partial v_j}{\partial x_i}\right)$$

is the rate of deformation tensor. From (8.5.8) and (8.5.11)

$$\left(\sigma_{ij} - 2\rho\frac{\partial U}{\partial C_{rs}}\frac{\partial x_i}{\partial X_s}\frac{\partial x_j}{\partial X_r}\right) d_{ij} = 0 . \qquad (8.5.12)$$

In the scalar equation (8.5.12) d_{ij}, σ_{ij} and

$$\rho\frac{\partial U}{\partial C_{rs}}\frac{\partial x_i}{\partial X_s}\frac{\partial x_j}{\partial X_r} \qquad (8.5.13)$$

are symmetric second order tensors.† Thus the coefficient of d_{ij} in (8.5.12) is symmetric. It follows that if (8.5.12) is to hold for an arbitrary choice of the six independent components of the symmetric tensor d_{ij} then

$$\sigma_{ij} = 2\rho \frac{\partial x_i}{\partial X_s} \frac{\partial x_j}{\partial X_r} \frac{\partial U}{\partial C_{rs}} \tag{8.5.14}$$

which are the constitutive equations of a Green elastic material.

Equations (8.5.14) may be re-written in matrix form

$$\boldsymbol{\sigma} = \mathbf{F}\mathbf{H}\mathbf{F}^T \tag{8.5.15}$$

where

$$\mathbf{H} = \left(2\rho \frac{\partial U}{\partial C_{rs}}\right) \equiv \mathbf{H}(\mathbf{U})$$

is a matrix function of the components of $\mathbf{C}$ (and hence of the components of $\mathbf{U}$).†† On writing $\mathbf{F} = \mathbf{R}\mathbf{U}$, (8.5.15) becomes

$$\boldsymbol{\sigma} = \mathbf{R}\mathbf{U}\mathbf{H}(\mathbf{U})\mathbf{U}\mathbf{R}^T$$

which is of the same form as (8.2.6). There is slightly less generality here since the quantity equivalent to **f** in (8.2.6) is expressed in terms of the partial derivatives of a single scalar function U. It is in this sense that Green materials form a sub-class of the more general Cauchy elastic materials.

The result (8.5.14) makes no appeal to the isotropy assumption. If the material is assumed further to be isotropic then U, a scalar function of $\mathbf{C}$, must be a function of the invariants of $\mathbf{C}$, which also are the invariants of $\mathbf{B}$

i.e.

$$U = U(\mathrm{I}_C, \mathrm{II}_C, \mathrm{III}_C) \ . \tag{8.5.16}$$

Inserting (8.5.16) into (8.5.14) gives

$$\sigma_{ij} = 2\rho \frac{\partial x_i}{\partial X_s} \frac{\partial x_j}{\partial X_r} \left(\frac{\partial U}{\partial \mathrm{I}_C} \frac{\partial \mathrm{I}_C}{\partial C_{rs}} + \frac{\partial U}{\partial \mathrm{II}_C} \frac{\partial \mathrm{II}_C}{\partial C_{rs}} + \frac{\partial U}{\partial \mathrm{III}_C} \frac{\partial \mathrm{III}_C}{\partial C_{rs}}\right) . \tag{8.5.17}$$

The invariants I, II and III of a 3 × 3 matrix are defined by equations (4.5.9)-(4.5.11). A slight re-arrangement of these expressions is more convenient for present purposes,

$$\mathrm{I}_C = C_{rr}$$

$$\mathrm{II}_C = \tfrac{1}{2}(\mathrm{I}_C{}^2 - C_{kl}C_{lk})$$

$$\mathrm{III}_C = \frac{1}{6}\left[-2\mathrm{I}_C{}^3 + 6\mathrm{I}_C\mathrm{II}_C + 2C_{kl}C_{lm}C_{mk}\right]$$

† The symmetry of (8.5.13) derives from $U(C_{rs}) = U(C_{sr})$ together with $C_{rs} = C_{sr}$ so that $\partial U/\partial C_{rs} = \partial U/\partial C_{sr}$, enabling the roles of r and s to be interchanged.

†† Because $\rho_0 = \rho|\mathbf{F}| = \rho|\mathbf{U}|$, ρ may be regarded also as a function of $\mathbf{U}$.

From these equations we derive easily (for $C_{rs} = C_{sr}$)

$$\frac{\partial \mathrm{I}_C}{\partial C_{rs}} = \delta_{rs}$$

$$\frac{\partial \mathrm{II}_C}{\partial C_{rs}} = \mathrm{I}_C \delta_{rs} - C_{rs}$$

$$\frac{\partial \mathrm{III}_C}{\partial C_{rs}} = \mathrm{II}_C \delta_{rs} - \mathrm{I}_C C_{rs} + C_{rk} C_{ks} .$$

Substituting these results into (8.5.17) and making use of the further results

$$\frac{\partial x_i}{\partial X_s}\frac{\partial x_j}{\partial X_r}\delta_{rs} = \frac{\partial x_i}{\partial X_s}\frac{\partial x_j}{\partial X_s} = (\mathbf{F}\mathbf{F}^T)_{ij} = B_{ij}$$

$$\frac{\partial x_i}{\partial X_s}\frac{\partial x_j}{\partial X_r} C_{rs} = \frac{\partial x_i}{\partial X_s}\frac{\partial x_k}{\partial X_s}\frac{\partial x_j}{\partial X_r}\frac{\partial x_k}{\partial X_r} = B_{ik}B_{kj} = (\mathbf{B}^2)_{ij}$$

$$\frac{\partial x_i}{\partial X_s}\frac{\partial x_j}{\partial X_r} C_{rk}C_{ks} = \frac{\partial x_i}{\partial X_s}\frac{\partial x_m}{\partial X_s}\frac{\partial x_m}{\partial X_k}\frac{\partial x_p}{\partial X_k}\frac{\partial x_p}{\partial X_r}\frac{\partial x_j}{\partial X_r}$$

$$= B_{im}B_{mp}B_{pj} = (\mathbf{B}^3)_{ij}$$

we obtain, in matrix notation

$$\boldsymbol{\sigma} = 2\rho\left[\mathbf{B}\left(\frac{\partial U}{\partial \mathrm{I}_C} + \mathrm{I}_C\frac{\partial U}{\partial \mathrm{II}_C} + \mathrm{II}_C\frac{\partial U}{\partial \mathrm{III}_C}\right) - \mathbf{B}^2\left(\frac{\partial U}{\partial \mathrm{II}_C} + \mathrm{I}_C\frac{\partial U}{\partial \mathrm{III}_C}\right) + \mathbf{B}^3\frac{\partial U}{\partial \mathrm{III}_C}\right] . \tag{8.5.18}$$

On remembering that the invariants of **B** and **C** are identical and using the Cayley-Hamilton identity (8.3.18) to eliminate $\mathbf{B}^3$, there results finally

$$\boldsymbol{\sigma} = 2\rho\left[\mathrm{III}_B\frac{\partial U}{\partial \mathrm{III}_B}\mathbf{I} + \left(\frac{\partial U}{\partial \mathrm{I}_B} + \mathrm{I}_B\frac{\partial U}{\partial \mathrm{II}_B}\right)\mathbf{B} - \frac{\partial U}{\partial \mathrm{II}_B}\mathbf{B}^2\right] . \tag{8.5.19}$$

Alternatively using the Cayley-Hamilton identity again to express $\mathbf{B}^2$ in terms of $\mathbf{B}$ and $\mathbf{B}^{-1}$ [by multiplying (8.3.18) by $\mathbf{B}^{-1}$] leads to the constitutive equations first derived by Finger in 1894

$$\boldsymbol{\sigma} = h_0\mathbf{I} + h_1\mathbf{B} + h_{-1}\mathbf{B}^{-1} \tag{8.5.20}$$

where h_0, h_1 and h_{-1} are expressed in terms of the partial derivatives of U with respect to the invariants. There obtains

$$\left.\begin{aligned} h_0 &= 2\rho\left[\mathrm{III}_B \frac{\partial U}{\partial \mathrm{III}_B} + \mathrm{II}_B \frac{\partial U}{\partial \mathrm{II}_B}\right] \\ h_1 &= 2\rho \frac{\partial U}{\partial \mathrm{I}_B} \\ h_{-1} &= -2\rho\,\mathrm{III}_B \frac{\partial U}{\partial \mathrm{II}_B} \quad . \end{aligned}\right\} \tag{8.5.21}$$

In these results ρ may be replaced, if required by

$$\rho = \rho_0/\mathrm{III}_B{}^{\frac{1}{2}}$$

which follows from remembering

$$\rho_0 = \rho\,|\mathbf{F}|$$

and noting $\qquad \mathrm{III}_B = |\mathbf{C}| = |\mathbf{F}^T\mathbf{F}| = |\mathbf{F}|^2 \ .$

Equations (8.5.19) and (8.5.20) are identical in form respectively with equations (8.3.15) and (8.3.16) but with the restriction that here the coefficients are expressed in terms of the derivatives of a single function U.

Some modifications to the argument and results are necessary for the incompressible state. For an incompressible solid

$$\mathrm{III}_B = 1 \tag{8.5.22}$$

and U is essentially a function only of I_B and II_B. It is adequate to insert these results into the second and third of (8.5.21) obtaining for the incompressible case

$$h_1 = 2\rho_0 \frac{\partial U}{\partial \mathrm{I}_B}\ , \quad h_{-1} = -2\rho_0 \frac{\partial U}{\partial \mathrm{II}_B} \tag{8.5.23}$$

where $\qquad U = U(\mathrm{I}_B, \mathrm{II}_B)\ . \qquad (8.5.24)$

This procedure fails for h_0 since (8.5.22) is to be interpreted as a constraint on the deformation as well as merely providing a particular value of III_B to insert into h_0. Physically it is evident, as in the case of an incompressible liquid, that the quantity h_0, which plays the role of a negative pressure, is now to be regarded as arbitrary and determined by the motion. Thus the constitutive equations of an incompressible elastic material are given by

$$\sigma = -P\mathbf{I} + h_1\mathbf{B} + h_{-1}\mathbf{B}^{-1} \tag{8.5.25}$$

where P is an arbitrary pressure, while h_1 and h_{-1} are given by (8.5.23). Mathematically this result is obtainable as follows. We notice that the argument leading from (8.5.12) to (8.5.14) fails for the incompressible case; for this case the restraint (8.5.22) is equivalent to the restraint div $\mathbf{v} = 0$

i.e. $\qquad d_{mm} = 0\ ,$

and the latter may be incorporated formally into (8.5.12) by the addition of a term χd_{ii}, involving an arbitrary scalar Lagrange multiplier χ. The resulting equation reads

$$\left(\sigma_{ij} + \chi\delta_{ij} - 2\rho \frac{\partial U}{\partial C_{rs}} \frac{\partial x_i}{\partial X_s} \frac{\partial x_j}{\partial X_r}\right) d_{ij} = 0 \ .$$

Clearly the effect of the extra term is equivalent to writing $\boldsymbol{\sigma}$ in the form (8.5.25), where P, which contains χ, is arbitrary. A similar modification replaces the term in $\mathbf{I}$ by $-P\mathbf{I}$ in (8.5.19) where again P is arbitrary.

The incompressible material

$$U = A(\mathrm{I}_B - 3) \ , \tag{8.5.26}$$

where A is a constant is predicted by the kinetic theory of rubber elasticity.† From these molecular calculations, somewhat similar to calculations in the kinetic theory of gases, A is given by

$$\rho_0 A = \Gamma N k \theta \ . \tag{8.5.27}$$

Here Γ is a number of order one, k is Boltzmann's constant, θ is absolute temperature and N is the density of cross links per unit volume.

The model provided by (8.5.26) and (8.5.27) gives a reasonable account of the behaviour of natural rubber for modest strains, i.e. for stretches up to about 10%. For larger deformations the *Mooney* model

$$U = A(\mathrm{I}_B - 3) + B(\mathrm{II}_B - 3) \ ,$$

where A and B are constants, has been used with some success to correlate experimental measurements in plane extension (see Problem 2 of this Chapter), although these is no moleculer basis for the coefficient B.

As indicated, with specific forms for the function U, the constitutive equations (8.5.19) accounts for the static mechanical behaviour of rubber; more generally equations (8.5.19) almost certainly describe the static behaviour of most polymeric and elastomer solids whose susceptibility to large recoverable deformations is due physically to the uncoiling of molecular chains.

Theories of elasticity of the type generally described in this chapter, except for the limiting linear case of Section 8.4, are known as finite elasticity theories. The word finite derives from the fact that, unlike the linear theory (see below), the quantities E_{ij} defined by equations (5.2.8) are not necessarily of order e where $e \ll 1$.

All isotropic finite elasticity theories degenerate to the theory of Section 8.4 for the case where the displacement gradients are of order $e \ll 1$. The simplified

† See L. R. G. Treloar 'The Physics of Rubber Elasticity' (*Oxford University Press* 2nd Edition 1974).In these calculations the 'free' energy (8.5.26) derives entirely from variations in entropy due to re-arrangement of molecular chains. There are no contributions to U from what is termed 'internal' energy in the full thermodynamic sense of the word. In this connexion the reader may recall the formula *Free energy* (here U) = *Internal energy* $-$ θS where S is entropy and θ is absolute temperature.

theory of Section 8.4 could, for example, be applied to small deformations in rubber. In fact the study of small static deformations in rubber and elastomer materials is of little interest, and the real signigicance of the theory of linear elasticity of Section 8.4 is in the application to structural materials, primarily metals, but also including concrete and to a lesser extent timber.

In particular for metals, equations (8.4.3) are valid for values of the E_{ij} at most of order 10^{-2}. For larger strains and stresses equations (8.4.3) cease to be valid, not immediately because of the failure of the equations to adequately approximate a finite elastic theory, but because metals cease to behave elastically. For stresses outside the limited linear elastic range, metals behave *plastically* and, with relatively small increases in stress beyond the elastic range, acquire large permanent strains, ultimately fracturing for extension ratios of about 1.1 - 1.5, depending on the material. Plasticity is discussed at greater length in Chapter 16. In the present context the important point is that for most structural and engineering purposes, it is highly undesirable to allow metals to enter the plastic region. It is for this reason that the linear elasticity theory is of great technological importance.†

The equations of linear elasticity are valid for metals in the appropriate stress range both under static and dynamic conditions. In contrast, the equations of finite elasticity are not usually adequate to describe the dynamical behaviour of polymers, elastomers and rubbers.

Although the strain magnitudes permitted in metals by the linear theory are very small (at most of order 10^{-2}), the associated stresses may be much much larger than those experienced by rubber. This is because the magnitude of the Lamé constants in metals is very large. Typically for steel λ and μ are of order 10^{11} N/m^2 so that strains of order 10^{-3} are associated with stresses of order 10^8 $N/m^2 = 6$ tons/in^2. Such large values of λ and μ enable steel(s) and some other metals to support large loads without significant deformation, and it is for this reason primarily that metals are important engineering materials.

The physical origin of the elastic behaviour of metals derives predominantly from internal energy changes consequent on the redistribution of the free conduction electrons with strain. Fundamental calculations of elastic constants based on this model entails the solution of a complex many body problem in quantum mechanics. The magnitude of the resulting theoretical elastic constants are of the right order.

As well as their application in technology, the linear equations of elasticity are valuable in the analysis of seismic phenomena such as earthquakes, and the propagation of mechanical waves generated explosively in the earth.

† It should be added that in the event structural engineering calculations based on the linear theory predicted stresses in the plastic region, it would be necessary to modify the design.

8.6 Strain energy and strain rate in the theory of linear elasticity

With the assumption that displacement gradients are of order $e \ll 1$, the strain E_{ij} is adequately approximated by equation (8.4.2)

$$E_{ij} = \frac{1}{2}\left(\frac{\partial U_i}{\partial X_j} + \frac{\partial U_j}{\partial X_i}\right) .$$

On differentiating with respect to time we obtain

$$\frac{\partial E_{ij}}{\partial t} = \frac{1}{2}\left(\frac{\partial V_i}{\partial X_j} + \frac{\partial V_j}{\partial X_i}\right) = \frac{1}{2}\left(\frac{\partial v_i}{\partial x_s}\frac{\partial x_s}{\partial X_j} + \frac{\partial v_j}{\partial x_s}\frac{\partial x_s}{\partial X_i}\right) . \quad (8.6.1)$$

With neglect of terms of order e times that retained, the first term on the right-side of (8.6.1) becomes

$$\frac{1}{2}\frac{\partial v_i}{\partial x_s}\left[\delta_{sj} + O(e)\right] \simeq \frac{1}{2}\frac{\partial v_i}{\partial x_j} .$$

Using a similar approximation for the second term leads to

$$\frac{\partial E_{ij}}{\partial t} \simeq \frac{1}{2}\left(\frac{\partial v_i}{\partial x_j} + \frac{\partial v_j}{\partial x_i}\right) = d_{ij}$$

which justifies (in a small displacement gradient theory) the terminology 'rate of strain tensor' for the rate of deformation tensor d_{ij}.

The result

$$\frac{\partial E_{ij}}{\partial t} = d_{ij} \quad (8.6.2)$$

may be used to find easily the strain energy of an isotropic linear elastic solid.

We appeal in the first instance to equation (8.5.8). However in view of the result

$$2E_{rs} = C_{rs} - \delta_{rs}$$

it is convenient to re-write the equation in the form

$$\sigma_{ij}d_{ij} = \rho_0 \frac{\partial U}{\partial E_{ij}}\frac{\partial E_{ij}}{\partial t} , \quad (8.6.3)$$

regarding U as a function of the E_{ij} rather than of the C_{ij}. Also in writing down (8.6.3) ρ has been replaced by ρ_0, consistant with approximations already employed. Substituting from (8.6.2) and 'cancelling' the quantity d_{ij} leads to

$$\sigma_{ij} = \rho_0 \frac{\partial U}{\partial E_{ij}} .$$

The 'cancellation' is legitimate since σ_{ij} is symmetric and (8.6.3) is to hold for arbitrary symmetric d_{ij}. Also linear elasticity does not utilise the incompressible approximation, so there is no necessity to introduce an additional arbitrary pressure.

$\rho_0 U$ now follows from substituting for σ_{ij} from the constitutive equations. We have

$$\rho_0 \frac{\partial U}{\partial E_{ij}} = \lambda E_{mm}\delta_{ij} + 2\mu E_{ij} \ .$$

By inspection, the solution of these equations is

$$\rho_0 U = \tfrac{1}{2}\lambda(E_{mm})^2 + \mu E_{ij}E_{ij} \ . \tag{8.6.4}$$

The solution is unique to within an arbitrary constant, here chosen to vanish in the absence of strain.

On physical grounds we infer that provided not all the E_{ij} are zero, U must be a positive definite quantity. This is ensured by the choice

$$\lambda > 0 \quad \text{and} \quad \mu > 0 \ . \tag{8.6.5}$$

All real materials which conform to the linear isotropic elastic model satisfy these two inequalities, although from a mathematical viewpoint the restrictions are excessive since the quadratic form (8.6.4) may also be written

$$\rho_0 U = \tfrac{1}{2}(\lambda + \tfrac{2}{3}\mu)(E_{mm})^2 + \mu\hat{E}_{ij}\hat{E}_{ij}$$

where

$$\hat{E}_{ij} = E_{ij} - \tfrac{1}{3}E_{mm}\delta_{ij}$$

Clearly, sufficient restrictions for U to be positive definite are

$$\lambda + \tfrac{2}{3}\mu > 0\,, \quad \mu > 0 \ .$$

However for real materials which conform to the linear elastic isotropic model the more restrictive inequalities (8.6.5) are satisfied.

Summary

1. The constitutive equations of an isotropic thermoelastic solid may be written

$$\boldsymbol{\sigma} = \alpha\mathbf{I} + \beta\mathbf{B} + \gamma\mathbf{B}^2$$

where α, β, γ are functions of the invariants of $\mathbf{B}$ and also of the temperature; here $\mathbf{B}$ is the matrix

$$\mathbf{B} = \mathbf{F}\mathbf{F}^T$$

For an elastic solid described by a strain energy function $U(\mathrm{I}_B, \mathrm{II}_B, \mathrm{III}_B)$

$$\alpha = 2\rho\mathrm{III}_B \frac{\partial U}{\partial \mathrm{III}_B}\,, \quad \beta = 2\rho\left[\frac{\partial U}{\partial \mathrm{I}_B} + \mathrm{I}_B\frac{\partial U}{\partial \mathrm{II}_B}\right]\,, \quad \gamma = -2\rho\frac{\partial U}{\partial \mathrm{II}_B}$$

In the incompressible case $U = U(\mathrm{I}_B, \mathrm{II}_B)$ and

$$\boldsymbol{\sigma} = -P\mathbf{I} + 2\rho_0\left[\frac{\partial U}{\partial \mathrm{I}_B} + \mathrm{I}_B\frac{\partial U}{\partial \mathrm{II}_B}\right]\mathbf{B} - 2\rho_0\frac{\partial U}{\partial \mathrm{II}_B}\mathbf{B}^2$$

where P is arbitrary.

2. The constitutive equations of a linear isotropic elastic solid are

$$\sigma_{ij} = \lambda\Delta\delta_{ij} + 2\mu E_{ij}$$

where λ and μ are positive constants, $\Delta = E_{mm}$ and E_{ij} is approximated by

$$E_{ij} = \tfrac{1}{2}(\partial U_i/\partial X_j + \partial U_j/\partial X_i)$$

Problems, Chapter 8

1. The constitutive equations of an isotropic finite elastic solid may be taken in the form

$$\boldsymbol{\sigma} = \alpha\mathbf{I} + \beta\mathbf{B} + \gamma\mathbf{B}^2$$

where α, β and γ are functions of the three invariants I_B, II_B, III_B, and $\mathbf{B}$ is the matrix $\mathbf{B} = \mathbf{F}\mathbf{F}^T$.

For a compressible material with strain energy function $U(\mathrm{I}_B, \mathrm{II}_B, \mathrm{III}_B)$ show that

$$\alpha = 2\rho_0 \mathrm{III}_B^{\frac{1}{2}} \frac{\partial U}{\partial \mathrm{III}_B}, \quad \beta = 2\rho_0 \mathrm{III}_B^{-\frac{1}{2}} \left[\frac{\partial U}{\partial \mathrm{I}_B} + \mathrm{I}_B \frac{\partial U}{\partial \mathrm{II}_B}\right]$$

$$\gamma = -2\rho_0 \mathrm{III}_B^{-\frac{1}{2}} \frac{\partial U}{\partial \mathrm{II}_B}$$

where ρ_0 is the initial density.

For the incompressible case $[\mathrm{III}_B = 1,\ U = U(\mathrm{I}_B, \mathrm{II}_B)]$ the corresponding equations are

$$\boldsymbol{\sigma} = -P\mathbf{I} + \beta\mathbf{B} + \gamma\mathbf{B}^2$$

where P is arbitrary and

$$\beta = 2\rho_0 \left(\frac{\partial U}{\partial \mathrm{I}_B} + \mathrm{I}_B \frac{\partial U}{\partial \mathrm{II}_B}\right), \quad \gamma = -2\rho_0 \frac{\partial U}{\partial \mathrm{II}_B}.$$

Deduce that the existence of $U(\mathrm{I}_B, \mathrm{II}_B)$ in the incompressible case implies that β and γ satisfy

$$\frac{\partial \beta}{\partial \mathrm{II}_B} + \frac{\partial \gamma}{\partial \mathrm{I}_B} + \mathrm{I}_B \frac{\partial \gamma}{\partial \mathrm{II}_B} = 0 \qquad (1)$$

and hence show that if $\gamma = 0$ then β is a function solely of I_B.

Does the existence of a strain energy function imply restrictions similar to (1) on α, β and γ for the compressible case. Without going into algebraic detail show that there are now three relations connecting α, β and γ. [Hint: The following analogy may be helpful. A necessary and sufficient condition for a vector $\mathbf{A}$ to be of the form $\mathbf{A} = \operatorname{grad}\phi$ is $\operatorname{curl}\mathbf{A} = 0$.]

2. (i) Show that for the homogeneous stretch

$$x_1 = \lambda_1 X_1, \quad x_2 = \lambda_2 X_2, \quad x_3 = \lambda_3 X_3 \tag{1}$$

where $\lambda_1, \lambda_2, \lambda_3$ are constants, that $\mathbf{B}$ is given by

$$\mathbf{B} = \text{diag}\,[\lambda_1^2, \lambda_2^2, \lambda_3^2]$$

and that the resulting stresses are constant with $\sigma_{12} = \sigma_{23} = \sigma_{31} = 0$. Verify that in the absence of body forces, the deformation (1) is admissable in a finite elastic material.

(ii) If the deformation (1) occurs in an incompressible material show that

$$\lambda_3 = (\lambda_1\lambda_2)^{-1} .$$

If further the stress σ_{33} vanishes everywhere show that σ_{11} and σ_{22} are related to the corresponding stretches λ_1 and λ_2 by

$$\sigma_{11} = \beta[\lambda_1^2 - (\lambda_1\lambda_2)^{-2}] + \gamma[\lambda_1^4 - (\lambda_1\lambda_2)^{-4}]$$

$$\sigma_{22} = \beta[\lambda_2^2 - (\lambda_1\lambda_2)^{-2}] + \gamma[\lambda_2^4 - (\lambda_1\lambda_2)^{-4}]$$

where α and β are functions of I_B and II_B and where

$$\mathrm{I}_B = \lambda_1^2 + \lambda_2^2 + (\lambda_1\lambda_2)^{-2}, \quad \mathrm{II}_B = \lambda_1^2\lambda_2^2 + \lambda_2^{-2} + \lambda_1^{-2} .$$

If β and γ are given by the formulae

$$\beta = 2\rho_0\left(\frac{\partial U}{\partial \mathrm{I}_B} + \mathrm{I}_B\frac{\partial U}{\partial \mathrm{II}_B}\right), \quad \gamma = -2\rho_0\frac{\partial U}{\partial \mathrm{II}_B}$$

show that σ_{11}, σ_{22} may be written in the form

$$\sigma_{11} = 2\rho_0[\lambda_1^2 - (\lambda_1\lambda_2)^{-2}]\left[\frac{\partial U}{\partial \mathrm{I}_B} + \lambda_2^2\frac{\partial U}{\partial \mathrm{II}_B}\right]$$

$$\sigma_{22} = 2\rho_0[\lambda_2^2 - (\lambda_1\lambda_2)^{-2}]\left[\frac{\partial U}{\partial \mathrm{II}_B} + \lambda_1^2\frac{\partial U}{\partial \mathrm{II}_B}\right] .$$

(These results have been used to determine $U(\mathrm{I}_B, \mathrm{II}_B)$ for vulcanised rubber – see note in the Answer.)

3. An incompressible elastic band of uniform cross sectional area is stretched in the ratio $\lambda : 1$ in the x direction by a stress σ_{xx}. All other stress components vanish. Show from symmetry and incompressibility arguments that

$$\lambda_2 = \lambda_3 = \lambda^{-\frac{1}{2}}$$

and that σ_{xx} is given by

$$\sigma_{xx} = \beta[\lambda^2 - \lambda^{-1}] + \gamma[\lambda^4 - \lambda^{-2}]$$

and that the associated axial force is

$$T_x = A_0[\beta(\lambda - \lambda^{-2}) + \gamma(\lambda^3 - \lambda^{-3})]$$

where A_0 is the initial cross sectional area.

4. An elastic material is subject to the homogeneous shear deformation

$$x = X + kY, \quad y = Y, \quad z = Z$$

where k is a constant. Show that

$$\mathbf{B} = \begin{bmatrix} 1+k^2 & k & 0 \\ k & 1 & 0 \\ 0 & 0 & 1 \end{bmatrix}, \quad \mathbf{B}^2 = \begin{bmatrix} 1+k^4+3k^2 & k(2+k^2) & 0 \\ k(2+k^2) & 1+k^2 & 0 \\ 0 & 0 & 1 \end{bmatrix}$$

and that the invariants of $\mathbf{B}$ are given by

$$\mathrm{I}_B = \mathrm{II}_B = 3 + k^2, \quad \mathrm{III}_B = 1 .$$

Deduce that $\sigma_{13} = \sigma_{23} = 0$ and that σ_{11}, σ_{22} and σ_{33} are even functions of k while σ_{12} is an odd function.

Finally show that σ_{12}, σ_{11} and σ_{22} satisfy the relation

$$(\sigma_{11} - \sigma_{22}) = k\sigma_{12}$$

independent of the nature of the functions α, β and γ.

5. A hollow cylindrical tube of circular cross section possesses initially internal and external radii a_0 and b_0 respectively. The material of the tube is incompressible and with constitutive equation

$$\boldsymbol{\sigma} = -P\mathbf{I} + \psi\mathbf{B}$$

where P is arbitrary and ψ is a constant.

The tube is subject to uniform internal pressure P_0 at the inner surface causing the latter to expand to a new radius a. If no expansion or contraction takes place in the axial (z) direction deduce from incompressibility arguments that material particles originally located at a radius R and axial position Z move to r, z where

$$r = (R^2 + a^2 - a_0{}^2)^{\frac{1}{2}}, \quad z = Z .$$

Show that principal stretches lie in the radial (r), circumferential (θ) and axial (z) directions and are given by

$$\lambda_r = R/r, \quad \lambda_\theta = r/R, \quad \lambda_z = 1 .$$

If σ_{rr}, $\sigma_{\theta\theta}$ and σ_{zz} denote the corresponding (principal) stresses (here σ_{rr} and $\sigma_{\theta\theta}$ are the normal components of stress on surfaces with normals respectively in the r and θ directions), show that

$$\left.\begin{aligned} \sigma_{rr} &= -P + \psi R^2/r^2 \\ \sigma_{\theta\theta} &= -P + \psi r^2/R^2 \\ \sigma_{zz} &= -P + \psi \end{aligned}\right\} \qquad (1)$$

In terms of cylindrical polar coordinates the radial equation of equilibrium (not discussed in the text)† is for the case of cylindrical symmetry

$$\frac{\partial \sigma_{rr}}{\partial r} + \frac{\sigma_{rr} - \sigma_{\theta\theta}}{r} = 0 \qquad (2)'$$

where σ_{rr}, $\sigma_{\theta\theta}$ are regarded as functions of r rather than as functions of R. The remaining equilibrium equations are satisfied identically if $P = P(r)$ is independent of z and θ.

Deduce from (1) and (2) that $P(r)$ satisfies the equation

$$\frac{d}{dr}[-P + \psi(R^2/r^2)] = \psi r^{-1}[(r^2/R^2) - (R^2/r^2)]$$

with solution

$$P = D + \psi[-(a^2 - a_0{}^2)/2r^2 + \log\{r/[r^2 - (a^2 - a_0{}^2)]^{\frac{1}{2}}\}$$

where D is a constant of integration.

If D is chosen to satisfy the boundary condition

$$(\sigma_{rr})_{r=a} = -P_0$$

derive the formula

$$\sigma_{rr} = -P_0 - \psi\left[-\frac{1}{2} + \frac{1}{2}\frac{a_0{}^2}{a^2} - \log\left(\frac{a}{a_0}\right) + \frac{(a^2 - a_0{}^2)}{2r^2} + \log\left(\frac{r}{[r^2 - (a^2 - a_0{}^2)]^{\frac{1}{2}}}\right)\right]$$

By using the boundary condition

$$(\sigma_{rr})_{r=b} = 0$$

where b is the current radius of the outer boundary [so that $b^2 = (a^2 + b_0{}^2 - a_0{}^2)$] deduce that the pressure P_0 required to inflate the tube is given by $P_0 = P_0(x, m)$ where

$$P_0(x, m) = \psi[-\tfrac{1}{2}x^{-2} + \log x - \tfrac{1}{2}\log[(x^2 + m^2 - 1)/m^2] + \tfrac{1}{2}m^2/(x^2 + m^2 - 1)]$$

and where $x = (a/a_0)$, $m = (b_0/a_0)$.

Finally show (i) for small expansions $[(x - 1) \ll 1]$

$$P_0 = 2\psi(m^2 - 1)m^{-2}(x - 1) + O(x - 1)^2$$

(ii) for large expansions $(x \to \infty)$

$$P_0 = \psi\left[\log m - \frac{1}{4}\frac{(m^4 - 1)}{x^4} + O(x^{-6})\right]$$

(iii) there is no value of x in the range $1 \leqslant x \leqslant \infty$ for which $\partial P_0/\partial x$ vanishes.

From these results deduce that the pressure is a monotonic increasing function of x and that the largest pressure the tube is able to sustain is $\psi \log(b_0/a_0)$.

† See Problem 10 of Chapter 4.

6. An initially rectangular block ($0 \leqslant X \leqslant L$, $0 \leqslant Y \leqslant M$, $0 \leqslant Z \leqslant N$) of a compressible finite elastic material is subject to the homogeneous deformation of Problem 6, Chapter 2, which is maintained by uniformly distributed tractions over the six faces.

Show that the stress field is uniform and that

$$\sigma_{xx} - \sigma_{yy} = (a - b)\sigma_{xy} \qquad *$$

independent of the nature of the response functions α, β and γ.

Show that the tractions on $Z = 0$ and $Z = N$ are normal. Also show from * and the result $ab < 1$, that it is not possible to maintain the deformation by purely normal tractions on the remaining four faces.

Chapter 9

Shear Flow Solutions for Reiner-Rivlin Fluids

9.1 Introduction

Until recently mathematical accounts of fluid flow situations have been based almost exclusively on the assumption that the mechanical behaviour of the fluid was modelled, at its most complex, by a compressible Stokes' fluid; for many purposes the further simplifications of an incompressible and/or inviscid fluid were adopted. The assumption that the behaviour of a liquid is modelled by a compressible Stokes' fluid leads to the subject conventionally called fluid dynamics or classical fluid dynamics.†

Selected topics from classical fluid dynamics are given in Chapters 10, 11, 12 and 13; the present chapter is concerned with some of the few known exact solutions for the flow of the more general Reiner-Rivlin fluids.

The importance of classical fluid dynamics stems from the fact that the behaviour of the two most common fluids, air and water, is modelled adequately by the assumption of a Stokes' fluid, or in some situations by the simpler incompressible and/or inviscid model. However in recent years it has become apparent that some fluids like polymer solutions, blood, printers ink, etc., do not conform to the assumption of a Stokes' fluid. Since a Stokes' fluid is a particular case of a Reiner-Rivlin fluid, the question arises as to whether the general form of a Reiner-Rivlin fluid, with general forms of the coefficients α, β and γ, provides a better description of real non-Stokesian materials.

It is in the latter context that exact solutions of the present chapter are of importance. Experimental checks of some of the predictions of these solutions suggest that the behaviour of real non-Stokesian materials is not accounted for adequately by the more general Reiner-Rivlin model. The implications of these results are that real non-Stokesian fluids exhibit memory effects characteristic of viscoelastic materials. These are discussed in a linear context in Chapter 15.

The constitutive equations for Reiner-Rivlin fluids, together with the equations of motion and mass conservation, admit a number of exact isothermal steady flow solutions of particularly simple character. In these solutions, discussed below, the spatial velocity field is time independent and such that (i) the vector

† Some writers reserve the description 'classical' for the most simplified model, i.e. for an inviscid, incompressible fluid.

v possesses a single non-vanishing component in one of three mutually orthogonal directions (ii) variations of the non-vanishing component with position occurs in one of the remaining two orthogonal directions. The three mutually orthogonal directions are either Cartesian (x, y and z) or cylindrical polar directions (r, θ and z). However the calculations given here are carried out primarily in Cartesian coordinates.†

Motions whose flow patterns conform to the above descriptions are readily visualised as relative motions of successive layers or sheets of material sliding over one another. Such motions are commonly called shear flows.

Shear flows are particular examples of flows for which the density is unchanged. The word **isochoric** is used to describe motions for which the density remains constant. Evidently all flows of an incompressible material are necessarily isochoric. The converse is not true. However most, but not all, of the shear flows discussed below are exact solutions only in an incompressible material.

In the solutions obtained here, body forces are neglected.

The motion of a Reiner-Rivlin fluid is governed by the equations of motion

$$\frac{\partial \sigma_{ij}}{\partial x_j} + \rho b_i = \rho\left[\frac{\partial v_i}{\partial t} + (\mathbf{v}\,.\,\mathrm{grad})v_i\right] \tag{9.1.1}$$

the equation of mass conservation

$$\frac{\partial \rho}{\partial t} + \mathrm{div}\,(\rho\mathbf{v}) = 0 \tag{9.1.2}$$

and the constitutive equations

$$\sigma_{ij} = -(P - \alpha)\delta_{ij} + \beta d_{ij} + \gamma d_{ik} d_{kj}\ . \tag{9.1.3}$$

The equivalent matrix form of (9.1.3)

$$\boldsymbol{\sigma} = -(P - \alpha)\mathbf{I} + \beta\mathbf{d} + \gamma\mathbf{d}^2 \tag{9.1.4}$$

is also useful.

For compressible isothermal materials, the coefficients α, β and γ are functions of ρ and of the invariants I_d, II_d and III_d, while P is a function of ρ. The case of an incompressible material for which $\rho = \rho_0$ and $\mathrm{I}_d(\equiv \mathrm{div}\,\mathbf{v}) = 0$, is met by regarding α, β and γ as functions of II_d and III_d, obtained from substituting $\mathrm{I}_d = 0$ in the compressible expressions for α, β and γ, and treating P as an arbitrary quantity to be determined in any particular problem concomitantly with the velocity field. Usually for incompressible materials, it is common to acknowledge the arbitrariness of the sum quantity $\mathrm{P}' = \mathrm{P} - \alpha$ and to replace $\mathrm{P} - \alpha$ in (9.1.4) by P'. However in the present chapter we wish to consider solutions to isochoric problems for both compressible and incompressible fluids, and there is some economy in retaining the common form (9.1.3) for the constitutive equation.

† In solving specific problems, we revert from suffix notation (x_1, x_2, x_3) to the conventional x, y and z.

9.2 Rectilinear shear flows between parallel plates

We seek shear flow solutions of equations (9.1.1), (9.1.2) and (9.1.3) of the form

$$v_z = f(x)\,, \quad v_x = v_y = 0\,, \quad \rho = \rho_0 = \text{const.} \tag{9.2.1}$$

for which equation (9.1.2) is automatically satisfied. For the velocity field (9.2.1) the matrices **d** and $\mathbf{d}^2$ are given by

$$\mathbf{d} = \tfrac{1}{2}f'(x)\begin{bmatrix} 0 & 0 & 1 \\ 0 & 0 & 0 \\ 1 & 0 & 0 \end{bmatrix} \tag{9.2.2}$$

$$\mathbf{d}^2 = \tfrac{1}{4}[f'(x)]^2\begin{bmatrix} 1 & 0 & 0 \\ 0 & 0 & 0 \\ 0 & 0 & 1 \end{bmatrix} . \tag{9.2.3}$$

Evidently the invariants I_d, II_d and III_d depend only on $f'(x)$; in fact from (9.2.2), (9.2.3) and equations (4.5.9) - (4.5.11)

$$\mathrm{I}_d = \mathrm{III}_d = 0\,, \quad \mathrm{II}_d = -\tfrac{1}{4}[f'(x)]^2$$

so that we may write

$$\begin{aligned} \alpha(\mathrm{I}_d, \mathrm{II}_d, \mathrm{III}_d, \rho) &= \alpha(0, -\tfrac{1}{4}(f')^2, 0, \rho_0) \\ &\equiv A[(f')^2] \qquad \text{(say)} \end{aligned} \tag{9.2.4}$$

and similarly $\qquad \beta = B[(f')^2]\,, \quad \gamma = C[(f')^2]\,. \qquad (9.2.5)$

Susbstituting these results together with (9.2.2) and (9.2.3) into the constitutive equations yields for the stresses

$$\left.\begin{aligned} \sigma_{xx} = \sigma_{zz} &= -P + A[(f')^2] + \tfrac{1}{4}C[(f')^2](f')^2 \\ \sigma_{yy} &= -P + A[(f')^2] \\ \sigma_{xz} = \sigma_{zx} &= \tfrac{1}{2}B[(f')^2]\,f' \\ \sigma_{xy} = \sigma_{yx} = \sigma_{zy} &= \sigma_{yz} = 0\,. \end{aligned}\right\} \tag{9.2.6}$$

With the velocity field of the form (9.2.1) the terms $\partial v_i/\partial t$ and $(\mathbf{v}\,.\,\mathrm{grad})v_i$ appearing in the equations of motion are identically zero for all components i. In the absence of body forces the equations of motion reduce to

$$\begin{aligned} \frac{\partial\sigma_{xx}}{\partial x} + \frac{\partial\sigma_{xy}}{\partial y} + \frac{\partial\sigma_{xz}}{\partial z} &= 0 \\ \frac{\partial\sigma_{yx}}{\partial x} + \frac{\partial\sigma_{yy}}{\partial y} + \frac{\partial\sigma_{yz}}{\partial z} &= 0\,. \\ \frac{\partial\sigma_{zx}}{\partial x} + \frac{\partial\sigma_{zy}}{\partial y} + \frac{\partial\sigma_{zz}}{\partial z} &= 0 \end{aligned} \tag{9.2.7}$$

Substituting from (9.2.6) into (9.2.7) now yields the three equations

$$\frac{\partial}{\partial x}[-P + A + \tfrac{1}{4}(f')^2 C] = 0 \tag{9.2.8}$$

$$-\frac{\partial P}{\partial y} = 0 \tag{9.2.9}$$

$$-\frac{\partial P}{\partial z} + \frac{d}{dx}[\tfrac{1}{2}f'B] = 0 \; . \tag{9.2.10}$$

For a compressible material behaving isochorically $P = P(\rho_0)$ is constant and equations (9.2.8), (9.2.9) and (9.2.10) admit a solution for which $f'(x)$ is constant i.e.

$$f'(x) = k \quad \text{(say)} \; . \tag{9.2.11}$$

Entering this result into equations (9.2.6) shows that a uniform stress state prevails; the actual stress magnitudes depend in detail on the manner in which α, β and γ depend on II_d.

The associated velocity field obtained from (9.2.1) and (9.2.11) is

$$v_z = kx + m \tag{9.2.12}$$

where m is a second constant of integration.

The particular choices $m = 0$, $k = v_0/h$ yielding

$$v_z = v_0 x/h \, , \quad (v_x = v_y = 0) \tag{9.2.13}$$

solves the boundary value problem for which a uniform infinite layer of liquid $0 \leqslant x \leqslant h$ adheres to a fixed surface $x = 0$ and to a surface $x = h$ which moves transversely in the z direction at a speed v_0.

Non-zero choice of m leads to the trivial generalisation where both surfaces $x = 0$, $x = h$, are in uniform motion in the z direction at different speeds.

The flow field (9.2.13) is realisable approximately in an experiment in which a plate of finite area glides over a larger plate; for sufficiently small h the field (9.2.13) would obtain at points between the plates and away from the periphery of the smaller plate.

The solution given above, in the absence of pressure gradients, is the only possible rectilinear shear flow solution of a compressible material.

The solution is also valid for incompressible materials; however for incompressible for which P is a dependent variable determined from (9.2.8) -(9.2.10) along with $f'(x)$, more general rectilinear shear flows are possible. Evidently equations (9.2.8) and (9.2.9) admit a solution

$$P = A + \tfrac{1}{4}f'^2 C + Q(z) \tag{9.2.14}$$

and this result is compatible with (9.2.10) provided

$$Q'(z) = \frac{d}{dx}(\tfrac{1}{2}Bf') \; . \tag{9.2.15}$$

In equation (9.2.15) the left side depends only on z, the right side only of x. It follows that both sides are constant (say $-m$) and that

$$\tfrac{1}{2}Bf' = -mx + p \qquad Q(z) = -mz + q$$

where p, q are constants of integration.

Thus incompressible fluids admit flows for which

$$P = A + \tfrac{1}{4}(f')^2C - mz + q \tag{9.2.16}$$

where $f'(x)$ is a solution of the equation

$$\tfrac{1}{2}Bf'(x) = -mx + p\,. \tag{9.2.17}$$

In this equation the quantity B is in general a function of $[f'(x)]^2$ so that the left side of (9.2.17) is an odd function of $f'(x)$. The choice $p = 0$ in (9.2.17) makes $f'(x)$ an odd function of x and hence $f(x)$ an even function of x. This result solves the problem of liquid flowing in the z direction under the action of a constant pressure gradient, $\partial P/\partial z = -m$, between two stationary parallel plates $x = \pm h$. For this problem the constant of integration associated with the integration of $f'(x)$ is determined by the adhesian condition $f(h) = 0$. The condition $f(-h) = 0$ is met automatically since $f(x)$ is even. For the particular case where B is constant, the velocity profile is given by

$$v_z(x)[\equiv f(x)] = m(h^2 - x^2)/B\,. \tag{9.2.18}$$

The parabolic profile (9.2.18) for flow between two parallel plates is a well known result in the theory of incompressible Stokesian liquids for which $\alpha = -\tfrac{2}{3}\mu$, $\beta(\equiv B) = 2\mu, \gamma = 0$, – see Chapter 10.

It should be emphasised that the solution found immediately above is not valid for a compressible material. This is because the motion is one for which the density is unchanged while at the same time P, given by (9.2.16), varies with x, Such behaviour is inconsistent for a compressible barotropic fluid for which $P = P(\rho)$. Of course for a nearly incompressible Reiner-Rivlin material, the solution found will be approximately valid.

9.3 Rectilinear shear flows with axial symmetry

In this section we seek solutions of the form

$$v_z = f(r)\,, \quad v_x = v_y = 0\,, \quad \rho = \rho_0 \tag{9.3.1}$$

where r is the cylindrical polar coordinate

$$r = (x^2 + y^2)^{\frac{1}{2}}$$

and where the z axis is the axis of symmetry. Equation (9.1.2) is satisfied identically.

From (9.3.1) we find for the only non-vanishing components of **d**

$$d_{zx} = d_{xz} = \frac{1}{2}\frac{\partial}{\partial x}[f(r)] = \tfrac{1}{2}xf'(r)/r$$

$$d_{zy} = d_{yz} = \frac{1}{2}\frac{\partial}{\partial y}[f(r)] = \tfrac{1}{2}yf'(r)/r\ .$$

Accordingly the matrices for **d** and $\mathbf{d}^2$ become

$$\mathbf{d} = \tfrac{1}{2}[f'(r)/r]\begin{bmatrix} 0 & 0 & x \\ 0 & 0 & y \\ x & y & 0 \end{bmatrix} \tag{9.3.2}$$

$$\mathbf{d}^2 = \tfrac{1}{4}[f'(r)/r]^2\begin{bmatrix} x^2 & xy & 0 \\ yx & y^2 & 0 \\ 0 & 0 & x^2+y^2 \end{bmatrix} \tag{9.3.3}$$

Evidently trace **d** $= 0$ and trace $(\mathbf{d}^2) = \tfrac{1}{2}[f'(r)]^2$ so that I_d is identically zero while from (4.5.10) II_d is a function solely of $[f'(r)]^2$. From (4.5.11), $\mathrm{III}_d \equiv |\mathbf{d}| = 0$ so that the coefficients α, β and γ appearing in the constitutive equations may be written

$$\alpha(\mathrm{I}_d, \mathrm{II}_d, \mathrm{III}_d) = A[(f')^2]$$
$$\beta(\mathrm{I}_d, \mathrm{II}_d, \mathrm{III}_d) = B[(f')^2]$$
$$\gamma(\mathrm{I}_d, \mathrm{II}_d, \mathrm{III}_d) = C[(f')^2]$$

where A, B and C are functions solely of $(f')^2$.

From these results and equations (9.3.2) and (9.3.3) the constitutive equations (9.1.4) yield

$$\boldsymbol{\sigma} = -(P-A)\mathbf{I} + (f'/2r)B\begin{bmatrix} 0 & 0 & x \\ 0 & 0 & y \\ x & y & 0 \end{bmatrix} + \tfrac{1}{4}(f'/r)^2 C\begin{bmatrix} x^2 & xy & 0 \\ yx & y^2 & 0 \\ 0 & 0 & x^2+y^2 \end{bmatrix}. \tag{9.3.4}$$

It remains to discuss whether or not equations (9.3.1) and (9.3.4) are compatible with the equations of motion. From (9.3.1) we find easily

$$\partial\mathbf{v}/\partial t = 0\,, \qquad (\mathbf{v}\,.\,\mathrm{grad})\mathbf{v} = 0$$

so that in the absence of body forces the equations of motion reduce to

$$\begin{aligned} \frac{\partial\sigma_{xx}}{\partial x} + \frac{\partial\sigma_{xy}}{\partial y} + \frac{\partial\sigma_{xz}}{\partial z} &= 0 \quad \text{(a)} \\ \frac{\partial\sigma_{yz}}{\partial x} + \frac{\partial\sigma_{yy}}{\partial y} + \frac{\partial\sigma_{yz}}{\partial z} &= 0 \quad \text{(b)}\ . \\ \frac{\partial\sigma_{zx}}{\partial x} + \frac{\partial\sigma_{zy}}{\partial y} + \frac{\partial\sigma_{zz}}{\partial z} &= 0 \quad \text{(c)} \end{aligned} \tag{9.3.5}$$

In view of the symmetry, P will depend on r and z and is independent of the polar angle $\theta = \tan^{-1}(y/x)$. On writing $P = P(r, z)$ and remembering that A, B and C depend on r [through $f'(r)$], substitution for the stresses from (9.3.4) into the first of (9.3.5) leads to

$$-\frac{x}{r}\frac{\partial}{\partial r}(P - A) + \frac{1}{4}\left[2x(f'/r)^2 C + (x^3/r)\frac{d}{dr}\{(f'/r)^2 C\}\right]$$

$$+\frac{1}{4}\left[x(f'/r)^2 C + (xy^2/r)\frac{d}{dr}\{(f'/r)^2 C\}\right] = 0 \ . \tag{9.3.6}$$

On combining the third and fifth terms, cancelling a common factor (x/r) and slightly re-arranging, we find

$$-\frac{\partial P}{\partial r} + \frac{dA}{dr} + \frac{1}{4}\left[(f')^2 C/r + \frac{d}{dr}\{(f')^2 C\}\right] = 0 \ . \tag{9.3.7}$$

[On substituting from (9.3.4) into the second of (9.3.5) there results an equation similar to (9.3.6) but with x and y interchanged. It follows that if (9.3.7) holds both (9.3.5a) and (9.3.5b) are satisfied.]

Substituting from (9.3.4) into the last of (9.3.5) yields a second equation

$$-\frac{\partial P}{\partial z} + Bf'/r + \frac{1}{2}r\frac{d}{dr}(Bf'/r) = 0$$

or on re-arranging
$$-\frac{\partial P}{\partial z} + \frac{1}{2}\{Bf'/r + \frac{d}{dr}(Bf')\} = 0 \ . \tag{9.3.8}$$

Equations (9.3.7) and (9.3.8) must both be satisfied if (9.3.1) is to be an admissible velocity field. In these equations A, B and C are specified functions of $[f'(r)]^2$.

The motion (9.3.1) is not in general an admissible motion for compressible materials. For, the motion (9.3.1) is isochoric and for a compressible material for which $P = P(\rho)$, the derivatives $\partial P/\partial r$ and $\partial P/\partial z$ vanish. In these circumstances since A, B and C are known functions of $f'(r)$, equations (9.3.7) and (9.3.8) provide two distinct differential equations for $f'(r)$. Evidently unless special relations hold between A, B and C these two equations are incompatible.† We do not discuss further the compressible case.

For incompressible materials the motion specified by (9.3.1) is admissible, for then equations (9.3.7) and (9.3.8) provide two equations for $P(r, z)$ and $f'(r)$. It is convenient to write

$$P' = P - A$$

† Examples of relations which make the equations compatible are (i) $B = 0$, (ii) $C = 0$, $A = $ const. The second of these is pertinent to the case of a Stokes' fluid for which $C = 0$ while both A and B are constants. The problem of the motion (9.3.1) of a compressible Stokes fluid in the absence of pressure gradients is discussed in Chapter 10.

so that the equations become

$$\frac{\partial P'}{\partial r} = \frac{1}{4}\left[\frac{d}{dr}\,(f')^2 C\} + (f')^2 C/r\right] \tag{9.3.9a}$$

$$\frac{\partial P'}{\partial z} = \frac{1}{2}\left[\frac{d}{dr}\,(Bf') + \frac{Bf'}{r}\right] . \tag{9.3.9b}$$

The consistency condition that $\partial^2 P'/\partial z \partial r$ [calculated from (9.3.9a)] is the same as $\partial^2 P'/\partial r \partial z$ [calculated from (9.3.9b)] shows that

$$\frac{\partial}{\partial r}\left(\frac{d}{dr}\,(Bf') + \frac{Bf'}{r}\right) = 0$$

i.e.

$$\frac{d}{dr}\,(Bf') + \frac{Bf'}{r} = 2Q'(z) \quad \text{(say)} \tag{9.3.10}$$

where $Q'(z)$ is solely a function of z. However the left side of (9.3.10) is independent of z so that $Q'(z)$ is at most a constant (say $-Q_0$). Accordingly

$$\frac{d}{dr}\,(Bf') + \frac{Bf'}{r} = -2Q_0 \tag{9.3.11}$$

and

$$\partial P'/\partial z = -Q_0 \quad . \tag{9.3.12}$$

Equation (9.3.11) may be integrated to yield

$$Bf' = -Q_0 r + E/r \tag{9.3.13}$$

where E is a second integration constant.

Equation (9.3.12) may be integrated also to yield

$$P' = -Q_0 z + \chi(r)$$

where $\chi(r)$ is obtained, to within an arbitrary integration constant from (9.3.9a)

i.e.

$$\chi(r) = \tfrac{1}{4}\,Cf'^2 + \tfrac{1}{4}\int r^{-1}(f')^2\,C dr \ . \tag{9.3.14}$$

Particular choices of constants solve particular problems. Once Q_0 and E are assigned, (9.3.13) is an (algebraic) equation for $f'(r)$. Solving for $f'(r)$ and integrating determines the velocity field; substituting the same value of $f'(r)$ in (9.3.14) determines χ and hence the pressure.

Problems solved by the present analysis are essentially those of axisymmetric axial flow of a fluid under the action of a pressure gradient in the z direction of $-Q_0$. For the particular case of a Stokes fluid ($C = 0, B =$ constant) the problems are easily solved explicitly. For example, for the case of Poiseuille flow of a liquid along a tube of circular cross section $0 \leqslant r \leqslant a$, we have, on writing $E = 0$ to avoid a singularity at $r = 0$ in f' and hence in the stresses, $f' = -(Q_0/B)r$, with integral

$$f = (Q_0/2B)(a^2 - r^2) \ . \tag{9.3.15}$$

In writing down (9.3.15), we have assumed the boundary condition $f = 0$ at $r = a$, i.e. that the material adheres to the cylindrical wall. This problem, together with related ones for a Stokes' fluid is considered in more detail in Chapter 10.

We conclude this section by noting that a somewhat simpler representation of the stresses (9.3.4) is given using cylindrical polar coordinates rather then Cartesian coordinates. If we define 'physical' components of stress σ_{rr}, $\sigma_{\theta\theta}$, σ_{zz}, $\sigma_{r\theta}$, σ_{rz}, $\sigma_{\theta z}$ in a manner analogous to the Cartesian components, so that the first suffix refers to the direction of the surface normal, while the second suffix refers to the direction of the resolved component of force per unit area, then the proper orthogonal matrix describing the rotation between the two sets of axes (Fig. 9.1) is given by

$$\mathbf{Q} = \begin{bmatrix} \cos\theta & \sin\theta & 0 \\ -\sin\theta & \cos\theta & 0 \\ 0 & 0 & 1 \end{bmatrix} \equiv \begin{bmatrix} x/r & y/r & 0 \\ -y/r & x/r & 0 \\ 0 & 0 & 1 \end{bmatrix} \tag{9.3.16}$$

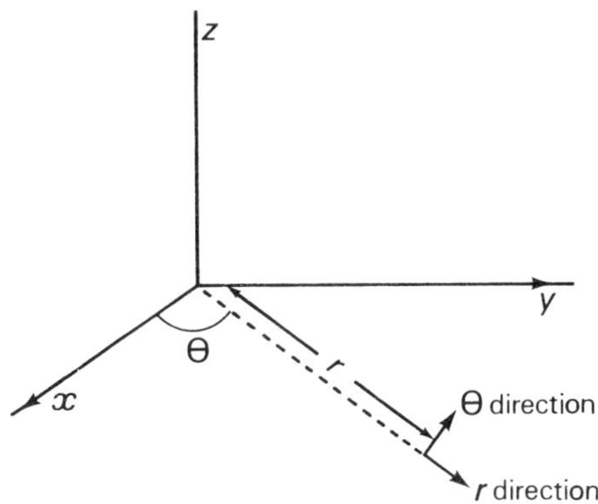

Fig. 9.1 The cylindrical polar r and θ directions

Pre-multiplying (9.3.4) by $\mathbf{Q}$ and post multiplying by $\mathbf{Q}^T$ leads to

$$\mathbf{Q}\sigma\mathbf{Q}^T \equiv \begin{bmatrix} \sigma_{rr} & \sigma_{r\theta} & \sigma_{rz} \\ \sigma_{\theta r} & \sigma_{\theta\theta} & \sigma_{\theta z} \\ \sigma_{zr} & \sigma_{z\theta} & \sigma_{zz} \end{bmatrix} \quad \text{(by definition)}$$

$$= -(P - A)\mathbf{I} + \tfrac{1}{2}Bf' \begin{bmatrix} 0 & 0 & 1 \\ 0 & 0 & 0 \\ 1 & 0 & 0 \end{bmatrix} + \tfrac{1}{4}C(f')^2 \begin{bmatrix} 1 & 0 & 0 \\ 0 & 0 & 0 \\ 0 & 0 & 1 \end{bmatrix} \tag{9.3.17}$$

In component form (9.3.17) reads

$$\sigma_{rr} = \sigma_{zz} = -(P - A) + \tfrac{1}{4}C(f')^2, \quad \sigma_{\theta\theta} = -(P - A)$$
$$\sigma_{rz} = \sigma_{zr} = \tfrac{1}{2}Bf', \quad \sigma_{r\theta} = \sigma_{\theta r} = \sigma_{z\theta} = \sigma_{\theta z} = 0 \tag{9.3.18}$$

9.4 Azimuthal (Couette) shear flows with axial symmetry

Here we seek solutions of the form

$$v_r = v_z = 0, \quad v_\theta = f(r), \quad \rho = \rho_0 \tag{9.4.1}$$

where r, θ and z are cylindrical polar coordinates, and where the motion is circular about the z axis. Equation (9.1.2) is satisfied identically. In Cartesian components (9.4.1) becomes

$$v_x = -yf(r)/r, \quad v_y = xf(r)/r, \quad v_z = 0 \tag{9.4.2}$$

so that

$$d_{xx} = -d_{yy} = -(xy/r)\frac{d}{dr}(f/r), \quad d_{zz} = 0$$

$$d_{xy} = d_{yx} = \tfrac{1}{2}[(x^2 - y^2)/r]\frac{d}{dr}(f/r)$$

$$d_{xz} = d_{zx} = d_{yz} = d_{zy} = 0\,.$$

Accordingly

$$\mathbf{d} = r^{-1}\frac{d}{dr}(f/r)\begin{bmatrix} -xy & \tfrac{1}{2}(x^2 - y^2) & 0 \\ \tfrac{1}{2}(x^2 - y^2) & xy & 0 \\ 0 & 0 & 0 \end{bmatrix} \tag{9.4.3}$$

and

$$\mathbf{d}^2 = \tfrac{1}{4}r^2\left(\frac{d}{dr}(f/r)\right)^2\begin{bmatrix} 1 & 0 & 0 \\ 0 & 1 & 0 \\ 0 & 0 & 0 \end{bmatrix}. \tag{9.4.4}$$

As in the previous example $\mathrm{I}_d = \mathrm{III}_d = 0$; here $\mathrm{II}_d = -\xi^2$ where

$$\xi = \tfrac{1}{2}r\frac{d}{dr}(f/r) \equiv \tfrac{1}{2}(f' - f/r) \tag{9.4.5}$$

Accordingly we write for the coefficients α, β, γ

$$\alpha = A(\xi^2), \quad \beta = B(\xi^2), \quad \gamma = C(\xi^2) \tag{9.4.6}$$

which are all (in principle) known functions of ξ^2.

In the present example $\partial \mathbf{v}/\partial t = 0$ but there exist non-vanishing components of $(\mathbf{v}\,.\,\mathrm{grad})\mathbf{v}$. We find from (9.4.2).†

$$(\mathbf{v}\,.\,\mathrm{grad})v_x = -(x/r^2)f^2\,, \quad (\mathbf{v}\,.\,\mathrm{grad})v_y = -(y/r^2)f^2$$
$$(\mathbf{v}\,.\,\mathrm{grad})v_z = 0\,. \tag{9.4.7}$$

Accordingly the equations of motion in the x and z directions become

$$\left.\begin{aligned}\frac{\partial\sigma_{xx}}{\partial x} + \frac{\partial\sigma_{xy}}{\partial y} + \frac{\partial\sigma_{xz}}{\partial z} &= -\rho(x/r^2)f^2 \qquad \text{(a)}\\ \frac{\partial\sigma_{zx}}{\partial x} + \frac{\partial\sigma_{zy}}{\partial y} + \frac{\partial\sigma_{zz}}{\partial z} &= 0 \qquad \text{(b)}\end{aligned}\right\} \tag{9.4.8}$$

The equation in the y direction is omitted; the situation is similar to that of the previous section where the equations of motion in the x and y directions lead to a common result.

From (9.4.3), (9.4.4) and the constitutive equations (9.1.4), the non-vanishing components of stress are

$$\left.\begin{aligned}\sigma_{xx} &= -(P-A) - Bxyr^{-1}\frac{d}{dr}(f/r) + \tfrac{1}{4}Cr^2\left[\frac{d}{dr}(f/r)\right]^2\\ \sigma_{yy} &= -(P-A) + Bxyr^{-1}\frac{d}{dr}(f/r) + \tfrac{1}{4}Cr^2\left[\frac{d}{dr}(f/r)\right]^2\\ \sigma_{xy} &= \tfrac{1}{2}B(x^2-y^2)r^{-1}\frac{d}{dr}(f/r)\\ \sigma_{zz} &= -(P-A)\,.\end{aligned}\right\} \tag{9.4.9}$$

From symmetry arguments P depends only on r. In these circumstances the stresses (9.4.9) satisfy equation (9.4.8b) identically. Substituting equations (9.4.9) into (9.4.8a) leads directly to

$$\begin{aligned}&-xr^{-1}\frac{d}{dr}(P-A) - Byr^{-1}\frac{d}{dr}(f/r) - (x^2y/r)\frac{d}{dr}\left[Br^{-1}\frac{d}{dr}(f/r)\right]\\ &+\tfrac{1}{4}xr^{-1}\frac{d}{dr}\left\{Cr^2\left[\frac{d}{dr}(f/r)\right]^2\right\} - Byr^{-1}\frac{d}{dr}(f/r)\\ &+\tfrac{1}{2}y(x^2-y^2)r^{-1}\frac{d}{dr}\left[Br^{-1}\frac{d}{dr}(f/r)\right] = -\rho xr^{-1}f^2\,.\end{aligned}$$

† If the components of $(\mathbf{v}\,.\,\mathrm{grad})\mathbf{v}$ are resolved in the r and θ direction, the terms (9.4.7) are seen to be equivalent to an acceleration in the radial direction of $-f^2/r$ which is a familiar result for circular motion.

After some simplication this equation may be re-written

$$\frac{x}{r}\left\{\frac{d}{dr}\left[P - A - \tfrac{1}{4}Cr^2\left[\frac{d}{dr}(f/r)\right]^2\right] - \frac{\rho f^2}{r}\right\} =$$

$$-\frac{y}{r}\left\{2B\frac{d}{dr}(f/r) + \tfrac{1}{2}r^2\frac{d}{dr}\left(Br^{-1}\frac{d}{dr}(f/r)\right)\right\} . \qquad (9.4.10)$$

Evidently a solution along the proposed lines is possible only if the two sets of terms in parentheses { } vanish independently; under these circumstances the pressure is given by

$$P = A + \tfrac{1}{4}Cr^2\left[\frac{d}{dr}(f/r)\right]^2 + \rho_0\int r^{-1}f^2dr \qquad (9.4.11)$$

while $f(r)$ is determined by the non-linear equation

$$4B\frac{d}{dr}(f/r) + r^2\frac{d}{dr}\left[Br^{-1}\frac{d}{dr}(f/r)\right] = 0 \qquad (9.4.12)$$

in which B is a (known) function of ξ given by (9.4.5).† Equation (9.4.12) admits a first integral

$$B\frac{d}{dr}(f/r) = G_1/r^3, \quad (G_1 = \text{arbitrary constant}) \qquad (9.4.13)$$

and for the particular case where B is constant, this last equation is integrable also to yield

$$f = Er + Gr^{-1}$$

where E and $G(= -G_1/2B)$ are arbitrary constants. The result is well known in the theory of Couette flow, i.e. azimuthal flow in the annulus of coaxial rotating cylinders, of Stokesian fluids. For this problem, discussed in Chapter 10, the constants E and G are determined by the requirement that the material adheres to the surfaces of the rotating cylinders.

When B is not constant (9.4.13) may be written in the form

$$\xi B(\xi^2) = \tfrac{1}{2}G_1r^{-2}$$

which is an algebraic equation for ξ. On solving for ξ, and conceivably there is a lack of uniqueness here, there results from (9.4.5) a first order linear differential equation for f. The solution of this equation entails a second constant of integration. This new constant, along with G_1, may be assigned as E and G in the previous paragraph for the Couette flow problem.

The pressure P, given by (9.4.11), varies with r; since the motion is isochoric, the velocity field (9.4.1) is admissible only for an incompressible material.

† It is left to the reader to verify that also the equation of motion in the y direction is satisfied by (9.4.11) and (9.4.12).

As in the previous problem the stress field assumes a simpler form when referred to cylindrical polar coordinates. Using the same transformation matrix **Q** given by (9.3.16), the physical components of stress in cylindrical polar coordinates are

$$\sigma_{rr} = \sigma_{\theta\theta} = -(P - A) + \tfrac{1}{4}Cr^2\left[\frac{d}{dr}(f/r)\right]^2 ; \quad \sigma_{zz} = -(P - A) ;$$

$$\sigma_{r\theta} = \tfrac{1}{2}Br\frac{d}{dr}(f/r) ; \quad \sigma_{rz} = \sigma_{\theta z} = 0 . \tag{9.4.14}$$

A number of interesting features attach to the three sets of stresses obtained in equations (9.2.6), (9.3.18) and (9.4.14) for the three shear flow situations. The three sets of equations are of similar structure and possess related features. For example for the case of a Reiner-Rivlin fluid for which $C = 0$ (e.g. a Stokesian fluid) all three sets of stresses yield equal normal stresses $[-(P - A)]$.† In these circumstances the state of stress is one of 'hydrostatic' pressure together with superimposed shear stresses.

For $C \neq 0$ two of the normal stresses are equal $[\sigma_{zz} = \sigma_{xx}$ in (9.2.6), $\sigma_{zz} = \sigma_{rr}$ in (9.3.18) and $\sigma_{\theta\theta} = \sigma_{rr}$ in (9.4.14)$]$. The indices involved in these equalities are those referring to the direction of the velocity field (respectively z, z and θ) and the direction in which the magnitude of the velocity field varies (respectively x, r and r). The normal stress in the third direction (respectively y, θ and z) differs by a term linear in the coefficient C.

The existence of one normal stress, different from the other two (equal) normal stresses in the shear flow situations discussed, is commonly called the *normal stress effect.* The reason for the terminology stems from the fact that for the much studied case of Stokesian fluids, for which A and B are constants while $C = 0$, all three normal stresses are the same for each of the above problems. Thus, historically, the existence of equal normal stresses was regarded as the norm for shear flows of fluids. Experimentally many fluids appear to conform to this norm.

However some fluids are known to display normal stress effects in shear flows. We omit the details; the interested reader should see, for example, the discussion given by Truesdall of the Weissenberg effect and Merrington's phenomena [p 161-163 *Principles of Continuum Mechanics* – Socony Mobil Oil Company (1961)]. It is natural to ask whether the full Reiner-Rivlin theory with $C \neq 0$ could account adequately for the normal stress effects in real materials.

The geometrical situations discussed in this Chapter provide in principle a partial answer to this question. For in each of the examples equality of two of the normal stresses occurs. This result, completely independent of the fine details as to how α, β and γ depend on the invariants of **d**, is in principle susceptible to an experimental check. Should the check fail then the Reiner-Rivlin model is necessarily inadequate.

† Of course P and A vary from point to point in each solution and are different in the different solutions.

According to Truesdall (loc. cit. p 177) it seems that results of measurements indicate that relations like $\sigma_{zz} = \sigma_{xx}$ in (9.2.6) are not fulfilled by real materials which display normal stress effects. The situation is complicated in that none of the geometrical situations discussed in this Chapter are exactly realisable experimentally, and also the necessary instrumentation usually interferes with the flow. The totality of evidence is small, but appears to suggest that the Reiner-Rivlin model is not adequate to account for the behaviour of real fluids which display normal stress effects. Also the model seems unnecessarily complicated for the simpler materials, which are described adequately by the Stokes' model.

Chapter 10

Some Solutions of the Navier-Stokes Equations

10.1 Introduction

Until recently most theoretical investigations in fluid dynamics were based on the assumption that the mechanical behaviour of the fluid was modelled by a compressible viscous (Stokes') fluid, or by simplified versions of the model. In recent years knowledge of fluids exhibiting distinctly non-Stokesian characteristics has led to an increasing number of investigations based on much more general constitutive equations; even so most current work in fluid mechanics is still directed towards finding solutions of the Navier-Stokes equations.

Here we give a number of well known exact solutions of the Navier-Stokes equations. The relevant equations for barotropic fluids with constant viscosity are summarised below.

(a) COMPRESSIBLE

The constitutive equations (7.2.23)

$$\sigma_{ij} = -P(\rho)\delta_{ij} + 2\mu d_{ij} - \tfrac{2}{3}\mu d_{mm}\delta_{ij} \tag{10.1.1}$$

where

$$d_{ij} = \tfrac{1}{2}(\partial v_i/\partial x_j + \partial v_j/\partial x_i) \ . \tag{10.1.2}$$

The equation of mass conservation in spatial form (2.3.18)

$$\frac{\partial \rho}{\partial t} + \operatorname{div}(\rho \mathbf{v}) = 0 \ . \tag{10.1.3}$$

The Navier-Stokes equations (7.2.24), obtained from substituting (10.1.1) into the general equations of motion of a continuum are

$$-\operatorname{grad} P + \rho \mathbf{b} + \mu\nabla^2\mathbf{v} + \tfrac{1}{3}\mu \operatorname{grad}(\operatorname{div}\mathbf{v}) = \rho[\partial \mathbf{v}/\partial t + (\mathbf{v}\,.\operatorname{grad})\mathbf{v}] \ . \tag{10.1.4}$$

With $P = P(\rho)$, a given function of the density, equations (10.1.3) and (10.1.4) are the governing equations for ρ and the three components of $\mathbf{v}$. With ρ and $\mathbf{v}$ found, equations (10.1.1) then determine the stresses. We also recall the formula (7.3.19) for the volume rate of energy dissipation

$$D = 2\mu\hat{d}_{ij}\hat{d}_{ij} \tag{10.1.5}$$

where

$$\hat{d}_{ij} = d_{ij} - \tfrac{1}{3}d_{mm}\delta_{ij} \ . \tag{10.1.6}$$

(b) INCOMPRESSIBLE

Considerable simplifications occur with the adoption of the incompressibility assumption. For this model $\rho = \rho_0$ and

$$d_{mm}(= \text{div } \mathbf{v}) = 0 .$$

Equations (10.1.1), (10.1.3), (10.1.4) and (10.1.5) become

$$\sigma_{ij} = -P\delta_{ij} + 2\mu d_{ij} \tag{10.1.7}$$

$$\text{div } \mathbf{v} = 0 \tag{10.1.8}$$

$$-\text{grad } P + \rho_0 \mathbf{b} + \mu\nabla^2\mathbf{v} = \rho_0 [\partial\mathbf{v}/\partial t + (\mathbf{v} \,.\, \text{grad})\mathbf{v}] \tag{10.1.9}$$

$$D = 2\mu d_{ij} d_{ij} . \tag{10.1.10}$$

In equations (10.1.8) and (10.1.9), governing the motion, the quantity P is here regarded as a dependent variable, determined by the equations concomitantly with $\mathbf{v}$.

10.2 Boundary conditions

Equations (10.1.3) and (10.1.4), or (10.1.8) and (10.1.9), are partial differential equations governing all possible motions of a Stokes' fluid. The feature which distinguishes one flow situation from another is the nature of the 'boundary conditions' satisfied by the velocity field $\mathbf{v}$.† Boundary conditions have been introduced already in Chapter 9, where it was assumed that the fluid adhered to rigid, but possibly moving, surfaces bounding the fluid.

This *adhesian* condition is adopted in finding solutions of the Navier-Stokes equations. Evidently at a rigid surface the normal component of fluid velocity must be the same as for the rigid surfaces; the assumption of adhesian implies further that the tangential velocity component is likewise the same as that of the rigid surface (*no slip condition*). The assumption of adhesian accords with experimental observations on real materials. For a fixed rigid surface the adhesian condition reduces simply to $\mathbf{v} = 0$ at the fixed surface.

Adoption of the adhesian condition is sufficient to determine uniquely solutions of the Navier-Stokes equations in the examples following.

Some of the solutions are particular cases of those of the previous chapter for more general constitutive equations.

10.3 Rectilinear shear flow solutions

Here we are concerned with solutions of the incompressible Navier-Stokes equation of the form

$$v_z = v_z(x, y) , \quad v_x = v_y = 0 \tag{10.3.1}$$

which is the most general form of a steady rectilinear shear flow.

† For the simpler problems of fluid dynamics, boundary conditions are imposed on the velocity field rather than the stress field, or the pressure. In more complicated examples such as the free streamline problems of Section (12.9), boundary conditions involve the pressure.

For the velocity field (10.3.1) the equation

$$\operatorname{div} \mathbf{v} = 0$$

is satisfied identically. Also $\partial \mathbf{v}/\partial t$ and $(\mathbf{v} \,.\, \operatorname{grad})\mathbf{v}$ vanish identically so that in the absence of body forces, the Navier-Stokes equations for an incompressible fluid reduce to

$$\left.\begin{aligned} -\partial P/\partial x &= 0 && \text{(a)} \\ -\partial P/\partial y &= 0 && \text{(b)} \\ -\partial P/\partial z + \mu \nabla^2 v_z &= 0 && \text{(c)} \end{aligned}\right\} \qquad (10.3.2)$$

where ∇^2 is here the two dimensional Laplacian operator

$$\nabla^2 = \partial^2/\partial x^2 + \partial^2/\partial y^2 \;.$$

[For compressible barotropic fluids for which $P = P(\rho)$ the pressure gradient terms in (10.3.2) vanish. Two of the equations are then satisfied identically and (10.3.2c) reduces to Laplace's equation

$$\nabla^2 v_z = 0 \;. \qquad (10.3.3)$$

Rectilinear shear flows, occuring in the absence of pressure gradients, satisfy (10.3.3) and are possible flows for both compressible and incompressible fluids.]

We return to equations (10.3.2) for an incompressible fluid. Equations (10.3.2a) and (10.3.2b) show that P is at most a function of z, i.e. $P = P(z)$. Inserting this result into (10.3.2c) yields the equation

$$dP/dz = \mu \nabla^2 v_z(x, y)$$

in which the left side is independent of x and y and the right side is independent of z. Evidently a solution is possible only if each side is equal to a constant ($-m$ say) so that

$$dP/dz = -m \;, \qquad (10.3.4a)$$

$$\mu \nabla^2 v_z = -m \;. \qquad (10.3.4b)$$

The first of these equations integrates immediately in the form

$$P = P_1 - mz \qquad (10.3.5)$$

where P_1 is a constant of integration. The integration of (10.3.4b) requires the boundary conditions satisfied by $v_z(x, y)$ to be specified for a particular problem.

Perhaps the simplest problem is that for which v_z depends only on one space variable, say x, for which (10.3.4b) becomes

$$\frac{d^2 v_z}{dx^2} = -\frac{m}{\mu}$$

with solution $$v_z = a + bx - \tfrac{1}{2}(m/\mu)x^2 \;.$$

10.3.1 Steady Flow Between Parallel Fixed Plates

The particular choices $b = 0$, $a = \frac{1}{2}(m/\mu)h^2$ in the equation above yielding

$$v_z = \frac{1}{2}(m/\mu)(h^2 - x^2) \tag{10.3.6}$$

solves the problem of the flow in the z direction of a layer of liquid $|x| \leqslant h$ confined between two fixed plates $x = \pm h$. For the two dimensional flow situation, i.e. flow confined to the xz plane independent of y, implied by the solution (10.3.6), the layer of liquid is unbounded in the y direction. Such a situation is not practically realisable, but the flow pattern (10.3.6) would be approximated in the experiment of figure (10.1).

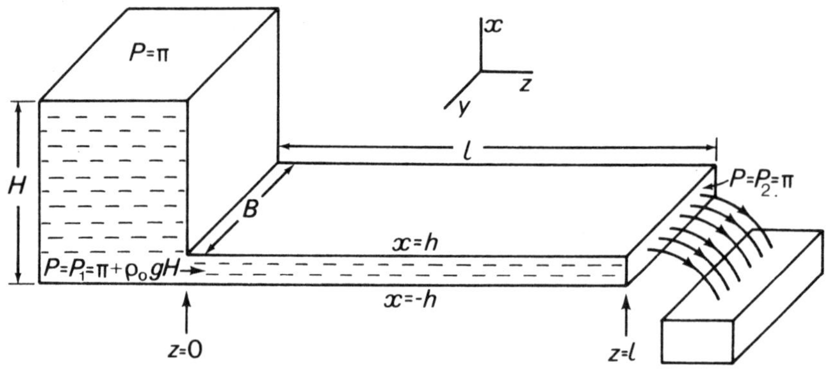

Fig. 10.1 Figure illustrating hypotherical experiment associated with problem of Section 10.3.1

Here the width of the plates in the y direction is sufficient to guarantee that the flow pattern (10.3.6) is realisable over most of the area between the plates. (In practise it would be necessary for the width in the y direction to be much larger than the separation distance $2h$.) The fluid is driven in the z direction by the pressure gradient $-m$. The latter quantity is at the disposal of the experimentalist who can arrange for arbitrary pressures differences $P_1 - P_2$ ($\equiv \rho_0 gH$); since P is a linear function of z [equation (10.3.5)], m is given by

$$m = (P_1 - P_2)/l \ .$$

The simplest arrangement is as shown in Fig. 10.1. The plates $x = \pm h$ lie in a horizontal plane. Liquid is supplied to the inlet $z = 0, |x| < h$ from a gravity feed tank which is kept topped up to a height H above the point $x = 0$; the liquid emerges at $z = l$ into the atmosphere for which the pressure is atmospheric ($= \Pi$). For such an arrangement

$$P_1 = \Pi + \rho_0 g H \text{ (a)}, \quad P_2 = \Pi \text{ (b)} \tag{10.3.7}$$

and
$$m = \rho_0 g H / l .$$

Here ρ_0 is the fluid density and g the acceleration due to gravity.

In writing down the first of (10.3.7) the quantity P_1 at the inlet $z = 0$ is calculated from the hydrostatics of the tank on the assumption that liquid motion in the tank is of negligible influence in determining the pressure P_1. Also it is assumed that the pressure across the inlet $|x| < h$ is sensibly constant. Both assumptions are physically realistic for plate separations which are small compared with the tank dimensions.

It may not have escaped attention that in the hydrostatic tank calculation (10.3.7a), gravitational body forces play a crucial role in determining P_1. Since body forces are neglected in the main flow calculation leading to (10.3.6) this might seem an inconsistent procedure. However for sufficiently small plate separation the neglect of body forces in the region between the plates is legitimate. This may be seen by noting that with the inclusion of gravity forces, variations in pressure across the tube will be of maximum order $\rho_0 g h$ (see below). These variations are small compared with the smallest value of P obtained in the absence of body forces if

$$\rho_0 g h \ll \Pi \tag{10.3.8}$$

providing an upper limit for h. With values of ρ_0 typical of liquids, $\Pi/\rho_0 g$ is about 10 m. and the inequality (10.3.8) is well satisfied by any practically conceivable arrangement of the type of Fig. 10.1. For any practical arrangement of this type, unstable turbulent flow due to instabilities caused by wall roughness would be initiated long before the inequality (10.3.8) is violated. With the onset of turbulent flow, the motion is no longer described by a velocity field of the form (10.3.1). A discussion of the onset of turbulence is beyond the scope of this book; however an indication of the range of the validity of the present analysis is that practical limits on plate separation for the flow of (say) a motor oil would be of the order of a millimetre.

It is perhaps of note that for the case of horizontal plates with gravitational forces acting in the x direction, an exact solution of the incompressible Navier-Stokes equation is given by a velocity field of the form (10.3.1) and (10.3.6) but with the pressure given by

$$P = P_1 - mz - \rho_0 g x \tag{10.3.9}$$

rather than by (10.3.5).† Here P_1 is now the pressure at $z = 0, x = 0$, i.e. at the midpoint of the inlet section $z = 0$. Unfortunately while (10.3.9) and

† The verification of this result is left to the reader.

(10.3.6) satisfy the Navier-Stokes equations and the boundary conditions on $x = \pm h$, the proposed solution fails to satisfy, to the accuracy of the additional term in (10.3.9) the boundary conditions at the emergent section $z = l$. For exact boundary conditions here are that at $z = l$ each surface $x = \pm h$ of the emergent liquid sheet is exposed to atmospheric pressure

i.e. $$P = \Pi \text{ at } z = l,\ x = \pm h\ .$$

Evidently the pressure field (10.3.9) will not meet these boundary conditions and the correct solution to the exact boundary value problem in the presence of body forces is of a much more complicated nature. However the 'solution' (10.3.9) does indicate that the maximum magnitude of additional terms in P due to body forces is or order $\rho_0 gh$, as used in the previous discussion leading to the inequality (10.3.8) above.

The forces acting on the plates $x = \pm h$ are normal forces due to the pressure (10.3.5) and drag forces due to the non-vanishing shear stress σ_{xz}.

We have from (10.1.7) $$\sigma_{xz} = 2\mu d_{xz} \tag{10.3.10}$$

while from (10.3.1) and (10.3.6)

$$d_{xz} = \tfrac{1}{2}(\partial v_x/\partial z + \partial v_z/\partial x) = -(mx/\mu)\ . \tag{10.3.11}$$

Combining (10.3.10) and (10.3.11) leads to

$$(\sigma_{xz})_{x = \pm h} = \mp 2mh \tag{10.3.12}$$

so that the drag force/unit area exerted by the liquid on each plate is uniform, independent of x and y, and of magnitude $2mh$ in the positive z direction. To maintain the plates stationary requires the experimenter to apply counteracting forces of the same magnitude but in the opposite direction.

The volume rate of fluid discharge (efflux) at $z = l$ is given by

$$Q = B\int_{-h}^{h} v_z(x)dx \tag{10.3.13}$$

where B is the width of the plates in the y direction. Inserting (10.3.6) into (10.3.13) and evaluating the integral leads to

$$Q = 2mh^3B/3\mu\ . \tag{10.3.14}$$

We conclude discussion of this solution by considering the energy balance question. Energy is supplied to the material between the planes at a rate

$$\frac{dW}{dt} = \iint_{(z=0)} -\sigma_{zz}v_z dxdy + \iint_{(z=l)} \sigma_{zz}v_z dxdy\ . \tag{10.3.15}$$

No energy is supplied from the plates $x = \pm h$ since on these surfaces the velocity field vanishes. From the constitutive equations $\sigma_{zz} = -P$ and (10.3.15) becomes

$$\frac{dW}{dt} = (P_1 - P_2) \iint\limits_{(z=0)} v_z dxdy = (P_1 - P_2)Q = 2m^2 h^3 lB/3\mu \,. \tag{10.3.16}$$

In obtaining this last result ml has been substituted for $(P_1 - P_2)$.

Since the motion is that of an incompressible fluid there are no changes in internal energy. Thus the quantity dW/dt must be balanced by the total rate of energy dissipation

$$\iiint\limits_{\Omega} Ddxdydz \tag{10.3.17}$$

where the integral is taken over the volume of material between the plates.

From (10.1.10), (10.3.1) and (10.3.6)

$$D = 2\mu(d_{xz}d_{xz} + d_{zx}d_{zx}) = 4\mu(d_{xz})^2 = \mu(\partial v_z/\partial x)^2 = m^2 x^2/\mu \,.$$

Inserting this result into (10.3.17) yields

$$\iiint\limits_{\Omega} Ddxdydz = m^2 lB\mu^{-1} \int_{-h}^{h} x^2 dx = 2m^2 h^3 lB/3\mu$$

in agreement with (10.3.16).

10.3.2 Shearing of Liquid Between Two Parallel Plates in Relative Motion

In the absence of pressure gradients $v_z(x, y)$ satisfies the two dimensional Laplace equation

$$\left(\frac{\partial^2}{\partial x^2} + \frac{\partial^2}{\partial y^2}\right) v_z = 0 \,. \tag{10.3.18}$$

If v_z depends only on x, (10.3.18) becomes $d^2 v_z/dx^2 = 0$ with solution $v_z = ax + b$ where a and b are constant of integration. The choice $b = 0$, $a = v_0/h$ solves the problem of the motion of a liquid layer $0 \leqslant x \leqslant h$ which adheres to a fixed infinite plate $x = 0$ and to an inifinite plate $x = h$ which moves transversely in the z direction at speed v_0. Details of stress and energy calculations are left to the reader.

10.3.3 RECTILINEAR SHEAR FLOW ALONG THE ANNULUS BETWEEN TWO PARALLEL CYLINDRICAL WALLS $C_1(x, y) = 0$, $C_2(x, y) = 0$

For this problem, depicted in Fig. 10.2, $v_z(x, y)$ depends on both x and y and it is now necessary to solve the partial differential equation (10.3.4b)

i.e.
$$\left(\frac{\partial^2}{\partial x^2} + \frac{\partial^2}{\partial y^2}\right) v_z = -m/\mu \tag{10.3.19}$$

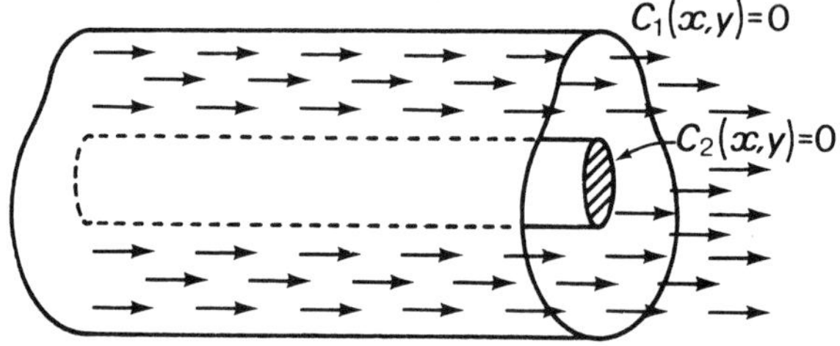

Fig. 10.2 Rectilinear shear flow along an annulus

For the case of fixed cylinders the problem specified by solving (10.3.19) subject to the boundary conditions

$$v_z = 0 \text{ on } C_1 = 0 \text{ and } C_2 = 0 \tag{10.3.20}$$

may be reduced to the solution of Laplace's equation. We write†

$$v_z = \phi - mx^2/2\mu$$

so that the problem specified by (10.3.19) and (10.3.20) reduces to the solution of

$$\left(\frac{\partial^2}{\partial x^2} + \frac{\partial^2}{\partial y^2}\right) \phi = 0, \tag{10.3.21}$$

subject to $\phi = \Phi_0(\equiv mx^2/2\mu)$ on $C_1 = 0$ and $C_2 = 0$. (10.3.22)

The problem of finding a function ϕ with continuous derivatives which satisfies $\nabla^2\phi = 0$ within a region Ω and subject to given conditions $\phi = \Phi_0$ on the boundaries S is known to possess a unique solution. A proof for the three dimensional case is as follows. Suppose there exist two functions ϕ_1 and ϕ_2 which satisfy the conditions

	$\nabla^2\phi_1 = 0$	for all (x, y, z) contained inside Ω bounded by S
	$\phi_1 = \Phi_0$	(x, y, z) contained in S
and	$\nabla^2\phi_2 = 0$	(x, y, z) contained in Ω
	$\phi_2 = \Phi_0$	(x, y, z) contained in S.

† The choice of $-mx^2/2\mu$ as a particular integral is not unique. Evidently many other choices are possible, e.g. $-m(x^2 + y^2)/4\mu$.

Consider now the function $\psi = \phi_1 - \phi_2$.

Evidently ψ satisfies Laplace's equation

$$\nabla^2 \psi = 0 \qquad (x, y, z) \text{ contained inside } \Omega \qquad (10.3.23)$$

and the boundary conditions

$$\psi = 0 \qquad (x, y, z) \text{ contained in } S. \qquad (10.3.24)$$

From the vector identity

$$\iint_S \psi \text{ grad } \psi \, . \, \mathbf{dS} = \iiint_\Omega (\text{grad } \psi)^2 d\Omega + \iiint_\Omega \psi \nabla^2 \psi d\Omega \quad (10.3.25)$$

and equation (10.3.23) together with the boundary condition (10.3.24) we derive

$$\iiint_\Omega (\text{grad } \psi)^2 \, d\Omega = 0 \qquad (10.3.26)$$

and for continuous positive definite integrands $(\text{grad } \psi)^2$, this is possible only if everywhere

$$\text{grad } \psi = 0 \, . \qquad (10.3.27)$$

In turn (10.3.27) implies that ψ is at most a constant and since from (10.3.24) $\psi = 0$ at the boundary it is seen that ψ is everywhere identically zero. In other words

$$\phi_1 = \phi_2$$

and the solution of the proposed potential problem is unique.†

The proof given above is readily modified to cover the two dimensional case; the details are left to the reader.

The existence of a solution of the rectilinear flow problem for arbitrary boundaries C_1 and C_2 devolves explicitly on the assumed form of the constitutive equations. Usually for the more general Reiner-Rivlin fluids it is *not* possible to find a rectilinear shear flow solution except for the case of circular boundaries.

Obtaining solutions of (10.3.21) subject to (10.3.22) for given boundaries $C_1 = 0$ and $C_2 = 0$ is, in general, a problem for numerical analysis. Exceptionally for a few simple cases it is possible to obtain an analytic solution. Some of these are considered in Sections 10.3d and 10.3e below.

Before discussing these analytic solutions we consider in general terms the energy balance aspects of the problems. Consider the material within the region contained between $C_1 = 0$ and $C_2 = 0$ and between the planes $z = 0$ and $z = l$.

† What has been proved is that if the proposed problem possesses a solution at all, then the solution is unique. Proving the existence of the solution is a rather more complicated problem [e.g. see Garabedian's *Partial Differential Equations* Chapter 8, (1964) Wiley].

Let the pressures at $z = 0$ and $z = l$ be respectively P_1 and P_2; then as in Section 10.3.1 the constant m in (10.3.19) is given by

$$m = (P_1 - P_2)/l . \tag{10.3.28}$$

Also as in Section 10.3.1 the rate of supply of energy to the material is given by

$$\frac{dW}{dt} = (P_1 - P_2)Q \tag{10.3.29}$$

where Q is the volume efflux

$$Q = \iint\limits_{(z = \text{const.})} v_z dxdy \tag{10.3.30}$$

and the surface integral is taken over the region bounded by the curves $C_1 = 0$ and $C_2 = 0$.

The volume rate of energy dissipation $D = 2\mu d_{ij}d_{ij}$ becomes for the assumed rectilinear flow

$$D = \mu[(\partial v_z/\partial x)^2 + (\partial v_z/\partial y)^2]$$

so that the total rate of energy dissipation is

$$\mu \iiint\limits_{\Omega} (\text{grad } v_z)^2 d\Omega \tag{10.3.31}$$

where Ω is the region bounded by $C_1 = 0, C_2 = 0, z = 0, z = l$. With the aid of the identity (10.3.25) (with ψ replaced by v_z) together with the differential equation (10.3.19), the expression (10.3.31) may be re-written

$$\mu \iint\limits_{S} (v_z \text{ grad } v_z) \,.\, \mathbf{dS} + m \iiint\limits_{\Omega} v_z d\Omega \,. \tag{10.3.32}$$

However the surface integral vanishes since either grad v_z is perpendicular to $\mathbf{dS}$ (surfaces $z = 0, z = l$) or $v_z = 0$ (surfaces $C_1 = 0, C_2 = 0$). On remembering that v_z is independent of z, the second integral in (10.3.31) becomes

$$ml \iint\limits_{(z = \text{const.})} v_z dxdy \equiv mlQ \tag{10.3.33}$$

or on recalling (10.3.28) we derive

$$\iiint\limits_{\Omega} D d\Omega = (P_1 - P_2)Q$$

so that, as expected, the total rate of energy dissipation is balanced by the rate (10.3.29) at which energy is supplied.

10.3.4 RECTILINEAR FLOW WITH AXIAL SYMMETRY

For concentric circular boundaries

$$C_1 = x^2 + y^2 - a^2 , \quad C_2 = x^2 + y^2 - b^2$$

equation (10.3.19) admits a solution for which v_z depends only on the radial coordinate $r = (x^2 + y^2)^{\frac{1}{2}}$. On writing $v_z = v_z(r)$ in (10.3.19) there results the ordinary differential equation†

$$\left(\frac{d^2}{dr^2} + \frac{1}{r}\frac{d}{dr}\right) v_z = -m/\mu \tag{10.3.34}$$

with general solution $v_z = A + B \log r - mr^2/4\mu$. (10.3.35)

On choosing the constants A and B to satisfy the no slip condition $v_z = 0$ on $r = a$ and $r = b$ there results

$$v_z = (\tfrac{1}{4}m/\mu)[(a^2 - r^2) - (a^2 - b^2)\log(a/r)/\log(a/b)] \tag{10.3.36}$$

for the rectilinear shear flow under pressure gradient m in the annulus $a \geqslant r \geqslant b$.

The particular case of flow along the inside of a tube $r \leqslant a$ given by choosing the limit $b \to 0$ in (10.3.36) yields

$$v_z = (\tfrac{1}{4}m/\mu)(a^2 - r^2) . \tag{10.3.37}$$

The flow (10.3.37) is commonly known as Poiseuille flow.††

For the flow (10.3.37) the volume efflux is given by

$$Q = 2\pi\int_0^a v_z r dr = \frac{\pi m a^4}{8\mu} = \frac{\pi(P_1 - P_2)a^4}{8\mu l} . \tag{10.3.38}$$

This last result, Poiseuille's formula, is well known in elementary physics. The formula readily provides one method for the determination of μ. A typical apparatus is sketched in Fig. 10.3. The horizontal tube of length l and internal radius a is gravity fed by the liquid contained in the large reservoir. The pressure P_2 at the end $z = l$ is atmospheric while at $z = 0$ the pressure is atmospheric plus the pressure head $\rho_0 gH$ where ρ_0 is the liquid density and H is the height of the liquid free surface above the horizontal tube. In terms of H (10.3.38) may be re-written

$$Q = \pi\rho_0 gHa^4/8\mu l . \tag{10.3.39}$$

† Alternately (10.3.34) may be obtained from appealing to the radially symmetric part of the Laplacian operator expressed in plane polar coordinates.

†† A simpler derivation of (10.3.37) directly from (10.3.35) is as follows. Because the axis $r = 0$ is here contained in the field of flow the constant B in (10.3.35) must be chosen to be zero, to avoid the logarithmic singularity. The remaining constant A is then determined by writing $v_z = 0$ on $r = a$.

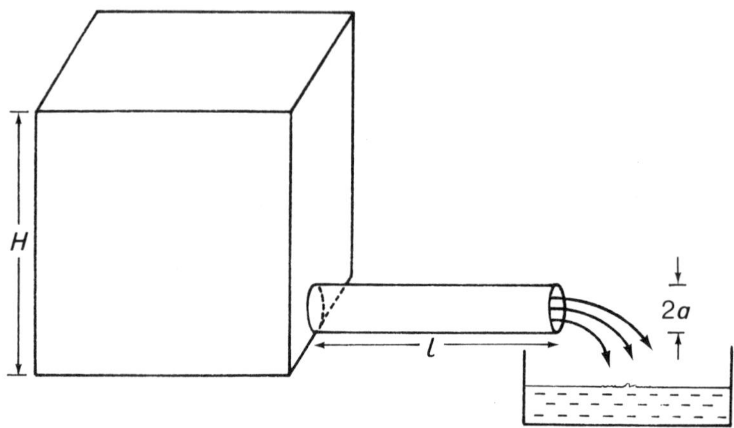

Fig. 10.3 Apparatus for determination of μ

In this formula ρ_0 is determinable, g is known while H, a and l are at the disposal of the experimentalist. The quantity Q is determined from $Q = L/t$ where L is the volume of liquid collected by the cup in time t.

Verification of (10.3.39) for a liquid in respect of the dependence of Q on h, a and l lends considerable, though not conclusive, support to the hypothesis that the liquid behaviour is adequately described by the Stokes' model; if the predicted dependence of Q on h, a and l is so verified then the formula (10.3.39) determines μ.†

In any practical apparatus an upper limit is placed on the radius of the tube in order to avoid instabilities of flow and turbulence. Typically for measurements on materials like glycerol and motor oils the tube is a capillary with radius of the order of a millimetre.

We conclude by determining the drag force F_z exerted by the liquid on the internal cylindrical surface of the tube. In terms of the shear stress σ_{rz}††, which by symmetry is independent of θ,

$$F_z = -\int_0^{2\pi}\int_0^l (\sigma_{rz})_{r=a}\, a\,d\theta dz = -2\pi a \int_0^l (\sigma_{rz})_{r=a} dz\,. \quad (10.3.40)$$

The non-Cartesian physical component of stress σ_{rz} may be obtained from the Cartesian components using the tensor transformation laws as in the previous chapter. However it is quicker to utilise the symmetry result that σ_{rz} is independent of θ, so that in particular

$$(\sigma_{rz})_{r=a} = (\sigma_{xz})_{x=a,\ y=0} = \mu(\partial v_z/\partial x)_{x=a,\ y=0} = -\tfrac{1}{2}ma$$

[from (10.3.37)] (10.3.41)

† However, see the discussion p. 18-19.

†† σ_{rz} is the component of shear stress acting in the z direction on an element of surface whose normal is in the positive r direction.

Combining (10.3.40) and (10.3.41) now yields

$$F_z = \pi m a^2 l = \pi a^2 (P_1 - P_2) . \tag{10.3.42}$$

Of course, in order that the tube remain stationary, it is necessary to exert a counter force, of the same magnitude in the negative z direction.

The result that the force component F_z is equal to the pressure difference $(P_1 - P_2)$ times the cross sectional area of the tube is more generally true for rectilinear shear flow along the bore of tubes of arbitrary cross section – see below.

10.3.5 Rectilinear Flow Along a Tube with Elliptical Cross Section

A particular solution of equation (10.3.4b) is

$$v_z = A[1 - x^2/a^2 - y^2/b^2] \tag{10.3.43}$$

provided the constant A is given by

$$A = ma^2 b^2 / 2\mu(a^2 + b^2) . \tag{10.3.44}$$

The quantity v_z vanishes on the ellipse $x^2/a^2 + y^2/b^2 = 1$ so that the solution found solves the problem of rectilinear shear flow under pressure gradient m along a tube of elliptical cross section.

The efflux calculation is the subject of Problem 2 at the end of the Chapter.

Evaluation of the drag force F_z exerted by the liquid on the tube is complicated by the necessity of determining the shear stress $\sigma_{\xi z}$ where ξ is the normal direction to the ellipse (Fig. 10.4).

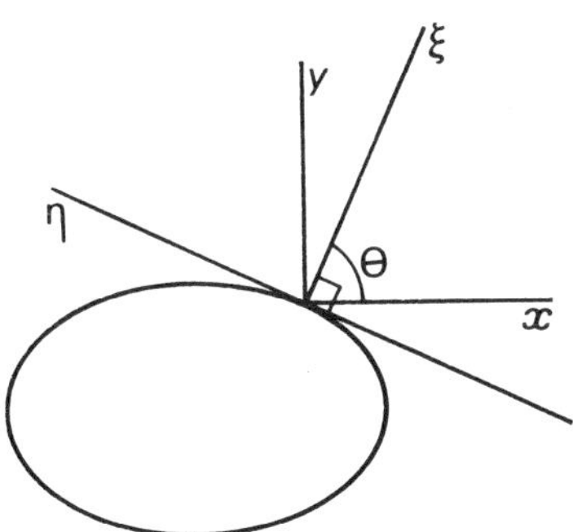

Fig. 10.4 Relation between directions of the x, y and ξ, η axes.

Referred to x, y, z coordinates, for a tube of arbitrary cross section, **d** is given by

$$\mathbf{d} = \tfrac{1}{2}\begin{bmatrix} 0 & 0 & \partial v_z/\partial x \\ 0 & 0 & \partial v_z/\partial y \\ \partial v_z/\partial x & \partial v_z/\partial y & 0 \end{bmatrix} \tag{10.3.45}$$

and $d_{\xi z}$ is given by the (1, 3) [or (3, 1)] component of

$$\mathbf{d}' = \mathbf{A}\mathbf{d}\mathbf{A}^T \tag{10.3.46}$$

where $\mathbf{A}$ is the proper orthogonal matrix

$$\mathbf{A} = \begin{bmatrix} \cos\theta & \sin\theta & 0 \\ -\sin\theta & \cos\theta & 0 \\ 0 & 0 & 1 \end{bmatrix} \tag{10.3.47}$$

relating the coordinates ξ, η, z, to x, y, z, and where θ is the angle between the x and ξ axes.

From (10.3.45), (10.3.46) and (10.3.47) we find

$$d_{\xi z} \equiv d'_{13} = \tfrac{1}{2}[\cos\theta(\partial v_z/\partial x) + \sin\theta(\partial v_z/\partial y)] \tag{10.3.48}$$

so that from the constitutive equations the shear stress $\sigma_{\xi z}$ is given by

$$\sigma_{\xi z} = \mu[\cos\theta(\partial v_z/\partial x) + \sin\theta(\partial v_z/\partial y)] \;. \tag{10.3.49}$$

For the ellipse $x^2/a^2 + y^2/b^2 = 1$

$$\frac{\partial v_z}{\partial x} = -2Ax/a^2 \,, \qquad \frac{\partial v_z}{\partial y} = -\frac{2Ay}{b^2}$$

$$\cos\theta = \frac{x/a^2}{(x^2/a^4 + y^2/b^4)^{\frac{1}{2}}} \,, \qquad \sin\theta = \frac{y/b^2}{(x^2/a^4 + y^2/b^4)^{\frac{1}{2}}}$$

so that

$$\sigma_{\xi z} = -2\mu A(x^2/a^4 + y^2/b^4)^{\frac{1}{2}} = -\frac{ma^2b^2}{(a^2+b^2)}(x^2/a^4 + y^2/b^4)^{\frac{1}{2}}. \tag{10.3.50}$$

Finally the drag force on the tube is in the z direction and of magnitude

$$F_z = -l \oint_{\text{ellipse}} \sigma_{\xi z} ds$$

where the line integral is taken around the ellipse. With the help of the parametric representation $x = a\cos\phi$, $y = b\sin\phi$ together with (10.3.50) we find

$$F_z = \frac{mlab}{(a^2+b^2)} \int_0^{2\pi} (a^2\sin^2\phi + b^2\cos^2\phi)d\phi = \frac{\pi mlab}{a^2+b^2}(a^2+b^2)$$

$$= ml\pi ab = (P_1 - P_2)\pi ab \tag{10.3.51}$$

which result illustrates again the general result that for rectilinear flow along the bore of a tube of arbitrary cross section with area A

$$F_z = (P_1 - P_2)A \;.$$

The proof of the general result for an arbitrary cross section makes use of (10.3.49). Following the arguments given above leads to

$$F_z = -\mu l \oint_C [\cos\theta(\partial v_z/\partial x) + \sin\theta(\partial v_z/\partial y)]\,ds \qquad (10.3.52)$$

where C is the perimeter of the cross section. Equation (10.3.52) may be written as

$$F_z = -\mu \iint_S \text{grad } v_z \,.\, \mathbf{dS} \qquad (10.3.53)$$

where S is the entire surface of the liquid in the tube. In writing down (10.3.53) use has been made of the fact that no contributions to the integral derive from the end surfaces $z = 0$ and $z = l$. [Also the argument leading to (10.3.53) is easily generalised to cover flow along an anuulus bounded by $C_1 = 0$ and $C_2 = 0$].

From the divergence theorem (10.3.53) becomes

$$F_z = -\mu \iiint_\Omega \nabla^2 v_z d\Omega = m \iiint_\Omega d\Omega = m\Omega = (P_1 - P_2)\Omega/l = (P_1 - P_2)A\,.$$

Here us has been made of (10.3.19).

The physical origin of the result is as follows. The material contained within the tube or annulus is in unaccelerated motion since $D\mathbf{v}/Dt = 0$. It follows that the sum of the external forces vanishes. In the z direction the pressures P_1 at $z = 0$ and P_2 at $z = l$ provide a force $(P_1 - P_2)A$ on the liquid. Evidently therefore the surface(s) of the tube (annulus) provided a counter force $-(P_1 - P_2)A$. By the action and reaction law the effect of the liquid on the tube is a force $(P_1 - P_2)A$ in the positive z direction.

10.4 Couette flow

Here we seek solutions of the incompressible Navier-Stokes equations (10.1.8) and (10.1.9) of the form

$$v_\theta = v_\theta(r)\,, \qquad v_r = v_z = 0 \qquad (10.4.1)$$

where r, θ and z are cylindrical polar coordinates.

Equation (10.1.8) is satisfied automatically. Also since $\mathbf{v}$ is independent of t the term $\partial\mathbf{v}/\partial t = 0$. However

$$(\mathbf{v}\,.\,\text{grad})\mathbf{v} = \frac{v_\theta}{r}\frac{\partial}{\partial\theta}[v_\theta(r)\mathbf{i}_\theta] \qquad (10.4.2)$$

where $\mathbf{i}_\theta$ is the unit vector in the θ direction. Using the first of the well known results

$$\frac{d}{d\theta}(\mathbf{i}_\theta) = -\mathbf{i}_r\,, \qquad \frac{d}{d\theta}(\mathbf{i}_r) = \mathbf{i}_\theta \qquad (10.4.3)$$

yields†

$$(\mathbf{v}\,.\,\text{grad})\mathbf{v} = -[v_\theta^2(r)/r]\,\mathbf{i}_r\,. \qquad (10.4.4)$$

† We give here a slightly different derivation of the result (10.4.4) to that of Chapter 9.

The vector $\nabla^2 \mathbf{v}$ can be found similarly.† We have, using equations (10.4.3), and noting that $\mathbf{v}$ is independent of z,

$$\nabla^2 \mathbf{v} = \left(\frac{\partial^2}{\partial r^2} + \frac{1}{r}\frac{\partial}{\partial r} + \frac{1}{r^2}\frac{\partial^2}{\partial \theta^2}\right)[v_\theta(r)\mathbf{i}_\theta(\theta)]$$

$$= \mathbf{i}_\theta \left(\frac{d^2}{dr^2} + \frac{1}{r}\frac{d}{dr} - \frac{1}{r^2}\right) v_\theta(r) \;. \tag{10.4.5}$$

In the absence of body forces and using the result that P is independent of θ (from symmetry arguments), the Navier-Stokes equations (10.1.9) become in component form

$$-\frac{\partial P}{\partial r} = \rho_0 {v_\theta}^2/r \text{ (a)}; \quad \mu\left(\frac{d^2}{dr^2} + \frac{1}{r}\frac{d}{dr} - \frac{1}{r^2}\right) v_\theta = 0 \text{ (b)}; \quad -\frac{\partial P}{\partial z} = 0 \text{ (c)} \tag{10.4.6}$$

The last of these equations shows that P depends only on r. The second equation is an ordinary differential equation for v_θ with solution

$$v_\theta = Ar + B/r \tag{10.4.7}$$

where A and B are constants of integration. Finally with the latter determined from appropriate boundary conditions the first equation determines $P(r)$ to within an arbitrary constant. The constant can be assigned only from knowledge of P at some point (or rather radius) in the flow. In what follows we are not concerned further with the pressure.

The result (10.4.7) solves the following problem. A viscous liquid occupies the annular region $a \leqslant r \leqslant b$ between a cylindrical shaft $0 \leqslant r \leqslant a$ rotating at constant angular velocity Ω_1 and a concentric circular tube of internal radius b rotating at constant angular velocity Ω_2 (Fig. 10.5)

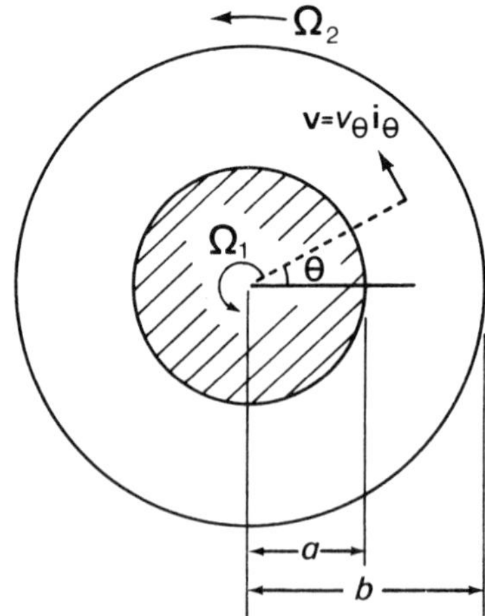

Fig. 10.5 Couette flow

† Alternately $\nabla^2 \mathbf{v}$ can be obtained using, in cylindrical polar coordinates, the vector identity curl (curl $\mathbf{v}$) $=$ grad (div $\mathbf{v}$) $- \nabla^2 \mathbf{v}$.

The boundary conditions

$$v_\theta = a\Omega_1\,, \quad (r = a)\,; \quad v_\theta = b\Omega_2\,, \quad (r = b)$$

determine the constants A and B in (10.4.7). There results

$$A = (b^2\Omega_2 - a^2\Omega_1)/(b^2 - a^2)\,, \quad B = (\Omega_1 - \Omega_2)a^2b^2/(b^2 - a^2)\,. \tag{10.4.8}$$

From the second of these we note the result that for equal angular velocities $\Omega_1 = \Omega_2 = \Omega$ (say), $A = \Omega$ and $B = 0$ so that (10.4.7) becomes $v_\theta = \Omega r$ and the motion of the liquid is that of a rigid body motion without deformation.

It is illuminating to investigate the energy balance aspects of the motion and we give the detailed calculations. Since no volume or kinetic energy changes occur we anticipate that the energy dissipated in the fluid is balanced by the energy supplied to the rotating shaft and outer cylinder. To determine D we write (from Fig. 10.5)

$$v_x = -v_\theta \sin\theta = -yv_\theta(r)/r$$
$$v_y = v_\theta \cos\theta = xv_\theta(r)/r$$

whence

$$d_{xx} = -(xy/r)\frac{d}{dr}(v_\theta/r)\,, \quad d_{yy} = (xy/r)\frac{d}{dr}(v_\theta/r)\,,$$

$$d_{xy} = d_{yx} = \tfrac{1}{2}(x^2 - y^2)\frac{1}{r}\frac{d}{dr}(v_\theta/r)\,, \quad d_{xz} = d_{zx} = d_{yz} = d_{zy} = d_{zz} = 0$$

and $D = 2\mu d_{ij}d_{ij} = 2\mu[d_{xx}{}^2 + d_{yy}{}^2 + 2d_{xy}{}^2] = \mu r^2\left[\frac{d}{dr}(v_\theta/r)\right]^2 = 4\mu B^2/r^4$.

As expected D does not depend on the constant A in (10.4.7) associated with the rigid body motion.

The total rate of energy dissipation per unit length in the z direction is

$$2\pi\int_a^b Drdr = 4\pi\mu B^2\left(\frac{1}{a^2} - \frac{1}{b^2}\right) = 4\pi\mu a^2b^2(\Omega_1 - \Omega_2)^2/(b^2 - a^2) \tag{10.4.9}$$

where we have substituted for B from the second of (10.4.8).

To calculate the rate of supply of energy entails evaluating the couple exerted on each of the rotating cylinders. Evidently because of the symmetry, the stress $\sigma_{r\theta}$ which determines the couple is independent of θ. Accordingly $\sigma_{r\theta}$ may be determined from

$$\sigma_{r\theta} = (\sigma_{xy})_{x=r,\,y=0} = 2\mu(d_{xy})_{x=r,\,y=0}$$
$$= \mu\left[(x^2 - y^2)r^{-1}\frac{d}{dr}(v_\theta/r)\right]_{x=r,\,y=0} = -2\mu B/r^2\,.$$

The couple exerted by the fluid per unit length in the z direction on the cylinder $0 \leqslant r \leqslant a$ is

$$\int_0^{2\pi} (\sigma_{r\theta})_{r=a}(a) a d\theta = -4\pi\mu B .$$

The counter couple per unit length required to maintain the motion is

$$M = 4\pi\mu B$$

and this couple works at a rate $M\Omega_1$. A similar calculation shows that the counter couple required to maintain the motion of the outer cylinder is $-M$ working at a rate of $-M\Omega_2$. The total rate of working of the external forces per unit length in the z direction is therefore

$$M(\Omega_1 - \Omega_2) = 4\pi\mu B(\Omega_1 - \Omega_2) = 4\pi\mu(\Omega_1 - \Omega_2)^2 a^2 b^2/(b^2 - a^2)$$

in agreement with (10.4.9).

10.5 Unsteady rectilinear flow

All the solutions discussed in the preceeding sections of this chapter have been concerned with stationary flow situations for which $\partial \mathbf{v}/\partial t = 0$. We conclude with an account of unsteady rectilinear shear flow with a velocity field of the form

$$v_z = v_z(x, y, t), \quad v_x = v_y = 0 . \tag{10.5.1}$$

Since div $\mathbf{v} = 0$, the motion is isochoric. Also although $\partial \mathbf{v}/\partial t \neq 0$, the quantity ($\mathbf{v}$. grad)$\mathbf{v}$ is zero as in Section 10.3. Accordingly, in the absence of body forces, the Navier-Stokes equations assume the form

$$\left.\begin{aligned} -\partial P/\partial x = 0 \text{ (a)}, \quad -\partial P/\partial y = 0 \text{ (b)}, \\ -\partial P/\partial z + \mu\nabla^2 v_z = \rho_0 \partial v_z/\partial t \text{ (c)} \end{aligned}\right\} \tag{10.5.2}$$

where ∇^2 is the two dimensional Laplacian $\partial^2/\partial x^2 + \partial^2/\partial y^2$. These equations differ from equations (10.3.2) only in the appearance of the additional term $\rho_0 \partial v_z/\partial t$ in the third equation. Also in (10.5.2) P in general depends on time as well as the space variables.

For compressible materials undergoing isochoric deformations grad $P = 0$ and the first two of equations (10.5.2) are satisfied identically while (10.5.2c) becomes

$$\nabla^2 v_z = (\rho_0/\mu)\partial v_z/\partial t . \tag{10.5.3}$$

In the absence of pressure gradients in incompressible materials v_z satisfies the same equation.

Slightly more generally for incompressible materials, equations (10.5.2) admit a solution for which $P = P(z, t)$ and where

$$\frac{\partial P}{\partial z} = -m(t) \qquad (10.5.4)$$

$$\mu \nabla^2 v_z = \rho_0 \partial v_z / \partial t - m(t) \qquad (10.5.5)$$

where m is an arbitrary time dependent pressure gradient in the z direction.

Equation (10.5.4) integrates in the form

$$P = P_1(t) - m(t)z$$

determining the pressure, while (10.5.5) provides an equation for v_z.

Equation (10.5.5) is reduced easily to (10.5.3) by the substitution

$$v_z = u + \rho_0{}^{-1} \int m(t')dt' \qquad (10.5.6)$$

where u is a new dependent variable. Thus for either case the basic mathematical problem of unsteady rectilinear flow is concerned with the solution of the homogeneous equation (10.5.3).†

Here we concern ourselves solely with problems governed directly by (10.5.3), i.e. problems for which there are no pressure gradients.

[Equation (10.5.3) is one of the archetypal equations of classical applied mathematics and mathematical physics. With ∇^2 symbolising in general the three dimensional Laplacian the equation

$$k \nabla^2 \psi = \partial \psi / \partial t \qquad (10.5.7)$$

is known as the **heat conduction** or **diffusion** equation. The origin of the terminology is that based on the simplest models, the equation governs both the variation of temperature and the variations of concentration of one substance diffusing through another. Of course the constant k is defined by very different physical parameters in the two cases.

The other archetypal equations of classical applied mathematics are the wave equation

$$\nabla^2 \psi = \frac{1}{c^2} \frac{\partial^2 \psi}{\partial t^2}, \qquad (10.5.8)$$

where c plays the role of a velocity of propagation, and Laplace's equation

$$\nabla^2 \psi = 0 \, . \qquad (10.5.9)$$

† In practice the substitution (10.5.6) may not be the simplest way of attempting a solution of (10.5.5) for a specific problem. For example see the method of solution given by Batchelor (*An Introduction to Fluid Dynamics* C.U.P 1967, p. 193-195) for a solution of the problem of the motion of a liquid, contained in a circular tube, which is at rest for $t < 0$ and subject to constant pressure gradient for $t > 0$.

In particular the latter equation arises from either of (10.5.7) or (10.5.8) for the case where ψ is independent of time. In certain circumstances Laplace's equation arises even when ψ is time dependent as for example in incompressible inviscid potential flow theory discussed in Chapters 11 and 12.

Equations (10.5.7), (10.5.8) and (10.5.9) arise in all branches of classical physics, and methods for solution of these and closely related equations† have been much studied. In the present text all three equations arise naturally and some of their practical methods of solution are discussed.]

We return to (10.5.3). The simplest problems in the present context are those for which v_z depends on the single space variable x when the equation becomes

$$\nu \frac{\partial^2 v_z}{\partial x^2} = \frac{\partial v_z}{\partial t} \tag{10.5.10}$$

where

$$\nu = \mu/\rho_0 \tag{10.5.11}$$

is called the kinematic viscosity.

Equation (10.5.10) describes the motion of an infinite layer of liquid $0 \leqslant x \leqslant h$ which adheres to planes $x = 0$ and $x = h$ both of which are in transverse motion in the z direction. If the plane $x = 0$ moves at a velocity $v_1(t)$ and the plane $x = h$ moves at $v_2(t)$, then evident boundary conditions for the solution of (10.5.10) are

$$\left.\begin{aligned} v_z(0, t) &= v_1(t) \\ v_z(h, t) &= v_2(t) \end{aligned}\right\} \tag{10.5.12}$$

and these conditions are sufficient to provide a unique solution.

A somewhat simpler problem concerns an infinitely thick layer $x > 0$ bounded below by the plane $x = 0$ which is in transverse motion in the x direction with velocity $v_1(t)$. One boundary condition is

$$v_z(0, t) = v_1(t) \tag{10.5.13}$$

as before, but the second condition, dictated by the form of the solution, is that the velocity field v_z behaves in a reasonable physical fashion as $x \to \infty$. We consider the particular case where $v_1(t)$ is oscillatory. In the first instance we assume that v_1 is complex so that

$$v_z(0, t) = v_0 e^{i\omega t} \qquad (-\infty \leqslant t \leqslant \infty) \tag{10.5.14}$$

where ω is a positive constant; also v_0 is a constant. There is no loss of generality in this seemingly artificial procedure for reasons which will become apparent.

A solution of (10.5.10) which satisfies (10.5.14) is obtained by looking for periodic solutions of the form

$$v_z = f(x) e^{i\omega t} \tag{10.5.15}$$

† A closely related equation is Poisson's equation $\nabla^2 \phi = Q$ where Q is known. One simple example of Poisson's equation is (10.3.19).

where
$$f(0) = v_0 \ . \tag{10.5.16}$$
Substituting (10.5.15) into (10.5.10) and cancelling a common factor $e^{i\omega t}$ leads to the ordinary differential equation
$$f''(x) - \frac{i\omega}{\nu} f(x) = 0 \tag{10.5.17}$$
with solution
$$f(x) = Ae^{(1 + i)\Omega x} + Be^{-(1 + i)\Omega x} \tag{10.5.18}$$
where
$$\Omega(\omega) = (\omega/2\nu)^{\frac{1}{2}} \ . \tag{10.5.19}$$
Combining (10.5.15) and (10.5.18) now leads to
$$v_z = (Ae^{(1 + i)\Omega x} + Be^{-(1 + i)\Omega x})e^{i\omega t}. \tag{10.5.20}$$

It appears that there is only one boundary condition [namely (10.5.16)] to determine the two constants A and B. However the first term in (10.5.20) contains an exponentially increasing factor $e^{\Omega x}$ and this implies unacceptable physical behaviour for large x. This behaviour is suppressed with the choice $A = 0$ and the condition (10.5.16) then determines $B = v_0$. Thus the complex solution of the original equation (10.5.10) which satisfies the complex boundary condition (10.5.14) is
$$v_z = v_0 e^{-\Omega x} e^{i(\omega t - \Omega x)} \ . \tag{10.5.21}$$
The solution of (10.5.10) which satisfies the real boundary condition
$$v_z(0, t) = v_0 \cos(\omega t)$$
is given by taking the real part of the right side of (10.5.21)

viz.
$$v_z = v_0 e^{-\Omega x} \cos(\omega t - \Omega x) \ . \tag{10.5.22}$$
Similarly the imaginary part of the right side of (10.5.21)
$$v_z = v_0 e^{-\Omega x} \sin(\omega t - \Omega x) \tag{10.5.23}$$
satisfies the boundary condition
$$v_z(0, t) = v_0 \sin(\omega t) \ .$$
Of course these two solutions are essentially the same, the second being obtained from the first by the transformation
$$\omega t \to \omega t - \tfrac{1}{2}\pi \ .$$
It is important to realise that steady state oscillatory solutions like (10.5.21), (10.5.23) are valid solutions only for the entire time range $-\infty < t < \infty$. It is incorrect, for example, to believe that the solution to the problem specified by
$$v_z(0, t) = \begin{cases} v_0 \cos(\omega t) & (t \geqslant 0) \\ 0 & (t < 0) \end{cases} \tag{10.5.24}$$
$$v_z(x, t) = 0 \quad t < 0,$$

is given by
$$v_z(x, t) = \begin{cases} v_0 e^{-\Omega x}\cos(\omega t - \Omega x) & (t \geqslant 0) \\ 0 & (t < 0) \end{cases} \tag{10.5.25}$$

The reason is that for almost all x, the function $v_z(x, t)$ specified by (10.5.25) fails to satisfy the physically necessary continuity condition $v_z(x, 0) = 0$ for $x > 0$.

It appears therefore that we have managed to solve what is at best a rather artificial problem for which the boundary condition is that for all t in the range $-\infty < t < \infty$, $v_z(0, t) = v_0 \cos(\omega t)$. However the real value of periodic steady state solutions of linear partial differential equations lies in being able to use a sum of such solutions, with different values of ω, to synthesise the solution of transient problems for which the boundary conditions are not periodic.

Consider for example the infinitely thick layer problem with *initial* conditions

$$v_z(x, t) = 0, \quad t < 0 \tag{10.5.26}$$

and with boundary condition $v_z(0, t) = v_1(t)$ (10.5.27)

where $v_1(t)$ is a specified function which vanishes for $t < 0$.

To find solutions of the problem specified above we make use of the complex periodic solutions proportional to $e^{i\omega t}$. For $\omega > 0$, physically acceptable solutions are [from (10.5.21)]

$$v_z = F(x, t, \omega) = e^{-\Omega(\omega)x + i[\omega t - \Omega(\omega)x]} \quad (\omega > 0) \tag{10.5.28}$$

where $\Omega(|\omega|)$ is given by (10.5.19) and where although in the first instance ω is a fixed parameter it is convenient to render explicit the dependence of F on ω.†
For $\omega < 0$ the acceptable solutions are

$$\begin{aligned} v_z = F(x, t, \omega) &= e^{-\Omega(|\omega|)x + i[\omega t + \Omega(|\omega|)x]} \quad (\omega < 0) \\ &= \bar{F}(x, t, -\omega)\ . \end{aligned} \tag{10.5.29}$$

The functions $F(x, t, \omega)$, regarded as a basic set of functions in terms of which more general solutions can be obtained are often called *eigenfunctions*; the quantity ω is then known as an *eigenvalue*. The word eigenvalue is used here in a somewhat different sense from that of Chapter 4 in the context of matrices.

Because (10.5.10) is a linear homogeneous equation it is possible to add distinct solutions, characterised by different values of ω, to obtain a more general solution. Such a general solution, utilising all the eigenfunctions $F(x, t, \omega)$ for every value of ω in the range $-\infty < \omega < \infty$, is given by the complex Fourier integral

$$v_z(x, t) = \int_{-\infty}^{\infty} B(\omega)F(x, t, \omega)d\omega \tag{10.5.30}$$

where the quantity $B(\omega)$ is a function of ω alone. When, as here, ω ranges continuously in the general solution, the eigenvalues are said to form a continuous

† It is convenient now to regard $\Omega = (|\omega|/2\nu)^{1/2}$ as a function of $|\omega|$ for positive and negative ω.

spectrum, and the general solution involves an integral over the spectrum. In other circumstances for other problems and other linear partial differential equations, allowable eigenvalues may form a discrete spectrum $\omega_1, \omega_2 \ldots \omega_n \ldots$ and the general solution then entails a sum instead of an integral. An example of such a problem is that of the transverse vibrations of a cantiliver beam considered in Section 14.6.

Usually the quantity $B(\omega)$ is complex, but, in order for v_z to be real, $B(\omega)$ necessarily satisfies the relation $B(-\omega) = \overline{B(\omega)}$. $B(\omega)$ is determined by the boundary condition (10.5.27). For $x = 0$, $F(x, t, \omega)$ degenerates to $e^{i\omega t}$ and the condition (10.5.27) is

$$\int_{-\infty}^{\infty} B(\omega)e^{i\omega t} = v_1(t)$$

which is an integral equation for $B(\omega)$. From the basic result in the theory of Fourier integrals

$$B(\omega) = (2\pi)^{-1} \int_{-\infty}^{\infty} v_1(t)e^{-i\omega t}dt \qquad (10.5.31)$$

determining $B(\omega)$. With $B(\omega)$ determined the integral (10.5.30) then yields the desired solution.

The solution for arbitrary $v_1(t)$ may be put in more explicit form. Consider first the *impulsive* problem

$$v_1(t) = \delta(t) \qquad (10.5.32)$$

where $\delta(t)$ is the improper Dirac delta function defined by

$$\delta(t) = 0 \qquad t \neq 0$$

$$\int \delta(t')dt' = \begin{cases} 1 & \text{when range of integration includes origin} \\ 0 & \text{otherwise.} \end{cases}$$

The delta function also possesses the important property

$$\int \chi(t')\delta(t')dt' = \begin{cases} \chi(0) & \text{when range of integration includes origin} \\ 0 & \text{otherwise} \end{cases}$$

where χ is any continuous function. An equivalent result used subsequently is

$$\int \chi(t')\delta(t - t')dt' = \chi(t)$$

when the range of t' includes t.

For the impulse (10.5.32) the Fourier spectrum $B(\omega)$ is well known

$$B(\omega) = (2\pi)^{-1} \int_{-\infty}^{\infty} \delta(t) e^{-i\omega t} dt = (2\pi)^{-1}$$

so that (10.5.31) becomes

$$v_z(x, t) \equiv (2\pi)^{-1} \int_{-\infty}^{\infty} F(x, t, \omega) d\omega \tag{10.5.33}$$

$$= \pi^{-1} \operatorname{Re} \int_{0}^{\infty} F(x, t, \omega) d\omega$$

where, in obtaining the last line we have taken $t > 0$ and made use of the property $F(x, t, -\omega) = \bar{F}(x, t, \omega)$.† Substituting from (10.5.28) for F and from (10.5.19) for Ω yields

$$v_z = \pi^{-1} \operatorname{Re} \int_{0}^{\infty} e^{-(\omega/2\nu)^{1/2} x + i[\omega t - (\omega/2\nu)^{1/2} x]} d\omega .$$

The real part of the definite integral may be evaluated using complex variable methods and the well known result

$$\int_{0}^{\infty} e^{-\xi^2} d\xi = \tfrac{1}{2}\pi^{\frac{1}{2}} . \tag{10.5.34}$$

The details, including proof of (10.5.34), are to be found in the Appendix to the chapter. There results

$$v_z = G(x, t)$$

where

$$G(x, t) = \tfrac{1}{2}(\pi\nu)^{-\frac{1}{2}} x t^{-\frac{3}{2}} e^{-x^2/4\nu t}. \tag{10.5.35}$$

Evidently $G(x, t) \to 0$ as $t \to 0+$ so that the solution (10.5.35) matches on to the initial condition that the material is undisturbed. Also $G(0, t) = 0$ except for $t = 0$ and this satisfies (10.5.32) for t other than zero. To verify that for $x = 0$, G in (10.5.35) behaves like a delta function, it is sufficient to consider the integral of $G(x, t)$ with respect to t over the range $0 \leqslant t \leqslant \infty$. It is left to the reader to show [using (10.5.34)]

$$\int_{0}^{\infty} G(x, t) dt = 1$$

† With $t < 0$, the Fourier integral (10.5.33) vanishes.

for all $x > 0$ and hence in particular for $x \to 0$,

$$Lt_{x \to 0} \int_0^\infty G(x, t)dt = 1$$

Regarded as a function of t, the quantity $G(x, t)$ is a positive quantity with a single maximum of $3(6/\pi e^3)^{\frac{1}{2}} \nu x^{-2}$ occurring at $t = x^2/6\nu$. The width of the profile, measured for example as the distance between the two points where the amplitude is (say) one half the maximum, increases proportionally with x^2. Thus $G(x, t)$ represents a pulse, initially sharp – in fact a delta function in time located at the origin – which propagates into the region $x > 0$ with increasing width and decreasing amplitude. This *diffusive* behaviour, together with the result that disturbances initiated in a region are felt instantaneously in infinitely remote regions, is characteristic of solutions of the diffusion equation.

Generalisation of the impulse solution (10.5.35) to the more general case of arbitrary $v_1(t)$ is accomplished by analysing the input data $v_1(t)$ into a sequence of delta functions of different amplitudes. In mathematical terms this is accomplished by writing

$$v_1(t) = \int_{-\infty}^{T} v_1(t')\delta(t - t')dt' \qquad (10.5.36)$$

where T the upper limit of the integral, is chosen to be greater than t. Choosing the lower limit to be $-\infty$ allows the possibility of $v_1(t')$ being non-zero for $t' < 0$. Of course if $v_1(t') = 0$ for $t' < 0$ it is possible to replace the lower limit in (10.5.36) by zero.

Each component $\qquad v_1(t')\delta(t - t')dt'$

of the integral (10.5.36) gives rise to the contribution

$$v_1(t')G(x, t - t')dt'$$

for $t > t'$. There is no contribution from $v_1(t')\delta(t - t')dt'$ for $t < t'$ since these times occur prior to the delta function pulse. Adding up all contributions to $v_z(x, t)$ from $t' = -\infty$ to the current time $t' = t$ then yields the solution

$$v_z = \int_{-\infty}^{t} v_1(t')G(x, t - t')dt' \qquad (10.5.37)$$

for the problem specified by the boundary condition (10.5.13).

Of course to proceed further requires $v_1(t')$ to be specified explicitly. However the character of the solution is dominated by the diffusive character of the function G.

[A function like $G(x, t)$ which solves a linear problem for an impulsive boundary condition, and then yields a solution for a more general boundary condition through an integral such as (10.5.37), may be termed, somewhat loosely, a Green's function. The significance of the slight reservation is that the term Green's function is commonly reserved to describe the solutions of problems where the delta function impulsive behaviour is associated with space variables rather than with time, as in the above example. However the ideas which permeate the concept of a Green's function are equally realised whether the impulsive behaviour is in space or time.

Perhaps the best known example of a Green's function is provided by the formal solution of the Poisson equation

$$\nabla^2 \phi = \rho/\epsilon_0 \tag{10.5.38}$$

in electrostatics. Here ϕ is the electrostatic potential, ϵ_0 is the (constant) specific permittivity and $\rho(\mathbf{r})$ is the charge density. We consider the case where it is required to solve (10.5.38) for an infinite space which is empty save for the presence of the space charge specified by ρ. A formal solution to (10.5.38) is readily constructed from a knowledge of the point charge solution. For a point charge q located at the origin of an otherwise empty space, the potential is $q/4\pi\epsilon_0|\mathbf{r}|$. For a point charge located at $\mathbf{r}'$ therefore, the potential is $q/4\pi\epsilon_0|\mathbf{r} - \mathbf{r}'|$. Thus in the case of a distribution of charge with density ρ, the contribution to ϕ from the element of charge $\rho(\mathbf{r}')d\Omega(\mathbf{r}')$, located in the volume element $d\Omega(\mathbf{r}')$, is $\rho(\mathbf{r}')d\Omega(\mathbf{r}')/4\pi\epsilon_0|\mathbf{r} - \mathbf{r}'|$. Finally adding up all such contributions yields

$$\phi = \iiint_{\Omega} \rho(\mathbf{r}')G(\mathbf{r} - \mathbf{r}')d\Omega(\mathbf{r}')$$

where

$$G(\mathbf{r}) = [4\pi\epsilon_0|\mathbf{r}|]^{-1}$$

plays the role of a Green's function. In fact G is the solution of (10.5.38) for the case when ρ assumes the delta function character

$$\rho = \delta(x - x')\delta(y - y')\delta(z - z') \text{ .]}$$

The Fourier integral technique used above to derive the impulse solution (10.5.35) is a common method for the solution of linear partial differential equations which contain overt reference to time only through the appearance of time derivatives – usually only the first and second derivatives $\partial/\partial t$ and $\partial^2/\partial t^2$ are involved. Most linear equations of classical mathematical physics and applied mathematics are of this character. In fact many of the partial differential equations studied in this book are linear equations. In general the equations of continuum mechanics are non-linear. The linear equations that occur in the subject are either a consequence of simplified geometrical situations, as in the present chapter, or are derived after various approximations have been invoked. For problems defined by non-linear equations the Fourier integral method, which devolves crucially on the possibility of superposing periodic solutions, necessarily fails.† Perhaps it should also be

† Because for non-linear equations it is not possible to obtain a solution by superposition of other solutions.

mentioned that the concept of a Green's function is also limited to linear equations, for again the technique relies explicitly on the superposition principle.

10.6 Appendix

DERIVATION OF (10.5.35)

It is required to evaluate

$$G(x, t) = \pi^{-1} \operatorname{Re} \int_0^\infty e^{-(\omega/2\nu)^{\frac{1}{2}}x + i[\omega t - (\omega/2\nu)^{\frac{1}{2}}]} d\omega .$$

The change of variable from ω to z where $\omega = 2\nu z^2$ leads to

$$G(x, t) = \pi^{-1} \operatorname{Re} \int_0^\infty 4\nu z e^{-z(1+i)x + 2i\nu t z^2} dz$$

$$= \pi^{-1} \operatorname{Re} \int_0^\infty \left[\frac{4i\nu t z - (1+i)x}{it}\right] e^{-z(1+i)x + 2i\nu t z^2} dz$$

$$+ \pi^{-1} \operatorname{Re} \int_0^\infty \frac{1+i}{i} x t^{-1} e^{-z(1+i) + 2i\nu t z^2} dz .$$

The first integral is evaluated trivially to yield

$$-\pi^{-1} \operatorname{Re} (it)^{-1} = 0$$

so that $$G(x, t) = \pi^{-1} \operatorname{Re} (1-i)xt^{-1} \int_0^\infty e^{-z(1+i)x + 2i\nu t z^2} dz . \tag{10.6.1}$$

We now regard $z = q_1 + iq_2$ as a complex variable and transform the path of integration in (10.6.1) from the real axis to the line $q_1 = q_2$. This is permissable because in the sector $0 \leqslant \arg z \leqslant \frac{1}{4}\pi$ the integrand in (10.6.1) is a regular function. Also in this sector the integrand tends to zero exponentially as $|z| \to \infty$. Thus writing $z = (1+i)q$ in (10.6.1) yields

$$G(x, t) = 2\pi^{-1} x t^{-1} \operatorname{Re} \int_0^\infty e^{-2iqx - 4\nu t q^2} dq .$$

The substitution $q = \xi/2(\nu t)^{\frac{1}{2}}$ now yields

$$G(x, t) = \pi^{-1}\nu^{-\frac{1}{2}}xt^{-\frac{3}{2}} \text{ Re} \int_0^\infty e^{-\xi^2 - ix\xi/(\nu t)^{1/2}} d\xi$$

$$= \pi^{-1}\nu^{-\frac{1}{2}}xt^{-\frac{3}{2}}e^{-x^2/4\nu t} \text{ Re} \int_0^\infty e^{-(\xi + ix/2(\nu t)^{1/2})^2} d\xi$$

$$= \pi^{-1}\nu^{-\frac{1}{2}}xt^{-\frac{3}{2}}e^{-x^2/4\nu t} \text{ Re} \int_{i\alpha}^{\infty + i\alpha} e^{-\zeta^2} d\zeta \qquad (10.6.2)$$

where $\alpha = x/2(\nu t)^{\frac{1}{2}}$ and where the path of the integral is along the straight line $\text{Im}(\zeta) = i\alpha$, $0 \leqslant \text{Re}\,\zeta \leqslant \infty$. The integrand (10.6.2) is regular in the region $\text{Re}\,\zeta \geqslant 0$, $0 \leqslant \text{Im}(\zeta) \leqslant \alpha$ and also the integrand tends to zero exponentially as $|\zeta| \to \infty$ within this region. If follows that the integral in (10.6.2) can be replaced by a line integral along the real axis together with an integral along the imaginary axis from $\zeta = i\alpha$ to $\zeta = 0$

viz.
$$\int_{i\alpha}^{\infty + i\alpha} e^{-\zeta^2} d\zeta = \int_{i\alpha}^{0} e^{-\zeta^2} d\zeta + \int_0^\infty e^{-\zeta^2} d\zeta \; .$$

This substitution $\zeta = ip$ where p is real shows the first integral to be purely imaginary. The second integral is $\frac{1}{2}\sqrt{\pi}$ so that finally

$$G(x, t) = \tfrac{1}{2}(\pi\nu)^{-\frac{1}{2}}xt^{-\frac{3}{2}}e^{-x^2/4\nu t} \; . \qquad (10.6.3)$$

An elementary proof of the result $\int_0^\infty e^{-\zeta^2} d\zeta = \frac{1}{2}\pi^{\frac{1}{2}}$ is as follows.

Consider the double integral

$$\int_0^\infty\int_0^\infty e^{-(x^2 + y^2)} dx dy = \int_0^\infty e^{-x^2} dx \int_0^\infty e^{-y^2} dy = \left(\int_0^\infty e^{-x^2} dx\right)^2 .$$

On changing from Cartesian to plane polar coordinates r and θ the double integral becomes

$$\int_0^{\pi/2} d\theta \int_0^\infty re^{-r^2} dr = \tfrac{1}{4}\pi$$

from which follows the desired result.

Problems, Chapter 10

1. An incompressible Stokes' liquid is driven through the annulus $a \leqslant (x^2 + y^2)^{\frac{1}{2}} \leqslant b$ in the z direction by a uniform pressure gradient

$$\frac{dP}{dz} = -m \, .$$

If the velocity field is assumed to be of the form

$$v_x = v_y = 0 \, , \quad v_z = v_z(r)$$

where $r = (x^2 + y^2)^{\frac{1}{2}}$, and if body forces are neglected, show that v_z satisfies

$$\frac{d^2 v_z}{dr^2} + \frac{1}{r}\frac{dv_z}{dr} = -\frac{m}{\mu}$$

with general solution $v_z = A + B \log r - \frac{1}{4}(m/\mu)r^2$.

If A and B are chosen to satisfy the non-slip conditions

$$v_z = 0 \quad \text{for} \quad r = a \quad \text{and} \quad r = b$$

derive $v_z = \frac{1}{4}(m/\mu)[(a^2 - r^2) + (b^2 - a^2)\{\log(r/a)/\log(b/a)\}]$

and deduce that the volume efflux through any section $z = $ const. is

$$Q \equiv 2\pi \int_a^b r v_z dr$$

$$= \frac{\pi m b^4}{8\mu}[1 - q^4 + (1 - q^2)^2(\log q)^{-1}]$$

where $q = (a/b)$.

2. An incompressible Stokes' liquid is driven through a pipe of elliptical cross section bounded by

$$\frac{x^2}{a^2} + \frac{y^2}{b^2} = 1 \, .$$

The pressure gradient is uniform and given by

$$\frac{dP}{dz} = -m \, .$$

Show that in the absence of body forces the Navier-Stokes equations are satisfied by the assumption of a velocity field

$$v_x = v_y = 0 \, , \quad v_z = v_z(x, y)$$

where

$$\frac{\partial^2 v_z}{\partial x^2} + \frac{\partial^2 v_z}{\partial y^2} = -\frac{m}{\mu} \, .$$

Verify that a solution of this equation which satisfies the non-slip condition $v_z = 0$ on the elliptical boundary is

$$v_z = \frac{ma^2 b^2}{2\mu(a^2 + b^2)} [1 - (x/a)^2 - (y/b)^2]$$

and show that the volume efflux through any section z = const. is given by

$$Q \equiv \iint_S v_z dxdy = \frac{1}{4} \frac{\pi m a^3 b^3}{\mu(a^2 + b^2)}$$

where
$$S \equiv \left\{ \frac{x^2}{a^2} + \frac{y^2}{b^2} \leqslant 1 \right\}$$

[Hint. – The double integral is most readily evaluated by choosing new integration variables (r, θ) where $x = ar\cos\theta$, $y = br\sin\theta$.]

3. A Stokes' liquid is confined to the annulus $a \leqslant r \leqslant b$ between a (long) solid cylinder $0 \leqslant r \leqslant a$ and a rigid cylindrical wall $r = b$. The solid cylinder moves in the axial (positive z) direction at a uniform speed v_0 while the rigid wall remains at rest. Show that in the absence of body forces and pressure gradients the Navier-Stokes equations are satisfied by the assumption of a velocity field

$$v_x = v_y = 0\,, \quad v_z = v_z(r)$$

where
$$\left(\frac{d^2}{dr^2} + \frac{1}{r}\frac{d}{dr} \right) v_z = 0\,.$$

Determine the solution of this equation satisfying the boundary conditions

$$v_z = 0\,, \quad r = b$$

$$v_z = v_0\,, \quad r = a$$

and show that the force on the solid cylinder required to maintain the motion is in the positive z direction and of magnitude

$$2\pi\mu v_0 [\log(b/a)]^{-1}$$

per unit length of the solid cylinder.

The rate of working of this force is

$$2\pi\mu v_0^2 [\log(b/a)]^{-1} \qquad *$$

per unit length. Show that (*) is also the rate at which energy is dissipated by the fluid per unit length in the z direction.

4. Show that in the Couette flow problem of Section 10.4 the pressure is given by

$$P = -\rho_0 [\tfrac{1}{2}A^2r^2 - \tfrac{1}{2}B^2r^{-2} + 2AB \log r + C]$$

where C is a constant of integration while A and B are the constants entering the solution for v_θ

i.e. $$v_\theta = Ar + Br^{-1} .$$

5. In the slow viscous flow approximation of a Stokes' fluid the acceleration terms $\rho D\mathbf{v}/Dt$ are ignored. Show that in this approximation $\mathbf{v}$ and P satisfy

$$\operatorname{div} \mathbf{v} = 0 \quad \text{(i)}, \quad -\operatorname{grad} P + \mu\nabla^2\mathbf{v} = 0 \text{ (ii)}$$

for an incompressible fluid.

Deduce that P is harmonic

i.e. $$\nabla^2 P = 0 .$$

Show that for plane two dimensional flow $[v_x = v_x(x, y), v_y = v_y(x, y), v_z = 0]$, equation (i) is satisfied by assuming the existence of an $\psi(x, y)$ such that

$$v_x = -\frac{\partial \psi}{\partial y}, \quad v_y = \frac{\partial \psi}{\partial x} .$$

The function $\psi(x, y)$ is called the stream function. The origin of the terminology is simply that in plane two dimensional flows the curves $\psi(x, y) = \text{const.}$ are streamlines. Prove this result by showing that $\mathbf{v} \cdot \operatorname{grad} \psi = 0$ and recalling that $\operatorname{grad} \psi$ is a vector normal to the curve $\psi = \text{const.}$

The stream function plays an important role in two dimensional flows. In inviscid irrotational flow theory ψ satisfies the two dimensional Laplace equation (see Chapter 12). In contrast, in the present slow viscous flow approximation, ψ satisfies the biharmonic equation. Prove this result by noting that (ii) may be written

$$\frac{\partial P}{\partial x} = -\mu \frac{\partial}{\partial y}(\nabla^2 \psi), \quad \frac{\partial P}{\partial y} = \mu \frac{\partial}{\partial x}(\nabla^2 \psi)$$

and hence deduce $$\nabla^2(\nabla^2 \psi) = 0 \text{ (iii)}$$

where $$\nabla^2 = \frac{\partial^2}{\partial x^2} + \frac{\partial^2}{\partial y^2}$$

is the two dimensional Laplacian operator.

The biharmonic equation (iii) appears also in the theory of plane strain [see Problem 17 of Chapter 14.].

6. AXISYMMETRIC INCOMPRESSIBLE SLOW VISCOUS FLOW IN CYLINDRICAL POLAR COORDINATES

In cylindrical polar coordinates (r, θ, z) axisymmetric flow is defined by

$$v_r = v_r(r, z), \quad v_z = v_z(r, z), \quad v_\theta = 0 .$$

The incompressibility equation in these coordinates reduces to

$$\frac{\partial v_r}{\partial r} + \frac{v_r}{r} + \frac{\partial v_z}{\partial z} = 0 .$$

(i) Show that this equation is satisfied by

$$v_r = -\frac{1}{r}\frac{\partial \psi}{\partial z}, \quad v_z = \frac{1}{r}\frac{\partial \psi}{\partial r}$$

where $\psi = \psi(r, z)$ is the axisymmetric stream function.

(ii) Deduce from the slow viscous flow equation

$$-\operatorname{grad} P + \mu\nabla^2 \mathbf{v} = 0$$

that

$$\nabla^2(\operatorname{curl} \mathbf{v}) = 0 \tag{i}$$

or equivalently

$$\nabla^2\left[r^{-1}\left\{\frac{\partial^2\psi}{\partial r^2} - \frac{1}{r}\frac{\partial\psi}{\partial r} + \frac{\partial^2\psi}{\partial z^2}\right\}\mathbf{i}_\theta\right] = 0 \tag{ii}$$

where $\mathbf{i}_\theta$ is a unit vector in the θ direction.

(iii) Deduce from (ii) that ψ satisfies the fourth order equation

$$\left(\frac{\partial^2}{\partial r^2} + \frac{1}{r}\frac{\partial}{\partial r} + \frac{\partial^2}{\partial z^2} - \frac{1}{r^2}\right)\left[r^{-1}\left\{\frac{\partial^2\psi}{\partial r^2} - \frac{1}{r}\frac{\partial\psi}{\partial r} + \frac{\partial^2\psi}{\partial z^2}\right\}\right] = 0 \tag{iii}$$

or equivalently

$$\left(\frac{\partial^2}{\partial r^2} - \frac{1}{r}\frac{\partial}{\partial r} + \frac{\partial^2}{\partial z^2}\right)\left(\frac{\partial^2\psi}{\partial r^2} - \frac{1}{r}\frac{\partial\psi}{\partial r} + \frac{\partial^2\psi}{\partial z^2}\right) = 0 . \tag{iv}$$

NOTES: (a) In proceeding from (ii) to (iii) it is important to retain the full expression for the Laplacian operator,

i.e.
$$\nabla^2 = \frac{\partial^2}{\partial r^2} + \frac{1}{r}\frac{\partial}{\partial r} + \frac{\partial^2}{\partial z^2} + \frac{1}{r^2}\frac{\partial^2}{\partial\theta^2}$$

and to recall that $\mathbf{i}_\theta$ depends on θ.

(b) It is of some interest to note that equation (iv) is <u>not</u> of the form

$$\nabla^2(\nabla^2\psi) = 0$$

where ∇^2 is the axisymmetric Laplacian operator in cylindrical polar coordinates

$$\nabla^2 = \frac{\partial^2}{\partial r^2} + \frac{1}{r}\frac{\partial}{\partial r} + \frac{\partial^2}{\partial z^2}$$

so there is not a precise parallel with equation (iii) of Problem 5.

7. AXISYMMETRIC INCOMPRESSIBLE SLOW VISCOUS FLOW IN SPHERICAL POLAR COORDINATES

In spherical polar coordinates (r, θ, ϕ) axisymmetric flow is defined by

$$v_r = v_r(r, \theta), \quad v_\theta = v_\theta(r, \theta), \quad v_\phi = 0 .$$

The incompressibility equation in these coordinates reduces to

$$\frac{\partial v_r}{\partial r} + \frac{2v_r}{r} + \frac{1}{r \sin\theta}\frac{\partial}{\partial \theta}(\sin\theta\, v_\theta) = 0 .$$

(i) Show that this equation is satisfied by choosing

$$v_r = -\frac{1}{r^2 \sin\theta}\frac{\partial \psi}{\partial \theta}, \quad v_\theta = \frac{1}{r\sin\theta}\frac{\partial \psi}{\partial r}$$

where $\psi = \psi(r, \theta)$ is known as the Stokes' stream function.

(ii) In terms of ψ show that curl $\mathbf{v}$ is given by

$$\operatorname{curl}\mathbf{v} = (r\sin\theta)^{-1}\left[\frac{\partial^2\psi}{\partial r^2} + \frac{1}{r^2}\frac{\partial^2\psi}{\partial \theta^2} - \frac{\cot\theta}{r^2}\frac{\partial\psi}{\partial\theta}\right]\mathbf{i}_\phi \qquad \text{(i)}$$

where $\mathbf{i}_\phi$ is a unit vector in the ϕ direction.

(iii) From the slow viscous flow equation

$$-\operatorname{grad} P + \mu\nabla^2\mathbf{v} = 0$$

deduce that

$$\operatorname{curl}\operatorname{curl}(\operatorname{curl}\mathbf{v}) = 0$$

and hence from (i) (and results in vector analysis for spherical polar coordinates) show that ψ satisfies the fourth order equation

$$\left(\frac{\partial^2}{\partial r^2} + \frac{1}{r^2}\frac{\partial^2}{\partial\theta^2} - \frac{\cot\theta}{r^2}\frac{\partial}{\partial\theta}\right)\left(\frac{\partial^2\psi}{\partial r^2} + \frac{1}{r^2}\frac{\partial^2\psi}{\partial\theta^2} - \frac{\cot\theta}{r^2}\frac{\partial\psi}{\partial\theta}\right) = 0 . \qquad \text{(ii)}$$

NOTE: Again it is of note that this equation is <u>not</u> of the form

$$\nabla^2(\nabla^2\psi) = 0$$

where ∇^2 is the axisymmetric Laplacian operator in spherical polar coordinates

$$\nabla^2 = \frac{\partial^2}{\partial r^2} + \frac{2}{r}\frac{\partial}{\partial r} + \frac{1}{r^2\sin\theta}\frac{\partial}{\partial\theta}\left(\sin\theta\frac{\partial}{\partial\theta}\right) .$$

8. Stokes' law for the force on a moving sphere

(i) Show that equation (ii) of Problem 7 possesses solutions of the form

$$\psi = f(r)\sin^2\theta$$

where

$$\left(\frac{d^2}{dr^2} - \frac{2}{r^2}\right)\left(\frac{d^2}{dr^2} - \frac{2}{r^2}\right) f(r) = 0$$

with general solution

$$f(r) = Ar + Br^{-1} + Cr^2 + Dr^4$$

where A, B, C, D are arbitrary constants.

(ii) Show that the velocity field for the stream function

$$\psi = \sin^2\theta\,[Ar + Br^{-1} + Cr^2]$$

is given by

$$v_r = -2\cos\theta\,[C + Ar^{-1} + Br^{-3}]\,, \qquad v_\theta = \sin\theta\,[2C + Ar^{-1} - Br^{-3}]$$

and that for sufficiently large r the field represents a stream with uniform velocity of magnitude $2C$ in the negative z direction.

(iii) Write $2C = v_0$ in the above solution (so that v_0 represents the velocity of a stream in the negative z direction) and show that A and B may then be determined to make $\mathbf{v} = 0$ on $r = a$. Show that the resulting velocity field is

$$v_r = -v_0\cos\theta\left[1 - \frac{3}{2}\left(\frac{a}{r}\right) + \frac{1}{2}\left(\frac{a}{r}\right)^3\right]$$

$$v_\theta = v_0\sin\theta\left[1 - \frac{3}{4}\left(\frac{a}{r}\right) - \frac{1}{4}\left(\frac{a}{r}\right)^3\right]. \qquad \text{(i)}$$

This field satisfies the boundary conditions

$$\mathbf{v} \to -v_0\mathbf{k} \qquad (r \to \infty)$$

$$\mathbf{v} = 0 \qquad (r = a)$$

and therefore provides the solution of the flow of an incompressible Stokes' fluid past a rigid sphere $r = a$ in the slow viscous flow approximation. By imposing a uniform velocity field

$$\mathbf{v} = v_0\mathbf{k}$$

on to (i) there obtains the velocity field associated with a sphere of radius a which is moving with velocity $v_0\mathbf{k}$ and whose centre is instantaneously located at $r = 0$. The resultant field is

$$v_r = v_0\cos\theta\left[\frac{3}{2}\left(\frac{a}{r}\right) - \frac{1}{2}\left(\frac{a}{r}\right)^3\right]$$

$$v_\theta = -v_0\sin\theta\left[\frac{3}{4}\left(\frac{a}{r}\right) + \frac{1}{4}\left(\frac{a}{r}\right)^3\right]. \qquad \text{(ii)}$$

(iv) Use equations (ii) and the vector equation

$$\operatorname{grad} P = -\mu\operatorname{curl}\operatorname{curl}\mathbf{v}$$

to obtain $$\frac{\partial P}{\partial r} = -\frac{3\mu a v_0\cos\theta}{r^3}\,, \qquad \frac{\partial P}{\partial\theta} = -\frac{3\mu a v_0\sin\theta}{2r^2}$$

with solution $$P = \frac{3}{2}\frac{\mu a v_0 \cos\theta}{r^2} + P_0$$

where P_0 is the pressure as $r \to \infty$.

(v) In spherical polar coordinates non-vanishing components of stress acting on the sphere are σ_{rr} (normally) and $\sigma_{r\theta}$ (tangentially). Show that the total force resisting the motion of the sphere is in the z direction and given by

$$F_z = \iint_S (\sigma_{rr}\cos\theta - \sigma_{r\theta}\sin\theta)dS$$

where S is the surface of the sphere. If σ_{rr} and $\sigma_{r\theta}$ are given by

$$\sigma_{rr} = -P + 2\mu d_{rr} \qquad \sigma_{r\theta} = 2\mu d_{r\theta}$$

and it is given that the physical components of rate of strain d_{rr} and $d_{r\theta}$ are related to v_r and v_θ by the following formulae.†

$$d_{rr} = \partial v_r/\partial r\,, \quad d_{r\theta} = \frac{1}{2}\left[\frac{\partial v_\theta}{\partial r} + \frac{1}{r}\frac{\partial v_r}{\partial \theta} - \frac{v_\theta}{r}\right]$$

show that $$F_z = -6\pi\mu a v_0$$

which is the well known law of Stokes for the resistance of a rigid sphere moving in an incompressible viscous fluid.

9. Show that the equation

$$\frac{\partial v_z}{\partial t} = \nu\frac{\partial^2 v_z}{\partial x^2}\,,$$

associated with time-dependent laminar flow in the absence of pressure gradients and body forces, admits similarity solutions of the form

$$v_z = t^n f(x^2/\nu t)$$

where n is a constant and where $f(\eta)$, a function of the variable η

$$\eta = (x^2/\nu t)\,,$$

satisfies the ordinary differential equation

$$4\eta f'' + (2+\eta)f' - nf = 0$$

and verify that for $n = -1$ one solution is given by

$$f = A\eta^{\frac{1}{2}}e^{-\frac{1}{4}\eta}$$

where A is an arbitrary constant.

NOTE: The choice $A = \frac{1}{2}\pi^{-\frac{1}{2}}$ leads to the function $G(x, t)$ specified by (10.6.3).

† The formulae relating physical components of **d** to physical components of **v** in curvilinear coordinate systems (here spherical polar) are not studied in the present text. The appropriate formulae are given in the present example because alternative ways of evaluating F_z (e.g. by using Cartesian components of **d** and **v**) are extremely tedious.

Chapter 11

General Theorems in Inviscid Hydrodynamics

11.1 Introduction

Under certain circumstances some problems in fluid dynamics are adequately solved in an approximation in which viscous forces are neglected. With the assumption of an inviscid barotropic fluid, the basic equations of hydrodynamics are the Euler equation of motion

$$-\operatorname{grad} P + \rho \mathbf{b} = \rho \mathbf{f} \tag{11.1.1}$$

and the equation of mass conservation

$$\frac{\partial \rho}{\partial t} + \operatorname{div}(\rho \mathbf{v}) = 0 . \tag{11.1.2}$$

In (11.1.1) the acceleration $\mathbf{f}$ is given in terms of $\mathbf{v}$ by

$$\mathbf{f} = \frac{\partial \mathbf{v}}{\partial t} + (\mathbf{v} \cdot \operatorname{grad})\mathbf{v} \tag{11.1.3}$$

while for a compressible barotropic fluid P in (11.1.1) is assumed to be a known function of ρ

$$P = P(\rho) . \tag{11.1.4}$$

For an incompressible fluid $\rho = \rho_0$ and (11.1.2) reduces to

$$\operatorname{div} \mathbf{v} = 0 . \tag{11.1.5}$$

For the compressible case equations (11.1.1), (11.1.2) and (11.1.4) provide five scalar equations for P, ρ and $\mathbf{v}$; for the incompressible case equations (11.1.1) (with $\rho = \rho_0$) and (11.1.5) provide 4 scalar equations for P and $\mathbf{v}$.

For either case (11.1.1) may be written in the slightly different form

$$-\operatorname{grad}[\psi(\rho)] + \mathbf{b} = \mathbf{f} \tag{11.1.6}$$

where

$$\left.\begin{aligned} \psi &= \int^{\rho} \frac{1}{\rho'} \frac{dP(\rho')}{d\rho'} d\rho' \qquad &\text{(compressible)} \\ &= P/\rho_0 . &\text{(incompressible)} \end{aligned}\right\} \tag{11.1.7}$$

Equation (11.1.6) is used below in the proof of Kelvin's circulation theorem.

11.2 Kelvin's circulation theorem

An important kinematic result not proved previously is the following. Consider a closed line circuit c which always links the same fluid particles. In general the geometrical configuration of the loop c depends on time and we denote this symbolically by the symbol c_t. The circulation circ c_t is defined to be the line integral

$$\text{circ } c_t = \oint_{c_t} \mathbf{v} \, . \, d\mathbf{r} \tag{11.2.1}$$

where $\mathbf{v}$ is the spatial velocity field. The kinematic theorem is

$$\frac{d}{dt}[\text{circ } c_t] = \oint_{c_t} \mathbf{f} \, . \, d\mathbf{r} \, . \tag{11.2.2}$$

The simplest proof of (11.2.2) is as follows. We re-write (11.2.1) in the form

$$\text{circ } c_t = \oint_{c_t} v_i \, dx_i$$

and transform from spatial coordinates x_i to material coordinates X_i obtaining

$$\text{circ } c_t = \oint_{c_0} V_i \frac{\partial x_i}{\partial X_j} \, dX_j \tag{11.2.3}$$

where c_0 is the time independent closed circuit in the reference configuration. Differentiating (11.2.3) with respect to time now presents no difficulties associated with moving circuits. On remembering

$$\frac{\partial}{\partial t}\left(\frac{\partial x_i}{\partial X_j}\right) = \frac{\partial}{\partial X_j}\left(\frac{\partial x_i}{\partial t}\right) = \frac{\partial V_i}{\partial X_j}$$

we find

$$\frac{d}{dt}(\text{circ } c_t) = \oint_{c_0} F_i \frac{\partial x_i}{\partial X_j} \, dX_j + \oint_{c_0} V_i \frac{\partial V_i}{\partial X_j} dX_j \, .$$

If $\mathbf{V}$ is single valued the second of these integrals vanishes since

$$\oint_{c_0} V_i \frac{\partial V_i}{\partial X_j} dX_j = \frac{1}{2} \oint_{c_0} \frac{\partial}{\partial X_j}(\mathbf{V}^2) dX_j = \tfrac{1}{2}[\mathbf{V}^2]_{\mathbf{R}_1}^{\mathbf{R}_1} = 0 \, .$$

Here $\mathbf{R}_1$ is any point of c_0. Finally, on changing the integration variables in the first integral from X_i to x_i there results

$$\frac{d}{dt}(\text{circ } c_t) = \oint_{c_t} f_i dx_i = \oint_{c_t} \mathbf{f} \, . \, d\mathbf{r} \tag{11.2.4}$$

which completes the proof.

Kelvin's circulation theorem for an inviscid barotropic fluid now follows on substituting for **f** from (11.1.6) into (11.2.4) and assuming that the body force **b** is expressible as the gradient of a potential function†

i.e.
$$\mathbf{b} = -\operatorname{grad} \Omega \ . \qquad (11.2.5)$$

From (11.1.6), (11.2.2) and (11.2.5) we find

$$\frac{d}{dt}[\text{circ } c_t] = -\oint_{c_t} \operatorname{grad}[\psi + \Omega] \cdot d\mathbf{r} = -[\psi + \Omega]_{\mathbf{R}_1}^{\mathbf{R}_1} = 0$$

where in the last step it has been assumed that ψ and Ω are single valued functions of position††.

The result that
$$\frac{d}{dt}[\text{circ } c_t] = 0 \ ,$$

known as Kelvin's circulation theorem, leads generally to

$$\text{circ } c_t = \text{constant} \qquad (11.2.6)$$

where the 'constant' may vary from one circuit of given particles to another. Of particular interest is the case of motion generated from rest. If the motion of an inviscid fluid is generated from rest so that at a particular time, say $t = 0$,

$$\text{circ } c_0 = 0$$

then the constant in (11.2.6) is determined as zero and there results

$$\text{circ } c_t = 0 \ . \qquad (11.2.7)$$

However in these circumstances (11.2.7) is valid for any choice of circuit c_t so that

$$\oint_c \mathbf{v} \cdot d\mathbf{r} = 0 \qquad (11.2.8)$$

for arbitrary choice of c.

In order for (11.2.8) to hold for any choice of c a necessary and sufficient condition, for continuous **v**, is that **v** be expressible as the gradient of a potential function. We write

$$\mathbf{v} = -\operatorname{grad} \phi \qquad (11.2.9)$$

where ϕ is known as the 'velocity potential'.

Flows described by a velocity potential are known as **potential flows** or **irrotational flows.** The latter terminology derives from the incidental result that for potential flows curl $\mathbf{v} = 0$ everywhere, so that everywhere the instantaneous rigid body angular velocity $\frac{1}{2}$ curl **v** is zero.

† This is almost always possible. For example if **b** is due to the earth's gravitational field then $\Omega = gz$ where the z axis is directed normally (upward) from the earth's surface.

†† The assumption that ψ is single valued fails if the circuit c_t intersects a surface across which the velocity and density are discontinuous, as in a shock. We exclude from present considerations the possibility of shock discontinuities.

The result that the velocity field is given by (11.2.9) leads to enormous simplifications in the theory of inviscid flow. If additionally the flow is incompressible as well as irrotational so that

$$\operatorname{div} \mathbf{v} = 0$$

the quantity ϕ in (11.2.9) satisfies Laplace's equation

$$\nabla^2 \phi = 0 \; . \tag{11.2.10}$$

The subject of irrotational inviscid flow, governed by (11.2.10), is discussed in Chapter 12. It is worth noting here that for most problems governed by Laplace's equation the equations of motion (11.1.1) play no role in determining the flow, which is governed entirely by the geometry of the problem in question. The significance of the equations of motion, in an integrated version, is discussed below in Section 11.3.

We have seen that for motion generated from rest in an ideal barotropic fluid the vorticity $\boldsymbol{\xi}$ defined by

$$\boldsymbol{\xi} = \operatorname{curl} \mathbf{v}$$

vanishes [since curl (grad ϕ) $= 0$]. In spite of this result there are occasions when it is of some interest to discuss the behaviour of a non-zero vorticity vector in an inviscid flow situation. The reasons are as follows. All real fluids are to some extent viscous and over a long period of time significant generation of vorticity may take place. Given the presence of vorticity $\boldsymbol{\xi}(\mathbf{r}, 0)$ at a time $t = 0$ (say) in the flow of a nearly inviscid fluid, the question arises as to the subsequent distribution of $\boldsymbol{\xi}(\mathbf{r}, t)$ at a later time t. If in the interval $[0, t]$ vorticity *generation* is neglected, an equation determining $\boldsymbol{\xi}(\mathbf{r}, t)$ is obtained from taking the curl of the equation of motion (11.1.6). For the case of a conservative body force ($\mathbf{b} = -\operatorname{grad} \Omega$), there results

$$\operatorname{curl}\left[\frac{\partial \mathbf{v}}{\partial t} + (\mathbf{v} \,.\, \operatorname{grad})\mathbf{v}\right] = 0 \; . \tag{11.2.11}$$

On interchanging the operators curl and $\partial/\partial t$, utilising the vector identity

$$\operatorname{grad}(\tfrac{1}{2}\mathbf{v}^2) = (\mathbf{v} \,.\, \operatorname{grad})\mathbf{v} + \mathbf{v} \wedge \operatorname{curl} \mathbf{v} \tag{11.2.12}$$

and writing $\boldsymbol{\xi} = \operatorname{curl} \mathbf{v}$, (11.2.11) yields

$$\frac{\partial \boldsymbol{\xi}}{\partial t} = \operatorname{curl}(\mathbf{v} \wedge \boldsymbol{\xi}) \; . \tag{11.2.13}$$

In turn curl $(\mathbf{v} \wedge \boldsymbol{\xi})$ may be expanded using the identity,

$$\operatorname{curl}(\mathbf{v} \wedge \boldsymbol{\xi}) = (\boldsymbol{\xi} \,.\, \operatorname{grad})\mathbf{v} - (\mathbf{v} \,.\, \operatorname{grad})\boldsymbol{\xi} + \boldsymbol{\xi} \operatorname{div} \mathbf{v} - \mathbf{v} \operatorname{div} \boldsymbol{\xi} \; .$$

However since here div $\boldsymbol{\xi} =$ div (curl $\mathbf{v}$) $= 0$, the last term on the right hand side of this last equation is identically zero. Also for the case of an incompressible fluid the third term is zero and (11.2.13) then becomes

$$\frac{\partial \boldsymbol{\xi}}{\partial t} + (\mathbf{v} \,.\, \operatorname{grad})\boldsymbol{\xi} = (\boldsymbol{\xi} \,.\, \operatorname{grad})\mathbf{v}$$

or
$$\frac{D\boldsymbol{\xi}}{Dt} = (\boldsymbol{\xi} \,.\, \mathrm{grad})\mathbf{v} \;. \tag{11.2.14}$$

Equation (11.2.14) is a predictive equation for $\boldsymbol{\xi}$ since given $\mathbf{v}$ as a function of $\mathbf{r}$ at (say) $t = 0$, and hence $\boldsymbol{\xi} \equiv \mathrm{curl}\,\mathbf{v}$ at $t = 0$, the equation calculates the convective derivative $D\boldsymbol{\xi}/Dt$. For the particular case of two dimensional flows where $v_x = v_x(x, y, t)$, $v_y = v_y(x, y, t)$ and $v_z = 0$, $\boldsymbol{\xi}$ is directed in the z direction and the quantity $(\boldsymbol{\xi} \,.\, \mathrm{grad})\mathbf{v}$ vanishes identically. Under these circumstances (11.2.13) simplifies further to

$$D\boldsymbol{\xi}/Dt = 0 \tag{11.2.15}$$

so that vorticity is conserved and convected with the fluid, i.e. the vorticity associated with any particular fluid particle is constant. This result is of importance subsequently in the analysis of the motion of rectilinear vortices in two dimensional flows of an inviscid incompressible liquid (Section 12.6).

[For an incompressible Stokes' fluid, the analogue of (11.2.14) is found by taking the curl of the Navier-Stokes equation and writing $\rho = \rho_0$. There results

$$\frac{D\boldsymbol{\xi}}{Dt} = (\boldsymbol{\xi} \,.\, \mathrm{grad})\mathbf{v} + (\mu/\rho_0)\nabla^2\boldsymbol{\xi} \tag{11.2.16}$$

for a fluid with constant viscosity. For a compressible Stokes' fluid the resulting equation is more complicated.

The character of the solutions of partial differential equations is dominated by the highest order derivatives. In the above equation these derivatives are $\partial\boldsymbol{\xi}/\partial t$ (appearing in $D\boldsymbol{\xi}/Dt$) for time, while the spatial derivatives are of second order (appearing in ∇^2). Thus equation (11.2.16) is basically of the diffusion type (10.5.7) and the spread of vorticity in a viscous fluid is essentially a diffusion process.]

11.3 The Bernoulli equations

The Euler equations, here written

$$-\,\mathrm{grad}[\psi + \Omega] = \partial\mathbf{v}/\partial t + (\mathbf{v} \,.\, \mathrm{grad})\mathbf{v} \tag{11.3.1}$$

admit a number of scalar integrals known variously as the Bernoulli equations. As a preliminary to obtaining these integrals we make use of the vector identity (11.2.12) to re-write (11.3.1) in the form

$$-\,\mathrm{grad}[\psi + \Omega + \tfrac{1}{2}\mathbf{v}^2] = \partial\mathbf{v}/\partial t - \mathbf{v} \wedge \mathrm{curl}\,\mathbf{v} \;. \tag{11.3.2}$$

The Bernoulli integrals of (11.3.2) take various closely related forms which depend in detail on the specific assumptions employed, i.e. incompressible and/or irrotational and/or steady flow. The most general form, in the sense of the least restrictive on the nature of the material and the flow, is obtained directly from integrating (11.3.2) along a streamline. If $\mathbf{dr}$ is a vector element of a streamline then since $\mathbf{dr}$ is parallel to $\mathbf{v}$ while $\mathbf{v} \wedge \operatorname{curl} \mathbf{v}$ is perpendicular to $\mathbf{v}$ the integral

$$\int_{\mathbf{r}_1}^{\mathbf{r}_2} \mathbf{dr} \cdot (\mathbf{v} \wedge \operatorname{curl} \mathbf{v}) = 0$$

where $\mathbf{r}_1$ and $\mathbf{r}_2$ are any points on the streamline. It follows that

$$\int_{\mathbf{r}_1}^{\mathbf{r}_2} \mathbf{dr} \cdot \operatorname{grad}(\psi + \Omega + \tfrac{1}{2}\mathbf{v}^2) = -\int_{\mathbf{r}_1}^{\mathbf{r}_2} (\partial \mathbf{v}/\partial t) \cdot \mathbf{dr} \qquad (11.3.3)$$

where the integrals are taken along a streamline. The integral on the left side of (11.3.3) is readily evaluated so that (11.3.3) becomes

GENERAL FORM OF BERNOULLI EQUATION

$$[\psi + \Omega + \tfrac{1}{2}\mathbf{v}^2]_{\mathbf{r}_1}^{\mathbf{r}_2} = -\int_{\mathbf{r}_1}^{\mathbf{r}_2} (\partial \mathbf{v}/\partial t) \cdot \mathbf{dr} \qquad (11.3.4)$$

which is the most general form of the Bernoulli equation. For steady flow, where $\mathbf{v} = \mathbf{v}(\mathbf{r})$ independent of time so that $\partial \mathbf{v}/\partial t = 0$, the general form simplifies to

STEADY FLOW

$$[\psi + \Omega + \tfrac{1}{2}\mathbf{v}^2]_{\mathbf{r}_1}^{\mathbf{r}_2} = 0\,. \qquad (11.3.5a)$$

Alternately this may be written

STEADY FLOW

$$\psi + \Omega + \tfrac{1}{2}\mathbf{v}^2 = f \text{ (along a streamline)} \qquad (11.3.5b)$$

where f is a constant which varies from streamline to streamline.

Another simplification is the assumption of incompressibility. With the latter assumption $\psi = P/\rho_0$ and (11.3.4) becomes

INCOMPRESSIBLE FLOW

$$[P/\rho_0 + \Omega + \tfrac{1}{2}\mathbf{v}^2]_{\mathbf{r}_1}^{\mathbf{r}_2} = -\int_{\mathbf{r}_1}^{\mathbf{r}_2} (\partial \mathbf{v}/\partial t) \cdot \mathbf{dr}\,. \qquad (11.3.6)$$

Combining the cases of steady and incompressible flow leads to

STEADY INCOMPRESSIBLE FLOW

$$[P/\rho_0 + \Omega + \tfrac{1}{2}\mathbf{v}^2]_{\mathbf{r}_1}^{\mathbf{r}_2} = 0 \qquad (11.3.7a)$$

or

$$P/\rho_0 + \Omega + \tfrac{1}{2}\mathbf{v}^2 = f \quad \text{(along a streamline).} \qquad (11.3.7b)$$

The simplest forms of the Bernoulli equation are obtained with the assumption of irrotational flow. In this case $\mathbf{v} = -\operatorname{grad}\phi$, $\operatorname{curl}\mathbf{v} = 0$ and (11.3.2) may be written

$$-\operatorname{grad}[\psi + \Omega + \tfrac{1}{2}\mathbf{v}^2 - \partial\phi/\partial t] = 0$$

with general integral.

IRROTATIONAL FLOW

$$\psi + \Omega + \tfrac{1}{2}\mathbf{v}^2 - \partial\phi/\partial t = f(t) \quad \text{(everywhere)} \qquad (11.3.8)$$

where $f(t)$ is an arbitrary function of time. It is important to realise that (11.3.8) holds everywhere and that unlike all the previous versions of the equation there is no restriction that the result holds only on a streamline.

Even simpler forms of the equation are derived from (11.3.8) with the additional assumptions of incompressibility and/or steady flow. We leave the reader to derive

INCOMPRESSIBLE IRROTATIONAL FLOW

$$P/\rho_0 + \Omega + \tfrac{1}{2}\mathbf{v}^2 - \partial\phi/\partial t = f(t) \quad \text{(everywhere)} \qquad (11.3.9)$$

STEADY IRROTATIONAL FLOW

$$\psi + \Omega + \tfrac{1}{2}\mathbf{v}^2 = f \quad \text{(everywhere)} \qquad (11.3.10)$$

where f is a constant independent of time

INCOMPRESSIBLE, STEADY IRROTATIONAL FLOW

$$P/\rho_0 + \Omega + \tfrac{1}{2}\mathbf{v}^2 = f \quad \text{(everywhere)} \qquad (11.3.11)$$

where f is a constant independent of time.

In applications of those equations immediately above which contain either arbitrary functions of time or arbitrary constants, the quantity $f(t)$ (or f) is commonly determined from knowing the velocity, pressure and body force potential at a particular point. In particular for (11.3.11), if P_0 and Ω_0 are the values of P and Ω at a *stagnation* point where $\mathbf{v} = 0$, the equations may be re-written

$$P/\rho_0 + \Omega + \tfrac{1}{2}\mathbf{v}^2 = P_0/\rho_0 + \Omega_0 \,. \qquad (11.3.12)$$

With the further approximation that body forces are neglected there results the very simplest form of the equation

$$P = P_0 - \tfrac{1}{2}\rho_0\mathbf{v}^2 \qquad (11.3.13)$$

implying that in this case the pressure is a maximum at a stagnation point.

11.4 Applications in hydraulic engineering of the Bernoulli equation

In the present text we encounter a number of applications of the Bernoulli equations, particularly the forms (11.3.9) and (11.3.11) for incompressible irrotational flow (Chapter 12).

One form of Bernoulli's equation is of technological importance in the theory of hydraulic engineering. The form of the equation utilised in hydraulics assumes that the fluid (water) is incompressible and that the term on the right side of (11.3.6)

i.e.
$$\int_{\mathbf{r}_1}^{\mathbf{r}_2} \partial\mathbf{v}/\partial t \,.\, d\mathbf{r}$$

is negligible even for unsteady flow. The resulting equation becomes

$$[P/\rho_0 + \Omega + \tfrac{1}{2}\mathbf{v}^2]_{\mathbf{r}_1}^{\mathbf{r}_2} = 0 \tag{11.4.1}$$

where $\mathbf{r}_1$ and $\mathbf{r}_2$ are points on the same streamline.

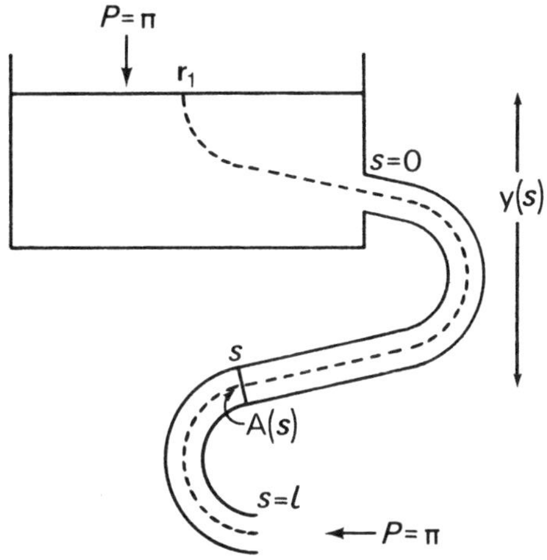

Fig. 11.1 Flow of water from reservoir

In a typical hydraulics situation water flows out from a reservoir through a network of pipes. Consider the situation depicted in Fig. 11.1, where water is fed into one pipe emerging from the base of a reservoir. To find the pressure at any point in the pipe, let s be the distance coordinate along the pipe, $A(s)$ the (variable) cross sectional area, Q the volume efflux at $s = l$, and $y(s)$ the height of the reservoir surface level above the cross section at s. Assuming that the flow is parallel to the pipe generators and writing $q(s) \equiv |\mathbf{v}|$ for the speed at s, conservation of mass leads to†

$$q(s)A(s) = Q, \quad \text{i.e. } q = Q/A(s) \;. \tag{11.4.2}$$

Also the potential Ω due to the earth's gravitational field is $\Omega = -gy$.

Consider now a streamline passing through the pipe and emerging at $s = l$. The other end of the streamline necessarily terminates at the reservoir surface since streamlines terminate either at boundaries of the flow or are closed. Let $\mathbf{r}_1$ be the point on the reservoir surface where the streamline emerges and assume that $\mathbf{v}(\mathbf{r}_1)$ is negligible; also at $\mathbf{r} = \mathbf{r}_1, P = \Pi$, where Π is the atmospheric pressure. If $\mathbf{r}_2$ is chosen in the first instance to be the point on the streamline where the latter emerges at $s = l$ and where also $P = \Pi$, application of (11.4.1) leads immediately to

$$\rho_0 g y(l) = \tfrac{1}{2}\rho_0 q^2(l), \quad \text{i.e. } q(l) = [2gy(l)]^{\frac{1}{2}}$$

so that from (11.4.2) Q is determined to be $A(l)\sqrt{2gy(l)}$ and

$$q(s) = \frac{A(l)}{A(s)}\sqrt{2gy(l)} \;.$$

Choosing now $\mathbf{r}_1$ as previously and $\mathbf{r}_2$ as the point where the streamline intersects the pipe section at s yields

$$\Pi + \rho_0 g y(s) = P(s) + \rho_0 g y(l) A^2(l)/A^2(s)$$

determining

$$P(s) = \Pi + \rho_0 g[y(s) - y(l)A^2(l)/A^2(s)] \;. \tag{11.4.3}$$

The approximations employed in deriving results like (11.4.3) are mostly adequate for hydraulic engineering purposes. However (11.4.3) is unlikely to provide a viable model near the pipe entry $s = 0$ where fluid speeds vary very rapidly and where the flow situation is not one dimensional.

Assuming that in a reservoir and pipe distribution system **all** streamlines in the pipes emerge at the reservoir free surface, equation (11.4.1) gives

$$\Pi + \rho_0 g y = P + \tfrac{1}{2}\rho_0 q^2 \tag{11.4.4}$$

where P and q are respectively the pressure and speed at any pipe section lying a distance y below the reservoir surface. Commonly used terminology in the present context would describe $\rho_0 g y$ as a static pressure head and $\frac{1}{2}\rho_0 q^2$ as a dynamic pressure head.

It is of interest that the assumptions leading to (11.4.4) in the hydraulic context do not include that of irrotational flow; this is perhaps as well since commonly the equation is used in hydraulic situations which are unlikely to be irrotational, e.g. as in turbulent flows.

† The pipes are of sufficiently large cross section for viscosity effects to be ignored. In these circumstances, and for sufficiently slowly varying $A(s)$, the assumptions leading to (11.4.2) are realistic.

11.5 The role played by the Bernoulli equation in irrotational inviscid incompressible flow

The theory of irrotational flow of an inviscid and incompressible fluid is governed by the equations (11.2.9), (11.2.10) and (11.3.9):

$$\mathbf{v} = -\operatorname{grad}\phi \tag{11.5.1}$$

$$\nabla^2\phi = 0 \tag{11.5.2}$$

$$P/\rho_0 + \Omega + \tfrac{1}{2}\mathbf{v}^2 - \partial\phi/\partial t = f(t)\ . \tag{11.5.3}$$

Usually the problems posed – see Chapter 12 – are such that the velocity field is obtained from (11.5.1) with ϕ determined by (11.5.2) and the boundary conditions. In this context the velocity field is determined entirely by the geometry of the situation and without any appeal whatsoever to the equations of motion. The latter, in the integrated scalar form of the Bernoulli equation (11.5.3), serve to determine the pressure in terms of the velocity potential when the latter has been found

This is quite an exceptional situation in the mechanics of continuous media; for almost all problems in other branches of continuum mechanics, the equations of motion play a crucial role in determining the motion, as for example in the flow problems of Chapters 9 and 10.

However because the assumptions of incompressible, inviscid and irrotational flow provide an adequate model of some flow situations, many problems have been solved in this approximation.

Even in the theory of incompressible, inviscid and irrotational flow, problems occur where the Bernoulli equation plays a marginally more significant role than is indicated in the second paragraph above. The problems in question involve flows bounded by unkown surfaces along which the pressure is constant. Examples of such problems are found in Section 12.9.

11.6 Cavitation in inviscid, incompressible and irrotational flow

As indicated in Section 4.5 the onset of cavitation in fluids occurs because these materials are unable to withstand positive normal tensile stresses, so that for all real flow situations the pressure necessarily satisfies $P \geqslant 0$. Cavitation occurs in fluids when the pressure is reduced to zero. More precisely in the case of liquids, cavitation occurs when the pressure is less than the saturated vapour pressure. For, under these conditions, the formation of cavities is accompanied by the evaporation of liquid into gaseous vapour at the cavity surface and this occurs when the pressure is reduced to the saturated vapour pressure. However except near the liquid/gas transition temperature, the condition for cavitation is adequately represented by $P = 0$.

Evidently the Bernoulli equation (11.5.3) carries no mathematical warranty that P is necessarily positive for the case of an inviscid incompressible fluid flowing irrotationally.

For the simplest case of the Bernoulli equation (11.3.13) the condition $P \geqslant 0$ is readily translated into the requirement that the magnitude $q(= |\mathbf{v}|)$ of the velocity be restricted by the inequality $q \leqslant q_{max}$ where

$$q_{\max} = (2P_0/\rho_0)^{\frac{1}{2}}$$

and P_0 is the pressure at a stagnation point. Incipient cavitation occurs in these circumstances in any region of flow where q approaches $q_{\max}$. With the inclusion of body forces, compressibility and unsteady flow situations the analagous restriction on the velocity field is obtained only by inserting the appropriate expressions for ϕ and Ω into the relevant Bernoulli equation.

It may happen that in the (apparent) solution to a particular flow problem there occur regions where the Bernoulli equation implies that the pressure is negative. In such circumstances the apparent solution is false, and a valid solution is obtained only by allowing for the existence of internal cavities in the flow. The resulting problems, involving cavities of unkown shape a priori, are much more complex than flow problems involving known boundaries, and fall into the category of those discussed in the last paragraph of Section 11.5.

Problems involving cavitation are not discussed further in this text. The interested reader should consult *Jets, Wakes and Cavities* by G. Birkhoff and E. H. Zaranonello, Academic Press (1957).†

11.7 Some simple (unsteady) hydraulics examples solved by appeal to the Bernoulli equation

We conclude this chapter with some examples of the unsteady flow of incompressible liquids in pipes which are solved by appeal solely to the Bernoulli equation (11.3.6),

We consider first the problem of the flow of liquid into a horizontal pipe of uniform cross section which is gravity fed by a large reservoir of liquid, the surface of which is of height h, which remains sensibly constant, above the pipe.

The problem is illustrated schematically in Fig. 11.2. At $t = 0$ the tap is turned on and the liquid flows into the pipe. Let $s(t)$ be the distance travelled along the pipe by the liquid and consider the steamline illustrated whose ends terminate at s and the reservoir surface.

† Alternately solutions may be found for which there exist 'dead' regions (rather than cavities) inside which there is no motion. Across the boundary of these dead regions the velocity field is necessarily tangentially discontinuous and this is possible only if the bounding surfaces contains a highly concentrated vorticity distribution. In fact in the direction normal to the surface the vorticity distribution must be of a delta function character. While flows of this type could not arise in an ideal inviscid fluid, some of these solutions are important in that they correlate with observed flow patterns of nearly inviscid materials. One example of such a solution is given in Section 12.9.3.

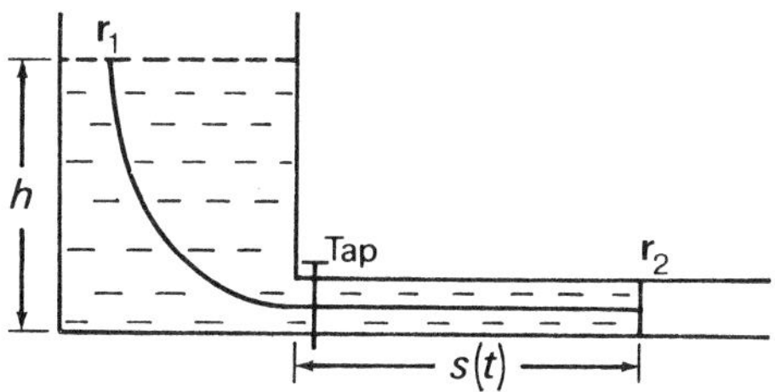

Fig. 11.2 Unsteady gravity fed flow into a horizontal pipe

Application of equation (11.3.6) with $\mathbf{r}_1$ and $\mathbf{r}_2$ endpoints of the streamline leads to

$$-gh + \tfrac{1}{2}\dot{s}^2 = -\int_{\mathbf{r}_1}^{\mathbf{r}_2} \frac{\partial \mathbf{v}}{\partial t} \cdot \mathbf{dr}\,. \tag{11.7.1}$$

In writing down (11.7.1) use has been made of the result that the pressures at $\mathbf{r}_1$ and $\mathbf{r}_2$ are equal, i.e. atmospheric pressure. Also $\dot{s}$ is the speed of the liquid at $\mathbf{r}_2$ while at $\mathbf{r}_1$ the speed is assumed negligible. The integral in (11.7.1) is now approximated by assuming that the contribution from that part of the streamline within the reservoir is negligible. In the tube $\partial\mathbf{v}/\partial t$ is of magnitude $\ddot{s}$ and directed along the streamline; accordingly the integral becomes $s\ddot{s}$ and (11.7.1) yields the non-linear differential equation

$$s\ddot{s} + \tfrac{1}{2}\dot{s}^2 = gh\,. \tag{11.7.2}$$

A first integral is easily obtained by writing $\dot{s} = q(s)$ and $\ddot{s} = q(dq/ds)$ to yield

$$\tfrac{1}{2}s\frac{d}{ds}(q^2) + \tfrac{1}{2}q^2 = gh, \qquad \text{i.e.} \quad \frac{d}{ds}(sq^2) = 2gh$$

whence

$$q^2 = 2gh + Cs^{-1} \tag{11.7.3}$$

where C is an arbitrary constant.

With the tap at the reservoir end of the tube, we must choose $C = 0$ in order to avoid infinite speeds initially. It follows that $q(\equiv \dot{s})$ is constant and given by $q = (2gh)^{\frac{1}{2}}$ whence $s = (2gh)^{\frac{1}{2}}t$, and the liquid flows into the pipe at a constant rate.

A slightly more complicated problem arises if initially the pipe is partly filled, i.e. if the tap is located part way along the pipe, say at $s = l$. In these circumstances C is determined by $q(\equiv \dot{s}) = 0$ at $s = l$ and (11.7.3) yields

$$\dot{s}^2 = 2gh(1 - l/s) \tag{11.7.4}$$

which is to be integrated subject to $s = l$ at $t = 0$.

We leave the reader to take the square root of each side of (11.7.4) and make the substitution $s = l\cosh^2\theta$ to obtain the result

$$t = \frac{l}{(2gh)^{\frac{1}{2}}}\left[\cosh^{-1}(s/l)^{\frac{1}{2}} + (s/l)^{\frac{1}{2}}\,[(s/l) - 1]^{\frac{1}{2}}\right].$$

We turn attention to the simpler problem of the oscillations of an incompressible inviscid liquid column contained in a bent tube as illustrated in Fig. 11.3. The ends of the tube are vertical and the tube is of uniform cross section.

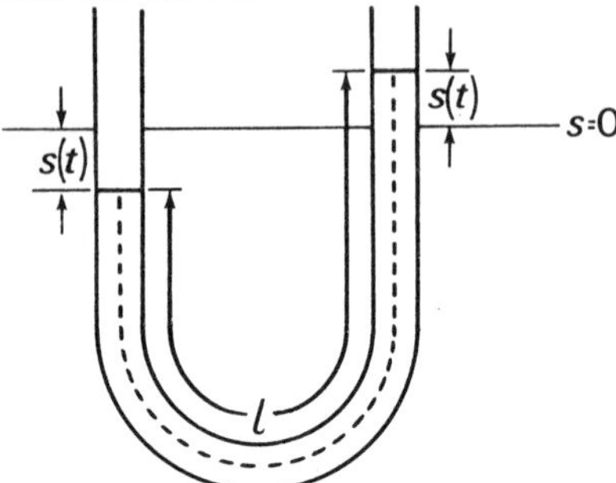

Fig. 11.3 Oscillations of a liquid column in a tube

Let $s = 0$ denote the equilibrium level of the two liquid surfaces and let $s(t)$ denote the displacement from equilibrium of the liquid surface in one of the vertical tubes. Since the tube is of uniform cross section and the liquid is incompressible the displacement of the other surface is $-s(t)$. At each surface the speed of the liquid is $\dot{s}(t)$ and the pressure is atmospheric. Hence if l is the (constant) length of the liquid column, application of the Bernoulli theorem (11.3.6) to a streamline connecting the two surfaces leads to

$$l\ddot{s} = -2gs . \tag{11.7.5}$$

In deriving this result it is assumed that the displacements $s(t)$ are sufficiently small for the liquid surfaces to remain in the vertical sections of the tube.

Equation (11.7.5) is of simple harmonic type and leads to the result that the period T of oscillations is given by the formula $T = \pi(2l/g)^{\frac{1}{2}}$.

The last problem considered is the steady flow of liquid along a pipe of finite length but variable cross section. Let P_1, A_1, h_1 denote the pressure, area of cross section and height of pipe above ground at one end of the pipe, while P_2, A_2, h_2 denote the corresponding quantity at the other end (Fig. 11.4).

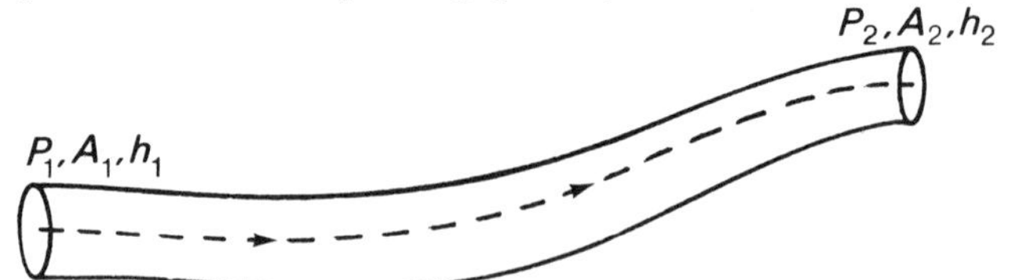

Fig. 11.4 Steady flow in pipe of variable cross section

Also let Q denote the volume efflux and assume that the sections A_1 and A_2 are located in regions of the pipe of reasonably uniform cross section so that at A_1 and A_2 the velocities $\mathbf{v}_1$ and $\mathbf{v}_2$ are approximately uniform over the cross section and parallel to the pipe. Let $q_1 = |\mathbf{v}_1|$, $q_2 = |\mathbf{v}_2|$ denote the associated speeds. Then conservation of mass implies

$$Q = A_1 q_1 = A_2 q_2, \quad \text{i.e.} \quad q_1 = Q/A_1, \; q_2 = Q/A_2 .$$

Application of the steady Bernoulli theorem (11.4.1) to a streamline connecting A_1 and A_2 now yields

$$P_1/\rho_0 + gh_1 + \tfrac{1}{2}Q^2/A_1^2 = P_2/\rho_0 + gh_2 + \tfrac{1}{2}Q^2/A_2^2$$

with solution
$$Q^2 = \frac{2A_1^2A_2^2}{\rho_0(A_1^2 - A_2^2)}[P_1 - P_2 + \rho_0 g(h_1 - h_2)] \qquad (11.7.6)$$

determining Q in terms of the parameters appertaining to the ends of the tube.

We suppose $A_1 > A_2$. Then in order that flow be possible it is necessary that

$$P_1 - P_2 + \rho_0 g(h_1 - h_2) > 0 \qquad (11.7.7)$$

since otherwise (11.7.6) yields $Q^2 < 0$ which is manifestly absurd. Provided (11.7.7) holds, then flow is possible in either direction along the tube; the calculation (11.7.6) makes no prediction about which direction the flow occurs. Since initiating flow in a tube is essentially a non-steady problem, the question as to which way the flow occurs depends on the details of the initiating motion.

Suppose flow is occurring in the direction indicated in Fig. 11.4; then even if $h_1 < h_2$, the flow is possible provided

$$P_1 - P_2 > \rho_0 g(h_2 - h_1)$$

which is a well known result in the theory of syphoning.

For the limiting case $A_1 = A_2$, Q is undetermined by (11.7.6), but a necessary condition to maintain *steady* flow is

$$P_1 - P_2 = \rho_0 g(h_2 - h_1)\,.$$

Summary

The important results of the present chapter are as follows.

For the motion of an inviscid fluid Kelvin's theorem states

$$\frac{d}{dt}\oint_{c_t} \mathbf{v}\cdot d\mathbf{r} = 0$$

where c_t is a time dependent closed contour formed from the same particles. For motion starting from rest in an inviscid fluid curl $\mathbf{v} = 0$ and

$$\mathbf{v} = -\operatorname{grad}\phi\,.$$

If additionally, the fluid is incompressible (div $\mathbf{v} = 0$) the velocity potential ϕ satisfies Laplace's equation

$$\nabla^2\phi = 0$$

and the vector equation of motion degenerates to the scalar Bernoulli equation (11.3.9)

$$P/\rho_0 + \tfrac{1}{2}\mathbf{v}^2 + \Omega - \partial\phi/\partial t = f(t)$$

where $f(t)$ is an arbitrary function of time.

Problems, Chapter 11

1. A pipe of variable cross sectional area $A(s)$ and of variable height $h(s)$, where s is distance along the pipe, is attached at $s = 0$ to the bottom of a large tank containing an incompressible liquid. The height of the liquid in the tank is maintained at a constant level H. A tap is located in the pipe at $s = s_0$ and initially fluid is at rest in the tank and along the section $0 \leqslant s \leqslant s_0$ of the pipe. At $t = 0$ the tap is opened and liquid flows into the region $s > s_0$ to occupy $0 < s < S(t)$, Both of the liquid surfaces at the top of the tank and at $S(t)$ are exposed to atmospheric pressure.

If $A(s)$ is sufficiently slowly varying to permit the assumption that everywhere in the pipe the fluid velocity is uniform across the cross section and directed normally to the cross section, and if also fluid motion in the tank is neglected, derive the equation

$$\tfrac{1}{2}\dot{S}^2 + [\ddot{S}A(S) + \dot{S}^2 A'(S)] \int_0^S A^{-1}(s)\,ds = g[H - h(S)]$$

for the determination of $S(t)$.

What boundary conditions required to be imposed on the solution of the equation. By writing

$$\dot{S}(t) \equiv q(S)$$

obtain the first integral

$$q^2(S) = \frac{2ge^{-\phi(S)}}{A^2(S)} \int_{s_0}^{S} \frac{A(s)\,[H - h(s)]\,e^{\phi(s)}}{\int_0^s A^{-1}(\xi)\,d\xi}\,ds$$

where
$$\phi(S) = \int_{\alpha}^{S} \frac{ds}{A(s) \int_0^s A^{-1}(\xi)\,d\xi}$$

and where α may be chosen arbitrarily.

Investigate in detail the cases (i) $h(s) = h = \text{const.}$, $A = \text{const.}$ (ii) $h(s) = h = \text{const.}$, $A = A_0 e^{\gamma s} (\gamma = \text{const.} > 0)$.

2. An incompressible inviscid liquid contained in a vertical cylindrical vessel is rotating as a rigid body at constant angular velocity about a central vertical axis in the earth's gravitational field. It is required to obtain the equation of the free surface of the liquid. Which of the following calculations is correct and which is fallacious and why?

(i) Since the liquid rotates as a rigid body the only non-vanishing velocity component is in the θ direction, where r, θ, z are cylindrical polar coordinates, and given by $v_\theta = \omega r$ where ω is the angular velocity. Thus $\mathbf{v}^2 = \omega^2 r^2$ and the Bernoulli equation may be written

$$P + \rho_0 g z + \tfrac{1}{2}\rho_0 \omega^2 r^2 = \text{const.}$$

where the positive z direction is upward.

The free surface is exposed to constant atmospheric pressure so measuring z from an origin at the free surface on the axis of symmetry $r = 0$, the equation of the free surface is

$$z = -\tfrac{1}{2}\omega^2 r^2/g\,.$$

(ii) From the equation of motion in the r direction and the equations of equilibrium in the θ and z directions

$$-\frac{\partial P}{\partial r} = -\frac{\rho_0 v_\theta^2}{r}\,, \qquad \frac{\partial P}{\partial \theta} = 0, \qquad -\frac{\partial P}{\partial z} - \rho_0 g = 0\,.$$

The second equation leads to $P = P(r,z)$ independent of θ while the first and third equations are consistent (since $\partial^2 P/\partial r \partial z = \partial^2 P/\partial z \partial r$) and lead to

$$P = -\rho_0 g z + \tfrac{1}{2}\rho_0 \omega^2 r^2 + \text{const.}$$

so that the equation of the free surface is

$$z = \tfrac{1}{2}\omega^2 r^2/g$$

which differs from the first calculation in sign.

3. An axially symmetric motion is specified to be of the form

$$\mathbf{v} = v_r(r, t)\mathbf{i}_r + v_z(z, t)\mathbf{k}$$

where $r = (x^2 + y^2)^{\frac{1}{2}}$ and $\mathbf{i}_r$ is a unit vector in the r direction. If the motion takes place in an incompressible medium, show that v_r and v_z are of the form

$$v_r = q(t)r^{-1} - \tfrac{1}{2}h(t)r\,, \qquad v_z = h(t)z + m(t)$$

where q, h and m are arbitrary functions of time. Verify further that the motion is irrotational and find the velocity potential. Hence deduce that in the absence of body forces the pressure is of the form

$$P/\rho_0 = \chi(t) - \tfrac{1}{2}(\dot{h} + h^2)z^2 + \tfrac{1}{4}(\dot{h} - \tfrac{1}{2}h^2)r^2 - (\dot{m} + mh)z$$
$$- \dot{q}\log r - \tfrac{1}{2}q^2 r^{-2}$$

where χ is arbitrary.

4. Use the solution given in Problem 3 above to solve the following problem. A long cylindrical annulus $a(t) \leqslant r \leqslant b(t)$ of incompressible inviscid fluid is inhibited from expansion in the axial (z) direction. The internal cavity $r \leqslant a(t)$ contains gas at a pressure $\Pi(t)$ (which results in the time dependent motion), while the external radius $r = b(t)$ is free from external pressure. Derive the equation relating $\Pi(t)$ to $a(t)$;

$$\Pi/\rho_0 = \tfrac{1}{2}(a\ddot{a} + \dot{a}^2)\log\left[1 + (b_0^2 - a_0^2)a^{-2}\right] - \tfrac{1}{2}\dot{a}^2(b_0^2 - a_0^2)/(a^2 + b_0^2 - a_0^2)$$

where a_0 and b_0 are initial values of a and b.

Show that cavitation will not occur if $\ddot{a} \geqslant 0$.

5. Show that in the absence of body forces the only possible steady flow of a compressible barotropic fluid of the form

$$\mathbf{v} = v(x)\mathbf{i}\,, \qquad \rho = \rho(x)\,, \qquad P = P(x)$$

is one for which v, ρ and P are all constant, independent of x.

6. Show that for the steady state flow

$$\mathbf{v} = v(r)\mathbf{i}_r$$

of a compressible fluid, where $r = (x^2 + y^2 + z^2)^{\frac{1}{2}}$ and $\mathbf{i}_r$ is a unit vector in the r direction

$$\rho v = Ar^{-2}$$

where A is a constant.

The flow of an ideal compressible fluid characterised by

$$P = K\rho^{\gamma}$$

where K and γ are positive constants with $\gamma > 1$, is steady and spherically symmetric. Deduce that with neglect of body forces

$$F(\rho, r) = C$$

where

$$F = \frac{\gamma K \rho^{\gamma - 1}}{\gamma - 1} + \frac{1}{2}\frac{A^2}{\rho^2 r^4}$$

and where C is constant.

The equation $F(\rho, r) = C$ defines $\rho = \rho(r)$. It is required to investigate the nature of this solution (or solutions).

Note that for fixed finite r the function F tends to infinity for both $\rho \to 0$ and $\rho \to \infty$. It follows that F possesses at least one turning point in the range $0 < \rho < \infty$. Show that there is just one such point for which F is a minimum, given by

$$\rho = (A^2/\gamma K r^4)^{\left(\frac{1}{1+\gamma}\right)}, \qquad F_{\min} = \tfrac{1}{2}\left(\frac{\gamma + 1}{\gamma - 1}\right)(\gamma K)^{\left(\frac{2}{1+\gamma}\right)}(A^2/r^4)^{\left(\frac{\gamma - 1}{\gamma + 1}\right)}.$$

Evidently for finite $r, F \geqslant F_{\min}$ so that no (real) solution of $F = C$ is possible if $C < F_{\min}$. For $C > F_{\min}$ there are two distinct solutions for ρ [say $\rho_1(r)$, $\rho_2(r)$]. A sketch of the situation is given below. [The reader should verify the essential features of this sketch.]

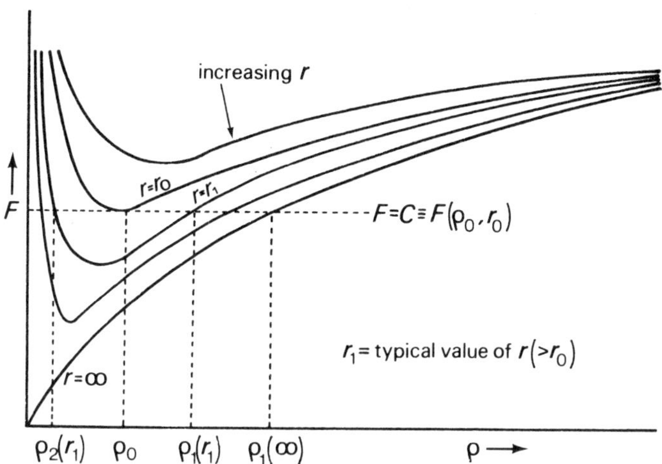

Sketch of F as a function of ρ for different values of r.

For any finite (positive) choice of C the relevant members of the family $F = F(\rho, r)$ are those whose minima are less than or equal to C. Therefore for given C there exists a finite r_0 and ρ_0 such that

$$C = F(\rho_0, r_0)$$

where $F(\rho_0, r_0) = F_{\min}(\rho, r_0)$, and the equation

$$F(\rho, r) = F(\rho_0, r_0)$$

possesses the single solution $\rho = \rho_0$ (at the minimum) for $r = r_0$. For $r > r_0$ there exist two solutions $\rho_1(r)$, $\rho_2(r)$ where ρ_1 tends to a finite limit for $r \to \infty$, while $\rho_2 \to 0$ for $r \to \infty$. For $r < r_0$, no solutions for ρ are possible. Therefore the steady state spherically symmetric flow of the compressible fluid is only possible for $r > r_0$ where $r_0 > 0$ may be chosen arbitrarily. Different values of r_0 lead to solutions which differ in detail but not in general structure. This last statement becomes clear if the equation determining ρ is put in dimensionless form.

Define
$$\sigma = \rho/\rho_0, \qquad x = r/r_0$$

and show that the equation
$$F(\rho, r) = C$$

becomes
$$\frac{\sigma^{\gamma-1}}{\gamma - 1} + \tfrac{1}{2}\sigma^{-2}x^{-4} = \tfrac{1}{2}\left(\frac{\gamma+1}{\gamma-1}\right)$$

with unique solution $\sigma = 1$ for $x = 1$ and two solution $\sigma_1(x)$ and $\sigma_2(x)$ [corresponding to $\rho_1(r)$ and $\rho_2(r)$] for $x > 1$.

Show that

$$\sigma_1(\infty) = [\tfrac{1}{2}(\gamma+1)]^{\frac{1}{\gamma-1}}, \qquad \sigma_2(x) \simeq \left(\frac{\gamma-1}{\gamma+1}\right)^{\frac{1}{2}} x^{-2} \qquad (x \to \infty).$$

Finally deduce that if $v = v_0$ at $r = r_0$ and

$$\nu = v(r)/v_0 ,$$

the dimensionless velocity distributions ν_1, ν_2 corresponding to σ_1 and σ_2 exhibit the asymptotic behaviour

$$\nu_1(x) \simeq \left(\frac{2}{1+\gamma}\right)^{\frac{1}{\gamma-1}} x^{-2}, \quad (x \to \infty); \qquad \nu_2(\infty) = \left(\frac{\gamma+1}{\gamma-1}\right)^{\frac{1}{2}}.$$

Chapter 12

Potential Theory of Classical Hydrodynamics

12.1 Basic equations

Classical hydrodynamics is concerned with the study of the irrotational flow of incompressible inviscid fluids. The basic equations, listed in the Summary of Chapter 11, are

$$\mathbf{v} = -\operatorname{grad}\phi \tag{12.1.1}$$

$$\nabla^2\phi = 0 \tag{12.1.2}$$

and the Bernoulli equation

$$P/\rho_0 + \Omega + \tfrac{1}{2}\mathrm{v}^2 - \partial\phi/\partial t = f(t)\,. \tag{12.1.3}$$

Before proceeding further we note for future purposes the simplest solution of Laplace's equation which leads to a non-vanishing velocity field. Equation (12.1.1), is satisfied by a linear function of the Cartesian coordinates; we write the solution

$$\phi = -\mathbf{u}\cdot\mathbf{r} \tag{12.1.4}$$

where $\mathbf{u}$ is a constant vector. From (12.1.1)

$$\mathbf{v} = \mathbf{u} \tag{12.1.5}$$

so that (12.1.4) is the velocity potential of a *uniform stream* flowing with velocity $\mathbf{u}$. Commonly, and without loss of generality, we take $\mathbf{u} = u\mathbf{k}$ so that the stream flows in the z direction. Accordingly for this case (12.1.4) becomes

$$\phi = -uz$$

and this result is used subsequently in the solution of more complicated problems.

In this chapter we are concerned almost exclusively with problems where ϕ is determined from (12.1.2) and pertinent boundary conditions occurring at known boundaries. However in Section 12.9, we deal briefly with a class of more complicated problems in which the flow is bounded by surfaces which are unknown a priori.

The problems of the present chapter provide a broad elementary introduction to the theory of the potential, i.e. solutions of $\nabla^2\phi = 0$, which are encountered in many branches of theoretical physics and applied mathematics. The principal topics are (a) the use of separation of variable solutions of Laplace's equation (i) in plane polar coordinates (ii) in spherical polar coordinates for the axisymmetric case, (b) solutions of the two dimensional Laplace equation in the plane by the powerful but highly specialised complex variable method.

12.2 Spherically symmetric flows and the underwater explosion problem.

The second simplest non-trivial solution of Laplace's equation is

$$\phi = m(t)/r \tag{12.2.1}$$

where $r = (x^2 + y^2 + z^2)^{\frac{1}{2}}$ is a spherical polar coordinate and where in general the quantity m appearing in (12.2.1) may be time dependent. (Proof that (12.2.1) satisfies $\nabla^2\phi = 0$ is left to the reader.)

The velocity field associated with (12.2.1) is spherically symmetric and given by

$$\mathbf{v} = -\operatorname{grad}\phi = \frac{m}{r^2}\mathbf{i}_r \tag{12.2.2}$$

where $\mathbf{i}_r$ is a unit vector in the radial direction. Equivalently if v_r, v_θ and v_ϕ denote components of $\mathbf{v}$ respectively in the r, θ, ϕ directions

$$v_r = m/r^2 , \quad v_\theta = v_\phi = 0 . \tag{12.2.3}$$

Conventionally the potential (12.2.1) is known as the potential due to a *point source* located at the origin. Proof that (12.2.1) may be associated with a point source, amongst other physical problems, is provided by the calculation of the volume efflux of fluid through any spherical surface of radius $r = b$ (say). The velocity at $r = b$ is normal to the surface and everywhere on the surface is of magnitude m/b^2. The volume efflux Q is given by multiplying m/b^2 by the surface area $4\pi b^2$ yielding

$$Q(t) = 4\pi m(t) . \tag{12.2.4}$$

independent of the choice of b.

If the flow is continuous up to the origin $r = 0$ then the calculation (12.2.4), which is valid for a surface of arbitrary small radius, implies that fluid is being created at $r = 0$ by a *point source.* Of course at the point $r = 0$ the equation of mass conservation

$$\operatorname{div}\mathbf{v} \equiv \left(\frac{\partial v_r}{\partial r} + \frac{2v_r}{r}\right) = 0$$

must be interpreted as failing since mass is not conserved at $r = 0$. In fact at $r = 0$ both $\partial v/\partial r$ and $2v_r/r$ are infinite in character. A more precise equation for $\operatorname{div}\mathbf{v}$ which acknowledges the existence of a source at $r = 0$ is

$$\operatorname{div}\mathbf{v} = 4\pi m(t)\,\delta(x)\,\delta(y)\,\delta(z)$$

where δ is the Dirac delta function.

For the case $m(t) < 0$, the potential (12.2.1) is known sometimes as the potential function of a *sink* rather than source.

Isolated point sources (or sinks) are rather artificial hydrodynamic concepts. However the potential function (12.2.1) and resultant velocity field (12.2.2) is pertinent to the important physical problem of an underwater explosion. Suppose at $t = 0$ an explosive charge, initially of radius a_0 is detonated under water. Subsequently the explosive gaseous debris from the charge creates a cavity of radius $a(t)$. It is required to determine $a(t)$ as a function of time.

With the neglect of body forces the velocity field lies in the radial direction and is described by (12.2.2). To determine $m(t)$ we make use of the physical condition that at $r = a$, v_r is necessarily the cavity surface velocity $\dot{a}(t)$. Accordingly $m(t) = a^2\dot{a}$ and

$$\phi = a^2\dot{a}/r \tag{12.2.5}$$

$$v_r = a^2\dot{a}/r^2 \,. \tag{12.2.6}$$

The Bernoulli equation (12.1.3) now yields, in the absence of body forces,

$$P/\rho_0 + \tfrac{1}{2}a^4\dot{a}^2/r^4 - (a^2\ddot{a} + 2a\dot{a}^2)/r = f(t) \tag{12.2.7}$$

where $f(t)$ is, as yet, an arbitrary function of time. To determine $f(t)$ we suppose that as $r \to \infty$ where the fluid is at rest, $P \to P_0$ (a constant pressure). This determines $f(t) = P_0/\rho_0$ and (12.2.7) becomes

$$P(r, t) = P_0 + \rho_0[(a^2\ddot{a} + 2a\dot{a}^2)/r - \tfrac{1}{2}a^4\dot{a}^2/r^4] \,.$$

Substituting $r = a$ into this result now yields

$$\bar{P} = P_0 + \rho_0(a\ddot{a} + \tfrac{3}{2}\dot{a}^2) \,. \tag{12.2.8}$$

where $\bar{P} = P(a, t)$ is the pressure at the cavity interface.

To proceed further we need some assumption about the pressure $\bar{P}$ at the cavity surface. Various possible models may be adopted, e.g. the gaseous debris from the explosive may be assumed to obey a highly compressible constitutive equation of the form†

$$\frac{P}{P_1} = \left(\frac{\rho}{\rho_1}\right)^\gamma$$

where P_1 and ρ_1 are the initially (very high) gas pressure and density immediately following the explosion. With this model and the assumption that the pressure is uniform in the cavity we have

$$\bar{P} = P_1(a_0/a)^{3\gamma} \tag{12.2.9}$$

which result follows from noting that ρ is inversely proportional to a^3. On substituting (12.2.9) into (12.2.8) there results a non-linear differential equation for $a(t)$.

The model (12.2.9) is exploited further in Problem 1 at the end of the chapter. Here we consider a slightly simpler model based on the assumption that all the energy in the explosion is released instantaneously.

† The model is realistic for real explosives with γ chosen to be of the order $\gamma = 10$.

As a preliminary we note that $a^2\dot{a}$ is an integrating factor of the right side of (12.2.8) and that the equation may be re-written

$$\bar{P}\,a^2\dot{a} = P_0 a^2\dot{a} + \tfrac{1}{2}\rho_0 \frac{d}{dt}(a^3\dot{a}^2)$$

i.e.
$$\int_{a_0}^{a} \bar{P}\,a^2 da = \tfrac{1}{3}P_0(a^3 - a_0^3) + \tfrac{1}{2}\rho_0 a^3\dot{a}^2\,. \tag{12.2.10}$$

In deriving (12.2.10), we have utilised the boundary condition

$$\dot{a} = 0 \quad \text{for} \quad t = 0\,. \tag{12.2.11}$$

After multiplication by the numerical factor 4π, the left side of (12.2.10) is the work done by the explosive gases on the fluid. If we assume that the energy release from the explosive E is instantaneous then for $t > 0$

$$\int_{a_0}^{a} \bar{P}\,a^2 da = \mathrm{E}/4\pi = \text{const.} \qquad (t > 0)\,.$$

Inserting this result into (12.2.10) now yields the equation

$$\tfrac{1}{2}\rho_0 a^3\dot{a}^2 + \tfrac{1}{3}P_0(a^3 - a_0^3) = \mathrm{E}/4\pi\,. \tag{12.2.12}$$

The expansion comes to rest when $a = a_f$, given by substituting $\dot{a} = 0$ in (12.2.12). This yields the formula

$$a_f^3 - a_0^3 = 3\mathrm{E}/4\pi P_0 \tag{12.2.13}$$

or since in practice $a_f \gg a_0$

$$a_f \simeq (3\mathrm{E}/4\pi P_0)^{\frac{1}{3}} \tag{12.2.14}$$

determining the final cavity size in terms of the available energy of the explosion and the ambient pressure in the fluid.

A further integration of (12.2.12) leads to an estimate of the time T for the formation of the final cavity of radius a_f. On substituting for E from (12.2.13) into (12.2.12), we derive easily

$$\int_{a_0}^{a_f} \frac{a^{\frac{3}{2}}\,da}{(a_f^3 - a^3)^{\frac{1}{2}}} = \sqrt{\frac{2\,P_0}{3\,\rho_0}}\;T\,.$$

Writing $a = a_f x$ in the integral and again approximating a_0 to the value zero then leads to

$$T = N(3\rho_0/2P_0)^{\frac{1}{2}} a_f \tag{12.2.15}$$

where N is the number.

$$N = \int_0^1 \frac{x^{\frac{3}{2}}\,dx}{(1 - x^3)^{\frac{1}{2}}} = \frac{2}{3}\int_0^{\frac{1}{2}\pi} \sin^{\frac{2}{3}}\theta\; d\theta\,.$$

The latter integral may be evaluated numerically or analytically in terms of gamma functions, with the result $N = 0.746$. Combining this figure with (12.2.14) and (12.2.15) leads finally to

$$T = 0.566(\rho_0/P_0)^{\frac{1}{2}}(\mathrm{E}/P_0)^{\frac{1}{3}} \qquad (12.2.16)$$

Provided E is taken to be about one half of the available energy in the explosive the formulae (12.2.14) and (12.2.16) are in reasonable agreement with experimental data for underwater explosions carried out at depths h which are large compared with a_f. In these circumstances the pressure P_0, given by $P_0 = \rho_0 gh$, is the ambient pressure in the vicinity of the explosion.

Defects of the calculation are the assumption that the energy of the explosive is released instantaneously, and also the neglect of compressibility effects in the fluid. In reality the energy is released over a finite period of time and this effect is readily incorporated in the model by use of equation (12.2.9) to describe the cavity pressure. However the calculations are still in question because the pressure in the cavity is not uniform; neither for short times is the velocity field described by (12.2.6). The latter expression, based on the assumption of incompressibility, predicts non-zero values of v_r everywhere in the fluid. For a compressible fluid v_r is only non-zero in a region bounded by an expanding wave front diverging from the charge with a speed determined predominantly by the bulk modulus K and to a lesser extent by the magnitude of the explosion. This wave propagation phenomena, completely neglected in the incompressible approximation, is important in the initial stages of cavity formation. Similar behaviour occurs also in the gaseous explosive debris in the cavity and leads to a non-uniform pressure for $r < a(t)$. Calculations incorporating these effects are very involved. Some account of the effects of compressibility in the fluid are given in Cole's *Underwater Explosions* Princeton University Press (1948).

One feature of the real behaviour of an underwater cavity resulting from an explosion is that the motion is non 'dead beat'. In other words the flow does not everywhere come to rest at the same instant as the cavity. The 'dead beat' nature of the incompressible problem is clear from (12.2.6) which predicts $v_r = 0$ everywhere when $\dot{a}(t) = 0$. In reality, following the first expansion of the cavity to an initial maximum radius a_f, the cavity then contracts and thereafter the radius $a(t)$ oscillates – for details see Cole (loc. cit.).

12.3 Potential due to a point dipole

A second seemingly artificial solution of Laplace's equation is obtained from the association of a source of strength m located near a sink of the same strength.

Consider the problem in which a sink of strength m is located at the origin together with a source of the same strength located at the point $\delta\mathbf{b}$. Superimposing the two relevant solutions of the (linear) Laplace equation yields

$$\phi = -m\left(\frac{1}{r} - \frac{1}{|\mathbf{r} - \delta\mathbf{b}|}\right) \qquad (12.3.1)$$

which is the potential of a 'finite dipole'. For points **r** for which $|\delta \mathbf{b}| \ll r$ the second term in (12.3.1) may be expanded by Taylor's theorem in the form

$$\frac{1}{|\mathbf{r} - \delta\mathbf{b}|} = \frac{1}{r} - \delta\mathbf{b} \cdot \operatorname{grad}\left(\frac{1}{r}\right) + O(\delta\mathbf{b}^2)$$

so that ϕ becomes

$$\phi = -m\delta\mathbf{b} \cdot \operatorname{grad}\left(\frac{1}{r}\right) + O(\delta\mathbf{b}^2).$$

We now define a *point dipole* by the limiting process $m \to \infty$, $\delta\mathbf{b} \to 0$ such that

$$\operatorname{Lt}(m\,\delta\mathbf{b}) \to \boldsymbol{\mu}(t)$$

where $\boldsymbol{\mu}$ is a vector, possibly time dependent. The resulting potential for a point dipole located at the origin is then

$$\phi = -\boldsymbol{\mu}(t) \cdot \operatorname{grad}\left(\frac{1}{r}\right) \qquad (12.3.2)$$

where $\boldsymbol{\mu}$ defines both the strength and orientation of the dipole.

With the dipole lying in the z direction so that μ may be written

$$\boldsymbol{\mu}(t) = \mu(t)\mathbf{k}$$

the potential (12.3.2) may be written

$$\phi = \mu(t)\cos\theta/r^2 \qquad (12.3.3)$$

where θ is the polar angle between the z axis and the radius vector.

It may be verified directly that (12.3.3) satisfies Laplace's equation. Subsequently we shall derive (12.3.3) as one term in a general series solution of Laplace's equation obtained by the method of separation of variables (Section 12.7.1). Here it is sufficient to note that the potential (12.3.3) arises in the quite unexceptional problem of the motion of a sphere through a fluid (Section 12.7.2).†

12.4 Boundary conditions in classical hydrodynamics

Almost all flow problems encountered in any branch of fluid mechanics entail consideration of the behaviour of the fluid at various boundaries occurring in the flow. There arises, as in Chapter 10, the question as to the nature of the boundary conditions obtaining at the interface of a fluid and a rigid body. For viscous fluids, adoption of the adhesian condition, motivated by experimental observations on real fluids, was seen to lead, at least for simple problems, to unique solutions of the basic equations.

† Many of the seemingly absurd 'simple' solutions of Laplace's equation like (12.2.1) and (12.3.3) which lead to infinite velocity fields at singular points, such as $r = 0$ in the above cases, arise in unexceptional real physical problems like the underwater explosion problem and the motion of a sphere through a liquid. Of course for these physical problems the singular points are excluded from the real flow field, e.g. in the underwater explosion problem the point $r = 0$ lies in the cavity.

The adhesian condition is specified mathematically by the boundary conditions that the components of fluid velocity v_n and v_s, respectively normal and tangential to the interface, shall be the same as the corresponding quantities for the surface of the rigid body.

In the present simplified theory it is no longer possible to specify both v_n and v_s to be the same as the rigid body. Attempts to impose both conditions lead to potential problems where the boundary conditions are 'overspecified' with the result that it is not possible to obtain solutions. To obtain solutions in the theory of inviscid, incompressible and irrotational flow it is necessary to compromise in the matter of boundary conditions and to retain only the condition that at an interface v_n be the same as that of the surface of the rigid body.

An indication of why it is necessary to abandon one of the boundary conditions is found by comparing the Euler equations of motion (11.1.1) with the Navier-Stokes equations (10.1.4). In the latter there occur spatial derivatives of **v** of second degree (in $\nabla^2 \mathbf{v}$), while for the Euler equations the highest order derivatives present are of first degree. By analogy with the theory of ordinary differential equations we expect the number of boundary conditions required to determine solutions of first degree equations to be less than in the case of equations of the second degree. This is indeed the case and explains why in the present theory we retain only the condition for the normal component of the velocity.

In mathematical terms the above considerations imply that solutions of Laplace's equation (12.1.2) are to be sought subject to boundary conditions at an interface

$$-\frac{\partial \phi}{\partial n} = \overline{v}_n \tag{12.4.1}$$

where $\overline{v}_n$ is the normal component of the velocity of the surface of the rigid body and $\partial/\partial n$ denotes the derivative in the normal direction to the interface. For the case where the rigid body is at rest, $\overline{v}_n = 0$ and (12.4.1) reduces to

$$\frac{\partial \phi}{\partial n} = 0 \tag{12.4.2}$$

Because the boundary conditions appertaining to the Navier-Stokes equations and the present theory are different it is not true that the velocity field obtained by writing $\mu \to 0$ in solutions of the Navier-Stokes equations necessarily leads to the corresponding inviscid solutions in the present theory. One counter example suffices. The velocity field obtained in the solution of the Couette flow problem (Section 10.4) happens to be independent of μ. Evidently therefore the velocity field is unchanged by the limiting process. On the other hand the solution to the same problem in the present theory for motion starting from rest yields a velocity field which is everywhere zero, unaffected by the rotating cylinders which are unable to exert a couple on the fluid.

Since the present approximation fails to take account of the boundary conditions satisfied by the tangential component of the velocity field at a fluid/solid interface, the inviscid theory tends to yield incorrect solutions in the vicinity of rigid body surfaces. Elsewhere in the flow the inviscid solution is often adequate and much current research in fluid mechanics is concerned with finding solutions to the Navier-Stokes equations in the vicinity of rigid bodies which 'match' on to solutions of the simpler inviscid equations in the rest of the flow field. The techniques involved are beyond the scope of the present text. In the remainder of this chapter we are concerned mostly with methods for obtaining solutions of (12.1.2) subject to either (12.4.1) or (more commonly) (12.4.2), at appropriate boundaries.

12.5 Point sources, dipoles and planes. Method of Images.

The problems of this Section are somewhat artificial in character and concern the flows arising from sources or dipoles located in the vicinity of plane surfaces.

The simplest problem, sketched in Fig. 12.1, involves a source of strength m located at $x = y = 0$, $z = h$ above the rigid plane surface $z = 0$.

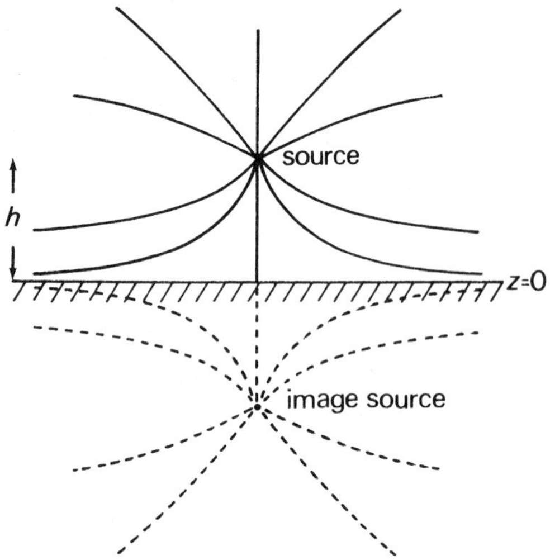

Fig. 12.1 Streamline pattern due to point source located near rigid surface $z = 0$

For the isolated source, the potential is

$$\phi_1 = \frac{m}{|\mathbf{r} - h\mathbf{k}|} = \frac{m}{[x^2 + y^2 + (z-h)^2]^{\frac{1}{2}}} \tag{12.5.1}$$

which result is obtained from (12.2.1) together with the fact that the source is located at $x = y = 0$, $z = h$ rather than at the origin. However the potential (12.5.1) fails to satisfy the boundary condition

$$-(v_z)_{z=0} \equiv (\partial\phi/\partial z)_{z=0} = 0 \tag{12.5.2}$$

demanded by the presence of the boundary $z = 0$.

The correct solution to the problem posed may be written in the form

$$\phi = \phi_1 + \phi_2$$

where the augmenting potential ϕ_2 is such that (12.5.2) is satisfied. It is also necessary that ϕ_2 should be unexceptional in the sense that ϕ_2 introduces no additional singularities, like sources or dipoles, into the flow region $z > 0$.

The method of *images* attempts to determine ϕ_2 by introducing 'image' singularities outside the region of flow (here $z < o$). For the present problem it seems intuitive that a solution is given by locating an image source of the same strength at the image point $x = y = 0$, $z = -h$. The resulting potential ϕ_2 is

$$\phi_2 = m/[x^2 + y^2 + (z+h)^2]^{\frac{1}{2}} .$$

It remains to verify that the boundary conditions (12.5.2) is satisfied. We have easily (on $z = 0$)

$$\left(\frac{\partial \phi_1}{\partial z}\right)_{z=0} = -\left(\frac{\partial \phi_2}{\partial z}\right)_{z=0} = \frac{mh}{(x^2 + y^2 + h^2)^{3/2}} .$$

Thus the potential $$\phi = \phi_1 + \phi_2 \qquad (12.5.3)$$
satisifes the conditions, (i) near $x = y = 0$, $z = h$, ϕ is dominated by the term ϕ_1 and hence 'looks' like a point source, (ii) it contains no other singularities in $z > 0$, (iii) it satisfies the boundary condition (12.5.2). It follows that ϕ solves the problem posed.

Having found ϕ, Bernoulli's equation determines the pressure. We assume m is constant, the pressure as $|\mathbf{r}| \to \infty$ is P_0 and that body forces are neglected. Accordingly (12.1.3) becomes

$$P = P_0 - \tfrac{1}{2}\rho_0 \mathrm{v}^2$$

since, because the velocity field tends to zero at large distances, $f(t) = P_0/\rho_0$.

We leave the reader to show that on $z = 0$, $\mathrm{v}^2 = 4m^2r^2/(r^2 + h^2)^3$ and $P = P_0 - 2m^2r^2\rho_0/(r^2 + h^2)^3$ where $r = (x^2 + y^2)^{1/2}$ is a cylindrical polar coordinate.

The problem is academic in that point sources are an artificial concept; also near the source the velocity field is very large and this implies cavitation (see Section 11.6).

A slightly more complicated problem is that of Fig. 12.2 where a point source is located at $x = h$, $y = k$, $z = 0$ near the two planes $x = 0$, $y = 0$. Here an intuitive solution is supplied using three image sources of the same strength located at the three image points ($x = -h$, $y = k$, $z = 0$), ($x = h$, $y = -k$, $z = 0$), and ($x = -h$, $y = -k$, $z = 0$).

We leave the reader to verify that the proposed image system satisfies the two boundary conditions

$$\left(\frac{\partial \phi}{\partial x}\right)_{x=0} = \left(\frac{\partial \phi}{\partial y}\right)_{y=0} = 0 .$$

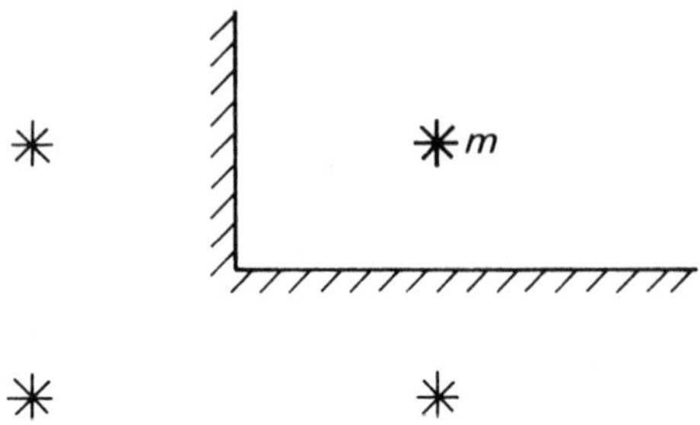

Fig. 12.2 Image system for point source located near a rectangular corner

The problem of a point dipole outside a rigid plane surface is most easily solved by considering the image systems due to the source and sink of a finite dipole and proceeding to the limiting case. The resulting image system is also a dipole with different orientation; The solution is depicted in Fig. 12.3.

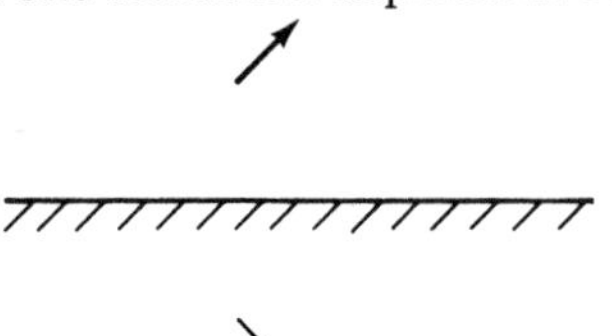

Fig. 12.3 Image system for a dipole located above a plane

12.6 Plane two dimensional flow solutions using plane polar coordinates – line sources and rectilinear vortices

12.6.1 General solution of Laplace equation for plane polar coordinates

Plane two dimensional flow situations occur when the velocity vector is both independent of z and contains no component in the z direction. For these problems $\phi(x, y, t)$ satisfies the two dimensional Laplace equation

$$\frac{\partial^2 \phi}{\partial x^2} + \frac{\partial^2 \phi}{\partial y^2} = 0 . \tag{12.6.1}$$

In this section, which is intended primarily as an introduction to the method of separation of variables for the solution of some linear partial differential equations, we seek solutions in terms of the polar variables r, θ defined by

$$x = r\cos\theta \ , \quad y = r\sin\theta \ . \tag{12.6.2}$$

Expressed in terms of r and θ (12.6.1) becomes

$$\frac{\partial^2 \phi}{\partial r^2} + \frac{1}{r}\frac{\partial \phi}{\partial r} + \frac{1}{r^2}\frac{\partial^2 \phi}{\partial \theta^2} = 0 . \tag{12.6.3}$$

This result may be obtained directly using (12.6.2) or alternately by expressing the operators div and grad in plane polar coordinates and using the identity div (grad) $= \nabla^2$.

A routine method of solving *linear* partial differential equations is the method of separation of variables. In the context of (12.6.3) we seek possible solutions of the form

$$\phi = g(r)f(\theta) \tag{12.6.4}$$

where g is solely a function of the variable r while f is solely a function of θ. On substituting (12.6.4) into (12.6.3) there results

$$f(\theta)\,[g''(r) + g'(r)/r] + [g(r)/r^2]\,f''(\theta) = 0\,. \tag{12.6.5}$$

For the method to work successfully it is necessary that equation (12.6.5) be rearrangable into two groups of terms, one group dependent only on r while the second group depends only on θ. This is accomplished in the present case by multiplying the equation by r^2/gf which, after transposition of one group of terms to the right side of the equation, leads to

$$[r^2g''(r) + rg']\,/g = -f''(\theta)/f(\theta)\,. \tag{12.6.6}$$

The argument proceeds as follows. In (12.6.6) the left side depends only on r while the right side depends only on θ. Evidently for both statements to be true simultaneously, it is necessary that each side of (12.6.6) be independent of both r and θ, i.e. to be constant.

Defining the 'constant' to be λ, there results the ordinary differential equations

$$f''(\theta) + \lambda f(\theta) = 0 \tag{12.6.7a}$$

$$r^2g''(r) + rg'(r) - \lambda g(r) = 0\,. \tag{12.6.7b}$$

The next stage in the argument seeks information about possible choices of the *eigenvalue* λ. The type of information sought concerns possible values of λ. It may happen that all possible values of λ are allowed in some range $\lambda_1 \leqslant \lambda \leqslant \lambda_2$, or only that certain discrete values are permitted.† In the former case we say there is a continuous spectrum of eigenvalues while otherwise the spectrum is discrete. It is possible that both a continuous spectrum and a discrete set of eigenvalues are permitted.

Each allowable eigenvalue leads to specific functions $g(r)f(\theta)$ which satisfy the original equation; by superimposing all such solutions for different λ values we obtained more general solutions of the equation.

The arguments determining the eigenvalues λ are different in different circumstances. However in all cases information is obtained only by initiating enquiries into the solutions of the ordinary differential equations.

Here we commence with the simpler equation (12.6.7a) with solutions which are periodic for $\lambda > 0$ and exponential for $\lambda < 0$. In the context of a flow problem in plane polar coordinates where θ may be taken to range over $0 \leqslant \theta \leqslant 2\pi$, imposition of the physical requirement that $\mathbf{v}$ is single valued in the x, y plane leads to the result that grad ϕ must be periodic in θ with period 2π. It follows

† We have encountered already the case of a continuous spectrum of eigenvalues in connexion with the diffusion equation in Section 10.5.

that the exponential solutions for $f(\theta)$ must be rejected, since then grad ϕ is also exponential in character and therefore non-periodic. It follows that λ is positive. Also for positive values of λ, periodic solutions of (12.6.7a), leading to periodic expressions for grad ϕ with period 2π, are possible if and only if λ is restricted to be one of the eigenvalues λ_n given by

$$\lambda_n = n^2 \qquad (n = 1, 2, 3 \ldots .)$$

The argument for $n = 0$ is more subtle and considered separately.

With $\lambda_n = n^2$ solutions of the second order equation (12.6.7a) are of the form

$$f(\theta) = \begin{matrix}\cos\\ \sin\end{matrix}(n\theta) \tag{12.6.8}$$

where the notation implies that either or both solutions are possible. Also of course each solution may be multiplied by an arbitrary constant; however as transpires subsequently the arbitrary constants are best incorporated at a later stage.

With $\lambda = \lambda_n$, the equation for g now becomes

$$r^2 g''(r) + rg'(r) - n^2 g(r) = 0 \tag{12.6.9}$$

which is homogeneous in the sense that derivatives of degree 0, 1 and 2 appearing in the equation are multiplied by the corresponding powers of r, i.e. r°, r^1, r^2. For a homogeneous equation which is also homogeneous in the sense that the right side is zero, a solution of the form $g = r^\alpha$, where α is a constant exponent, is always possible. Here substitution of $g = r^\alpha$ into the equation leads to

$$r^\alpha[\alpha(\alpha - 1) + \alpha - n^2] = 0$$

so that α satisfies the quadratic equation $\alpha^2 = n^2$ with roots $\alpha = \pm n$. Thus solutions of (12.6.9) are

$$g = r^{\pm n} \qquad (n = 1, 2, 3 \ldots .)$$

which, together with (12.6.8) leads to solutions of (12.6.3) given by

$$\phi = r^{\pm n} \begin{matrix}\cos\\ \sin\end{matrix}(n\theta) \qquad (n = 1, 2, 3 \ldots .) \tag{12.6.10}$$

Each of these solutions is a possible solution of (12.6.3) and a more general solution is obtained by multiplying each solution by an arbitrary constant and superimposing all such solutions.

However before writing down this general solution it remains to consider the slightly exceptional case $n = 0$ for which $\lambda = 0$. For this case

$$f''(\theta) = 0 \text{ and } r^2 g''(r) + rg'(r) = 0 .$$

The first of these has solution

$$f(\theta) = A\theta + B \tag{12.6.11}$$

while the solution of the second equation is

$$g(r) = C \log r + D \tag{12.6.12}$$

so that possible solutions for ϕ obtained from multiplying (12.6.11) and (12.6.12) are

$$fg = \text{const.},\ \log r,\ \theta,\ \theta \log r\,. \tag{12.6.13}$$

It might appear that both the terms θ and $\theta \log r$ are inadmissible because of the periodicity requirement. However the precise periodicity requirement is imposed on grad ϕ rather than on ϕ itself (which has no physical significance). Now in terms of r and θ the equation $\mathbf{v} = -\text{grad}\,\phi$ leads to components,

$$v_r = -\,\partial\phi/\partial r\ ,\quad v_\theta = -\frac{1}{r}\,(\partial\phi/\partial\theta) \tag{12.6.14}$$

and these are both periodic in θ for the first three solutions written down in (12.6.13).

Combining these results with (12.6.10), and incorporating arbitrary constants now leads to the general solution

$$\phi = a_0 + a_1 \log r + a_2\theta + \sum_{n=1}^{\infty}\left[(b_n r^n + c_n r^{-n})\cos(n\theta) + (d_n r^n + e_n r^{-n})\sin(n\theta)\right]. \tag{12.6.15}$$

This general solution is useful for solutions of two dimensional flow problems involving circular boundaries. Determination of the coefficients a_0, a_1 etc. for any particular problem devolves on the specific boundary conditions pertinent to the problem in question. For steady flow situations a_0, a_1 etc. are constant while for unsteady flow the coefficients are time-dependent.

Before seeking specific solutions of boundary value problems we consider the flows arising from some of the individual terms appearing in (12.6.15).

The solution $\phi = a_0$ is trivial since the resulting velocity field is everywhere zero. In fact in the present context of irrotational fluid mechanics ϕ is undetermined to with an arbitrary additive constant (or function of time).

The solutions $\phi = (b_1\, r\cos\theta,\ d_1 r\sin\theta)$ may also be written

$$\phi = (b_1 x,\ d_1 y)$$

and represent uniform streams with velocities $-b_1\,\mathbf{i}$ and $-d_1\,\mathbf{j}$ respectively.

The solution

$$\phi = a_1 \log r \tag{12.6.16}$$

is the potential due to a *line* source which is located at $r = 0$ and whose volume efflux is $-2\pi a_1$ per unit length. To prove this consider the efflux, per unit length in the z direction, through the cylindrical surface $r = a$, From (12.6.14) and (12.6.16)

$$v_r = -\,a_1/r\,,\quad v_\theta = 0$$

and $\mathbf{v}$ is both normal to the surface $r = a$ and independent of θ. The volume efflux through $r = a$ is therefore

$$Q = 2\pi a(v_r)_{r=a} = -\,2\pi a_1 \qquad \text{(per unit length in } z \text{ direction)}$$

independent of the radius a. If now we choose a to be the radius of a very small circle surrounding $r = 0$, the interpretation follows analagous to the spherically symmetric case discussed previously (Section 12.2). The more conventional expression for a line source potential is

$$\phi = -m \log r \tag{12.6.17}$$

where m is called the strength of the source. Sometimes, with negative values of m, (12.6.17) is known as the potential due to a line sink.

We leave the reader to show that a line sink of strength m located at the origin together with a parallel line source located at $x = \delta h$, $y = 0$, leads (to first order in δh) to the velocity potential

$$\phi = m\delta hx/r^2 + O(\delta h^2) \tag{12.6.18}$$

In the limit $m \to \infty$, $\delta h \to 0$ such that $\mathrm{Lt}(m\,\delta h) \to \mu$ there results the potential

$$\phi = \mu x/r^2 \tag{12.6.19}$$

due to a line dipole oriented in the x direction lying at the origin. Since $x = r \cos\theta$, (12.6.19) is essentially the term $c_1 \cos\theta/r$ appearing in (12.6.15). The term $e_1 \sin\theta/r$ is the potential due to a line dipole also located at the origin but oriented in the y direction. More generally for a line dipole located at the origin with arbitrary orientation in the (x, y) plane, ϕ is given by

$$\phi = \boldsymbol{\mu} \cdot \mathbf{r}/r^2 \tag{12.6.20}$$

where the vector $\boldsymbol{\mu}$ characterises both the strength and orientation of the dipole. Both terms involving c_1 and e_1 in (12.6.15) appear in (12.6.20). Of more interest is the flow due to the non-single valued term $a_2\theta$ considered below.

12.6.2 THE RECTILINEAR VORTEX

The potential given by the third term in (12.6.15) is of much importance subsequently. In the notation conventionally employed we write $a_2 = -\kappa$ obtaining

$$\phi = -\kappa\theta\,. \tag{12.6.21}$$

The resulting velocity field determined by (12.6.14) is

$$v_\theta = \kappa/r\,, \quad v_r = 0 \tag{12.6.22}$$

and the motion is in the θ direction with circular streamlines as illustrated below.

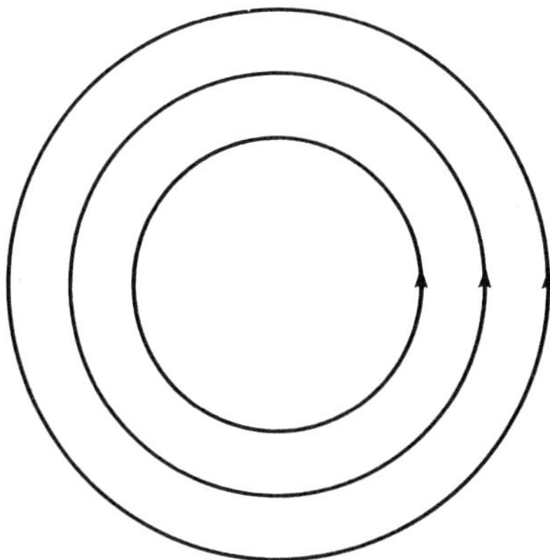

Fig. 12.4 Streamlines for an isolated rectilinear vortex

The velocity potential (12.6.22) is sometimes known as the potential due to a rectilinear vortex located at the origin. The reason for the terminology (with the implication of the presence of vorticity in spite of the fact that for all r, except possibly for $r = 0$, (12.6.22) is necessarily an irrotational velocity field) is as follows.

Consider the circulation integral

$$\text{circ } c = \oint_c \mathbf{v} \cdot \mathbf{dr} \qquad (12.6.23)$$

where c is any closed contour in the (x, y) plane enclosing $r = 0$. It is readily verified directly from (12.6.22) that for all circular contours centred on $r = 0$ that

$$\text{circ } c = 2\pi\kappa\ . \qquad (12.6.24)$$

The result is more generally true for any contour enclosing the origin. A general proof is obtained by writing $\mathbf{v} = -\text{grad}\,\phi$ in (12.6.23) and substituting for ϕ from (12.6.21). We obtain

$$\text{circ } c = \kappa\,[\theta]_{\mathbf{r}_1}^{\mathbf{r}_2} \qquad (12.6.25)$$

where $\mathbf{r}_1$ and $\mathbf{r}_2$ are any pair of neighbouring points on the contour and where in evaluating the indefinite integral θ between the limits $\mathbf{r}_1$ and $\mathbf{r}_2$ the path c is traversed completely. As the path is traversed, θ increases from its initial value (say α) at $\mathbf{r}_1$ to the value $\alpha + 2\pi$ at $\mathbf{r}_2$ (see Fig. 12.5a). Thus for all closed circuits c enclosing $r = 0$ the result (12.6.24) follows.

On the other hand for circuits not enclosing the origin, θ at $\mathbf{r}_2$ returns to the value α so that $[\theta]_{\mathbf{r}_1}^{\mathbf{r}_2} = 0$ and $\text{circ } c = 0$ (Fig. 12.5b).

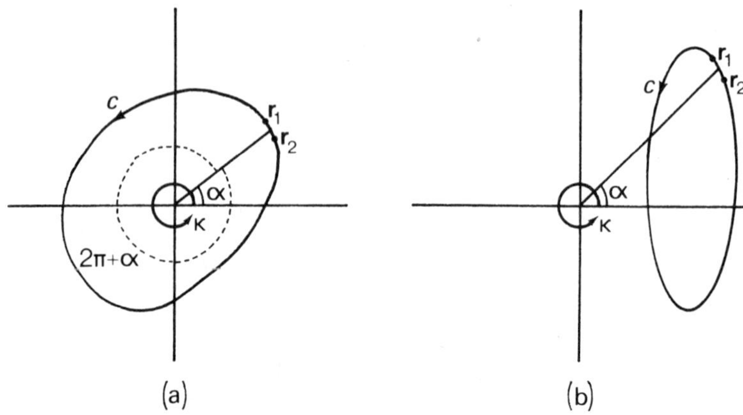

Fig. 12.5 Evaluation of circ c for rectilinear vortex for the cases where (a) c encloses origin (b) c does not enclose origin

The result that circ $c = 0$ for all circuits not enclosing $\mathbf{r} = 0$ is readily reconciled with Stokes' theorem

i.e.
$$\oint_c \mathbf{v} \cdot d\mathbf{r} = \iint_S \text{curl}\, \mathbf{v} \cdot d\mathbf{S}$$

(where S is an area spanning c), on remembering that the flow is irrotational (curl $\mathbf{v} = 0$) for all r, other possibly than for $r = 0$. For all circuits enclosing the origin the result (12.6.24) is reconcilable with Stokes' theorem provided we assume that at the origin $r = 0$, curl $\mathbf{v}$ is infinite in such a manner that the surface integral $\iint_S \text{curl}\, \mathbf{v} \cdot d\mathbf{S}$ is finite. This is possible if we associate a vorticity vector of delta function character with the velocity field (12.6.22),

i.e. we write
$$\boldsymbol{\xi}(\equiv \text{curl}\, \mathbf{v}) = 2\pi\kappa\, \delta(x)\, \delta(y)\mathbf{k} = [\kappa\, \delta(r)/r]\, \mathbf{k} \qquad (12.6.26)$$

An alternative viewpoint may be helpful. Consider the two dimensional velocity field of a *finite* rectilinear vortex specified by

$$v_\theta = \begin{cases} \kappa r/a^2 & (r \leqslant a)\ \text{(a)} \\ \kappa/r & (r \geqslant a)\ \text{(b)} \end{cases} \qquad (12.6.27)$$

$$v_r = 0$$

For $r \geqslant a$, the motion is the same as for the rectilinear vortex while for $r \leqslant a$ the motion is that of a rigid body (v_θ proportional to r), and is geometrically possible for any material. Also v is continuous at $r = a$. The associated angular velocity for $r \leqslant a$ is κ/a^2.

For $r \geqslant a$, curl $\mathbf{v} = 0$ as expected; for $r < a$ it is a simple calculation, left to the reader, to derive curl $\mathbf{v} = (2\kappa/a^2)\mathbf{k}$ and hence

$$\boldsymbol{\xi} = \text{curl } \mathbf{v} = \begin{cases} 2(\kappa/a^2)\mathbf{k} & (0 \leqslant r \leqslant a) \\ 0 & (r > a) \end{cases} \tag{12.6.28}$$

[In passing we note that this is in agreement with the result that $\frac{1}{2}$ curl $\mathbf{v}$ measures the instantaneous angular velocity associated with a rigid body motion.] For any circuit c enclosing the 'core' region $r \leqslant a$, circ c is $2\pi\kappa$ as before. Also for the finite vortex with core of radius a, equation (12.6.28) show that $\boldsymbol{\xi}$ is finite and uniform in the core with a value $2(\kappa/a^2)\mathbf{k}$. As $a \to 0$, $\boldsymbol{\xi} \to \infty$ but the integral

$$\iint_{(r \leqslant a)} \boldsymbol{\xi} \cdot d\mathbf{S} = 2\pi\kappa$$

remains finite. Thus in the limiting case $a \to 0$, $\boldsymbol{\xi}$ assumes the delta function character indicated in (12.6.26).

The rectilinear vortex obtained by the limiting process is for many purposes an adequate model of a finite rectilinear vortex.

Vortices are of some importance in aerodynamics since the vorticity generated by viscous forces in the vicinity of aircraft wings is shed by a succession of line vortices at the trailing edge of the wing. The finite vortex also provides a crude model for tornadoes and hurricanes. In this context the core $r \leqslant a$ may be identified as the tornado or hurricane 'eye'. From (12.6.27) wind velocities are evidently at a maximum at $r = a$ and decay to zero at the centre $r = 0$, in accordance with observations. In tornadoes a is of the order of a few feet while for hurricanes a may be several tens of miles. For tornadoes the model is crude in that it neglects vertical funneling of wind which is characteristic of real tornadoes. The visible realisation of tornadoes is due commonly to the dust gathered up by the vertical funneling.

In the tornado/hurricane context it is of some interest to calculate the radial pressure distribution.

We assume that as $r \to \infty$, $P \to P_0$ and that body forces may be neglected. For $r > a$ Bernoulli's theorem (12.1.3) together with (12.6.27b) yields

$$P = P_0 - \tfrac{1}{2}\rho_0\kappa^2/r^2 \tag{12.6.29}$$

so that at $r = a$

$$(P)_{r=a} = P_0 - \tfrac{1}{2}\rho_0\kappa^2/a^2 \,. \tag{12.6.30}$$

For $r < a$, because the motion is not irrotational, it is not possible to use Bernoulli's theorem (12.1.3). Nor are the more general forms of the theorem involving integrals along streamlines of any help since the streamlines are here circular while we wish to find P as a function of r. It is necessary to return to the Euler equation of motion,

$$-\text{ grad } P = \rho_0 \left(\frac{\partial \mathbf{v}}{\partial t} + (\mathbf{v} \cdot \text{grad})\mathbf{v}\right) .$$

The only component of this vector equation not satisfied identically is in the radial direction. The radial component reads

$$-\frac{dP}{dr} = -\rho_0 v_\theta^2/r$$

or on substituting from (12.6.27a)

$$\frac{dP}{dr} = \rho_0\kappa^2 r/a^4 .$$

Integrating this last equation subject to (12.6.30) gives

$$P = P_0 + \tfrac{1}{2}\rho_0 \frac{\kappa^2}{a^2}\left(\frac{r^2}{a^2} - 2\right)$$

so that the pressure takes the minimum value

$$(P)_{r=0} = P_0 - \rho_0\kappa^2/a^2$$

at $r = 0$. The result that P is a minimum at the centre of the eye of a hurricane is well known.

Because for $r \leqslant a$ there exists a non-vanishing vorticity vector, it would be impossible to generate rectilinear vortices in motion starting from rest in a completely inviscid fluid. However the behaviour of vortices is of importance in inviscid hydrodynamics since vortices, readily created in regions of flow where viscous forces are important, may escape into regions where the flow is adequately modelled in an inviscid approximation. For example, as indicated above, the trailing edges of aircraft wings shed vortices into the region away from the wing surfaces.

In two dimensional inviscid incompressible motion the vorticity vector $\boldsymbol{\xi}$ is governed by the equation (11.2.15)

$$\frac{D\boldsymbol{\xi}}{Dt} = 0 \qquad (12.6.31)$$

and this result is important in dealing with the motion of rectilinear vortices.

We show first that in the absence of any additional velocity field, due for example to line sources, dipoles, other vortices or the presence of boundaries, an isolated rectilinear vortex is not disturbed. For let $\mathbf{v}_s$ be the *self* field of a finite rectilinear vortex located at the origin

$$\mathbf{v}_s = \begin{cases} (\kappa/r)\,\mathbf{i}_\theta & (r > a) \\ (\kappa r/a^2)\,\mathbf{i}_\theta & (r < a) \end{cases}$$

The vorticity vector $\boldsymbol{\xi}$ is non-zero only for $r < a$. In this region $\boldsymbol{\xi}$ is constant and given by

$$\boldsymbol{\xi} = 2(\kappa/a^2)\mathbf{k} \qquad (r < a). \qquad (12.6.32)$$

According to (12.6.31) ξ remains constant for each particle. In a context where (i) initially ξ is zero for $r > a$ and uniform for $r < a$ and (ii) the motion takes place in circles around the origin, it is clear that the vorticity distribution remains as in (12.6.32) and is undisturbed by $\mathbf{v}_s$. (There is some analogy with the result in electrostatics where a charged particle does not experience a force due to its own electric field.)

The argument is valid for arbitrary values of a and is not destroyed by the limiting process $a \to 0$. It follows that an isolated rectilinear vortex is also undisturbed by its own velocity field.

However the situation is modified by the presence of additional velocity fields due, for example, to other vortices or sources, or boundaries in the flow. We confine attention to the case where the additional flow is also two dimensional and in the same plane as $\mathbf{v}_s$.

Initially we consider the case of a finite rectilinear vortex together with a uniform stream of velocity $\mathbf{u}_1$ lying in the (x, y) plane. The sum velocity field is

$$\mathbf{v} = \mathbf{v}_s + \mathbf{u}_1$$

and because of (12.6.31) ξ is convected with the velocity $\mathbf{v}$. We have already seen that $\mathbf{v}_s$ does not alter the distribution of ξ; however the additional (uniform) velocity $\mathbf{u}_1$ results in every particle in the vortex core moving with the additional velocity $\mathbf{u}_1$, and carrying with it its own vorticity. Evidently the net result is equivalent to the entire vortex core (and associated flow field for $r > a$) moving as a whole with the velocity $\mathbf{u}_1$. In other words the vortex drifts with the stream velocity $\mathbf{u}_1$. Observations on tornadoes bear this result out. Tornadoes are observed to drift with capricious changes in direction due to changes in the local wind velocity.

For the more general case of a non-uniform additional velocity field $\mathbf{u}_1(\mathbf{r}, t)$, the behaviour of the finite vortex core is more complicated, for the core will distort in shape and within the distorted core the vorticity distribution will vary with time and position. However this problem does not arise in the limiting case $a \to 0$ since the instantaneous velocity of the core is simply that of $\mathbf{u}_1$ evaluated at the current core position.

These last considerations lead to an equation of motion of the vortex core. Let $\mathbf{s}(t)$ denote the position of a rectilinear vortex and write the total velocity field in the form

$$\mathbf{v} = \mathbf{v}_s + \mathbf{u}_1 \tag{12.6.33}$$

where $\mathbf{v}_s$ is the self field due to the vortex itself. $\mathbf{v}_s$ is given by (12.6.22) provided it is remembered to refer the coordinates r and θ to an origin located at $\mathbf{s}(t)$. Also in general $\mathbf{u}_1 = \mathbf{u}_1(\mathbf{r}, t)$ is a function of both $\mathbf{r}$ and t. The resulting motion $\mathbf{s}(t)$ is now determined by the ordinary (vector) differential equation

$$\frac{d\mathbf{s}}{dt} = \mathbf{u}_1[\mathbf{s}(t), t] \, . \tag{12.6.34}$$

A simple example is provided by a line source of constant strength m located at the origin and a parallel rectilinear vortex located initially at $x = A,\ y = 0$ (Fig. 12.6).

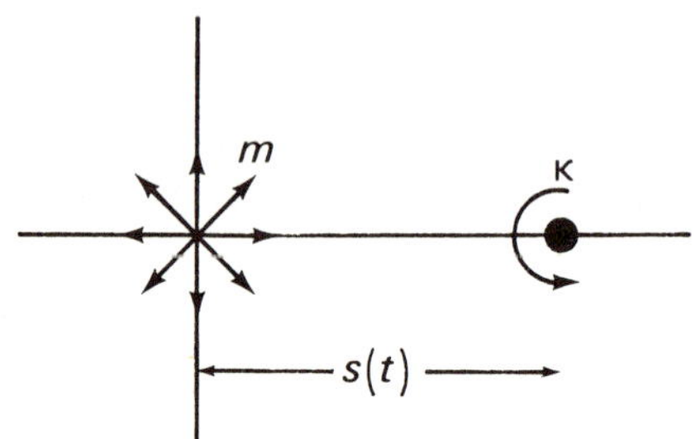

Fig. 12.6 Vortex in presence of a source

Here
$$\mathbf{u}_1 = (m/r)\,\mathbf{i}_r \tag{12.6.35}$$

Evidently the vortex, located initially on the x axis, remains on the x axis. Let the x coordinate of the vortex core be $s(t)$. Then from (12.6.34) and (12.6.35)

$$\frac{ds}{dt} = \frac{m}{s}$$

or on integrating subject to $s = A$ at $t = 0$

$$s = (A^2 + 2mt)^{\frac{1}{2}}. \tag{12.6.36}$$

The velocity field resulting from the source and the moving vortex may be constructed now from the time-dependent potential

$$\phi = -\,m \log r + \kappa \tan^{-1}\big[y/[x - s(t)]\big] \tag{12.6.37}$$

where s is given by (12.6.36).†

Other examples involving vortices are considered in Section 12.8 which is concerned with complex variable methods for the solution of two dimensional flow problems (see also Problems 10, 11, 13–16). For evident reasons it is exceptional to find steady flow problems involving vortices.

12.6.3 CIRCULATION AROUND A CYLINDER

The velocity field
$$v_\theta = \kappa/r\,, \qquad v_r = 0 \tag{12.6.38}$$

associated with a rectilinear vortex may arise in a context divorced from vortices.

† For $m < 0$, the vortex is sucked into the sink after a finite time $A^2/2|m|$ and remains there indefinitely, since any attempt to escape is frustrated by the inwardly directed (infinite) radial velocity.

It is evident for example that (12.6.38) describes a possible flow of fluid around a circular cylinder of radius a centered on the origin (Fig. 12.7).

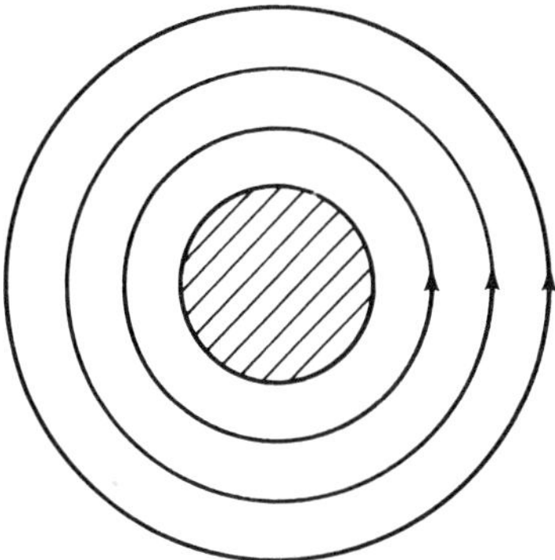

Fig. 12.7 Circulation around a solid circular cylinder.

For the flow depicted in Fig. 12.7 there is no 'real' vortex whose core is located at $r = 0$. However it remains true that the circulation for any contour enclosing the cylinder is $2\pi\kappa$ and in the present context the field (12.6.38) is said to describe a circulation flow around the cylinder. Of course because of Kelvin's theorem this circulation flow could not arise in motion initiated from rest in an inviscid fluid. However for reasons which will become clearer subsequently the circulation flow described by (12.6.38) is important in the theory of lift on an aircraft wing, (Section 12.8).

A further important point is as follows. Suppose we have solved some problem concerning flow past a cylinder, e.g. the problem of a cylinder placed in a uniform stream which is the subject of the next section; once the possibility of circulation is admitted, the solution obtained may be augmented with the velocity field (12.6.38) without in any sense disturbing the boundary conditions on the cylindrical surface $r = a$. In other words if circulation is admissible there is a lack of uniqueness in the solutions of flow problems around circular cylinders. In order to obtain a unique solution for these problems it is important to specify the strength of the circulation. Usually, in view of Kelvin's theorem there exist grounds for assuming that $\kappa = 0$.

A similar situation arises in problems of flow past cylinders of non-circular cross section, e.g. see Section (12.8) for the case of an elliptical cylinder.

12.6.4 TRANSLATORY MOTION OF RIGID CYLINDER IMMERSED IN A FLUID

We consider first the problem of a stream flowing in the negative x direction past a cylinder of radius a whose centre is at the origin, illustrated below.

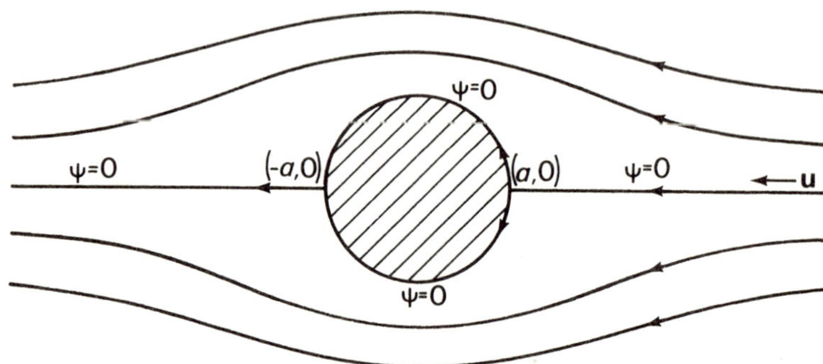

Fig. 12.8 Streaming motion past a circular cylinder

We suppose that at points remote from the cylinder

$$\mathbf{v} = -u\mathbf{i} \tag{12.6.39}$$

where u is the constant stream velocity. The velocity field (12.6.39) derives from the potential

$$\phi = ux \equiv ur\cos\theta\,.$$

Accordingly we seek a solution to the flow problem specified by Fig. 12.8 of the form

$$\phi = ur\cos\theta + \phi_1\,, \tag{12.6.40}$$

where

(i) $\partial\phi/\partial r = 0$ on $r = a$,

(ii) grad $\phi_1 \to 0,\ r \to \infty$,

(iii) ϕ_1 satisfies Laplace's equation.

In particular the second of these conditions states that at large distances the stream is undisturbed by the presence of the cylinder.

In view of (i) and the form of the term $ur\cos\theta$ appearing in (12.6.40) we seek a solution ϕ_1 also proportional to $\cos\theta$. Inspection of (12.6.15) in conjunction with (ii) shows that the appropriate term is $c_1\cos\theta/r$. With this choice ϕ may be written

$$\phi = [ur + c_1 r^{-1}]\cos\theta$$

and condition (i) is met by choosing $c_1 = ua^2$. Accordingly the flow problem is solved (in the absence of circulation) by

$$\phi = u(r + a^2 r^{-1}) \cos\theta \qquad (12.6.41)$$

which, formally, is a combination of a uniform stream together with a line dipole.

The pressure distribution on $r = a$ determines the force experienced by the cylinder. From Bernoulli's theorem (12.1.3) and the result that on $r = a$

$$v_r = 0, \qquad v_\theta = -\left(\frac{1}{r}\frac{\partial\phi}{\partial\theta}\right)_{r=a} = 2\,u \sin\theta$$

we find for the pressure distribution on the cylinder in the absence of body forces

$$P(a, \theta) = \text{const.} - 2\,\rho_0 u^2 \sin^2\theta\,. \qquad (12.6.42)$$

[The constant is of no importance in determining the force experienced by the cylinder.]

From Fig. 12.9 the forces F_x and F_y per unit length of cylinder are given by

$$\left.\begin{aligned} F_x &= -\int_0^{2\pi} aP(a,\theta)\cos\theta\,d\theta \qquad &(a)\\ F_y &= -\int_0^{2\pi} aP(a,\theta)\sin\theta\,d\theta \qquad &(b)\end{aligned}\right\} \qquad (12.6.43)$$

On substituting from (12.6.42) for P, there are no contributions to F_x and F_y from the constant, as might be expected from symmetry. More surprisingly, since the integrals of $\sin^2\theta\cos\theta$ and $\sin^3\theta$ over the range $0 \leqslant \theta \leqslant 2\pi$ also vanish, there are no contributions from the term involving u^2 in (12.6.42). The present result is one example of d'Alembert's paradox which states that the force experienced by an immobilised rigid body immersed in a stream is zero. A general proof of d'Alembert's paradox, which is true for an inviscid fluid in the absence of circulation, is given by Milne-Thomson in *Theoretical Hydrodynamics,* Macmillan (1949).

The paradox contradicts experimental findings and arises from neglect of viscosity effects in the fluid.

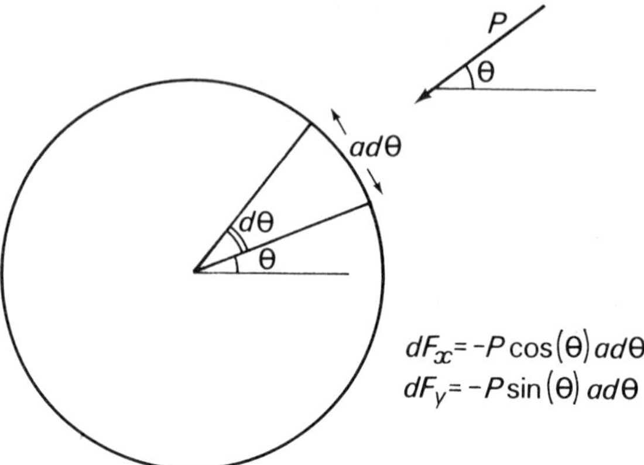

Fig. 12.9 Calculation of forces acting on a cylinder.

Superimposing the uniform stream potential $-ur\cos\theta$ on (12.6.41) leads to

$$\phi = \frac{ua^2\cos\theta}{r} \equiv \frac{ua^2x}{r^2} \qquad (12.6.44)$$

which, from the method of construction, is the potential of a cylinder whose centre is *currently* located at $r = 0$ and whose velocity is u in the positive x direction. To obtain the potential for a cylinder moving in the x direction at an arbitrary velocity we denote the position of the centre of the cylinder by $x = \alpha(t)$, $y = 0$. The velocity of the cylinder is then $u(t) \equiv \dot{\alpha}(t)$ in the x direction and the acceleration is $\dot{u} = \ddot{\alpha}(t)$. Taking account of the fact that the cylinder centre is no longer at the origin, the unsteady motion of the cylinder through the fluid, which is at rest at large distances, is obtained by replacing (12.6.44) by

$$\phi = \frac{ua^2x'}{(r')^2} \qquad (12.6.45)$$

where x' and r' are coordinates referred to the moving origin. With respect to a fixed origin

$$x' = x - \alpha(t), \quad r'^2 = [x-\alpha(t)]^2 + y^2$$

and (12.6.45) becomes

$$\phi = \frac{u(t)a^2[x-\alpha(t)]}{[x-\alpha(t)]^2+y^2}. \qquad (12.6.46)$$

Equation (12.6.46) is required for calculating the quantity $\partial\phi/\partial t$ which enters the unsteady Bernoulli equation. We find (on remembering $\dot{\alpha} = u$)

$$\left(\frac{\partial\phi}{\partial t}\right) = \frac{\dot{u}a^2\cos\theta'}{r'} - \frac{u^2a^2}{r'^2} + \frac{2u^2a^2\cos^2\theta'}{r'^2}$$

$$= \frac{\dot{u}a^2\cos\theta'}{r'} + \frac{u^2a^2\cos(2\theta')}{r'^2} \qquad (12.6.47)$$

where r' and θ' are polar coordinates referred to the centre of the sphere. On the cylinder $r' = a$, (12.6.47) yields

$$\left(\frac{\partial\phi}{\partial t}\right)_{(r'=a)} = \dot{u}a\cos\theta' + u^2\cos(2\theta')\,. \tag{12.6.48}$$

To find $\mathbf{v}^2$ on $r' = a$ it is more convenient to express (12.6.46) in the form

$$\phi = \frac{ua^2\cos\theta'}{r'}$$

whence on $r' = a$

$$v_{r'} = -\left(\frac{\partial\phi}{\partial r'}\right)_{(r'=a)} = u\cos\theta'\,\cdot$$

$$v_{\theta'} = -\left(\frac{1}{r'}\frac{\partial\phi}{\partial\theta'}\right)_{(r'=a)} = u\sin\theta'$$

so that on the surface of the cylinder

$$\mathbf{v}^2 = u^2\,, \qquad (r' = a)\,. \tag{12.6.49}$$

Combining Bernoulli's theorem (12.1.3) (in the absence of body forces) with (12.6.48) and (12.6.49) now yields for the pressure distribution on the cylinder

$$P = \rho_0 f(t) - \tfrac{1}{2}\rho_0 u^2 + \rho_0\dot{u}a\cos\theta' + \rho_0 u^2\cos(2\theta')\,. \tag{12.6.50}$$

Substituting (12.6.50) into (12.6.43), with θ replaced by θ' in the latter, now yields $F_y = 0$ and

$$F_x = -\rho_0\dot{u}a^2\int_0^{2\pi}\cos^2\theta'\,d\theta'$$

$$= -\pi a^2\rho_0\dot{u}\,, \tag{12.6.51}$$

the only non-zero contribution to F_x from (12.6.50) arising from the term involving the acceleration $\dot{u}$.

Thus for accelerated motion of the cylinder in the positive x direction the liquid exerts per unit length of cylinder a force $-\pi a^2\rho_0\dot{u}$ in the x direction. To oppose the liquid motion it is necessary to impose a force $\pi a^2\rho_0\dot{u}$ in the positive x direction on the cylinder. For a cylinder in arbitrary motion with velocity $\mathbf{u}$ and acceleration $\dot{\mathbf{u}}$ this last result generalises to the following. In order to overcome the resistive inertia of the liquid, the force per unit length required to maintain the motion is

$$\mathbf{F} = M'\dot{\mathbf{u}}$$

where M' is the mass of fluid displaced by the cylinder $[M' = \pi a^2\rho_0]$. If additionally the cylinder itself possesses inertial mass M per unit length, Newton's second law requires a further force $M\dot{\mathbf{u}}$. Combining these two results shows that the total force per unit length required to accelerate the immersed cylinder is

$$\mathbf{F} = (M+M')\dot{\mathbf{u}} \tag{12.6.52}$$

and the cylinder behaves as if it possessed an effective mass $(M+M')$.

The result that there is no fluid resistance for unaccelerated motion is a consequence of d'Alembert's paradox. The result is easily obtained from the previous one (that no force is exerted by a stream flowing past a stationary cylinder) by noting that with the imposition of a *uniform* velocity the relationships between a stationary coordinate system and one centred on the cylinder is Galilean. It is well known that for Galilean transformations in Newtonian physics forces are invariant so that no force is required to maintain a constant cylinder velocity.

In the above calculation, body forces have been neglected. Body forces due to a uniform earth's gravitational field are easily taken into account as follows. For a body of *arbitrary* shape bounded by a closed surface S the term $\rho_0 \Omega$ contributing to the pressure in Bernoulli's euqation (12.1.3) leads to a total force $\mathbf{F}_b$

$$\mathbf{F}_b = \iint_S (-\rho_0\Omega)(-\mathbf{dS}) = \iint_S \rho_0\Omega\,\mathbf{dS}\,. \tag{12.6.53}$$

For the earth's gravitational field we write

$$\Omega = gy$$

where y is in the outward normal direction to the earth's surface. The integral (12.6.53) becomes

$$\mathbf{F}_b = \rho_0 g \iint_S y\,\mathbf{dS}\,. \qquad (12.6.54$$

It is a simple exercise to evaluate the vector integral above to obtain finally

$$\mathbf{F}_b = -\rho_0\overline{\Omega}\,\mathbf{g} \tag{12.6.55}$$

where $\mathbf{g}(=-g\mathbf{j})$ denotes the vector acceleration due to gravity, and $\overline{\Omega}$ is the volume.†

Equation (12.6.55) shows that the effect of the body force is a vertical upthrust opposing the gravitational field and of magnitude equal to the weight of fluid displaced. This result is of course the famous one attributed to Archimedes.

In the above analysis it is pre-supposed that the density is constant. This is a necessary restriction for the precise validity of Archimedes principle in hydrodynamics. However the incompressibility restriction may be lifted for the hydrostatic case. Here the equilibrium equation

$$-\operatorname{grad} P + \rho\,\mathbf{g} = 0 \tag{12.6.56}$$

shows that with $\mathbf{g}$ directed along the (negative) y axis, $\partial P/\partial x = \partial P/\partial z = 0$ so that P depends only on y. If now we compare the two problems (presence or absence of the body) for a reservoir whose free surface is maintained at the same vertical height, then $P = P(y)$, $\rho = \rho(y)$ are the same functions in each case.

† Here $\overline{\Omega}$, the volume of the body, is not to be confused with Ω the gravitational potential.

It follows from (12.6.56) that $\mathbf{F}_b$ given by

$$\mathbf{F}_b = -\iint_S P\,d\mathbf{S} = -\iiint_{\bar{\Omega}} \text{grad}\,P d\bar{\Omega}$$

may be written

$$\mathbf{F}_b = -\mathbf{g}\iiint \rho d\bar{\Omega}$$

and the upthrust is of the same magnitude as the weight of displaced fluid. Notice that the derivation of Archimedes' principle entails converting a surface integral into a volume integral, assuming in the region occupied by the body a hypothetical fluid density $\rho = \rho(y)$ identical to that of the real fluid outside the body. This latter step has no parallel in a compressible hydro*dynamic* situation and in these circumstances Archimedes' principle fails. Of course, for nearly incompressible flows, the principle holds approximately.

A simple application of (12.6.52) together with Archimedes' principle concerns the acceleration of a long cylinder moving freely in a fluid. The force **F** in (12.6.52) is given by the weight of the body acting downwards (Mg per unit length) together with the upthrust ($-$M$'$g per unit length). The resulting equation leads to the constant acceleration

$$\dot{\mathbf{u}} = \frac{(M - M')\mathbf{g}}{M + M'} \tag{12.6.57}$$

which is in the direction of g for $M > M'$ and contrary if the cylinder is less dense than the displaced fluid $(M' > M)$.

Many other two dimensional flow problems may be solved using the expansion (12.6.15). However these problems also yield readily to the complex variable method discussed subsequently (Section 12.8).

12.7 Axisymmetric flow problems in spherical polar coordinates

12.7.1 General solution of axisymmetric Laplace equation

Spherical polar coordinates (r, θ, ψ) are defined by (see Fig. 12.10)

$$x = r\sin\theta\cos\psi\,, \quad y = r\sin\theta\sin\psi\,, \quad z = r\cos\theta\,. \tag{12.7.1}$$

In terms of r, θ, ψ Laplace's equation assumes the form

$$\frac{\partial^2\phi}{\partial r^2} + \frac{2}{r}\frac{\partial\phi}{\partial r} + \frac{1}{r^2\sin\theta}\frac{\partial}{\partial\theta}\left(\sin\theta\,\frac{\partial\phi}{\partial\theta}\right) + \frac{1}{r^2\sin^2\theta}\frac{\partial^2\phi}{\partial\psi^2} = 0\,. \tag{12.7.2}$$

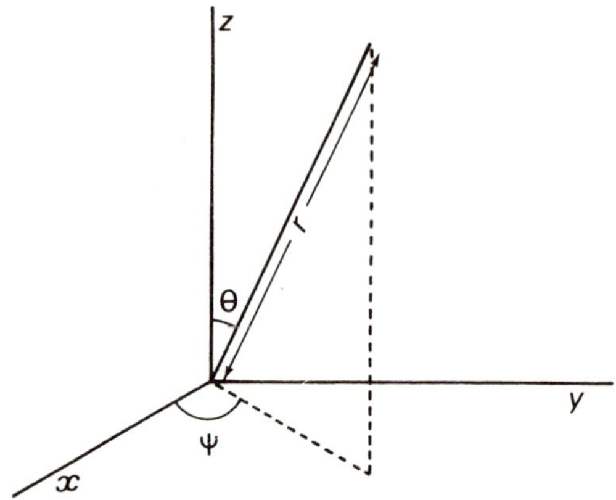

Fig. 12.10 Spherical polar coordinates (r, θ, ψ)

Here we shall be concerned only with solutions of the form $\phi = \phi(r, \theta)$ of (12.7.2) independent of the second polar angle ψ. In these circumstances (12.7.2) reduces to

$$\frac{\partial^2\phi}{\partial r^2} + \frac{2}{r}\frac{\partial\phi}{\partial r} + \frac{1}{r^2\sin\theta}\frac{\partial}{\partial\theta}\left(\sin\theta\,\frac{\partial\phi}{\partial\theta}\right) = 0\,. \tag{12.7.3}$$

Also the component of $\mathbf{v}$ in the ψ direction is zero while v_r and v_θ, both independent of ψ, are given by

$$v_r = -\,\partial\phi/\partial r\,, \qquad v_\theta = -\frac{1}{r}\,(\partial\phi/\partial\theta)\,. \tag{12.7.4}$$

Flows described by (12.7.3) and (12.7.4) with non-zero components of velocity only in the r and θ directions, and for which v_r and v_θ are independent of ψ, are termed axisymmetric.

Separation of variable solutions of (12.7.3) are obtained by writing

$$\phi = g(r)\,f(\theta)\,. \tag{12.7.5}$$

The resulting ordinary differential equations are

$$\frac{1}{\sin\theta}\frac{d}{d\theta}\left(\sin\theta\,\frac{df}{d\theta}\right) + \lambda f = 0 \tag{12.7.6}$$

$$\frac{d^2g}{dr^2} + \frac{2}{r}\frac{dg}{dr} - \frac{\lambda g}{r^2} = 0 \tag{12.7.7}$$

where λ is the separation constant.

As in the previous section there arises an eigenvalue problem for λ. However here the arguments determining possible eigenvalues of λ are much more subtle than those needed previously. The arguments employed devolve on the fine detail of the forms of the solution of equation (12.7.6) for $f(\theta)$.

To solve (12.7.6) we introduce a new independent variable

$$\eta = \cos\theta \tag{12.7.8}$$

where, since θ ranges over the values $0 \leqslant \theta \leqslant \pi$, the corresponding range of η is

$$1 \geqslant \eta \geqslant -1\,. \tag{12.7.9}$$

From (12.7.8)
$$\frac{d}{d\theta} = \frac{d\eta}{d\theta}\frac{d}{d\eta} = -\sin\theta\,\frac{d}{d\eta}$$

so that
$$\sin\theta\,\frac{d}{d\theta} = -\sin^2\theta\,\frac{d}{d\eta} = -(1-\eta^2)\,\frac{d}{d\eta}\,.$$

Also
$$\frac{1}{\sin\theta}\,\frac{d}{d\theta} = -\frac{d}{d\eta}$$

and on inserting these results into (12.7.6) we obtain

$$\frac{d}{d\eta}\left[(1-\eta^2)\,\frac{dP}{d\eta}\right] + \lambda P = 0 \tag{12.7.10}$$

where
$$P(\eta) \equiv g(\theta)\,.$$

Equation (12.7.10) is known as Legendre's equation and solutions of the equation for arbitrary values of λ are known as Legendre functions. Here we shall be concerned with a restricted class of polynomial solutions, known as the Legendre Polynomials $P_n(\eta)$ associated with the eigenvalues $\lambda_n = n(n+1)$, $(n = 0, 1, 2 \ldots)$.

The arguments determining these discrete eigenvalues λ_n devolve on the series solutions of (12.7.10). The point $\eta = 0$ is an ordinary point of (12.7.10) so there are two power series solutions in η which may be obtained by the Frobenius method. We leave the details to the reader and state the results

$$P(\eta) = a_0\left[1 - \frac{\lambda}{2!}\,\eta^2 + \frac{\lambda[\lambda-(2.3)]}{4!}\,\eta^4 - \frac{\lambda[\lambda-(2.3)]\,[\lambda-(4.5)]}{6!}\,\eta^6 \ldots\right]$$

$$+\; a_1\left[\eta - \frac{[\lambda-(1.2)]}{3!}\,\eta^3 + \frac{[\lambda-(1.2)]\,[\lambda-(3.4)]}{5!}\,\eta^5 - \frac{[\lambda-(1.2)]\,[\lambda-(3.4)]\,[\lambda-(5.6)]}{7!}\,\eta^7 \ldots\right]$$

where a_0 and a_1 are the two arbitrary multiplicative constants associated with the solution of a second order ordinary linear homogeneous differential equation. The first series terminates for $\lambda = 0, 2.3, 4.5, \ldots$ and the second series terminates for $\lambda = 1.2, 3.4 \ldots$. For values of λ other than these neither series terminates and there then arises the question of convergence. In the latter case the ratio R_m of the $(m + 1)$th to mth term in both series is given by

$$R_m = -\frac{[\lambda - n(n+1)]}{(n+1)(n+2)}\eta^2 \qquad \begin{aligned} n &= 2m-2 \text{ (first series)} \\ n &= 2m-1 \text{ (second series).} \end{aligned}$$

In the limit $m \to \infty$ $$\operatorname*{Lt}_{m\to\infty} |R_m| = \eta^2$$

so that by the ratio test both series converge absolutely for $\eta^2 < 1$. The ratio test is inadequate for the case $\eta^2 = 1$. It is to be remembered that the end-points $\eta = \pm 1$ of the range (12.7.9) are in the field of legitimate interest. In fact both series diverge logarithmically for $\eta^2 = 1$ for most values of λ.†

These divergent series are not acceptable solutions since there is no reason why ϕ and its derivatives should diverge on the z axis specified by $\theta = 0$, $\theta = \pi$. The way out of the difficulty has been indicated. It is to note that any of the choices $\lambda = 0,\ 2.3,\ 4.5\ \ldots$ result in the first series terminating (the second series remaining divergent at $\eta^2 = 1$) while the choices $\lambda = 1.2, 3.4, \ldots$ results in the second series terminating (the first series remaining divergent). It follows that acceptable solutions of (12.7.10) are polynomials obtained either by

(i) choosing $\lambda = 0, 2.3, 4.5, \ldots$ with $a_1 = 0$

or by (ii) choosing $\lambda = 1.2, 3.4 \ldots$ with $a_0 = 0$.

The resulting totality of acceptable solutions corresponding to $\lambda = n(n+1)$, $n = 0, 1, 2 \ldots$ and obtaining alternately from the two series are

$$\begin{aligned}
n &= 0 & \lambda &= 0 & P(\eta) &= a_0 \\
n &= 1 & \lambda &= 1.2 & P(\eta) &= a_1\eta \\
n &= 2 & \lambda &= 2.3 & P(\eta) &= a_0(1 - 3\eta^2) \\
n &= 3 & \lambda &= 3.4 & P(\eta) &= a_1(\eta - 5\eta^3/3) \\
n &= 4 & \lambda &= 4.5 & P(\eta) &= a_0(1 - 10\eta^2 + 35\eta^4/3)\,. \quad \text{etc.}
\end{aligned}$$

The Legendre polynomials $P_n(\eta)$, $n = 0, 1, 2 \ldots$ are defined to be the above solutions but with a_0 or a_1 chosen according to the prescription $P_n(1) = 1$.

† Proof that the series diverge requires the Gauss test. If we write $n = (2m - 2)$ (first series) or $n = (2m - 1)$ (second series) we have in each case $1/R_m = 1 + 1/m + O(1/m^2)$ (m large). The Gauss test states that a series converges if in the expansion $1/R_m = 1 + \mu/m + O(1/m^2)$ the coefficient μ is greater than unity. For $\mu \leqslant 1$ the series diverges and for $\mu = 1$ the series diverges logarithmically. It is arguable that although both series diverge at $\eta = \pm 1$, that nevertheless, by judicious choice of the ratio a_1/a_0, it may be possible to cancel the (logarithmic) singularities. This is indeed possible at one end of the range (say $\eta = 1$), but since one series is odd in η while the other series is even, the singularities then fail to cancel at $\eta = -1$.

Thus for $n = 1$, $a_1 = 1$ and $P_1(\eta) = \eta$, etc. The resulting polynomials are

$$\begin{array}{lll} n = 0 & \lambda = 0 & P_0(\eta) = 1 \\ n = 1 & \lambda = 1.2 & P_1(\eta) = \eta \\ n = 2 & \lambda = 2.3 & P_2(\eta) = \tfrac{1}{2}(3\eta^2 - 1) \\ n = 3 & \lambda = 3.4 & P_3(\eta) = \tfrac{1}{2}(5\eta^3 - 3\eta) \\ n = 4 & \lambda = 4.5 & P_4(\eta) = \tfrac{1}{8}(35\eta^4 - 30\eta^2 + 3) \quad \text{etc.} \end{array} \tag{12.7.11}$$

Substituting $\eta = \cos\theta$ now yields admissible solutions for $f(\theta)$. We note in particular the results

$$P_0(\cos\theta) = 1\,, \qquad P_1(\cos\theta) = \cos\theta\,, \qquad P_2(\cos\theta) = \tfrac{1}{2}(3\cos^2\theta - 1)\,.$$

The functions $P_n(\eta)$ are orthogonal in the sense

$$\int_{-1}^{1} P_n(\eta)\, P_m(\eta) d\eta = 0 \qquad n \neq m\,. \tag{12.7.12}$$

This result is easily proved from the differential equations

$$\frac{d}{d\eta}\,[(1-\eta^2)\,P_n'(\eta)] + n(n+1)\,P_n = 0$$

$$\frac{d}{d\eta}\,[(1-\eta^2)\,P_m'(\eta)] + m(m+1)\,P_m = 0\,.$$

On multiplying the first by P_m, the second by P_n, subtracting and integrating from $\eta = -1$ to $\eta = 1$ there results

$$- [n(n+1) - m(m+1)] \int_{-1}^{1} P_n(\eta) P_m(\eta)\, d\eta$$

$$= \int_{-1}^{1} \left[P_m \frac{d}{d\eta}[(1-\eta^2)P_n'(\eta)] - P_n \frac{d}{d\eta}[(1-\eta^2)P_m'(\eta)] \right] d\eta$$

$$= [(1-\eta^2)(P_m P_n' - P_n P_m')]_{-1}^{1} = 0$$

In obtaining the last result we have used the fact that the polynomials P_n are finite at $\eta = \pm 1$. Since n and m are positive and $n \neq m$, the factor $[n(n+1) - m(m+1)]$ is non-zero and the result (12.7.12) follows.

The case $n = m$ is more difficult and is not discussed.in the main body of the text (see however Problem 6). The result is

$$\int_{-1}^{1} P_n^2(\eta)\, d\eta = 2/(2n+1)\,. \tag{12.7.13}$$

The results (12.7.12) and (12.7.13) may be summarised in the formula

$$\int_{-1}^{1} P_n(\eta)\, P_m(\eta)\, d\eta \;=\; (2/(2n+1))\, \delta_{nm} \qquad (12.7.14)$$

where δ_{nm} is the Kronecker delta

$$\delta_{nm} \;=\; \begin{cases} 1 & n = m \\ 0 & n \neq m . \end{cases}$$

Alternately in terms of θ

$$\int_0^{\pi} P_n(\cos\theta)\, P_m(\cos\theta)\; \sin\theta\; d\theta \;=\; [2/(2n+1)]\, \delta_{nm} \, . \qquad (12.7.15)$$

One final result, not proved here (also covered in Problem 6) is that for $0 \leqslant h < 1$ the polynomials $P_n(\eta)$ are given by the *generating* function

$$(1 - 2h\eta + h^2)^{-\frac{1}{2}} \;=\; \sum_{n=0}^{\infty} h^n P_n(\eta) \qquad (0 \leqslant h < 1)\, . \qquad (12.7.16)$$

where the series converges absolutely in the indicated range.

We return to equation (12.7.7) for $g(r)$. With λ one of the eigenvalues $\lambda_n \;=\; n(n+1)$ equation (12.7.7) becomes

$$r^2 g'' \;+\; 2rg' \;-\; n(n+1)g \;=\; 0\, . \qquad (12.7.17)$$

The equation is homogeneous in the same sense as equation (12.6.9). Solutions are of the form r^{α} where α is a constant. The equation yields

$$\alpha(\alpha - 1) \;+\; 2\alpha \;-\; n(n+1) \;=\; 0$$

which factories into

$$(\alpha - n)\,(\alpha + n + 1) \;=\; 0$$

with solutions $\alpha = n,\; -(n+1)$. Therefore two independent solutions of (12.7.17) are r^n and $r^{-(n+1)}$. Combining these results with the Legendre polynomials leads to the general solution of the axisymmetric Laplace equation in the form

$$\phi \;=\; \sum_{n=0}^{\infty} \left[a_n r^n \;+\; b_n r^{-(n+1)} \right] P_n(\cos\theta) \qquad (12.7.18)$$

where the quantities a_n and b_n are two sets of arbitrary constants, whose determination devolves on the boundary conditions of the specific problem under consideration.

Of the various terms appearing in (12.7.18) we have already encountered the point source potential (b_0/r), the uniform stream in the z direction $[a_1 r P_1(\cos\theta) \;\equiv\; a_1 r\cos\theta \;\equiv\; a_1 z]$, the point dipole oriented in the z direction $[b_1 P_1(\cos\theta)/r^2 \;\equiv\; b_1 \cos\theta/r^2]$ and the constant term a_0 which is indeterminate in the present context.

We consider two problems the first of which concerns the flow of a stream past a sphere and the second the flow due to a source located outside a sphere.

12.7.2 TRANSLATORY MOTION OF A RIGID SPHERE IMMERSED IN A FLUID

We consider first the simple problem of a sphere of radius a, whose centre is at the origin, immersed in a uniform stream flowing with speed u in the negative z direction. For large r the velocity potential satisfies

$$\phi \;\to\; uz \;\equiv\; u\,r\,P_1(\cos\theta)\,. \tag{12.7.19}$$

Accordingly we seek a solution of the form

$$\phi \;=\; u\,r\,P_1(\cos\theta) \;+\; \phi_1$$

where

(i) $\partial\phi/\partial r \;=\; 0$ on $r \;=\; a$, (ii) grad $\phi_1 \to 0$ for $r \to \infty$,

(iii) ϕ_1 satisfies Laplace's equation.

The argument is similar to that used in the derivation of (12.6.41) for the analogous problem of a cylinder. Evidently in order to satisfy (i) for all possible θ and also to ensure (ii) we choose from the general expansion (12.7.18) the dipole term $b_1 r^{-2} P_1(\cos\theta)$ and write (on absorbing u into the definition of b_1)

$$\phi \;=\; u\cos\theta(r + b_1/r^2)\,.$$

Condition (ii) is now satisfied with the choice $b_1 \;=\; \frac{1}{2}a^3$ so that the solution of the problem posed is given by

$$\phi \;=\; u\cos\theta(r + a^3/2r^2)\,. \tag{12.7.20}$$

On the sphere $r = a$, $v_r = 0$ and v_θ is given by

$$v_\theta \;=\; -\left(\frac{1}{r}\frac{\partial\phi}{\partial\theta}\right)_{r=a} \;=.\; \frac{3}{2}\,u\sin\theta\,.$$

From Bernoulli's theorem (12.1.3) in the absence of body forces

$$P(a,\theta) \;=\; \text{const.} \;-\; \frac{9}{8}\,u^2\sin^2\theta\,. \tag{12.7.21}$$

[As in the case of the cylinder, the constant does not contribute to the force on the sphere.] From symmetry the force on the sphere due to (12.7.21) is in the z direction and of magnitude

$$F_z \;=\; -\int_0^{\pi}\int_0^{2\pi} P(a,\theta)\cos\theta\; a^2\sin\theta\; d\theta\; d\psi \tag{12.7.22}$$

$$\;=\; -\,2\pi\,a^2\int_0^{\pi} P(a,\theta)\cos\theta\,\sin\theta\; d\theta\,. \tag{12.7.23}$$

[In (12.7.22) the element of surface area dS is $a^2 \sin\theta\, d\theta\, d\psi$ (see Fig. 12.11) and the resolved components of the force on dS is $-P\, dS \cos\theta$]. Substituting from (12.7.21) for P into (12.7.23) leads to $F_z = 0$, providing a further example of d'Alembert's paradox.

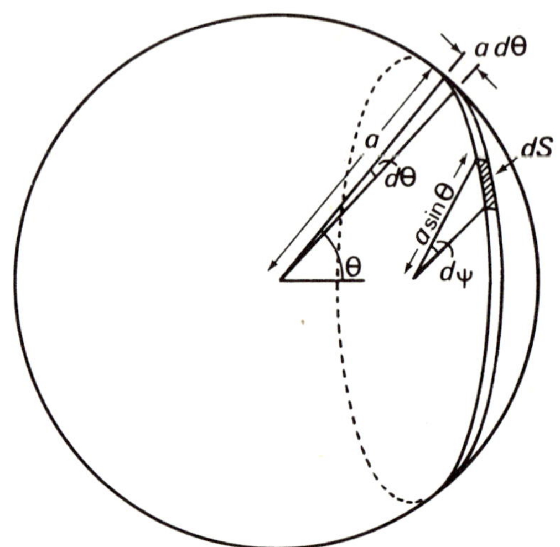

Fig. 12.11 Element of area of a sphere $dS = a^2 \sin\theta\, d\theta\, d\psi$

A more interesting problem is provided by the accelerated motion of a sphere. Following the prescription adopted for the cylinder in obtaining (12.6.44) the potential of a sphere *currently* located with centre at the origin and moving in the positive z direction with velocity u is obtained by superimposing on (12.7.20). the uniform stream potential $-ur\cos\theta$. There results

$$\phi = u\, a^3 \cos\theta / 2r^2 = ua^3 z / 2r^3\,.$$

Finally to adapt this potential to the problem of a sphere whose centre is located at $z = \alpha(t)$ moving with speed $u(t) \equiv \dot{\alpha}(t)$ we write in place of z and r

$$z' = z - \alpha(t)\,, \quad r' = (x^2 + y^2 + z'^2)^{\frac{1}{2}}$$

obtaining

$$\phi = \frac{u(t)\, a^3\, [z - \alpha(t)]}{2\,[x^2 + y^2 + [z - \alpha(t)]^2]^{\frac{3}{2}}}\,. \tag{12.7.24}$$

From (12.7.24) we find on the sphere $r' = a$ for $\partial\phi/\partial t$

$$\left(\frac{\partial\phi}{\partial t}\right)_{r'=a} = \tfrac{1}{2}\,\dot{u}\, a \cos\theta' - \tfrac{1}{2}\, u^2 + \tfrac{3}{2}\, u^2 \cos^2\theta'$$

$$= \tfrac{1}{2}\,\dot{u}\, a\, P_1(\cos\theta') + u^2 P_2(\cos\theta') \tag{12.7.25}$$

where θ' is the polar angle, defined by $z' = r'\cos\theta'$, referred to the centre of the sphere.

The velocity components are most easily obtained by writing ϕ in the form

$$\phi = \frac{u\,a^3 \cos\theta'}{2r'^2}$$

whence from (12.7.4)

$$(v_{r'})_{(r=a)} = u\cos\theta' , \qquad (v_{\theta'})_{(r=a)} = \tfrac{1}{2}u\sin\theta'$$

$$(\mathrm{v}^2)_{(r=a)} = u^2(\cos^2\theta' + \tfrac{1}{4}\sin^2\theta')$$

$$= \tfrac{1}{2}u^2[P_2(\cos\theta') + P_0(\cos\theta')] . \qquad (12.7.26)$$

From (12.7.25), (12.7.26) and Bernoulli's equation (12.1.3) in the absence of body forces

$$P(a,\theta) = \rho_0 f(t)P_0 - \tfrac{1}{4}\rho_0 u^2[P_2 + P_0] + \rho_0 u^2 P_2 + \tfrac{1}{2}\rho_0 \dot{u}\,a\,P_1 \qquad (12.7.27)$$

where for subsequent convenience we have inserted a factor $P_0(\equiv 1)$ in the first term. On inserting (12.7.27) into (12.7.23) and recalling that the quantity $\cos\theta$ appearing in the latter may be replaced by P_1 we obtain integrals of the form

$$\int_0^{\pi} P_1(\cos\theta)\,P_n(\cos\theta)\sin\theta\,d\theta \qquad (n = 0, 1, 2) .$$

In view of the orthogonality relations (12.7.15), the integrals for $n = 0$ and $n = 2$ are zero so that the only contribution from (12.7.27) to F_z is the term involving the acceleration $\dot{u}$. We find

$$F_z = -2\pi a^2(\tfrac{1}{2}\rho_0 \dot{u} a) \int_0^{\pi} P_1^2(\cos\theta)\sin\theta\,d\theta = -\frac{2}{3}\pi a^3 \rho_0 \dot{u} = -\frac{1}{2}M'\dot{u}$$

where M' is the mass of fluid displaced by the sphere.

Following the arguments used previously in Section 12.6 the effective mass of the sphere is $M + \frac{1}{2}M'$ where M is the mass of the sphere. The acceleration of a sphere immersed in a fluid in the earth's gravitational field is

$$\dot{\mathrm{u}} = \frac{M - M'}{M + \frac{1}{2}M'}\,\mathrm{g} \qquad (12.7.28)$$

which result is to be compared with (12.6.57) for the case of a cylinder. Equation (12.7.28) is relevant to the motion of a balloon in the atmosphere.

12.7.3 FLOW DUE TO A POINT SOURCE LOCATED OUTSIDE A SPHERE

The second axisymmetric problem considered makes use of all the terms in the general expansion (12.7.18) and also uses the generating function result (12.7.16).

The problem in question is to determine the flow arising from the location of a point source at $z = f,\ x = y = 0,$ outside a sphere of radius a (Fig. 12.12).

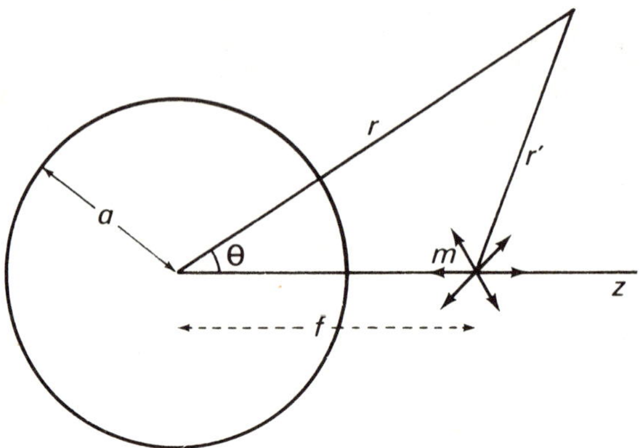

Fig. 12.12 Point source located outside a sphere

For the isolated point source in the absence of the sphere the potential at **r** is given by

$$\phi = \frac{m}{r'}$$

where r' is the distance of the point **r** from the source.

We expect the velocity field to tend to zero as r (and r') tend to infinity. Accordingly we look for a solution

$$\phi = \frac{m}{r'} + \phi_1 \tag{12.7.29}$$

where

(i) grad $\phi_1 \to 0$ as $r \to \infty$, (ii) $\left(\dfrac{\partial\phi}{\partial r}\right)_{r=a} = 0$,

(iii) ϕ_1 satisfies Laplace's equation .

As a preliminary we express the point source potention m/r' in terms of Legendre polynomials. From Fig. 12.12

$$\frac{1}{r'} = (r^2 + f^2 - 2rf\cos\theta)^{-\frac{1}{2}}. \tag{12.7.30}$$

For $r > f$, (12.7.30) may be expanded in the form [from (12.7.16)]

$$\frac{1}{r'} = (r^2 + f^2 - 2rf\cos\theta)^{-\frac{1}{2}} = \frac{1}{r}\left(1 - \frac{2f}{r}\cos\theta + \frac{f^2}{r^2}\right)^{-\frac{1}{2}}$$

$$= \sum_{n=0}^{\infty} \frac{f^n}{r^{n+1}} P_n(\cos\theta) \qquad (r > f), \tag{12.7.31}$$

while for $r<f$

$$\frac{1}{r'} = \frac{1}{f}\left(1-\frac{2r}{f}\cos\theta+\frac{r^2}{f^2}\right)^{-\frac{1}{2}} = \sum_{n=0}^{\infty}\frac{r^n}{f^{n+1}}P_n(\cos\theta) \qquad (r<f)\,. \quad (12.7.32)$$

We observe that (12.7.31) and (12.7.32) are particular examples of the general expansion (12.7.18). In particular for $r \gg f$ the leading term $(n=0)$ in (12.7.31) is $1/r$ and this concurs with physical expectation.

Since the boundary condition (ii) refers to the radius a which is less than f, the expansion of immediate importance is (12.7.32).

We write for $r<f$

$$\phi = m\sum_{n=0}^{\infty}\frac{r^n}{f^{n+1}}P_n(\cos\theta) + \phi_1 \qquad (r<f)$$

where in the first instance ϕ_1 may be of the general form (12.7.18). However because of (i) we reject the terms involving positive powers of r and write

$$\phi_1 = \sum_{n=0}^{\infty}\frac{b_n}{r^{n+1}}P_n(\cos\theta) \qquad (a \leqslant r < \infty)\,.$$

The boundary condition (ii) is now satisfied if

$$\sum_{n=0}^{\infty}\left(\frac{mna^{n-1}}{f^{n+1}} - \frac{(n+1)b_n}{a^{n+2}}\right)P_n(\cos\theta) = 0$$

and this is true for all θ if

$$b_n = \frac{mn}{(n+1)}\frac{a^{2n+1}}{f^{n+1}}\,.$$

Thus finally ϕ may be written

$$\phi = m\left[\frac{1}{r'} + \sum_{n=0}^{\infty}\frac{n}{n+1}\frac{a^{2n+1}}{(fr)^{n+1}}P_n(\cos\theta)\right]. \qquad (12.7.33)$$

Since $(n/n+1) < 1$ while $f>a$ and $r \geqslant a$ the series converges for all relevant values of r [by comparison with (12.7.16) with $h=a/f$].

Direct calculation of the thrust on the sphere is tedious. From symmetry the thrust is in the z direction. On the sphere $v_r=0$ and $v_\theta = -\left(\frac{1}{r}\frac{\partial\phi}{\partial\theta}\right)_{r=a}$.

Substituting for $\frac{1}{r'}$ from (12.7.32) into (12.7.33) and differentiating yields

$$(v_\theta)_{r=a} = -\frac{m}{af}\sum_{0}^{\infty}\left(\frac{2n+1}{n+1}\right)h^n\frac{d}{d\theta}P_n(\cos\theta) \qquad (12.7.34)$$

where $$h = a/f\,.$$

From Bernoulli's theorem $$P(a, \theta) = \text{const.} - \tfrac{1}{2}\rho_0 v_\theta^2$$

and from (12.7.34) and (12.7.23)

$$F_z = \frac{\pi\rho_0 m^2}{f^2} \int_0^\pi \cos\theta \sin\theta \sum_{n,q=0}^{\infty} \frac{(2n+1)(2q+1)}{(n+1)\ (q+1)} h^{n+q} \frac{d}{d\theta} P_n(\cos\theta) \times \frac{d}{d\theta} P_q(\cos\theta) d\theta \, .$$

On interchanging the order of the (double) summation and integration, and changing the integration variable to $\eta = \cos\theta$ we obtain

$$F_z = \frac{\pi\rho_0 m^2}{f^2} \sum_{n,q=0}^{\infty} \frac{(2n+1)(2q+1)}{(n+1)(q+1)} h^{n+q} \int_{-1}^{1} \eta(1-\eta^2) P_n'(\eta)\, P_q'(\eta) d\eta \, . \tag{12.7.35}$$

It is a non-trivial exercise in integration by parts together with the help of various properties of the Legendre polynomials [equations (12.7.14), the differential equation for P_n and equation (iv) of Problem 6] to show that (12.7.35) reduces to

$$F_z = \frac{4\pi\rho_0 m^2 h}{f^2} \sum_1^{\infty} n\, h^{2n} = \frac{4\pi\rho_0 m^2 a^3}{f(f^2 - a^2)^2}$$

so that the sphere is attracted towards the source.

12.8 Solution of plane two dimensional flow problems by complex variable theory

12.8.1 General results

Some two dimensional incompressible irrotational flow problems were encountered in Section 12.6. There, solutions were obtained by the method of separation of variables of Laplace's equation using plane polar coordinates.

Other separation of variable solutions are useful on occasion but the most versatile tool for the solution of the plane two dimensional Laplace equation

$$\frac{\partial^2\phi}{\partial x^2} + \frac{\partial^2\phi}{\partial y^2} = 0 \tag{12.8.1}$$

is based on complex variable theory. The basic result on which the technique devolves is that the general solution of (12.8.1) is expressible as the real part of an analytic (i.e. regular) function f of the complex variable $x + iy$. Proof of this result follows almost immediately.

In complex variable theory it is common to denote the independent variable $(x + iy)$ by z. We adopt this notation here writing

$$z = x + iy\,, \qquad \bar{z} = x - iy \tag{12.8.2}$$

where the bar implies change of sign of any quantities i appearing in the function. Of course z defined by (12.8.2) does not refer to the Cartesian coordinate z, which plays no role in two dimensional problems. For the remainder of this chapter, unless otherwise explicitly stated, the symbol z is defined by the first of (12.8.2).

To show that the general solution of (12.8.1) is given by

$$\phi = \operatorname{Re} f(z) \tag{12.8.3}$$

where f is an analytic function of z, consider a change of independent variables from x, y to $z, \bar{z}$. By the usual rules for change of variable

$$\frac{\partial}{\partial x} = \frac{\partial z}{\partial x}\frac{\partial}{\partial z} + \frac{\partial \bar{z}}{\partial x}\frac{\partial}{\partial \bar{z}} = \left(\frac{\partial}{\partial z} + \frac{\partial}{\partial \bar{z}}\right)$$

Similarly

$$\frac{\partial}{\partial y} = i\left(\frac{\partial}{\partial z} - \frac{\partial}{\partial \bar{z}}\right)$$

so that

$$\frac{\partial^2}{\partial x^2} + \frac{\partial^2}{\partial y^2} = \left(\frac{\partial}{\partial z} + \frac{\partial}{\partial \bar{z}}\right)^2 - \left(\frac{\partial}{\partial z} - \frac{\partial}{\partial \bar{z}}\right)^2 = 4\frac{\partial^2}{\partial z \partial \bar{z}}\,.$$

Accordingly (12.8.1) reduces to the simpler equation

$$\frac{\partial^2 \phi}{\partial \bar{z}\,\partial z} = 0 \quad \text{or} \quad \frac{\partial}{\partial \bar{z}}\left(\frac{\partial \phi}{\partial z}\right) = 0\,. \tag{12.8.4}$$

From the second version immediately above it follows that $\partial\phi/\partial z$ is independent of $\bar{z}$ and is therefore solely a function of z. For convenience we write this arbitrary function as $\frac{1}{2}f'(z)$ where the prime denotes differentiation and where f is also arbitrary, obtaining

$$\frac{\partial \phi}{\partial z} = \tfrac{1}{2}f'(z)\,.$$

A further integration shows that ϕ is the sum of $\frac{1}{2}f(z)$ together with any arbitrary function of $\bar{z}$ [say $\frac{1}{2}\bar{g}(\bar{z})$]. Thus

$$\phi = \tfrac{1}{2}[f(z) + \bar{g}(\bar{z})] \tag{12.8.5}$$

and this result is symmetrical in z and $\bar{z}$ as might be expected from the form of the equation (12.8.4).

If now we remember that the solution to (12.8.1) is necessarily real, it follows that the imaginary part of $\bar{g}(\bar{z})$ is minus the imaginary part of $f(z)$. However since a function of a complex variable is determined uniquely to within an arbitrary additive real constant C by its imaginary part, we must have $\bar{g}(\bar{z}) = \overline{f(z)} + C$. Further since $\frac{1}{2}C$ can be absorbed into the definition of $f(z)$ and also

$$\overline{f(z)} = \bar{f}(\bar{z}) \tag{12.8.6}$$

then without loss of generality $g = f$ and

$$\phi = \tfrac{1}{2}[f(z) + \overline{f(z)}] = \operatorname{Re} f(z)\,. \tag{12.8.7}$$ †

Equally ϕ may be taken to be the imaginary part of an analytic function, say $F(z)$, but this result is included in (12.8.6) if f is replaced by $-iF$.
In the hydrodynamic context if we write

$$f(z) = \phi + i\psi \tag{12.8.8}$$

where ψ is the imaginary part of $f(z)$, then ψ also plays a significant role in the theory – indeed the quantity ψ is of more direct physical significance than ϕ.††
Before interpreting the role of ψ we recall the Cauchy-Riemann relations

$$\frac{\partial\phi}{\partial x} = \frac{\partial\psi}{\partial y}\,, \quad \frac{\partial\phi}{\partial y} = -\frac{\partial\psi}{\partial x} \tag{12.8.9}$$

Eliminating ϕ and ψ in turn from (12.8.9) shows that both ψ and ϕ satisfy Laplace's equation

$$\nabla^2\phi = 0\,, \qquad \nabla^2\psi = 0$$

as would be expected from previous considerations.

The significance of ψ in hydrodynamical problems follows from noticing that the mass conservation equation

$$\frac{\partial v_x}{\partial x} + \frac{\partial v_y}{\partial y} = 0 \tag{12.8.10}$$

is satisfied automatically by writing

$$v_x = -\partial\hat{\psi}/\partial y\,, \quad v_y = \partial\hat{\psi}/\partial x \tag{12.8.11}$$

where $\hat{\psi}$ is called the stream function. However also $v = -\operatorname{grad}\phi$ so that

$$\frac{\partial\phi}{\partial x} = \frac{\partial\hat{\psi}}{\partial y}\,, \quad \frac{\partial\phi}{\partial y} = -\frac{\partial\hat{\psi}}{\partial x}$$

and these equations are of exactly the same form as (12.8.9). It follows that $\hat{\psi}$ is identifiable with ψ, at least to within an irrelevant constant. For, it may be shown that if ϕ is given then ψ is determined by equations (12.8.9) to within a constant. A constructive proof of this result leading to the determination of $f(z)$ is the subject of Problem 19.

† The bar notation is sometimes found confusing in the function context. $\bar{f}(z)$ is obtained by changing the sign of every i appearing in the function $f(z)$ except the i appearing in z itself. Thus, for example if $f(z) = az - 3ib \sin z$, $\bar{f}(z) = \bar{a}z + 3i\bar{b} \sin z$ and $\overline{f(z)} = \bar{a}\,\bar{z} + 3i\bar{b} \sin \bar{z}$. Since $\bar{f}(\bar{z})$ is obtained by changing the sign of every i, including the one associated with z, the result (12.8.6) follows.

†† ψ is not to be confused with the third polar coordinate of (r, θ, ψ), which does not appear in the rest of this chapter.

The physical role played by ψ follows from recalling in complex variable theory that the one parameter family of curves ϕ = const. intersect the one parameter family ψ = const. orthogonally. Since $\mathbf{v} = -\text{grad}\,\phi$ is necessarily normal to the curves ϕ = const., the curves ψ = const. are everywhere parallel to **v**; in other words the curves ψ = const. are streamlines.

The origin of the terminology stream function is now apparent. It is also worth noting that for *fixed* boundaries in two dimensional flow the boundary condition $\partial\phi/\partial n = 0$ is replaceable by the simpler condition that the boundary be a streamline, i.e. along a fixed boundary

$$\psi = \text{const.}$$

One final general result concerning ψ is of importance subsequently in Section 12.9. In Figure 12.13, $\psi = \beta_1$ and $\psi = \beta_2$ denote two streamlines. No flow occurs across these lines instantaneously so that from mass conservation arguments the volume flux of fluid across any curve e.g. c_1 or c_2 connecting the streamlines is the same. We shall now show that this volume flux (per unit distance in the direction normal to the plane of the paper) is $-(\beta_1 - \beta_2)$.

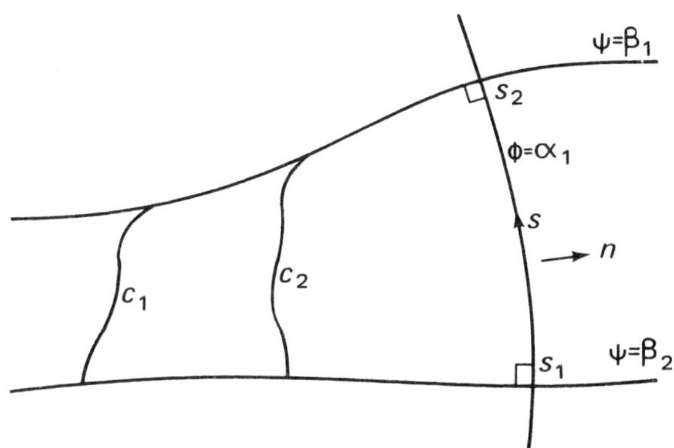

Fig. 12.13 Calculation of flux between two streamlines

Since the flux is independent of whichever path is chosen connecting the streamlines, we evaluate the flux for an equipotential say $\phi = \alpha_1$. Let n denote the normal direction to $\phi = \alpha_1$ and s a distance coordinate along $\phi = \alpha_1$. Then the flux of fluid through $\phi = \alpha_1$ per unit length in the (coordinate) z direction is

$$Q = -\int_{s_1}^{s_2} \frac{\partial\phi}{\partial n}\, ds \qquad (12.8.12)$$

where s_1 and s_2 denote the points of intersection of $\phi = \alpha_1$ with the two streamlines. However if at every point along $\phi = \alpha_1$ we erect a local Cartesian coordinate system with x' and y' axes directed respectively in the normal n and tangential s directions we have from the Cauchy-Riemann equations

$$\frac{\partial \phi'}{\partial x'} = \frac{\partial \psi}{\partial y'}$$

or more generally
$$\frac{\partial \phi}{\partial n} = \frac{\partial \psi}{\partial s} .$$

Substituting this result into (12.8.12) now leads to

$$Q = -\int_{s_1}^{s_2} \frac{\partial \psi}{\partial s}\, ds = -[\psi(s_2) - \psi(s_1)] = -(\beta_1 - \beta_2) \qquad (12.8.13)$$

which completes the proof.

The combined quantity $\phi + i\psi$ is usually denoted by $\omega(z)$ rather than $f(z)$ and is known as the complex potential. On differentiating the complex potential

$$\omega(z) = \phi + i\psi$$

with respect to x we obtain the complex velocity

$$\omega'(z) \equiv v = \frac{\partial \phi}{\partial x} + i\,\frac{\partial \psi}{\partial x} \qquad (12.8.14)$$

i.e.
$$v = -v_x + i\,v_y\,. \qquad (12.8.15)$$

A useful result in the context of Bernoulli's theorem is

$$\mathbf{v}^2 = v\bar{v} = \frac{d\omega}{dz}\,\frac{d\bar{\omega}}{d\bar{z}}\,. \qquad (12.8.16)$$

This completes the general results concerning the use of complex variable theory for two dimensional potential flows. Before discussing some simple examples we re-iterate that the method is restricted to plane flow problems.

On the other hand the stream function concept itself is not so limited. For example for irrotational incompressible axisymmetric flows expressed in terms of spherical coordinates, the equation of mass conservation is

$$\frac{\partial v_r}{\partial r} + \frac{2v_r}{r} + \frac{1}{r\sin\theta}\,\frac{\partial}{\partial\theta}(\sin\theta\, v_\theta) = 0\,. \qquad (12.8.17)$$

By writing this in the form

$$\frac{\partial}{\partial r}(r^2\sin\theta\, v_r) + \frac{\partial}{\partial\theta}(r\sin\theta\, v_\theta) = 0 \qquad (12.8.18)$$

it is evident that v_r and v_θ may be expressed in terms of the *Stokes'* stream function ψ_S

$$v_r = -\frac{1}{r^2\sin\theta}\,\frac{\partial\psi_S}{\partial\theta}\,, \qquad v_\theta = \frac{1}{r\sin\theta}\,\frac{\partial\psi_S}{\partial r}\,. \qquad (12.8.19)$$

However ψ_S does *not* satisfy Laplace's equation,† nor are the dimensions of $\psi_S\,[L^3T^{-1}]$ in this case the same as the dimensions of $\phi\,[L^2T^{-1}]$. In these circumstances it is hardly sensible to attempt to construct a complex potential like $\phi + i\psi$.

No use is made of the Stokes' stream function in the text; however see Problem 9 of this chapter and Problems 7 and 8 of Chapter 10.

It is not generally possible to introduce a stream function for flows in three dimensions. First in order to be able to cast the mass conservation equation into a form susceptible to an integral like (12.8.11) or (12.8.19), it is imperative that there exist only two non-vanishing components of velocity. Secondly in general it is not possible to represent a stream*line* by an equation like $\psi(x, y, z) = \text{const.}$†† for such an equation represents a surface and not a line. In fact a little consideraation will show that even the 'streamlines' represented by $\psi(x, y) = \beta$ in the theory of plane flows are not really streamlines at all but stream sheets formed from the family of streamlines obtained from intersections of the three dimensional sheet $\psi(x, y) = \beta$ with the planes $z = \text{const.}$†† Similarly in the axisymmetric case the equation $\psi_S(r, \theta) = \beta$ defines a stream sheet. The streamlines here are formed from the intersection of the surface $\psi_S = \beta$ with the planes $\Psi = \text{const.}$ where Ψ is the third spherical polar coordinate of r, θ, Ψ.*

12.8.2 Some simple examples of complex potentials

Simple examples of $\omega(z)$ are as follows

Uniform stream with velocity u in the positive x direction

Here
$$\omega = -uz \tag{12.8.20}$$

and
$$\frac{d\omega}{dz} = v = -v_x + i\,v_y = -u\,.$$

Hence
$$v_x = u\,, \quad v_y = 0$$

as indicated.

Uniform stream with velocity u making an angle α with the x axis

$$\omega = -ue^{-i\alpha}z \tag{12.8.21}$$

$$v = -v_x + i\,v_y = -ue^{-i\alpha}\,.$$

Hence $v_x = u\cos\alpha\,, \quad v_y = u\sin\alpha$ as indicated.

† See Problem 9.

†† Here z is the z coordinate.

* Here we have written Ψ for the polar angle to avoid confusion with the stream function.

A possible flow in a rectangular corner

$$\omega = Az^2 \qquad (A = \text{real constant}) . \tag{12.8.22}$$

Here
$$\phi + i\psi = A(x^2 - y^2) + 2iAxy$$

so that the streamlines are given by $2Axy = \text{const.}$ i.e. by the family of rectangular hyperbolae $xy = c$. In particular the choice $c = 0$ shows that $x = 0$, $y = 0$ are streamlines and (12.8.22) is a possible flow in the vicinity of a rectangular corner (Fig. 12.14).

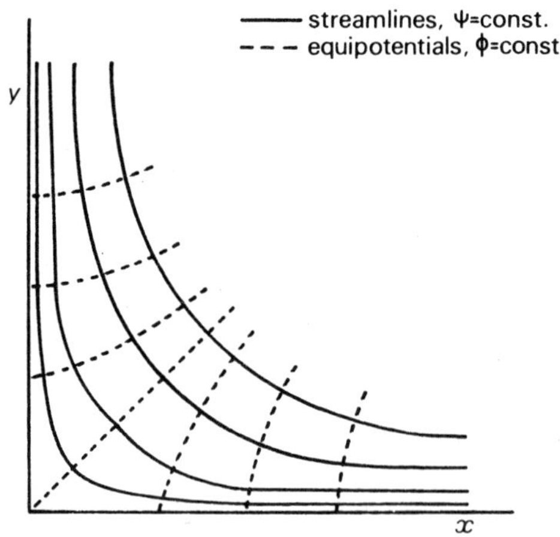

Fig. 12.14 Possible flow in a rectangular corner.

Line source

The complex potential for a line source is given by

$$\omega = -m \log z \tag{12.8.23}$$

for on writing $z = re^{i\theta}$ where r and θ are plane polar coordinates

$$\phi + i\psi = -m(\log r + i\theta)$$

and
$$\phi = -m \log r, \qquad \psi = -m\theta .$$

The potential $\phi = -m \log r$ is that of a line source [equation (12.6.17)] and the streamlines given by $\theta = \alpha$ lie in the radial direction as expected.

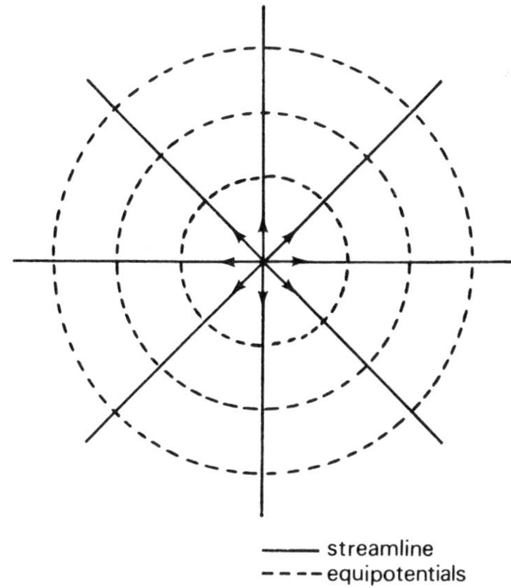

Fig. 12.15 Line source

Rectilinear vortex and circulation flow around a cylinder

Multiplying the complex potential (12.8.23) by $-i$ interchanges the roles of streamlines and equipotentials. The resulting flow is that of a rectilinear vortex. For the notation to be consistent with that of Section 12.6.2 we also replace m by κ obtaining

$$\omega = i\kappa \log z \qquad (12.8.24)$$

so that $\qquad \phi = -\kappa\theta, \qquad \psi = \kappa \log r.$

The potential $-\kappa\theta$ is identical with (12.6.21) while the streamlines, as expected, are given by $r = \text{const.}$ (Fig. 12.16).

The reader may like to verify that the detailed example of Section 2.4 used to illustrate the relation between spatial and material quantities is that of a rectilinear vortex.

Line dipole with orientation in the x direction

Here $\omega = \mu/z = (\mu/r)e^{-i\theta}$ so that $\phi = \mu \cos\theta/r$ in agreement with equation (12.6.19). For a line dipole with orientation α with respect to the x axis, the complex potential is

$$\omega = \mu e^{i\alpha}/z .$$

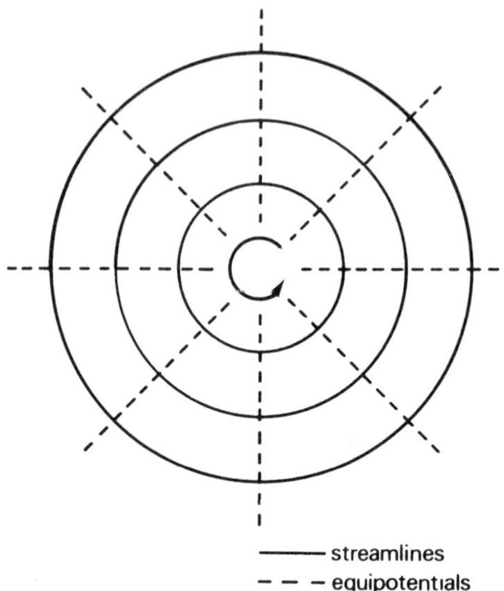

Fig. 12.16 Rectilinear vortex

Singular flows centred on points other than the origin

For a line source located at $z = z_0$ $(= x_0 + i\,y_0)$ rather than at the origin a trivial modification of (12.8.23) leads to

$$\omega = -m\log(z - z_0) \qquad \text{(source located at } z_0\text{)} \qquad (12.8.25)$$

Similarly for a rectilinear vortex located at z_0 the complex potential is given by

$$\omega = i\kappa\log(z - z_0) \qquad \text{(rectilinear vortex located at } z_0\text{)} \quad (12.8.26)$$

while for a line dipole with orientation α located at z_0

$$\omega = \mu e^{i\alpha}/(z - z_0)\,.$$

12.8.3 SUPERPOSITION OF LINE SOURCES, RECTILINEAR VORTICES, STREAMS IN ABSENCE OF BOUNDARIES

It is not entirely trivial that the complex potential due to (say) a stream $[\omega = -ue^{-i\alpha}z]$, a rectilinear vortex located at z_0 $[\omega = i\kappa\log(z - z_0)]$, a line source located at z_1 $[\omega = -m\log(z - z_1)]$, etc. is given by the superposition

$$\omega = -ue^{-i\alpha}z + i\kappa\log(z - z_0) - m\log(z - z_1), \text{ etc.}.$$

However in the absence of boundaries, this result is correct. To prove this it is necessary to show (i) at large distance the flow is dominated by the stream [and this is clearly so since $d\omega/dz = -ue^{-i\alpha} + O(z^{-1})$], (ii) that the efflux per unit length in the (coordinate) z direction through all closed circuits enclosing z_1 is $2\pi m$, (iii) that for all closed circuits surrounding z_0 the circulation is $2\pi\kappa$. We prove (iii) and leave (ii) to the reader. The circulation is given by

$$\text{circ}\ c = \oint_c \mathbf{v}\,.\,d\mathbf{r} = \oint_c v_x dx + v_y dy = -\text{Re}\oint_c \upsilon\, dz\,.$$

The latter result is easily obtained from $\upsilon = -v_x + i\,v_y$ and $dz = dx + i\,dy$.

Now $$\upsilon = \frac{d\omega}{dz} = -ue^{-i\alpha} + \frac{i\kappa}{z - z_0} - \frac{m}{z - z_1} + \text{etc.}$$

so that from the residue theorem, for all circuits enclosing z_0 and z_1 (say)

$$\text{circ}\ c = -\text{Re}\ 2\pi i\,(i\,\kappa - m) = 2\pi\kappa\,.$$

Evidently the result is unaffected by the location of z_1, whether it be inside, outside or indeed on the contour.

More generally, in the absence of boundaries, the flow due to a stream together with any distribution of rectilinear vortices, line dipoles, sources etc. is obtained by the superposition of the complex potentials of the individual flows.

Of course if rectilinear vortices are present the complex potential is nearly always time dependent since usually for a vortex located at a point $z = z_0$ (say), the velocity induced at z_0 by the stream and all the remaining singular flows is non-zero. In what follows we consider a number of problems which contain rectilinear vortices in motion.

12.8.4 IMAGES IN PLANE BOUNDARIES

The method of images is sometimes useful for two dimensional flow problems involving plane boundaries. Consider for example the problem of a line source located outside the plane $y = 0$ at $x = 0$, $y = h$. From previous experience of the point source problem it might be expected that a solution is effected by locating an image source of the same strength at the image point $x = 0$, $y = -h$. The resulting complex potential is

$$\begin{aligned}\omega &= -m\log(z - ih) - m\log(z + ih)\\ &= -m\log(z^2 + h^2)\,.\end{aligned}$$

On $z = x$ $$\omega = -m\log(x^2 + h^2)$$

which is entirely real. It follows that on the boundary $y = 0$ the stream function is zero. In other words the line $y = 0$ is a streamline as required.

The problem of a line source located at $z_0 = x_0 + iy_0$ outside two rigid walls $x = 0$, $y = 0$ is solved by using three images as in the point source case (Fig. 12.2). We leave the details to the reader. The solution is

$$\omega = -m\log\left[(z^2 - z_0^2)(z^2 - \bar{z}_0^2)\right]\,. \tag{12.8.27}$$

Of more interest is the problem of a rectilinear vortex located at some instant of time (say $t = 0$) at $x = 0$, $y = h$, outside the plane boundary $y = 0$. It is intuitive that the problem is solved by locating at the image point $x = 0$, $y = -h$ a rectilinear vortex of the same strength but *opposite* sign (Fig. 12.17).

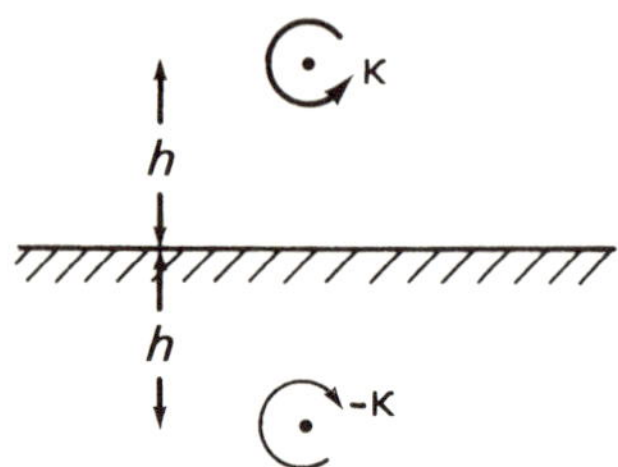

Fig. 12.17 Image solution for rectilinear vortex lying parallel to rigid plane.

The resulting potential is

$$\omega = i\kappa \log(z - ih) - i\kappa \log(z + ih).$$

Proof that $y = 0$ is a streamline follows if we can show that for $z = x$, ψ is constant. On $z = x$

$$\omega = i\kappa \log\left(\frac{x - ih}{x + ih}\right)$$

and on writing $(x + ih)$ in polar form, say $(x + ih) = Qe^{i\eta}$, we find $(\omega)_{z=x} = i\kappa \log(e^{-2i\eta}) = 2\eta\kappa$ which is real.

The vortex does not remain at $x = 0$, $y = h$ because the velocity induced at this point by the image vortex with potential

$$\omega_1 = -i\kappa \log(z + ih)$$

is $\kappa/2h$ in the positive x direction. It follows that the vortex moves parallel to the plane $y = 0$ at a speed $\kappa/2h$ carrying with it the image vortex. Thus at time t the complex potential is

$$\omega = i[\kappa \log(z - ih - \kappa t/2h) - \kappa \log(z + ih - \kappa t/2h)]. \quad (12.8.28)$$

For this problem we calculate the pressure on the plane $y = 0$. On $y = 0$, v is given by

$$v = \frac{d\omega}{dz} = i\kappa\left[\frac{1}{(x - \kappa t/2h - ih)} - \frac{1}{(x - \kappa t/2h + ih)}\right]$$
$$= -2\kappa h/[(x - \kappa t/2h)^2 + h^2] \quad (12.8.29)$$

which is real, i.e. $v_y = 0$ as expected. Also on $z = x$, i.e. $y = 0$,

$$\frac{\partial\phi}{\partial t} = \text{Re}\left(\frac{\partial\omega}{\partial t}\right) = \kappa^2/[(x - \kappa t/2h)^2 + h^2]. \quad (12.8.30)$$

From these results and the Bernoulli theorem (12.1.3)

$$P = P_0 - \rho_0 \kappa^2 \left[\frac{2h^2}{[(x - \kappa t/2h)^2 + h^2]^2} - \frac{1}{[(x - \kappa t/2h)^2 + h^2]} \right] \quad (12.8.31)$$

where P_0 is the pressure (assumed constant) as $|x| \to \infty$.

The problem of a rectilinear vortex located near the corner formed by the planes $x = 0$, $y = 0$ is the subject of Problem 11.

Problems containing singular flows in the velocity field, like those above, are largely artificial since near the singularities the fluid speeds are very large and tend to infinity at the singularities themselves. This behaviour is of course physically unrealistic. Also for steady state flow situations for which $\partial\phi/\partial t = 0$, ther Bernoulli equation then implies further infinitely negative pressures at the singularities – this is also true in many non-steady situations. Again this is physically unrealistic, for any attempt to realise such flows experimentally would lead to cavitation. However there is some value in studying problems about rectilinear vortices, since the latter provide a tractable model for the real vortices with cores of finite radius which are encountered, for example, in the flow behind accelerating aircraft.

In general therefore problems involving line sources, dipoles etc. in the real flow field are academic and are included only with the object of familiarizing the reader with methods of solving flow problems.

12.8.5 THE MILNE-THOMSON CIRCLE THEOREM

The circle theorem due to Milne-Thomson (p 149 *Theoretical Hydrodynamics* (1949) Macmillan) is concerned with the following problem. Suppose that flow due to a stream together with singular flows occurs in the absence of any boundaries and with complex potential $f(z)$. Suppose further that all the singular flows are located outside the circle $|z| = a$. Then with the introduction of a rigid circular cylinder bounded by $|z| = a$ the potential

$$\omega = f(z) + \bar{f}(a^2/z) \quad (12.8.32)$$

makes $z = ae^{i\theta}$ a streamline and, in the absence of any other boundaries, provides a solution to the flow problem in the presence of the cylinder.

It is important that no other boundaries are present since although (12.8.32) makes $|z| = a$ a streamline, the additional term $\bar{f}(a^2/z)$ would not in general make ψ constant on other boundaries which originally were streamlines for $\omega = f(z)$.

Proof that (12.8.32) makes $z = ae^{i\theta}$ a streamline is trivial for on the circle, $a^2/z = ae^{-i\theta}$ and

$$\omega = f(ae^{i\theta}) + \bar{f}(ae^{-i\theta})$$

is entirely real, i.e. $z = ae^{i\theta}$ is (part of) the streamline $\psi = 0$. Also if $f(z)$ in singular for $|z| > a$, $\bar{f}(a^2/z)$ is singular for $|z| < a$ so that the additional potential $\bar{f}$ does not introduce any new singularities in the real flow region.

A simple application concerns the flow of a stream past a cylinder. If the stream is flowing in the negative x direction at speed u then in the absence of the cylinder

$$f(z) = uz .$$

Application of (12.8.32) now yields

$$\omega = u(z + a^2/z) \tag{12.8.33}$$

which is the complex potential for a stream together with a line dipole oriented in the x direction. This is in agreement with previous calculations (Section 12.6.4). Indeed if we split ω into its real and imaginary parts we find

$$\phi = u(r + a^2/r)\cos\theta \tag{12.8.34}$$

in agreement with equation (12.6.41) while ψ is given by

$$\psi = u(r - a^2/r)\sin\theta . \tag{12.8.35}$$

The consequences of (12.8.34) have already been fully exploited while from (12.8.35) we see that the streamline $\psi = 0$ is given by $r = a$ as expected and also by $\theta = \pm\pi$. The last equations, equivalent to $y = 0$, show that $\psi = 0$ is a dividing streamline Fig. (12.8), with stagnation points at $x = \pm a$, $y = 0$.

The complex potential due to a stream flowing past a cylinder together with circulation is of some subsequent interest. The complex potential is

$$\omega = u(z + a^2/z) + i\kappa \log z . \tag{12.8.36}$$

Another application of the circle theorem concerns the problem of a rectilinear vortex located outside and parallel to the rigid cylinder $|z| = a$. (Fig. 12.18). We shall show that the vortex revolves around the cylinder at a constant angular velocity. Let $z = f$ (f real) denote the position of the vortex core at $t = 0$.

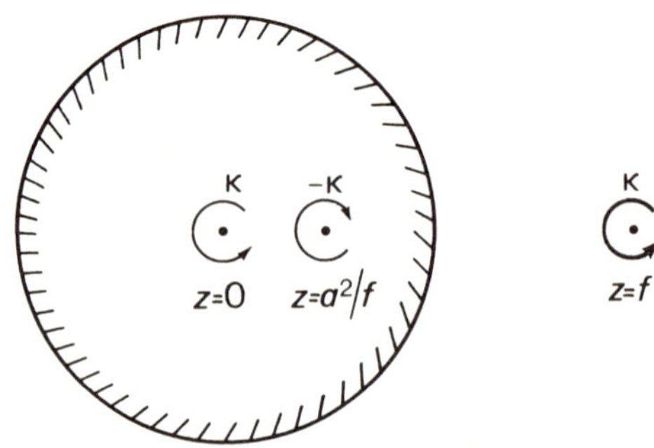

Fig. 12.18 Image solution for a rectilinear vortex lying externally parallel to a cylinder

Then in the absence of the cylinder the vortex potential at $t = 0$ is $i\kappa \log(z - f)$. From the circle theorem the potential in the presence of the cylinder is

$$\omega = i\kappa \log(z - f) - i\kappa \log\left(\frac{a^2}{z} - f\right). \qquad (12.8.37)$$

After discarding an irrelevant constant $[-i\kappa \log(-f)]$, (12.8.37) may be re-written

$$\omega = i\kappa \log(z - f) + i\kappa \log z - i\kappa \log(z - a^2/f) \qquad (12.8.38)$$

showing that at $t = 0$ the problem is solved by locating an image vortex of strength κ at the origin $z = 0$ together with a vortex of the same strength but opposite sign at the *image* point $z = a^2/f$. Since $a/f < 1$, the image point $z = a^2/f$ lies inside $|z| = a$. †

The image vortex at the origin induces a velocity κ/f in the positive y direction at $z = f$ while the velocity induced by the vortex at the image point is $\kappa/(f - a^2/f)$ in the negative y direction. The net induced velocity is $\kappa a^2/[f(f^2 - a^2)]$ in the negative y direction. It is more appropriate to think of the direction as the negative θ direction as is evident from considering the case when the original vortex does not lie on the real axis. Evidently for positive κ the vortex revolves clockwise around the cylinder in a circle of radius f with angular velocity $\kappa a^2/[f^2(f^2 - a^2)]$.

The circle theorem, if used with care, may also provide the solution for flows containing singularities within $|z| = a$ and bounded by $|z| = a$. For example suppose we have a line dipole with potential $\omega = \mu/(z - f)$ in isolation. If now the dipole is located *inside* the circle $|z| = a$, the circle theorem states that (for real μ and f)

$$\omega = \mu/(z - f) + \mu/[(a^2/z) - f]$$

makes $|z| = a$ a streamline. The second term may be rewritten as the sum of an (irrelevant) constant together with an image dipole at the point a^2/f, which is now outside the flow field. The resulting complex potential (ignoring the constant).

$$\omega = \mu/(z - f) - (\mu a^2/f^2)/(z - a^2/f)$$

is an acceptable solution to the problem posed.

However consider the problem of a line source located at $z = f < a$ with f real. The circle theorem yields

$$\begin{aligned} \omega &= -m \log(z - f) - m \log(a^2/z - f) \\ &= -m \log(z - f) - m \log(z - a^2/f) + m \log z - m \log(-f). \end{aligned}$$

The last term is an irrelevant constant while the second term is an acceptable line source lying outside $|z| = a$. The difficulties are associated with the term $m \log z$ which denotes a sink at the origin, so that a new singularity in the real flow field has been thrown up by the theorem.

† The point located at $z = a^2/f$ is commonly called the 'image' point for the circle.

Here it may be argued that a nonsensical problem is posed, for it is not possible to locate a source inside a *closed* boundary without arranging suitable sink(s) to compensate. In fact what the circle theorem has accomplished here is the provision of a precisely suitable sink at the origin.

Consider now the problem of a rectilinear vortex of strength κ located at $z = f < a$ inside $|z| = a$. In this case the circle theorem yields the potential

$$\omega = i\kappa \log(z - f) + i\kappa \log z - i\kappa \log(z - a^2/f) + \text{const.}$$

The third term is an acceptable image vortex located outside $z = a$, but the second term is a vortex of strength κ in the real flow field. Thus the problem solved by the circle theorem is not the problem posed, and there is no question here that the problem posed is nonsensical.

In fact here the correct solution is easily found simply by removing the term $i\kappa \log z$ from the complex potential (for the latter makes $|z| = a$ a streamline). Then the solution to the problem posed is

$$\omega = i\kappa \log(z - f) - i\kappa \log(z - a^2/f)\,.$$

12.8.6 THE BLASIUS THEOREMS

The Blasius theorems are concerned with the evaluation of the force and couple per unit length in the (coordinate) z direction exerted on a stationary cylinder with arbitrary bounding curve c. The cylinder is supposed immersed in an incompressible fluid whose motion is irrotational, steady and two dimensional in a plane whose normal lies in the direction of the cylinder generators.

We suppose the complex potential describing the motion to be $\omega(z)$ and ignore body forces. The latter are readily taken into account, as we have seen, by Archimedes' principle for the case where the body forces are due to the earth's gravitational field.

The pressure is given by the simplest of the Bernoulli equations

$$\begin{aligned} P &= P_0 - \tfrac{1}{2}\rho_0 \vec{v}^{\,2} \\ &= P_0 - \tfrac{1}{2}\rho_0 \frac{d\omega}{dz}\frac{d\bar{\omega}}{d\bar{z}}\,. \end{aligned} \tag{12.8.39}$$

Let ds be an element of length of the bounding curve c and consider the component of force dF_x exerted on the area formed by multiplying ds by unit length in the (coordinate) z direction. From Fig. 12.19

$$dF_x = P\,ds\,(-\sin\theta) = -P\,dy\,.$$

Similarly

$$dF_y = P dx$$

and it proves profitable to consider the complex combination

$$\begin{aligned} dF_y + i\,dF_x &= P(dx - i\,dy) \\ &= P\,d\bar{z}\,. \end{aligned} \tag{12.8.40}$$

The net force components per unit length of cylinder are given by integrating (12.8.40) around c obtaining

$$F_y + i F_x = \oint_c P \, d\bar{z} .$$

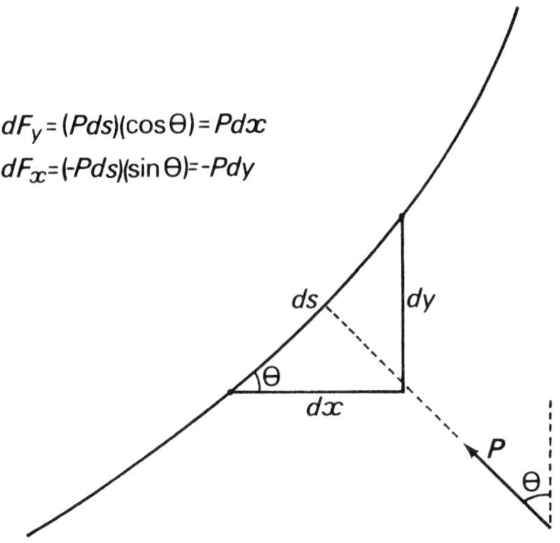

Fig. 12.19 Forces on cylinder in two dimensional flow

Substituting from (12.8.39) for P now gives

$$F_y + i F_x = \oint_c P_0 d\bar{z} - \tfrac{1}{2}\rho_0 \oint_c \left(\frac{d\omega}{dz}\right)\left(\frac{d\bar{\omega}}{d\bar{z}}\right) d\bar{z} .$$

Since c is a closed curve the first of these integrals vanishes leaving

$$F_y + i F_x = -\tfrac{1}{2}\rho_0 \oint_c \frac{d\omega}{dz}\frac{d\bar{\omega}}{d\bar{z}} d\bar{z}$$

$$= -\tfrac{1}{2}\rho_0 \oint_c \frac{d\omega}{dz} d\bar{\omega} .$$

We now make use of the fact that the cylinder is stationary so that along c, which is a streamline, ψ is constant and $d\psi = 0$. It follows that along c

$$d\bar{\omega} = d\phi - i\, d\psi = d\phi + i\, d\psi = d\omega$$

so that finally

$$F_y + i F_x = -\tfrac{1}{2}\rho_0 \oint_c \left(\frac{d\omega}{dz}\right)^2 dz \qquad (12.8.41)$$

which is the first of the Blasius theorems.

The theorem reduces the problem of finding F_y and F_x to the evaluation of a contour integral. It is worth noting that in (12.8.41), c may be replaced by any other contour c' which encloses c provided that within the region between c and c', $(d\omega/dz)^2$ is a regular function free from singularities.

A simple application of (12.8.41) is the case of a circular cylinder of radius a lying in a stream flowing in the negative x direction with speed u. We know from direct calculations that for a stationary cylinder the forces are zero (Section 12.6.4). To obtain this result using (12.8.41) we have from (12.8.33) for $\omega(z)$

$$\omega(z) = u(z + a^2/z)$$

so that

$$\left(\frac{d\omega}{dz}\right)^2 = u^2(1 - a^2/z^2)^2$$

Thus the integrand of (12.8.41) while containing a pole of order four at $z = 0$ does not contain any terms in z^{-1}. At $z = 0$ therefore the residue is zero and the integral (12.8.41) vanishes.

However wtih the inclusion of circulation of amount $2\pi\kappa$ the situation is different. The complex potential is now

$$\omega = u(z + a^2/z) + i\,\kappa \log z$$

and

$$\left(\frac{d\omega}{dz}\right)^2 = [u(1 - a^2/z^2) + i\,\kappa/z]^2 .$$

The integrand of (12.8.41), while still containing a pole of order four, also yields the term $2i\kappa u/z$ with residue $2i\kappa u$. Accordingly

$$F_y + i F_x = -\tfrac{1}{2}\rho_0(2\pi i)(2i\kappa u) = 2\pi\rho_0\kappa u$$

i.e.

$$F_y = 2\pi\rho_0\kappa u\,, \quad F_x = 0 \qquad (12.8.42)$$

so that in the presence of circulation the cylinder experiences a *lift* force in the vertical direction.

As we shall see in Section 12.8.11, a similar situation occurs with an aerofoil. In this case it is the function of the trailing edge to induce the circulation whose magnitude is related to the speed of the stream and the dimensions of the aerofoil.

The second Blasius theorem determines the couple exerted on the cylinder in terms of a contour integral. The element of force **dF** per unit length of cylinder gives rise to a couple **dM** (per unit length) about the origin.

$$\mathbf{dM} = \mathbf{r} \wedge \mathbf{dF}$$

which is directed along the (coordinate) z axis. The magnitude of **dM** is given by

$$\begin{aligned} dM &= x\,dF_y - y\,dF_x \\ &= P(xdx + ydy)\,. \end{aligned} \tag{12.8.43}$$

The total couple given by integrating (12.8.43) around c is

$$M = \oint_c P(xdx + ydy)$$

On substituting from (12.8.39) for P, the term P_0 yields a zero contribution since

$$P_0 \oint_c (x\,dx + y\,dy) = \tfrac{1}{2} P_0 \oint_c d(x^2 + y^2)$$

which vanishes since c is closed and $(x^2 + y^2)$ is single valued. It follows that

$$M = -\tfrac{1}{2}\rho_0 \oint_c \frac{d\omega}{dz}\frac{d\bar{\omega}}{d\bar{z}}(x\,dx + y\,dy)\,.$$

We note that

$$\begin{aligned} x\,dx + y\,dy &= \mathrm{Re}(x + iy)(dx - i\,dy) \\ &= \mathrm{Re}(z\,d\bar{z}). \end{aligned}$$

Following now the arguments used for the first theorem leads to

$$M = -\tfrac{1}{2}\rho_0 \,\mathrm{Re} \oint_c z\left(\frac{d\omega}{dz}\right)^2 dz \tag{12.8.44}$$

for the couple per unit length exerted on the cylinder.

Again if $z(d\omega/dz)^2$ is regular in the region outside the cylinder it is possible to replace c by any suitable contour enclosing c.

12.8.7 CONFORMAL MAPPING TECHNIQUE (BASIC RESULTS)

So far the complex variable method has been seen to yield solutions to straightforward problems involving streams, sources and vortices together with plane or circular boundaries. The real versatility of the method becomes apparent only with the introduction of conformal mapping into the theory.

The value of conformal mapping lies in the result that if $\zeta(\equiv \xi + i\eta) = \zeta(z)$ is a regular function of z and $\omega = \omega(\zeta)$ is a regular function of ζ, then

$$\omega = \omega[\zeta(z)] \tag{12.8.45}$$

is a regular function of z. [The proof is trivial. A necessary and sufficient condition for a function to be regular is the existence of the derivative. Thus $\omega'(\zeta)$ and $\zeta'(z)$ exist and hence by the chain rule $\omega'(z) = \omega'(\zeta)\zeta'(z)$ exists so that ω is a regular function of z.]

Suppose now a two dimensional flow problem is posed in the (x, y) plane with fixed boundaries and whose solution is not obviously obtainable by the elementary complex variable methods discussed previously. Since the boundaries are fixed, these boundaries are streamlines. The conformal transformation $\zeta = \zeta(z)$ transforms the streamlines in question to new streamline boundaries in the ζ plane. An appropriate choice of ζ may yield a simpler potential problem in the ζ plane whose solution $\omega(\zeta)$ is more easily found. [E.g. suppose the original problem is that of the flow of a stream past a fixed elliptic cylinder and a transformation $\zeta = \zeta(z)$ is known which transforms the elliptical boundary into a circle. The solution (12.8.33) for the problem of the flow of a stream past a circular cylinder is known, $[\omega = \omega(\zeta)$ (say)], and the resulting solution of the original problem is then $\omega = \omega[\zeta(z)]$. This particular problem is discussed in Section 12.8.10.]

Proof that streamlines in the ζ plane map into streamlines in the z plane follows merely from the observation that along a streamline in the ζ plane $\text{Im}(\omega) \equiv \psi(\xi, \eta)$ is constant, and that at corresponding points in the ζ and z planes values of ψ are equal. Thus the line in the z plane corresponding to a given streamline in the ζ plane is also a streamline.

It should be noted that the conformal transformation method deals readily with problems involving fixed boundaries; for moving boundaries, which are no longer streamlines, there are complications. Problems dealing with moving boundaries are not considered in the present text; some examples are given by Milne-Thomson (p 232–243 *Theoretical Hydrodynamics* (1949) Macmillan).

From a more formal viewpoint the applicability of conformal transformation theory in the solution of two dimensional potential problems may be founded on the following argument. Suppose $\phi = \phi[\xi(x, y), \eta(x, y)]$ is an arbitrary function of x and y determined via ξ and η, where the relations $\xi = \xi(x, y)$, $\eta = \eta(x, y)$ are given by the conformal transformation

$$\zeta(\equiv \xi + i\eta) = \zeta(z).$$

Then from the usual rules involving changes of independent variables it may be shown that

$$\frac{\partial^2\phi}{\partial x^2} + \frac{\partial^2\phi}{\partial y^2} = |\zeta'(z)|^2 \left(\frac{\partial^2\phi}{\partial \xi^2} + \frac{\partial^2\phi}{\partial \eta^2}\right),$$

e.g. see E.G. Phillips p 38–39, *Functions of a Complex Variable,* Oliver and Boyd (1940).

If now ϕ is restricted to satisfy Laplace's equation in the z plane then, provided $\zeta'(z) \neq 0$, ϕ also satisfies Laplace's equation in the ζ plane. Of course this result is largely a restatement that $\omega(z)$ defined by (12.8.45) is regular: however there is the slight additional emphasis that for points where the transformation $z \to \zeta$

ceases to be conformal, i.e. points where $\zeta'(z) = 0$, the equation $\partial^2\phi/\partial x^2 + \partial^2\phi/\partial y^2 = 0$ does not imply $\partial^2\phi/\partial\xi^2 + \partial^2\phi/\partial\eta^2 = 0$. We encounter subsequently examples containing points where $\zeta'(z) = 0$.

Some useful results in what follows are that except for those points where $\zeta'(z) = 0$, sources and vortices located in the ζ plane transform into sources and vortices of the same strength in the z plane (together usually with some non-singular terms). We consider the case of a source in the ζ plane located at $\zeta_0 = \zeta(z_0)$ where z_0 is the corresponding point in the z plane.

The complex potential of an isolated source at ζ_0 is

$$\omega = -m \log[\zeta(z) - \zeta_0(z_0)]$$

However near z_0, by Taylor's theorem

$$\zeta(z) = \zeta(z_0) + \zeta'(z_0)(z-z_0) + \tfrac{1}{2}\zeta''(z_0)(z-z_0)^2 + \ldots$$

so that in the z plane

$$\omega = -m\Big[\log\zeta'(z_0) + \log(z-z_0) + \log[1 + \tfrac{1}{2}\{\zeta''(z_0)/\zeta'(z_0)\}(z-z_0) + \cdots]\Big]$$

On disposing of the irrelevant constant $[-m\log\zeta'(z_0)]$ there results

$$\omega = -m\log(z-z_0) + \text{regular terms}$$

so that near z_0 the flow is dominantly that of a source of strength m.

It will be observed that the proof devolves on the non-vanishing of $\zeta'(z_0)$. Should $\zeta'(z_0) = 0$ a different situation arises; we leave the reader to show (for example) that should $\zeta'(z_0) = 0$ while $\zeta''(z_0) \neq 0$ then a source of strength m in the ζ plane transforms into a source of strength $2m$ in the z plane. There is little point in enunciating general rules to deal with such exceptional situations; it is simpler to examine each case separately.

In general, flow in the ζ plane which behaves as a stream for large $|\zeta|$ does not transform into a stream for large $|z|$ except for important specific transformations where $\zeta \to z$ for large $|z|$.

12.8.8 SOME SIMPLE CONFORMAL MAPPINGS

The transformation $\zeta = z^2$ maps the positive quadrant $x \geqslant 0$, $y \geqslant 0$ onto the half plane $\eta \geqslant 0$ (Fig. 12.20), as is readily shown by expressing z in polar form.

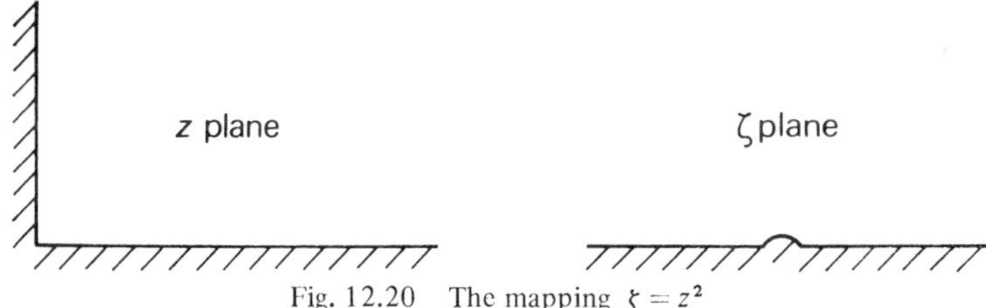

Fig. 12.20 The mapping $\zeta = z^2$

Thus the problem of a uniform stream flowing parallel to the ξ axis and solved by $\omega = -u\zeta$ transforms into the problem of flow in a corner with solution $\omega = -uz^2$ (Fig. 12.14).

The problem of a line source located in the positive quadrant at $x = x_0$, $y = y_0$ (i.e. at $z = z_0 \equiv x_0 + iy_0$) and lying parallel to the two plane walls $x = 0$, $y = 0$ is converted into the problem of a line source located at $\zeta_0 = z_0^2$ lying parallel to the plane $\eta = 0$. The solution of the latter problem is (from image theory)

$$\omega = -m\log(\zeta - \zeta_0) - m\log(\zeta - \bar{\zeta}_0)$$

so that the solution of the original problem is

$$\begin{aligned}\omega &= -m\log(z^2 - z_0^2) - m\log(z - \bar{z}_0^2) \\ &= -m\{\log(z - z_0) + \log(z - \bar{z}_0) + \log(z + z_0) + \log(z + \bar{z}_0)\} .\end{aligned}$$

Of course this latter result is also obtainable directly from the theory of images [equation (12.8.27) and Fig. 12.2.†].

We can find similarly the solution for the problem of a rectilinear vortex located at z_0 and lying parallel to the walls $x = 0$, $y = 0$. However,a word of warning; the path of the vortex in the z plane is not the path corresponding to the one that would be travelled by the corresponding vortex in the ζ plane. Again the problem is solvable by image theory (see Problem 11).

The transformation $\zeta = z^{\pi/\alpha}$ transforms the wedge $0 \leqslant |\text{Arg}\, z| \leqslant \alpha$ to the half plane $0 \leqslant \text{Arg}\, \zeta \leqslant \pi$ (Fig. 12.21).

Fig. 12.21 The mapping $\zeta = z^{\pi/\alpha}$

Thus for example the problem of a source located at z_0 is solved by

$$\omega = -m\log(z^{\pi/\alpha} - z_0^{\pi/\alpha}) - m\log(z^{\pi/\alpha} - \bar{z}_0^{\pi/\alpha}) .$$

Unless π/α is integral [so that $(z^{\pi/\alpha} - z_0^{\pi/\alpha})$ factorises into a product of π/α linear factors], it is no longer possible to obtain a solution by image theory.

A slightly more complicated transformation is $\zeta = e^{\alpha z}$ (α real and positive) which maps the strip $0 \leqslant y \leqslant \pi/\alpha$ on to the half plane $\eta > 0$ (Fig. 12.22). We leave the details to the reader.

† Fig. (12.2) refers originally to the problem of a *point* source; the same diagram is relevant to the present example.

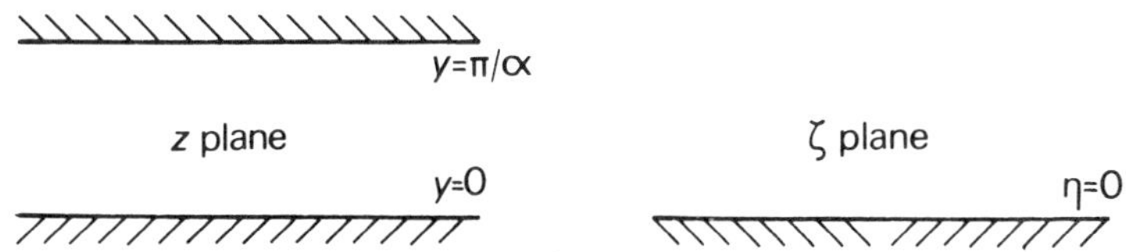

Fig. 12.22 The mapping $\zeta = e^{\alpha z}$

Thus for example the problem of a rectilinear vortex located between two parallel planes is transformed into the tractable problem of a vortex lying parallel to a single plane.

One further transformation which plays an important role in a variety of problems ranging from the flow of a stream past an elliptic cylinder to an elementary account of aerofoil theory, is the **Kutta-Joukowski transformation** discussed below.

12.8.9 THE KUTTA-JOUKOWSKI TRANSFORMATION

The Kutta Joukowski transformation

$$\zeta = z + a^2/z \tag{12.8.46}$$

transforms the circle $z = ae^{i\theta}$ into the straight line segment $\xi = 2a\cos\theta$

i.e.
$$-2a \leqslant \xi \leqslant 2a, \qquad \eta = 0. \tag{12.8.47}$$

There is a one to one correspondence between the points $|z| < a$ and the entire ζ plane. For, the circles $z = r\,e^{i\theta}$ (r constant) map into the ellipses

$$\frac{\xi^2}{(a^2r^{-1}+r)^2} + \frac{\eta^2}{(a^2r^{-1}-r)^2} = 1$$

with major axis $(a^2r^{-1}+r)$ and minor axis $(a^2r^{-1}-r)$; as r decreases from a to zero $(a^2r^{-1}-r)$ and $(a^2r^{-1}+r)$ increase respectively from 0 to ∞ and from $2a$ to ∞. With decreasing r therefore each ellipse encloses its predecessors and the sequence of ellipses cover the entire ζ plane. [For $r = a$ the ellipse degenerates to the line segment (12.8.47).] Similarly there is a one to one correspondence between the points $|z| > a$ and the points of the ζ plane. This proof follows that of E.G. Phillips, p.72, *Functions of a complex variable* (1949), Oliver and Boyd.

The inverse mappings obtained by solving (12.8.46) for z are

$$z = \tfrac{1}{2}\,[\zeta - (\zeta^2 - 4a^2)^{\frac{1}{2}}] \qquad (|z| < a \text{ maps into } \zeta \text{ plane})$$

and
$$z = \tfrac{1}{2}\,[\zeta + (\zeta^2 - 4a^2)^{\frac{1}{2}}] \qquad (|z| > a \text{ maps into } \zeta \text{ plane}).$$

The choice of sign is dictated by the result that for the second of these $z \to \zeta$ as $|\zeta| \to \infty$ and clearly this is associated with mapping the exterior of $|z| = a$ onto the ζ plane.†

† The radical is interpreted to be real and positive for $\eta = 0,\ \xi > 2a$.

We shall be concerned solely with the second of these, i.e. the bijective mapping of the exterior of the circle onto the ζ plane (Fig. 12.23) defined by

$$\zeta = z + a^2/z \tag{12.8.48}$$

with inverse

$$z = \tfrac{1}{2}\,[\zeta + (\zeta^2 - 4a^2)^{\frac{1}{2}}] \tag{12.8.49}$$

In what follows we also invert the roles of ζ and z.

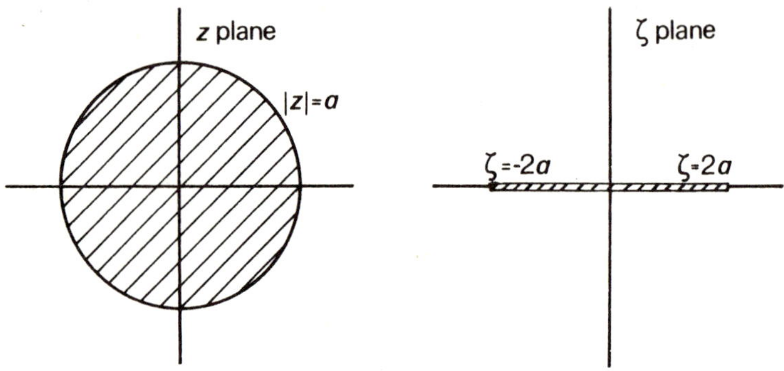

Fig. 12.23 The mapping $\zeta = z + a^2/z$ with inverse $z = \frac{1}{2}[\zeta + (\zeta^2 - 4a^2)^{\frac{1}{2}}]$

12.8.10 FLOWS PAST PLATES AND ELLIPTIC CYLINDERS

The simplest problem to be solved by the transformation (12.8.48) is the flow of a stream past a circular cylinder. In the ζ plane consider a stream flowing in the positive ζ direction without interference by the plate $|\xi| \leqslant 2a,\ \eta = 0$. The complex potential is $\omega = -u\zeta$ and the corresponding problem in the z plane is solved by

$$\omega = -u(z + a^2/z)$$

which is the complex potential for the flow of a stream past the cylinder $|z| = a$ [equation (12.8.33) but with a sign change]. (For the bijective mapping (12.8.48) and (12.8.49) streams in the z and ζ planes correspond since at large distances $\zeta \to z$).

Of rather more interest is the problem of a stream flowing obliquely past the plate at an angle α. We invert the roles of ζ and z, noting the transformation

$$\zeta = \tfrac{1}{2}[z + \sqrt{z^2 - 4a^2}\,] \tag{12.8.50}$$

maps the plate $|x| < 2a,\ y = 0$ onto the circle $|\zeta| = a$.

In the ζ plane a stream flowing obliquely making an angle α with the ξ axis is solved in the absence of the cylinder by the complex potential $\omega = -ue^{-i\alpha}\zeta$. From the circle theorem in the presence of the cylinder ω is given by

$$\omega = -u\,[e^{-i\alpha}\zeta + e^{i\alpha}a^2/\zeta]\ .$$

Finally substituting from (12.8.50) for ζ leads to

$$\omega = -\tfrac{1}{2}u\,[e^{-i\alpha}(z+\sqrt{z^2-4a^2}) + e^{i\alpha}(z-\sqrt{z^2-4a^2})] \qquad (12.8.51)$$

for the complex potential of a stream flowing obliquely past a plate of width $4a$. In obtaining this we have used the identity

$$(z-\sqrt{z^2-4a^2})\;(z+\sqrt{z^2-4a^2}) = 4a^2\,. \qquad (12.8.52)$$

The solution (12.8.51) is somewhat artificial since at the points $z = \pm 2a$, $|d\omega/dz|$ is infinite. In practice this would lead to cavitation; a more acceptable solution for the plate problem in the case $\alpha = \frac{1}{2}\pi$ is given in Section 12.9.3.

More realistic is the solution for the flow of a stream past an elliptic cylinder. For the transformation

$$z = \zeta + a^2/\zeta$$

inverse to (12.8.50), the circles $\zeta = be^{i\theta}$ $(b > a)$ map onto

$$x = (b + a^2/b)\cos\theta\,, \qquad y = (b - a^2/b)\sin\theta$$

i.e. onto the ellipses
$$\frac{x^2}{\gamma^2} + \frac{y^2}{\beta^2} = 1 \qquad (12.8.53)$$

where the semi-major and minor axes are given by

$$\gamma = b + a^2/b\,, \qquad \beta = b - a^2/b \qquad (12.8.54)$$

so that
$$b = \tfrac{1}{2}(\gamma+\beta)\,, \qquad a = \tfrac{1}{2}(\gamma^2-\beta^2)^{\frac{1}{2}} \qquad (12.8.55)$$

The oblique flow of a stream at angle α past the circle $|\zeta| = b$ is given by

$$\omega = -u(e^{-i\alpha}\zeta + b^2e^{i\alpha}/\zeta) \qquad (12.8.56)$$

and the solution of the corresponding problem for the ellipse is given by substituting from (12.8.50) for ζ leading to

$$\omega = -\tfrac{1}{2}u\,[e^{-i\alpha}(z+\sqrt{z^2-4a^2}) + (b^2/a^2)e^{i\alpha}(z-\sqrt{z^2-4a^2})\,] \qquad (12.8.57)$$

where b and a are given by (12.8.55). The points $z = \pm 2a$ are now excluded from the flow† and (12.8.57) is regular for all points outside the ellipse. Thus (12.8.57) leads everywhere to an acceptable finite velocity field for the problem of the flow of a stream past an elliptic cylinder. [Of course if velocities are too large, cavitation may still occur and invalidate the solution.]

† The semi-major axis of length $(b + a^2/b)$ is greater than $2a$ if $b > a$ (for $b + a^2/b$ is an increasing function of b for $b > a$ and is of value $2a$ for $b = a$).

If circulation around the cylinder of radius b is included in the ζ plane implying an additional term $i\kappa \log \zeta$ is (12.8.56) the resulting complex potential for an ellipse immersed in a stream together with circulation is

$$\omega = -\tfrac{1}{2}u\,[e^{-i\alpha}(z+\sqrt{z^2-4a^2}) + (b^2/a^2)e^{i\alpha}(z-\sqrt{z^2-4a^2})] + i\kappa \log[\tfrac{1}{2}(z+\sqrt{z^2-4a^2})]\,. \tag{12.8.58}$$

We calculate from (12.8.58) both the couple and force (per unit length) exerted on the elliptic cylinder. The insertion of (12.8.58) into the Blasius formulae lead to very tedious contour integrals. However, the only singularities of (12.8.58) in the finite part of the z plane occur at $z = \pm 2a$ and are located inside the ellipse. For values of z outside the ellipse it is possible to expand the integrands appearing in the Blasius integrals as a Laurent power series, and to calculate the integrals from the residue theorem.

As a preliminary we expand $z \pm \sqrt{z^2-4a^2}$ in a Laurent series valid for $|z| > 2a$. We have

$$\tfrac{1}{2}[z+\sqrt{z^2-4a^2}] = z - a^2/z + O(z^{-3}),\quad \tfrac{1}{2}[z-\sqrt{z^2-4a^2}] = a^2/z + O(z^{-3})$$

and also
$$\log[\tfrac{1}{2}(z+\sqrt{z^2-4a^2})] = \log z + O(z^{-2})$$

so that for large $|z|$, (12.8.58) becomes

$$\omega = -ue^{-i\alpha}z + i\kappa\log z - u(b^2e^{i\alpha} - a^2e^{-i\alpha})/z + O(z^{-2})\,. \tag{12.8.59}$$

This expansion confirms that at large distances the behaviour is that of a stream $(-ue^{-i\alpha}z)$ together with circulation $(i\kappa \log z)$.

On differentiating (12.8.59) there results

$$\frac{d\omega}{dz} = -ue^{-i\alpha} + i\kappa/z + u(b^2e^{i\alpha} - a^2e^{-i\alpha})/z^2 + O(z^{-3})$$

which leads to

$$\left(\frac{d\omega}{dz}\right)^2 = u^2e^{-2i\alpha} - 2i\kappa ue^{-i\alpha}/z - [2u^2(b^2 - a^2e^{-2i\alpha}) + \kappa^2]/z^2 + O(z^{-3})\,. \tag{12.8.60}$$

From the first Blasius theorem (12.8.41)

$$F_y + iF_x = -\tfrac{1}{2}\rho_0 \oint_c \left(\frac{d\omega}{dz}\right)^2 dz \tag{12.8.61}$$

where c may be taken now as any contour enclosing the ellipse. We choose c of sufficiently large radius to validate the expansion (12.8.60) and integrate term by term. The only non-vanishing contribution to the integral (12.8.61) derives from the coefficient of the terms in z^{-1}, and by the residue theorem

$$F_y + iF_x = -\tfrac{1}{2}\rho_0(2\pi i)(-2i\kappa u e^{-i\alpha}) = -2\pi\rho_0\kappa u e^{-i\alpha}$$

i.e. $$F_y = -2\pi\rho_0\kappa u\cos\alpha\,, \quad F_x = 2\pi\rho_0\kappa u\sin\alpha$$

and **F** is directed perpendicular to the stream (Fig. 12.2.4).

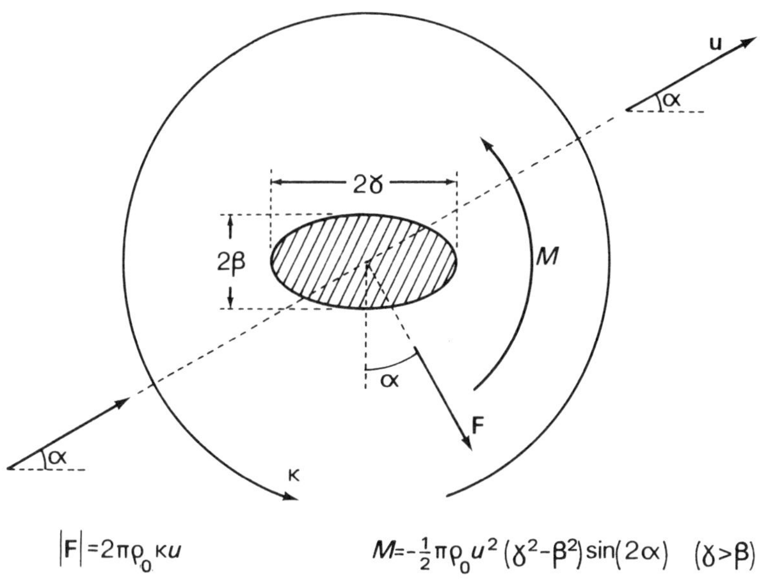

Fig. 12.24 Forces and couple acting on an elliptic cylinder located in a stream and with circulation

We find similarly for the couple from (12.8.44) and (12.8.60)

$$M = \tfrac{1}{2}\rho_0 \operatorname{Re}(2\pi i)\,[2u^2(b^2 - a^2 e^{-2i\alpha}) - \kappa^2] = -2\pi\rho_0 a^2 u^2\sin(2\alpha)$$
$$= -\tfrac{1}{2}\pi\rho_0 u^2(\gamma^2 - \beta^2)\sin(2\alpha)\,. \qquad (12.8.62)$$

The couple vanishes for $\alpha = 0$, $\alpha = \frac{1}{2}\pi$ (as might be expected from symmetry arguments), and for the limiting case of a circle ($\gamma = \beta$), and is independent of the circulation. On the other hand the existence of a non-zero force depends on the presence of circulation.

The introduction of circulation is artificial in the present context; in the elementary account of aerofoil theory given below circulation is introduced mathematically into the problem in order to avoid infinite fluid velocity occurring at the trailing edge of the aerofoil.

12.8.11 AEROFOIL THEORY

We conclude the discussion of two dimensional potential problems with known boundaries with a simplified account of the theory of the lift forces developed on an aircraft wing in flight. As might be expected from the d'Alembert paradox, for unaccelerated aircraft motion no force would be experienced by the wings unless there were some mechanism for inducing circulation. In physical terms it is the function of the trailing edge of the aerofoil to induce the circulation, whose strength is determined in the present theory by the wind velocity and the aerofoil geometry.

Typical aerofoil cross sections are similar in shape to that depicted in Fig. 12.25. Characteristically the leading edge is blunt, the trailing edge sharp while the ratio (l/t) of the chord to thickness is large.

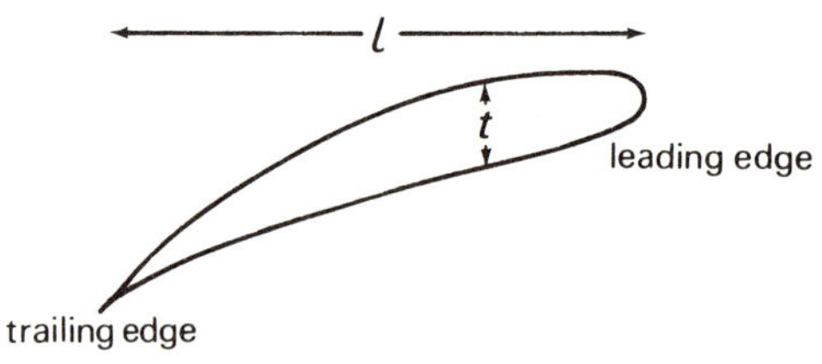

Fig. 12.25 Cross section of aerofoil

It happens that profiles of this general appearance are readily generated by use of the Kutta-Joukowski transformation. Thus with the further assumption that the flow is inviscid, incompressible and essentially two dimensional over each wing section, it is possible to give some account of the forces exerted on aircraft wings.

The problem of an aircraft in flight is one of unsteady motion. However if the wind velocity is constant and the motion of the aircraft unaccelerated, then the forces developed on the aircraft are identical with the forces developed in the steady problem of the flow of a stream past a stationary aircraft. [The two problems are related by a Galilean transformation.]

It follows that the forces developed in the steady problem posed in Fig. 12.26a are the same as those for the unsteady problem of Fig. 12.26b. For the steady problem the stream velocity is given by the difference $\mathbf{u} = \mathbf{v}_1 - \mathbf{u}_1$ where $\mathbf{u}_1$ is the aircraft velocity and $\mathbf{v}_1$ the wind velocity.

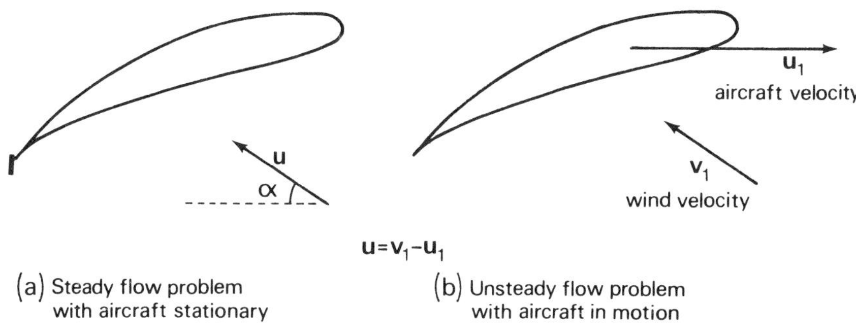

Fig. 12.26   Related steady and unsteady flow problems for airofoil

Characteristic aerofoil shapes may be obtained in the z plane by applying the Kutta-Joukowski transformation to a circle of radius R in the ζ plane (Fig. 12.27). The centre of the circle is chosen to lie at $\zeta = \delta e^{i\beta}$ $(0 < \beta < \frac{1}{2}\pi,\ \delta > 0)$ in the first quadrant and the radius is chosen so that the circle passes through one of the singular points $(\zeta = -a)$, where the transformation ceases to be conformal. Evidently the second singular point lies inside the circle. Thin profile airofoils are obtained by choosing $(\delta/a) \ll 1$.

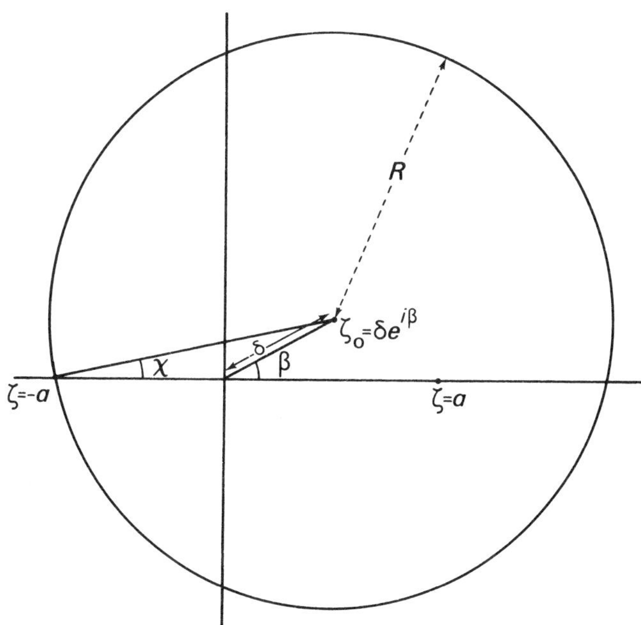

Fig. 12.27   Circle in ζ plane used to generate aerofoil profile

Because the circle is chosen to pass through $\zeta = -a$ the resulting figure in the z plane possesses a cusp representing the trailing edge of the airofoil. This is shown directly by Milne-Thomson from geometrical arguments [p 181-183 *Theoretical Hydrodynamics* Macmillan (1949)] and is also evident in the slender profile approximation discussed below.

In the ζ plane the circle is described by

$$\zeta = \delta e^{i\beta} + Re^{i\theta}$$

where, since the circle passes through $\zeta = -a$,

$$R = [(a + \delta \cos \beta)^2 + \delta^2 \sin^2 \beta]^{\frac{1}{2}}. \tag{12.8.63}$$

From the Kutta-Joukowski transformation the corresponding curve in the z plane is given by

$$z = \delta e^{i\beta} + Re^{i\theta} + a^2/(\delta e^{i\beta} + Re^{i\theta}) \tag{12.8.64}$$

and by separating this equation into real and imaginary parts the airofoil profile is obtained parametrically in the form $x = x(\theta)$, $y = y(\theta)$. Alternately the profile may be constructed geometrically (see Milne-Thomson, loc. cit. p 181-183).

For thin profiles $(\delta/a) \ll 1$ it is permissable to expand (12.8.64) in a power series in δ. With the help of (12.8.63) we derive, on separating into real and imaginary parts,

$$x = 2a \cos \theta + O(\delta),$$

$$y = 2\delta [\cos \beta \sin \theta (1 + \cos \theta) + \sin^2 \theta \sin \beta] + O(\delta^2).$$

To the approximation indicated, elimination of θ leads to

$$y = 2\delta [\pm \cos \beta (1 + x/2a)^{\frac{3}{2}} (1 - x/2a)^{\frac{1}{2}} + \sin \beta (1 - x^2/4a^2)] \tag{12.8.65}$$

where the $\pm$ signs, deriving respectively from $\sin \theta$ in the ranges $0 \leqslant \theta \leqslant \pi$ and $-\pi \leqslant \theta \leqslant 0$, refer to the upper and lower halves of the profile (Fig. 12.28).

In (12.8.65) the range of x is $-2a \leqslant x \leqslant 2a$. The point $x = -2a$, $y = 0$ $(\theta = \pi)$ is precisely associated with the singular point $\zeta = -a$ and leads to the cusp formed by the two branches. The existence of the cusp is demonstrated analytically by noting that at $x = -2a$ both branches possess a common derivative.

$$\left(\frac{dy}{dx}\right)_{x=-2a} = (2\delta/a) \sin \beta$$

From Fig. 12.27 this yields approximately for the (small) angle χ, $(dy/dx)_{-2a} \simeq \tan 2\chi$. (In fact it is exactly true that the trailing edge cusp makes an angle 2χ with the x axis, even for large values of δ/a.)

The point $x = 2a$, $y = 0$ which satisfies the approximation (12.8.65) would appear to correspond to the other singular point $\zeta = a$ which however, as noted earlier, lies inside the circle. What has happened is that for small values of δ/a the point $x = 2a, y = 0$ lies just inside the true aerofoil profile and this fine detail has been lost in the approximation (12.8.65). At the moment we are concerned with the approximate aerofoil profile; subsequently when we come to calculate the lift forces on the aerofoil it is to be recalled that the point $x = 2a$, $y = 0$ lies inside the areofoil profile. In the present approximation the point $x = 2a$, $y = 0$, for which both branches of (12.8.65) possess infinite slope, represents the leading edge.

From subtracting the two curves (12.8.65) the thickness of the profile is

$$t = 4\delta(1+m)^{\frac{3}{2}}(1-m)^{\frac{1}{2}}\cos\beta$$

where m is $x/2a$. The maximum thickness, occurring at $m = \frac{1}{2}$, is $3\sqrt{3}\,\delta\cos\beta$. In Fig. 12.28 we have chosen $\beta = \frac{1}{4}\pi$, $\delta/a = (6\sqrt{2})^{-1}$.

Fig. 12.28 Aerofoil profile obtained from (12.8.65) with $\beta = \frac{1}{4}\pi$, $\delta/a = (6\sqrt{2})^{-1}$

The problem and analogue problem with which we are concerned are depicted respectively in Figs. 12.29(a) and (b).

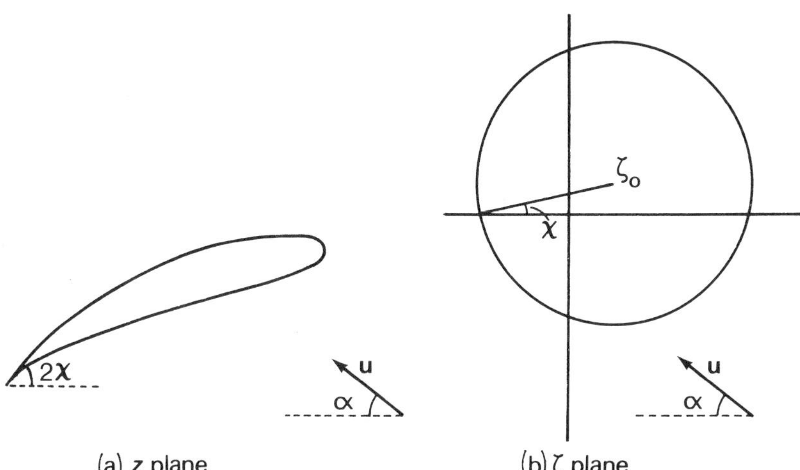

Fig. 12.29 Aerofoil problem and analogue problem in ζ plane

In both the z and ζ planes the stream velocity at large distances is given by

$$v_x = -u\cos\alpha \qquad v_y = u\sin\alpha$$

i.e.

$$v = -v_x + i\,v_y = ue^{i\alpha}$$

with corresponding complex potential (in the ζ plane)

$$\omega \to ue^{i\alpha}\zeta \qquad (|\zeta| \to \infty)\,.$$

For a circle centred on the origin in the ζ plane of radius R, the circle theorem yields an additional term $ue^{-i\alpha}a^2/\zeta$. However since the circle is centred on $\zeta_0 = \delta e^{i\beta}$ rather than the origin, the correct complex potential in the absence of circulation is

$$\omega = u\,[e^{i\alpha}\zeta + e^{-i\alpha}R^2/(\zeta - \zeta_0)]\ .$$

With the inclusion of circulation centred on ζ_0 we find finally for the complex potential in the ζ plane

$$\omega = u\,[e^{i\alpha}\zeta + e^{-i\alpha}R^2/(\zeta - \zeta_0)] + i\,\kappa\,\log(\zeta - \zeta_0)\,. \quad (12.8.66)$$

The reason for inclusion of circulation becomes apparent below.

The solution in the z plane now follows by writing

$$\zeta = \tfrac{1}{2}(z + \sqrt{z^2 - 4a^2}) \quad (12.8.67)$$

in (12.8.66).

The derivative $d\omega/dz$ may be calculated from the chain rule

$$\frac{d\omega}{dz} = \frac{d\omega}{d\zeta}\frac{d\zeta}{dz}$$

and we note that everywhere except $\zeta = \zeta_0$ (which lies inside the circle) $d\omega/d\zeta$ is regular. On the other hand $d\zeta/dz$ is infinite at $z = \pm 2a$. Of these two points $z = 2a$ lies inside the aerofoil and is of no further concern while $z = -2a$ is the trailing edge cusp. Thus in general the velocity is infinite at the trailing edge. However this infinity may be removed by choosing a (real) value of κ such that $d\omega/d\zeta$ is zero at the point $\zeta = -a$ (corresponding to $z = -2a$).

We have

$$\frac{d\omega}{d\zeta} = u\,[e^{i\alpha} - R^2e^{-i\alpha}/(\zeta - \zeta_0)^2] + i\,\kappa/(\zeta - \zeta_0)$$

and from Fig. 12.27, on writing for $\zeta = -a$

$$(\zeta - \zeta_0)_{\zeta=-a} = -Re^{i\chi}$$

there results

$$\left(\frac{d\omega}{d\zeta}\right)_{\zeta=-a} = u\,[e^{i\alpha} - e^{-i(\alpha+2\chi)}] - i\,\kappa\, e^{-i\chi}/R$$

This vanishes if

$$\kappa = 2Ru\sin(\alpha + \chi) \quad (12.8.68)$$

determining a unique real value of κ which annihilates the infinity in the velocity field at the trailing edge.

That the choice (12.8.68) for κ leads to a finite velocity at $z = -2a$ may be seen from the following argument. Near $\zeta = -a$, ω is a regular function of ζ and since (now) $d\omega/d\zeta$ vanishes at $\zeta = -a$ we must have near $\zeta = -a$ (from the Taylor series)

$$\frac{d\omega}{d\zeta} = (\zeta + a)\left(\frac{d^2\omega}{d\zeta^2}\right)_{\zeta=-a} + O(\zeta + a)^2\ .$$

On the other hand $$\frac{d\zeta}{dz} = \left(\frac{dz}{d\zeta}\right)^{-1} = [1 - a^2/\zeta^2]^{-1} = \zeta^2/(\zeta^2 - a^2)\,.$$

On multiplying these results together and proceeding to the limit $\zeta \to -a$ it is evident that $d\omega/dz$ is finite at the trailing edge.

The calculation of the lift forces is virtually identical to the elliptic cylinder case. We revert to (12.8.66) and write

$$\zeta = \tfrac{1}{2}[z + (z^2 - 4a^2)^{\frac{1}{2}}] = z - a^2/z + O(z^{-3})$$

obtaining for large $|z|$

$$\omega = u\,e^{i\alpha}z + i\,\kappa\,\log z + [u(R^2 e^{-i\alpha} - a^2 e^{i\alpha}) - i\,\kappa\,\zeta_0]/z + O(z^{-2})\,.$$

On calculating $(d\omega/dz)^2$ from this result and evaluating the contour integral (12.8.41) there obtains

$$F_y + i\,F_x = -\tfrac{1}{2}\,\rho_0(2\pi i)\,(2i\kappa u e^{i\alpha}) = 2\pi\rho_0\kappa u e^{i\alpha} \qquad (12.8.69)$$

so that $$F_y = 2\pi\rho_0\kappa u\cos\alpha\,, \quad F_x = 2\pi\rho_0\kappa\,u\sin\alpha$$

and the force, directed perpendicular to the stream velocity (Fig. 12.30) is of magnitude

$$|\mathbf{F}| = 2\pi\rho_0\kappa u\,. \qquad (12.8.70)$$

per unit length of wing.

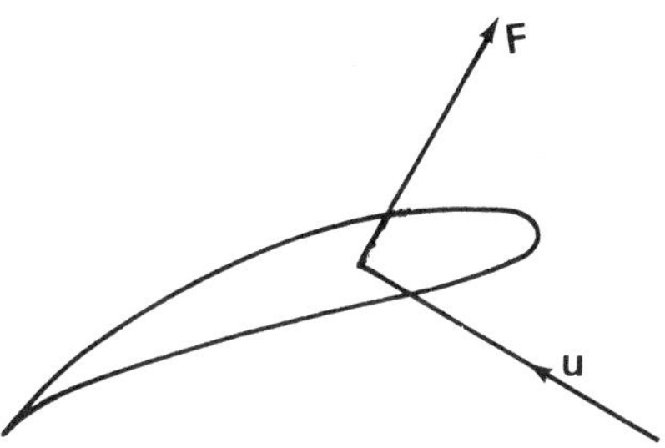

Fig. 12.30 Direction of force on aerofoil

We leave the reader to show that the couple per unit length of wing about the origin is given by

$$M = 2\pi\rho_0\,[u^2 a^2\sin(2\alpha) + \kappa u\delta\,\cos(\alpha + \beta)]\;.$$

With the couple determined and the direction and magnitude of the force known, the point at which the force acts is determined. Ideally this point (the centre of pressure) should coincide with the centre of mass of the entire aircraft.

With κ given by (12.8.68), equation (12.8.70) yields

$$|\mathbf{F}| = 4\pi\rho_0 R u^2\sin(\alpha + \chi) \qquad (12.8.71)$$

and depends on the square of the speed, the air density, the dimensions of the wing ($R \simeq a \simeq \frac{1}{4}$span) and $\sin(\alpha + \chi)$. This result gives an explanation for the existence of a critical speed for take-off of an aircraft. Suppose that with the aircraft stationary on the ground, the aerofoil trailing edge makes an angle $\alpha_0 + 2\chi$ with the horizontal. [In terms of Fig. 12.28 we are supposing that on attachment to the aircraft frame, the profile depicted is rotated through an angle α_0]. For simplicity we neglect the real wind velocity. Then for an aircraft travelling along the ground with a real speed u directed parallel to the ground, the force F is directed vertically upwards and is of magnitude $\pi\rho_0 L u^2 \sin\Gamma$ per unit thickness of wing in the (coordinate) z direction. Here $L \simeq 4R$ is the span while $\Gamma = \alpha_0 + \chi$. The total force due to both wings of length B is

$$2\pi\rho_0 BLu^2 \sin\Gamma \tag{12.8.72}$$

and in order for the plane to take off it is necessary to achieve a critical speed u_0 given by

$$2\pi\rho_0 BLu_0^2 \sin\Gamma = W$$

where W is the weight of the aircraft.

In-flight speeds are commonly much greater than take-off and landing speeds, but the implied increase of lift force due to the dependence on u^2 is largely counterbalanced by the much smaller air densities prevailing at operational altitudes. Adjustments of lift force, commonly employed by the pilot following take-off and before landing, are the raising and lowering of flaps, corresponding to decreasing and increasing χ (and hence Γ), and the varying of wing area.

The discussion above is qualitative since for real aircraft neither the take-off nor in-flight wing profile is likely to conform exactly to a Joukowski profile. Also in-flight speeds of modern jet aircraft are comparable with the speed of sound, and in these circumstances it is not realistic to assume that air is incompressible.

In recent years aerofoils have been used on racing cars to provide a downward thrust to augment wheel adhesion. Here α_0 is chosen to be sufficiently negative so that $\Gamma(= \alpha_0 + \chi)$ is negative and the thrust is vertically downward, see Fig. 12.31.

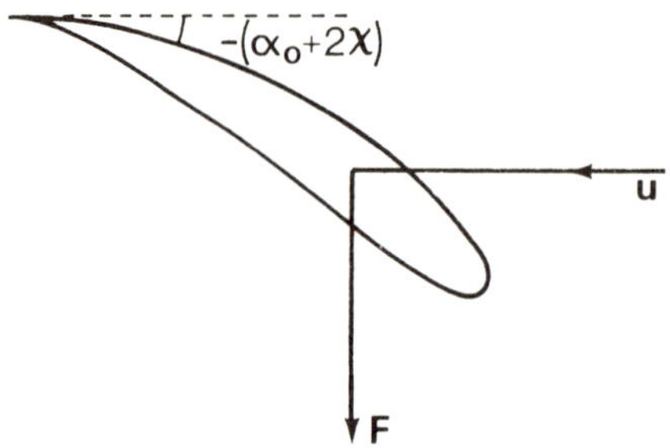

Fig. 12.31 Force on aerofoil attached to a racing car

Other well known examples, which make use of the *Magnus* effect, i.e. a force developed because of circulation induced around an 'aerofoil', are yacht sails and hydrofoils. It is well known to yachtsmen that the force developed on the sail is perpendicular to the wind direction, while in hydrofoil boats the submerged hydrofoil is used to elevate the hull from the water, thereby reducing the water resistance on the hull. The swerve of a spinning cricket ball has also been attributed to the Magnus effect.

It remains to give a physical explanation of the origin of the circulation which is introduced into the theory in a somewhat arbitrary fashion. In the theory κ is chosen so that the fluid velocity at the trailing edge is finite; this is of course an eminently sensible physical choice, which however does not account by itself for the origin of the circulation. In the initial accelerating motion of an aerofoil from rest, at first the flow is without circulation. In consequence the fluid speed at the trailing edge is very large. The speed is not infinite because surrounding the aerofoil there is a thin **boundary layer** in which viscous forces are important and in which vorticity is generated. Because of the large fluid velocity, and hence large velocity gradients and viscous forces, at the trailing edge the boundary layer separates from the aerofoil shedding vorticity into the main body of the virtually inviscid flow region. The vorticity is shed in the form of line vortices which pass downstream away from the aerofoil. Since there is conservation of circulation in the inviscid flow region the shedding of vortices leads to the establishment of circulation around the aerofoil. At constant velocity the circulation is constant and line vortices are shed only during acceleration (or decceleration) of the aerofoil. Photographs of vortices shed by accelerating and deccelerating aerofoils have been taken and accord with the arguments given above. A more extended account together with relevant photographs is given in F.K. Batchelor's *An Introduction to Fluid Dynamics* pp 438-440, Cambridge University Press (1967). There it is argued that while it is not yet mathematically possible to analyse in detail the vortex shedding process, the total amount of vorticity shed in attaining a uniform speed u must be such that the circulation is given by equation (12.8.68), for, if not, any other value of κ entails large velocity gradients at the trailing edge and in turn this leads to further boundary layer separation and shedding of vortices until the value (12.8.68) is established.

Evaluation of the drag forces on aerofoils requires detailed analysis of the behaviour of the boundary layer where viscous forces predominate. This is not encompassed by the present (essentially inviscid) theory.

12.9 Problems with free streamlines

12.9.1 PLANE JET INCIDENT NORMALLY ON RIGID WALL

Problems in hydrodynamics may occur with initially unknown boundaries. Consider for example the problem depicted in Fig. 12.32 where a jet of incompressible fluid, initially of velocity u in the positive x direction, impacts a rigid wall at normal incidence. Here the bounding curve c of the jet is unknown a priori and indeed is one of the principal items of interest in the solution. We consider the

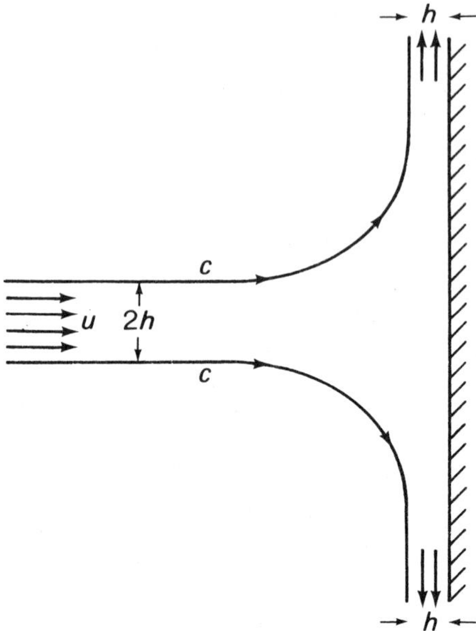

Fig. 12.32 Normal impact of a two dimensional jet on a plane wall

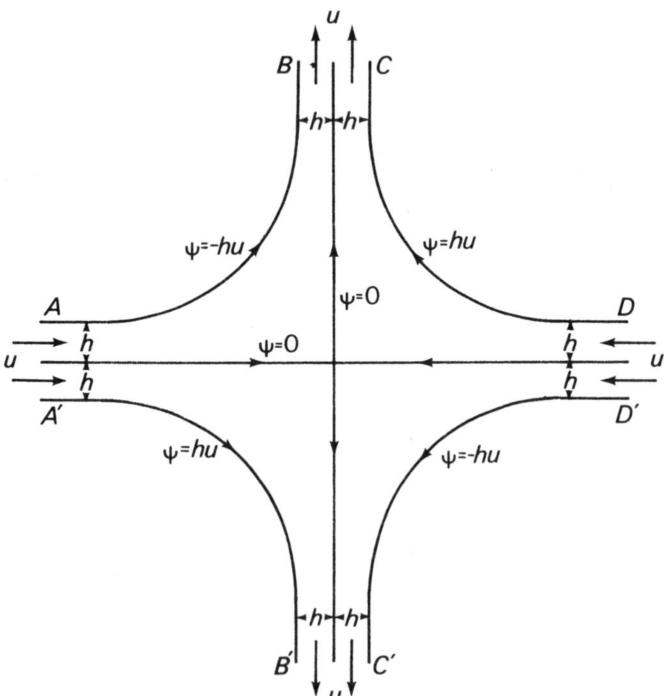

Fig. 12.33 Normal impact of two similar plane jets

case of a plane two dimensional jet of width $2h$ which extends indefinitely in the (coordinate) z direction. The problem posed is 'one half' of the problem depicted in Fig. 12.33 where two jets initially of equal width $2h$ and speed u impinge head on. From symmetry the emergent jets approach the same width and speed at large distances from the impact zone. We consider in detail this second problem.

The motion is assumed steady, so that the bounding curves are streamlines. Along these streamlines the pressure is constant, i.e. atmospheric, and if body forces are neglected Bernoulli's theorem leads immediately to the result that the speed along these bounding streamlines is everywhere the same and given by u.

Also because of the flux theorem (12.8.13), values of ψ may be attached to the bounding stream lines. We label the dividing streamline $(x=0,\ y=0)$ by $\psi=0$. By the flux result and from consideration of the jets $AB, A'B'$ at distances remote from the impact region, the streamline AB is $\psi=-hu$ and the streamline $A'B'$ is $\psi=hu$. Similarly the remaining pair of streamlines CD and $C'D'$ are respectively $\psi=hu,\ \psi=-hu$. [In this symmetric example, AB and $C'D'$ together form the streamline $\psi=-hu$. Similarly CD and $A'B''$ together comprise $\psi=hu$.]

If the complex velocity v is expressed in polar form

$$-v_x+i\,v_y = -qe^{-i\theta}$$

so that

$$v_x = q\cos\theta\,,\qquad v_y = q\sin\theta$$

then along AB, $q=u$ and $0\leqslant\theta\leqslant\frac{1}{2}\pi$. Similar results hold for $A'B'$, CD and $C'D'$. These results are summarised below

$$\left.\begin{array}{llll}\text{streamline } AB; & \psi=-hu, & q=u, & 0\leqslant\theta\leqslant\frac{1}{2}\pi.\\ \text{streamline } CD; & \psi=hu, & q=u, & \frac{1}{2}\pi\leqslant\theta\leqslant\pi.\\ \text{streamline } C'D'; & \psi=-hu, & q=u & -\pi\leqslant\theta\leqslant-\frac{1}{2}\pi.\\ \text{streamline } A'B'; & \psi=hu, & q=u, & -\frac{1}{2}\pi\leqslant\theta\leqslant 0.\end{array}\right\}\qquad(12.9.1)$$

It will be seen that $q=u$ for each of the four bounding streamlines, and for the entire free streamline system θ ranges over the complete range $-\pi\leqslant\theta\leqslant\pi$.

The key to the solution of the problem is to recognise that $v=d\omega/dz$ is a complex variable in its own right and that ω may be regarded as a function of v rather than z. Formally this is readily achieved in principle for known $\omega(z)$ by solving $v=v(z)$ for $z=z(v)$ and substituting the result in ω. Of course for known $\omega(z)$ there is no point in expressing ω in terms of v; however recognition that ω may be expressed in terms of v rather than z leads to a solution of steady flow two dimensional free streamline problems in the absence of body forces.†

† The absence (or rather neglect) of body forces is crucial, since otherwise the speed is not constant along the free streamlines and under these circumstances the method fails.

In what follows the stream function ψ is regarded as a function of q and θ and indeed since ω is a function of the complex variable v, ψ satisfies Laplace's equation in plane polar coordinates with independent variables $-q$ and $-\theta$, i.e.

$$\frac{\partial^2\psi}{\partial(-q)^2} + \frac{1}{(-q)}\frac{\partial\psi}{\partial(-q)} + \frac{1}{(-q)^2}\frac{\partial^2\psi}{\partial(-\theta)^2} = 0\,.$$

ψ therefore satisfies Laplace's equation in polar form in the variables q and θ

$$\frac{\partial^2\psi}{\partial q^2} + \frac{1}{q^2}\frac{\partial\psi}{\partial q} + \frac{1}{q^2}\frac{\partial^2\psi}{\partial\theta^2} = 0\,. \tag{12.9.2}$$

We proceed by mapping the free streamlines of Fig. (12.31) onto the **v** plane with v_x and v_y as axes and for which q and θ are polar coordinates.† The bounding streamlines along which $q = u$ map onto the perimeter of the circle $q = u$, and with the information provided by (12.9.1) values of ψ on $q = u$ are known (Fig. 12.34).

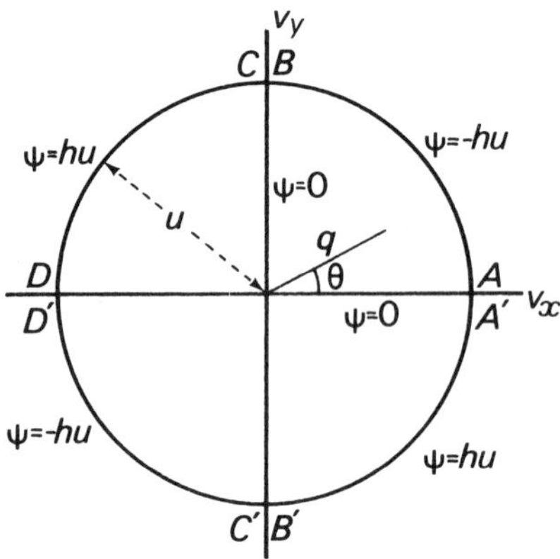

Fig. 12.34 Mapping of Fig. 12.33 onto **v** plane

The problem is now reduced to finding an ψ which satisfies (12.9.2) within the circle $q = u$ and with given values of ψ on the boundary.†† There is a unique solution for this problem (Section 10.3.3). The method adopted here to find it makes use of the separation of variables solutions developed in Section 12.6.1. This method of solution is taken from A.W. Mackie, *Proceedings of the Edinburgh Mathematical Society*, Vol. 11, Part 2, p 109 (1958).

† This is not a conformal mapping.

†† That we are concerned with solving for ψ *inside* the circle is evident from the fact that the dividing streamline $\psi = 0$ maps onto the v_x and v_y axes, passing through $\mathbf{v} = 0$.

For the present problem ψ is an odd function of θ, as is evident from the boundary conditions, and there are no singularities inside the circle. It follows that ψ can be represented by the Fourier series

$$\psi = \sum_{n=1}^{\infty} a_n q^n \sin(n\theta) \qquad (12.9.3)$$

where the coefficients a_n are to be determined from the values of ψ on the boundary. We have

$$\sum_{1}^{\infty} a_n u^n \sin(n\theta) = \begin{cases} -hu & (0 \leqslant \theta \leqslant \tfrac{1}{2}\pi) \\ hu & (\tfrac{1}{2}\pi \leqslant \theta \leqslant \pi) \end{cases}$$

whence from the usual formulae for Fourier coefficients

$$a_n u^n == (2hu/\pi)\left[\int_0^{\frac{1}{2}\pi} -\sin(n\theta)d\theta + \int_{\frac{1}{2}\pi}^{\pi} \sin(n\theta)d\theta\right]$$

$$= \begin{cases} 0 & (n \neq 4m+2) \\ -8hu/\pi(4m+2) & (n = 4m+2) \end{cases} \qquad (m = 0, 1, 2 \ldots)$$

With these results (12.9.3) becomes

$$\psi = -\frac{8uh}{\pi}\sum_{m=0}^{\infty}\frac{(q/u)^{4m+2}}{(4m+2)}\sin(4m+2)\theta = \operatorname{Im}\frac{8uh}{\pi}\sum_{m=0}^{\infty}\frac{(-qe^{-i\theta}/u)^{4m+2}}{4m+2}$$

$$= \operatorname{Im}\frac{8uh}{\pi}\sum_{m=0}^{\infty}\frac{(v/u)^{4m+2}}{(4m+2)}$$

Since to within an irrelevant constant ψ determines ω, the last result implies

$$\omega(v) = \frac{8uh}{\pi}\sum_{m=0}^{\infty}\frac{(v/u)^{4m+2}}{(4m+2)}$$

The series is readily summed to yield

$$\omega(v) = -\frac{2uh}{\pi}\log\left(\frac{u^2-v^2}{u^2+v^2}\right) \qquad (12.9.4)$$

determining $\omega(v)$.†

† Let $\xi = v/u$ and $g(\xi) = \sum_0^\infty \xi^{4m+2}/(4m+2)$. Then $g'(\xi) = \xi\sum_0^\infty \xi^{4m} = \xi/(1-\xi^4)$. On expressing this last result in partial fractions and integrating (and using $g(0) = 0$) there obtains $g = -\frac{1}{4}\log[(1-\xi^2)/(1+\xi^2)]$.

Equation (12.9.4) solves the problem posed but does not provide immediately the equations of the free streamlines. In fact from the derivation it is clear that on writing $\upsilon = -ue^{-i\theta}$ in (12.9.4) we merely reproduce the initial information about the ψ values on the circle of Fig. 12.34.

One method of obtaining the free streamlines is to write $\upsilon = d\omega/dz$ in (12.9.4), solve the resulting differential equation for ω, and then write $\psi(x,y) = \pm hu$. This is possible analytically here but it is simpler to obtain z in terms of υ from the formula

$$\upsilon = \left(\frac{d\omega}{dz}\right) = \frac{d\omega}{d\upsilon}\frac{d\upsilon}{dz}$$

so that

$$z = \int \upsilon^{-1}\frac{d\omega}{d\upsilon}\,d\upsilon\,.$$

Here

$$\frac{d\omega}{d\upsilon} = \frac{8u^3h}{\pi}\frac{\upsilon}{(u^2-\upsilon^2)(u^2+\upsilon^2)}$$

and

$$z = \frac{2h}{\pi}\left[\log\left(\frac{u+\upsilon}{u-\upsilon}\right) + i\log\left(\frac{u-i\upsilon}{u+i\upsilon}\right)\right] \tag{12.9.5}$$

We have disposed of the constant of integration by noting that at $z=0$ the streamline $\psi=0$ divides and hence $z=0$ is a stagnation point for which $\upsilon=0$.

The bounding streamline AB is now found in parametric form by writing $\upsilon = -ue^{-i\theta}$ (where $0 \leqslant \theta \leqslant \frac{1}{2}\pi$) in (12.9.5). There results for the streamline AB

$$\begin{aligned} x + iy &= \frac{2h}{\pi}\left[\log\left(\frac{1-e^{-i\theta}}{1+e^{-i\theta}}\right) + i\log\left(\frac{1+ie^{-i\theta}}{1-ie^{-i\theta}}\right)\right] \\ &= \frac{2h}{\pi}\left[\log[i\tan(\tfrac{1}{2}\theta)] + i\log[i\cot(\tfrac{1}{4}\pi-\tfrac{1}{2}\theta)]\right]. \end{aligned}$$

For the range of θ in question $\tan(\frac{1}{2}\theta)$ and $\cot(\frac{1}{4}\pi-\frac{1}{2}\theta)$ are both positive and the i appearing in the arguments of the logarithms is to be interpreted as $e^{\frac{1}{2}i\pi}$. Accordingly parametric representation of AB is given by

$$\left.\begin{aligned} x(\theta) &= \frac{2h}{\pi}\left[\log[\tan(\tfrac{1}{2}\theta)] - \tfrac{1}{2}\pi\right] \\ y(\theta) &= \frac{2h}{\pi}\left[\tfrac{1}{2}\pi - \log[\tan(\tfrac{1}{4}\pi-\tfrac{1}{2}\theta)]\right] \end{aligned}\right\} \tag{12.9.6}$$

As $\theta \to 0$, $x \to -\infty$ and $y \to h$ and as $\theta \to \frac{1}{2}\pi$, $x \to -h$ and $y \to \infty$ in accordance with Fig. 12.33.

Detailed knowledge of the velocity field is not needed in order to evaluate the force exerted on the plate in the original problem of Fig. 12.32. We suppose first that the pressure on the bounding streamlines is zero. Then the force on the plate is due entirely to the destruction of momentum in the incident stream. Per unit thickness in the (coordinate) z direction the momentum destroyed per unit time is the momentum contained in a mass $2\rho_0 hu$. Initially the linear momentum in the x direction associated with this mass when located at $x = -\infty$ is $2\rho_0 hu^2$. Ultimately this momentum is destroyed (when the mass, having split into the two streams is at $y = \pm\infty$). Since the problem is one of steady flow the momentum destroyed per second by the plate is also $2\rho_0 hu^2$, which is therefore the force exerted in the x direction on unit width of the plate in the (coordinate) z direction.

If the pressure on the bounding streamlines is $P_0 \neq 0$ the pressure in the flow is everywhere increased by P_0 and there is an additional (infinite) force of uniform intensity P_0 per unit area of the plate. In reality P_0 would be atmospheric pressure, which would be exerted even in the absence of the jet. In any practical situation not only would the plate be finite in area but the force due to P_0 would be counteracted by atmospheric pressure exerted on the rear of the plate.

12.9.2 Emergence of a plane jet from a slit

The problem is defined by Fig. 12.35. A rigid wall $x = 0$ contains fluid in $x < 0$. Fluid escapes through the slit $|y| < H$ and forms a plane jet in the region $x > 0$. Ultimately the jet is of width $2h$ and travels with speed u in the positive x direction.

We assume the free streamlines BC, $B'C'$ are continuations respectively of the streamlines AB and $A'B'$ lying on the inside of the wall. From symmetry $y = 0$ is a streamline which we label $\psi = 0$. With this labelling, and from consideration of the behaviour of the jet as $x \to \infty$, the streamlines ABC, $A'B'C'$ become respectively $\psi = -hu$ and $\psi = hu$.

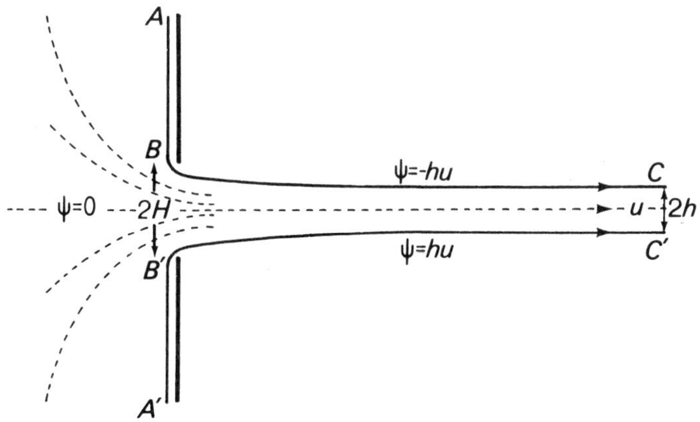

Fig. 12.35 Plane jet issuing from a slit in a wall

On the free portions of $\psi = \pm hu$, $q = u$. At B', $\theta = \frac{1}{2}\pi$ while at C', $\theta = 0$, so that along $B'C'$, θ varies from $\frac{1}{2}\pi$ to zero. Similarly along BC, $q = u$ and θ varies from $-\frac{1}{2}\pi$ to 0. Along $A'B'$, $\theta = \frac{1}{2}\pi$ and we assume q varies from zero (at $y = -\infty$) to u at B'. Similarly along AB, $\theta = -\frac{1}{2}\pi$ and q varies from 0 to u.

Accordingly the corresponding map in the **v** plane is the semi-circle. $q = u$, $|\theta| < \frac{1}{2}\pi$ with values of ψ determined on the boundaries as in Fig. 12.36.

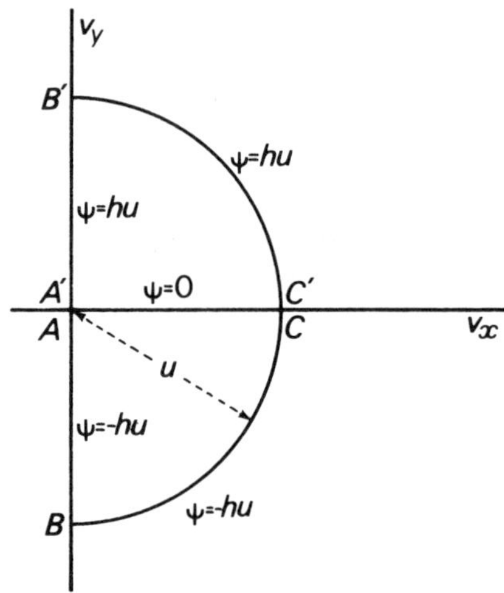

Fig. 12.36 Mapping of Fig. 12.35 onto **v** plane

Again we are concerned with finding a solution of (12.92) but inside a semi-circle. As in the previous example the boundary conditions imply that ψ is an odd function of θ, and since only the range $-\frac{1}{2}\pi \leqslant \theta \leqslant \frac{1}{2}\pi$ is involved, this suggests a Fourier series solution of the form

$$\psi = \sum_{n=1}^{\infty} a_{2n} q^{2n} \sin(2n\theta). \tag{12.9.7}$$

However this solution yields $\psi = 0$ on $\theta = \pm\frac{1}{2}\pi$ which does not meet the boundary condition on $B'A'AB$ in Fig. 12.36.

We recall that $\psi = \theta$ is also a solution of Laplace's equation and augment (12.9.7) with a term $2uh\theta/\pi$ obtaining

$$\psi = 2hu\theta/\pi + \sum_{n=1}^{\infty} a_{2n} q^{2n} \sin(2n\theta) \tag{12.9.8}$$

which now satisfies the condition $\psi = \pm hu$ on $\theta = \pm\frac{1}{2}\pi$. The remaining conditions on $q = u$ lead to

$$\sum_{n=1}^{\infty} a_{2n} u^{2n} \sin(2n\theta) = \begin{cases} -2uh\theta/\pi + hu & (0 < \theta \leqslant \frac{1}{2}\pi) \\ -2uh\theta/\pi - hu & (-\frac{1}{2}\pi \leqslant \theta < 0) \end{cases}$$

whence a_{2n} is given by

$$u^{2n} a_{2n} = (4uh/\pi) \int_0^{\frac{1}{2}\pi} [1 - 2\theta/\pi] \sin(2n\theta) d\theta = 4uh/2n\pi$$

The stream function may now be written

$$\psi = \frac{2uh}{\pi} \left[\theta + 2 \sum_{n=1}^{\infty} \frac{(q/u)^{2n} \sin(2n\theta)}{2n} \right]$$

$$= \frac{2uh}{\pi} \text{ Im } \left[-\log(-qe^{-i\theta}) - \log(-1) - 2 \sum_{n=1}^{\infty} \frac{(-qe^{-i\theta}/u)^{2n}}{2n} \right]$$

Apart from an irrelevant constant $[-\log(-1)]$ this gives, on summing the series

$$\omega = -(2uh/\pi) [\log v - \log(1 + v/u) - \log(1 - v/u)] .$$

Following the analysis of the previous problem

$$z = \int v^{-1} \frac{d\omega}{dv} dv = (2uh/\pi) \left[v^{-1} - u^{-1} \log\left(\frac{u+v}{u-v} \right) \right] + C$$

where C is a complex constant of integration. While there is no mechanism in this problem immediately for determining C, the precise value of C is of no consequence since it merely relates to the position of the origin in the x, y plane. Accordingly we choose $C = 0$.

On the free streamlines $B'C'$, $v = -ue^{-i\theta}$ $(0 \leqslant \theta \leqslant \frac{1}{2}\pi)$ and

$$z = -(2h/\pi) \left[e^{i\theta} + \log\left(\frac{1 - e^{-i\theta}}{1 + e^{-i\theta}} \right) \right] = -(2h/\pi) \left[e^{i\theta} + \log[i \tan(\theta/2)] \right]$$

or on separating into real and imaginary parts

$$x = -(2h/\pi) [\cos\theta + \log\tan(\tfrac{1}{2}\theta)] , \qquad y = -(2h/\pi) [\sin\theta + \tfrac{1}{2}\pi] .$$

The point B' is given by $\theta = \frac{1}{2}\pi$

i.e. $$x_{B'} = 0, \quad y_{B'} = -(2h/\pi)(1 + \tfrac{1}{2}\pi) .$$

Similarly

$$x_B = 0, \quad y_B = (2h/\pi)(1 + \tfrac{1}{2}\pi) .$$

Thus in fact the choice $C = 0$ corresponds to a choice of origin midway between B and B'.

The gap BB' is given by

$$2H = y_B - y_{B'} = (4h/\pi)(1 + \tfrac{1}{2}\pi)$$

and the jet contracts in width from $2H$ to $2h$ where the contraction ratio is

$$(h/H) = \pi/(\pi + 2) = 0.611 .$$

12.9.3 STREAM INCIDENT NORMALLY ON A PLATE OF FINITE WIDTH

Finally we consider the problem of Fig. 12.37 where a plate CBC' of finite width $2L$ in the y direction is placed perpendicular to a stream with velocity u in the positive x direction. A solution of this problem given by choosing $\alpha = \frac{1}{2}\pi$ in equation (12.8.51) assumes that the flow takes place on both sides of the plate. The resulting complex potential leads to infinite fluid velocities at the plate edges and this is not acceptable physically.

Here we assume that behind the plate there is a region of motionless *dead* fluid separated by the streamlines CD, $C'D'$ from the region in which flow occurs. In the dead region the pressure is supposed constant so that along CD and $C'D'$ the speed q is constant.

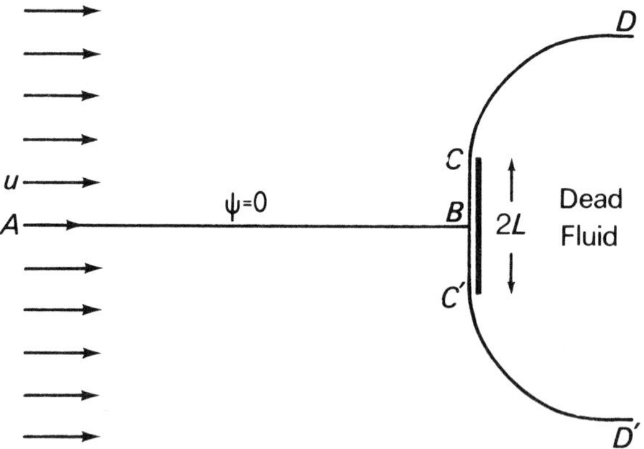

Fig. 12.37 Stream incident normally on a plate of finite width

The streamlines CD and $C'D'$ are part of the dividing streamline $ABCD$, $ABC'D'$ which we label $\psi = 0$. Along CD, $q = u$ (say) and θ varies from $\frac{1}{2}\pi$ at C to 0 as $x \to \infty$. Along BC, $\theta = \frac{1}{2}\pi$ and q varies from zero at the stagnation point B to u at C. Finally along AB, $\theta = 0$ and q varies from $u(x \to -\infty)$ to zero at B. Thus the streamline $ABCD$ with $\psi = 0$ maps onto a quarter circle of the $\mathbf{v}$ plane (Fig. 12.38).

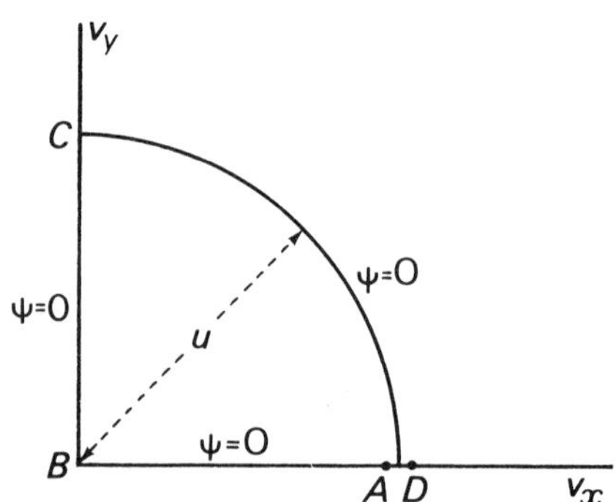

Fig. 12.38 Mapping of streamline $ABCD$ of Fig. 12.37 onto $\mathbf{v}$ plane

An immediate difficulty arises in that it would appear that if ψ vanishes all around the boundary then ψ must be identically zero inside the semicircle (since the solution is unique and $\psi = 0$ satisfies the Laplace equation and the boundary conditions). The resolution of the difficulty lies in the result that at the common point A,D of Fig. 12.38 there is a (double pole) singularity.

To see this we consider initially the more complicated problem with a stream of finite width (Fig. 12.39).

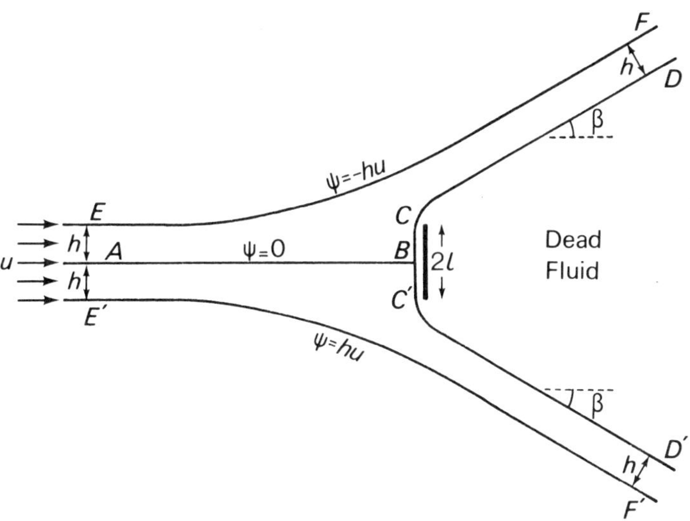

Fig. 12.39 Symmetric normal impact of a stream of finite width on a plate of finite width

We suppose the incident stream is divided by the plate of width $2h$ into two streams whose speeds are ultimately u in the directions $\pm\beta$ and whose widths are also h. (This last result follows from either mass conservation or the flux result.) At the moment β is an undetermined parameter; ultimately we find an equation relating h, l and β.

The top part of Fig. 12.39 now maps into a quarter circle in the **v** plane, but CD only maps into part of the circular boundary $\frac{1}{2}\pi \geqslant \theta \geqslant \beta$, $q = u$. The remaining part of the circular boundary is associated with the streamline EF for which $\psi = -hu$ (Fig. 12.40).

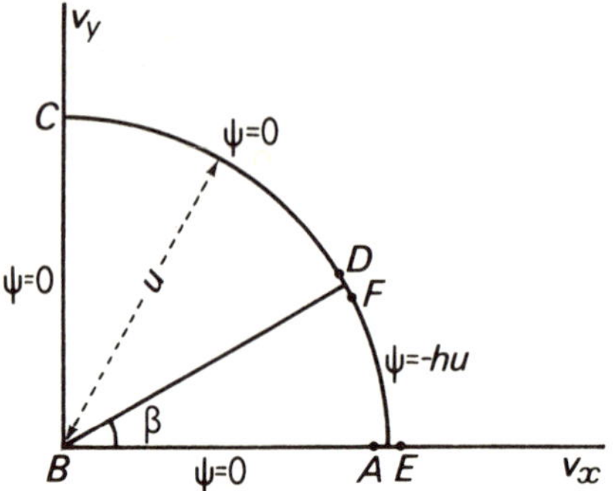

Fig. 12.40 Mapping of streamlines *ABCD* and *EF* of Fig. 12.39

ψ is finite inside the quarter circle and a suitable half range Fourier series which satisfies $\psi = 0$ on the axes is

$$\psi = \sum_{n=1}^{\infty} a_{2n} q^{2n} \sin(2n\theta) .$$

From the boundary conditions on $q = u$

$$\sum_{n=1}^{\infty} a_{2n} u^{2n} \sin(2n\theta) = \begin{cases} -hu & (0 \leqslant \theta \leqslant \beta) \\ 0 & (\beta \leqslant \theta \leqslant \frac{1}{2}\pi) \end{cases}$$

and therefore

$$a_{2n} u^{2n} = -(4uh/\pi) \int_0^{\beta} \sin(2n\theta) d\theta = -(2hu/n\pi)\,[1 - \cos(2n\beta)]$$

$$\text{whence } \psi = -(2hu/\pi \sum_{n=1}^{\infty} [1 - \cos(2n\beta)]\,(q/u)^{2n} \sin(2n\theta)/n .$$

We leave the reader to supply the details which yield

$$\omega(v) = -(2uh/\pi) \Big[\log[1 - v^2/u^2] - \tfrac{1}{2}\log[1 - v^2 e^{2i\beta}/u^2] - \tfrac{1}{2}\log[1 - v^2 e^{-2i\beta}/u^2]\Big] \tag{12.9.9}$$

and

$$z = (2h/\pi)\left[\log\left(\frac{u+v}{u-v}\right) - \tfrac{1}{2}e^{i\beta}\log\left(\frac{ue^{-i\beta}+v}{ue^{-i\beta}-v}\right) - \tfrac{1}{2}e^{-i\beta}\log\left(\frac{ue^{i\beta}+v}{ue^{i\beta}-v}\right)\right] . \tag{12.9.10}$$

In obtaining this last result the integration constant has been chosen so that $v = 0$ at the stagnation point $z = 0$.

For the streamline CD, $q = u$ and $\frac{1}{2}\pi \geqslant \theta \geqslant \beta$. On substituting $v = -ue^{-i\theta}$ into (12.9.10) and separating into real and imaginary parts there results

$$x = \frac{2h}{\pi}\left[\log\tan\tfrac{1}{2}\theta - \tfrac{1}{2}\cos\beta\,[\log\tan\tfrac{1}{2}(\theta-\beta) + \log\tan\tfrac{1}{2}(\theta+\beta)]\right]$$

$$y = \frac{2h}{\pi}\left[\tfrac{1}{2}\pi(1-\cos\beta) - \tfrac{1}{2}\sin\beta\,[\log\tan\tfrac{1}{2}(\theta-\beta) - \log\tan\tfrac{1}{2}(\theta+\beta)]\right]$$

(12.9.11)

A parametric representation of the streamline EF may be found similarly.

The point $C(x = 0, y = l)$ is given by substituting $\theta = \frac{1}{2}\pi$ in equations (12.9.11). With $\theta = \frac{1}{2}\pi$, $\log\tan\frac{1}{2}\theta = 0$ and $\log[\tan(\frac{1}{4}\pi - \frac{1}{2}\beta)\tan(\frac{1}{4}\pi + \frac{1}{2}\beta)] = 0$ so that the coordinate value $x = 0$ at the edge of the plate is confirmed. For the y coordinate we find

$$l = h(1-\cos\beta) + [2h\sin\beta/\pi]\,\log\left[\frac{1+\tan(\frac{1}{2}\beta)}{1-\tan(\frac{1}{2}\beta)}\right] \qquad (12.9.12)$$

which is the relation determining β in terms of l and h. It is readily verified that in the range $0 \leqslant \beta \leqslant \frac{1}{2}\pi$, l is an increasing function of β $(dl/d\beta \geqslant 0)$ and that for $\beta = 0$, $l/h = 0$ while for $\beta \to \frac{1}{2}\pi$, $l/h \to \infty$.

The case of a very wide stream requires $h \to \infty$; however, in order to keep the plate dimension l finite in (12.9.12) we must suppose that $\beta \to 0$ in such a manner that the limit of the right side of (12.9.12) is finite (say L). It is easily verified that (12.9.12) may be written (for small β)

$$l = h\left[\beta^2\left(\tfrac{1}{2} + \frac{2}{\pi}\right) + O(\beta^4)\right]$$

so that

$$L = \frac{\pi + 4}{2\pi} \underset{\substack{h\to\infty \\ \beta\to 0}}{\text{Lt}} (\beta^2 h)\,. \qquad (12.9.13)$$

If use is made of the limit (12.9.13) in the complex potential (12.9.9), there results, for the case of an infinite stream flowing past a plate of width $2L$ (Fig. 12.37),

$$\omega = \frac{8u^3L}{(\pi+4)}\,\frac{v^2}{(u^2-v^2)^2} \qquad (12.9.14)$$

and it is readily verified that the imaginary part of (12.9.14) satisfies the boundary conditions of Fig. 12.38. However as stated earlier there is a (double pole) singularity at $v = -u$ (i.e. at the common point A, D).

It is possible to obtain the streamline CD of Fig. (12.37) direct from (12.9.14). However it is simpler to proceed to the limit defined by (12.9.13) in equations (12.9.11). There results (after some algebra)

$$x = \frac{2L}{4+\pi}\,[\log\tan(\tfrac{1}{2}\theta) + \cos\theta/\sin^2\theta], \quad y = \frac{L}{4+\pi}\left(\pi + \frac{4}{\sin\theta}\right).$$

For $\theta \to 0$ only the terms involving $\sin\theta$ are important and

$$x \sim \frac{2L}{(4+\pi)\theta^2}, \quad y \sim \frac{4L}{(4+\pi)\theta}$$

and the streamline is asymptotic to the parabola

$$y^2 = \frac{8}{(4+\pi)}\,Lx\,.$$

The force on the plate per unit length in the (coordinate) z direction is most simply evaluated from momentum arguments. For the case where the stream is of finite width, the incident momentum per unit time in the positive x direction is $2\rho_0 hu^2$ and the momentum carried away per unit time by the two streams at angle β is $2(\rho_0 hu^2)\cos\beta$. Thus the force per unit length on the plate is in the positive x direction and of magnitude

$$F = 2\rho_0 hu^2\,(1 - \cos\beta)\,.$$

With the help of (12.9.12) F may be expressed in terms of h and l. For the limiting case $\beta \to 0$, $h \to \infty$ there results, after using (12.9.13), the simple formula

$$F = \frac{2\pi}{\pi + 4}\,2L(\tfrac{1}{2}\rho_0 u^2)\,.$$

It is usual to express the total force on the plate in terms of a drag coefficient C_D defined by

$$C_D = (F/2L)/(\tfrac{1}{2}\rho_0 u^2)$$

where $F/2L$ is the force per unit area of the plate. For the present problem the dimensionless coefficient C_D is given by

$$C_D = \frac{2\pi}{\pi+4} = 0.880 \tag{12.9.15}$$

The assumption that a streamline divides a region of dead fluid from fluid in motion is not compatible with the assumption of an inviscid fluid, since the existence of a discontinuous tangential velocity necessarily implies that the streamline (or rather stream sheet) contains a highly concentrated distribution of vorticity. However experimental behaviour is in accord, at least qualitatively, with (12.9.15). To achieve the motion described, it would be necessary in the transient stage, prior to establishing the steady state, that the necessary vorticity be shed by boundary layer effects at the edges of the plate. [In fact the sheet across which the velocity is discontinuous may be shown to be equivalent to a continuous distribution of line vortices of infinitesimal strength.]

Problems, Chapter 10

1. Underwater explosion

A spherical explosive capsule of initial radius a_0 is immersed in a sea of incompressible inviscid liquid. At large distances from the explosive the pressure in the liquid is P_0. At $t = 0$ the explosive is detonated and the explosive debris is described by a constitutive equation

$$\bar{P}/P_1 = (\sigma/\sigma_0)^\gamma$$

where $\gamma > 1$ is a constant, $\bar{P}$ is the pressure and σ is the density of the debri, σ_0 is the initial density and P_1 is the initial pressure. If the spherical cavity formed by the explosive debris is of radius $a(t)$ and the pressure of the debris is uniform for $0 \leqslant r \leqslant a(t)$, derive the equation of motion of the cavity

$$P_1 (a_0/a)^{3\gamma} = P_0 + \rho_0 (a\ddot{a} + \tfrac{3}{2}\dot{a}^2)$$

where ρ_0 is the (constant) density of the liquid.

Show that a first integral of the equation is

$$\frac{P_1 a_0^{3\gamma}}{3(\gamma - 1)} [a_0^{-3(\gamma-1)} - a^{-3(\gamma-1)}] = \tfrac{1}{3}P_0(a^3 - a_0^3) + \tfrac{1}{2}\rho_0 a^3 \dot{a}^2$$

Deduce that the cavity first comes to rest at a radius a_f where

$$\frac{P_1}{P_0(\gamma - 1)} [1 - (a_0/a_f)^{3(\gamma-1)}] = (a_f/a_0)^3 - 1$$

and that this is achieved in time T given by

$$\left(\frac{2P_0}{3\rho_0 a_0^2}\right)^{\frac{1}{2}} T =$$

$$[1 - X^{-3(\gamma-1)}]^{\frac{1}{2}} \int_1^X \frac{\theta^{3/2}\, d\theta}{[(X^3 - 1)(1 - \theta^{-3(\gamma-1)}) - (\theta^3 - 1)(1 - X^{-3(\gamma-1)})]^{\frac{1}{2}}}$$

where $X = a_f/a_0$.

2. Explosion in a cylindrical tube

A hollow cylindrical tube of initial cross section $a_0 \leqslant r \leqslant b_0$ is at rest for $t < 0$. The cavity is filled with explosive which is detonated at $t = 0$ so that for $t > 0$ the cylinder expands radially and occupies the annulus $a(t) \leqslant r \leqslant b(t)$. If the material of the tube is assumed to behave as an incompressible inviscid fluid of density ρ_0 subject to a constant external pressure P_0 at $r = b(t)$, and an internal pressure at $r = a(t)$ given by $P = \bar{P}(a)$, where $\bar{P}(a)$ is the pressure due to the explosive debris, deduce the equation of motion of the cavity

$$\bar{P}/\rho_0 = (P_0/\rho_0) + \tfrac{1}{2}(a\ddot{a} + \dot{a}^2)\log Q + \tfrac{1}{4}a\dot{a}\frac{d}{dt}(\log Q)$$

where
$$Q = (b_0^2 - a_0^2 + a^2)/a^2$$

and show that a first integral is given by

$$\int_{a_0}^{a} \overline{P}(a)a\,da = \tfrac{1}{2}P_0(a^2 - a_0^2) + \tfrac{1}{4}\rho_0 a^2 \dot{a}^2 \log Q \,.$$

3. Point dipole outside a plane

A point dipole of constant strength $\boldsymbol{\mu} = \mu(\sin\alpha, 0, \cos\alpha)$ is located at $x = y = 0$, $z = h$ above a rigid barrier $z = 0$. Show that the boundary condition $v_z = 0$ is satisfied by locating an image dipole of strength $\boldsymbol{\mu}' = \mu(\sin\alpha, 0, -\cos\alpha)$ at $x = y = 0, z = -h$. Also show that if the pressure at points remote from the dipole is P_0, and body forces are neglected, the pressure distribution on the plane $z = 0$ is given by

$$P = P_0 - 2\rho_0\mu^2\left\{\frac{\sin^2\alpha}{R^6} + \frac{3\xi\eta}{R^8} - \frac{9\xi^2 h^2}{R^{10}}\right\}$$

where $\xi = x\sin\alpha - h\cos\alpha$, $\eta = x\sin\alpha - 3h\cos\alpha$, $R^2 = (x^2 + y^2 + h^2)$

and ρ_0 is the (constant) density of the fluid.

4. General theorem on kinetic energy of incompressible, inviscid and irrotational fluid

(a) A number of solid objects with bounding surfaces denoted by S_m are in motion in an otherwise unbounded incompressible inviscid liquid. Show that if S_R denotes a sphere of large radius R and ϕ is such that

$$\underset{R\to\infty}{\mathrm{Lt}} \iint_{S_R} \phi\,\frac{\partial\phi}{\partial r}\,dS = 0$$

then the kinetic energy of the fluid is

$$T = -\tfrac{1}{2}\rho_0 \sum_m \iint_{S_m} \phi\,\frac{\partial\phi}{\partial n}\,dS$$

where $\partial\phi/\partial n$ denotes the outward normal derivative to S_m and ρ_0 is the fluid density.

(b) Show that if a sphere of radius a moves at a velocity $\mathbf{u}$ in an unbounded fluid which is at rest at large distances from the sphere, the kinetic energy of the fluid is

$$T = \tfrac{1}{4}M'\mathbf{u}^2$$

where M' is the mass of fluid displaced by the sphere. Hence deduce that the resistive forces of the fluid on the sphere is given by $\mathbf{F} = -\frac{1}{2}M'\dot{\mathbf{u}}$ in agreement with the results of Section 12.7.2.

5. Perturbation analysis of nearly circular cylinder in a stream

(a) Show that if $z = x + iy$, the stream function associated with the flow of a stream past a stationary circular cylinder with boundary $|z| = a$ is given by

$$\psi = \psi_0 = -u[r - a^2/r]\sin(\theta - \alpha)$$

where u is the speed of the stream which makes an angle α with the x axis.

(b) A nearly circular cylinder with boundary

$$r = a[1 + \epsilon\cos(p\theta)] \qquad (\epsilon \ll 1;\ p \text{ integer},\ p \geqslant 2)$$

is located at rest in a stream of speed u which makes an angle α with the x axis. By writing the stream function in the form $\psi = \psi_0 + \epsilon\psi_1$ determine ψ_1 so that to first order in ϵ, ψ vanishes on the boundary of the cylinder.

Hence determine the associated complex potential $\omega(z)$ and show that to order ϵ and for all choices of p, the force on the cylinder vanishes while there is a non-vanishing couple only for the case $p = 2$.

6. Generating function of Legendre polynomials and associated results

(a) It may be shown (see below) that the Legendre Polynomials $P_n(\eta)$ are generated by expanding the *generating function*

$$V(h, \eta) = (1 - 2h\eta + h^2)^{-\frac{1}{2}}$$

as a power series in h; that is for sufficiently small h and $|\eta| \leqslant 1$

$$V = (1 - 2h\eta + h^2)^{-\frac{1}{2}} = \sum_{n=0}^{\infty} h^n P_n(\eta)\,. \tag{i}$$

By writing the generating function in the form

$$V = (1 - 2h\eta + h^2)^{-\frac{1}{2}} = [(1 - he^{i\theta})(1 - he^{-i\theta})]^{-\frac{1}{2}}$$

where $\eta = \cos\theta$ show that the expansion (i) is absolutely convergent for $0 \leqslant h < 1$ provided $|\eta| \leqslant 1$.

Verify that if $P_n(\eta)$ is defined by (i) then

$$P_n(1) = 1\,. \tag{ii}$$

To verify that $P_n(\eta)$ defined by (i) are the Legendre polynomials, note first that the $P_n(\eta)$ so defined are polynomials of degree n in η. Also from (ii) $P_n(1) = 1$. Further the Legendre Polynomials defined by (12.7.11) also satisfy $P_n(1) = 1$ and are the only polynomial solutions of

$$(1-\eta^2)P_n''(\eta) - 2\eta P_n'(\eta) + n(n+1)P_n(\eta) = 0 \qquad \text{(iii)}$$

(for the second solution of this equation is not a polynomial). It is therefore sufficient to show that the polynomials generated by the generating function also satisfy equation (iii). This is proved in (c) below.

(b) (Recurrence relation, etc.)

From the definition of V show that the following identities hold

$$(\eta-h)V = (1-2h\eta+h^2)(\partial V/\partial h), \qquad h(\partial V/\partial h) = (\eta-h)(\partial V/\partial\eta)$$

and by substituting into these results from the power series (i) for V and comparing coefficients of h^n deduce the results

$$nP_n - (2n-1)\eta P_{n-1} + (n-1)P_{n-2} = 0 \qquad \text{(iv)}$$

$$\eta P_n' - P_{n-1}' = nP_n \qquad \text{(v)}$$

In particular, the first of these connects three successive Legendre Polynomials and is usually known as the recurrence relation.

(c) To show that the P_n satisfying the differential-difference formulae (iv) and (v) also satisfy Legendre's equation, write (iv) in the form

$$n[P_n - 2\eta P_{n-1} + P_{n-2}] = P_{n-2} - \eta P_{n-1},$$

differentiate, and make use of (v) to obtain

$$P_n' = 2\eta P_{n-1}' + P_{n-1} - P_{n-2}'. \qquad \text{(vi)}$$

Multiply (v) by η and subtract from (vi) to obtain

$$\begin{aligned}(1-\eta^2)P_n' &= \eta P_{n-1}' + P_{n-1} - P_{n-2}' - n\eta P_n \\ &= n[P_{n-1} - \eta P_n] \qquad \text{(vii)}\end{aligned}$$

where in the last step use has been made again of (v). Finally differentiate (vii) and again use (v) to show that P_n satisfies

$$\frac{d}{d\eta}[(1-\eta^2)P_n'(\eta)] + n(n+1)P_n(\eta) = 0$$

which is Legendre's equation.

(d) To prove (12.7.13) write

$$(1-2h\eta+h^2)^{-1} = \left(\sum_{n=0}^{\infty} h^n P_n(\eta)\right)\left(\sum_{m=0}^{\infty} h^m P_m(\eta)\right) = \sum_n \sum_m h^{n+m} P_n(\eta) P_m(\eta)$$

and integrate with respect to η from -1 to 1. On making use of (12.7.12) there results

$$\sum_n h^{2n} \int_{-1}^{1} P_n^2(\eta)d\eta = \int_{-1}^{1} \frac{d\eta}{(1-2h\eta+h^2)} . \tag{viii}$$

Show that the integral on the right hand side of (viii) is $h^{-1}\log[(1+h)/(1-h)]$ and by expanding the latter as a power series in h deduce that

$$\int_{-1}^{1} P_n^2(\eta)d\eta = 2/(2n+1) .$$

A further discussion on Legendre's polynomials is found in F. Chorlton's *Vector and Tensor Methods,* Ellis Horwood Ltd. (1976).

7. Point dipole outside sphere

Show that a point dipole of strength μ lying along the z axis and located at $x = y = 0$, $z = \xi$ exerts a velocity potential

$$\phi = \phi_1 \equiv \mu \frac{\partial}{\partial \xi} (r^2 + \xi^2 - 2r\xi \cos\theta)^{-\frac{1}{2}} \tag{i}$$

where r and θ are spherical polar coordinates. For $r < \xi$ show that ϕ_1 may be expanded in the form

$$\phi_1 = -\mu \sum_0^\infty \frac{(n+1)r^n}{\xi^{n+2}} P_n(\cos\theta) .$$

A solid sphere $r \leqslant a$ where $a < \xi$ is introduced into the flow given by $\phi = \phi_1$. Show that the modified velocity potential is $\phi = \phi_1 + \phi_2$ where

$$\phi_2 = -\mu \sum_0^\infty \frac{na^{2n+1}}{r^{n+1}\xi^{n+2}} P_n(\cos\theta) .$$

Further show that ϕ_2 is the potential of an image dipole of strength $-\mu(a/\xi)^3$ located at the image point $z = \xi' \equiv a^2/\xi$, $x = y = 0$.

Hence obtain the closed form solution of the problem

$$\begin{aligned}\phi = \;& \mu[-(\xi - r\cos\theta)\,[r^2 + \xi^2 - 2r\xi\cos\theta]^{-3/2} \\ &+ (a/\xi)^3(\xi' - r\cos\theta)\,[r^2 + \xi'^2 - 2r\xi'\cos\theta]^{-3/2}]\end{aligned}$$

and verify directly from this closed form that $(\partial\phi/\partial r)_{r=a} = 0$.

Calculate $\mathbf{v}^2$ on the surface $r = a$ and show that the force on the sphere is of magnitude $24\pi\rho_0 a^3\mu^2\xi/(\xi^2 - a^2)^4$ and points in the positive z direction.

8. Two spheres moving along line of centres

(a) Show that the velocity potential of a sphere of radius a travelling in the z direction with centre located at $[0, 0, q_1(t)]$ is

$$\phi = \phi_0 \equiv \tfrac{1}{2}u_1(t)a^3 \cos\theta/R^2$$

where $R^2 = x^2+y^2+[z-q_1(t)]^2, \quad u_1 = \dot{q}_1(t), \quad R\cos\theta = z-q_1$.

(b) A second sphere of radius b is centred on $x=y=0,\ z=q_2(t)$ and travelling with velocity $u_2 = \dot{q}_2$. It may be assumed that $q_2 > q_1$ (and then $q_2-q_1 > a+b$ so that the spheres are separated). In the absence of the first sphere, the velocity potential of this second sphere is that of a dipole of strength $\frac{1}{2}u_2b^3$ lying at the centre of the second sphere. Use the results of Problem 7 to show that the boundary condition on the first sphere is satisfied by writing

$$\phi = \phi_0 + \hat{\phi} \tag{i}$$

where near $R=a$

$$\hat{\phi} = -\tfrac{1}{2}u_2b^3\left\{\sum_0^\infty \frac{(n+1)R^n}{\xi^{n+2}}P_n(\cos\theta) + \sum_0^\infty \frac{na^{2n+1}}{R^{n+1}\xi^{n+2}}P_n(\cos\theta)\right\}$$

and where $\xi = (q_2-q_1)$.

Equation (i) does not constitute an exact solution to the problem posed since it fails to satisfy the boundary condition on the second sphere. However if the spheres are sufficiently far apart equation (i) will be approximately correct near the surface of the first sphere. A similar (but in detail different) approximation will hold near the surface of the second sphere.

Use (i) to show that in this approximation

$$T_1 \equiv -\tfrac{1}{2}\rho_0 \iint_{S_a} \phi\frac{\partial\phi}{\partial R}\,dS = \tfrac{1}{4}M_1'u_1^2 - \pi\rho_0a^3b^3u_1u_2\xi^{-3}$$

where S_a is the surface $R=a$ and $M_1' = \frac{4}{3}\pi a^3\rho_0$.

From symmetry arguments and the results of Problem 4 deduce that in the present approximation the kinetic energy of the fluid and spheres is given by

$$T = \tfrac{1}{4}(M_1'\dot{q}_1^2 + M_2'\dot{q}_2^2) - \frac{2\pi\rho_0a^3b^3\dot{q}_1\dot{q}_2}{(q_2-q_1)^3} + \tfrac{1}{2}(M_1\dot{q}_1^2 + M_2\dot{q}_2^2)$$

where $M_2' = \frac{4}{3}\pi b^3\rho_0$ and where M_1 and M_2 are respectively the actual masses of the spheres of radius a and b.

(c) If no other forces are present, use Lagrange's equations to deduce that the force acting on the sphere of radius a is

$$\mathbf{F} = \left(-\tfrac{1}{2}M_1'\ddot{q}_1 + \frac{2\pi\rho_0a^3b^3\ddot{q}_2}{(q_2-q_1)^3} - \frac{6\pi\rho_0a^3b^3\dot{q}_2^2}{(q_2-q_1)^4}\right)\mathbf{k}.$$

In this result, the first term is identical with that obtained in the text (and in Problem 4) for the force on an isolated accelerating sphere. The remaining terms are due to the presence of the other moving sphere. There is a curiosity here. It might be supposed that the term in $\dot{q}_2^2$ is the same as that obtained in Problem 7 for a stationary point dipole outside a stationary sphere. In fact the two terms are of different sign as well as of different magnitude. For large separation distance ξ the force due to the stationary dipole of Problem 7 varies as ξ^{-7} whereas the term appearing above is of magnitude ξ^{-4}. The contribution above is due essentially to the fact that the second sphere is a dipole *in motion* while the force obtaining in Problem 7 (which is ignored by the present approximation) is a higher order term altogether.

9. Axisymmetric stream function (spherical polar coordinates)

(a) Show that for axisymmetric flows $[v_r = v_r(r, \theta),\ v_\theta = v_\theta(r, \theta),\ v_\phi = 0]$ the incompressibility equation $\operatorname{div} \mathbf{v} = 0$ becomes

$$\frac{\partial v_r}{\partial r} + \frac{2v_r}{r} + \frac{1}{r \sin \theta} \frac{\partial}{\partial \theta} (\sin \theta\, v_\theta) = 0$$

where r, θ and ϕ are spherical polar coordinates. Prove that this equation is satisfied by the choice

$$v_r = -\frac{1}{r^2 \sin \theta} \frac{\partial \psi}{\partial \theta}, \qquad v_\theta = \frac{1}{r \sin \theta} \frac{\partial \psi}{\partial r}$$

where $\psi(r, \theta)$ is known as the Stokes' stream function [see also Problem 7, Chapter 10].

If additionally the flow is irrotational, deduce that ψ satisfies

$$\frac{\partial^2 \psi}{\partial r^2} + \frac{1}{r^2} \frac{\partial^2 \psi}{\partial \theta^2} - \frac{\cot \theta}{r^2} \frac{\partial \psi}{\partial \theta} = 0 . \qquad \text{(i)}$$

Note, this is not Laplace's equation. Also, there is no possibility of combining ϕ and ψ into a complex form $\phi + i\psi$ (which is exploited extensively in the theory of plane flow) since, amongst other difficulties, even the dimensions of ϕ and ψ are different.

Equation (i) may be compared with the corresponding result in the slow viscous flow approximation (Problem 7, Chapter 10).

(b) Show that

$$\mathbf{v} \cdot \operatorname{grad} \psi = 0$$

and deduce that $\mathbf{v}$ lies parallel everywhere to the surfaces $\psi(r, \theta) = \text{const.}$ Deduce that in steady flow situations about a stationary axisymmetric body, ψ is constant on the surface of the body.

The surfaces $\psi(r, \theta) = \text{const.}$ are 'stream sheets'; however the usual terminology is *streamline* in conformity with the terminology used in plane flows.

(c) Show that three specific solutions of equation (i) are

$$\psi = m\cos\theta, \qquad \psi = -\tfrac{1}{2}ur^2\sin^2\theta$$

$$\psi = -\tfrac{1}{2}u\sin^2\theta\,[r^2 - a^3/r]$$

and that these solutions represent respectively a point source of strength m at the origin, a stream travelling in the positive z direction with speed u and the flow of a stream of speed u in the positive z direction past the stationary sphere $r = a$.

Discuss in detail why the last of these is different to the solution of Problem 8 of Chapter 10 for the same physical situation.

10. Pair of parallel rectilinear vortices

(i) At time $t = 0$ parallel rectilinear vortices of strength κ_1 and κ_2 lie respectively at $(a, 0)$ and $(-b, 0)$ where $a > 0$, $b > 0$ and where $\kappa_1 a = \kappa_2 b$. Show that the subsequent motion of the vortices is one in which both vortices revolve around the origin at the common angular velocity Ω where

$$\Omega = \frac{\kappa_1}{b(a+b)} \equiv \frac{\kappa_2}{a(a+b)} .$$

(ii) If the pressure at points remote from the vortices is P_0, and body forces are neglected, show that the pressure at the origin is constant with value

$$P = P_0 - \tfrac{1}{2}\rho_0[(\kappa_2/b)^2 + (\kappa_1/a)^2] .$$

(iii) Use the results of part (i) to discuss the subsequent motion of two arbitrarily located parallel rectilinear vortices of arbitrary strengths.

11. Rectilinear vortex in a rectangular corner

A rectilinear vortex of strength κ lies at $[x(t), y(t)]$ in the first quadrant outside the two rigid planes $x = y = 0$. Construct the image system and show that $x(t)$ and $y(t)$ satisfy the equations

$$\dot{x} = \tfrac{1}{2}\kappa x^2 y^{-1}(x^2+y^2)^{-1}, \quad \dot{y} = -\tfrac{1}{2}\kappa y^2 x^{-1}(x^2+y^2)^{-1}$$

and that the vortex path is given by

$$x^{-2} + y^{-2} = c^{-2} \qquad *$$

where c^{-2} is a constant of integration.

Equation * has the parametric representation

$$x = c(\cos\gamma)^{-1}, \quad y = c(\sin\gamma)^{-1} .$$

Show that

$$\tan 2\gamma = [A - (\kappa t/4c^2)]^{-1}$$

where A is a second constant of integration.

12. Extension of Blasius' first theorem to unsteady flow around a stationary cylinder

Show that if the unsteady flow of an inviscid incompressible irrotational fluid around a cylinder is described by the velocity potential $\omega(z, t)$, then with neglect of body forces, the force exerted on the cylinder per unit length is given by

$$\mathbf{F} = (F_x + F_x')\mathbf{i} + (F_y + F_y')\mathbf{j}$$

where $F_y + iF_x = -\frac{1}{2}\rho_0 \oint_c \left(\frac{\partial \omega}{\partial z}\right)^2 dz, \quad F_y' - iF_x' = \rho_0 \oint_c \frac{\partial \omega}{\partial t}\, dz$

and where c denotes the bounding contour of the cylinder.

13. Force on cylinder due to vortex

At $t = 0$ a line vortex of strength κ is located at $z = f$ (f real and positive) outside the stationary circular cylinder $|z| = a$ (so that $f > a$).

Show that the resulting unsteady flow problem is solved by

$$\omega = i\kappa \log\left[z - fe^{-i\Omega t}\right] + i\kappa \log z - i\kappa \log\left[z - (a^2/f)e^{-i\Omega t}\right]$$

where $$\Omega = \kappa a^2/[f^2(f^2 - a^2)]\ .$$

Use the results of Problem 12 to show that the force on the cylinder lies in the outward direction of the radius vector from the origin to the current vortex position, and is of magnitude

$$2\pi\rho_0\kappa^2 a^2/f^3\ .$$

14. Equilibrium of vortex in a stream in the presence of stationary cylinder

(i) A rectilinear line vortex of strength κ is located on the positive x axis at $(f, 0)$ outside a circular cylinder with boundary $|z| = a$ where $a < f$. Also a stream of speed u flows past the cylinder; in regions remote from the cylinder the stream flows in the positive y direction. Show that the velocity potential is given by

$$\omega = i\kappa \log(z - f) + i\kappa \log z - i\kappa \log[z - (a^2/f)] + iu[z - a^2/z]$$

and deduce that the vortex remains stationary provided u, κ, a and f are related by

$$u = \kappa a^2 f/(f^4 - a^4)\ .$$

(ii) Show that the equilibrium situation described above is unstable for small disturbances of the position of the vortex. [Consider the case of a vortex located at $z = f + Z(t)$ where $Z = X + iY$ and $|Z| \ll [f - a^2/f]$; write down the equations of motion for Z in a linearised approximation and show that the equations possess solutions with exponential time dependence.]

15. Rectilinear vortex inside a circular boundary

A rectilinear vortex of strength κ lies inside the circular boundary $|z| = a$. If at $t = 0$ the vortex is located at $(f, 0)$ where $a > f > 0$, deduce that at time t the velocity potential is

$$\omega = i\kappa \log[z - fe^{i\Omega t}] - i\kappa \log[z - (a^2/f)e^{i\Omega t}]$$

where

$$\Omega = \kappa/(a^2 - f^2)$$

and calculate the total thrust per unit length on the inside wall of the circular boundary.

16. Rectilinear vortex outside a flat plate

(i) A rectilinear vortex of strength κ lies at $\zeta_0(t)$ outside the circular cylinder $|\zeta| = a$. Solve the flow problem in the ζ plane and show that the vortex moves in a circle $|\zeta_0(t)| = $ const. at angular velocity

$$\Omega = -\kappa a^2 |\zeta_0|^{-2} [\,|\zeta_0|^2 - a^2]^{-1}.$$

(ii) Use the solution of the potential problem above in conjunction with the Kutta-Joukowski transformation

$$\zeta = \tfrac{1}{2}[z + (z^2 - 4a^2)^{\frac{1}{2}}] \qquad \text{with inverse} \qquad z = \zeta + a^2\zeta^{-1}$$

to solve the problem of a rectilinear vortex lying at $z_0(t)$ outside the flat plate $|x| \leqslant 2a$, $y = 0$. Show that the image path $\zeta_0(t) \equiv \zeta[z_0(t)]$ in the ζ plane of the vortex path $z = z_0(t)$ satisfies the differential equation

$$-\frac{d\bar{\zeta}_0}{dt} = \frac{i\kappa a^2 \bar{\zeta}_0^2 \zeta_0 [2a^2 - \zeta_0(\zeta_0 + \bar{\zeta}_0)]}{(\zeta_0^2 - a^2)^2(\bar{\zeta}_0^2 - a^2)(\zeta_0\bar{\zeta}_0 - a^2]}$$

Show that this equation is *not* satisfied by the circular path $\zeta_0 = fe^{-i\Omega t}$ where $\Omega = \kappa a^2/f^2(f^2 - a^2)$ (f real) so that while conformal mapping techniques map one problem into another, there is no parallel correspondence between vortex motions.

17. Force on cylinder due to line source in the presence of circulation and a stream

A cylinder of arbitrary cross section possesses a boundary which encloses the origin. There is a circulation of strength κ around the cylinder together with a line source of strength m located at z_0 (outside the cylinder). Also a stream of speed u in the positive x direction flows past the cylinder.

The total complex potential is therefore of the form

$$\omega = i\kappa \log z - m \log(z - z_0) - uz + \omega_1(z)$$

where $\omega_1(z)$ is a function of z, regular in the region outside the cylinder, which is chosen to satisfy the boundary conditions. Since ω is regular for $|z| \to \infty$ it may be assumed that for large z (after discarding perhaps an irrelevant constant)

$$\omega(z) = O(z^{-1}).$$

Use this result together with the first Blasius theorem and the theory of contour integration to show that the complex force $(F_y + iF_x)$ acting on unit length of the cylinder is given by

$$F_y + iF_x = 2\pi\rho_0[-\kappa u + \kappa m z_0^{-1} - im(d\omega_1/dz)_{z=z_0}] .$$

18. Force on circular and elliptical cylinder due to line source

(i) Use the results of Problem 17 to show that in the absence of circulation a line source of strength m located at $z = f$ (f real and $f > 0$) outside a circular cylinder $|z| = a$ exerts a force in the positive x direction on the cylinder of magnitude $2\pi\rho_0 m^2 a^2 f^{-1}(f^2 - a^2)^{-1}$ per unit length.

(ii) A line source of strength m lies at $z = z_0$ (z_0 complex) outside the elliptical cylinder

$$\frac{x^2}{\gamma^2} + \frac{y^2}{\beta^2} = 1 \qquad \text{where} \quad \gamma > \beta .$$

Solve the potential problem and by using the results of Problem 17, show that the (complex) force $F_y + iF_x$ exerted per unit length of the cylinder is

$$F_y + iF_x = \frac{\pi\rho_0 m^2 i[(\gamma+\beta)^2\zeta_0 - (\gamma^2-\beta^2)\bar{\zeta}_0]\,\zeta_0^2}{2\,[\zeta_0^2 - \frac{1}{4}(\gamma^2-\beta^2)]^2\,[\zeta_0\bar{\zeta}_0 - \frac{1}{4}(\gamma+\beta)^2]}$$

where ζ_0 is the complex number

$$\zeta_0 = \tfrac{1}{2}[z_0 + (z_0^2 - 4a^2)^{\frac{1}{2}}]$$

and where

$$a = \tfrac{1}{2}(\gamma^2 - \beta^2)^{\frac{1}{2}} .$$

19. Milne-Thomson's construction of $f(z)$ given either $\phi(x, y)$ or $\psi(x, y)$

It is required to find $f(z)$ such that

$$f(z) = \phi(x, y) + i\psi(x, y)$$

where ϕ and ψ are real and either ϕ or ψ is given. Consider first the case where ϕ is given.

(i) Show formally that $f(z)$ may be written

$$f(z) = \phi\left[\frac{1}{2}(z+\bar{z}), \frac{1}{2i}(z-\bar{z})\right] + i\psi\left[\frac{1}{2}(z+\bar{z}), \frac{1}{2i}(z-\bar{z})\right] . \qquad *$$

This may be regarded as an identity involving two independent variables z and $\bar{z}$. Since the left side of $*$ is independent of $\bar{z}$, the right side is likewise independent of $\bar{z}$. In other words with any choice of $\bar{z}$, $*$ remains valid. The particular choice $\bar{z} = z$ leads to

$$f(z) = \phi[z, 0] + i\psi[z, 0]$$

so that
$$f'(z) = \left[\frac{\partial\phi(x,y)}{\partial x}\right]_{x=z,\ y=0} + i\left[\frac{\partial\psi(x,y)}{\partial x}\right]_{x=z,\ y=0}$$

(ii) Use the Cauchy-Riemann equations to show that

$$f'(z) = \left[\frac{\partial\phi(x,y)}{\partial x}\right]_{x=z,\ y=0} - i\left[\frac{\partial\phi(x,y)}{\partial y}\right]_{x=z,\ y=0}$$

which determines $f'(z)$ for given $\phi(x, y)$. Hence $f(z)$ is determined to within an imaginary constant.

(iii) Alternatively if $\psi(x, y)$ is given, show that

$$f'(z) = \left[\frac{\partial\psi(x,y)}{\partial y}\right]_{x=z,\ y=0} + i\left[\frac{\partial\psi(x,y)}{\partial x}\right]_{x=z,\ y=0}.$$

which determines $f(z)$ to within a real constant.

(iv) Find the function $f(z)$ whose real part is

$$\phi(x, y) = x^3 - 3xy^2$$

Verify that the real part of the resulting $f(z)$ is as given and find $\psi(x, y)$.

Explain the inconsistencies that seemingly arise if the function ϕ is replaced by

$$\phi = x^3 - \lambda xy^2$$

where λ is a real constant of any value except $\lambda = 3$.

20. Reconciliation of separation of variables solution of Laplace's equation in plane polar coordinates with complex variable solution

If $f'(z)$ is regular for $a \leqslant |z| \leqslant b$ $(\infty > b > a > 0)$ state the form of the Laurent expansion of $f'(z)$ and show that on integrating, $f(z)$ assumes the form

$$f(z) = \sum_{-\infty}^{\infty} a_n z^n + A \log z \qquad *$$

where A and the a_n are constants (usually complex).

By writing $z = re^{i\theta}$ reconcile $*$ with the separation of variable solution (12.6.15) of Laplace's equation.

21. Rectilinear vortex lying between two parallel planes

(i) Show that the transformation

$$\zeta = e^{\frac{1}{2}\pi z/h}$$

where $\zeta = \xi + i\eta$, $z = x + iy$ and $h > 0$, maps the region $|y| < h$ onto $\xi > 0$ and that the straight lines $y = \pm h$ map respectively into the positive and negative η axes. Note in particular that the point $\zeta = 0$ corresponds to the region $x \to -\infty$, $|y| \leqslant h$. This is of some importance in Problem 22.

(ii) A rectilinear vortex of strength κ lies at $z = il$ (l real, $|l| < h$) between two walls $y = \pm h$. Show that the vortex remains on $y = l$ moving in the x direction with velocity

$$-\tfrac{1}{4}\kappa\pi h^{-1}\tan(\tfrac{1}{2}\pi l/h).$$

22. Line source located between two parallel planes

Use the transformation of Problem 21 (i) to show that the problem of a line source of strength m located at $z = il$ (l real, $|l| < h$) between two planes $y = \pm h$ is solved by the complex potential

$$\omega(z) = -m\log[\sinh(\tfrac{1}{2}\pi z/h) - i\sin(\tfrac{1}{2}\pi l/h)].$$

Hint. Note that although the source in the z plane maps into a source of the same strength in the ζ plane, there is an additional sink of strength m required at the origin in the ζ plane. The reason is because at large distances from the source in the z plane the fluid created at the source is diverted into one of two streams which flow in the positive x direction for $x \to \infty$ and the negative x direction for $x \to -\infty$. In particular since the region $x \to -\infty$ maps into the immediate neighbourhood of $\zeta = 0$ the efflux of the stream in the z plane requires a sink at $\zeta = 0$ in the ζ plane. Evidently the strength of the sink is m.† There is no sink associated with the stream located in the region $x \to \infty$ since this region maps into remote parts of the ζ plane and not to the immediate vicinity of a single point. In fact the region $x > A$ (A real, $A > 0$) maps into $|\zeta| > \exp(\tfrac{1}{2}\pi A/h)$, $\mathrm{Re}\,\zeta > 0$.

23. Two dimensional plane jet emerging from finite width slit between two semi-infinite inclined plates

Two semi-infinite plates, inclined symmetrically to each other at an angle $2\alpha(2\alpha < \pi)$, bound a sea of incompressible inviscid irrotational fluid. The plates are not joined so that a plane jet of fluid emerges from the gap as shown in the Figure. If body forces are neglected show that the complex potential is given by

$$\omega(v) = (hu/\alpha)\left[-\log v + (2\alpha/\pi)\log\left[1 - (ve^{i\pi}/u)^{\pi/\alpha}\right]\right]$$

† The sink is of strength m in the ζ plane (and not $\frac{1}{2}m$ as may appear at first sight) since the sink is located on the wall $\xi = 0$. In these circumstances the influx into the sink is only one half of what would occur if the sink was completely surrounded by fluid.

where u is the ultimate speed of the jet of width $2h$ and $v = -v_x + iv_y$ is the complex velocity.

Deduce for the particular case $\alpha = \frac{1}{4}\pi$ that the contraction ratio of the jet, i.e. the ratio of the ultimate width ($2h$) to the initial gap ($2H$) between the plates, is given by

$$h/H = \left[1 + \frac{2\sqrt{2}}{\pi} - \frac{2}{\pi}\log(1 + \sqrt{2})\right]^{-1}$$

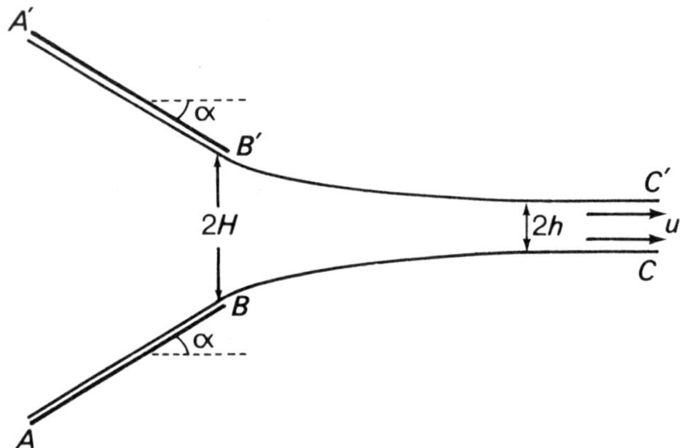

24. Collision of two coplanar collinear jets of the same speed

Two plane jets travelling in a common plane but in opposite directions at a common speed u collide. Body forces are negligible and the external pressure is constant. The resulting flow pattern is as depicted in the figure with two emergent jets making angles of α and β respectively with the incident direction. If $2h_1$, $2h_2$, $2h_3$ and $2h_4$ denote the widths of the jets, at large distances from the collision region and it is assumed that a steady state flow pattern is possible, deduce from momentum and mass conservation considerations that

$$h_1 + h_2 = h_3 + h_4$$

$$h_3 \sin\alpha = h_4 \sin\beta$$

$$h_3 \cos\alpha + h_4 \cos\beta = h_1 - h_2 .$$

Note that these equations provide three equations for four unknowns $(\alpha, \beta, h_3, h_4)$ so that there is a lack of uniqueness to the solution.

Find the complex potential as a function of complex velocity for the particular case of a symmetrical collision for which $\alpha = \beta$ and $h_3 = h_4$.

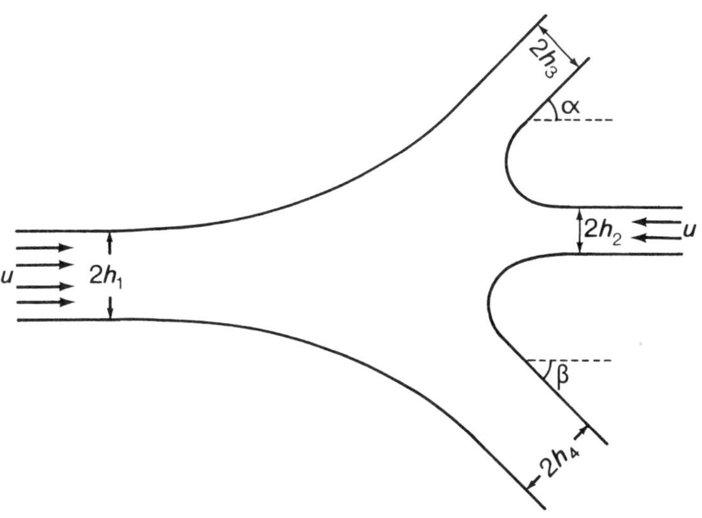

25. Invariance of Bernoulli equation with respect to Galilean transformations

Show that the Bernoulli equation

$$P + \tfrac{1}{2}\rho_0 \mathbf{v}^2 - \rho_0(\partial\phi/\partial t) = f(t),$$

valid for incompressible, irrotational and inviscid flow, is invariant with respect to Galilean transformations

$$\mathbf{r} = \mathbf{r}' + \mathbf{U}t$$

where $\mathbf{r}, \mathbf{r}'$ are coordinates referred respectively to two Cartesian coordinate systems whose relative motion is one of translation at (constant) velocity $\mathbf{U}$.

Hint. Let $\mathbf{v}$ and $\mathbf{v}'$ be the velocity vector of the fluid in the two systems and note that $\mathbf{v}' = \mathbf{v} - \mathbf{U}$. Also if ϕ and ϕ' are the two corresponding velocity potentials then

$$\phi'(\mathbf{r}', t) = \phi(\mathbf{r}, t) + \mathbf{U}\cdot\mathbf{r}'.$$

26. Collision of two planar collinear jets of different speeds

Two coplanar parallel plane jets travelling at speeds u_1 and $-u_2$ ($u_1 > 0$, $u_2 > 0$) collide. Body forces are negligible and the external pressure is constant.

Show that unless $u_1 = u_2$ it is not possible to obtain a stationary flow situation as in Problem 24. On the other hand use the result of Problem 25 to show that it is possible to obtain an unsteady flow solution which is stationary with respect to a coordinate system travelling at speed $s = \tfrac{1}{2}(u_1 - u_2)$, in the direction of u_1. Deduce that there exist solutions of the original problem for which the collision region travels at speed s in the direction of u_1.

Chapter 13

Compressible Flow of Inviscid Fluids: The Acoustic Approximation

13.1 Introduction and basic equations

None of the fluid mechanics problems studied so far encompass the possibility of compressible flow, although the solutions obtained in Chapter 10 in the absence of pressure gradients are valid isochoric flows for compressible viscous fluids.

In this chapter we study compressible flow based on the simplest possible model in the so called *acoustic approximation.* The origin of the terminology derives from the fact that the model provides a reasonable valid description of the small amplitude sound waves encountered in everyday life, e.g. in speech, music, etc. With more intense sound waves, such as occur in the immediate vicinity of an explosion or the exhaust of a jet engine, the acoustic approximation fails.

The basic equations of the acoustic approximation may be derived by writing

$$\rho = \rho_0(1 + s) \qquad (13.1.1)$$

where ρ_0 is the ambient density and $|s|$ is assumed to be much less than unity.† In older texts s is sometimes known as the **condensation.** Inserting (13.1.1) into the relation $P = P(\rho)$ for a barotropic fluid and expanding by Taylor's theorem leads to

$$P(\rho) = P(\rho_0) + \rho_0 s P'(\rho_0) + \ldots \qquad (13.1.2)$$

where $P'(\rho_0)$ is the derivative of $P(\rho)$ evaluated at $\rho = \rho_0$, and terms in s^2 and higher have been neglected. For real fluids $P' > 0$ and we write the positive constant $P'(\rho_0)$ as††

$$P'(\rho_0) = c^2 \qquad (13.1.3)$$

where, ultimately, c will be identified as the speed of wave propagation.

In most acoustic problems body forces are negligible except inasmuch that they may lead to a medium of variable density in which c varies [as occurs for example in the earth's atmosphere]. Here we ignore body forces so that the equation of motion (11.1.1) becomes

$$-\operatorname{grad} P = \rho\left(\frac{\partial \mathbf{v}}{\partial t} + (\mathbf{v} \,.\, \operatorname{grad})\mathbf{v}\right).$$

† In the approximation in question $s = -\Delta$ where Δ is the dilatation.

†† For real fluids the pressure is an increasing function of density.

Inserting (13.1.1) and (13.1.2) into this equation leads to

$$-c^2 \operatorname{grad} s = (1 + s)\left(\frac{\partial \mathbf{v}}{\partial t} + \mathbf{v} \cdot \operatorname{grad} \mathbf{v}\right) . \tag{13.1.4}$$

Since $s \ll 1$ the right side of (13.1.4) is well approximated by neglecting completely the term s in the first pair of brackets. Also, in the acoustic approximation the term $(\mathbf{v} \cdot \operatorname{grad})\mathbf{v}$ is neglected on the grounds that $\mathbf{v}$ is in some sense 'small', and that terms involving products of $\mathbf{v}$ and its derivatives can be ignored. The latter approximation is hardly justifiable at this stage; however if we accept the validity of the argument and explore the consequences, it is possible to give some justification for the validity of the process at a later stage. Accordingly we approximate (13.1.4) by

$$-c^2 \operatorname{grad} s = \frac{\partial \mathbf{v}}{\partial t} . \tag{13.1.5}$$

A second relation connecting s and $\mathbf{v}$ follows from the mass conservation equation (2.3.18)

$$\frac{\partial \rho}{\partial t} + \operatorname{div}(\rho \mathbf{v}) = 0 .$$

Substituting from (13.1.1) into this equation leads to

$$\frac{\partial s}{\partial t} + (1 + s) \operatorname{div} \mathbf{v} + \mathbf{v} \cdot \operatorname{grad} s = 0$$

or on neglecting the last term, also justified subsequently, and again neglecting s compared with unity

$$\frac{\partial s}{\partial t} + \operatorname{div} \mathbf{v} = 0 . \tag{13.1.6}$$

From (13.1.5) we have $\quad \dfrac{\partial}{\partial t}(\operatorname{curl} \mathbf{v}) = 0$

(since $\operatorname{curl} \operatorname{grad} s = 0$) and it follows that if initially the fluid is undisturbed so that curl $\mathbf{v}$ is initially zero **everywhere**, then curl $\mathbf{v}$ vanishes for all time, whence $\mathbf{v}$ may be written

$$\mathbf{v} = -\operatorname{grad} \phi \tag{13.1.7}$$

where ϕ is the velocity potential. From (13.1.5) and (13.1.7)

$$\operatorname{grad}\left(c^2 s - \frac{\partial \phi}{\partial t}\right) = 0$$

so that $c^2 s - \partial\phi/\partial t$ is solely a function of time, say $\dot{q}(t)$, and

$$c^2 s = \frac{\partial \phi}{\partial t} + \dot{q}(t)$$

where $\dot{q}(t)$ is arbitrary. Inserting this result and (13.1.7) into (13.1.6) yields

$$\nabla^2\phi - \frac{1}{c^2}\frac{\partial^2\phi}{\partial t^2} = \frac{\ddot{q}(t)}{c^2} . \qquad (13.1.8)$$

It is a trivial matter to dispose of the term in $\ddot{q}$ by noting that a particular integral of (13.1.8) is $\phi = -q(t)$. Since for this ϕ, grad $\phi = 0$ there is no contribution from q to the velocity field. Therefore without loss of generality q may be taken to vanish and ϕ to satisfy the homogeneous equation

$$\nabla^2\phi - \frac{1}{c^2}\frac{\partial^2\phi}{\partial t^2} = 0 . \qquad (13.1.9)$$

In terms of ϕ the basic equations of the acoustic approximation are

$$\mathbf{v} = -\operatorname{grad}\phi \qquad (13.1.10)$$

$$P = P(\rho_0) + \rho_0(\partial\phi/\partial t) \qquad (13.1.11)$$

where ϕ is a solution of the **wave equation**

$$\nabla^2\varphi - \frac{1}{c^2}\frac{\partial^2\phi}{\partial t^2} = 0 \qquad (13.1.12)$$

13.2 General solutions of the wave equation

Like Laplace's equation and the Diffusion equation, equation (13.1.12) is one of the common equations of classical applied mathematics and physics. The equation arises in many different fields including the present context and also electromagnetic theory, where c^2 is defined by $c^2 = (\epsilon\mu)^{-1}$ in which ϵ and μ are respectively permittivity and permeability. The reader may be familiar also with the one dimensional version

$$\frac{\partial^2\phi}{\partial x^2} - \frac{1}{c^2}\frac{\partial^2\phi}{\partial t^2} = 0 \qquad (13.2.1)$$

for the transverse vibrations of a stretched string. Here ϕ is the transverse displacement and $c^2 = T/m$ where T is the tension in the string and m the mass per unit length.

Many different solutions of (13.1.12) have been found and applied to a variety of different problems. Perhaps the best known solution of the one dimensional equation (13.2.1), due to d'Alembert, expresses ϕ in terms of arbitrary functions. This solution is obtainable formally from the general solution of Laplace's equation in two dimensions. In Section 12.8 it was shown that the general solution of

$$\frac{\partial^2\phi}{\partial x^2} + \frac{\partial^2\phi}{\partial y^2} = 0$$

could be written as $\phi = f(x + iy) + g(x - iy)$

where f and g are arbitrary functions. The present g is to identified with $\frac{1}{2}\bar{g}$ and the present f with $\frac{1}{2}\bar{f}$ in (12.8.5). Writing formally $y = ict$ in Laplace's equation produces the one dimensional wave equation (13.2.1) so that the general solution of the latter is

$$\phi = f(x - ct) + g(x + ct)$$

where f and g are ***arbitrary*** functions of the combined variables $x - ct$, $x + ct$. Since an arbitrary function of $x \pm ct$ may also be regarded as an arbitrary function of $t \pm x/c$, an alternative, and often more useful, form is

$$\phi = f(t - x/c) + g(t + x/c) \ . \tag{13.2.2}$$

Each of the terms on the right side of (13.2.2) is susceptible to an interpretation as a transient pulse. We consider in detail the solution

$$\phi = f(t - x/c) \ .$$

Then for $x = 0$, $\phi = f(t)$ which, as a function of t, we suppose to be as shown on Fig. 13.1. In the figure it has been supposed, without loss of generality, that $f = 0$ for $t < 0$. For non-vanishing x, say $x = x_1$, $\phi = f(t - x_1/c)$. Evidently the *function* f is the same *function* as previously and therefore the graph of ϕ against t possesses the same profile as for $x = 0$ but with a different time origin. In other words the profile of ϕ is undisturbed in shape but translated along the t axis a distance x_1/c (Fig. 13.1). It is this result which justifies the description wave equation for (13.1.12) and also provides the interpretation of c as the speed of propagation of the wave.

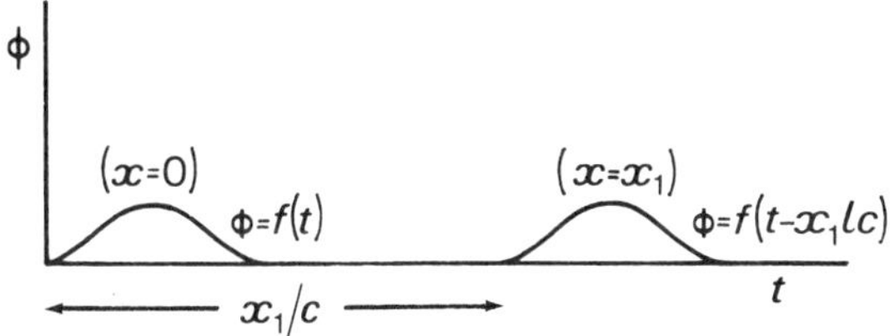

Fig. 13.1 Interpretation of $f(t - x/c)$ as a travelling pulse.

The term $g(t + x/c)$ represents the propagation of a pulse in the negative x direction.

A general solution of (13.1.12) in terms of arbitrary functions is also possible for the case of spherical symmetry where $\phi = \phi(r, t)$ and $r^2 = x^2 + y^2 + z^2$. In this case (13.1.12) reduces to

$$\frac{\partial^2 \phi}{\partial r^2} + \frac{2}{r}\frac{\partial \phi}{\partial r} = \frac{1}{c^2}\frac{\partial^2 \phi}{\partial t^2} \ . \tag{13.2.3}$$

The reader is left to show that the substitution $\phi = \Phi(r, t)/r$ results in

$$\frac{\partial^2 \Phi}{\partial r^2} - \frac{1}{c^2}\frac{\partial^2 \Phi}{\partial t^2} = 0$$

with solution $\Phi = f(t - r/c) + g(t + r/c)$

so that (13.2.3) is solved generally by

$$\phi = \frac{f(t - r/c)}{r} + \frac{g(t + r/c)}{r} \tag{13.2.4}$$

where f and g are arbitrary functions.

In the light of the earlier discussion the physical interpretation of the solution (13.2.4) is clear; e.g. the term $f(t - r/c)/r$ represents a wave travelling out in the outward radial direction but with a decreasing amplitude caused by the factor r^{-1}. This arises physically from the expanding region of space into which the function ϕ is propagating. Some books describe $f(t - r/c)/r$ as a **spherically diverging** wave, while $g(t + r/c)/r$ is called a **converging** (or occasionally **imploding**) spherical wave.

Physical intuition suggests that the cylindrically symmetric wave equation

$$\frac{\partial^2 \phi}{\partial r^2} + \frac{1}{r}\frac{\partial \phi}{\partial r} = \frac{1}{c^2}\frac{\partial^2 \phi}{\partial t^2}, \tag{13.2.5}$$

where $r^2 = x^2 + y^2$, represents a situation intermediate between the **plane** wave equation (13.2.1) and the **spherical** equation (13.2.3); in turn this suggests the possibility of solutions of the form $\phi = r^{-\frac{1}{2}}[f(t - r/c) + g(t + r/c)]$ but in fact the latter expressions do not satisfy (13.2.5), although for large r they are approximate solutions.†

A generalisation of the solution (13.2.2) is provided by

$$\phi = f\left(t \pm \frac{\mathbf{n}\,.\,\mathbf{r}}{c}\right) \tag{13.2.6}$$

where $\mathbf{n}$ is a constant unit vector. These solutions represent plane waves travelling in the directions $\pm\mathbf{n}$. For such waves

$$\mathbf{v} \equiv -\operatorname{grad}\phi = \mp\frac{\mathbf{n}}{c}\dot{f}\left(t \pm \frac{\mathbf{n}\,.\,\mathbf{r}}{c}\right) \tag{13.2.7}$$

and $\mathbf{v}$ is in the direction of propagation $\mathbf{n}$. In (13.2.7) the dot denotes differentiation either with respect to time or with respect to the entire argument $[t \pm (\mathbf{n}\,.\,\mathbf{r})/c]$.

It is possible to provide now some justification for the approximations entailed in deriving (13.1.5) and (13.1.6), at least in the context of a plane wave solution. For the velocity potential $\phi = f[t - (\mathbf{n}\,.\,\mathbf{r})/c]$, the assumption that $(\mathbf{v}\,.\,\operatorname{grad})\mathbf{v}$ is much smaller in magnitude than $\partial\mathbf{v}/\partial t$ becomes

$$\frac{1}{c^3}|\dot{f}\ddot{f}| \ll \left|\frac{\ddot{f}}{c}\right| \quad \text{i.e.} \quad |\dot{f}/c^2| \ll 1 .$$

† For an exact solution of $\dfrac{\partial^2\phi}{\partial r^2} + \dfrac{1}{r}\dfrac{\partial\phi}{\partial r} + \dfrac{1}{r^2}\dfrac{\partial^2\phi}{\partial\theta^2} = \dfrac{1}{c^2}\dfrac{\partial^2\phi}{\partial t^2}$ involving arbitrary functions see Problem 12.

However since $|\mathbf{v}| = |\dot{f}|/c$, the inequality implies $|\mathbf{v}| \ll c$ and it is in this sense that $|\mathbf{v}|$ is small. The neglect of $(\mathbf{v} \,.\, \text{grad})s$ compared with div $\mathbf{v}$ in the mass conservation equation requires the same inequality, so that in physical terms the acoustic approximation seems to be validated provided the velocity is much less than the sound speed c. These conclusions are not entirely correct because it is possible to find solutions of the *full* equations for which $|\mathbf{v}| \ll c$ everywhere but for which there exist localised regions where $\mathbf{v}$ is so rapidly changing with position that $|(\mathbf{v} \,.\, \text{grad})\mathbf{v}|$ is much larger than $|\partial\mathbf{v}/\partial t|$.† However excluding these very localised regions, which do not affect significantly the solution elsewhere, it is largely correct to say that the validity of the acoustic approximation devolves on the inequality $|\mathbf{v}| \ll c$.

13.3 Magnitudes of c and applications of the acoustic approximation

For most common liquids, $P(\rho)$ is of the form (7.1.15) with the bulk modulus in the range $10^3 - 10^4$ N/mm^2. The corresponding speed of wave propagation, or speed *of sound,* is given by (7.1.16) and is of order 10^3 m/sec. For ideal gases c is given by (7.1.11): in particular for air at sea level at normal ambient temperatures c is 3.45×10^2 m/sec.

In terms of pressure variations the range of validity of the acoustic approximation is quite different in gases and liquids. There are few situations in compressible liquids for which the acoustic approximation is inadequate. This is because the acoustic approximation, valid generally for density variations of up to a few per cent, encompasses pressure variations from zero up to about 20 N/mm^2 in most liquids, and this is an enormous pressure compared with those encountered in almost all practical situations. In contrast gases may behave in a highly compressible fashion; a mere increase of pressure of 10^{-1} N/mm^2 doubles the ambient pressure in air at sea level and leads to density changes far too large to be encompassed by the acoustic approximation. Even so there are not all that many practical earthbound situations in gas dynamics where the acoustic approximation is completely inadequate – e.g. even in dealing with the extreme problem of blast waves generated by an explosion, at points distant from the explosive the acoustic approximation is valid. Perhaps the most important areas of failure of the acoustic approximation in air are near intense sources of disturbance such as the immediate vicinities of explosions or the exhausts of jet engines. It is worth recalling that many problems in the flow of gases do not call even for the acoustic approximation, e.g. the flow of air around the wing of a subsonic aircraft is adequately approximated by an incompressible model.

One well-known example of the use of wave propagation effects is the use of underwater submarine detection equipment. Sound waves generated by the equipment on the surface vessel are propagated into the ocean. The waves in

† The statement is not true for plane wave solutions of the *approximate* equations, as has been shown.

question are essentially short bursts, i.e. pulsed, spherical waves whose amplitude is attenuated by spherical divergence as in (13.2.4). In normal circumstances the pulses propagate into the ocean and are dispersed by both spherical divergence and dissipative processes. Exceptionally if a portion of the expanding pulse surface is intercepted by a relatively solid object such as a submarine, locally there occurs wave reflection and the submarine itself acts then as a source of spherical waves which propagate in various directions. Those waves which propagate back towards the surface vessel are detectable by appropriate receiving equipment. The time lapse between emission of the initial pulse and receipt of the reflected pulse determines the distance of the submarine; orientation of the receiving equipment to maximum signal strength determines the direction.

Measurement of the depth of the ocean bed is also carried out by the same technique.

Seismological investigations of the thickness of the earth's crust and the detection of distant nuclear explosions also depend on wave propagation phenomena. Wave propagation in elastic solids, which is rather more complicated than in fluids, is discussed in Section 14.6.

13.4 Some simple wave propagation problems

13.4.1 Piston Problem for Plane Waves

We defer to Section 14.6 a number of transient wave problems which are there considered in the context of classical elasticity. Here, in the context of the acoustic approximation for fluids, only the simplest problems are considered, such as the generation of plane, cylindrically symmetric (Section 13.4.3) and spherically symmetric (Section 13.4.2) waves; in Section (13.4.4) the problem of non-normal reflection of a plane wave at a plane interface is solved. Finally in Section 13.4.5 we consider the standing wave problems associated with organ pipes and room acoustics.

Generation of a plane wave is realised in the following **piston** problem. A close fitting rigid piston of arbitrary cross section is located initially at the accessible end $x = 0$ of a semi-infinite tube $0 \leqslant x \leqslant \infty$ of the same (uniform) cross section. Initially, say at $t = 0$ the tube is filled with stationary fluid at uniform pressure $P_0 \equiv P(\rho_0)$ and density ρ_0. For $t > 0$ the piston is set in motion with a given velocity $v_0(t)$. The problem is depicted in Fig. (13.2).

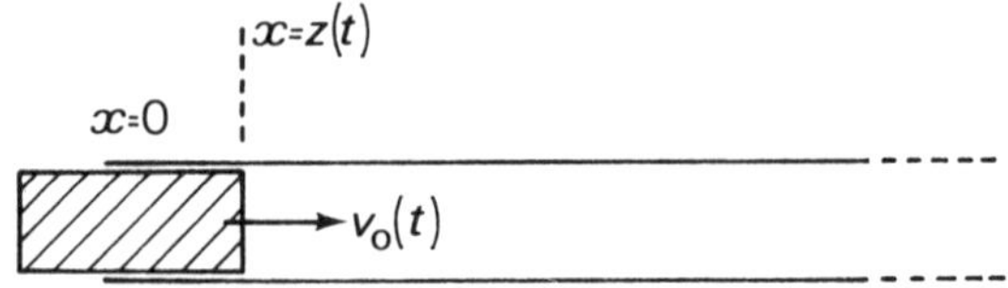

Fig. 13.2 Generation of a plane wave by a piston motion

Evidently only **outgoing** waves, travelling in the positive x direction, are generated so that in the fluid the velocity potential is

$$\phi = f(t - x/c) \qquad (13.4.1)$$

where the function f is determined by the condition that the fluid velocity on the front face of the piston is given by $v_0(t)$

i.e.
$$v_0(t) = \frac{1}{c} f'\left(t - \frac{z(t)}{c}\right) . \qquad (13.4.2)$$

Here $z(t)$ is the location of the front face of the piston, so that

$$\dot{z}(t) = v_0(t) . \qquad (13.4.3)$$

With $z(t)$ given, equations (13.4.2), (13.4.3) determine the unkown function f. Except for the case of uniform piston velocity v_0, ($z = v_0 t$), equations (13.4.2) and (13.4.3) only determine f implicitly; however for $|v_0| \ll c$ it is possible to make the further approximation that in the argument of f' in (13.4.2) the term $z(t)/c$ is small compared with t. In these circumstances (13.4.2) becomes $f'(t) = cv_0(t)$ with solution, continuous at $t = 0$,

$$f = \begin{cases} c\displaystyle\int_0^t v_0(t')dt' & (t \geqslant 0) \\ 0 & (t \leqslant 0) \end{cases}$$

It follows that

$$\phi = \begin{cases} c\displaystyle\int_0^{[t - x/c]} v_0(t')dt' & (t \geqslant x/c) \\ 0 & (t \leqslant x/c) \end{cases} \qquad (13.4.4)$$

Evidently the fluid is undisturbed for $x > ct$ so that the **wave front** $x = ct$ represents the limit of disturbance. The velocity and pressure fields derived from (13.1.10), (13.1.11) and (13.4.4) are, for $x < ct$,

$$v = v_0(t - x/c), \quad P = P_0 + \rho_0 c_0 v_0(t - x/c) ,$$

and the relation $(P - P_0) = \rho_0 c_0 v_0$ relating the overpressure $(P - P_0)$ to particle velocity v is a fundamental result in the theory of plane waves.

The case of a **square pulse** for which

$$v_0(t) = \begin{cases} v_0 = \text{const.} & (0 < t < T) \\ 0 & (t > T) \end{cases}$$

is illustrated in Fig. 13.3 which depicts the pressure distribution for $t = 0$, $0 < t < T$, $t = T$ and $t > T$.

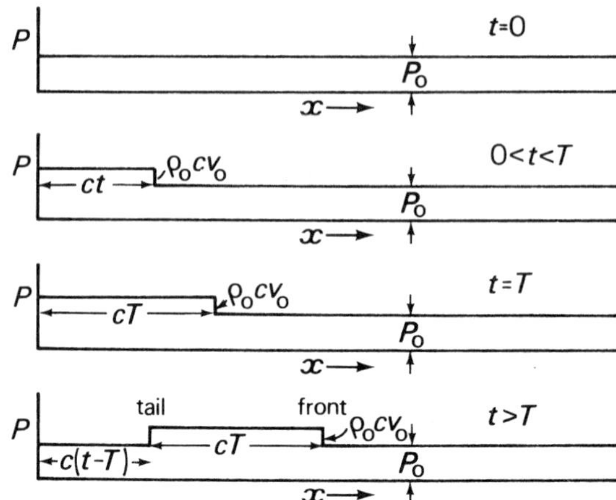

Fig. 13.3 Pressure distribution due to square pulse

The square overpressure pulse propagates along the tube unchanged in shape at speed c. Here the acoustic approximation is seen to lead to discontinuities in pressure and velocity. The approximation fails of course if v_0 is comparable with c, when we would need a full non-linear treatment of the wave problem. Even for $v_0/c \ll 1$ the acoustic approximation fails in the immediate vicinity of the wave front for at the wave front itself the velocity gradient is infinite.† (In the present example a similar situation occurs at the tail). In the immediate vicinities of the wave front and tail large viscous forces, dependent on velocity gradients, are called into play. As a result the wave front and tail are in practice smoothed out, as in Fig. 13.4, into transition regions.

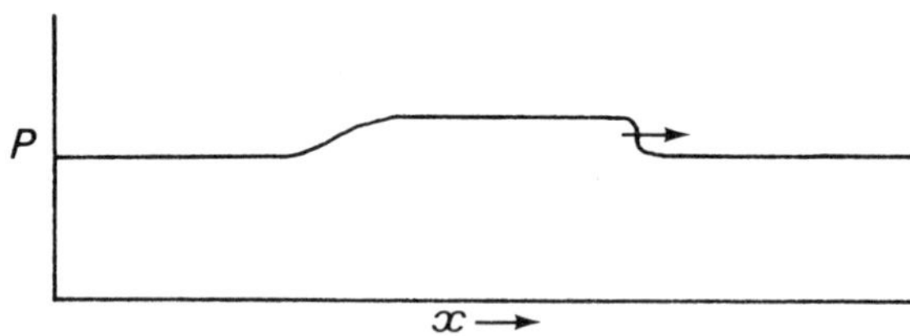

Fig. 13.4 Smoothing of wave front and tail due to viscous forces

The mathematics of the smoothing process is too complicated to discuss in the present text; even the simplest model entails the compressible Navier-Stokes equation.

† See discussion at end of Section 13.2.

The square pulse displayed in Fig. 13.3 continues to propagate along the hypothetically semi-infinite tube indefinitely. For a real tube of finite length L which is either *closed* at $x = L$ or *open* (i.e. open to atmospheric pressure), the solution (13.4.4) of the piston problem is valid until the wave front reaches $x = L$, in other words, for a time period $0 \leqslant t \leqslant L/c$. For $t \geqslant L/c$, a reflection process occurs at $x = L$ which generates a wave propagating in the negative x direction back towards the piston. Further reflections occur at the piston after times $2L/c$, $4L/c$, $6L/c \ldots$ and at $x = L$ after times $3L/c$, $5L/c \ldots$. A simple reflection problem is considered in Section 13.4.4.

For problems involving multiple reflections, the transient wave solutions $f(t \pm x/c)$ are less profitable usually then a separation of variables solution of the wave equation (see Section 13.4.5).

13.4.2 EXPANDING CYLINDRICALLY SYMMETRIC WAVES – A SIMILARITY SOLUTION

Because the cylindrical wave equation

$$\frac{\partial^2 \phi}{\partial r^2} + \frac{1}{r}\frac{\partial \phi}{\partial r} = \frac{1}{c^2}\frac{\partial^2 \phi}{\partial t^2} \tag{13.4.5}$$

fails to admit solutions like (13.2.2) and (13.2.4), problems involving cylindrically symmetric waves are of a slightly different character to the plane and spherically symmetric cases. However it is possible to generate solutions of (13.4.5) from a basic **similarity** solution which is assumed to be of the form

$$\phi(r, t) = rf(r/ct) . \tag{13.4.6}$$

This assumed form of solution leads to a radial velocity field

$$-\frac{\partial \phi}{\partial r} \equiv v_r(r, t) = -[f(\eta) + \eta f'(\eta)] \tag{13.4.7}$$

where $\eta = r/ct$ is called the similarity variable.† Thus the velocity field depends only on the combined variable r/ct so that for fixed r/t, v_r is constant. Solutions of this type are called similarity solutions. Similarity solutions have proved immensely valuable in continuum mechanics in indicating the nature of solutions of quite complicated (sometimes non-linear) partial differential equations. The mathematical advantage of a similarity solution is that the partial differential equation is reduced to an ordinary differential equation which is much easier to solve; however this has to be balanced against the disadvantage that the solution obtained is only applicable to a particular boundary value problem, which is established *a posteriori.*

† It is convenient to regard f as a function of the (dimensionless) variable r/ct rather than as a function of r/t.

From (13.4.6) the partial derivatives of ϕ are given by

$$\frac{\partial\phi}{\partial r} = f(\eta) + \eta f'(\eta), \quad \frac{\partial^2\phi}{\partial r^2} = [\eta f''(\eta) + 2f'(\eta)]/ct$$

$$\frac{\partial\phi}{\partial t} = -\frac{r^2}{ct^2}f'(\eta), \quad \frac{\partial^2\phi}{\partial t^2} = \frac{2r^2}{ct^3}f'(\eta) + \frac{r^3}{c^2t^4}f''(\eta) .$$

Substituting for $\partial\phi/\partial r$, $\partial^2\phi/\partial r^2$ and $\partial^2\phi/\partial t^2$ into (13.4.5) and making use of $\eta = r/ct$ leads to the ordinary differential equation†

$$\eta(1 - \eta^2)f''(\eta) + (3 - 2\eta^2)f' + f/\eta = 0 . \qquad (13.4.8)$$

A readily verifiable solution of (13.4.8) is $f(\eta) = \eta^{-1}$ so that the general solution may be found by the usual methods. We write $f = q(\eta)/\eta$ obtaining

$$\eta(1 - \eta^2)q''(\eta) + q'(\eta) = 0 \qquad (13.4.9)$$

which is readily solved for $q'(\eta)$ by writing

$$\frac{q''}{q'} = -\frac{1}{\eta(1 - \eta^2)}$$

or on integration $\log q' = -\log\eta + \frac{1}{2}\log(1 - \eta^2) + \log A_1$

where A_1 is a constant of integration. The last equation may be written $q'(\eta) = A_1\eta^{-1}\sqrt{1-\eta^2}$ from which $q(\eta)$ may be found by quadrature. However since $q'(\eta) = f + \eta f'$, the velocity field of the similarity solution is given directly by $-q'(\eta)$, (c.f. 13.4.7), so that

$$v_r(\eta) = A\sqrt{1 - \eta^2}/\eta \quad (\eta \leqslant 1) \qquad (13.4.10)$$

where $A = -A_1$.

The solution (13.4.10) is valid for $\eta \leqslant 1$; for $\eta \geqslant 1$, (13.4.9) admits the solution

$$v_r(\eta)[= -q'(\eta)] = B\sqrt{\eta^2 - 1}/\eta$$

where B is a constant of integration. This solution is not used here.

The velocity field

$$v_r = \begin{cases} A\sqrt{1 - \eta^2}/\eta & (\eta \leqslant 1) \\ 0 & (\eta \geqslant 1) \end{cases} \qquad (13.4.11)$$

† Whether or not the method works depends crucially on whether or not the resulting equation [here (13.4.8)], is solely an equation involving η, and not involving η and r (or η and t) separately. Different partial diffential equations admit slightly different forms of similarity solution. The most general type of solution assumed is of the form $r^\alpha\psi(r/t^\beta)$ where (hopefully) one can find an α and β so that an ordinary differential equation results for $\psi(\eta)$ with $\eta = r/t^\beta$. In the present example $\alpha = \beta = 1$. It is not always possible to find similarity solutions, although for (13.4.5) there are an infinite number of solutions with $\beta = 1$ and α arbitrary.

which is continuous at $\eta = 1$, solves the problem of the growth of a cylindrical cavity at uniform rate v_0 for $t > 0$ in a fluid initially at rest at $t = 0$. The boundary condition for this problem is

$$v_r = v_0 \quad \text{for} \quad r = v_0 t \qquad \text{i.e.} \quad v_r = v_0 \quad \text{for} \quad \eta = v_0/c$$

and this is satisfied by (13.4.11) provided $v_0 < c$ and A is given by

$$A = v_0{}^2 c^{-1}(1 - v_0{}^2/c^2)^{-\frac{1}{2}} .$$

The solution $B\sqrt{\eta^2 - 1}/\eta$ for $\eta > 1$ is rejected because otherwise for $r \to \infty$, the velocity field tends to the finite value B and this is incompatible with the initial quiescent condition that for the cavity problem the fluid is everywhere at rest for $t = 0$; the latter condition is met automatically by equation (13.4.11).

Evidently for the solution (13.4.11) the wave front is located at $r = ct$ and this concurs with physical expectations of the nature of the solution for $v_0 < c$. For values of v_0 comparable with or greater than c the acoustic approximation fails so that the problem of solving (13.4.9) for $v_0 > c$ is not a valid quest.

Somewhat more general solutions of (13.4.5) may be obtained from (13.4.11) by noting that (13.4.5) is invariant with respect to time translation so that a solution of (13.4.5) is given by (13.4.11) but with η replaced by

$$\eta = r/c(t - t')$$

where for the moment t' is a fixed parameter. If A is also regarded as a function of t', then for $r < ct$ (13.4.11) becomes

$$v_r = \begin{cases} A(t')r^{-1}[c^2(t - t')^2 - r^2]^{\frac{1}{2}} & (t' < t - r/c) \\ 0 & (t' > t - r/c) \end{cases}$$

Because (13.4.5) is linear we can superpose solutions with different values of t'. For example, treating t' as a continuous parameter which varies from $t' = -\infty$ to $t' = t$, leads to

$$v_r = r^{-1} \int_{-\infty}^{(t - r/c)} A(t')[c^2(t - t')^2 - r^2]^{\frac{1}{2}} dt' \tag{13.4.12}$$

Different choices of $A(t')$ lead to different solutions; the solution (13.4.11) is obtained by choosing $A(t') = A\delta(t')$ where δ is the Dirac delta function.

13.4.3 SPHERICALLY SYMMETRIC WAVES – OSCILLATING SPHERE

Here we consider the problem of a sphere immersed in an infinite sea of fluid and with a surface whose radius a varies with time so that

$$a(t) = a_0[1 + \gamma(t)] \qquad (\gamma = 0 \quad \text{for} \quad t \leqslant 0)$$

where a_0 is constant and where we assume the dimensionless quantity γ to satisfy $\gamma(t) \ll 1$. Only outgoing waves are generated by the sphere so that from (13.1.10) and the first term of (13.2.4) the (radial) velocity is given by

$$v_r = -\frac{\partial}{\partial r}\left[\frac{f\left(t - \dfrac{r - a_0}{c}\right)}{r}\right] . \qquad (13.4.13)$$

In writing down this formula the constant a_0/c has been absorbed into the argument of f, this is convenient subsequently and may be justified by a change of origin of t. From (13.4.13)

$$v_r = \frac{f\left(t - \dfrac{r - a_0}{c}\right)}{r^2} + \frac{\dot{f}\left(t - \dfrac{r - a_0}{c}\right)}{cr} . \qquad (13.4.14)$$

The boundary condition $v_r = a_0\dot{\gamma}$ at $r = a_0(1 + \gamma)$ is met by choosing ϕ to satisfy the equation

$$a_0\dot{\gamma} = \frac{f\left(t - \dfrac{a_0\gamma}{c}\right)}{a_0^2(1+\gamma)^2} + \frac{\dot{f}\left(t - \dfrac{a_0\gamma}{c}\right)}{ca_0(1 + \gamma)} . \qquad (13.4.15)$$

With the assumption $|\gamma| \ll 1$, the factors $(1 + \gamma)^2$ and $(1 + \gamma)$ are adequetely approximated by unity. Further, for the validity of the acoustic approximation, $|v_r|/c \ll 1$; in particular at the boundary $v_r = a_0\dot{\gamma}$ and therefore

$$|a_0\gamma| = \left|\int_0^t a_0\dot{\gamma}\,dt\right| \ll ct$$

suggesting that in the arguments of f and $\dot{f}$ the term $a_0\gamma/c$ may also be neglected. With these approximations (13.4.15) takes the much simpler form

$$\dot{f}(t) + \frac{c}{a_0}f(t) = ca_0^2\dot{\gamma}(t) \qquad (13.4.16)$$

which is a linear first order differential equation determining f; if for $t < 0$ the fluid is at rest so that $\gamma(t) = 0$ for $t \leq 0$, the appropriate solution of (13.4.16) for $t > 0$ is

$$f = ca_0^2 e^{-t/\tau}\int_0^t \dot{\gamma}(t')e^{t'/\tau}dt' \qquad (13.4.17)$$

where $\tau = a_0/c$. With $\gamma(t)$ given, equation (13.4.17) solves the proposed problem; to proceed further necessitates specifying in detail the quantity $\gamma(t)$.

13.4.4 REFLECTION OF A PLANE WAVE AT NON-NORMAL INCIDENCE BY A RIGID PLANE BARRIER

The generation of plane waves by piston motions was discussed in Section 13.4.1. There the lateral dimensions of the wave were confined by the cross sectional area of the tube. In this section we consider the somewhat hypothetical problem of a plane wave of infinite cross sectional dimensions impinging on a rigid barrier $x = 0$ (Fig. 13.5). Without loss of generality the direction of propagation may be assumed to lie in the xy plane. If this direction is $\mathbf{n} = (n_1, n_2, 0)$, the incident wave may be represented by

$$\phi = f[t - (\mathbf{n} \, . \, \mathbf{r})/c] \tag{13.4.18}$$

Evidently in general (13.4.18) fails to satisfy the boundary condition

$$v_x = 0$$

at the rigid barrier. It is therefore necessary to postulate a reflected wave in the direction $\mathbf{n}'$, as yet unkown.

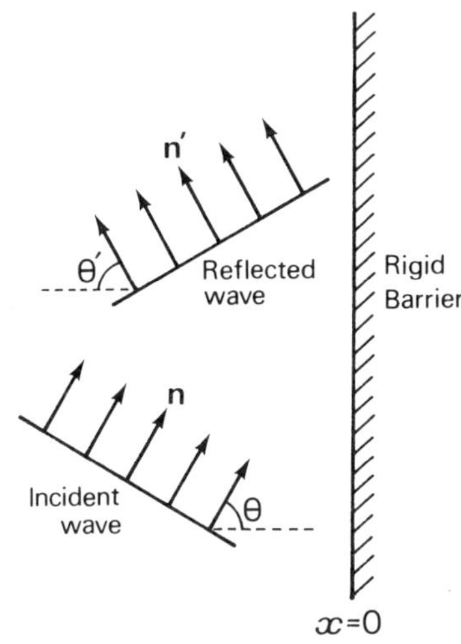

Fig. 13.5 Reflection of a plane wave at a rigid barrier

If the reflected wave velocity potential is denoted by g then the sum velocity potential of the incident and reflected waves is

$$\phi = f[t - (\mathbf{n} \, . \, \mathbf{r})/c] + g[t - (\mathbf{n}' \, . \, \mathbf{r})/c] \tag{13.4.19}$$

where $\mathbf{n}' = (n_1', n_2', n_3')$. The boundary condition $v_x = 0$ on $x = 0$ is met by

$$\left(\frac{\partial \phi}{\partial x}\right)_{x = 0} = 0$$

or from (13.4.19) $\quad - n_1 \dot{f}(t - n_2 y/c) - n_1' \dot{g}(t - n_2' y/c - n_3' z/c) = 0$

and this is possible only if the arguments of $\dot{f}$ and $\dot{g}$ are identical so that $n_3' = 0$ and $n_2 = n_2'$. However since **n** and **n**$'$ are unit vectors this last result implies also $n_1'^2 = n_1^2$ so that

$$n_1' = \pm n_1 \ .$$

The choice $n_1' = n_1$ leads to $\dot{g}(t) = -\dot{f}(t)$ and hence $g = -f$ thus obliterating the incident wave, since the arguments of f and g are then identical everywhere. Therefore only the choice $n_1' = -n_1$ is of any physical significance; for this case $\dot{g} = \dot{f}$ and hence to within an irrelevant constant $g = f$ and ϕ is given by

$$\phi = f\left(t - \frac{\mathbf{n}\,.\,\mathbf{r}}{c}\right) + f\left(t - \frac{\mathbf{n}'\,.\,\mathbf{r}}{c}\right)$$

with

$$\mathbf{n}' = (-n_1, n_2, 0) \ .$$

The directions of the incident and reflected wave are in accordance with the well known priciples of reflection so that in Fig. 13.5 the angles θ and θ' are equal. The slightly more complicated problem of reflection and refraction of a plane wave at a plane interface between two fluids is the subject of Problem 7.

13.4.5 Room Acoustics, Musical and Wind Instruments

Sound waves generated by orchestral instruments or other sources are transmitted to the audience by acoustic waves propagating through the air. The waves emanating from various instruments are best pictured at source as long trains of periodic waves with slowly varying amplitude, e.g. see the discussion following on organ pipes. These waves spread out in all directions but with an intensity diminished by spherical divergence as in the solution (13.2.4). In the open air or in an acoustically 'dead' auditorium lined with sound absorbing material, most of the sound reaching the listener is direct. In a concert hall a small proportion of the sound reaches the listener after one or more reflections; in the case of a large hall not many reflections are important because the wave amplitude will have been rendered inaudible by spherical divergence, natural dissipative processes and absorption.

The presence of reflected sound is said to give **ambience** to an auditorium and for many people an ambient hall is preferable to an acoustically dead one.† In very large auditoriums, such as the Albert Hall in London, some of the reflected sound travels so far that a perceptible time delay results in an echo effect which ruins the performance for some unfortunately located members of the audience. In cases like this usually some focussing of the sound in specific directions must have occurred, perhaps by reflection from a convex surface, to counteract the spherical divergence effect. In recent years the famous Albert Hall problem has been overcome successfully by acoustic engineers.

† The advantages of quadraphonic sound to the listener of classical music is said to lie in using the two additional speakers, located behind the listener, in order to reproduce the ambience of a real concert hall.

In normal circumstances sound waves in air are extinguished very rapidly once the source is switched off, so that usually the listener receives an attenuated but otherwise faithful reproduction of the original. Of course if the listener is located very near the orchestra the balance will be wrong because the attenuation will be much smaller for the nearer instruments than for those further away.

Each instrument sets into motion the surrounding air. For example the air in the vicinity of a drum or gong is excited directly by the vibrations of the drum skin or the metal. In the case of a string instrument the vibrations of the air caused directly by the strings are very feeble, and an essential amplifying mechanism is provided by the cavity resonator formed by the body of the instrument. For an organ pipe the vibrations of the column of air in the pipe are communicated directly to the external air via the mouth of the pipe. The sound emanating from a simple instrument like a gong depends on the free vibrations of the gong excited by the hammer blow. In a more complicated instrument like a violin, vibrations of the string may be caused to vary in frequency by the player whose fingering alters the length of the strings; the prime purpose of the bow is merely to provide the force which excites the string motion.

Instruments which use vibrating air columns directly are the organ and members of the woodwind and brass families. A somewhat crude discussion of the characteristic sound emanating from any of these instruments may be based on the theory of the motion of air columns contained in tubes which are open at one end and either closed or open at the other end. How the air is excited is another question, e.g. by an external bellows and sometimes with an internally located reed as in the case of an organ pipe, and directly by the performer's lips and lungs in the case of a trumpet, horn, trombone, flute and piccolo.

The behaviour of a column of air in a tube is assumed described by the one dimensional wave equation

$$\frac{\partial^2 \phi}{\partial x^2} = \frac{1}{c^2}\frac{\partial^2 \phi}{\partial t^2} \qquad (13.4.20)$$

where ϕ is the velocity potential. Although a general solution of this equation has already been found [equation (13.2.2)] this transient wave solution is not very convenient in the present context. One reason is that for the practical case of a musical wind instrument the length of the column of air in question is at most a few metres; even in the extreme case of an organ the longest pipe is conventionally 32 feet although exceptionally some organs possess a 64 foot pipe. Usually as in a trumpet the air column is rather less than one metre. With a column of length (say) 1 metre, the time for a wave to traverse the column is about 0·003 seconds, and this is to be compared with the shortest sensible note duration of about 0·25 seconds. Therefore during the time the note is emitted about 80 wave passages occur with 40 reflections at each end of the air column. While the solution (13.2.2) is well adapted to deal with problems involving one reflection, as in Section 13.4.4, and is suitable for a small number of reflections, for the musical instrument problem where many reflections are involved, a much better approach is to use a separation of variables solution of (13.4.20). Also the latter description provides a much more suitable basis for discussing the characteristics of a musical instrument.

We write
$$\phi = F(t)G(x)$$
obtaining from (13.4.20)
$$\frac{F''(t)}{F(t)} = \frac{c^2 G''(x)}{G(x)} = -\omega^2 \quad \text{(say)}$$
where $-\omega^2$ is the separation constant, assumed negative.† Solutions of the equation for F and G lead to trigonometric expressions of the form

$$\phi = \begin{matrix}\cos\\ \sin\end{matrix}\{\omega t\}\begin{matrix}\cos\\ \sin\end{matrix}\left\{\frac{\omega x}{c}\right\} \tag{13.4.21}$$

To proceed further it is necessary to determine allowable values of the eigenvalue ω. These are determined by the boundary conditions obtaining at the ends of the column. At a closed end the velocity is zero so that $\partial\phi/\partial x = 0$. At an end open to the atmosphere the pressure assumes the ambient value and the condensation vanishes so that $\partial\phi/\partial t = 0$. However since $\partial\phi/\partial t$ is zero for all t, then equivalently (on integration with respect to t at fixed x) ϕ is zero at an open end.

Summarising $\quad \phi = 0$ (open end), $\quad \dfrac{\partial\phi}{\partial x} = 0$ (closed end)

Consider for the moment a tube open at each end. With ϕ given by functions of the type (13.4.21), only those solutions involving sin $(\omega x/c)$ satisfy the boundary condition $\phi = 0$ at $x = 0$. Further in order for ϕ to be zero at the other end of the tube (say $x = l$), ω must satisfy the eigenvalue equation $\sin(\omega l/c) = 0$ so that the only admissible values of ω are the discrete set

$$\omega_n = n\pi c/l \qquad n = 1, 2, 3, \ldots \quad .$$

Thus subject to the stated boundary conditions the admissable (circular) frequencies of vibrations are the **fundamental** frequency

$$\omega_1 = \pi c/l$$

and the **overtones** or **higher harmonics** with frequency $n\omega_1$ $(n = 2, 3, \ldots)$. This feature that the overtone frequencies are integer multiples of the fundamental, also characteristic of string instruments, is supposedly the reason for the ear's acceptance of the sound as agreeable. The overtone frequencies of timpani are not integer multiples of the fundamental and the sound is less 'pure'.

Admissable solutions of the vibrating air column, open at each end are now given by

$$\phi = \sin(n\pi x/l)\begin{matrix}\sin\\ \cos\end{matrix}\left(\frac{n\pi ct}{l}\right)$$

† Separation constants of the opposite sign lead to exponential varying solutions which are not appropriate to the problem under consideration.

and the general solution is given by a superposition of all such eigenfunctions

$$\phi = \sum_{1}^{\infty} [A_n \sin(n\pi ct/l) + B_n \cos(n\pi ct/l)] \sin(n\pi x/l) \ . \quad (13.4.22)$$

This solution is pertinent, as will be seen, to the case of an (open) organ pipe.

Strictly speaking the solution (13.4.22), which would appear to persist for all time, is not absolutely correct. In reality almost as soon as the mechanism providing the disturbance ceases, e.g. the wind chest in the case of an organ, the vibrations are damped out by viscosity effects and by communication of energy to the surrounding atmosphere. The damping time is short compared with the note duration. During the time the note is being emitted (13.4.22) provides a valid description for the behaviour of the air throughout most of the pipe. At the ends there are complications because the air flow pattern is very complex.

Equation (13.4.22) may also be written

$$\phi = \sum_{1}^{\infty} C_n \cos[(n\pi ct/l) + \epsilon_n] \sin(n\pi x/l)$$

where ϵ_n is a phase angle given by $\tan \epsilon_n = -A_n/B_n$, and $C_n = (A_n{}^2 + B_n{}^2)^{\frac{1}{2}}$. The human ear is not over sensitive to phase angles and what is heard by the listener depends largely on the magnitudes of the amplitudes C_n. In most instruments the basic frequency dominates so that C_1 is rather larger than the other amplitudes. The relative magnitudes of C_n/C_1 determine the **colour** or **timbre** of the instrument; loudness depends on absolute magnitude. In most instruments only the first few harmonics are of importance.

Readers familiar with the theory of vibrating strings will recognise (13.4.22) as the pertinent solution of the wave equation for a string $0 \leqslant x \leqslant l$ held at $x = 0$ and $x = l$. [Here ϕ is the transverse displacement and $c^2 = T/m$ where T is the tension and m the mass/unit length of the string.] In the latter theory the usual problems encountered entail the determination of A_n and B_n from boundary conditions of the type

$$(\phi)_{t=0} = f(x) \, , \quad \left(\frac{\partial \phi}{\partial t}\right)_{t=0} = g(x)$$

where f and g are given. Determination of A_n and B_n is then an exercise in the theory of Fourier series.

It would appear that similar problems might arise in the wind instrument context, but such problems, like the violin string problem, are academic rather than practical. For example consider the case of an open organ pipe depicted in Fig. 13.6.† Near the base of the pipe a 'mouth' is cut into the cylindrical wall; from

† An *open* pipe is one open to the atmosphere at $x = l$. A *closed* pipe is sealed at $x = l$. All organ pipes are open at $x = 0$ by virtue of the mouth made in the cylindrical wall.

the present viewpoint the mouth is located at $x = 0$ and the working section of the pipe is $0 \leqslant x \leqslant l$. Below the mouth the pipe is continued into a wind chest which, when the pipe is activated, provides a stream of *wind* or air in the vicinity of the mouth. The air flow causes disturbances at the mouth which set up the solution (13.4.22) in the working section.

Any mathematical explanation of the tone colour of an organ pipe would necessitate determination of the coefficients A_n and B_n in (13.4.22). However the relative magnitudes of these coefficients will depend on the very complicated unsteady air flow patterns near the mouth and the end of the tube. A crude model is quite inappropriate. For example suppose we assumed that during the period T the pipe was activated, the steady stream of air provided by the wind chest resulted in a constant velocity (say v_0) of air at $x = 0$. The resulting boundary condition would be

$$\left(\frac{\partial \phi}{\partial x}\right)_{x=0} = -v_0$$

which from (13.4.22) implies

$$\sum_1^\infty (n\pi/l)[A_n \sin(n\pi ct/l) + B_n \cos(n\pi ct/l)] = \begin{cases} -v_0 & (0 < t \leqslant T) \\ & (t < 0, t > T) \end{cases} \tag{13.4.23}$$

At first sight (13.4.23) appears to provide an equation for the determination of A_n and B_n. However the Fourier series problem posed is nonsensical since the function on the right hand side of the equation is not periodic with period l/c. [Rather more subtly it might be argued that because $T \gg l/c$, while we know that (13.4.23) is valid for $0 < t < T$, then a legitimate problem is posed by assuming that the right side is v_0 for all t; since then the constant v_0 may be regarded as a periodic function, a proper Fourier series problem is posed. Unfortunately the solution involves $A_n = B_n = 0$ except for the case $n = 0$ and this particular term is already excluded from the Fourier sum on the left!]

Evidently the mechanism of energy transfer from the wind chest to the normal modes of vibration of the air in the pipe is very complicated and devolves in detail on the unsteady flow mechanism operating in the vicinity of the mouth and the open end of the pipe.

The theory of acoustics is not used commonly to determine the complete sound spectrum of an instrument (i.e. the relative magnitudes of the C_n), but only to determine the natural frequencies of vibration and hence the possible frequencies present in the note emitted. In any event in the case of most instruments, the relative magnitudes of C_n are to some extent under the control of the performer in the way he uses his lips and/or his tongue.

Instruments other than the organ which effectively utilise a column of air vibrating in a tube open at each end are members of the flute family. Here of course the length of the column of air is (discretely) variable by the use of holes in the instrument, so that different notes can be emitted.

In instruments like trombones a more appropriate model is of air vibrating in a tube with the end at the player's lips sealed. Effectively the player is providing variable pressure with his lips. Of course in an instrument like a trombone the 'pipe' is not straight, but it is probable that the curvature is not very important in the problem.

In the particular case of a trombone different notes are obtainable with continuous variation by altering the length of the pipe by means of the slide mechanism. In trumpets and horns depression of stops being channels of different length into play.

The theory of vibrations in a tube open at one end ($x = 0$) and sealed at the other ($x = l$) parallels that given above except that the admissable eigenfunctions must now satisfy

$$\phi = 0, \quad (x = 0)\,; \quad \frac{\partial\phi}{\partial x} = 0, \quad (x = l)\;.$$

The first of these is met by choosing from (13.4.21) eigenfunctions of the type

$$\phi = \begin{matrix}\cos\\ \sin\end{matrix}(\omega t)\sin\left(\frac{\omega x}{c}\right)$$

while the second condition requires $\cos(\omega l/c) = 0$ with solutions

$$\omega_n = (n + \tfrac{1}{2})\pi c/l, \quad (n = 0, 1, 2, 3\ldots)\;.$$

The fundamental (circular) frequency is $\omega_0 = \frac{1}{2}\pi c/l$ and the overtone frequencies are

$$3\omega_0, 5\omega_0 \;\ldots, \quad (2n + 1)\omega_0$$

so that here the even harmonics ($2\omega_0, 4\omega_0 \ldots$) are missing.

Even in the limited context of using solutions of the wave equation simply to find eigenvalues, the discussions given are only very approximate. For example the character of the note emerging from a trumpet is affected significantly by the bell-shape at the open end of the air column; similar considerations prevail for all instruments. Also the material from which the instrument is fabricated is evidently of great importance, and this factor is not remotely taken into account in the above analysis which assumes the material rigid.

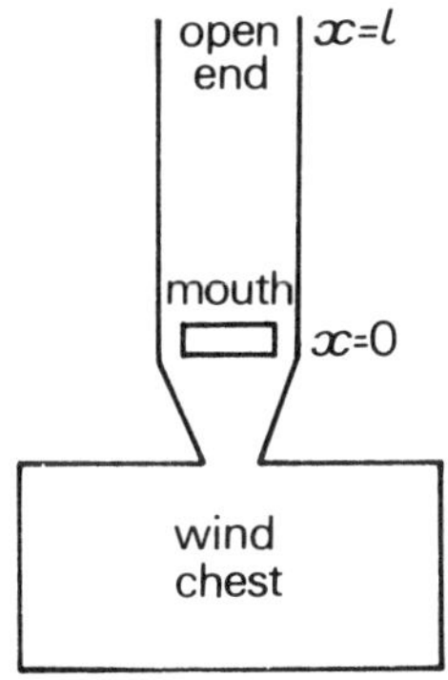

Fig. 13.6 Schematic diagram of the construction of an 'open' organ pipe

We conclude with a brief discussion of the related topic of room acoustics from the viewpoint of eigenfunction solutions of the wave equation. For very large rooms however it may be more appropriate to think in terms of transient wave solutions as indicated earlier.

We consider first an idealised rectangular parallelopiped of a room defined by $0 \leqslant x \leqslant h_1, 0 \leqslant y \leqslant h_2, 0 \leqslant z \leqslant h_3$ with rigid perfectly reflecting surfaces at which the normal components of velocity vanish. The reader will readily verify that eigenfunction solutions of

$$\frac{\partial^2 \phi}{\partial x^2} + \frac{\partial^2 \phi}{\partial y^2} + \frac{\partial^2 \phi}{\partial z^2} = \frac{1}{c^2}\frac{\partial^2 \phi}{\partial t^2}$$

which satisfy

$$\frac{\partial \phi}{\partial x} = 0,\ x = 0,\ x = h_1;\ \frac{\partial \phi}{\partial y} = 0,\ y = 0,\ y = h_2;\ \frac{\partial \phi}{\partial z} = 0,\ z = 0, z = h_3;$$

are $$\phi = \cos(l\pi x/h_1)\cos(m\pi y/h_2)\cos(n\pi z/h_3)\begin{matrix}\cos\\ \sin\end{matrix}(\omega_{l,m,n}t)$$

where the allowable (circular) frequencies $\omega_{i,m,n}$ are given by

$$\omega_{l,n,m} = c\pi\left[\frac{l^2}{h_1^2} + \frac{m^2}{h_2^2} + \frac{n^2}{h_3^2}\right]^{\frac{1}{2}}$$

and where l, m, n take integer values (0, 1, 2, . . .) but with $l = 0, m = 0, n = 0$ excluded.†

The 'natural' frequencies of vibration, given by multiplying the admissable ω values by $(2\pi)^{-1}$ are

$$f_{l,n,m} = \frac{1}{2}c\left[\frac{l^2}{h_1^2} + \frac{m^2}{h_2^2} + \frac{n^2}{h_3^2}\right]^{\frac{1}{2}}. \tag{13.4.24}$$

This formula, of much use in practical acoustic work, provides considerable insight into a number of important problems.

The lowest frequency of vibration is given by

$$\frac{1}{2}c\ \mathrm{Min}\left[\frac{1}{h_1}, \frac{1}{h_2}, \frac{1}{h_3}\right]$$

and is determined by the largest dimension of the room. For example for a cathedral (for which the model is somewhat idealised geometrically but is correct as regards boundary conditions – since stone is an excellent reflector), typically the longest dimension is about 100 metres and this leads to a lowest frequency of 1.72 Hz (say about 2 cycles per second).†† A frequency of 2 Hz is much less than the lowest frequency associated with even the longest (64′)

† The latter solution leads to $\phi = $ const. and is of no physical significance.

†† Hz is an abbreviation for 'Hertz' which is the conventional unit for frequency in the present context.

open organ pipe (of about 9Hz). In contrast, for a domestic living room of maximum dimension 5 metres, the lowest frequency is about 34Hz. So, for example, there is little point in demanding that a gramophone record of organ music be able to reproduce accurately the sound of frequency 18Hz from a 32′ pipe!

The range of frequencies to which the best of human ears respond is from about 20Hz to 20KHz.† Very good Hi-Fi equipment reproduces faithfully frequencies in the range 40Hz-20KHz; well engineered gramophone records cover a similar range with an extension down to 20Hz at the lower end of the range. As we have seen 40Hz is about the lowest frequency of vibration that can be sustained in a medium size domestic room.

The problems of reproducing sound accurately in a domestic room are rather more than buying good quality equipment sufficient to guarantee a good source. Room acoustic problems arise mostly at the lower end of the frequency range and are occasioned by the fact that for small values of l, m, n the formula (13.4.24) predicts quite large gaps between adjacent admissable frequency values. For example for a room of dimensions 15′ × 10′ × 10′ the five lowest admissable values are (approximately) 38Hz, 56Hz, 68Hz, 75Hz and 80Hz. With increasing l, m, n the spacings become much less. Consider now sound emanating from a loudspeaker which contains a wide spectrum of frequencies. In the idealised room of our model, sound with components of frequency other than the natural room frequencies will be converted into sounds with frequencies given by (13.4.24); the mechanism of conversion is the reflection process. For the higher frequencies given out by the source there are always admissable natural room frequencies which are very close, and little distortion occurs. At the lower end of the frequency scale the problems are more severe, as would occur for a 30Hz or a 50Hz sound wave in the numerical example quoted. Even more important for the case where the sound source contains significant components of exactly an admissable frequency, e.g. at 68Hz in the above example, the source and room may act as a coupled resonating system. The resulting low frequency 'boom' may be quite pronounced, and to the discriminating ear is quite disconcerting.

In a real domestic room the boom effect is considerably mitigated by soft furnishings which absorb sound and inhibit the reflection process. Even in this case there are some advantages in rooms whose linear dimensions are not in simple ratios, i.e. ratios of small integers like 1 : 1, 3 : 2 or 2 : 1, for in this case the same frequency value will occur in two different modes or more. This is called degeneracy. In these circumstances the boom effect is enhanced and also the gaps in the frequency spectrum are wider than would be the case where the ratios are none simple.

† This range excludes the fundamental note of a 64′ pipe which is not *audible* in the normal sense of the word.

Problems, Chapter 13

USE OF ARBITRARY FUNCTION SOLUTION OF ONE DIMENSIONAL WAVE EQUATION

1. A long tube $x \geqslant 0$ of uniform cross section A is filled with a fluid initially at rest. In this initial state the fluid is characterised by uniform values of pressure (P_0), density (ρ_0) and sound speed c. The accessible end of the tube $x = 0$ is sealed by a close fitting piston of mass m also at rest initially. [In these circumstances the piston must be subject to an external force of magnitude P_0A to counteract the force exerted by the fluid in the tube – in the subsequent analysis of the piston motion it is to be assumed that this external force is maintained for all time so that the motion is independent of P_0.]

At $t = 0$ the piston receives an impulsive force, additional to the steady force P_0A, which results in an initial piston motion of velocity V in the positive x direction. Assuming that the motion of the fluid is governed by the acoustic approximation show that the fluid velocity is given by

$$v_x = \begin{cases} Ve^{-k[t - (x/c)]} & (x \leqslant ct) \\ 0 & (x > ct) \end{cases}$$

where $k = \rho_0 cA/m$. It is to be assumed, as in Section 13.4.1 that the position of the piston $x = z(t)$ is small compared with ct.

Verify the conservation of linear momentum for the system.

2. A tube of finite length $0 \leqslant x \leqslant L$ and of uniform cross section is closed at $x = L$ and sealed at $x = 0$ by a close fitting piston free to move in the postive x direction. Initially the tube is filled with a fluid which is stationary and characterised by an ambient pressure P_0, density ρ_0 and sound speed c. Also initially the piston is at rest. For $t \geqslant 0$ the piston is set into motion with a velocity $v_0(t)$ in the positive x direction. Assuming that $v(t) \ll c$ (so that the approximation employed in Section 13.4.1 and in Problem 1 may be used here when discussing the boundary condition applicable at the piston), and also that motion of the fluid is governed by the acoustic approximation, show that for $0 \leqslant ct < L$ the velocity potential $\phi(x, t)$ is given by

$$\phi = \phi_1(t - x/c) \equiv \begin{cases} c\displaystyle\int_0^{(t - x/c)} v_0(t')dt' & (t \geqslant x/c) \\ 0 & (t \leqslant x/c) \end{cases} \tag{1}$$

and that following the first reflexion at $x = L$ the velocity potential for $L < ct < 2L$ is given by

$$\phi = \phi_1[t - x/c] + \phi_1[t - (2L - x)/c] \ .$$

Deduce that the overpressure, i.e. pressure in excess of P_0, at $x = L$ for the time $L < ct < 2L$ is twice that associated with the incident wave (1).

Extend the solution to the time period $2L < ct < 3L$.

3. A tube of finite length $0 \leqslant x \leqslant L$ and of uniform cross section is open at $x = L$ to the atmosphere, where the pressure is P_0, and sealed at $x = 0$ by a close fitting piston free to move in the x direction. Initially the open tube contains air at rest characterised by an ambient pressure P_0, density ρ_0 and sound speed c. Also initially the piston is at rest.

For $t \geqslant 0$ the piston is set into motion with a velocity $v_0(t)$ in the x direction. Using the same approximations as in the previous two questions calculate the velocity and pressure fields for $0 \leqslant t < 2L/c$. The motion of the air outside the tube is assumed negligible so that at $x = L$ the pressure is P_0.

USE OF LAPLACE TRANSFORMS

4. (For readers familiar with the application of Laplace transform methods to linear partial differential equations.) Solve Problem 2 for all $t > 0$ by the Laplace transform method and compare the solution with that obtained previously for $0 \leqslant t < L/c$.

5. (For readers familiar with the application of Laplace transform methods to linear partial differential equations.) A tube of finite length $0 \leqslant x \leqslant L$ and of uniform cross section of area α is sealed at each end by pistons of mass M_1 (at $x = 0$) and of mass M_2 (at $x = L$). Initially the pistons are at rest and the tube is filled with stationary air characterised by an ambient pressure P_0, density ρ_0 and sound speed c. Also the air surrounding the tube is at pressure P_0 and this is assumed maintained in the subsequent piston motion. For $t \geqslant 0$ the piston of mass M_1 is subject to a force $F(t)$ in the x direction. Show that the Laplace transform of the velocity of the piston of mass M_2 is given by

$$\frac{2\rho_0 \alpha c e^{-sL/c} \bar{F}(s)}{(M_1 s + \rho_0 \alpha c)(M_2 s + \rho_0 \alpha c) - (M_1 s - \rho_0 \alpha c)(M_2 s - \rho_0 \alpha c) e^{-2sL/c}}$$

where

$$\bar{F}(s) = \int_0^\infty F(t) e^{-st} dt$$

is the Laplace transform of $F(t)$, and find the piston velocity for $0 \leqslant t < 3L/c$ for the case $F(t) = F_0$ where F_0 is a constant. It may be assumed that the displacements of the pistons are sufficiently small so that boundary conditions obtaining at the pistons may be applied at $x = 0$ and $x = L$. It is also to be assumed that the tube overlaps the internal face of the pistons sufficiently so that the pistons do not fall out of the tube.

6. Verify that the one dimensional wave equation

$$\frac{\partial^2 \phi}{\partial x^2} = \frac{1}{c^2} \frac{\partial^2 \phi}{\partial t^2}$$

admits (complex) separation of variable solutions of the type

$$e^{i\omega[t - x/c]}, \text{ (a)}; \qquad e^{i\omega(t + x/c)}, \text{ (b)} \tag{1}$$

where ω is a constant.

If there are no restrictions on ω, the most general solution, obtained by superimposing solutions for all values of ω, is

$$\phi = \int_{-\infty}^{\infty} a(\omega)e^{i\omega[t - x/c]}\,d\omega + \int_{-\infty}^{\infty} b(\omega)e^{i\omega[t + x/c]}\,d\omega \tag{2}$$

where $a(\omega)$ and $b(\omega)$ are arbitrary. Of course these integrals must be real and this requires $a(-\omega) = \bar{a}(\omega)$, $b(-\omega) = \bar{b}(\omega)$. Verify that these are necessary and sufficient conditions for ϕ to be real. Since a and b are arbitrary, equation (2) represents the sums of two arbitrary functions respectively with arguments $(t - x/c)$ and $(t + x/c)$; thus we recover the wave function solution

$$\phi = f(t - x/c) + g(t + x/c)\ . \tag{3}$$

Steady state oscillatory solutions like (1) which persist for all time are sometimes known as progressive waves. The origin of the terminology is as follows. Consider for example the real part of (1a). The solution

$$\cos \omega(t - x/c)$$

is periodic in x with period $2\pi c/\omega$. At two different times say $t = 0$ and $t = t_1$ the solutions are identical except for a shift of phase ωt_1. When this latter phase difference is 2π (so that $t_1 = 2\pi/\omega$) the solutions are exactly the same, so that a ciné film of the function (3) would show that the sinusoidal pattern moves in the positive x direction. The speed with which the pattern appears to move is $(2\pi c/\omega)(2\pi/\omega)^{-1} = c$ and in the context of solutions of the type (1) c is called a **phase velocity**. When the individual solutions (1) are used to synthesise the arbitrary function solution (3), c is more properly called a pulse velocity. [See also the discussion on pp. 404 and 407 and Problem 36 of Chapter 14.]

7. Reflection of a Plane Wave at a Plane Interface

Two liquids characterised by ambient densities and sound speeds (ρ_1, c_1) and (ρ_2, c_2) occupy respectively the regions $x < 0$ and $x > 0$. Initially the liquids are at rest and the ambient pressures P_1 and P_2 are negligible compared with the pressures associated with the subsequent motion.

An incident plane wave of infinite extent, described by the velocity potential

$$\phi = f[t - (\mathbf{n}_I, \mathbf{r})/c_1]$$

is propagated in $x < 0$ in the direction $\mathbf{n}_I = (\cos\theta_1, \sin\theta_1, 0)$ towards the interface $x = 0$ (so that $0 < \theta_1 < \frac{1}{2}\pi$). Show that provided $c_2 \sin\theta_1/c_1 < 1$ the boundary conditions of continuity of pressure and normal component of velocity imply a reflected wave travelling in $x < 0$ in the direction $\mathbf{n}_R =$

$(-\cos\theta_1, \sin\theta_1, 0)$ together with a transmitted wave travelling in the direction $\mathbf{n}_T = (\cos\theta_2, \sin\theta_2, 0)$ where θ_1 and θ_2 are related through Snell's law

$$\frac{\sin\theta_1}{\sin\theta_2} = \frac{c_1}{c_2} \quad (0 < \theta_2 < \tfrac{1}{2}\pi), \tag{1}$$

and determine the relative amplitudes of these waves.

If $c_2 \sin\theta_1/c_2 > 1$ note that (1) fails to determine a real value of θ_2. In these circumstances, somethimes called total internal reflexion, there is no solution of the problem in the terms proposed. In this connection see part (iii) of Problem 30 of Chapter 14 where a solution to a problem with a similar difficulty in the theory of elastic waves is obtained using Fourier integrals.

SPHERICAL WAVE EQUATION

8. Show that the substition $\phi = \Phi(r, t)/r$ reduces the spherically symmetric wave equation

$$\frac{\partial^2\phi}{\partial r^2} + \frac{2}{r}\frac{\partial\phi}{\partial r} = \frac{1}{c^2}\frac{\partial^2\phi}{\partial t^2}$$

to the equation
$$\frac{\partial^2\Phi}{\partial r^2} = \frac{1}{c^2}\frac{\partial^2\Phi}{\partial t^2}$$

and hence deduce that the general solution of (1) is

$$\phi = r^{-1}[f(t - r/c) + g(t + r/c)]$$

where f and g are functions of the indicated variables.

A sphere of time dependent radius $a(t)$ is centred on the origin and surrounded by a compressible fluid characterised by sound speed c. For $t < 0$, $a(t) = a_0$ (constant) and the fluid is at rest. For $t > 0$ $a(t)$ is given by

$$a(t) = a_0[1 + \epsilon\sin(\omega t)]$$

where ϵ ($\ll 1$) and ω are constants. Show that the velocity potential of the fluid is

$$\phi = r^{-1}f[t - (r - a_0)/c]$$

where $f(t) = \omega\epsilon a_0^3[\cos(\omega t) + \omega\tau\sin(\omega t) - e^{-(t/\tau)}]/(1 + \omega^2\tau^2)$

and where $\tau = a_0/c$. Deduce that for sufficiently large times and sufficiently large radii $[(\omega r/c) \gg 1]$ the velocity of the fluid is periodic with circular frequency ω and with (velocity) amplitude

$$\frac{\omega^2\epsilon a_0^3}{cr[1 + \omega^2\tau^2]^{\frac{1}{2}}}.$$

9. Show that the spherical wave equation

$$\frac{\partial^2\phi}{\partial r^2} + \frac{2}{r}\frac{\partial\phi}{\partial r} = \frac{1}{c^2}\frac{\partial^2\phi}{\partial t^2}$$

admits a similarity solution of the form $\phi = rf(\eta)$ where f satisfies the equation

$$\eta^2(1 - \eta^2)f'' + \eta(4 - 2\eta^2)f' + 2f = 0$$

and $\eta = r/ct$.

Verify that $f = \eta^{-1}$ is a solution of the latter equation and find the velocity field associated with the similarity solution. Show that the velocity field may be used to solve the problem of the steady growth of a spherical cavity in a compressible fluid from zero radius at $t = 0$ to a radius $v_0 t$ at time t where $v_0 \ll c$.

10. Show that the axisymmetric wave equation expressed in spherical polar coordinates (r, θ) (where $z = r\cos\theta, x^2 + y^2 + z^2 = r^2$)

i.e.
$$\frac{\partial^2 \phi}{\partial r^2} + \frac{2}{r}\frac{\partial \phi}{\partial r} + \frac{1}{r^2 \sin\theta}\frac{\partial}{\partial \theta}\left(\sin\theta \frac{\partial \phi}{\partial \theta}\right) = \frac{1}{c^2}\frac{\partial^2 \phi}{\partial t^2} \tag{1}$$

admits separation of variable solutions of the form

$$\phi = F_n(r, t)P_n(\cos\theta) \qquad (n = 0, 1, 2, \ldots)$$

where the P_n are Legendre Polynomials and where the F_n satisfy the partial differential equations

$$\frac{\partial^2 F_n}{\partial r^2} + \frac{2}{r}\frac{\partial F_n}{\partial r} - \frac{n(n+1)F_n}{r^2} = \frac{1}{c^2}\frac{\partial^2 F_n}{\partial t^2}. \tag{2}$$

Consider in particular the case $n = 1$. Show that if $\phi(r, t)$ satisfies the spherical wave equation

$$\frac{\partial^2 \phi}{\partial r^2} + \frac{2}{r}\frac{\partial \phi}{\partial r} = \frac{1}{c^2}\frac{\partial^2 \phi}{\partial t^2}$$

then $\partial\phi/\partial r$ satisfies (2) for the case $n = 1$. Deduce that one solution of (1) is

$$\phi = \cos\theta \frac{\partial}{\partial r}[r^{-1}\{f(t - r/c) + g(t + r/c)\}] \tag{3}$$

where f and g are arbitrary functions of the indicated variables. [Alternately verify that if ϕ is a solution of the wave equation then so is $\partial\phi/\partial z$ and apply the result to the general solution of the spherical wave equation to obtain (3).]

Use the outgoing wave part of the solution (3) to solve Problem 11 below.

11. A rigid sphere of fixed radius a is immersed in a compressible fluid whose ambient pressure and density are respectively P_0 and ρ_0 and whose sound speed is c. For $t < 0$ both fluid and sphere are at rest. For $t \geqslant 0$ the sphere executes rectilinear motion in the z direction so that the centre of the sphere is located at $[0, 0, \xi(t)]$ where $|\xi| \ll a$.

Show that the boundary condition for the normal component of fluid velocity on the sphere is given in a first approximation by

$$-\left(\frac{\partial \phi}{\partial r}\right)_{r\,=\,a} = \dot{\xi}(t)\cos\theta$$

and that this condition together with the outgoing wave condition is met by the solution

$$\phi = \cos\theta\,\frac{\partial}{\partial r}\left[r^{-1}f\,[t - (r - a)/c]\right]$$

provided $f(t)$ satisfies the ordinary differential equation

$$\frac{\ddot{f}}{c^2 a} + \frac{2\dot{f}}{ca^2} + \frac{2f}{a^3} = -\dot{\xi}(t)\ .$$

Find the solution of this equation for the case $\xi = v_0 t$ where v_0 is constant. (Evidently the solution is valid only for times satisfying $v_0 t \ll a$.) In finding the solution, two boundary conditions are needed. Since the fluid is undisturbed beyond the wave front it is shown below that ϕ is continuous at $r = a + ct$. If this is to be true for all t then $f(0) = \dot{f}(0) = 0$. In general $\partial\phi/\partial r$ is then discontinuous at the wave front and hence so is the velocity. This is acceptable physical behaviour. On the other hand discontinuities in ϕ imply velocities of unacceptable delta function character at the wave front.

Calculate the pressure on the sphere and from this result show that the external force on the sphere required to impose the motion is

$$\mathbf{F} = \tfrac{4}{3}\pi\rho_0 a^2 v_0 c e^{-(ct/a)}\cos(ct/a)\mathbf{k}$$

together with the necessary instantaneous impulsive (delta function) force required to accelerate initially the sphere from rest to the velocity v_0.

CYLINDRICAL WAVE EQUATION

12. Show that the substitution $\phi = r^{-\frac{1}{2}}f$ in the cylindrically symmetric wave equation

$$\frac{\partial^2\phi}{\partial r^2} + \frac{1}{r}\frac{\partial\phi}{\partial r} = \frac{1}{c^2}\frac{\partial^2\phi}{\partial t^2}$$

yields

$$\frac{\partial^2 f}{\partial r^2} + \frac{1}{4}\frac{f}{r^2} = \frac{1}{c^2}\frac{\partial^2 f}{\partial t^2}$$

which differs from the one dimensional wave equation by the inclusion of the term $\frac{1}{4}r^{-2}f$. If the latter may be neglected, which is plausible for large r, the cylindrically symmetric wave equation possesses the approximate solution

$$\phi \simeq r^{-\frac{1}{2}}[f(t - r/c) + g(t + r/c)]$$

but this is inaccurate for small r.

On the other hand show that the plane polar wave equation

$$\frac{\partial^2 \phi}{\partial r^2} + \frac{1}{r}\frac{\partial \phi}{\partial r} + \frac{1}{r^2}\frac{\partial^2 \phi}{\partial \theta^2} = \frac{1}{c^2}\frac{\partial^2 \phi}{\partial t^2}$$

possesses exact solutions of the type

$$\phi = \begin{matrix}\cos\\ \sin\end{matrix}\,(\tfrac{1}{2}\theta) r^{-\frac{1}{2}}\,[f(t - r/c) + g(t + r/c)]$$

where f and g are arbitrary. [I am indebted to Professor Eshelby, University of Sheffield, for pointing this result out to me.]

Chapter **14**

The Theory of Classical Linear Elasticity for Isotropic Solids

14.1 Introduction and basic equations

The oldest established theory governing the behaviour of deformable solid materials is that of classical elasticity, founded in its present form in the early nineteenth century. In discussing the basic equations we assume in the first instance that the initial and final body configurations are virtually identical, as for example are the configurations of (say) a desk in (a) a situation in which the desk surface is clear; (b) a situation where the surface is subject to external loading caused by the presence of books or a paperweight, etc. Most technological applications of classical elasticity are encompassed by this type of situation, although there are examples of importance which entail additionally large rigid body motions, as for example in rotating machinery. We show subsequently how these may be incorporated into the theory.

The equations of classical elasticity are based on the equations of motion of a continuum (4.4.13)

$$\frac{\partial \sigma_{ij}}{\partial x_j} + \rho b_i = \rho f_i \qquad (14.1.1)$$

the strain-displacement gradient relations (8.4.2)

$$E_{ij} = \frac{1}{2}\left(\frac{\partial U_i}{\partial X_j} + \frac{\partial U_j}{\partial X_i}\right) \qquad (14.1.2)$$

and the constitutive equations (8.4.4)

$$\sigma_{ij} = \lambda \Delta \delta_{ij} + 2\mu E_{ij} \ . \qquad (14.1.3)$$

In the latter equations λ and μ are the Lamé constants, which are positive for real materials, while Δ is the dilatation given by

$$\Delta = \frac{\partial U_i}{\partial X_i} \equiv \operatorname{div} \mathbf{U} \equiv E_{mm} \ . \qquad (14.1.4)$$

In writing down (14.1.2) use has been made of the small displacement gradient assumption discussed in Section 5.3. Conventionally, and incorrectly, the approximation entailed is commonly described as the **small** or **infinitesimal strain** approximation. The assumption of small displacement gradients is more stringent than the assumption of small strains. Indeed there are some problems of elasticity (not considered here) where the displacement gradients may be large while the strains remain small. In these circumstances, notably in the theory of the large deflexions of elastic plates, equations (14.1.2) are inadequate even though the behaviour of the material may be small strain elastic; also the initial and final body configurations are not virtually identical.

Equations (14.1.1) are not quite in the form used in classical elasticity theory. As written the equations imply that the σ_{ij} are regarded as functions of the spatial coordinates x_i rather than of the material coordinates X_i. Evidently in (14.1.2) and (14.1.3) it is much more convenient to think of all field quantities (U_i, E_{ij}, σ_{ij}) as functions of the X_i. Accordingly we express (14.1.1) also in terms of the X_i. The term ρf_i is simply $\rho \partial^2 U_i/\partial t^2$ [from equations (2.2.4) and (2.3.29)] while b_i, which is given, is to be expressed in terms of the X_i. In almost all practical applications where the body force terms are retained, there is no difficulty in this step since usually b_i is identified with the body force due to gravity and is constant. The terms $\partial\sigma_{ij}/\partial x_j$ are approximated by $\partial\sigma_{ij}/\partial X_j$; this may be justified by the argument

$$\frac{\partial\sigma_{ij}}{\partial X_j} = \frac{\partial\sigma_{ij}}{\partial x_k}\frac{\partial x_k}{\partial X_j} = \frac{\partial\sigma_{ij}}{\partial x_k}\frac{\partial}{\partial X_j}(X_k + U_k) = \frac{\partial\sigma_{ij}}{\partial x_k}\left(\delta_{kj} + \frac{\partial U_k}{\partial X_j}\right)$$

$$= \frac{\partial\sigma_{ij}}{\partial x_j} + \text{terms of order } e \text{ times the stress gradients} \ .$$

Here $e(\ll 1)$ is the symbol introduced in Section 5.3 to denote the order of magnitude of the displacement gradients. It follows that with neglect of terms of order e times those retained

$$\frac{\partial\sigma_{ij}}{\partial x_j} \simeq \frac{\partial\sigma_{ij}}{\partial X_j} \ . \tag{14.1.5}$$

Finally we recall equation (5.3.11), derived on the basis of the small displacement gradient assumption,

$$\rho = \rho_0(1 - E_{mm}) = \rho_0[1 + O(e)]$$

which implies that to an accuracy consistent with (14.1.5), ρ in (14.1.1) may be approximated by ρ_0. Assembling all these approximations leads to

$$\frac{\partial\sigma_{ij}}{\partial X_j} + \rho_0 b_i = \rho_0\frac{\partial^2 U_i}{\partial t^2} \tag{14.1.6}$$

which together with (14.1.2), (14.1.3) provide 15 equations for the 15 unkowns σ_{ij}, E_{ij}, U_i.† The equations are **linear** with many resulting simplifications, e.g. such equations are much simpler to solve than non-linear equations, while also new solutions can be built up by the addition or **superposition** of independent solutions. The resultant theory, based on equations (14.1.2), (14.1.3), (14.1.6) is sometimes known as **linear elasticity theory**; other common descriptions are

† We recall that only 6 of the 9 components of E_{ij} and σ_{ij} are independent in view of the symmetry relations $E_{ij} = E_{ji}$, $\sigma_{ij} = \sigma_{ji}$.

classical, reflecting perhaps the historical fact that the theory provided the first complete description of the possible behaviour of a deformable solid, and **infinitesimal,** reflecting the small displacement gradient assumption.

Until recently classical elasticity theory was the only extensively developed theory of solid mechanics. Technologically it remains the most important branch of the subject, and the theory is used extensively in civil, mechanical and aeronautical engineering design. The present chapter is concerned with some of the simpler solutions of equations (14.1.2), (14.1.3) and (14.1.6).

14.2 Boundary conditions and notation

In Section 14.1 the small displacement gradient assumption was used to approximate the equations of motion (14.1.1) to equations (14.1.6). It is also possible to adopt a slightly different viewpoint and regard all the field quantities as functions of the spatial coordinates x_i rather than of the material coordinates X_i. The resulting equations are then

$$\frac{\partial \sigma_{ij}}{\partial x_j} + \rho_0 b_i = \rho_0 \frac{\partial^2 u_i}{\partial t^2} \tag{14.2.1}$$

$$\sigma_{ij} = \lambda \Delta \delta_{ij} + 2\mu e_{ij} \tag{14.2.2}$$

$$e_{ij} = \frac{1}{2}\left(\frac{\partial u_i}{\partial x_j} + \frac{\partial u_j}{\partial x_i}\right) \tag{14.2.3}$$

where

$$\Delta = e_{mm} \tag{14.2.4}$$

and where **u** is the displacement expressed in terms of the x_i.

Equations (14.2.1) - (14.2.4) are **identical in structure** respectively to equations (14.1.6), (14.1.3), (14.1.2) and (14.1.4) of the previous section. They are not of course identical equations, since the two sets differ from each other at most by terms of order e times terms retained. However since both sets of equations are founded on the small displacement gradient approximation, either set may be used and are equally valid. In other words the basic equations of classical elasticity fail to distinguish between the material and spatial coordinates in their roles as **independent** variables.†

Consistent with this last result, the theory of classical elasticity also fails to distinguish between initial and current configurations of a body in the application of boundary conditions. This is made clear in the following illustration. Fig. 14.1 depicts a body in the initial and current configurations. The current configuration is supposed a consequence of the application of given external forces **F** distributed over part of the bounding surface. Consider as an illustration the boundary conditions associated with the traction free part of the surface, not subjected to external forces. The precise boundary condition for a traction free surface is $\mathbf{T}_{(\mathbf{n})} = 0$ where $\mathbf{T}_{(\mathbf{n})}$ is the stress vector and **n** is the normal to the deformed surface in the current configuration. In classical elasticity theory

† Of course the situation is quite different in respect of the **dependent** variables U_i (or u_i) which represent precisely the difference between x_i and X_i.

this boundary condition is assumed to be $\mathbf{T}_{(\mathbf{n}')} = 0$ where $\mathbf{n}'$ is the normal to the undeformed surface in the initial configuration. This type of approximation is always employed in specifying boundary conditions in classical elasticity, i.e. boundary conditions are always supposed to apply to unperturbed surfaces. The approximation involved is consistent with the failure of the governing equations to distinguish between the x_i and X_i, and makes the solutions of problems much simpler than they would be otherwise.

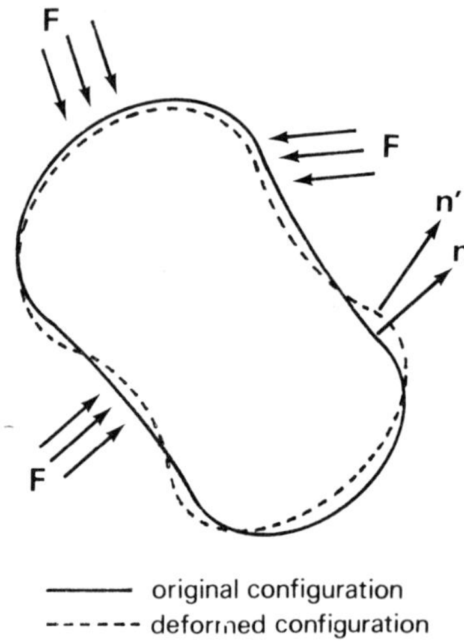

Fig. 14.1 Boundary conditions in classical elasticity

It remains to discuss the case where, in addition to any *small* elastic deformations, the body undergoes also *large* rigid body motions of translation and/or rotation. In all cases the rigid body motion will either be directly known, as in the case of rotating machinery, or alternately may be inferred from the (known) external forces in conjunction with Newton's laws of motion. Accordingly we decompose the particle displacement into two components

$$\mathbf{U} = \mathbf{U}^{(e)}(\mathbf{R}, t) + \mathbf{U}^{(R)}(\mathbf{R}, t) \tag{14.2.5}$$

where $\mathbf{U}^{(R)}$ is the known rigid body displacement and $\mathbf{U}^{(e)}$ the displacement associated with elastic deformation. In (14.2.5) $\mathbf{R}$ is the material coordinate which is referred to axes moving with the rigid body motion. From (14.2.5) the acceleration is readily calculated,

$$\frac{\partial^2 \mathbf{U}}{\partial t^2} = \frac{\partial^2 \mathbf{U}^{(e)}}{\partial t^2} + \frac{\partial^2 \mathbf{U}^{(R)}}{\partial t^2}$$

for which the second term is known. Accordingly, inserting this expression into the equations of motion (14.1.6) results in

$$\frac{\partial \sigma_{ij}}{\partial X_j} + \rho_0 \left(b_i - \frac{\partial^2 U_i^{(R)}}{\partial t^2} \right) = \rho_0 \frac{\partial^2 U_i^{(e)}}{\partial t^2} \tag{14.2.6}$$

which differ from the original equations only by the presence of a modified body force $[b_i - \ddot{U}_i^{(R)}]$.

We encounter one example in Section (14.4.5).

14.3 Young's modulus, Poisson's ratio: Engineering version of the theory

In the remainder of this chapter the notation of equations (14.2.1) - (14.2.4) which utilises lower case letters for displacement, strain and coordinates is adopted.

Equations (14.2.2) may be solved for e_{ij}. To effect the solution, note

$$\sigma_{ii} = 3\lambda\Delta + 2\mu e_{mm} = (3\lambda + 2\mu)\Delta$$

so that
$$\Delta = \frac{\sigma_{ii}}{(3\lambda + 2\mu)} .$$

Substituting this result back into (14.2.2) and re-arranging leads to

$$e_{ij} = \frac{\sigma_{ij}}{2\mu} - \frac{\lambda\sigma_{mm}}{2\mu(3\lambda + 2\mu)}\delta_{ij} . \tag{14.3.1}$$

In full, these equations become (diagonal component)

$$e_{11} = \frac{(\lambda + \mu)}{\mu(3\lambda + 2\mu)}\left[\sigma_{11} - \frac{\lambda}{2(\lambda + \mu)}(\sigma_{22} + \sigma_{33})\right] \tag{14.3.2}$$

with similar equations for e_{22}, e_{33}, together with (off diagonal components)

$$e_{12} = \sigma_{12}/2\mu \qquad \text{etc.} \tag{14.3.3}$$

With the following definitions of Young's modulus (E) and the dimensionless quantity Poisson's ratio (ν),

$$E = \frac{\mu(3\lambda + 2\mu)}{\lambda + \mu}, \quad \nu = \frac{\lambda}{2(\lambda + \mu)} \tag{14.3.4}$$

equations (14.3.2) and (14.3.3) may be written

$$e_{11} = \frac{1}{E}[\sigma_{11} - \nu(\sigma_{22} + \sigma_{33})]; \qquad e_{22} = \frac{1}{E}[\sigma_{22} - \nu(\sigma_{33} + \sigma_{11})];$$

$$e_{33} = \frac{1}{E}[\sigma_{33} - \nu(\sigma_{11} + \sigma_{22})]; \quad e_{12} = \frac{\sigma_{12}}{2\mu}; \quad e_{23} = \frac{\sigma_{23}}{2\mu}; \quad e_{13} = \frac{\sigma_{13}}{2\mu} \tag{14.3.5}$$

which are the forms of the elasticity constitutive equations familiar in most of the engineering literature.

From (14.3.4) the following relations connecting the **elastic constants** λ, μ, E, ν may be derived

$$\lambda = 2\mu\nu/(1 - 2\nu), \quad E = 2\mu(1 + \nu) . \tag{14.3.6}$$

Often in the engineering literature the second of these results is used to express the off diagonal components of (14.3.5) in the form

$$e_{12} = \frac{(1 + \nu)}{E}\sigma_{12} \qquad \text{etc.}$$

[It should also be mentioned that all the older (and some of the current) literature define off diagonal strain components differing from the present formulae by a factor of two. The strain-displacement gradient relations employed in the old literature are

$$e_{11} = \frac{\partial u_1}{\partial x_1}, \text{ etc.}, \qquad e_{12} = \frac{\partial u_1}{\partial x_2} + \frac{\partial u_2}{\partial x_1}, \text{ etc.},$$

with associated constitutive (or stress-strain) relations

$$e_{11} = \frac{1}{E}[\sigma_{11} - \nu(\sigma_{22} + \sigma_{33})], \text{ etc.}, \qquad e_{12} = \frac{2(1 + \nu)}{E}\sigma_{12}, \text{ etc.}$$

These definitions of e_{12}, e_{13} and e_{23}, which are gradually disappearing, and result in a non-tensor definition of strain, are *not* used here.]

Physical interpretations of E and ν provided by the well known laboratory experiment in which a weight W is attached to a long wire of uniform cross section are readily shown to be consistent with the diagonal components of equation (14.3.5). In Fig. 14.2 a wire of initial length l, and of uniform cross section A, is suspended from the fixed support x_3 $(\equiv z) = 0$ and carries a weight W at $z = l$.

The stress field is modelled adequately by the uniform field

$$\sigma_{33} = W/A, \quad \sigma_{11} = \sigma_{22} = \sigma_{12} = \sigma_{23} = \sigma_{13} = 0$$

except in the vicinities of $z = 0, z = l$ where the local field is critically dependent on methods of support, and likely to vary from one experiment to another. The resultant strains are also uniform: in particular the **longitudinal** strain e_{33} is uniform and is therefore represented by the ratio $\Delta l/l$ where Δl is the increase in length of l due to W. From these results and the third of (14.3.5) there follows the well known equation

$$\frac{W}{A} = E\frac{\Delta l}{l}$$

which is used to define Young's modulus in elementary physics. Near $z = 0$ and $z = l$ the strain e_{33} is neither given by $\Delta l/l$ nor is the strain uniform. However for sufficiently long wires, the contributions to Δl of the non-uniform regions are negligible compared with the uniform regions.

From the first and second of (14.3.5), the strains e_{11} and e_{22}, which are the proportional length changes for lines drawn parallel to the x_1 and x_2 axes are given by

$$e_{11} = e_{22} = -\nu e_{11}$$

and are also uniform. The lines in question, which lie in the plane of the cross section, are therefore shortened in the ratio $(1 - \nu e_{11}) : 1$. However since as yet it has not been necessary to prescribe either the origin of the x_1, x_2 coordinates in the cross section, or the directions of the axes, all lines lying in the cross section plane are reduced in length by the same proportion. In particular for a circular wire of initial radius a, the contracted radius under load is $a(1 - \nu e_{11})$. This result usually provides the definition of ν in elementary physics.

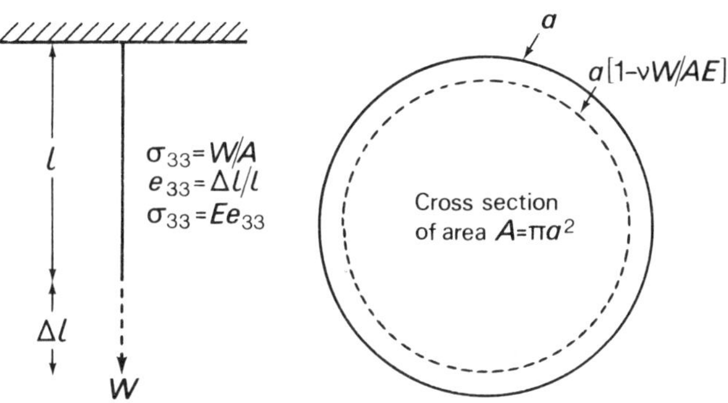

Fig. 14.2 Schematic diagram of Young's modulus experiment in elementary physics

From the sum of the first three equations in (14.3.5) there obtains

$$e_{mm} = \Delta = \frac{(1 - 2\nu)}{E}[\sigma_{11} + \sigma_{22} + \sigma_{33}]$$

so for $\nu = \frac{1}{2}$, the dilatation Δ, i.e. the relative volume change, vanishes. Therefore $\nu = \frac{1}{2}$ characterises incompressible classical elastic solids. For positive λ and μ possible values of ν lie in the range $0 \leqslant \nu \leqslant \frac{1}{2}$, for most metals ν lies in the range $0.25 < \nu < 0.35$ while small deformations of rubbers and elastomers are more nearly incompressible (typically $\nu = 0.48$ - 0.49).

Values of Young's modulus for metals are typically in the range from 2.10^5 N/mm^2 ($\simeq 1.25 \times 10^4$ tons/in^2) for steels to about one sixth of this for aluminium. For rubbers and elastomers, E is very sensitive to both temperature and conditions of testing (rapid or slow loading) and can vary from as low as 1 N/mm^2 to 10^4 N/mm^2.†

† Materials whose properties depend on the speed of testing are commonly called **strain-rate sensitive** – such materials are not really elastic at all, but possess time-dependent constitutive relations (e.g. as occurs in the theory of **viscoelasticity** discussed in Chapter 15).

14.4 Static problems

14.4.1 Prism of Uniform Cross Section Suspended in Earth's Gravitational Field

We consider a prism of length l, of uniform cross section A and density ρ_0 suspended at $z = 0$ in the earth's gravitational field (Fig. 14.3).

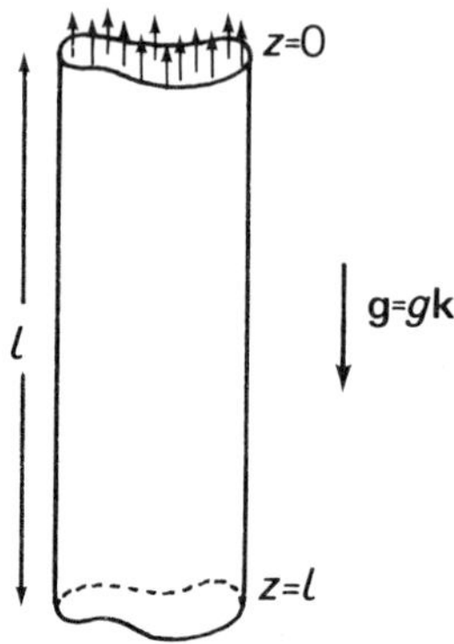

Fig. 14.3 Prism suspended in earth's field

A stress field is assumed for which the only non-vanishing component of stress is σ_{zz} dependent only on z. With the positive direction of the z axis chosen as indicated in Fig. 14.3 the only non-trivial equilibrium equation is

$$\frac{d\sigma_{zz}}{dz} = -\rho_0 g$$

with solution

$$\sigma_{zz} = \rho_0 g(l - z) \ . \tag{14.4.1}$$

In carrying out the integration, the constant of integration has been chosen to satisfy the boundary condition $\sigma_{zz} = 0$ on $z = l$ so that the surface $z = l$ is traction free. Since $\sigma_{xx} = \sigma_{yy} = \sigma_{xy} = \sigma_{xz} = \sigma_{yz} = 0$, the condition that the vertical walls of the cylinder remain stress free is automatically satisfied. This is most easily seen by noting that for $\mathbf{n}$ normal to the curved vertical wall $\mathbf{n} = (n_x, n_y, 0)$, and hence the components $\sigma_{ij}n_j$ of $\mathbf{T}_{(\mathbf{n})}$ always vanish.

On $z = 0$ the normal stress is $\sigma_{zz} = \rho_0 gl$ and is independent of x and y. The proposed solution is strictly correct only provided the method of suspension, indicated schematically by the vertical arrows in Fig. 14.3, permits the uniform stress distribution. In reality it would be very difficult, if not impossible, to arrange a uniform normal stress distribution over $z = 0$, so that the solution obtained is not valid in the immediate vicinity of the suspended end of the prism. The total load exerted on $z = 0$ is the integral of $\rho_0 gl$ over the cross section. If A is the cross section area – the integral is $\rho_0 gAl$ which, not surprisingly, is the total weight of the prism.

The stress-strain equations (14.3.5) in conjunction with the strain-displacement gradient equations (14.2.3) yield

$$\frac{\partial u_z}{\partial z} = \frac{\rho_0 g}{E}(l - z) \qquad (14.4.2)$$

$$\frac{\partial u_y}{\partial y} = -\frac{\nu\rho_0 g}{E}(l - z), \ (14.4.3); \qquad \frac{\partial u_x}{\partial x} = -\frac{\nu\rho_0 g}{E}(l - z), \ (14.4.4);$$

$$\frac{\partial u_x}{\partial y} + \frac{\partial u_y}{\partial x} = 0, \ (14.4.5); \qquad \frac{\partial u_y}{\partial z} + \frac{\partial u_z}{\partial y} = 0, \ (14.4.6);$$

$$\frac{\partial u_z}{\partial x} + \frac{\partial u_x}{\partial z} = 0, \ (14.4.7),$$

which are six partial differential equations for u_x, u_y, u_z. Conceivably there is no triplet of functions u_x, u_y, u_z which satisfy these equations. In fact here there are many possible solutions as is seen below.†

It is instructive to solve equations (14.4.2) - (14.4.7) directly. From (14.4.2), (14.4.3) and (14.4.4) we derive after integration

$$u_z = \frac{\rho_0 g}{E}(lz - \tfrac{1}{2}z^2) + f_3(x, y) \qquad (14.4.8)$$

$$u_x = -\frac{\nu\rho_0 g x}{E}(l - z) + f_1(y, z) \qquad (14.4.9)$$

$$u_y = -\frac{\nu\rho_0 g y}{E}(l - z) + f_2(x, z) \qquad (14.4.10)$$

where, as yet, f_1, f_2 and f_3 are arbitrary functions of the indicated variables. Substituting (14.4.8) - (14.4.10) into (14.4.5) - (14.4.7) now yield the following partial differential equations for f_1, f_2, f_3,

$$\frac{\partial f_1}{\partial y} + \frac{\partial f_2}{\partial x} = 0, \ (14.4.11); \qquad \frac{\partial f_2}{\partial z} + \frac{\partial f_3}{\partial y} + \frac{\nu\rho_0 g y}{E} = 0, \ (14.4.12);$$

$$\frac{\partial f_3}{\partial x} + \frac{\partial f_1}{\partial z} + \frac{\nu\rho_0 g x}{E} = 0, \ (14.4.13).$$

The equations are simplified by writing

$$f_3(x, y) = -\frac{1}{2}\frac{\nu\rho_0 g}{E}(x^2 + y^2) + F_3(x, y) \qquad (14.4.14)$$

† If the *assumption* of a stress field happens to yield inconsistencies in the equations for the displacement, the only conclusion to draw is that the inferred stress field does not provide a correct solution to the problem posed. The situation is rare, for except in the simplest of problems, as here and in Section 13.4.2, we do not infer stress fields, but rather solve for displacement and stress fields concomitantly.

obtaining

$$\frac{\partial f_1}{\partial y} + \frac{\partial f_2}{\partial x} = 0, \ (14.4.15); \quad \frac{\partial f_2}{\partial z} + \frac{\partial F_3}{\partial y} = 0, \ (14.4.16);$$

$$\frac{\partial F_3}{\partial x} + \frac{\partial f_1}{\partial z} = 0, \ (14.4.17) \ .$$

Because $\partial f_1/\partial y$ is independent of x and $\partial f_2/\partial z$ independent of y, (14.4.15) and (14.4.16) may be integrated in the form

$$f_1 = -y\frac{\partial f_2}{\partial x} + \psi(z), \quad F_3 = -y\frac{\partial f_2}{\partial z} + \phi(x) \ . \tag{14.4.18}$$

Substituting these results into (14.4.17) leads to

$$\phi'(x) + \psi'(z) = 2y\frac{\partial^2 f_2}{\partial z \partial x} \ .$$

However in this equation the left side is independent of y while the right side appears to be linear in y. This inconsistency is resolvable only if

$$\phi'(x) + \psi'(z) = 0, \ (14.4.19); \quad \text{and} \quad \frac{\partial^2 f_2}{\partial z \partial x} = 0, \ (14.4.20) \ .$$

Again, in (14.4.19), since $\phi'(x)$ is independent of z and $\psi'(z)$ independent of x we must have $\phi'(x) = -a_2$, $\psi'(z) = a_2$ where a_2 is constant. These equations integrate to give

$$\phi(x) = -a_2 x + c_3, \ (14.4.21); \quad \psi(z) = a_2 z + c_1, \ (14.4.22)$$

where c_1 and c_3 are two further arbitrary constants. Also equation (14.4.20) may be integrated to yield

$$f_2 = \theta(z) + \tau(x) \tag{14.4.23}$$

where $\theta(z)$, $\tau(x)$ are as yet arbitrary functions of x and z. From (14.4.18), (14.4.21), (14.4.22) and (14.4.23)

$$f_1 = -y\tau'(x) + a_2 z + c_1; \quad f_2 = \tau(x) + \theta(z); \quad F_3 = -y\theta'(z) - a_2 x + c_3 \ .$$

Finally recalling that f_1 is independent of x and also F_3 is independent of z shows that $\tau'(x)$ and $\theta'(z)$ are at most constants. Writing $\tau'(x) = a_3$, $\theta'(z) = -a_1$ now yields $\tau(x) = a_3 x + m$, $\theta(z) = -a_1 z + n$ and

$$f_1 = -a_3 y + a_2 z + c_1; \quad f_2 = a_3 x - a_1 z + c_2; \quad F_3 = a_1 y - a_2 x + c_3 \ ,$$

where $c_2 = m + n$ is arbitrary. (14.4.24)

The final expression for the displacement field found from (14.4.8), (14.4.9), (14.4.10), (14.4.14) and (14.4.24) is (in vector form)

$$\mathbf{u} = \frac{\rho_0 g}{E}\left[-\nu x(l-z), \ -\nu y(l-z), \ lz - \tfrac{1}{2}z^2 - \tfrac{1}{2}\nu(x^2+y^2)\right] + \mathbf{c} + (\mathbf{a} \wedge \mathbf{r}) \tag{14.4.25}$$

where $\mathbf{c}$ and $\mathbf{a}$ are arbitrary vectors.

The vector **c** represents an arbitrary rigid body translation and is determined by fixing one point in the body. For example if in the present case the point (0, 0, 0) is undisturbed then $\mathbf{c} = 0$. The origin of the $\mathbf{a} \wedge \mathbf{r}$ term is more interesting. We recall from Section 5.3 that in the small displacement gradient approximation the vector $\frac{1}{2}$ curl **u** represents locally the rigid body motion of a material element in the vicinity of **r**. For the displacement field $\mathbf{u} = (\mathbf{a} \wedge \mathbf{r})$

$$\tfrac{1}{2}\,\text{curl}(\mathbf{a} \wedge \mathbf{r}) = \mathbf{a}$$

for all **r**, so that $\mathbf{u} = \mathbf{a} \wedge \mathbf{r}$ represents a rigid body rotation of (small) magnitude $|\mathbf{a}|$ about the axis of **a** for the whole body. If, for the current problem, the z axis passes through the centre of mass of the prism, we expect $u_x = u_y = 0$ for $x = y = 0$, implying $\mathbf{a} = 0$.

With $\mathbf{a} = \mathbf{c} = 0$ in (14.4.25), the resultant displacement field is

$$\mathbf{u} = \frac{\rho_0 g}{E}\,[-\nu x(l - z), \quad -\nu y(l - z), \quad lz - \tfrac{1}{2}z^2 - \tfrac{1}{2}\nu(x^2 + y^2)] \;.$$

We consider in detail the case of a circular cylinder $x^2 + y^2 \leqslant a^2$. The centre point (0, 0, 0) of the suspended section $z = 0$ has been chosen to remain undisturbed. Elsewhere on $z = 0$ the displacement u_z is

$$(u_z)_{z=0} = -\frac{\frac{1}{2}\rho_0 g\nu}{E}\,r^2 \qquad (r^2 = x^2 + y^2)$$

so that the surface assumes a dished shape, which for realistic values of E would be imperceptible. The surface $z = l$ is also dished with normal displacement

$$(u_z)_{z=l} = \frac{\frac{1}{2}\rho_0 g}{E}\,[l^2 - \nu r^2] \;. \qquad (14.4.26)$$

For $l \gg a$, the dominant contribution to (14.4.26) derives from the term involving l^2. Even so for a steel rod of length one metre for which $E = 2.10^5$ N/mm^2, $l = 1$ m, $\rho_0 = 8.10^3$ K/m^3, $g \simeq 10$ m/s^2, the displacement $\frac{1}{2}\rho_0 g l^2/E$ is a mere 2.10^{-4} mm.

It is of interest to note that if $W = \rho_0 g l A$ is the weight of the rod (where A = area of cross section), then, ignoring the term νr^2,

$$(u_z)_{z=l} \simeq \frac{\frac{1}{2}\rho_0 g l^2}{E} = \frac{\frac{1}{2}Wl}{EA}$$

so that the displacement at $z = l$ is one half of what would occur if the weight were concentrated at $z = l$, cf. the discussion on the elementary Young's modulus experiment in Section 14.3. The perceptive reader will have noted that in this discussion, the body force of the wire was neglected and the wire extension attributed entirely to the external load. For the experiment to yield displacements Δl of measurable magnitude with simple instrumentation, i.e. $\Delta l \gg 2 \times 10^{-4}$ mm, the external load must be much greater than the weight of the wire; in these circumstances the body force terms are negligible.

A careful examination of the analysis leading to (14.4.25) shows that the contribution

$$\mathbf{u} = \mathbf{c} + \mathbf{a} \wedge \mathbf{r} \tag{14.4.27}$$

to the displacement field is quite generally the solution of the equations

$$e_{ij} \equiv \frac{1}{2}\left(\frac{\partial u_i}{\partial x_j} + \frac{\partial u_j}{\partial x_i}\right) = 0$$

and evidently therefore *must* represent the displacement of a rigid (i.e. undeformed) body; Indeed it has been shown that $\mathbf{c}$ and $\mathbf{a} \wedge \mathbf{r}$ represent respectively a rigid body translation and a (small) rotation.† Since the equations of classical elasticity are linear it is always possible to add to any displacement field additional terms like (14.4.27) without altering the stress field. In problems for which the boundary conditions take the form of specified surface tractions over the whole surface, the displacement field is indeterminate to within a rigid body displacement of the form (14.4.27). For other problems, where the boundary conditions specify surface displacements on part or the whole of the boundary, this indeterminacy is removed.

14.4.2 Bending of Beams by Terminal Couples

Here we exploit the feasibility of a solution of the elastostatic equations with a stress field

$$\sigma_{xx} = Ey/R, \quad \sigma_{yy} = \sigma_{zz} = \sigma_{xy} = \sigma_{yz} = \sigma_{zx} = 0 \tag{14.4.28}$$

where E is Young's modulus and R is a constant which will be interpreted subsequently. With neglect of body forces the stress field (14.4.28) satisfies the equilibrium equations $\partial\sigma_{ij}/\partial x_j = 0$. From (14.3.5) and (14.2.3) the displacement field satisfies the equations

$$\left.\begin{aligned} &\frac{\partial u_x}{\partial x} = \frac{y}{R}, \quad \frac{\partial u_y}{\partial y} = -\frac{\nu y}{R}, \quad \frac{\partial u_z}{\partial z} = -\frac{\nu y}{R} \\ &\frac{\partial u_y}{\partial x} + \frac{\partial u_x}{\partial y} = \frac{\partial u_z}{\partial y} + \frac{\partial u_y}{\partial z} = \frac{\partial u_z}{\partial x} + \frac{\partial u_x}{\partial z} = 0 \end{aligned}\right\} \tag{14.4.29}$$

The first three of (14.4.29) lead to

$$\left.\begin{aligned} &u_x = \frac{yx}{R} + f_1(y, z), \quad u_y = -\frac{1}{2}\frac{\nu y^2}{R} + f_2(x, z), \\ &u_z = -\frac{\nu yz}{R} + f_3(x, y) \end{aligned}\right\} \tag{14.4.30}$$

† Of course, the most general displacement of a rigid body is a combination of rotation and translation. The rotation may be large and the only reasons for the emphasis on small rotations in the present context are (a) large rotations are not encompassed by the small displacement gradient assumption, (b) large rotations are not expressible by $\mathbf{u} = \mathbf{a} \wedge \mathbf{r}$ (see discussion in Section 5.3).

and substituting these results into the remaining equations of (14.4.29) yields

$$\frac{\partial f_1}{\partial y} + \frac{\partial f_2}{\partial x} = -\frac{x}{R}, \quad \frac{\partial f_2}{\partial z} + \frac{\partial f_3}{\partial y} = \frac{\nu z}{R}, \quad \frac{\partial f_3}{\partial x} + \frac{\partial f_1}{\partial z} = 0 .$$

If now we write $f_2 = [(\nu z^2 - x^2)/2R] + F_2$ the equations for f_1, F_2 and f_3 assume precisely the homogeneous form (14.4.15), (14.4.16), (14.4.17). Following the arguments of Section 14.4.1 the displacement field (14.4.30) is therefore given by

$$\mathbf{u} = \frac{1}{R}[yx, \tfrac{1}{2}\nu(z^2 - y^2) - \tfrac{1}{2}x^2, -\nu yz] + \mathbf{c} + \mathbf{a} \wedge \mathbf{r} \qquad (14.4.31)$$

where **c** and **a** are arbitrary vectors.

So far the stress field (14.4.28) and associated displacement field (14.4.31) have not been identified with a problem. In Fig. 14.4 a horizontal cylinder of uniform cross section is depicted with the x axis lying parallel to the cylinder generators. The y axis is directed vertically (and the direction of the z axis is then specified uniquely). For convenience the y and z axes are chosen to intersect at the centroid of the cross sectional area. We consider the solution (14.4.28), (14.4.31) in the context of Fig. 14.4 for the case $\mathbf{a} = \mathbf{c} = 0$ so that

$$\mathbf{u} = \frac{1}{R}[yx, \tfrac{1}{2}\nu(z^2 - y^2) - \tfrac{1}{2}x^2, -\nu yz] . \qquad (14.4.32)$$

For a cylinder whose lateral dimensions in the cross section are small compared with the length l, the dominant displacement component of (14.4.32) is u_y which is well approximated by

$$u_y = -\tfrac{1}{2}x^2/R .$$

In any event for points on the line of centroids, i.e. on the x axis, we have exactly

$$\mathbf{u} = (0, -\tfrac{1}{2}x^2/R, 0)$$

so that the centroid line is deformed into a parabola (Fig. 14.5). Denoting by Y the deflection of the centroid line in the negative y direction so that $Y = -u_y$, the equation of the parabola is

$$Y = \tfrac{1}{2}x^2/R . \qquad (14.4.33)$$

Within the small displacement gradient approximation, which implies in particular $|dY/dx| \ll 1$, the parabola (14.4.33) is also the arc of a circle of radius R. For, from the well known formula for radius of curvature ρ†

$$\rho^{-1} = \frac{d^2Y/dx^2}{[1 + (dY/dx)^2]^{3/2}}$$

we find from (14.4.33)

$$\rho = R[1 + O(dY/dx)^2]$$

i.e.

$$\rho \simeq R$$

† Here the symbol ρ denotes radius of curvature and not density.

so that the displacement field (14.4.32) may be associated with the flexure of a cylinder whose centroid line is bent into the arc of a circle of radius R.

We examine now the forces necessary to maintain this deformation. From (14.4.28) the only non-vanishing component of stress σ_{xx} is a tensile (or compressive) stress directed along lines parallel to the line of centroids. Fig. 14.6 shows the nature of the stress field across any cross section. Evidently lines parallel to the x axis are extended by tensile stresses for $y > 0$ and compressed for $y < 0$.†

Because the z axis has been chosen to pass through the centroid, the net force over the section $x = l$, or indeed over any section, is zero. This is shown by the following argument. The only possible non-vanishing component of force on $x = l$ is in the x direction and is given by

$$F_x = \iint_S \sigma_{xx} dS = \frac{E}{R} \iint_S y \, dydz$$

where S is the area of cross section. Since y is measured from the centroid, the double integral vanishes so that $F_x = 0$.

However it is clear from Fig. 14.6 that the stress distribution σ_{xx} must exert a couple, usually known as the **bending couple**, on each section about the z axis. The magnitude of the couple is

$$M = \iint_S y\sigma_{xx} dS = \frac{E}{R} \iint_S y^2 dydz \tag{14.4.34}$$

The double integral, which is purely a geometrical property of the cross section, is known as the **second moment of area** and denoted (universally) by the symbol I. The dimensions of

$$I = \iint_S y^2 dydz \tag{14.4.35}$$

are evidently $[\mathrm{L}^4]$; for example for a rectangular cross section of depth $2a$ and width $2b$

$$I = \int_{-b}^{b} dz \int_{-a}^{a} y^2 dy = \tfrac{4}{3} ba^3$$

while for a circle of radius a [on writing $y = r \sin\theta$, $dydz \to rdrd\theta$ in (14.4.35)]

$$I = \int_0^{2\pi} \sin^2\theta \, d\theta \int_0^a r^3 dr = \tfrac{1}{4}\pi a^4$$

† In some engineering texts lines parallel to the line of centroids are termed **fibres.** Engineers usefully conceive a bent beam as an assembly of compressed and extended fibres, each of which obeys the elementary Hooke's law $\sigma_{xx} = Ee_{xx}$.

From (14.4.34), (14.4.35) we derive a well known result in the theory of beam flexure

$$M = EI/R \tag{14.4.36}$$

relating bending moment, Young's modulus and radius of curvature.

We have interpreted (14.4.28) and (14.4.32) as the solution of the problem of the bending of a uniform cylinder into the arc of a circle of radius R under the action of terminal couples M of equal magnitude. The conventional diagram indicating the presence of the couples is shown in Fig. 14.7.

Important results in the theory which are used extensively in technological mechanics are (14.4.36) and [from (14.4.33)]

$$\frac{d^2 Y}{dx^2} = \frac{1}{R} . \tag{14.4.37}$$

In engineering terminology the line of centroids is usually known as the *neutral axis* and the plane $y = 0$, for which $\sigma_{xx} = 0$, as the *neutral plane.*

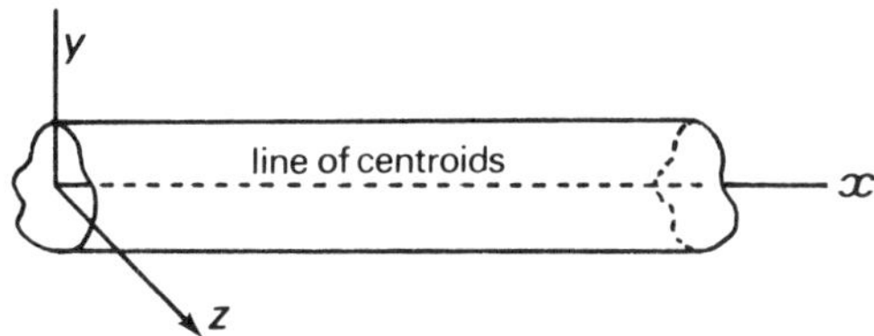

Fig. 14.4 Geometry of cylinder and choice of axes

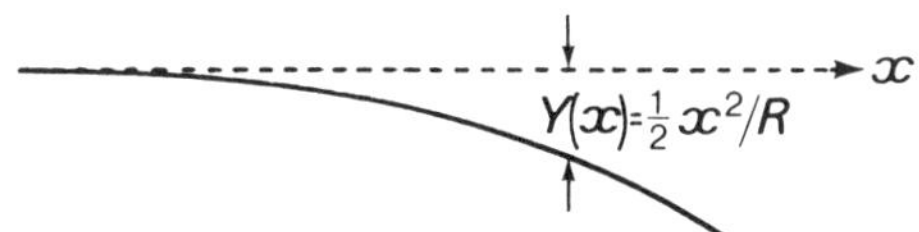

Fig. 14.5 Deflection of line of centroids

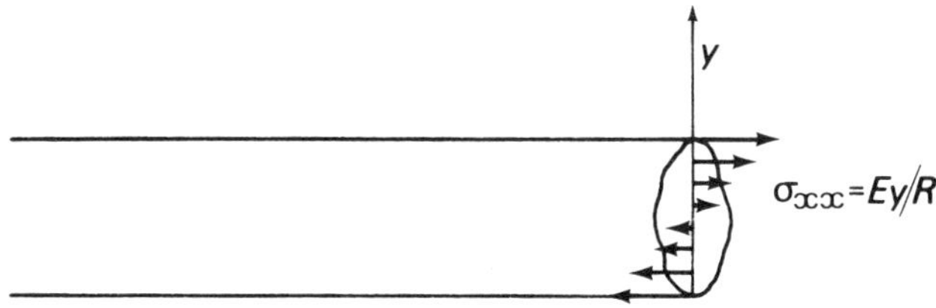

Fig. 14.6 Stress distribution over any section and (in particular) over the end faces

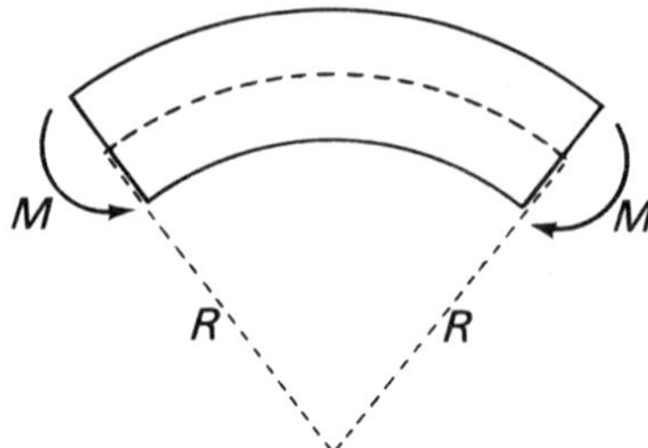

Fig. 14.7 Schematic diagram of beam bent under equal terminal couples M. (Curvature exaggerated)

The solution found above provides a basis for the important technological theory of the bending of beams. This elementary theory, discussed in more detail in Section 14.5, is concerned with the mechanics of the bending of beams subjected to external transverse forces, as for example in Fig. 1.1. Here the beam experiences a non-uniform bending moment due to the load W.

In the theory of beam bending the quantity $M(x)$ usually varies with x, and the stress distribution over any cross section x = const. is assumed to be essentially as depicted in Fig. 14.6, but with $R(x)$ also varying with x†. In consequence the relation (14.4.36) is preserved, and this equation together with (14.4.37) then provides a differential equation for the deflection $Y(x)$. $M(x)$ is derived from the (known) distribution of external transverse forces acting on the beam.

Evident deficiencies in the model are that external forces such as W in Fig. 1.1 must induce locally a stress field, with perhaps large components, which does not remotely resemble the assumed linear distribution of longitudinal stress of Fig. 14.6. In spite of these defects the elementary theory of beam bending works well, certainly in respect of the deflection of the neutral axis, and is one of the foundation stones of structural mechanics.

We have encountered previously the idea that a simple solution to a problem may be largely correct, even when the solution fails in some details. In the present context consider the example of a flat ruler bent by holding the ends between (say) the thumb and first finger of each hand. The forces exerted manually *are equivalent to* imposing a pair of couples of equal magnitude but opposite sign on the ends of the ruler, as in Fig. 14.7. However the method of applying the couples is quite different from that envisaged in the theory of beam bending (Fig. 14.6). The elementary theory fails to take account of the precise method of loading of the ruler, and the solution obtained in the theory is that given above for the bending of a beam under terminal couples of equal magnitude and opposite sign. Evidently the solution fails, certainly for the stress field, in the immediate vicinity of the ends of the ruler.

† For variable $M(x)$ there are also shear stresses acting on the planes x = const. These shear stresses do not contribute significantly to the bending.

This discussion leads to the practically important concept known as **St. Venant's principle**. It is rather hard to pin the 'principle' down, if only because it is not a principle at all, but rather a statement of belief which works satisfactorily in a wide variety of situations. The principle is concerned with the situation where an approximate and usually simple solution to a problem describes the grosser features of the problem quite well, but fails to meet in detail the exact boundary conditions over some limited areas of the bounding surface. Usually the failure is associated with traction boundary conditions, as in the example of the bent ruler. The principle then states that even though the solution, particularly for the stress field, may be gravely in error in the vicinity of the limited areas, elsewhere the solution is satisfactory. In most cases it is further inferred that some features of the displacement field, such as the transverse displacement of the neutral axis in the theory of beam bending, are correct, even in the vicinity of those boundaries where the boundary conditions are violated. If the former hypothesis is right, the latter will follow for those features of the displacement field which in some sense devolve on the distribution of stress throughout the body, and do not depend crucially on the local stress distribution.

In no sense is the principle 'provable' in general. For a particular case, conceivably it is possible to prove the principle, or rather to quantify the precise meanings of *vicinity* and *elsewhere,* by solving the exact problem with the correct boundary conditions. In practice this is likely to be a formidable task in the numerical solution of partial differential equations, which is only just within reach of present day computers. Perhaps it is of even more relevance to note that for many practical situations, such as occur in structural engineering, the exact boundary conditions may not be known.

It is difficult perhaps to leave the reader with any firm conviction of the value of the principle. Nevertheless the principle is used extensively with satisfactory results, particularly in the theory of the bending of beams. In many cases experimental confirmation of the principle is feasible, e.g. by measuring strains and deflections in loaded beams.

St. Venant's principle is almost an article of faith in the engineering world and in fact seems largely to work. We encounter the principle again in the theory of torsion (Section 14.4.6).

14.4.3 SPHERICAL PRESSURE VESSEL

Figure 14.8 depicts a spherical pressure vessel occupying $a \leqslant r \leqslant b$, (where $r = \sqrt{x^2 + y^2 + z^2}$ is a spherical polar coordinate), and subject to a uniform internal pressure P. In the following analysis body forces are neglected.

With neglect of body forces it is evident from symmetry arguments that the displacement field $\mathbf{u}$ is directed in the radial direction and depends only on r, i.e.

$$\mathbf{u} = u(r)\mathbf{i}_r \tag{14.4.38}$$

where $\mathbf{i}_r$ is a unit vector in the r direction.

An equation for $u(r)$ may be found from the general (vector) equation for $\mathbf{u}$ in the absence of body forces. The latter follows from (14.2.1) and (14.2.2). In the absence of body forces there results

$$\frac{\partial \sigma_{ij}}{\partial x_j} = \lambda \frac{\partial \Delta}{\partial x_i} + \mu \frac{\partial}{\partial x_j}\left(\frac{\partial u_i}{\partial x_j} + \frac{\partial u_j}{\partial x_i}\right)$$

$$= \lambda \frac{\partial \Delta}{\partial x_i} + \mu \nabla^2 u_i + \mu \frac{\partial}{\partial x_i}\left(\frac{\partial u_j}{\partial x_j}\right) = \rho_0 \frac{\partial^2 u_i}{\partial t^2}$$

i.e. $$(\lambda + \mu)\frac{\partial \Delta}{\partial x_i} + \mu \nabla^2 u_i = \rho_0 \frac{\partial^2 u_i}{\partial t^2} \quad (\text{since } \Delta = \operatorname{div} \mathbf{u} \equiv \partial u_j/\partial x_j).$$

In vector form this may be written

$$(\lambda + \mu)\operatorname{grad}(\operatorname{div} \mathbf{u}) + \mu \nabla^2 \mathbf{u} = \rho_0 \frac{\partial^2 \mathbf{u}}{\partial t^2} \tag{14.4.39}$$

For static problems, as here, the equation reduces to

$$(\lambda + \mu)\operatorname{grad}(\operatorname{div} \mathbf{u}) + \mu \nabla^2 \mathbf{u} = 0 . \tag{14.4.40}$$

For the spherically symmetric case (14.4.38), it is readily verified that curl $\mathbf{u} = 0$; utilising this result together with the vector identity curl curl $\mathbf{u} = \operatorname{grad}(\operatorname{div} \mathbf{u}) - \nabla^2 \mathbf{u}$ shows that for the present problem

$$\operatorname{grad}(\operatorname{div} \mathbf{u}) = 0 \tag{14.4.41}$$

so that div $\mathbf{u}$ is a constant.

We write
$$\operatorname{div} \mathbf{u} = 3A \tag{14.4.42}$$
where $3A$ is a constant of integration and the factor 3 is inserted for subsequent convenience. Since for the vector field (14.4.38), div $\mathbf{u} = u'(r) + 2u/r$, (14.4.42) becomes

$$\frac{d}{dr}(r^2 u) = 3Ar^2$$

with solution
$$u = Ar + \frac{B}{r^2} \tag{14.4.43}$$

where B is a second constant of integration.

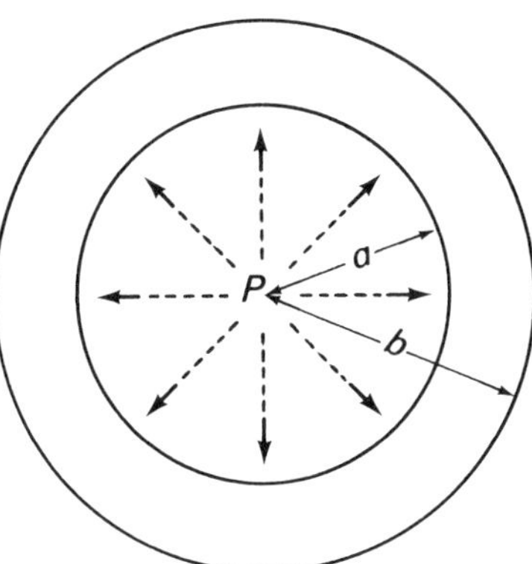

Fig. 14.8 Spherical pressure vessel subject to internal pressure P

It remains to find A and B which are determined by the conditions that the surfaces $r = a$, $r = b$ are subject respectively to normal traction $-P$ and zero traction.

We evaluate the stresses along the z axis, for which $z = r, x = y = 0$. The Cartesian components of displacement are

$$u_x = \frac{x}{r}u = Ax + Bx/r^3, \quad u_y = \frac{y}{r}u = Ay + By/r^3,$$

$$u_z = \frac{z}{r}u = Az + Bz/r^3 .$$

From these expresssions and (14.2.3)

$$e_{xx} = A + B\left(\frac{1}{r^3} - \frac{3x^2}{r^5}\right), \quad e_{yy} = A + B\left(\frac{1}{r^3} - \frac{3y^2}{r^5}\right),$$

$$e_{zz} = A + B\left(\frac{1}{r^3} - \frac{3z^2}{r^5}\right),$$

$$e_{xy} = -3Bxy/r^5, \quad e_{xz} = -3Bxz/r^5, \quad e_{yz} = -3Byz/r^5 .$$

From these results and equations (14.2.2), the corresponding stresses on $z = r, x = y = 0$ are given by

$$\left.\begin{aligned} \sigma_{xx} = \sigma_{yy} = 3\lambda A + 2\mu(A + B/r^3), \quad \sigma_{zz} &= 3\lambda A + 2\mu(A - 2B/r^3), \\ \sigma_{xy} = \sigma_{yz} = \sigma_{xz} &= 0 \end{aligned}\right\} \quad (14.4.44)$$

so that along the z axis, σ_{xx}, σ_{yy} and σ_{zz} are principal stresses. However because of the spherical symmetry of the problem any choice of direction for the z axis leads to a similar result. It follows that everywhere in the body the radial direction constitutes a principal axis. Denoting in particular by σ_{rr} the normal stress associated with an element of surface whose normal is in the r direction (i.e. following the conventions discussed in Section 9.3 on p. 173 in the context of cylindrical polar coordinates), we have

$$\sigma_{rr} = (\sigma_{zz})_{z = r,\, x = y = 0} = 3\lambda A + 2\mu(A - 2B/r^3). \tag{14.4.45}$$

Also in the same convention (see Fig. 14.9)

$$\sigma_{\theta\theta} = \sigma_{\phi\phi} = 3\lambda A + 2\mu(A + B/r^3), \quad \sigma_{r\theta} = \sigma_{r\phi} = \sigma_{\theta\phi} = 0. \tag{14.4.46}$$

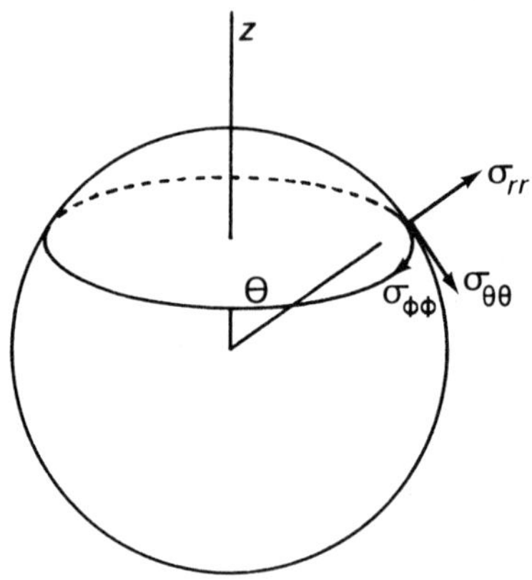

Fig. 14.9 Directions of principal axes and labelling of stress components for spherical pressure vessel

The boundary conditions are

$$(\sigma_{rr})_{r = a} = -P, \quad (\sigma_{rr})_{r = b} = 0$$

so that from (14.4.45)

$$3\lambda A + 2\mu(A - 2B/a^3) = -P, \quad 3\lambda A + 2\mu(A - 2B/b^3) = 0$$

with solution $B = \dfrac{Pa^3 b^3}{4\mu(b^3 - a^3)}, \quad A = \dfrac{Pa^3}{(3\lambda + 2\mu)(b^3 - a^3)}.$

Finally, inserting these results into (14.4.43), (14.4.45) and (14.4.46), and expressing λ and μ in terms of E and ν, yields

$$u(r) = \frac{Pa^3}{E(b^3 - a^3)} [(1 - 2\nu)r + \tfrac{1}{2}(1 + \nu)b^3/r^2] \tag{14.4.47}$$

$$\sigma_{rr} = -\frac{Pa^3}{(b^3 - a^3)}\left(\frac{b^3}{r^3} - 1\right), \tag{14.4.48}$$

$$\sigma_{\theta\theta} = \sigma_{\phi\phi} = \frac{Pa^3}{(b^3 - a^3)}\left(1 + \frac{1}{2}\frac{b^3}{r^3}\right), \tag{14.4.49}$$

14.4.4 CYLINDRICAL PRESSURE VESSEL

Here $r = (x^2 + y^2)^{\frac{1}{2}}$ denotes a cylindrical polar coordinate. The problem considered is that of a cylindrical tube $a \leqslant r \leqslant b$ subject to internal pressure P at $r = a$. We assume a displacement field

$$\mathbf{u} = u_r(r)\mathbf{i}_r + u_z(z)\mathbf{k} \qquad (14.4.50)$$

where $\mathbf{k}$ is the usual unit vector in the z direction and $\mathbf{i}_r$ a unit vector in the (cylindrical polar) r direction.

It is readily verified for the proposed displacement field (14.4.50) that curl $\mathbf{u} = 0$ so that equation (14.4.40) is equivalent to grad (div $\mathbf{u}$) $= 0$ or

$$\operatorname{div} \mathbf{u} = A' \qquad (14.4.51)$$

where A' is a constant of integration.

From (14.4.50) and the formula for div $\mathbf{u}$ in cylindrical polar coordinates, (14.4.51) may be written

$$\frac{du_r}{dr} + \frac{u_r}{r} = A' - \frac{du_z}{dz} \,. \qquad (14.4.52)$$

Since the left side of (14.4.52) is independent of z and the right side independent of r, each side is constant. Therefore in particular the axial strain

$$e_{zz} = \frac{du_z}{dz}$$

is constant while u_r satisfies

$$\frac{du_r}{dr} + \frac{u_r}{r} = 2A \qquad (14.4.53)$$

where A is a new constant defined by $2A = A' - e_{zz}$, and where the factor 2 is inserted for subsequent convenience. It will be seen subsequently that different values of the constant e_{zz} correspond to slightly different problems.

Equation (14.4.53) is readily integrated by multiplying by the integrating factor r, obtaining

$$\frac{d}{dr}(ru_r) = 2Ar$$

with solution $\qquad u_r = Ar + Br^{-1} \qquad (14.4.54)$

where B is an additional constant of integration.

Equation (14.4.54) is equivalent to

$$u_x = Ax + Bx/r^2, \quad u_y = Ay + By/r^2$$

from which we derive the strains

$$\left.\begin{aligned} e_{xx} &= A + B\left(\frac{1}{r^2} - \frac{2x^2}{r^4}\right), \quad e_{yy} = A + B\left(\frac{1}{r^2} - \frac{2y^2}{r^4}\right), \\ e_{xy} &= -2Bxy/r^3 \,. \end{aligned}\right\} \qquad (14.4.55)$$

Also since u_x and u_y are independent of z, while u_z is independent of x and y

$$e_{xz} = e_{zy} = 0 .$$

From (14.4.55) and (14.2.2) together with the result that e_{zz} is constant we find for $x = r, y = 0$

$$\left.\begin{aligned} \sigma_{xx} &= \lambda(2A + e_{zz}) + 2\mu[A - B/r^2] , \\ \sigma_{yy} &= \lambda(2A + e_{zz}) + 2\mu[A + B/r^2] , \\ \sigma_{zz} &= \lambda(2A + e_{zz}) + 2\mu e_{zz}, \\ \sigma_{xy} &= \sigma_{yz} = \sigma_{xz} = 0 . \end{aligned}\right\} \qquad (14.4.56)$$

Evidently along the x axis σ_{xx}, σ_{yy} and σ_{zz} are principal stresses. From the symmetry of the problem the radial (r) direction and the axial (z) direction always constitute principal stress directions; the third principal stress direction is in the θ direction and in the notation of page 173 equations (14.4.56) are equivalent to

$$\begin{aligned} \sigma_{rr} &= 2(\lambda + \mu)A + \lambda e_{zz} - 2\mu B/r^2 , \\ \sigma_{\theta\theta} &= 2(\lambda + \mu)A + \lambda e_{zz} + 2\mu B/r^2 , \\ \sigma_{zz} &= \lambda(2A + e_{zz}) + 2\mu e_{zz} . \end{aligned} \qquad (14.4.57)$$

The stresses σ_{rr} and $\sigma_{\theta\theta}$ are indicated in Fig. 14.10; the stress σ_{zz} is directed normal to the plane of the diagram.

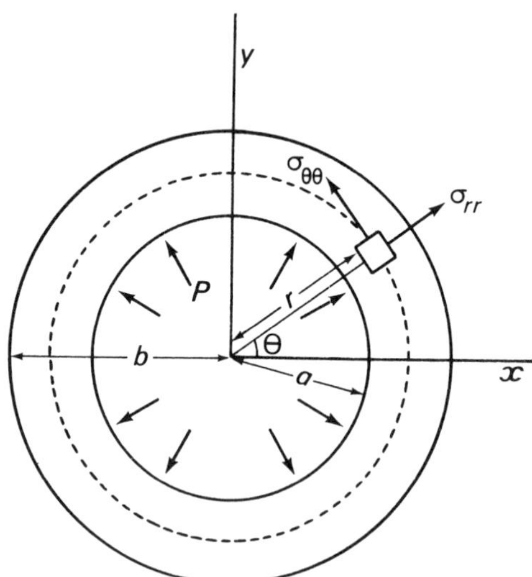

Fig. 14.10 Radial stress σ_{rr} and hoop stress $\sigma_{\theta\theta}$ in tube under internal pressure.

The boundary conditions

$$\sigma_{rr} = -P, \ (r = a); \quad \sigma_{rr} = 0, \ (r = b);$$

serve to determine completely the combined constant $2(\lambda + \mu)A + \lambda e_{zz}$ and also B and hence to determine σ_{rr} and $\sigma_{\theta\theta}$. We have

$$2(\lambda + \mu)A + \lambda e_{zz} - 2\mu B/a^2 = -P, \quad 2(\lambda + \mu)A + \lambda e_{zz} - 2\mu B/b^2 = 0$$

from which

$$B = \frac{a^2 b^2 P}{2\mu(b^2 - a^2)}, \quad 2(\lambda + \mu)A + \lambda e_{zz} = \frac{a^2 P}{(b^2 - a^2)} \tag{14.4.58}$$

so that

$$\sigma_{rr} = -\frac{a^2 P}{(b^2 - a^2)}\left(\frac{b^2}{r^2} - 1\right), \ (14.4.59); \quad \sigma_{\theta\theta} = \frac{a^2 P}{(b^2 - a^2)}\left(\frac{b^2}{r^2} + 1\right), \ (14.4.60)$$

Also, solving the second of (14.4.58) for A, substituting the result into the third of (14.4.57) and expressing λ and μ in terms of E leads to

$$\sigma_{zz} = \frac{2\nu a^2 P}{(b^2 - a^2)} + E e_{zz}. \tag{14.4.61}$$

At this stage three slightly different problems may be solved. We consider first the case of a tube whose ends are restrained to prevent axial expansion or contraction. For this problem $e_{zz} = 0$ and the axial stress is

$$\sigma_{zz} = \frac{2\nu a^2 P}{b^2 - a^2}.$$

A second problem concerns the case of a tube which is subject to no axial restraint, i.e. a tube which is allowed freely to expand (or contract) in the z direction. For this problem $\sigma_{zz} = 0$ and e_{zz} is given by

$$e_{zz} = \frac{-2\nu a^2 P}{E(b^2 - a^2)}$$

so that in fact the tube contracts.

Finally, and perhaps of greater practical engineering interest, there is the problem of a tube whose ends are sealed by rigid plates over whose surfaces of internal area πa^2 there is a pressure P. Due to the internal pressure therefore, each end plate is subject to a force $\pi a^2 P$ and this is balanced by the force exerted on the plate by the tube. Since σ_{zz} is uniform the latter force is σ_{zz} times the area of the annulus between $r = a$ and $r = b$ so that

$$\pi a^2 P = \pi(b^2 - a^2)\left[E e_{zz} + \frac{2\nu a^2 P}{(b^2 - a^2)}\right]$$

which implies $e_{zz} = \dfrac{(1 - 2\nu)a^2 P}{E(b^2 - a^2)}$ and $\sigma_{zz} = \dfrac{a^2 P}{b^2 - a^2}$

For the more realistic case where the ends are sealed by plates which are not rigid, the end plates deform and there is an interaction between the tube and the plates. Calculations of the true stress and displacement fields in the tube in the vicinity of the ends would be much more complicated than those presented here. Away from the ends of the tube St. Venant's principle suggests that the simple solution given above is satisfactory.

14.4.5 Stresses in Rotating Shafts

We consider here the problem of a cylindrical shaft of circular cross section $x^2 + y^2 \leqslant a^2$ rotating at constant angular velocity ω about the z axis. In this example the body is undergoing large rigid body rotations. Superposed on the rigid body rotation is a 'small' elastic displacement field. Due to the angular velocity each particle of the body experiences a radial acceleration of magnitude $r\omega^2$ directed towards $r = 0$. Accordingly with respect to axes fixed in the body the effect of the rotation is equivalent to a body force of magnitude $\rho_0 r\omega^2 \mathbf{i}_r$. The resultant equation of motion is (14.2.6) with $\mathbf{b} = 0$ (in the absence of true body forces) and with $\ddot{\mathbf{u}}^{(R)}$ given by $-\rho_0\omega^2 r\mathbf{i}_r$. There results for $\mathbf{u}$, after using (14.2.2),

$$(\lambda + \mu)\,\mathrm{grad}\,(\mathrm{div}\,\mathbf{u}) + \mu\nabla^2\mathbf{u} = -\rho_0\omega^2 r\mathbf{i}_r \quad .$$

Here $\mathbf{i}_r$ is the unit vector in the direction of the (plane) polar coordinate r.

We seek a solution of the form (14.4.50); curl $\mathbf{u} = 0$ as in the previous problem and hence

$$(\lambda + 2\mu)\mathrm{grad}(\mathrm{div}\,\mathbf{u}) = -\rho\omega^2 r\mathbf{i}_r$$

or in component form

$$\frac{\partial}{\partial z}(\mathrm{div}\,\mathbf{u}) = 0\,, \qquad \frac{\partial}{\partial r}(\mathrm{div}\,\mathbf{u}) = -\frac{\rho\omega^2 r}{\lambda + 2\mu}\,.$$

Since here

$$\mathrm{div}\,\mathbf{u} = \frac{du_r}{dr} + \frac{u_r}{r} + \frac{du_z}{dz}$$

the first equation implies $e_{zz} \equiv du_z/dz$ is constant whereupon the second equation becomes

$$\frac{d}{dr}\left(\frac{du_r}{dr} + \frac{ur}{r}\right) = -\frac{\rho_0\omega^2 r}{\lambda + 2\mu}$$

or on integration

$$\frac{1}{r}\frac{d}{dr}(ru_r) = -\frac{\rho_0\omega^2 r^2}{2(\lambda + 2\mu)} + 2A$$

with solution

$$u_r = -\frac{\rho_0\omega^2 r^3}{8(\lambda + 2\mu)} + Ar + B/r\,.$$

For a solid shaft continuous up to $r = 0$, B must be chosen to vanish so that

$$u_r = -\frac{\rho_0\omega^2 r^3}{8(\lambda + 2\mu)} + Ar\,.$$

Calculations of (principal) stresses σ_{rr}, $\sigma_{\theta\theta}$, σ_{zz} may be carried out as in the previous example. The details are left to the reader; in terms of A and the (as yet unkown) constant e_{zz} there results

$$\sigma_{rr} = \lambda(2A + e_{zz}) + 2\mu A - \frac{(2\lambda + 3\mu)}{4(\lambda + 2\mu)}\rho_0\omega^2 r^2 \qquad (14.4.62)$$

$$\sigma_{\theta\theta} = \lambda(2A + e_{zz}) + 2\mu A - \frac{(2\lambda + \mu)}{4(\lambda + 2\mu)}\rho_0\omega^2 r^2 \qquad (14.4.63)$$

$$\sigma_{zz} = \lambda(2A + e_{zz}) + 2\mu e_{zz} - \frac{\lambda}{2(\lambda + 2\mu)}\rho_0\omega^2 r^2 \ . \qquad (14.4.64)$$

The boundary condition $\sigma_{rr} = 0$ at $r = a$ determines $[\lambda(2A + e_{zz}) + 2\mu A]$ and hence determines σ_{rr} and $\sigma_{\theta\theta}$. We find

$$\lambda(2A + e_{zz}) + 2\mu A = \frac{(2\lambda + 3\mu)}{4(\lambda + 2\mu)}\rho_0\omega^2 a^2 \qquad (14.4.65)$$

$$\sigma_{rr} = \frac{(2\lambda + 3\mu)}{4(\lambda + 2\mu)}\rho_0\omega^2(a^2 - r^2) \equiv \frac{(3 - 2\nu)}{8(1 - \nu)}\rho_0\omega^2(a^2 - r^2)$$

$$\sigma_{\theta\theta} = \frac{(2\lambda + 3\mu)}{4(\lambda + 2\mu)}\rho_0\omega^2 a^2 - \frac{(2\lambda + \mu)}{4(\lambda + 2\mu)}\rho_0\omega^2 r^2$$

$$\equiv \frac{\rho_0\omega^2}{8(1 - \nu)}[(3 - 2\nu)a^2 - (1 + 2\nu)r^2] \ .$$

Eliminating A between (14.4.65) and (14.4.64) gives

$$\sigma_{zz} = Ee_{zz} + \frac{\rho_0\omega^2\nu}{4(1 - \nu)}[(3 - 2\nu)a^2 - 2r^2] \ . \qquad (14.4.66)$$

If the shaft is prevented from contracting by the end bearings ($e_{zz} = 0$) then the bearings must exert an axial tensile stress

$$\sigma_{zz} = \frac{\rho_0\omega^2\nu}{4(1 - \nu)}[(3 - 2\nu)a^2 - 2r^2]$$

on the ends of the shaft. The total (tensile) force exerted is

$$T = \int_0^a 2\pi r\sigma_{zz}dr = \tfrac{1}{2}\pi\rho_0\omega^2\nu a^4 \ .$$

On the other hand the ends of the shaft may be traction free (if for example the shaft is located in roller bearings). The traction free condition implies $\sigma_{zz} = 0$. Evidently it is not possible to satisfy this condition precisely with the present simple solution. However it is possible to approximate the end condition by choosing e_{zz} so that the total force exerted on the shaft is zero,† i.e.

$$2\pi\int_0^a \sigma_{zz}rdr = 0 \ .$$

† The solution found is therefore not correct near the ends of the shaft but is presumed satisfactory elsewhere by St. Venant's principle.

Substituting from (14.4.66) for σ_{zz} and evaluating the integrals leads to the value

$$e_{zz} = -\frac{\frac{1}{2}\rho_0 \omega^2 \nu a^2}{E} \qquad (14.4.67)$$

so that the shaft contracts. Presumably contraction is likely to occur even if the shaft is located between thrust bearings since the shaft will tend to pull away from the bearings. With e_{zz} given by (14.4.67), σ_{zz} becomes

$$\sigma_{zz} = \frac{\rho_0 \omega^2 \nu}{4(1 - \nu)} (a^2 - 2r^2) .$$

14.4.6 Torsion of a Cylinder of Arbitrary Cross Section

Figure 14.11 depicts a cylinder of arbitrary uniform cross section and of length l. The periphery of the surface $z = 0$ is presumed held while the face $z = l$, also gripped peripherally, is rotated about the z axis through an angle α. The essential geometrical features of this deformation are described by the assumption of a displacement field of the form

$$u_\theta = rz\alpha/l, \quad u_z = \alpha\phi(x, y)/l \qquad (14.4.68)$$

where r, θ, z are cylindrical polar coordinates. For each cross sectional plane $z =$ const. of the cylinder, the displacement component u_θ represents a twist of the entire cross section through an angle $\alpha z/l$ which varies linearly with z from zero at $z = 0$ to α at $z = l$. The twist/unit length is denoted by

$$\tau = \alpha/l$$

so that (14.4.68) becomes

$$u_\theta = \tau rz, \quad u_z = \tau\phi(x, y) . \qquad (14.4.69)$$

The proposed displacement function u_z is the same for each cross section and this seems intuitively correct for the problem posed.

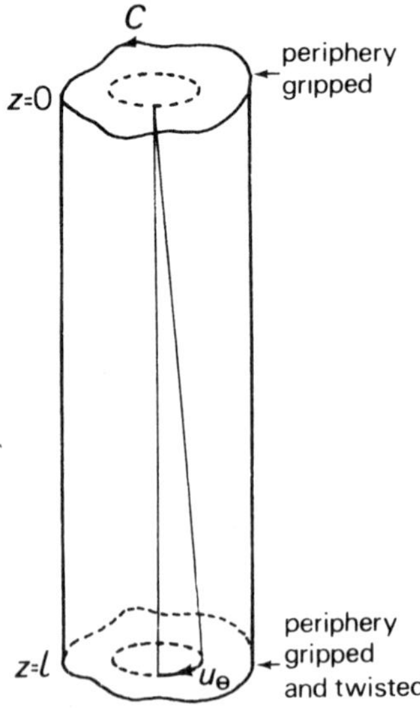

Fig. 14.11 Torsion of a cylinder

It is here convenient to work in terms of Cartesian components of displacement. Resolving u_θ into x and y components, the displacement field (14.4.69) may be written (see Fig. 14.12)

$$u_x = -\tau yz \qquad u_y = \tau xz \qquad u_z = \tau\phi(x, y) .$$

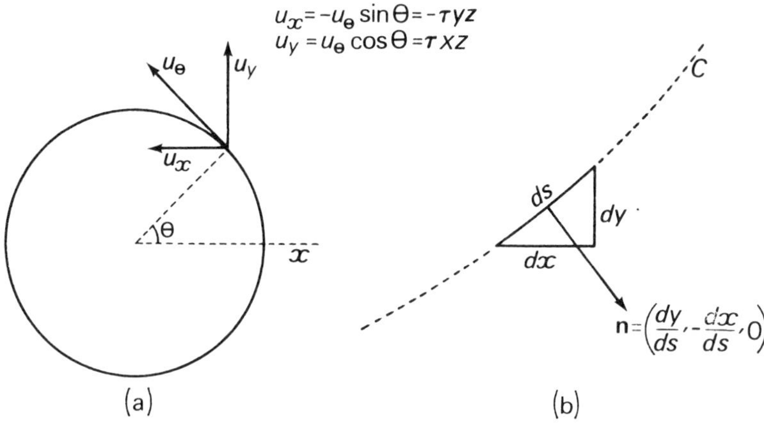

Fig. 14.12 Resolution of tangential displacement u_θ and outward normal unit vector **n** into Cartesian components

The strain components are given by

$$e_{xz} = \frac{1}{2}\tau\left[-y + \frac{\partial\phi}{\partial x}\right], \qquad e_{yz} = \frac{1}{2}\tau\left[x + \frac{\partial\phi}{\partial y}\right],$$

$$e_{xx} = e_{yy} = e_{zz} = e_{xy} = 0$$

so that the only non-vanishing stress components are

$$\sigma_{xz} = \mu\tau\left[-y + \frac{\partial\phi}{\partial x}\right], \qquad \sigma_{yz} = \mu\tau\left[x + \frac{\partial\phi}{\partial y}\right]. \tag{14.4.70}$$

In the absence of body forces the equations of equilibrium in the x and y direction are satisfied identically; in the z direction the equation reduces to

$$\frac{\partial\sigma_{xz}}{\partial x} + \frac{\partial\sigma_{yz}}{\partial y} = 0$$

or from (14.4.70)

$$\frac{\partial^2\phi}{\partial x^2} + \frac{\partial^2\phi}{\partial y^2} = 0 \tag{14.4.71}$$

so that the proposed displacement field is compatible with the equilibrium equation provided ϕ satisfies the two dimensional Laplace equation.

Boundary conditions for ϕ are established from consideration of the traction free nature of the curved surface of the cylinder. On this surface, which except near the ends is traction free, $\sigma_{ij}n_j = 0$ where $\mathbf{n} = (n_1, n_2, 0)$ is the outward normal vector whose z component vanishes. Evidently for $i = 1, 2, \sigma_{ij}n_j = 0$ is automatically satisfied because $\sigma_{11} = \sigma_{12} = \sigma_{22} = 0$. However for $i = 3$

$$\sigma_{3j}n_j = \sigma_{31}n_1 + \sigma_{32}n_2 \equiv \sigma_{xz}n_x + \sigma_{yz}n_y$$

and if this is to vanish on the bounding curve C of the cross section then from (14.4.70)

$$\left(-y + \frac{\partial\phi}{\partial x}\right)n_x + \left(x + \frac{\partial\phi}{\partial y}\right)n_y = 0$$

i.e. $$\mathbf{n}\,.\,\text{grad}\,\phi \equiv \frac{\partial\phi}{\partial n} = yn_x - xn_y \qquad (\text{on } C)$$

The condition $$\frac{\partial\phi}{\partial n} = yn_x - xn_y \tag{14.4.72}$$

is most easily handled by introducing the conjugate function $\psi(x, y)$, defined by the requirement that $\phi + i\psi = f(z)$ where f is a regular function of $z = x + iy$.† The functions ϕ and ψ are related by the Cauchy-Riemman equations

$$\frac{\partial\phi}{\partial x} = \frac{\partial\psi}{\partial y}, \qquad \frac{\partial\phi}{\partial y} = -\frac{\partial\psi}{\partial x}.$$

Also if s is the parameter describing distance along the bounding curve C

$$\frac{\partial\phi}{\partial n} = \frac{\partial\psi}{\partial s}$$

[see p. 274], so that (14.4.72) may be written

$$\frac{d\psi}{ds} = yn_x - xn_y$$

where, since the differentiation is along C, $\partial/\partial s$ has been replaced by d/ds. From Fig. 14.12, $n_y = -dx/ds$, $n_x = dy/ds$, so that

$$\frac{d\psi}{ds} = y\frac{dy}{ds} + x\frac{dx}{ds} = \frac{1}{2}\frac{d}{ds}(x^2 + y^2)$$

i.e. $$\psi = \tfrac{1}{2}(x^2 + y^2) + \psi_1 \qquad (\text{on } C) \tag{14.4.73}$$

where ψ_1 is a constant of integration.

We recall that ψ also satisfies Laplace's equation so that the problem is reduced to finding a function ψ which solves

$$\frac{\partial^2\psi}{\partial x^2} + \frac{\partial^2\psi}{\partial y^2} = 0$$

and is subject to (14.4.73) on the bounding curve C.

† The complex variable z is *not* to be confused with the third Cartesian coordinate.

We introduce a new function

$$\chi = \psi - \tfrac{1}{2}(x^2 + y^2) \qquad (14.4.74)$$

which satisfies the Poisson equation

$$\nabla^2\chi = -2 \qquad (14.4.75)$$

and the boundary condition $\chi = \psi_1$ (on C) . (14.4.76)

It transpires that for a solid cylinder the precise value of ψ_1 is irrelevant and that ψ_1 may be chosen to be zero.

Before considering the solution of (14.4.75) subject to (14.4.76) we discuss the nature of the forces exerted on the surfaces $z = l$ (and $z = 0$). On $z = l$ the non-vanishing components of stress are shear components σ_{zx}, σ_{zy} which lead to a net force

$$\mathbf{F} = \iint_S (\sigma_{zx}\mathbf{i} + \sigma_{zy}\mathbf{j})dxdy$$

where S is the surface bounded by C.

We show that $\mathbf{F} = 0$. To this end it is more convenient to consider the quantity

$$\hat{\mathbf{F}} = \iint_S (\sigma_{zx}\mathbf{j} - \sigma_{zy}\mathbf{i})dxdy \,.$$

Evidently if $\hat{\mathbf{F}}$ vanishes, so does $\mathbf{F}$. From (14.4.70), the Cauchy-Riemman equations and (14.4.74)

$$\hat{\mathbf{F}} = \mu\tau \iint_S (\text{grad}\ \chi)dxdy = \mu\tau \oint_C \chi\mathbf{n}ds \,,$$

the latter result following from an obvious modification of the divergence theorem applied to a two dimensional plane region. However on C, $\chi = \psi_1$, so that

$$\hat{\mathbf{F}} = \mu\tau\psi_1 \oint_C \mathbf{n}ds = \mu\tau\psi_1 \iint_S \text{grad}(1)dxdy = 0$$

and the net forces on $z = 0$, $z = l$ are zero.

However, as might be expected, the shear stresses σ_{zx} and σ_{zy} acting on the surfaces $z = 0$ and $z = l$ give rise to couples of equal magnitude but opposite sign. The magnitude of the couple acting in a counter clockwise sense in the xy plane about the z axis on the surface $z = l$ is given by

$$G = \iint_S (x\sigma_{yz} - y\sigma_{xz})dxdy = \mu\tau \iint_S \left(x^2 + y^2 + x\frac{\partial\phi}{\partial y} - y\frac{\partial\phi}{\partial x}\right)dxdy$$

$$= \mu\tau \iint_S \left(x^2 + y^2 - x\frac{\partial \psi}{\partial x} - y\frac{\partial \psi}{\partial y} \right) dxdy$$

$$= -\mu\tau \iint_S (\mathbf{r} \,.\, \text{grad}\, \chi) dxdy \tag{14.4.77}$$

where
$$\mathbf{r} = x\mathbf{i} + y\mathbf{j} \;.$$

The solution of (14.4.75) subject to (14.4.76) may be expressed in the form $\chi = \chi_0(x, y) + \psi_1$, where χ_0 satisfies

$$\nabla^2 \chi_0 = -2$$

and the boundary condition $\chi_0 = 0 \quad (\text{on } C)$.

The solution for χ_0 is unique since the problem is equivalent to determining an ψ which satisfies $\nabla^2 \psi = 0$ and $\psi = \frac{1}{2}(x^2 + y^2)$ on C; from p. 186-187 the latter problem possesses a unique solution. Therefore χ is determined to within an arbitrary constant of integration ψ_1. However the formula (14.4.77) involves grad χ rather than χ, and therefore is independent of ψ_1.

Choosing $\psi_1 = 0$, so that $\chi = \chi_0(x, y)$, we can simplify slightly (14.4.77). For the two dimensional vector $\mathbf{r} = (x, y, 0)$

$$\mathbf{r} \,.\, \text{grad}\, \chi = \text{div}(\mathbf{r}\chi) - 2\chi \;.$$

Substituting this result into (14.4.77) and using the divergence theorem to evaluate the integral

$$\iint_S (\text{div}(\mathbf{r}\chi) dxdy = \oint_C \chi \mathbf{r} \,.\, \mathbf{n} ds = 0 \;,$$

since $\chi = 0$ on C. In these circumstance

$$G = 2\mu\tau \iint_S \chi dxdy \tag{14.4.78}$$

where χ satisfies the equation $\quad \nabla^2 \chi = -2 \quad$ (14.4.79)

and boundary condition $\quad \chi = 0 \quad \text{on } C$. $\quad$ (14.4.80)

The discussion above is valid only for a solid cylinder. For a hollow cylinder bounded internally by a second contour, the torsion problem is *not* solved by choosing χ to satisfy (14.4.79) together with $\chi = 0$ on both boundaries – nor indeed is (14.4.78) correct in these circumstances. In fact the torsion problem for a hollow cylinder is, in general, considerably more complicated than that for a solid cylinder.

The problem posed by (14.4.79) and (14.4.80) is isomorphic with the problem of Poiseuille flow of a Stokes' viscous fluid along a tube of similar shape. For this problem we found simple solutions for the cases of an elliptical and circular boundary (Section 10.3). Here for the ellipse we note the parallel solution

$$\chi = \frac{a^2 b^2}{a^2 + b^2}\left[1 - \frac{x^2}{a^2} - \frac{y^2}{b^2}\right]$$

which evidently satisfies (14.4.79) and (14.4.80). The couple G is given by

$$G = \frac{2\mu\tau a^2 b^2}{a^2 + b^2}\iint_S \left[1 - \frac{x^2}{a^2} - \frac{y^2}{b^2}\right] dxdy$$

where S is now the region $x^2/a^2 + y^2/b^2 \leqslant 1$. The integral is readily evaluated by changing to coordinates r, θ where $x = ar\cos\theta$, $y = br\sin\theta$ for which the Jacobian is abr. We find

$$G = \frac{2\mu\tau a^3 b^3}{a^2 + b^2}\int_0^1 (1 - r^2) r dr \int_0^{2\pi} d\theta = \frac{\pi\mu\tau a^3 b^3}{a^2 + b^2} \qquad \text{(elliptical cylinder)}$$

from which the corresponding result for the circle ($b = a$) is easily derived

$$G = \tfrac{1}{2}\pi\mu\tau a^4 \qquad \text{(circular cylinder).}$$

The torsional problem also has a simple solution (curiously) for an equilateral triangle – see Problem 13, and somewhat more complicated but closed form solutions are known for a number of other boundary shapes. Some simple shapes, e.g. rectangles and squares, do not admit closed form solutions for χ, which then require an infinite series representation. For complicated boundaries the most practical method of solving (14.4.79) subject to (14.4.80) is by numerical methods.

We turn now to the case of hollow cylinders bounded externally and internally respectively by closed curves C_1 and C_2. The previous theory is valid up to (14.4.75) and requires modification only in that (14.4.76) needs augmenting to read

$$\left.\begin{aligned} \chi &= \psi_1 \quad \text{on} \quad C_1 \\ \chi &= \psi_2 \quad \text{on} \quad C_2 \end{aligned}\right\} \qquad (14.4.81)$$

where ψ_1 and ψ_2 are (different) constants.

However as will become very clear in the example of the torsion of a circular cylindrical shell only one of ψ_1 and ψ_2 may be assigned arbitrarily.

Once χ has been found G is given by (14.4.77), and *not* by (14.4.78) which is only applicable to the case of a solid cylinder.

Consider the case of a cylindrical shell $a \leqslant r \leqslant b$. Evidently by symmetry χ depends only on r and therefore from (14.4.79)

$$\frac{d^2\chi}{dr^2} + \frac{1}{r}\frac{d\chi}{dr} = -2$$

with solution

$$\chi = -\tfrac{1}{2}r^2 + A\log r + B\,. \qquad (14.4.82)$$

There are no grounds at present for rejecting the term $A \log r$ since $r = 0$ is excluded from the range of interest ($a \leqslant r \leqslant b$).

With χ given by (14.4.82), ψ is given by $\psi = A \log r + B$ and the function ϕ, conjugate to ψ, is (apart from an irrelevant constant) $\phi = -A\theta$. Hence the resulting displacement $\tau\phi$ in the z direction contains a multi-valued term which is physically unacceptable. Accordingly A must be chosen to vanish, so that for the torsion of a cylindrical shell

$$\chi = -\tfrac{1}{2}r^2 + B\,. \tag{14.4.83}$$

The constant B can be chosen arbitrarily and does not affect the calculation (14.4.77) of G.

We find from (14.4.77) and (14.4.83)

$$G = \tfrac{1}{2}\pi\mu\tau(b^4 - a^4) \qquad \text{(cylindrical shell).}$$

The above calculation is much over-elaborate for the posed problem. The displacement field is simly $u_r = u_z = 0$, $u_\theta = \tau rz$ from which the calculation of G may be made without recourse to the general theory. The purpose of the present calculation is to show that it is not possible to assign arbitrary values to χ on both C_1 and C_2. χ may be assigned an arbitrary value on C_1 (say) and the value of χ on C_2 is then determined by requirements that ϕ be a single valued function. For arbitrary C_1 and C_2 this is a rather complicated problem and we do not pursue the general matter further.

An example is given by the case where the tube is bounded by two similar ellipses, namely

$$\frac{x^2}{a^2} + \frac{y^2}{b^2} = 1 \quad (C_1), \qquad \frac{x^2}{a^2} + \frac{y^2}{b^2} = k^2 \quad (C_2)$$

where for $k < 1$, C_2 lies inside C_1. It is evident that

$$\chi = -\frac{a^2b^2}{a^2+b^2}\left(\frac{x^2}{a^2} + \frac{y^2}{b^2}\right) \tag{14.4.84}$$

satisfies $\nabla^2\chi = -2$ and the conditions $\chi = \text{const.}$ on C_1 and C_2. In order for (14.4.84) to be an acceptable solution to the problem it must be shown that ϕ is single valued. We have from (14.4.74)

$$\psi = \tfrac{1}{2}(x^2 + y^2) - \frac{a^2b^2}{a^2+b^2}\left(\frac{x^2}{a^2} + \frac{y^2}{b^2}\right) = \frac{\tfrac{1}{2}(a^2 - b^2)}{(a^2 + b^2)}(x^2 - y^2)$$

whose conjugate function is

$$\phi = -\left(\frac{a^2 - b^2}{a^2 + b^2}\right)xy$$

which is single valued. Accordingly (14.4.84) is the solution of the posed problem. From (14.4.77) and (14.4.84) the moment is given by

$$G = \frac{2\mu\tau a^2 b^2}{(a^2 + b^2)} \iint_S \left(\frac{x^2}{a^2} + \frac{y^2}{b^2}\right) dxdy = \frac{\pi\mu\tau a^3 b^3 (1 - k^4)}{(a^2 + b^2)}$$

where S is the area between the ellipses.

In the solution of the torsion problem the couple G is assumed applied through shearing stresses distributed over the flat surfaces $z = 0, z = l$. In many practical applications, as is implied by Fig. 14.11, the couple is applied through circumferential contact on the periphery of the cylinder. The validity of results obtained from the torsion theory at points distant from the ends is supposed justified by appeal to St. Venant's principle.

14.5 Approximate theories in classical elasticity; Bending of beams

The equations of classical elasticity, even though they provide the most tractable model for the behaviour of an elastic solid, have proved formidable for all except the simplest problems. For wide classes of problems of technological importance it is possible to set up approximate theories; the latter make use of simplifications deriving from particular geometric features, characteristic of the class of problems under consideration.

The simplest of the approximate theorics is concerned with the transverse displacements of beams caused by transverse forces. One example is provided by the problem of the deflection of a cantilever beam carrying a weight W (Fig. 1.1). The predominant geometric feature common to most beams used in structural engineering is that the lengths of the beams are much greater than the cross section dimensions. In these circumstances the engineer is less concerned with a detailed knowledge of the displacement and stress fields (which would be difficult to obtain in full, even using numerical solutions obtained with the help of computers), than with a more 'macroscopic' description which takes account of the grosser features of the fields. The theory of the bending of beams provides just such a description. For example, the only displacement that features in the theory is the transverse displacement of the neutral axis (or line of centroids). To the reader it will be readily evident from the common experience of flexing a ruler, that the transverse displacement is of much greater magnitude than the others, which are imperceptible to the naked eye.

In the theory of beams, forces in the beam are described solely by the bending moment $M(x)$ and transverse shear force $Q(x)$, defined subsequently. The bending moment $M(x)$ is *assumed* related to the local radius of curvature of the beam, i.e. the radius of curvature of the neutral axis, by the equation

$$M(x) = EI/R(x) \tag{14.5.1}$$

derived in Section 14.4.2 for the case of constant M and R.† Also with neglect of $(dY/dx)^2$ compared with unity the curvature R^{-1} is given by

$$R^{-1} = d^2 Y/dx^2 \tag{14.5.2}$$

where $Y(x)$ is the transverse deflection of the neutral axis. The basic equation for the theory of beams obtained by eliminating R^{-1} between (14.5.1) and (14.5.2) is

$$M(x) = EI\frac{d^2 Y}{dx^2} \tag{14.5.3}$$

In many engineering applications, beams of variable cross section are employed and then I is also a (known) function of x; equation (14.5.3) is presumed valid in this case – even, for example when I is discontinuous. In the present text only problems invoving constant I are discussed.

In most engineering applications $M(x)$ is derivable from a knowledge of the external forces acting on the beam. Consider for example the problem of a cantilever beam $0 \leqslant x \leqslant l$ carrying a weight W at $x = l$ and **built in** (or **encastré**) at $x = 0$ (Fig. 1.1). The problem is depicted schematically in Fig. 14.13a. Here $M(x)$ may be found from consideration of the rotational equilibrium of the section BC. With neglect of the weight of the beam and on taking moments about the neutral plane at the section x (Fig. 14.13b), there results

$$M(x) = W(l - x)$$

so that (14.5.3) yields the linear differential equation of second degree

$$\frac{d^2 Y}{dx^2} = \frac{W}{EI}(l - x) \tag{14.5.4}$$

with solution $$Y = \frac{W}{EI}\left(\tfrac{1}{2}lx^2 - \tfrac{1}{6}x^3 + Ax + B\right)$$

where A and B are constants of integration. One evident boundary condition for the determination of A and B is $Y(0) = 0$. The second condition is $Y'(0) = 0$, since otherwise the beam suffers a discontinuous change in $Y'(x)$ at $x = 0$ between the built in section $x < 0$ and the exposed part $x > 0$. In turn this leads to an infinite second derivative of delta function character for $Y''(0)$ and a corresponding infinite value for $M(0)$. With $Y(0) = Y'(0) = 0, A = B = 0$ and

$$Y = \frac{W}{EI}\left(\tfrac{1}{2}lx^2 - \tfrac{1}{6}x^3\right).$$

† Most engineering texts derive the result (14.5.1) on the assumption that the beam is supposed composed of a dense set of parallel fibres in each of which the longitudinal stress is assumed to obey Hooke's law. The relation between curvature and local strain in each fibre is derived geometrically assuming that fibres distance ζ from the (unstrained) neutral plane are bent into a circle of radius $R + \zeta$.

In particular at $x = l$, $$Y(l) = \frac{Wl^3}{3EI}$$

so that the maximum deflection varies with the cube of the length of the beam.

Body forces are readily taken into account in this theory. For the problem posed in Fig. 14.13, if w is the weight per unit length of the beam then for the section BC the weight is $w(l - x)$ acting at the centre of gravity of BC. The moment contribution of this at B is $\frac{1}{2}w(l - x)^2$. Accordingly (14.5.4) is now replaced by

$$\frac{d^2Y}{dx^2} = \frac{W}{EI}(l - x) + \frac{\frac{1}{2}w}{EI}(l - x)^2 \tag{14.5.5}$$

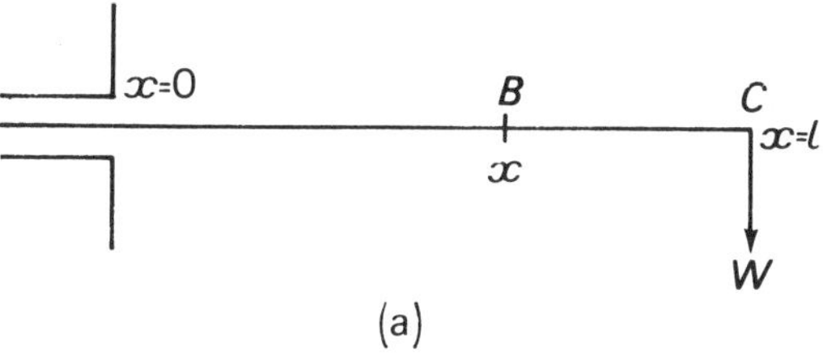

Fig. 14.13 (a) Schematic diagram of cantilever beam supporting weight W

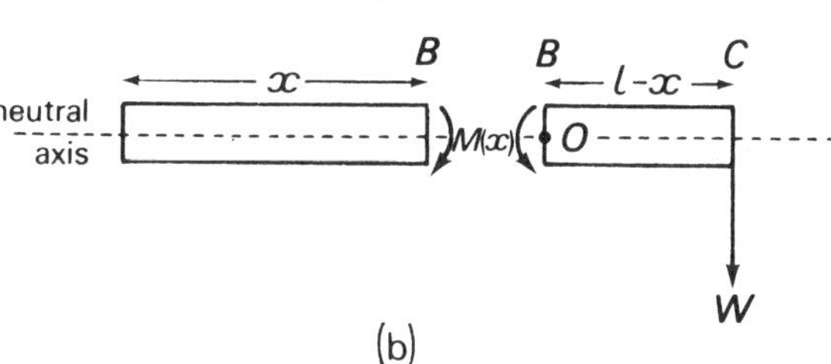

(b) Determination of M from external load. Moments about 0 yields $M - W(l - x) = 0$.

whose solution, subject to $Y(0) = Y'(0) = 0$, is

$$Y(x) = \frac{W}{EI}(\tfrac{1}{2}lx^2 - \tfrac{1}{6}x^3) + \frac{w}{2EI}(\tfrac{1}{2}l^2x^2 - \tfrac{1}{3}lx^3 + \tfrac{1}{12}x^4)$$

A slightly more complicated problem is shown in Fig. 14.14. Here a weightless beam $-L \leqslant x \leqslant L$ is **simply supported** at $x = \pm l$, $(l < L)$, and carries a weight W at $x = 0$. From symmetry and equilibrium arguments the reactions at the points of support are each $\frac{1}{2}W$. For $x > l$, $M(x) = 0$ while for $0 \leqslant x \leqslant l$ $M(x) = -\frac{1}{2}W(l - x)$. Accordingly

$$EI\frac{d^2Y}{dx^2} = \begin{cases} -\frac{1}{2}W(l - x) & (0 \leqslant x \leqslant l) \\ 0 & (L \geqslant x \geqslant l) \end{cases} \tag{14.5.6}$$

One boundary condition is evidently $Y(l) = 0$. Also $Y'(l)$ is continuous since otherwise $M(l)$ would be infinite. Finally from symmetry arguments and the requirement that M be finite at $x = 0$, we have $Y'(0) = 0$. The solution of (14.5.6) satisfying these conditions is

$$EIY(x) = \begin{cases} W[\frac{1}{6}l^3 - \frac{1}{4}lx^2 + \frac{1}{12}x^3] & (0 \leqslant x \leqslant l) \\ -\frac{1}{4}Wl^2(x - l) & (l \leqslant x \leqslant L) \end{cases} \qquad (14.5.7)$$

so that the central deflexion is $\frac{1}{6}Wl^3/EI$ while the edge deflexion is $-\frac{1}{4}Wl^2(L - l)/EI$.

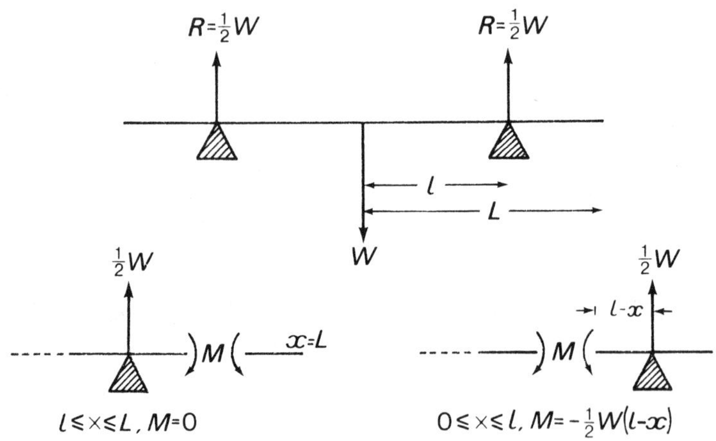

Fig. 14.14 Symmetrically supported weightless beam carrying a central load

For non-symmetrical problems involving freely supported beams the reactions of the supports are not as readily calculable as above. For example in the diagram of Fig. 14.15, which is an unsymmetric version of the problem just discussed, we have from static translational equilibrium

$$R_1 + R_2 = W\,. \qquad (14.5.8)$$

A second equation can be found by taking moments about any convenient point. Choosing the point to be where the load W acts leads to

$$R_1 l_1 = R_2 l_2\,,$$

which together with (14.5.8) determines R_1 and R_2 and hence $M(x)$.

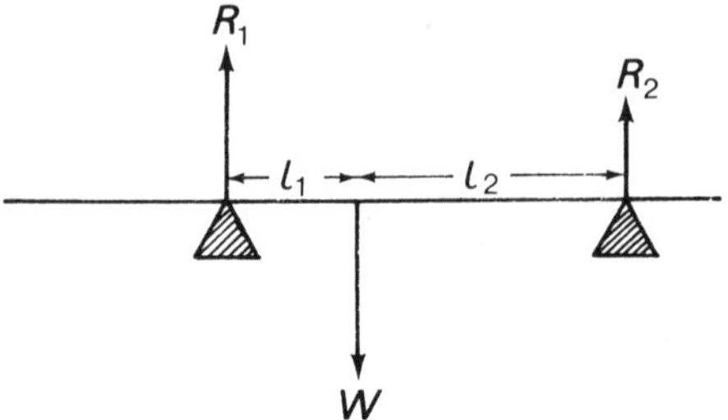

Fig. 14.15 Non-symmetric problem

Sometimes the overall translation and rotation equilibrium equations fail to determine reactions at supports. In these circumstances the problem is said to be statically indeterminate. A seemingly innocuous example is shown in Fig. 14.16 where a cantilever beam rests on a support and carries a weight W.

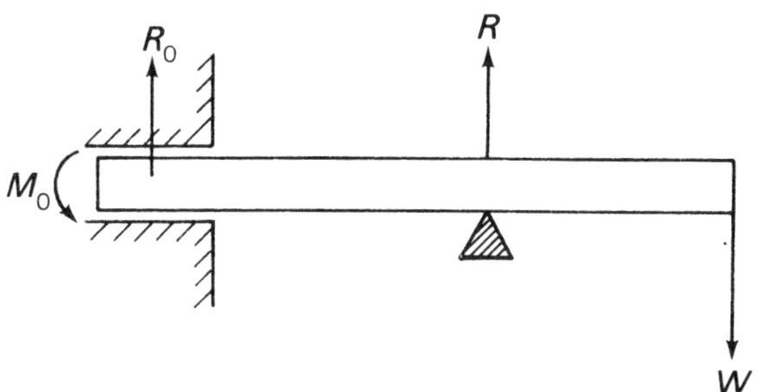

Fig. 14.16 Supported cantilever carrying a load

Here the overall equilibrium equations are inadequate to determine R since we know neither the reaction R_0 nor the bending moment M_0 at the built in end. In these circumstances R must be carried through the calculations to the end and then determined from some other condition. This particular example is discussed further in Problem 21.

Evidently there are many variants on the problems discussed above; some of these are posed as examples at the end of the chapter.

A variable bending moment $M(x)$ is accompanied by a shear force across the beam. For example consider the elementary section of width Δx shown in Fig. 14.17. On one side of this section the bending moment is $M(x + \Delta x)$ while on the other side the bending moment is $M(x)$. The net moment about O is

$$M(x + \Delta x) - M(x) = \frac{dM}{dx}\Delta x$$

where the derivative is evaluated at some point in the interval $[x, x + \Delta x]$. This is balanced by the moment $Q(x + \Delta x)\Delta x$ due to the shear force $Q(x + \Delta x)$ acting across the section at $x + \Delta x$. Since there must be rotational equilibrium of the section of width Δx, as well as of finite sections of the beam,

$$\frac{dM}{dx}\Delta x = -Q(x + \Delta x)\Delta x$$

which after division by Δx and proceeding to the limit $\Delta x \to 0$, yields

$$Q = -\frac{dM}{dx} . \tag{14.5.9}$$

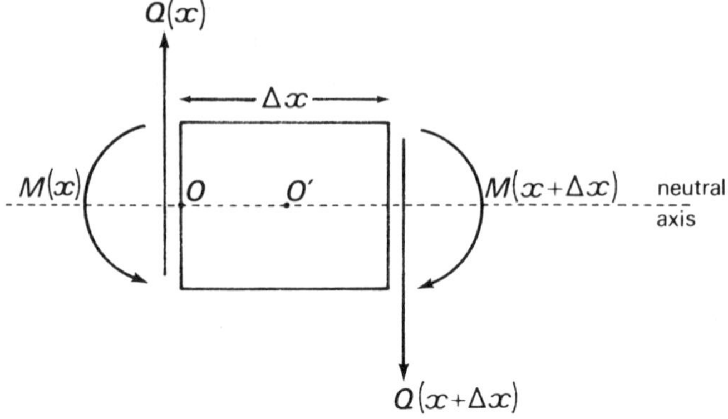

Fig. 14.17 Relation between shear force and bending moment

The shear force Q plays no role in determining deflexions in static beam problems and (14.5.9) serves merely to determine the Q that necessarily arises because of variable $M(x)$. For dynamic problems equation (14.5.9) plays a more basic role in the theory (see Section 14.6.4).

The shear force Q is given by the integral

$$Q = -\iint_S \sigma_{xy}\,dxdy$$

where S is the area of cross section. However, again, this relation plays no role whatsoever in the approximate bending theory. For although for non-zero Q the equation carries the implication of non-vanishing components σ_{xy} acting over at least part of the cross section, this is not taken into account in the derivation of (14.5.3).

The theory of beams is paralleled by a somewhat similar theory of plates. In the latter theory the simplifying geometrical characteristic is that the thickness of the plate is small compared with the lateral dimensions. While the theory of beams reduces elasticity to a one dimensional filament model and ordinary differential equations, the theory of plates is necessarily two dimensional and leads to partial differential equations in two variables. The theory of plates, and more generally the theory of curved shells, is not discussed in the present text.

14.6 Elastodynamics

14.6.1 Basic Equations and General Solutions

In the absence of body forces the equations of classical **elastodynamics**, may be taken in the form (14.4.39) for the displacement field $\mathbf{u}$

$$(\lambda + \mu)\operatorname{grad}(\operatorname{div}\mathbf{u}) + \mu\nabla^2\mathbf{u} = \rho_0\frac{\partial^2\mathbf{u}}{\partial t^2} \tag{14.6.1}$$

together with the constitutive relations (14.2.2) and the strain-displacement gradient equations (14.2.3).

Neglect of body forces is not a serious limitation for usually body forces are time independent as in the common case of the earth's gravitational field. In these circumstances it is possible to decompose the displacement field into a time independent portion which satisfies the equations of elastostatics (including the body force term) and a time dependent portion which satisfies (14.6.1). This decomposition of the displacement field is possible because the basic equations are linear. The time dependent solution then represents transient disturbances about a statically deformed configuration. A general solution of (14.6.1) may be obtained as follows. We write

$$\mathbf{u} = \operatorname{grad} \phi + \operatorname{curl} \mathbf{A} \tag{14.6.2}$$

where ϕ and $\mathbf{A}$ are respectively a scalar and vector potential.†

Substituting (14.6.2) into (14.6.1) and re-arranging leads to

$$\operatorname{grad}\left\{(\lambda + 2\mu)\nabla^2\phi - \rho_0\frac{\partial^2\phi}{\partial t^2}\right\} + \operatorname{curl}\left\{\mu\nabla^2\mathbf{A} - \rho_0\frac{\partial^2\mathbf{A}}{\partial t^2}\right\} = 0 \tag{14.6.3}$$

so that (14.6.1) is satisfied provided ϕ and $\mathbf{A}$ are respectively solutions of the wave equations

$$\nabla^2\phi - \frac{1}{c_1^2}\frac{\partial^2\phi}{\partial t^2} = 0, \text{ (14.6.4)}; \qquad \nabla^2\mathbf{A} - \frac{1}{c_2^2}\frac{\partial^2\mathbf{A}}{\partial t^2} = 0, \text{ (14.6.5)}$$

where the two wave speeds c_1 and c_2 are given by the formulae

$$c_1^2 = (\lambda + 2\mu)/\rho_0, \qquad c_2^2 = \mu/\rho_0 \, . \tag{14.6.6}$$

Evidently c_1 is larger than c_2; in terms of Young's modulus and Poisson's ratio the formulae (14.6.6) become

$$c_1 = \left[\frac{(1-\nu)}{(1+\nu)(1-2\nu)}\frac{E}{\rho_0}\right]^{\frac{1}{2}}, \qquad c_2 = \left[\frac{E}{2(1+\nu)\rho_0}\right]^{\frac{1}{2}} .$$

Typically for steel, $E = 2.10^5$ N/mm^2, $\rho_0 = 8.10^3$ K/m^3, $\nu = 0.3$ and these figures lead to

$$c_1 = 5.8 \times 10^3 \text{ m/s}, \quad c_2 = 3.1 \times 10^3 \text{ m/s}$$

so that wave propagation speeds in metals are an order of magnitude larger than the speed of sound in air at normal temperature and pressure.

† All continuous vector fields **u** may be decomposed in the manner indicated. Sometimes grad ϕ is termed the **irrotational** component and curl **A** the **solenoidal** component. The term irrotational has been introduced previously (Chapter 11); the description solenoidal stems from the theory of electromagnetism.

For a proof of (14.6.2) see A. Sommerfields *Mechanics of Deformable Bodies,* p. 147 Academic Press (1964).

In elastodynamics displacement fields of the type

$$\mathbf{u} = \operatorname{grad} \phi$$

are usually known as **dilatational**, indicating that such fields necessarily entail (time dependent) volume changes. Displacement fields for which

$$\mathbf{u} = \operatorname{curl} \mathbf{A}$$

are sometimes known as **equivoluminal**, since for these fields div $\mathbf{u} = 0$ and there are no volume changes.

Evidently on taking respectively the gradient and curl of (14.6.4) and (14.6.5), dilatational and equivoluminal displacement fields themselves directly satisfy vector wave equations respectively with wave speeds c_1 and c_2.

Simple solutions of (14.6.4) are of the type discussed in Chapter 13. For example for $\phi = \phi(x, t)$ independent of y and z, (14.6.4) reduces to

$$\frac{\partial^2 \phi}{\partial x^2} - \frac{1}{c_1^2}\frac{\partial^2 \phi}{\partial t^2} = 0$$

with general solution $\phi = f(t - x/c_1) + g(t + x/c_1)$ (14.6.7)

where f and g represent respectively waves propagating in the positive and negative x directions. For these waves, the displacement $\mathbf{u} = \operatorname{grad} \phi$ possesses non-zero components only in the x direction, that is in the direction of propagation.

In contrast simple solutions for equivoluminal waves possess displacement components which are perpendicular to the direction of propagation. For, consider solutions of (14.6.5) of the form (14.6.7) but where c_2 replaces c_1 and where f and g possess a vector character, as is necessary since (14.6.5) is a vector wave equation. There is no point in considering vectors **A** which lie in the x direction since solutions of (14.6.5) of the form

$$\mathbf{A} = [f(t - x/c_2) + g(t + x/c_2)]\,\mathbf{i}$$

result in curl $\mathbf{A} = 0$. However if we write

$$\mathbf{A} = [f(t - x/c_2) + g(t + x/c_2)]\,\mathbf{j}$$

then **u** is given by

$$\mathbf{u} = \operatorname{curl} \mathbf{A} = -\frac{1}{c_2}[f'(t - x/c_2) - g'(t + x/c_2)]\,\mathbf{k}$$

which lies at right angles to the direction of propagation.

For self-evident reasons simple solutions of (14.6.5) for which the displacement field is perpendicular to the direction of propagation are known as **shear** waves. Shear wave propagation in elastic solids is without parallel in ideal fluids, considered in Chapter 13, which are unable to sustain shear stresses.

Some of the simpler problems in dilatational wave propagation parallel closely those discussed in Chapter 13 and are not discussed here in detail. Plane dilatational waves may be realised experimentally (approximately) by detonating a sheet of explosive laid in contact with one plane face of a rectangular metallic block – though in the immediate vicinity of the explosion the stress levels are so high that the material does not behave elastically but deforms plastically and also fractures.

Shear waves are most simply generated in dynamic torsion experiments on circular cylinders; the description of these waves is undertaken in Section 14.6.2.

The reflection of a plane dilatational or shear wave at non-normal incidence by a traction free plane boundary is more complicated than in the case of acoustic waves. For example an incident plane dilatational wave results in both a reflected dilatational wave and a shear wave. The algebra is rather complicated (see Problems 28-31); the more general (and more complicated) problems of the reflection and transmission of plane waves by a plane boundary between two elastic media is discussed by H. Kolsky in *Stress waves in solids,* Oxford University Press (1954).

The theory of elastic waves is used in geological prospecting. Elastic waves are generated at the surface of the earth's crust by means of small explosions. Receiver devices (seismographs) also located on the surface detect direct disturbances from the explosion together with reflected waves generated by the presence of (say) cavities filled with oil.

The analysis of seismographic records has also enabled the pin pointing of the location of earthquake centres and nuclear explosions. In these practical applications of elastic wave theory it is necessary to analyse data recorded simultaneously by a number of seismographs located in different positions.

It has been inferred from experimental observations of wave motions in the earth's crust that the earth comprises an outer solid crust of several miles thickness together with an inner zone whose behaviour is that of a liquid. The inference is based on evidence from wave propagation experiments which suggests the existence of an inner zone through which it is impossible to propagate shear waves. The boundary between the two regions is known as the **Mohovoric discontinuity**; there have been proposals, not yet carried out, to bore a **Mohole** in the earth's crust to the depths of the discontinuity.

In the theory of elastic waves there arises the problem discussed briefly in the context of fluids in Section 7.1. For very rapidly varying motions in elastic solids the elastic moduli which should be used in the basic equations are the adiabatic rather than isothermal values. (The latter are measured in static experiments.) However in many elastodynamic situations, particularly for metals at normal ambient temperatures, there is no significant difference between the isothermal and adiabatic elastic constants. In these circumstances the (small-strain)† elastodynamic behaviour is that of a barotropic material, independent of temperature.

With the exception of problems which are closely paralleled by those discussed in Chapter 13, and are the subject of some of the examples at the end of this chapter, most elastic wave problems involve lengthy algebra and advanced

† Metals cease to behave elastically for large strains – see Chapter 16.

analysis. In the remaining subsections we discuss only some of the simpler problems. The generation of shear waves in dynamic torsion experiments is studied in Section 14.6.2, the theory of Rayleigh waves in Section 14.6.3, and the approximate theories of longitudinal waves and transverse waves in bars of uniform cross section in Section 14.6.4.

14.6.2 Torsional Waves

Let r, θ, z be cylindrical polar coordinates and consider the displacement field

$$\mathbf{u} = rF(t, z)\mathbf{i}_\theta \tag{14.6.8}$$

where $\mathbf{i}_\theta$ is a unit vector in the θ direction. In Cartesian coordinates (14.6.8) may be written

$$u_x = -yF(t, z), \quad u_y = xF(t, z), \quad u_z = 0$$

so that $\operatorname{div} \mathbf{u} = 0$ and (14.6.8) represents an equivoluminal disturbance. The equation (14.6.1) reduces to

$$\nabla^2 \mathbf{u} - \frac{1}{c_2^2}\frac{\partial^2 \mathbf{u}}{\partial t^2} = 0 \tag{14.6.9}$$

where c_2 is given by the second of (14.6.6). Recalling in cylindrical polar coordinates the result

$$\nabla^2 = \frac{\partial^2}{\partial r^2} + \frac{1}{r}\frac{\partial}{\partial r} + \frac{\partial^2}{\partial z^2} + \frac{1}{r^2}\frac{\partial^2}{\partial \theta^2}$$

and remembering that $\mathbf{i}_\theta$ is a function of θ which satisfies

$$\frac{d^2}{d\theta^2}\mathbf{i}_\theta = -\mathbf{i}_\theta$$

we find that (14.6.8) solves (14.6.9) provided F satisfies the one dimensional wave equation

$$\frac{\partial^2 F}{\partial z^2} - \frac{1}{c_2^2}\frac{\partial^2 F}{\partial t^2} = 0$$

with solution $$F = f(t - z/c_2) + g(t + z/c_2)$$

where f and g are arbitrary functions of the indicated variables $(t \pm z/c_2)$.

The solution $$\mathbf{u} = rf(t - z/c_2)\mathbf{i}_\theta \tag{14.6.10}$$

is pertinent to the following problem. A long cylinder $z > 0$ of uniform circular cross section $0 \leqslant r \leqslant a$ and with axis in the z direction is subject at $z = 0$ to a time-dependent twist through an angle $\alpha = f(t)$.

The reader is left to verify that the surface $r = a$ is traction free and that, referred to cylindrical polar coordinates, the only non-vanishing stress component $\sigma_{z\theta}$ is given by

$$\sigma_{z\theta} = -\frac{\mu r}{c_2} f'(t - z/c_2)$$

so that the couple exerted across the section $z = 0$ is

$$G = \frac{\mu f'(t)}{c_2} \int_0^{2\pi} d\theta \int_0^a r^3 dr = \frac{\pi \mu a^4}{2c_2} f'(t) . \qquad (14.6.11)$$

Thus if G is known as a function of time, the last equation can be integrated to provide $f(t) (= \alpha)$.

For a cylinder $0 \leqslant z \leqslant l$ of finite length l, the solution (14.6.10) persists only for a finite time. If the motion at $z = 0$ is initiated at $t = 0$, so that $f(t) = 0$ for $t < 0$, the solution (14.6.10) shows that the material is undisturbed for values of z satisfying $z > c_2 t$. At $t = l/c$ the disturbance reaches $z = l$ whereupon reflections occur. For times $t > l/c_2$ the solution involves waves travelling in both directions. The precise nature of the solution depends now on the boundary conditions obtaining at $z = l$ as well as at $z = 0$. Problems of this type are discussed in a longitudinal wave context in Section 14.6.4.

In practical experiments on torsional waves usually the couple is applied rather differently, and St. Venant's principle is invoked. The end of the cylinder is placed in a collar which grips the bar periphery. As shown in Fig. 14.18, flanges attached to the collar are struck by pendulums or bullets, or are subject to explosive impact.

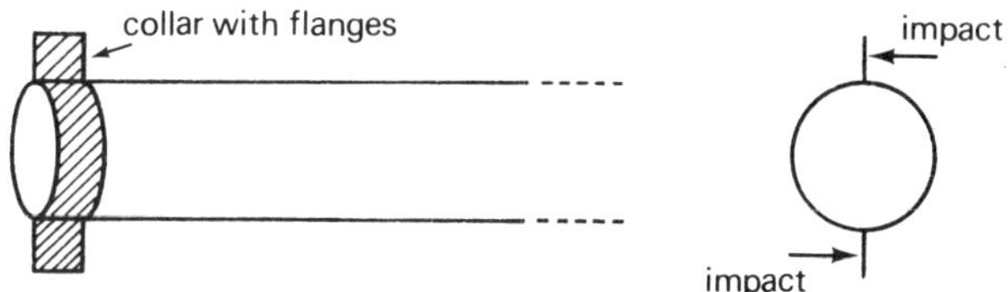

Fig. 14.18 Experimental generation of torsional waves

14.6.3 Rayleigh Waves

Earthquake tremors originate mostly from **fault** regions, e.g. the San Andreas fault in California, where from time to time sudden relative slippage of rock occurs, releasing energy into the surrounding earth's crust. In travelling over the earth's surface the resulting waves are characterised by large surface disturbances which decay with depth, and a well defined velocity of propagation which is somewhat less than c_2. An important solution of the equations of elastodynamics in this context is the two dimensional displacement field discussed below, which represents a steady state oscillatory motion with circular frequency ω. Initially we assume $\omega > 0$. The field in question is characterised by a scalar potential

$$\phi = P(\omega) e^{i\omega[t - x/c(\omega)] - \omega\gamma_1(\omega) z} \qquad (14.6.12)$$

and a vector potential

$$\mathbf{A} = \mathrm{j} Q(\omega) e^{i\omega[t - x/c(\omega)] - \omega\gamma_2(\omega) z} . \qquad (14.6.13)$$

The expressions (14.6.12) and (14.6.13) are separation of variables solutions of (14.6.4) and (14.6.5) chosen to have exactly the same periodic dependence on x and t. Here $P(\omega)$, $Q(\omega)$ are amplitudes to be determined from boundary and initial conditions, $\omega\gamma_1$ and $\omega\gamma_2$ are attenuation factors,† the earth is represented by the half space $z \geqslant 0$ with $z = 0$ a free surface, while finally $c(\omega)$ is the **phase velocity**. The terminology derives from the result that along the straight lines $x = c(\omega)t + \text{const.}$ (for which $dx/dt = c$) the phases $\omega[t - x/c(\omega)]$ appearing in (14.6.12), (14.6.13) are constant and equations (14.6.12), (14.6.13) represent disturbances which are periodic in x but appear to propagate in the positive x direction with speed $c(\omega)$: see also the discussion on p. 356.

The solutions (14.6.12) and (14.6.13) are complex and cannot represent real disturbances; we shall see subsequently how real disturbances are synthesised out of solutions of the type indicated, using all real values of ω.

In the analysis below it is assumed in the first instance that c depends on ω; in fact it transpires that for the present problem c is independent of ω and in these circumstances c represents a real speed of transient wave packages like those depicted in Fig. 13.1.††

Equation (14.6.12) is a solution of (14.6.4) provided

$$\gamma_1^2 = \frac{1}{c^2} - \frac{1}{c_1^2}. \tag{14.6.14}$$

Similarly (14.6.13) satisfies (14.6.5) provided

$$\gamma_2^2 = \frac{1}{c^2} - \frac{1}{c_2^2}. \tag{14.6.15}$$

Since $c_1 > c_2$, (14.6.15) admits real (positive) values of γ_2 and (14.6.14) admits real (positive) values of γ_1 if it can be shown that $c < c_2$. It will be seen subsequently that there exists a value of c which satisfies this inequality.

From (14.6.12) and (14.6.13) the displacement $\mathbf{u} = \operatorname{grad}\phi + \operatorname{curl}\mathbf{A}$ is given by

$$\mathbf{u} = e^{i\omega(t - x/c)}\left[\left(-\frac{i\omega}{c}, 0, -\omega\gamma_1\right)Pe^{-\omega\gamma_1 z} + \left(\omega\gamma_2, 0, -\frac{i\omega}{c}\right)Qe^{-\omega\gamma_2 z}\right]. \tag{14.6.16}$$

We now choose P, Q and c so that the boundary conditions

$$\sigma_{zz} = \sigma_{zx} = \sigma_{zy} = 0$$

are met on the free surface $z = 0$. Since u_z is independent of y and u_y vanishes, the condition

$$\sigma_{zy}\left[\equiv \mu\left(\frac{\partial u_y}{\partial z} + \frac{\partial u_z}{\partial y}\right)\right] = 0$$

† ω is inserted here for subsequent convenience; also we assume γ_1 and γ_2 positive.

†† When c depends on ω, $c(\omega)$ does *not* represent any physical speed but only the apparent speed of an infinite train of periodic disturbances.

is met automatically. The condition $\sigma_{zx} = 0$ on $z = 0$ may be written

$$\left(\frac{\partial u_z}{\partial x} + \frac{\partial u_x}{\partial z}\right)_{z=0} = 0$$

or from (14.6.16)

$$\frac{2i\gamma_1}{c} P - \left(\gamma_2{}^2 + \frac{1}{c^2}\right) Q = 0 \tag{14.6.17}$$

The condition $\sigma_{zz} = 0$ on $z = 0$ may be written

$$\left[\lambda(\mathrm{div}\,\mathbf{u}) + 2\mu \frac{\partial u_z}{\partial z}\right]_{z=0} = 0$$

or since $\mathrm{div}\,\mathbf{u} = \nabla^2\phi = \dfrac{1}{c_1{}^2}\dfrac{\partial^2\phi}{\partial t^2}$,

$$\left[\frac{\lambda}{c_1{}^2}\frac{\partial^2\phi}{\partial t^2} + 2\mu\frac{\partial u_z}{\partial z}\right]_{z=0} = 0\,.$$

From (14.6.12) and (14.6.16) this condition becomes

$$\left(2\mu\gamma_1{}^2 - \frac{\lambda}{c_1{}^2}\right) P + \frac{2i\mu\gamma_2}{c} Q = 0\,. \tag{14.6.18}$$

Equations (14.6.17) and (14.6.18) are consistent provided c satisfies

$$-\frac{4\mu\gamma_1\gamma_2}{c^2} + \left(\gamma_2{}^2 + \frac{1}{c^2}\right)\left(2\mu\gamma_1{}^2 - \frac{\lambda}{c_1{}^2}\right) = 0\,. \tag{14.6.19}$$

With c determined either of (14.6.17) or (14.6.18) then determines P/Q.

On recalling the definitions of $c_1{}^2, c_2{}^2, \gamma_1{}^2$ and $\gamma_2{}^2$ equation (14.6.19) may be put in the form

$$4(1 - \xi^2)^{\frac{1}{2}}(1 - k\xi^2)^{\frac{1}{2}} = (2 - \xi^2)^2 \tag{14.6.20}$$

where

$$k = \frac{(1 - 2\nu)}{2(1 - \nu)} \equiv \frac{c_2{}^2}{c_1{}^2}\,.$$

and

$$\xi = \frac{c}{c_2}\,. \tag{14.6.21}$$

For all values of ν in the range $0 \leqslant \nu \leqslant \frac{1}{2}$, it may be shown that (14.6.20) possesses one real root for which $\xi^2 < 1$, i.e. for which $c < c_2$.† Other roots are real or complex, depending on the value of ν, and lead to solutions physically unacceptable for a variety of reasons, e.g. solutions with pure imaginary values of γ_1 so that the disturbance does not decay with depth, or worse, solutions which grow exponentially with depth.

We consider the case of an incompressible material ($\nu = \frac{1}{2}$) for which $k = 0$. On substituting $k = 0$ into (14.6.20), squaring each side and re-arranging there results

$$\xi^2\{\xi^6 - 8\xi^4 + 24\xi^2 - 16\} = 0\,.$$

† See Problem 26.

The root $\xi = 0$ is of no physical significance; the expression in parentheses is a cubic ξ^2 with one real root

$$\xi^2 = 0.91262$$

and two (irrelevant) complex roots. For an incompressible material therefore

$$c = (0.91262)^{\frac{1}{2}} c_2 = 0.9553 c_2 \;.$$

Associated with this value of c we have from (14.6.14), $\gamma_1 = c^{-1}$, since for the incompressible case $c_1 = \infty$, and from (14.6.15), $\gamma_2 = 0.2956/c$.

Since the ratio P/Q is determined either by (14.6.17) or (14.6.18), only one of P and Q in (14.6.16) is unknown so that the solution found could be written

$$\mathbf{u} = P(\omega)e^{i\omega(t - x/c)}\left[\left(-\frac{i\omega}{c}, 0, -\omega\gamma_1\right) e^{-\omega\gamma_1 z} + h\left(\omega\gamma_2, 0, -\frac{i\omega}{c}\right) e^{-\omega\gamma_2 z}\right] \tag{14.6.22}$$

where h ($\equiv Q/P$) is a known constant [in fact from (14.6.17) h is independent of ω and pure imaginary]. The expression in square parentheses in (14.6.22) is now to be regarded as an eigenfunction with continuous eigenvalue ω so that more general solutions for Rayleigh waves may be synthesised from (14.6.22) by multiplying by $d\omega$ and integrating over the entire range of continuous eigenvalues $-\infty \leqslant \omega \leqslant \infty$.† In the resulting expression $P(\omega)$ is an arbitrary amplitude. These ideas, exploited previously in Section (10.5), provide an important tool in the solution of linear partial differential equations. There results for example for u_z on $z = 0$

$$(u_z)_{z=0} = \int_{-\infty}^{\infty} U(\omega)e^{i\omega(t - x/c)}d\omega \tag{14.6.23}$$

where for $\omega > 0$, $U(\omega) = P(\omega)[-\omega\gamma_1 - ih\omega/c]$. On substituting for $h(\equiv Q/P)$ from (14.6.17) and taking account of the modifications necessary for $\omega < 0$, † there results for all ω

$$U(\omega) = |\omega| P(\omega)\gamma_1(1 - \gamma_2{}^2 c^2)/(1 + \gamma_2{}^2 c^2)$$

where γ_1 is the positive root of (14.6.14).

If now, for example, the profile of the travelling pulse $(u_z)_{z=0}$ is known equation (14.6.23) in conjunction with Fourier's integral theorem determines $U(\omega)$ and hence $P(\omega)$. With P found, the entire solution is determined in terms of Fourier integrals. We do not pursue the matter in detail.

† For $\omega < 0$, in order to ensure an exponential decay with depth, it is necessary to choose the negative roots for γ_1 and γ_2 in (14.6.14) and (14.6.15); also then, from (14.6.17) the imaginary quantity h changes sign. Equivalently, if the original positive definitions for γ_1 and γ_2 are retained, the eigenfunction for $\omega < 0$ is obtained by replacing h by $-h$ in (14.6.22) and also replacing the ω appearing in the combinations $\omega\gamma_1$ and $\omega\gamma_2$ by $|\omega|$.

The Fourier integral solution justifies the earlier statement that because the phase velocity c is independent of ω, then c is also the actual speed of wave propagation of transient waves. To prove this it is sufficient to note that the integral of the right side of (14.6.22) with respect to ω represents a function of two variables, namely z and the combined variable $(t - x/c)$. As has already been demonstrated in discussion of the wave equation (Chapter 13), a function of $(t - x/c)$ represents a propagating transient disturbance which moves in the positive x direction with speed c. This interpretation is not disturbed by the presence of the other variable z.

The situation is somewhat different in situations where the phase velocity $c(\omega)$ varies with ω. In this case it is no longer possible to state that an integral of the form (14.6.23) (say) is solely a function of $(t - x/c)$, and while indeed the integral represents a travelling pulse, the pulse in question varies its profile usually by continuously spreading out and with a continuously diminishing maximum amplitude. This phenomena is called *phase dispersion* and is encountered in a number of elastodynamic (and other) situations. One example is discussed in Problem 35. Fig. 14.19 displays qualitatively the usual features of a dispersing pulse. These features are a lengthening of the pulse accompanied by a decrease in amplitude. It is not possible to give a more precise account in the absence of specific knowledge of the phase velocity $c(\omega)$ and the Fourier spectrum $U(\omega)$ of the pulse at $x = 0$.

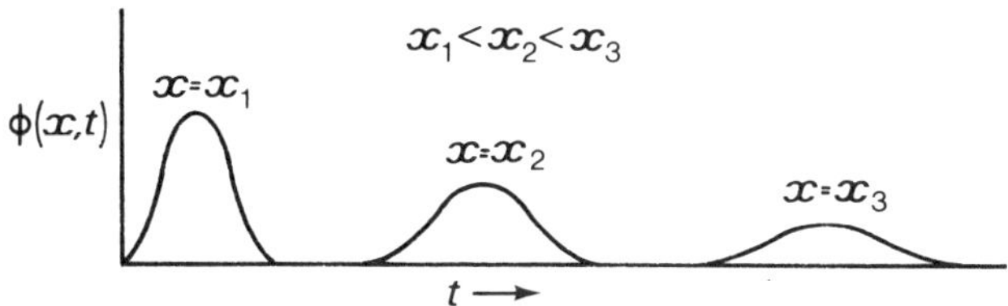

Fig. 14.19 Schematic diagram exhibiting phase dispersion for a pulse

$$\phi(x, t) = \int_{-\infty}^{\infty} U(\omega) e^{i\omega[t - x/c(\omega)]} d\omega \quad \text{where } c \text{ depends on } \omega.$$

Phase dispersion occurs commonly in other branches of the physical sciences and the evaluation of integrals of the form (14.6.23) is an important topic. An approximate method of evaluation leading to the concept of **group velocity** is due to Kelvin and is discussed in Problem 36.

Phase dispersion, accompanied also by attenuation in the direction of propagation, is a feature of longitudinal viscoelastic waves discussed in Chapter 15.

14.6.4 Approximate Elastodynamic Theories

In Section 14.5 we discussed the advantages of approximate elastostatic theories for bodies characterised by particular geometric features. A similar situation prevails in an elastodynamic context; for rods, plates and shells it is possible to set up approximate elastodynamic equations. Here we examine only problems of the longitudinal and transverse vibrations of rods.

Fig. 14.20 depicts a straight rod of uniform cross section subject to longitudinal motions, supposed described by the assumption of an axial displacement $u_x = u(x, t)$. The longitudinal strain e_{xx} is given by the usual formula

$$e_{xx} = \partial u/\partial x \,. \tag{14.6.24}$$

In the absence of motion and body forces e_{xx} is constant and for a rod subject to uniform tension σ_{xx}, the relation between longitudinal stress and longitudinal strain is that of the elementary Hooke's law

$$\sigma_{xx} = E e_{xx} \tag{14.6.25}$$

while all other stress components vanish. Again in the static case the lateral strains are related to e_{xx} by

$$e_{yy} = e_{zz} = -\nu e_{xx} \tag{14.6.26}$$

and the shear strains e_{xy}, e_{xz} and e_{yz} vanish – see Section 14.3.

In the simplest theory of longitudinal vibrations in rods, σ_{xx} is assumed still to be the only non-vanishing stress component and equations (14.6.25) and (14.6.26) are presumed to hold even though the behaviour is dynamic and e_{xx} and σ_{xx} vary with x as well as t. In fact the assumptions are approximate as indeed is the assumption that u depends only on x and t. In a more refined analysis account is taken of the warping of the section originally located at x and the displacement u depends on the other space variables; also displacements in the transverse directions enter the theory together with the stresses σ_{yy}, σ_{zz}, σ_{yz}, σ_{xz} and σ_{xy} which do not everywhere vanish.

With the assumption that σ_{xx} is the only non-vanishing stress component the equation of motion in the x direction is

$$\frac{\partial \sigma_{xx}}{\partial x} = \rho_0 \frac{\partial^2 u}{\partial t^2} \,. \tag{14.6.27}$$

Eliminating e_{xx} and σ_{xx} from (14.6.24), (14.6.25) and (14.6.27) leads to the wave equation

$$\frac{\partial^2 u}{\partial x^2} = \frac{1}{c^2}\frac{\partial^2 u}{\partial t^2} \tag{14.6.28}$$

where the propagation speed c is given by

$$c = (E/\rho_0)^{\frac{1}{2}} \,.$$

The relations (14.6.26) do not enter the basic theory and serve only to determine e_{yy} and e_{zz} once $u(x, t)$ has been found.

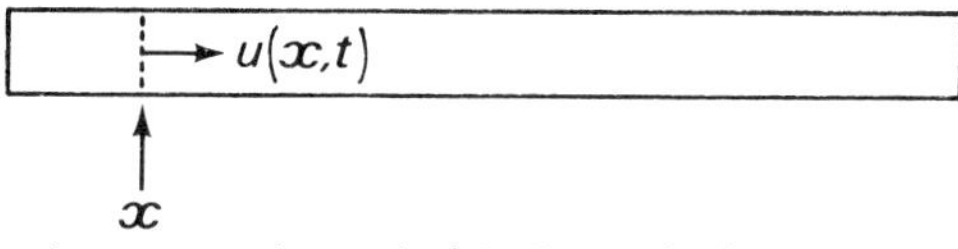

Fig. 14.20 Longitudinal motion of rod of uniform cross section

General solutions of (14.6.28) are

$$u = f(t - x/c) + g(t + x/c) \tag{14.6.29}$$

where f and g represent transient waves propagating respectively in the $\pm x$ directions at a speed c.

Since the theory is approximate the question arises as to whether it is realistic.† This question can only be answered by reference to more exact theories (see Problem 35) which show that (14.6.29) is an adequate representation of the longitudinal displacement provided that physical quantities like the displacement u are 'sufficiently smooth'. More precisely there should be no rapid variations of u for variations in x of order a where a is a length characterising the cross sectional dimensions of the rod, for example the radius of a circular rod.

We consider an example based on the elementary theory. A wire is suspended vertically and carries a light tray at $x = 0$. A mass M is allowed to travel along the wire and is caught by the tray, impacting the end $x = 0$ at a speed v_0 (Fig. 14.21). The experimental arrangement, with of course a wire of finite length, was used by Bertram Hopkinson in the late nineteenth century in studies on dynamic fracture.

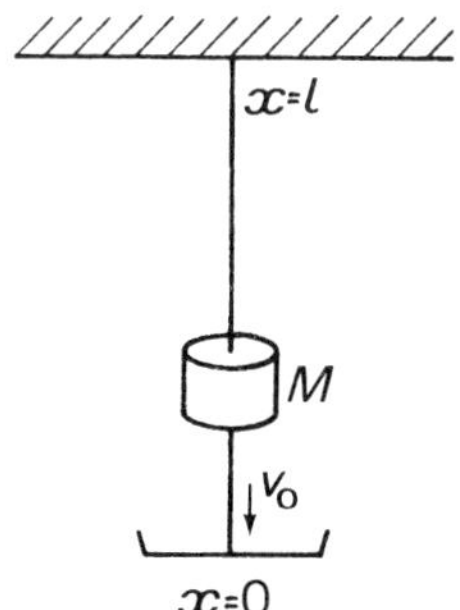

Fig. 14.21 Hopkinson's fracture experiment

† The theory is approximate since because of the Poisson ratio effect the transverse acceleration components are non-zero. However since it is also assumed that $\sigma_{yy} = \sigma_{yx} = \sigma_{yz} = 0$, the equation of motion in the y direction (and also in the z direction) is not satisfied.

Suppose the wire is of length l. From the moment of impact, say $t = 0$, until $t = l/c$ only outgoing waves are generated at $x = 0$ so that the displacement in the wire is given by

$$u = f(t - x/c) .$$

It is required to determine f. The stress is

$$\sigma_{xx} = E(\partial u/\partial x) = -\rho_0 c f'(t - x/c), \tag{14.6.30}$$

on remembering $E = \rho_0 c^2$, so that in particular at $x = 0$

$$\sigma_{xx} = -\rho_0 c f'(t)$$

and the total force acting on the mass is

$$F = -\rho_0 c A f'(t)$$

where A is the cross-sectional area of the wire. The displacement of the mass is that of the end of the wire, $f(t)$, so that the acceleration of M is $f''(t)$. Accordingly the equation of motion of M, ignoring gravity, is

$$Mf''(t) = -\rho_0 c A f'(t)$$

with first integral

$$f'(t) = Be^{-\rho_0 cAt/M}$$

where B is an arbitrary constant. However since $f'(t)$ is the velocity of M with initial value $-v_0$, $B = -v_0$ and

$$f'(t) = -v_0 e^{-\rho_0 cAt/M} . \tag{14.6.31}$$

A further integration, subject to $f(0) = 0$ (since the wire is undisturbed for $t \leqslant 0$) yields

$$f(t) = \begin{cases} -\dfrac{Mv_0}{\rho_0 cA}(1 - e^{-\rho_0 cAt/M}) & (t \geqslant 0) \\ 0 & (t \leqslant 0) \end{cases} . \tag{14.6.32}$$

Both f and f' are negative concurring with physical expectation.

Finally the stress in the wire given by (14.6.30) is tensile, again as would be expected physically,

$$\sigma_{xx} = \begin{cases} \rho_0 c v_0 e^{-\rho_0 cA(t - x/c)} & (t \geqslant x/c) \\ 0 & (t < x/c) \end{cases} \tag{14.6.33}$$

so that the pulse rises sharply at the wave front $x = ct$ and is followed by an exponential tail (Fig. 14.22).

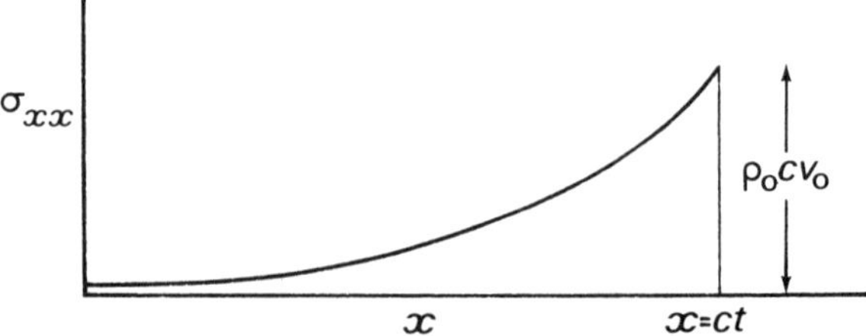

Fig. 14.22 Distribution of stress in long wire impacted at the end

The situation described by (14.6.32) and Fig. 14.22 persists until a time $t = T$ where $T = l/c$. At $t = T$ the tensile stress wave front of amplitude $\rho_0 c v_0$ reaches the end of the wire $x = l$ and a reflection process occurs. We suppose the end $x = l$ of the wire to be supported rigidly so that for $x = l$, $u(l, t) = 0$. For $T < t < 2T$ we write

$$u = f(t - x/c) + g(t + x/c) \tag{14.6.34}$$

where f is the function already derived [equation (14.6.32)] and g is to be determined. For times $t > 2T$ it is necessary to take account of reflections at $x = 0$; there is no difficulty in principal in the theory which merely becomes increasing cumbersone algebraically with each reflection, occurring at T, $2T$, $3T$, etc. Here we only investigate behaviour up until $t = 2T$.

Since $u = 0$ at $x = l$, (14.6.34) yields for $T < t < 2T$

$$f(t - l/c) + g(t + l/c) = 0$$

i.e.
$$g(t + x/c) = \begin{cases} -f[t - (2l - x)/c] & [t > (2l - x)/c] \\ 0 & [T < t < (2l - x)/c] \end{cases}$$

so that for $t > (2l - x)/c$

$$u = f[t - x/c] - f[t - (2l - x)/c] \ .$$

It is of particular interest to compute the stress at $x = l$. From the above and

$$\sigma_{xx} = E(\partial u/\partial x)$$

we derive for $T < t < 2T$ and $x = l$

$$\sigma_{xx} = -\rho_0 c\{f'[t - l/c] + f'[t - (2l - l)/c]\} = -2\rho_0 c f'(t - T)$$

so that during the time interval $T < t < 2T$, the stress experienced by $x = l$ is twice that of the incident pulse but initiated at $t = T$ rather than $t = 0$. In particular the maximum stress occurring at $t = T + 0$ is given by

$$\sigma_{xx} = 2\rho_0 c v_0 \ .$$

This analysis goes some way to explaining experimental results obtained and correctly interpreted by Hopkinson on the failure of brittle wires, i.e. wires which behave elastically for $\sigma_{xx} < \sigma_0$ and which fracture when σ_{xx} attains the value σ_0. For velocities v_0 less than v_1 no fractures occurred, while for velocities in a range $v_1 < v_0 < 2v_1$, failure occurred at the supported end of the wire ($x = l$). Here v_1 is an experimentally determined quantity. For velocities greater than $2v_1$, failure occurred near the impacted end $x = 0$. These experimental results, at first surprising, are consistent with the above theory provided v_1 is identified with $\sigma_0/2\rho_0 c$.

The sharp, indeed discontinuous, wave front depicted in Fig. 14.22 is an idealisation of the real stress pulse which propagates along the wire during the initial interval $0 \leqslant t \leqslant T$. This is because for pulses for which significant variations of stress, strain and displacement occur over distances of the order of the wire diameter, the simple longitudinal wave theory breaks down. More elaborate theories† show that all longitudinal wave propagation in elastic solids is accompanied by **phase dispersion** (sometimes known as **geometrical dispersion** since the phenomenom depends on geometric dimensions of the wire or bar cross section). Phase dispersion effects are most severe in the vicinity of initially sharp discontinuities which are smoothed out by the process. A more realistic picture of the wave profile for the Hopkinson experiment is sketched in Fig. 14.23. Here the discontinuous wave front has been spread out; also shown are some oscillations in the tail which are also a consequence of phase dispersion in elastic longitudinal wave propagation.

The maximum stress amplitude in Fig. 14.23 is about $\rho_0 c v_0$ as in Fig. 14.22, while the 'width' of the wave front, defined by w in the sketch varies rapidly in the first few wire diameters of pulse travel, and then somewhat insensitively with further travel. By the time the wave front of the incident pulse occurring for $0 \leqslant t < T$ in the Hopkinson experiment reaches $x = l$, w is of several wire diameters width and the behaviour of the dispersed profile is reasonably described by the simple theory. Thus in particular, Hopkinson's fracture arguments remain theoretically substantiated.

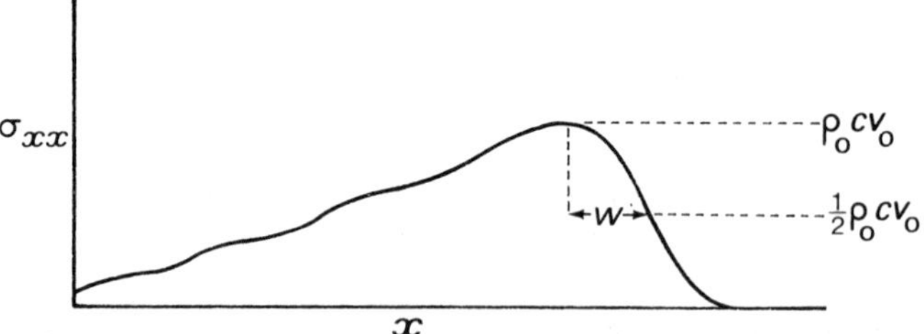

Fig. 14.23 The effect of geometric dispersion on the profile of Fig. 14.22

We turn now to the problem of transverse vibrations of a bar of uniform cross section. The simplified theory here is a straightforward extension of the static theory developed in Section 14.5. We consider the problem of the free vibrations of a cantilever beam of uniform cross section and of length l sketched in Fig. 14.24.

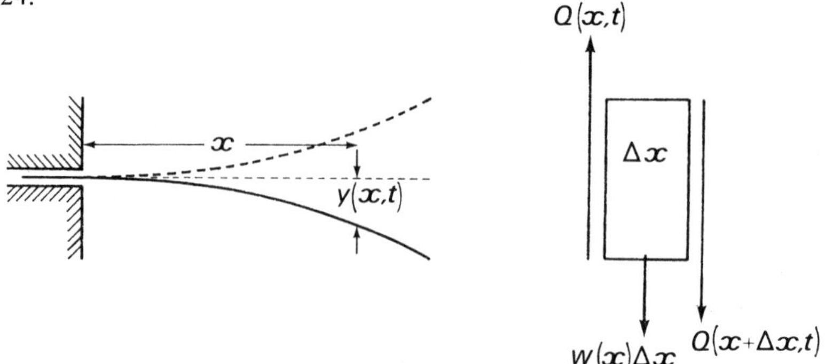

Fig. 14.24 Transverse vibrations of a cantilever beam

Fig. 14.25 Transverse forces on element of width Δx

† The simplest of these theories is the subject of Problem 35.

The transverse deflection of the beam is assumed described by $Y(x, t)$ and we retain from Section 14.5 the constitutive equation (14.5.3)

$$M = EI\frac{\partial^2 Y}{\partial x^2} \qquad (14.6.35)$$

where now a second order *partial* derivative is written since here Y also depends on t. The equation (14.5.9)

$$Q = -\partial M/\partial x \qquad (14.6.36)$$

(again with a partial derivative) also is retained since even though the bar is in motion inertial contributions to this equation are nil. This can be seen from the following argument. Referring to Fig. 14.17 and the associated analysis, we have in the case of angular motion

$$\frac{\partial M}{\partial x}\Delta x + \tfrac{1}{2}[Q(x + \Delta x) + Q(x)]\Delta x + O(\Delta x^3) = J\frac{d\Omega}{dt} \qquad (14.6.37)$$

where J is the moment of inertia of the section Δx about a line through the centre of mass O' and perpendicular to the figure, and where $d\Omega/dt$ is the corresponding angular acceleration. Also in (14.6.37), $\partial M/\partial x$ is evaluated at some point in the interval $[x, x + \Delta x]$, while finally the term not explicitly identified, and of order Δx^3, derives from any distribution of external transverse forces present including perhaps body forces, of intensity w per unit length. These latter, of magnitude $w\Delta x$, give rise at most to contributions of order Δx^3 for continuous w.

The contributions from the right side of (14.6.37) are also of order Δx^3 since J is given by the product of the mass of the element, of order Δx, and the square of the radius of gyration. Since the latter is also of order Δx, J is of order Δx^3. Thus dividing (14.6.37) by Δx and proceeding to the limit recovers (14.6.36), even for the dynamic case.

Referring to Fig. 14.24 contributions from Q and w to the transverse equation of motion lead to

$$Q(x + \Delta x) - Q(x) + w\Delta x = \rho_0 A_0 \Delta x\frac{\partial^2 Y}{\partial t^2} \qquad (14.6.38)$$

where A_0 is the cross sectional area. Following now familiar arguments leads to

$$\frac{\partial Q}{\partial x} + w = \rho_0 A_0\frac{\partial^2 Y}{\partial t^2}\,. \qquad (14.6.39)$$

From (14.6.35), (14.6.36) and (14.6.39) we obtain the partial differential equation

$$EI\frac{\partial^4 Y}{\partial x^4} + \rho_0 A_0\frac{\partial^2 Y}{\partial t^2} = w\,. \qquad (14.6.40)$$

In what follows w is neglected; for the common case of time independent w there is no loss of generality since it is always possible to find a time independent $Y(x)$ which satisfies

$$EI\frac{d^4 Y}{dx^4} = w(x) \tag{14.6.41}$$

which together with the time dependent solution of the homogeneous equation

$$EI\frac{\partial^4 Y}{\partial x^4} + \rho_0 A_0 \frac{\partial^2 Y}{\partial t^2} = 0 \tag{14.6.42}$$

provides the complete solution for Y. In any event the quantity of interest is the deflection measured from the static configuration and this is given directly by the solution of the homogeneous equation.

It is possible to find solutions of (14.6.42) in terms of arbitrary functions but the result is complicated and involves arbitrary functions appearing inside an integral. The usual method of solution is by separation of variables. These latter solutions are of direct interest in the theory of the tuning fork.

We seek solutions of (14.6.42) which are periodic in time and of the separation of variable form

$$Y = F(x)G(t) .$$

For periodic $G(t)$

$$G(t) = \begin{matrix}\cos\\ \sin\end{matrix}(\omega t)$$

the equation for $F(x)$ is easily found to be

$$\frac{d^4 F}{dx^4} - \alpha^4 F = 0 \tag{14.6.43}$$

where

$$\alpha^4 = \frac{\rho_0 A_0 \omega^2}{EI} . \tag{14.6.44}$$

Solutions of (14.6.43) are

$$F = A\cos(\alpha x) + B\sin(\alpha x) + C\cosh(\alpha x) + D\sinh(\alpha x) \tag{14.6.45}$$

where A, B, C, D are constants of integration. Boundary conditions for Y at the built in end are

$$Y = \frac{\partial Y}{\partial x} = 0 \quad (x = 0) \tag{14.6.46}$$

while the stress free end $x = l$ is characterised by

$$M = Q = 0 \quad (x = l) . \tag{14.6.47}$$

In terms of Y, the conditions (14.6.47) may be written

$$\frac{\partial^2 Y}{\partial x^2} = \frac{\partial^3 Y}{\partial x^3} = 0 \quad (x = l) .$$

In terms of F, the boundary conditions (14.6.46) and (14.6.47) yield

$$F(0) = F'(0) = 0, \text{ (a)}; \quad F''(l) = F'''(l) = 0, \text{ (b)} . \qquad (14.6.48)$$

The algebra is expedited if (14.6.48a) is first utilised to derive $C = -A$, $D = -B$ so that (14.6.45) may be written

$$F = A(\cos \alpha x - \cosh \alpha x) + B(\sin \alpha x - \sinh \alpha x) .$$

The conditions (14.6.48b) now become, after cancelling common factors α^2 and α^3,

$$\left.\begin{aligned} -A[\cos(\alpha l) + \cosh(\alpha l)] - B[\sin(\alpha l) + \sinh(\alpha l)] &= 0 \\ A[\sin(\alpha l) - \sinh(\alpha l)] - B[\cos \alpha l + \cosh(\alpha l)] &= 0 \end{aligned}\right\} \qquad (14.6.49)$$

These equations are consistent only if

$$[\cos(\alpha l) + \cosh(\alpha l)]^2 + [\sin(\alpha l) - \sinh(\alpha l)][\sin(\alpha l) + \sinh(\alpha l)] = 0$$

which is readily simplified to

$$\cos(\alpha l)\cosh(\alpha l) + 1 = 0 . \qquad (14.6.50)$$

Equation (14.6.50) provides a transcendental equation for the determination of the discrete set of allowable values (eigenvalues) of αl. Writing $\alpha l = \theta$ the equation may be written

$$\cos \theta = -\operatorname{sech} \theta$$

so that the roots are given by the intersection of the graphs $y = \cos \theta$ with $y = -\operatorname{sech} \theta$. From Fig. 14.26 it is clear that the roots θ_n $(n = 1, 2, 3, \ldots)$ rapidly approach the roots of $\cos \theta = 0$,

i.e. $$\theta_n \to (n - \tfrac{1}{2})\pi .$$

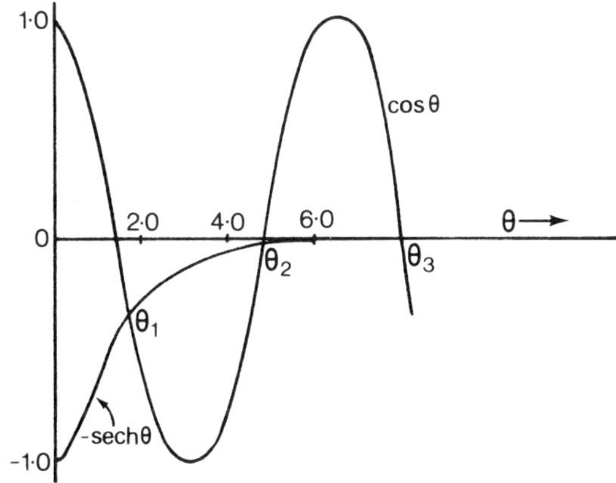

Fig. 14.26 Roots of $\cos \theta = -\operatorname{sech} \theta$

The important roots, which may readily be computed, e.g. by Newton's method using $(n - \frac{1}{2})\pi$ as a first approximation, are the first few which are

$$\theta_1 = 0.5969\pi \qquad \theta_2 = 1.4942\pi \qquad \theta_3 = 2.5003\pi$$
$$= 1.875 \qquad\qquad = 4.694 \qquad\qquad = 7.855$$

showing that the approach to $(n - \frac{1}{2})\pi$ is very rapid indeed.

With $\theta_1, \theta_2 \ldots \theta_n \ldots$ evaluated, the ratio of $B/A = k_n$ (say) is determined for each allowable θ_n by either of the equations (14.6.49). Choosing for the moment $A = 1$ there results the set of eigenfunctions

$$F_n = [\cos(\theta_n x/l) - \cosh(\theta_n x/l)] + k_n[\sin(\theta_n x/l) - \sinh(\theta_n x/l)]$$

so that a complete solution of the free vibration problem is given by

$$Y(x, t) = \sum_1^\infty [a_n \cos(\omega_n t) + b_n \sin(\omega_n t)] F_n(x)$$

where from (14.6.44) $$\omega_n = \left(\frac{EI}{\rho_0 A_0}\right)^{\frac{1}{2}} (\theta_n{}^2/l^2) ,$$

and where the a_n and b_n are coefficients to be determined.

These coefficients are readily determined in principle for initial conditions

$$Y(x, 0) = h(x) , \qquad \frac{\partial Y}{\partial t}(x, 0) = g(x)$$

with h and g given. The calculation is trivial formally since it is readily shown that the eigenfunctions $F_n(x)$ are orthogonal

i.e. $$\int_0^l F_n(x) F_m(x) dx = 0 \qquad (n \neq m) .$$

For a proof of this see Problem 33. For the case where the bar is released from rest from a non-equilibrium configuration, $g = 0$ and hence $b_n = 0$. The remaining condition leads to

$$\sum_1^\infty a_n F_n(x) = h(x)$$

so that by virtue of the orthogonality relation

$$a_n = \int_0^l h(x) F_n(x) dx \Bigg/ \int_0^l F_n{}^2(x) dx .$$

The algebra is tedious for any specific $h(x)$. For the particular case where the initial beam displacement is

$$h(x) = \frac{W}{EI}\left(\tfrac{1}{2}lx^2 - \tfrac{1}{6}x^3\right)$$

caused by a terminal load W (see Section 14.5) which is suddenly removed at $t = 0$, the coefficients a_n may be shown to be

$$a_n = \frac{4Wl^3}{EI}\frac{[\cos(\theta_n) + \cosh(\theta_n)]}{[\sin(\theta_n) + \sinh(\theta_n)]}\frac{1}{\theta_n{}^4}$$

and it is of note that the coefficients decay so rapidly with increasing θ_n that the entire expansion for Y is dominated by the first term with (circular) frequency ω_1.

For the case where $h(x) = 0$ but $g(x)$ is prescribed, e.g. by tapping the bar with a hammer, the coefficients a_n vanish while the b_n are given by

$$\omega_n b_n = \int_0^l g(x)F_n(x)dx \bigg/ \int_0^l F_n{}^2(x)dx\ .$$

If $g(x)$ is reasonably smooth then again the fundamental circular frequency ω_1 will dominate.

The result that the free vibrations of a cantilever beam are largely dominated by the fundamental frequency finds application in the theory of the tuning fork. The fork is bent into a U shape and the base of the U attached to a stem; the assumption that each prong of the fork vibrates independently with a built-in end at the base of the U is evidently idealised.

"The fundamental mode of a tuning fork may be further reinforced relative to the other modes if the stem be screwed into the upper face of a resonance box of suitable dimensions" (H. Lamb, p. 134, *Dynamical Theory of Sound* – Dover Reprint 1960). Lamb also states that "the first overtone (circular frequency ω_2) may be elicited in considerable intensity by bowing one of the prongs near the bend," resulting, for conventional forks, in a "very shrill" note. In practical forks slightly different constructional details in the two prongs may lead to slightly different fundamental modes, resulting in the familiar experience of "beats". The beats usually decay rapidly because of the coupling between the two prongs; after a short time the fork vibrates with a single frequency near the fundamental one for either prong.

Problems, Chapter 14

1. Relations connecting elastic constants

(a) Derive the relations

$$\mu = \frac{E}{2(1+\nu)}, \quad \lambda = \frac{\nu E}{(1+\nu)(1-2\nu)}, \quad 3\lambda + 2\mu = \frac{E}{(1-2\nu)} .$$

(b) The compressible behaviour of many liquids is modelled adequately by the assumption that $\sigma_{ij} = -P(\rho)\delta_{ij}$ where $P(\rho)$ is given by

$$P(\rho) = K(\rho - \rho_0)/\rho_0 . \tag{1}$$

Here ρ is density, ρ_0 is density in the absence of pressure and the constant K is the Bulk Modulus. Show that for an elastic solid, provided P is interpreted as the "mean" hydrostatic pressure, i.e. $P = -\frac{1}{3}\sigma_{mm}$, (so that under hydrostatic stress conditions $\sigma_{ij} = -P\delta_{ij}$), a pressure-density relation similar to (1) holds with K given by

$$K = \lambda + \tfrac{2}{3}\mu = \frac{E}{3(1-2\nu)} .$$

(c) Deduce the relation $\quad \dfrac{3}{\mu} + \dfrac{1}{K} = \dfrac{9}{E}$

connecting the shear modulus, bulk modulus and Young's modulus.

2. An elastic slab $0 \leqslant z \leqslant h$ whose lateral dimensions are sufficiently large to ensure $u_x = u_y = 0$ everywhere is subject to uniform normal tractions $\sigma_{zz} = T$ on $z = 0$ and $z = h$. Deduce that in the absence of body forces the state of stress is everywhere uniform and given by

$$\sigma_{zz} = T, \quad \sigma_{xx} = \sigma_{yy} = \nu T/(1-\nu), \quad \sigma_{xy} = \sigma_{yz} = \sigma_{zx} = 0$$

and that the only non-vanishing component of strain is also uniform and given by

$$e_{zz} = \frac{(1+\nu)(1-2\nu)T}{(1-\nu)E} .$$

3. Verification of a solution

(i) Calculate all components of stress for the displacement field

$$u_x = A[\tfrac{1}{2}lz^2 - \tfrac{1}{6}z^3 + \tfrac{1}{2}\nu(l-z)(x^2-y^2)]$$

$$u_y = A\nu(l-z)xy$$

$$u_z = -A[x(lz - \tfrac{1}{2}z^2) + xy^2]$$

where A and l are constants.

(ii) Verify that in the absence of body forces the equilibrium equations are satisified.

4. Verification of a solution

(i) Calculate the components of stress for the displacement field

$$u_x = u_y = 0 , \quad u_z = A\theta$$

where A is a constant and θ is the plane polar angle

$$\theta = \tan^{-1}(y/x) .$$

(ii) Verify that in the absence of body forces the equations of equilibrium are satisfied.

5. Verification of a general solution

(i) Show that in the absence of body forces the displacement vector of a classical elastic solid satisfies the equation

$$(\lambda + \mu)\text{grad}(\text{div } \mathbf{u}) + \mu\nabla^2 \mathbf{u} = \rho_0(\partial^2 \mathbf{u}/\partial t^2) . \qquad (1)$$

(ii) Verify that an elastostatic (i.e. time-independent) solution of (1) is given by

$$\mathbf{u} = 4(1 - \nu)\mathbf{A} - \text{grad}(\phi + \mathbf{r} . \mathbf{A}) \qquad (2)$$

where ϕ and $\mathbf{A}$ are scalar and vector potentials which satisfy the Laplace equations

$$\nabla^2 \phi = 0 , \quad \nabla^2 \mathbf{A} = 0 .$$

6. Spherically symmetric problems in electrostatics

(i) Show that if $r = (x^2 + y^2 + z^2)^{\frac{1}{2}}$ and $\mathbf{i}_r$ is a unit vector in the radial direction, the displacement field

$$\mathbf{u} = u(r)\mathbf{i}_r \qquad (1)$$

is irrotational (i.e. curl $\mathbf{u} = 0$). Hence or otherwise deduce that if (1) is a solution of the elastostatic equation

$$(\lambda + \mu)\text{grad}(\text{div } \mathbf{u}) + \mu\nabla^2 \mathbf{u} = 0$$

then

$$u(r) = Ar + Br^{-2} \qquad (2)$$

where A and B are constants.

(ii) Show that (1) and (2) lead to (principal) stresses

$$\sigma_{rr} = (3\lambda + 2\mu)A - 4\mu Br^{-3}$$

$$\sigma_{\theta\theta} = \sigma_{\phi\phi} = (3\lambda + 2\mu)A + 2\mu Br^{-3}$$

where (r, θ, ϕ) are spherical polar coordinates. Use these results to solve Problems (7), (8) and (9).

7. (i) Derive the formulae

$$\sigma_{rr} = \frac{Pa^3}{(b^3 - a^3)}\left(1 - \frac{b^3}{r^3}\right), \qquad \sigma_{\theta\theta} = \sigma_{\phi\phi} = \frac{Pa^3}{(b^3 - a^3)}\left(1 + \frac{b^3}{2r^3}\right)$$

$$u(r) = \frac{Pa^3}{E(b^3 - a^3)}\left[(1 - 2\nu)r + \tfrac{1}{2}(1 + \nu)\frac{b^3}{r^2}\right]$$

for the principal stresses and radial displacement of an elastic spherical shell $a \leqslant r \leqslant b$ subject to internal pressure P at $r = a$ and zero external pressure.

(ii) From the above derive the approximate "thin shell' formulae

$$\sigma_{rr} \simeq P(r - b)/h, \qquad \sigma_{\theta\theta} = \sigma_{\phi\phi} \simeq \tfrac{1}{2}Pa/h, \qquad u(r) \simeq \tfrac{1}{2}(1 - \nu)Pa^2/Eh$$

valid for $(h/a) \ll 1$ where $h = (b - a)$ is the shell thickness.

(iii) If the material remains elastic up to fracture and if fracture occurs as soon as the largest principal stress attains the value F, deduce that the maximum safe pressure of a thin pressure vessel is $2hF/a$.

8. (i) A solid elastic sphere $0 \leqslant r \leqslant b$ is subject to uniform external pressure P at $r = b$. If body forces are neglected show that the state of stress throughout the sphere is one of uniform hydrostatic pressure. Show for a solid body of arbitrary shape subject to uniform external pressure that the same result holds.

(ii) An elastic sphere of outer radius b contains a concentric internal spherical cavity of radius a. The surface $r = b$ is subject to uniform pressure P. Deduce that in the absence of body forces the stress field is given by

$$\sigma_{rr} = -\frac{Pb^3}{(b^3 - a^3)}\left(1 - \frac{a^3}{r^3}\right), \qquad \sigma_{\theta\theta} = \sigma_{\phi\phi} = -\frac{Pb^3}{(b^3 - a^3)}\left(1 + \frac{a^3}{2r^3}\right).$$

(iii) Show that for $a < r \leqslant b$ the limiting form of the latter stress field as $a \to 0$ is the same as the stress field in Part (i). Exceptionally, however, show that for $r = a$ the limiting stress field is

$$\sigma_{rr} = 0, \quad \sigma_{\theta\theta} = \sigma_{\phi\phi} = -\tfrac{3}{2}P.$$

Thus in the immediate vicinity of a cavity (no matter how small) in a body subject to uniform external pressure, the stress field is different to that which occurs in the absence of the cavity.

9. An elastic spherical shell $a \leqslant r \leqslant b$ is subject to internal pressure P on the inner surface $r = a$, while the outer surface $(r = b)$ is restrained from expansion by a surrounding rigid (spherical) wall. Show that the internal radial displacement is

$$u(a) = \frac{Pa[(b/a)^3 - 1](1 - 2\nu)}{E[1 + \{2(1 - 2\nu)/(1 + \nu)\}(b/a)^3]}$$

while the principal stresses at $r = b$ are given by

$$(\sigma_{rr})_b = -\frac{3(1-\nu)P/(1+\nu)}{[1+\{2(1-2\nu)/(1+\nu)\}(b/a)^3]},$$

$$(\sigma_{\theta\theta})_b = (\sigma_{\phi\phi})_b = [\nu/(1-\nu)](\sigma_{rr})_b .$$

10. Gravitating elastic sphere

The body force due to self gravity in a solid sphere of uniform density ρ_0 is $\mathbf{b} = -(gr/a)\mathbf{i}_r$ where g is the gravitational constant on the surface $r = a$, r is the distance from the centre of the sphere and $\mathbf{i}_r$ a unit vector in the r direction. Show that the elastostatic equations, including the self induced body force, admit a spherically symmetric solution of the form

$$\mathbf{u} = u(r)\mathbf{i}_r$$

where $u(r)$ satisfies $$\frac{d}{dr}\left(\frac{du}{dr} + \frac{2u}{r}\right) = \frac{\rho_0 g r}{(\lambda + 2\mu)a} .$$

Further show that the relevant solution of this equation which is finite at the origin is

$$u = Ar + \frac{\rho_0 g r^3}{10(\lambda + 2\mu)a}$$

and determine the constant A so that the surface $r = a$ is free of traction. Hence determine the stress field within the sphere.

11. Rotating hollow shaft

(i) A hollow cylindrical elastic shaft $a \leqslant r \leqslant b$, where $r = (x^2 + y^2)^{\frac{1}{2}}$ rotates at constant angular velocity ω about the axis of symmetry (z axis). Show that the assumption of a displacement field of the form

$$\mathbf{u} = u_r(r)\mathbf{i}_r + u_z(z)\mathbf{k}$$

leads to $$e_{zz} \equiv \frac{du_z}{dz} = \text{const.}$$

and $$u_r = Ar + Br^{-1} - \frac{\rho_0 \omega^2 r^3}{8(\lambda + 2\mu)}$$

where A and B are arbitrary constants.

(ii) Evaluate in terms of A, B and e_{zz} the principal stresses σ_{rr}, $\sigma_{\theta\theta}$, σ_{zz} and determine A, B and e_{zz} (and hence σ_{rr}, $\sigma_{\theta\theta}$, σ_{zz}) for the case of a shaft whose inner and outer curved surfaces $r = a$ and $r = b$ are free of traction, and with a zero mean axial load.

Torsion problems

12. The displacement field associated with the torsion of a prism about the z axis is given by

$$u_x = -\tau yz, \quad u_y = \tau xz, \quad u_z = \tau\phi(x, y) \tag{1}$$

where ϕ is harmonic (i.e. satisfies Laplaces equation) and certain boundary conditions.

Verify that the displacement field

$$u_x = \tau y_1 z, \quad u_y = -\tau x_1 z, \quad u_z = -\tau(y_1 x - x_1 y) \tag{2}$$

where y_1 and x_1 are constants, represents a rigid body rotation in the small displacement gradient approximation. By superimposing the fields (1) and (2) deduce that the sum field, (a) represents torsion of the prism about an axis parallel to the z axis but located at (x_1, y_1), (b) leads to the same stresses as the field (1).

13. **(i)** Verify that with an appropriate choice of the constant A, the function

$$\chi = A(x - a)(x + 2a - \sqrt{3}y)(x + 2a + \sqrt{3}y)$$

satisfies

$$\nabla^2\chi = -2$$

and provides the solution of the torsion problem for an equilateral triangle.

(ii) Find the warping function ϕ conjugate to $\chi + \frac{1}{2}(x^2 + y^2)$.

(iii) Show that the couple required to provide a twist of τ per unit length of the prism is $9\sqrt{3}\,\mu\tau a^4/5$.

14. **(i)** Verify that the function

$$\chi = -\frac{(x^2 + y^2 - 2ay)(x^2 + y^2 - b^2)}{2(x^2 + y^2)} \tag{1}$$

satisfies

$$\nabla^2\chi = -2\,.$$

HINT, the calculation is somewhat easier if use is made of polar coordinates and it is recalled that $r^{\pm n} \begin{smallmatrix}\cos\\ \sin\end{smallmatrix} (n\theta)$ are separation of variable solutions of the two dimensional Laplace equation.

(ii) Show that χ vanishes on the circles $x^2 + y^2 = b^2$, $x^2 + (y - a)^2 = a^2$. Consider the case $b < 2a$ and sketch the two circles.

(iii) Calculate the warping function ϕ which is conjugate to $\chi + \frac{1}{2}(x^2 + y^2)$. From this result show that the origin $r = 0$ must be excluded from the domain of χ and hence deduce that (1) solves the torsion problem for a cylinder whose boundary is given by $r = b$, $r = 2a \sin\theta$ where $1 \geqslant \sin\theta \geqslant b/2a$.

(iv) Calculate the couple required to twist the cylinder an amount τ per unit length.

15. The torsion problem for a solid cylinder of arbitrary cross-section bounded by the surface $C(x, y) = 0$ may be specified by the equation

$$\nabla^2 \chi = -2,$$

where $\chi = 0$ on $C = 0$. With χ so determined, the torsional rigidity of the cylinder is given by

$$G = 2\mu\tau \iint_S \chi dx dy,$$

where S is the plane surface bounded by $C = 0$ and μ is the shear modulus.

(i) Verify that for a circular cylinder of radius a

$$\chi = \tfrac{1}{2}(a^2 - r^2),$$

where $r = (x^2 + y^2)^{\frac{1}{2}}$, satisfies the differential equation and boundary condition and that

$$(G/\mu\tau) = \tfrac{1}{2}\pi a^4.$$

(ii) The cross-section of a nearly circular cylinder is specified in plane polar coordinates by the boundary curve

$$r = a[1 + \epsilon \cos(n\theta)],$$

where n is a positive integer and ϵ ($\ll 1$) is a constant. Show that the function

$$\chi = \tfrac{1}{2}(a^2 - r^2) + a^2 \left[\epsilon(r/a)^n \cos(n\theta) - \tfrac{1}{2}\epsilon^2(n - \tfrac{1}{2})[1 + (r/a)^{2n} \cos(2n\theta)]\right]$$

satisfies the differential equation exactly and satisfies the boundary condition provided terms of order ϵ^3 and higher are neglected. Deduce that G is given by

$$(G/\mu\tau) \equiv 2\int_0^{2\pi} d\theta \int_0^{a[1 + \epsilon\cos(n\theta)]} r\chi(r, \theta) dr$$

$$= \tfrac{1}{2}\pi a^4 [1 + \epsilon^2(3 - 2n) + O(\epsilon^3)].$$

16. For a simply connected region S bounded by the closed curve C the torsion problem may be specified by

$$\chi = -\tfrac{1}{2}(x^2 + y^2) + \text{Im} f(z)$$

where f is a regular function of $z = x + iy$. Also $\chi = 0$ on C while the warping function ϕ is given by $\text{Re}\, f(z)$. Finally the couple is

$$G = 2\mu\tau \iint_S \chi dx dy$$

where μ is the shear modulus and τ is the twist per unit length.

By choosing $$f(z) = 4ib^{\frac{3}{2}} z^{\frac{1}{2}}$$

where b is a positive constant and where $z^{\frac{1}{2}}$, defined in the plane cut along the entire negative real axis, is chosen to be the branch which is positive for $(x > 0, y = 0)$, show that the resulting χ solves the torsion problem for the region bounded by

$$r = 4b[\cos(\tfrac{1}{2}\theta)]^{\frac{2}{3}} \qquad -\pi \leqslant \theta \leqslant \pi$$

where r and θ are plane polar coordinates.

Sketch the curve and verify that ϕ is single valued inside and on C. Show also that

$$G = N\mu\tau b^4$$

where $$N = (768/5)\int_0^{\pi/2} (\cos\gamma)^{\frac{8}{3}} d\gamma \simeq 107$$

17. Two dimensional theory – plane strain

(i) Introduction and the Airy Function

A state of plane strain is said to occur when the displacement field is of the form

$$u_x = u_x(x, y), \qquad u_y = u_y(x, y), \qquad u_z = 0\,.$$

Show that under these circumstances the non-vanishing components of stress in an elastic medium are σ_{xx}, σ_{yy}, σ_{xy} and σ_{zz}. Show further that these non-vanishing stress components depend only on x and y and also that $\sigma_{zz} = \nu(\sigma_{xx} + \sigma_{yy})$.

In the absence of body forces deduce that one equation of equilibrium is satisfied identically while the remaining pair reduce to

$$\frac{\partial\sigma_{xx}}{\partial x} + \frac{\partial\sigma_{xy}}{\partial y} = 0\,, \qquad \frac{\partial\sigma_{xy}}{\partial x} + \frac{\partial\sigma_{yy}}{\partial y} = 0$$

Show that these latter are satisfied by the choice

$$\sigma_{xx} = \frac{\partial^2\chi}{\partial y^2}\,, \qquad \sigma_{yy} = \frac{\partial^2\chi}{\partial x^2}\,, \qquad \sigma_{xy} = -\frac{\partial^2\chi}{\partial x\partial y}$$

where $\chi = \chi(x, y)$, is called the Airy stress function.

[The present χ is not to be confused with the function χ which appears in the torsion problem.]

(ii) Derivation of Equation for χ

(a) Show from the equation

$$(\lambda + \mu)\,\mathrm{grad}(\mathrm{div}\,\mathbf{u}) + \mu\nabla^2\mathbf{u} = 0$$

that the dilatation satisfies Laplace's equation, i.e. $\nabla^2(e_{mm}) = 0$. This result is independent of whether or not plane strain conditions prevail.

(b) For plane strain show that

$$(1 + \nu)(\sigma_{xx} + \sigma_{yy}) = (3\lambda + 2\mu)e_{mm}$$

and hence deduce that χ satisfies the bi-harmonic equation

$$\nabla^2(\nabla^2\chi) = 0$$

where ∇^2 is the two dimensional Laplacian operator

$$\nabla^2 = \frac{\partial^2}{\partial x^2} + \frac{\partial^2}{\partial y^2} .$$

(iii) SOLUTION OF THE BI-HARMONIC EQUATION IN COMPLEX VARIABLE TERMS

By writing $z = x + iy \quad \bar{z} = x - iy$

show that the bi-harmonic equation reduces to

$$\frac{\partial^4\chi}{\partial z^2\partial\bar{z}^2} = 0$$

and that the general solution of this latter equation may be written

$$\chi = z\bar{\phi}(\bar{z}) + \bar{z}\phi(z) + \psi(z) + \bar{\psi}(\bar{z}) \tag{1}$$

where $\phi(z)$ and $\psi(z)$ are arbitrary functions. In deriving (1), make use of the fact that χ is real.

The solution (1) is the starting point for the investigation of many two dimensional problems in elastostatics. Use of complex variable theory in two-dimensional elastostatics is akin to the use in two dimensional potential theory, but the elastostatic case is more complicated in that two complex variable functions rather than one appear in the analysis. Some of the basic results stemming from (1) are discussed in (iv) and (v) below, and in the corresponding parts of the answer.

(iv) FORMULAE FOR STRESSES

Utilise the solution (1) to calculate σ_{xx}, σ_{yy} and σ_{xy} and deduce in particular the formulae

$$\sigma_{xx} + \sigma_{yy} = 4[\phi'(z) + \bar{\phi}'(\bar{z})] ,$$

$$\sigma_{xx} - \sigma_{yy} + 2i\sigma_{xy} = -4[z\bar{\phi}''(\bar{z}) + \bar{\psi}''(\bar{z})] .$$

(v) FORMULAE FOR THE DISPLACEMENTS

From the constitutive equations deduce

$$e_{xx} \equiv \frac{\partial u_x}{\partial x} = \frac{(1-\nu^2)}{E}\left[\sigma_{xx} - \frac{\nu}{(1-\nu)}\sigma_{yy}\right] \tag{2}$$

$$e_{yy} \equiv \frac{\partial u_y}{\partial y} = \frac{(1-\nu^2)}{E}\left[\sigma_{yy} - \frac{\nu}{(1-\nu)}\sigma_{xx}\right] \tag{3}$$

$$e_{xy} \equiv \frac{1}{2}\left(\frac{\partial u_x}{\partial y} + \frac{\partial u_y}{\partial x}\right) = \frac{1+\nu}{E}\sigma_{xy} \tag{4}$$

By substituting into (2) and (3) from the results obtained in section (iv) for σ_{xx} and σ_{yy} derive the equations

$$\frac{\partial u_x}{\partial x}, \frac{\partial u_y}{\partial y} = \frac{2(1-2\nu)(1+\nu)}{E}[\phi'(z) + \bar{\phi}'(\bar{z})]$$
$$\mp \frac{(1+\nu)}{E}[\bar{z}\phi''(z) + z\bar{\phi}''(\bar{z}) + \psi''(z) + \bar{\psi}''(\bar{z})] \quad (5)$$

where the $\mp$ signs are associated respectively with $(\partial u_x/\partial x)$ and $(\partial u_y/\partial y)$.

Integrate equations (5) to obtain

$$u_x = \frac{(1+\nu)}{E}\Big\{(3-4\nu)[\phi(z) + \bar{\phi}(\bar{z})]$$
$$- [\bar{z}\phi'(z) + z\bar{\phi}'(\bar{z}) + \psi'(z) + \bar{\psi}'(\bar{z})]\Big\} + M(y) \quad (6)$$

$$u_y = \frac{i(1+\nu)}{E}\Big\{-(3-4\nu)[\phi(z) - \bar{\phi}(\bar{z})]$$
$$+ [-\bar{z}\phi'(z) + z\bar{\phi}'(\bar{z}) - \psi'(z) + \bar{\psi}'(\bar{z})]\Big\} + N(x) \quad (7)$$

where at this stage M and N are arbitrary.

Substitute from (6) and (7) into (4) for u_x and u_y, and from the results obtained in section (iv) for σ_{xy}, show that

$$M'(y) + N'(x) = 0 .$$

Deduce that $M(y) = ky + a$, $N(x) = -kx + b$ where k, a, b are arbitrary constants and that $M(x)$ and $N(y)$ represent merely a rigid body displacement, i.e. rotation plus translation in the xy plane.

Show that by an appropriate modification to ϕ the terms $M(y)$ and $N(x)$ may be incorporated into the terms within the brace parentheses { } in (6) and (7) of this problem so that effectively M and N may be taken to vanish. In these circumstances deduce that without loss of generality $(u_x + iu_y)$ may be expressed in the form

$$u_x + iu_y = \frac{2(1+\nu)}{E}[(3-4\nu)\phi(z) - z\bar{\phi}'(\bar{z}) - \bar{\psi}'(\bar{z})] .$$

Bending of beams

18. **(i)** Show that for a beam of uniform rectangular cross section $-b \leqslant z \leqslant b$, $-a \leqslant y \leqslant a$ bent in the xy plane, the second moment of area is $I = \frac{4}{3}ba^3$.

(ii) A massive beam of uniform rectangular cross section with breadth $2b$ and depth $2a$ is of weight w per unit length. The beam of length l is horizontal and supported freely at the ends $x = 0$ and $x = l$. Show that the bending moment $M(x)$ is given by

$$M(x) = -\tfrac{1}{2}wx(l - x)$$

and that the deflexion $Y(x)$ is

$$Y = \frac{w}{32a^3bE} x(l - x)(l^2 + lx - x^2) .$$

(iii) If the beam is made from a material whose maximum safe tensile stress is σ_0, and if shear stresses across the cross section are neglected, show that the maximum length of beam which can be supported safely in the indicated manner is $4a[2b\sigma_0/3w]^{1/2}$.

19. A light horizontal beam $0 \leqslant |x| \leqslant (l + L)$, of second moment of area I, is simply supported at $x = \pm l$ and carries identical weights W at $x = \pm(l + L)$. Show that for $0 \leqslant x \leqslant (l + L)$

$$M(x) = \begin{cases} WL & (0 \leqslant x \leqslant L) \\ W(L + l - x) & (l \leqslant x \leqslant l + L) \end{cases}$$

and that the deflection for $x > 0$ is given by

$$EIY = \begin{cases} \frac{1}{2}WL(x^2 - l^2) & (0 \leqslant x \leqslant l) \\ W[\frac{1}{2}(L + l)x^2 - \frac{1}{6}x^3 - \frac{1}{2}l^2x + \frac{1}{6}l^3 - \frac{1}{2}Ll^2] & (l \leqslant x \leqslant l + L) \end{cases}$$

while for negative values of x, $Y(-x) = Y(x)$.

20. A light horizontal beam, of second moment of area I, occupies $0 \leqslant x \leqslant l + L$. The beam is simply supported at the end points and carries a single weight W at $x = L$. Deduce that $M(x)$ is given by

$$M(x) = \begin{cases} -\dfrac{Wlx}{(L + l)} & (0 \leqslant x \leqslant l) \\ -\dfrac{WL}{(L + l)}(L + l - x) & (l \leqslant x \leqslant l + L) \end{cases}$$

and hence derive the associated $Y(x)$.

21. A light horizontal beam of uniform cross section and second moment of area I occupies $0 \leqslant x \leqslant (l + L)$. The beam is built in at $x = 0$, i.e. is supported at $x = 0$ as a cantilever so that $Y(0) = Y'(0) = 0$, and is simply supported at $x = l$. At $x = l + L$ the beam carries a weight W.

Let R denote the (upward) reaction at $x = l$ so that at $x = 0$ the (upward) reaction due to the cantilever support is $(W - R)$. Also at $x = 0$ let the (initially unknown) bending moment be M_0. The problem contains two unkown constants (R and M_0) and while these are related $[M_0 = W(L + l) - Rl]$, elementary statics yields no other relation enabling the determination of M_0 and R.

In what follows R is carried through the deflexion calculation and determined ultimately from the condition $Y(l) = 0$.

Show that the bending moment $M(x)$ is given by

$$M(x) = \begin{cases} W(L + l - x) - R(l - x) & (0 \leqslant x \leqslant l) \\ W(L + l - x) & (l \leqslant x \leqslant L + l) \end{cases}$$

Deduce that for $0 \leqslant x \leqslant l$

$$EIY(x) = \tfrac{1}{2}[W(L + l) - Rl]x^2 - \tfrac{1}{6}(W - R)x^3$$

and hence show from $Y(l) = 0$ that $R = W(1 + 3L/2l)$ and that also for $0 \leqslant x \leqslant l$,

$$EIY(x) = -\tfrac{1}{4}WLx^2 + \tfrac{1}{4}WLx^3/l \ .$$

Finally using the results that $Y(l) = 0$ and $Y'(l)$ is continuous, derive for $l \leqslant x \leqslant l + L$

$$EIY(x) = W[\tfrac{1}{2}(L + l)x^2 - \tfrac{1}{6}x^3 - (\tfrac{3}{4}L + \tfrac{1}{2}l)lx + \tfrac{1}{4}l^2L + \tfrac{1}{6}l^3] \ .$$

Dynamic (wave) problems

22. The region $x \geqslant 0$ is occupied by an elastic material. The surface $x = 0$ is subjected to a pressure pulse

$$P(t) = \begin{cases} P_0 e^{-(t/\tau)} & (t \geqslant 0) \\ 0 & (t < 0) \end{cases}$$

generated explosively. Here P_0 and τ are constants.

If motion is assumed to occur only in the x direction and depends only on x show that $u_x(x, t)$ satisfies

$$\frac{\partial^2 u_x}{\partial x^2} - \frac{1}{c^2}\frac{\partial^2 u_x}{\partial t^2} = 0 \quad \text{where} \quad c^2 = (\lambda + 2\mu)/\rho_0$$

and that the solution of this equation satisfying the boundary conditions is

$$u_x(x, t) = \begin{cases} \dfrac{P_0 \tau}{\rho_0 c}[1 - e^{-(t - x/c)/\tau}] & (x \leqslant ct) \\ 0 & (x \geqslant ct) \end{cases}$$

and calculate the stress components.

23. An elastic rod of uniform cross section occupies $0 \leqslant x \leqslant l$. If the axial stress σ_{xx} is assumed uniform over the cross section and related to the axial strain e_{xx} by $\sigma_{xx} = Ee_{xx}$ show that the longitudinal displacement $u_x(x, t)$ satisfies

$$\frac{\partial^2 u_x}{\partial x^2} - \frac{1}{c^2}\frac{\partial^2 u_x}{\partial t^2} = 0$$

where

$$c^2 = E/\rho_0 \ .$$

A pulse $$u_x = f[t - x/c]$$
is generated at $x = 0$ by an explosion while $x = l$ is free of traction. Show that for $0 \leqslant t \leqslant l/c$ the end $x = l$ is undisturbed and that for $l/c < t < 3l/c$ the displacement is given by
$$u(l, t) = 2f[t - l/c] \ .$$

24. In the split Hopkinson pressure bar experiment, two elastic bars of identical uniform cross section and material are placed in contact with a specimen of (soft) material whose uniaxial (compressive) stress-strain properties are to be investigated under impact loading conditions (see diagram).

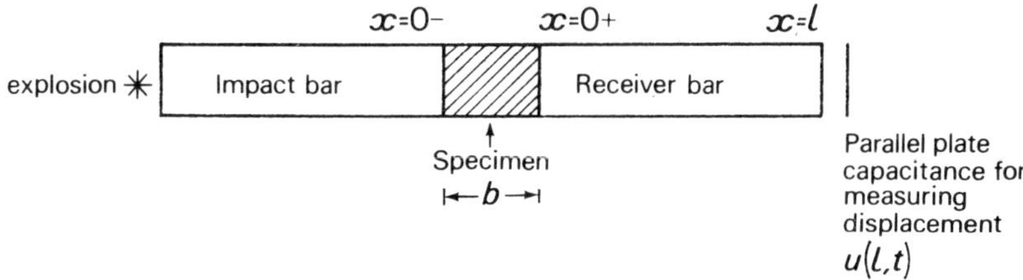

An explosion at the end of the impact bar generates a displacement pulse $u_x = f(t - x/c)$ which encounters the specimen at $x = 0-$ On encountering the specimen a reflected wave [say $g(t + x/c)$] is generated in the impact bar ($x < 0$) and a transmitted wave [say $h(t - x/c)$] in the receiver bar ($x > 0$). It is possible to measure $f(t \quad l/c)$ and $h(t - l/c)$ by carrying out experiments in the absence and presence of the specimen. [Measurements are made by the parallel plate capacitance device of the movement of the free surface $x = l$. From the results of Problem 23 these displacements are $2f(t - l/c)$ and $2h(t - l/c)$.] It is desired to express the compressive stress $\sigma = -\sigma_{xx}$ and compressive strain $e = -e_{xx}$ in the specimen in terms of the measured quantities f and h.

Show that if σ and e are respectively longitudinal (compressive) stress and strain in the specimen (assumed uniform), then
$$\sigma(t) = \rho c h'(t) , \quad e(t) = 2[f(t) - h(t)]/b \ .$$
[In the experiment the pulse widths are short compared with the lengths of the bars, so that there is no interference in the measurements from multiple reflections. Also b is small compared with the pulse widths so that for all but the initial period of stress, the assumption of uniform stress and strain in the specimen is realistic.]

25. **(i)** Show that if $\mathbf{u} = \text{grad}\,\phi$ then the equation
$$(\lambda + \mu)\text{grad}\,(\text{div}\,\mathbf{u}) + \mu\nabla^2\mathbf{u} = \rho_0 \frac{\partial^2 \mathbf{u}}{\partial t^2}$$

is satisfied provided ϕ is a solution of

$$\nabla^2 \phi - \frac{1}{c^2}\frac{\partial^2 \phi}{\partial t^2} = 0$$

where $$c^2 = (\lambda + 2\mu)/\rho_0 \ .$$

Show that if $\phi = \phi(r, t)$ where $r = (x^2 + y^2 + z^2)^{1/2}$ the wave equation reduces to

$$\frac{\partial^2 \phi}{\partial r^2} + \frac{2}{r}\frac{\partial \phi}{\partial r} = \frac{1}{c^2}\frac{\partial^2 \phi}{\partial t^2}$$

and that the solution for outgoing spherical waves is

$$\phi = \frac{f(t - r/c)}{r}$$

where f is an arbitrary function of the variable $(t - r/c)$.

(ii) A spherical cavity $0 \leqslant r \leqslant a$ in an elastic solid is filled with explosive which is detonated at $t = 0$ to yield a pressure $P(t)$ acting on the internal surface $r = a$. As a consequence of the explosion, for $t > 0$, the pressure pulse $P(t)$ generates an outgoing wave

$$\phi = f[t - (r - a)/c]/r$$

in $r > a$. [It is convenient here to absorb the constant (a/c) into the argument of f.]

Deduce that the displacement is purely radial and given by

$$u_r = -(f/r^2) - (f'/cr) \ .$$

Show that (principal) stresses σ_{rr}, $\sigma_{\theta\theta}$, $\sigma_{\phi\phi}$ are given by

$$\sigma_{rr} = (\lambda + 2\mu)(f''/c^2 r) + 4\mu[(f'/cr^2) + (f/r^3)]$$

$$\sigma_{\theta\theta} = \sigma_{\phi\phi} = \lambda(f''/c^2 r) - 2\mu[(f'/cr^2) + (f/r^3)] \ .$$

From the first of these deduce that $f(t)$ satisfies the equation

$$\frac{(\lambda + 2\mu)f''(t)}{c^2 a} + \frac{4\mu f'(t)}{ca^2} + \frac{4\mu f(t)}{a^3} = -P(t) \ .$$

Show that for the case

$$P = \begin{cases} P_0 & (t \geqslant 0) \\ 0 & (t < 0) \end{cases}$$

$f(t)$ is given by $f = 0$ for $t < 0$ and for $t \geqslant 0$

$$f = -\frac{a(1 - \nu)\tau^2 P_0}{2\rho_0(1 - 2\nu)} + e^{-(\beta t/\tau)}\,[A \sin(\gamma t/\tau) + B \cos(\gamma t/\tau)]$$

where $\beta = (1 - 2\nu)/(1 - \nu)$, $\gamma = (1 - 2\nu)^{\frac{1}{2}}/(1 - \nu)$, $\tau = a/c$ and where A and B are constants of integration.

Finally from the condition that at the wave front ($r = a + ct$) the material is continuous (so that $u_r = 0$), deduce that

$$B = \frac{a(1 - \nu)\tau^2 P_0}{2\rho_0(1 - 2\nu)}, \qquad A = (\beta B/\gamma) .$$

26. Show that one solution of (14.6.20) is $\xi = 0$ and that the remaining solutions satisfy

$$\xi^6 - 8\xi^4 + 8(3 - 2k)\xi^2 - 16(1 - k) = 0 .$$

Prove that the latter possesses a real root $0 < \xi^2 < 1$ for $0 \leqslant k < 1$.

Relevant values of k for $0 \leqslant \nu \leqslant \frac{1}{2}$ are $0 \leqslant k \leqslant \frac{1}{2}$, so that the Rayleigh equation always possesses one real root $(c/c_2) < 1$.

27. **(i)** Show that $\qquad \phi = f[t - (\mathbf{n} \cdot \mathbf{r})/c_1]$

(where $\mathbf{n}$ is a constant unit vector) is a solution of

$$\nabla^2 \phi - \frac{1}{c_1^2} \frac{\partial^2 \phi}{\partial t^2} = 0$$

and that the associated displacement $\mathbf{u} = \text{grad}\, \phi$ represents a plane dilatational wave travelling in the direction $\mathbf{n}$.

Show similarly that $\qquad \mathbf{A} = \mathbf{m} g[t - (\mathbf{n} \cdot \mathbf{r})/c_2]$

(where $\mathbf{n}$ is a constant unit vector and $\mathbf{m}$ a constant vector) is a solution of

$$\nabla^2 \mathbf{A} - \frac{1}{c_2^2} \frac{\partial^2 \mathbf{A}}{\partial t^2} = 0$$

and that the associated displacement $\mathbf{u} = \text{curl}\, \mathbf{A}$ represents a plane shear wave travelling in the direction $\mathbf{n}$, with $\mathbf{u}$ in the direction $\mathbf{m} \wedge \mathbf{n}$.

(ii) If $\phi = f(x, y, t)$ is independent of z and $\mathbf{A} = A(x, y, t)\mathbf{k}$ lies in the z direction but is independent of z, the resulting sum displacement field is two dimensional given by

$$u_x = \frac{\partial \phi}{\partial x} + \frac{\partial A}{\partial y}, \qquad u_y = \frac{\partial \phi}{\partial y} - \frac{\partial A}{\partial x}, \qquad u_z = 0 .$$

These results are useful in formulating reflection problem for plane waves (Problems 28-31 below).

28. An elastic half space $y \geqslant 0$ is characterised by dilatational and shear wave speeds c_1 and c_2 respectively. The surface $y = 0$ is bonded to a rigid barrier. A plane dilatational wave ϕ_I is incident on the barrier at an angle θ to the normal so that ϕ_I is of the form

$$\phi_I = f[t - (x \sin \theta/c_1) + (y \cos \theta/c_1)] . \qquad (1)$$

Evidently (1) does not satisfy the boundary condition $u_x = u_y = 0$ on $y = 0$, so that on colliding with the barrier, the incident wave must generate reflected waves. Further since *two* boundary conditions must be met, it seems that in general both a dilatational and shear wave are necessary to solve the problem.

Assume a two-dimensional displacement field

$$u_x = \frac{\partial \phi}{\partial x} + \frac{\partial A}{\partial y}, \quad u_y = \frac{\partial \phi}{\partial y} - \frac{\partial A}{\partial x}$$

where
$$\phi = \phi_I + \phi_R, \quad A = A_R$$

and where ϕ_R and A_R represent respectively reflected dilatational and shear waves. ϕ_R and A_R may be written

$$\phi_R = g[t - (x \sin \psi/c_1) - (y \cos \psi/c_1)]$$
$$A_R = h[t - (x \sin \omega/c_2) - (y \cos \omega/c_2)]$$

where g and h are functions to be determined, while ψ and ω are the angles of reflection of the two waves. A schematic diagram of the reflection process is shown below.

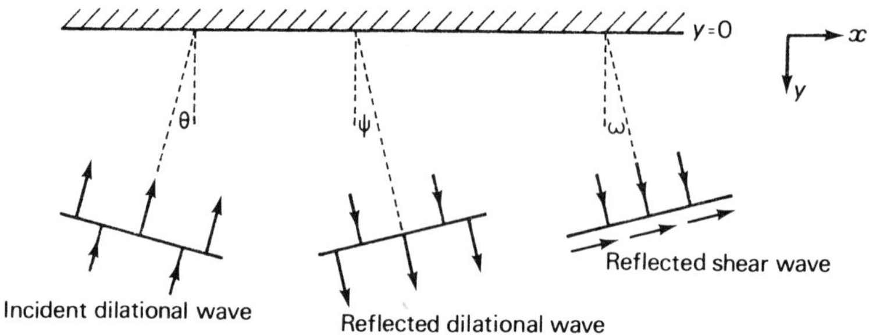

(i) Show that the boundary condition $u_x = u_y = 0$ at $y = 0$ can be met for all values of x only if the arguments of f, g and h are identical for $y = 0$,

i.e.
$$\frac{\sin \theta}{c_1} = \frac{\sin \psi}{c_1} = \frac{\sin \omega}{c_2}.$$

Deduce that $\psi = \theta$ and $\sin \omega = q \sin \theta \quad (q = c_2/c_1)$

and, since $q < 1$, that the second equation always possesses a real solution for ω.

(ii) Derive the equations

$$q \sin \theta\,[f'(t) + g'(t)] + [1 - q^2 \sin^2 \theta]^{\frac{1}{2}} h'(t) = 0$$
$$\cos \theta\,[f'(t) - g'(t)] + \sin \theta\, h'(t) = 0$$

and deduce that

$$h = -\frac{(2q \sin\theta \cos\theta)f}{[q \sin^2\theta + \cos\theta(1 - q^2 \sin^2\theta)^{\frac{1}{2}}]},$$

$$g = \frac{[\cos\theta(1 - q^2 \sin^2\theta)^{\frac{1}{2}} - q \sin^2\theta]\, f}{[q \sin^2\theta + \cos\theta\,(1 - q^2 \sin^2\theta)^{\frac{1}{2}}]}$$

determining h and g in terms of the incident wave.

(iii) Show that there is no reflected dilatational wave for an incident angle θ given by $\tan\theta = q^{-1}$.

29. Consider Problem 28 with the modification that the surface $y = 0$ is free of traction (i.e. $\sigma_{yy} = \sigma_{xy} = 0$). In the notation of Problem 28 show that

$$\psi = \theta\,, \qquad \sin\omega = q \sin\theta$$

as previously, but that the relations connecting f, g and h are now

$$h = \frac{[4q^2 \cos\theta \sin\theta(1 - 2q^2 \sin^2\theta)]\, f}{4q^3 \sin^2\theta \cos\theta(1 - q^2 \sin^2\theta)^{\frac{1}{2}} + (1 - 2q^2 \sin^2\theta)^2}$$

$$g = \frac{[4q^3 \cos\theta \sin^2\theta(1 - q^2 \sin^2\theta)^{\frac{1}{2}} - (1 - 2q^2 \sin^2\theta)^2]\, f}{[4q^3 \sin^2\theta \cos\theta(1 - q^2 \sin^2\theta)^{\frac{1}{2}} + (1 - 2q^2 \sin^2\theta)^{\frac{1}{2}}]}\,.$$

HINT, show first that the boundary conditions on $y = 0$ are equivalent to

$$\frac{\partial u_x}{\partial y} + \frac{\partial u_y}{\partial x} = 0\,, \qquad \frac{\partial u_y}{\partial y} = (2q^2 - 1)\frac{\partial u_x}{\partial x}\,.$$

30. **(i)** A plane shear wave

$$A_I = f[t - (x \sin\theta/c_2) + (y \cos\theta/c_2)]$$

is incident on the rigid barrier $y = 0$. Show that the boundary conditions $u_x = u_y = 0$ are met by the assumption of a reflected shear wave

$$A_R = g[t - (x \sin\psi/c_2) - (y \cos\psi/c_2)]$$

together with a reflected dilatational wave

$$\phi_R = h[t - (x \sin\omega/c_1) - (y \cos\omega/c_1)]$$

provided $\qquad \psi = \theta \quad$ and $\quad \sin\omega = q^{-1} \sin\theta$

where $\qquad q = \dfrac{c_2}{c_1} = \left[\dfrac{1 - 2\nu}{2(1 - \nu)}\right]^{\frac{1}{2}}.$

Since $q < 1$ deduce that the proposed form of solution is valid only for angles of incidence θ satisfying

$$\sin\theta < q\,.$$

(ii) For these angles of incidence derive the relations

$$h = \frac{(2\sin\theta\cos\theta)f}{[\sin^2\theta + \cos\theta(q^2 - \sin^2\theta)^{\frac{1}{2}}]},$$

$$g = \frac{[\cos\theta(q^2 - \sin^2\theta)^{\frac{1}{2}} - \sin^2\theta]f}{[\sin^2\theta + \cos\theta(q^2 - \sin^2\theta)^{\frac{1}{2}}]}$$

and deduce that there is no reflected shear wave for the case

$$\tan\theta = q .$$

Verify that this possibility lies within the valid range $0 < \theta < \sin^{-1}(q)$.

(iii) For $\theta > \sin^{-1} q$ there is no general wave function solution.

Show however that for the particular case of an incident (complex) harmonic wave

$$f = f_0(\omega)e^{i\omega[t - (x\sin\theta/c_2) + (y\cos\theta/c_2)]}$$

that the problem admits a (complex) solution

$$h = h_0(\omega)e^{i\omega[t - (x\sin\theta/c_2)] - |\omega| y(\sin^2\theta - q^2)^{1/2}/qc_1}$$

$$g = g_0(\omega)e^{i\omega[t - (x\sin\theta/c_2) - (y\cos\theta/c_2)]}$$

where
$$h_0 = \frac{(2\sin\theta\cos\theta)f_0}{[\sin^2\theta \mp i\cos\theta(\sin^2\theta - q^2)^{\frac{1}{2}}]},$$

$$g_0 = -\frac{[\sin^2\theta \pm i\cos\theta(\sin^2\theta - q^2)^{\frac{1}{2}}]f_0}{[\sin^2\theta \mp i\cos\theta(\sin^2\theta - q^2)^{\frac{1}{2}}]}$$

and where the upper sign is valid for $\omega > 0$ and the lower sign for $\omega < 0$.

[Thus for $\theta > \sin^{-1} q$ the reflected dilatational "wave" is damped exponentially in the positive y direction. Exponentially increasing solutions also satisfy the mathematical requirements of the problem on $y = 0$, but must be rejected on physical grounds.]

Discuss how the above solution may be generalised by using the theory of Fourier integrals and show formally how to solve the problem for an incident plane wave

i.e.
$$A_I = f[t - (x\sin\theta/c_2) + (y\sin\theta/c_2)]$$

where f is an arbitrary function.

31. **(i)** Show that the equations of elastodynamics

$$(\lambda + \mu)\,\mathrm{grad}\,(\mathrm{div}\,\mathbf{u}) + \mu\nabla^2\mathbf{u} = \rho_0\partial^2\mathbf{u}/\partial t^2$$

possess solutions of the form

$$u_x = \frac{\partial\Phi}{\partial x} + \frac{\partial\psi}{\partial y}, \quad u_y = \frac{\partial\Phi}{\partial y} - \frac{\partial\psi}{\partial x}, \quad u_z = 0,$$

where $\Phi(x, y, t)$ and $\psi(x, y, t)$ satisfy

$$c_1^2\left(\frac{\partial^2 \Phi}{\partial x^2} + \frac{\partial^2 \Phi}{\partial y^2}\right) = \frac{\partial^2 \Phi}{\partial t^2}, \quad c_2^2\left(\frac{\partial^2 \psi}{\partial x^2} + \frac{\partial^2 \psi}{\partial y^2}\right) = \frac{\partial^2 \psi}{\partial t^2}$$

and where $$c_1^2 = (\lambda + 2\mu)/\rho_0, \quad c_2^2 = \mu/\rho_0 .$$

(ii) A plane shear wave propagating in the half plane $y \geqslant 0$ is specified by

$$\psi_{\text{incident}} = f[t - (x \sin\theta)/c_2 + (y \cos\theta)/c_2] ,$$

where $f(\tau)$ is known. The wave is incident on the plane boundary $y = 0$ from $y > 0$ and the angle between the direction of propagation and the positive y direction is θ, where $0 < \theta < \frac{1}{2}\pi$.

The boundary $y = 0$ adjoins a plane rigid barrier inhibiting any normal displacement so that $u_y = 0$ on $y = 0$. Also the surfaces of the half plane and the barrier are lubricated so that the half plane surface is unable to sustain shear tractions.

Show how one might expect to satisfy the boundary conditions on $y = 0$ by assuming a reflected shear wave specified by

$$\psi_{\text{reflected}} = g[t - (x \sin\theta)/c_2 - (y \cos\theta)/c_2]$$

together with a reflected dilatational wave specified by

$$\Phi_{\text{reflected}} = h[t - (x \sin\gamma)/c_1 - (y \cos\gamma)/c_1] ,$$

where γ is given by $$\sin\gamma = (c_1/c_2)\sin\theta .$$

(it may be supposed for the moment that $(c_1/c_2)\sin\theta \leqslant 1$, so that a real angle γ exists.)

Show from detailed calculations that in fact there is no reflected dilatational wave and find the relation between g and f. Is the solution obtained restricted to angles of incidence $0 < \theta \leqslant \sin^{-1}(c_2/c_1)$?

The solution obtained is valid only if the normal stress supported by the lubricated contact between the half plane and the rigid barrier is not tensile; show that this is the case provided $f''(\tau) \leqslant 0$ for all relevant τ.

Transverse vibrations of a free beam

32. A free beam of uniform cross section is set into transverse motion by an impact at $t = 0$. For $t > 0$ the transverse motion satisfies the beam equation

$$\frac{\partial^4 Y(x, t)}{\partial x^4} + \frac{\rho_0 A_0}{EI}\frac{\partial^2 Y}{\partial t^2} = 0$$

and the boundary conditions $M = Q = 0 \quad (x = 0, \quad x = l)$

where l is the length of the beam.

Show that the equation admits periodic solutions

$$Y = f(x)\,{}^{\cos}_{\sin}(\omega t)$$

where f is of the form

$$f(x) = A \sin \alpha x + B \cos \alpha x + C \sinh \alpha x + D \cosh \alpha x$$

and where $$\alpha^4 = \rho_0 A_0 \omega^2 / EI\ .$$

Show that the boundary conditions are met provided αl satisfies the equation

$$\cos(\alpha l)\cosh(\alpha l) = 1$$

and indicate how to determine approximately the admissable (circular) frequencies ω_1, ω_2, etc. Show that the higher frequencies are given approximately by

$$\omega_n \simeq \left(\frac{EI}{\rho_0 A_0}\right)^{\frac{1}{2}} \pi^2 l^{-2} (n - \tfrac{1}{2})^2\ .$$

33. **(i)** The set of eigenfunctions $F_n(x)$ for the vibrating cantilever beam problem satisfy the equations

$$F_n^{(iv)}(x) = \alpha_n^4 F_n(x)$$

where the α_n form a discrete and distinct set of eigenvalues.

From the differential equations satisfied by $F_n(x)$ and $F_m(x)$ deduce

$$(\alpha_n^4 - \alpha_m^4)\int_0^l F_n F_m\, dx = \int_0^l (F_m F_n^{(iv)} - F_n F_m^{(iv)})dx$$

Deduce that if the F_n satisfy the boundary conditions

$$F_n(0) = F_n'(0) = 0\,, \quad F'''(l) = F''(l) = 0$$

then for $n \neq m$

$$\int_0^l F_n(x)F_m(x)dx = 0\ .$$

(ii) Show similarly that the eigenfunctions of the freely vibrating beam problem (Problem 32) are also orthogonal.

It is of interest that orthogonality is provable from the differential equation and boundary conditions without appeal to the specific form of the eigenfunctions.

34. Uniqueness theorem for classical elastostatics

It is shown on p 186-187 that the solution of Laplace's equation $\nabla^2\phi = 0$ in a region Ω bounded by a surface S on which ϕ is prescribed is unique. A related problem in classical elastostatics concerns uniqueness for stress and strain in the region Ω where the surface S is subject to the following mixed boundary conditions. The surface S is supposed divided into regions S_u and the complementary region $S_T = S - S_u$. On S_u the displacement is prescribed while on S_T the surface traction is prescribed.

Suppose there are two solutions $(\sigma_{ij}{}^{(1)}, u_i{}^{(1)}, e_{ij}{}^{(1)})$ and $(\sigma_{ij}{}^{(2)}, u_i{}^{(2)}, e_{ij}{}^{(2)})$ of the elastostatic problem which satisfy the equilibrium equations (including body force), the constitutive equations, the strain-displacement gradient relations and the boundary conditions. Show that the "difference" solution

$$\sigma_{ij} = \sigma_{ij}{}^{(1)} - \sigma_{ij}{}^{(2)}, \quad u_i = u_i{}^{(1)} - u_i{}^{(2)}, \quad e_{ij} = e_{ij}{}^{(1)} - e_{ij}{}^{(2)}$$

satisfies

$$\frac{\partial \sigma_{ij}}{\partial x_j} = 0, \quad e_{ij} = \frac{1}{2}\left(\frac{\partial u_i}{\partial x_j} + \frac{\partial u_j}{\partial x_i}\right), \quad \sigma_{ij} = \lambda e_{mm}\delta_{ij} + 2\mu e_{ij}$$

and also $u_i = 0$ on S_u, $\sigma_{ij}n_j = 0$ on S_T.

Derive the identity

$$\iint_S \sigma_{ij} n_j u_i dS = \iiint_\Omega u_i \frac{\partial \sigma_{ij}}{\partial x_j} d\Omega + \iiint_\Omega \sigma_{ij} e_{ij} d\Omega$$

and with the help of the above results and [from equation (8.6.4)]

$$\sigma_{ij}e_{ij} = 2\rho_0 U,$$

where U is the stored energy per unit mass, deduce that

$$\rho_0 \iiint_\Omega U d\Omega = 0 \tag{1}$$

Since $\rho_0 U$ is a positive definite quadratic form in the components of strain, the only way in which (1) can be fulfilled is if $e_{ij} = 0$ throughout Ω. It follows that

$$e_{ij}{}^{(1)} = e_{ij}{}^{(2)}$$

so that the solution of the mixed boundary problem is unique for strain, and therefore also for stress. However the displacement is not uniquely specified in the particular case of the simpler boundary value problem where for the whole of S the surface tractions are prescribed. For evidently for any solution of the latter problem, it is possible to add to the displacement field the terms $\boldsymbol{\alpha} + \boldsymbol{\omega} \wedge \mathbf{r}$ where $\boldsymbol{\alpha}$ and $\boldsymbol{\omega}$ are arbitrary vectors. The additional terms representing a rigid body displacement, does not alter the prescribed boundary conditions.

35. The Rayleigh equation for longitudinal wave propagation

The simple one-dimensional longitudinal wave propagation theory of Section 14.6.4, culminating in equation (14.6.28), fails for waves whose characteristic lengths are comparable with the linear dimensions of the cross section of the bar. An improved, but still approximate, one-dimensional theory, which takes some account fo the lateral dimensions of the bar, may be obtained as follows.

The assumption that the only non-vanishing component of stress is $\sigma_{xx}(x, t)$, independent of y and z, is retained along with the assumption that the longitudinal displacement $u_x(x, t)$ is also independent of y and z. Displacement components in the y and z direction are estimated on the assumption that the strains e_{yy} and e_{zz} are as occur in the static case

i.e.
$$e_{zz} = e_{yy} = -\nu e$$

where $e = \partial u_x/\partial x \equiv e_{xx}$ and ν is Poisson's ratio. In particular for a bar of circular cross section this leads to displacements (in plane polar coordinates) $u_\theta = 0$ and

$$u_r = -\nu r e(x, t) \tag{1}$$

These assumptions are not quite correct for the dynamic case, but the resulting theory represents an improvement over the simple theory of Section (14.6.4).

With the assumption (1) show that the total kinetic energy of a disc of thickness Δx to first order in Δx is given by

$$K = \tfrac{1}{2}\rho_0 A\Delta x\left[\left(\frac{\partial u_x}{\partial t}\right)^2 + \tfrac{1}{2}\nu^2 a^2 \dot{e}^2\right]$$

where $A = \pi a^2$ is the radius of the disc. The corresponding elastic stored energy is

$$V = \tfrac{1}{2}AEe^2\,\Delta x$$

where E is Young's modulus. To first order in Δx the rate of change of $(K + V)$ is

$$A\Delta x\left[\rho_0\,\frac{\partial u_x}{\partial t}\,\frac{\partial^2 u_x}{\partial t^2} + \tfrac{1}{2}\rho_0\nu^2 a^2\dot{e}\,\ddot{e} + Ee\dot{e}\right] \tag{2}$$

The rate of working of the external forces on the disc is

$$A\left[\sigma_{xx}(x + \Delta x, t)\,\frac{\partial u_x}{\partial t}\,(x + \Delta x, t) - \sigma_{xx}(x, t)\,\frac{\partial u_x}{\partial t}\,(x, t)\right]$$

$$= A\left[\frac{\partial \sigma_{xx}}{\partial x}\,\frac{\partial u_x}{\partial t} + \sigma_{xx}\dot{e}\right]\Delta x \tag{3}$$

Equating (2) and (3) and recalling the equation of motion

$$\frac{\partial \sigma_{xx}}{\partial x} = \rho_0\,\frac{\partial^2 u_x}{\partial t^2} \tag{4}$$

leads to
$$\sigma_{xx} = Ee + \tfrac{1}{2}\rho_0\nu^2 a^2\ddot{e} \tag{5}$$

which differs from the elementary theory by the additional *lateral inertia* correction term involving a^2 and $\ddot{e}$. (Similar corrections are possible for bars of non-circular cross section).

Derive from (4) and (5) the Rayleigh equation

$$c_0{}^2 \frac{\partial^2 u_x}{\partial x^2} + \tfrac{1}{2}\nu^2 a^2 \frac{\partial^4 u_x}{\partial x^2 \partial t^2} = \frac{\partial^2 u_x}{\partial t^2} \tag{6}$$

where $c_0{}^2 = E/\rho_0$. Equation (6) is to be compared with (14.6.28) in the elementary theory.

Show that (6) is satisfied by separation of variable solutions of the form

$$u_x = e^{i\omega[t \pm x/c(\omega)]}$$

where
$$c(\omega)/c_0 = [1 - \tfrac{1}{2}\omega^2 a^2 \nu^2/c_0{}^2]^{\frac{1}{2}} .$$

These solutions yield real $c(\omega)$ for sufficiently small ω. In fact the entire theory is only valid provided the correction term $\frac{1}{2}\omega^2 a^2 \nu^2/c_0{}^2$ is small compared to unity. This means in the context of Fourier integral solutions of the form

$$u_x = \int_{-\infty}^{\infty} A(\omega) e^{i\omega[t \pm x/c(\omega)]} d\omega \tag{7}$$

which represent pulses propagating in the negative and positive x directions with phase distortion, that the theory is reasonably valid provided the Fourier spectrum $A(\omega)$ of the pulse is effectively zero for values of ω for which $\frac{1}{4}\omega^2 a^2 \nu^2/c_0{}^2$ is no greater than (say) about 0.2. In these circumstances it is adequate to approximate $c(\omega)$ by

$$c(\omega) = c_0 [1 - \tfrac{1}{4}\omega^2 a^2 \nu^2/c_0{}^2] \tag{8}$$

The wavelength $2\pi c/\omega$ of waves travelling with the maximum allowable frequency is $\pi a\nu/(0.2)^{1/2} \simeq 2.3a$ (for $\nu = \frac{1}{3}$), so that the Rayleigh approximation is adequate for pulses whose profiles do not change significantly over distances much shorter than a bar diameter.

36. Kelvin method of stationary phase

(i) Basic Analysis

The mathematics of Problem 35 provides one of many examples in applied mathematics which lead to Fourier integrals of the type

$$u = \int_{-\infty}^{\infty} A(\omega) e^{i\omega[t - x/c(\omega)]} d\omega \tag{1}$$

representing a pulse $u(x, t)$ propagating in the positive x direction. For the simplest case where $c(\omega)$ is constant, independent of ω equation (1) shows that u is an arbitrary function of $(t - x/c)$ which, from the discussion of Section 13.2, represents a pulse which propagates without distortion in the positive x direction at speed c.

However if $c(\omega)$ depends on ω the interpretation given immediately above fails, and the question arises as to how the pulse propagates and changes profile. Some answers to these questions are provided by the Kelvin method of stationary phase which evaluates the integral (1) approximately for large values of t. The argument presented here is without mathematical rigor and based on physical intuition; for a more precise account see E. T. Copson, *Asymptotic Expansions,* Chapter 4 C.U.P. (1965) and A. Erdélyi *Asymptotic Expansions,* Chapter 2, Dover Publications (1965).

In what follows the analysis is simplified if t and x are regarded as positive (independent) variables. Accordingly we assume that the Fourier integral (1) refers to a pulse propagating along $x \geqslant 0$ for $t \geqslant 0$. In particular we assume, without loss of generality that the pulse is initiated at $x = 0$ and for $t < 0$ there is no disturbance so that

$$u(0, t) \equiv \int_{-\infty}^{\infty} A(\omega)e^{i\omega t} = f(t)H(t)$$

where H is the Heaviside function and $f(t)$ is a given function of t.†

From Fourier's integral theorem

$$2\pi A(\omega) = \int_{0}^{\infty} f(t)e^{-i\omega t}dt$$

determining the Fourier spectrum (from here on assumed known) of $f(t)$.

Kelvin argues that the phase

$$\Phi(\omega, x, t) = \omega[t - x/c(\omega)]$$

appearing in the exponential in the integrand of (1) is, for most values of ω, a rapidly changing quantity. In these circumstances, and providing $A(\omega)$ is reasonably smooth, one might expect that individual wavelets

$$A(\omega)e^{i\omega[t - x/c(\omega)]}d\omega,$$

contributing to the integral, destructively interfere with each other for neighbouring values of ω. Exceptionally Φ will not vary rapidly in the vicinity of a stationary point $\omega = \omega_1$ (say) for which $(\partial\Phi/\partial\omega)_{\omega_1} = 0$.

To assess contributions to u from the vicinity of $\omega = \omega_1$ we write

$$\Phi = \gamma(\omega, x/t)t \tag{2}$$

† $H(t)$ is defined on p. 445.

where $$\gamma = \omega[1 - (x/t)/c(\omega)] \ . \tag{3}$$

The reason for this is that ultimately we are looking for an approximation to (1) valid for sufficiently large t while (x/t) remains finite. In turn this is motivated by physical arguments. For if $u(0, t)$ is a pulse of finite width we expect the pulse to change shape as it propagates but to remain recognisably a travelling pulse. In some sense the *centre* of the travelling pulse is expected to move at a reasonably well defined velocity, and this implies that for those values of x for which the pulse is of significant amplitude, the quantity (x/t) remains finite, somewhere about the velocity of the centre of the pulse.

The phase Φ is stationary for $\partial\gamma/\partial\omega = 0$, i.e. for points $\omega = \omega_1(x/t)$ (say). Near $\omega = \omega_1$ we have from Taylor's theorem

$$\gamma = \gamma(\omega_1) + \chi(\omega_1)(\omega - \omega_1)^2 + \ldots \tag{4}$$

where $\chi(\omega_1) = \frac{1}{2}(\partial^2\gamma/\partial\omega^2)_{\omega_1}$ is assumed non-zero. [If $\chi(\omega_1) = 0$, the method may be elaborated, see Copson and Erdélyi – loc. cit.] The principal contribution to (1) from $\omega = \omega_1$ arises from some interval – say $\omega_1 - \alpha < \omega < \omega_1 + \alpha$ where α is chosen small enough to justify approximating γ by the explicitly identified terms in (4). Accordingly, from the stationary point $\omega = \omega_1$ there arises the contribution

$$u_1 = \int_{-\alpha}^{\alpha} A(\omega_1 + q)e^{i[\gamma(\omega_1)t + \chi(\omega_1)q^2 t]}dq$$

For sufficiently smooth $A(\omega)$, i.e. for a sufficiently sharp initial pulse $u(0, t)$, $A(\omega_1 + q) \simeq A(\omega_1)$ so that

$$u_1 \simeq A(\omega_1)e^{i\gamma(\omega_1)t}\int_{-\alpha}^{\alpha} e^{i\chi(\omega_1)q^2 t}dq$$

To evaluate the integral for large t, we change the integration variable from q to $\eta = |\chi(\omega_1)|^{1/2}t^{1/2}q$ obtaining

$$u_1 = A(\omega_1)e^{i\gamma(\omega_1)t}|\chi(\omega_1)|^{-\frac{1}{2}}t^{-\frac{1}{2}}\int_{-|\chi t|^{1/2}\alpha}^{|\chi t|^{1/2}\alpha} e^{\pm i\eta^2}d\eta$$

where the $\pm$ signs are associated respectively with $\chi \gtrless 0$. For sufficiently large t the limits on the integrand may be taken to be $\pm\infty$.

Show from contour integration

$$\int_{-\infty}^{\infty} e^{\pm ix^2}dx = (1 \pm i)(\pi/2)^{\frac{1}{2}}$$

and hence derive for sufficiently large t

$$u_1 = A(\omega_1)e^{i\gamma(\omega_1)t}|\chi(\omega_1)|^{-\frac{1}{2}}t^{-\frac{1}{2}}(\pi/2)^{\frac{1}{2}}(1 \pm i) . \tag{5}$$

The sum of all such contributions from the various stationary points then yields the Kelvin approximation to (1).

What is meant by sufficiently smooth, sufficiently sharp and sufficiently large t, depends in detail on the nature of the functions $A(\omega)$ and $\gamma(\omega)$.

(ii) GROUP VELOCITY

The stationary points, located at values of ω where $\partial\gamma/\partial\omega = 0$ are given by

$$\frac{\partial}{\partial\omega}\left\{\omega\left[1 - \frac{x/t}{c(\omega)}\right]\right\} = 0$$

The *group velocity* $c_g(\omega)$ is defined by

$$c_g^{-1}(\omega) = \frac{d}{d\omega}\left[\frac{\omega}{c(\omega)}\right] \tag{6}$$

so that stationary points occur at the solution of

$$c_g(\omega) = (x/t) . \tag{7}$$

Equation (7) above, carries the physical implication that if the Fourier spectrum $A(\omega)$ is dominated by contributions for which ω is in the vicinity of Ω (say), then the propagating pulse travels at a speed of about $c_g(\Omega)$, since the only values of (x/t) for which $A(\omega)$ is appreciably non-zero occur near $x = c_g(\Omega)t$. This particular viewpoint provides the interpretation favoured by physicists, i.e. that a *group* of wavelets with frequencies in the vicinity of Ω travels at the group velocity $c_g(\Omega)$.

(iii) We consider here integrals of the type (7) of Problem 35

$$u = \int_{-\infty}^{\infty} A(\omega)e^{i\omega[t - x/c(\omega)]}\,d\omega$$

where $c^{-1}(\omega)$ is approximated by

$$c^{-1}(\omega) = c_0^{-1}(1 + \tfrac{1}{4}\nu^2a^2\omega^2/c_0^2) .$$

Show that the corresponding group velocity is given by

$$c_g^{-1}(\omega) = c_0^{-1}(1 + \tfrac{3}{4}\nu^2a^2\omega^2/c_0^2)$$

and that stationary points occur for $\omega = \pm\,\omega_1$ where

$$\omega_1 = \frac{2c_0}{\sqrt{3}}\left(\frac{c_0t}{x} - 1\right)^{\frac{1}{2}}\Big/\nu a .$$

This result suggests that only values of x less than $c_0 t$ are relevant, i.e. that no disturbances propagate at a speed greater than c_0. This is hardly surprising since in the present model $c_g \leqslant c_0$ for all ω.

Calculate the corresponding values of $\chi(\pm\,\omega_1)$ and $\gamma(\pm\,\omega_1)$ and show that the stationary phase approximation leads to

$$u = (2\pi/\nu a x)^{\frac{1}{2}} c_0 \left[3\left(\frac{c_0 t}{x} - 1\right)\right]^{-\frac{1}{4}} \operatorname{Re} A\left[\frac{2c_0}{\sqrt{3}\,\nu a}\left(\frac{c_0 t}{x} - 1\right)^{\frac{1}{2}}\right] e^{i\left[\frac{4x}{3\sqrt{3\nu a}}\left(\frac{c_0 t}{x} - 1\right)^{\frac{1}{2}} - \frac{1}{4}\pi\right]}$$

Because the approximations to the group velocity and phase velocity are only accurate for small values of ω, this result is only accurate for values of $c_0 t/x$ not too far removed from unity.

Chapter 15

The Theory of Linear Viscoelasticity

15.1 Introduction

In the final two chapters of this book the reader is introduced to two non-classical theories which are used widely to describe two disparate types of mechanical behaviour in materials. The theory of linear viscoelasticity discussed in the present chapter is largely pertinent to the small deformations of rubber, polymer and elastomer types of materials, i.e. materials whose fundamental structure is one of a tangled web of long chain hydrocarbon molecules, possibly cross linked. The theory of plasticity discussed in the final chapter is pertinent to the deformation of metals at stress levels in excess of a **yield stress**, beyond which permanent deformation ensues.

15.2 Basic behaviour and one dimensional constitutive equations for linear viscoelastic materials

Many materials exhibit time dependent mechanical properties so that necessarily time enters the constitutive equations. A familiar example occurs on stretching a rubber band. On release of the tension the band contracts at first rather rapidly and then more slowly, resuming its initial configuration only after a lapse time. The behaviour is more marked in some rubbery materials than others and is remarkably sensitive to temperature.

To introduce the constitutive equations of linear viscoelasticity it is simplest to work initially in one dimension and to discuss the relation between longitudinal stress and strain as in the elementary Hooke's law experiment.

For a classical elastic material the constitutive equation is

$$\sigma = Ee \tag{15.2.1}$$

where σ and e are respectively the longitudinal stress and strain and E is Young's modulus. Because time enters as an essential ingredient into the constitutive equation of a viscoelastic material, the equation analogous to (15.2.1) relates stress and strain cycles $\sigma(t)$, $\epsilon(t)$.† We consider first the case of a step function stress cycle defined by

$$\sigma(t) = \sigma_0 H(t) \tag{15.2.2}$$

† In the discussion on constitutive equations we suppress explicit reference to dependence of stress and strain on space variables. Of course in the solution of stress analysis problems stress, strain and displacement depend on position as well as time.

where $H(t)$ is the Heaviside step function

$$H(t) = \begin{cases} 1 & (t \geqslant 0) \\ 0 & (t < 0) . \end{cases}$$

For a linear viscoelastic material the resulting strain for $t \geqslant 0$ is given by

$$e(t) = \frac{\sigma_0}{E_D} [1 + \psi(t)] . \qquad (15.2.3)$$

Here E_D is a constant which is known as the **dynamic** (Young's) **modulus** and σ_0/E_D is an instantaneous elastic response. The term $\sigma_0 \psi(t)/E_D$ represents a delayed strain where $\psi(t)$ is a dimensionless function of time for which $\psi(0) = 0$. This last result is a consequence of the definition of $\psi(t)$. The function $\psi(t)$ may be termed the **creep function,** since extension of materials under constant load is commonly called creep. For those materials for which the present theory is applicable, experimental evidence suggests that $\psi(t)$ is a monotonically increasing function with monotonically decreasing slope. For materials like rubber whose long time behaviour in the present context is that of a solid, $\psi(t)$ approaches a constant value as $t \to \infty$ while for materials like bitumens and tars whose ultimate behaviour is that of a (viscous) liquid $\psi(t)$ increases without limit. A schematic representation of the creep behaviour of linear viscoelastic solids and fluids is depicted in Fig. 15.1. Here we concern ourselves primarily with the solid case.

In (15.2.3) the strain $e(t)$ is directly proportional to the stress amplitude σ_0; this is the characteristic assumption for a *linear* viscoelastic material. The validity or otherwise of the assumption is a matter for experiment; for many rubber, polymer and elastomer materials the assumption (15.2.3) appears to be reasonably valid up to strains of about 10^{-1}.

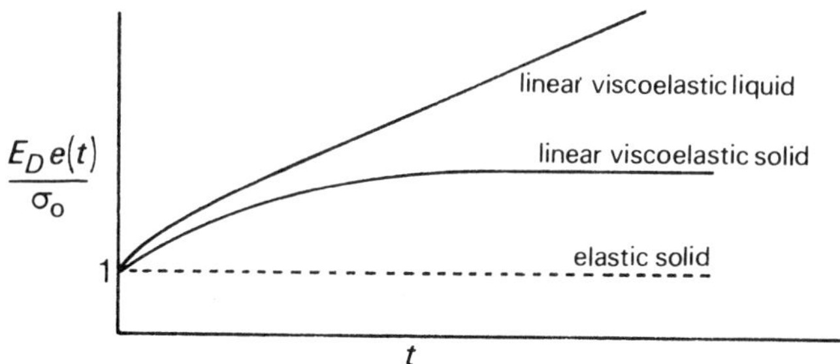

Fig. 15.1 Schematic diagram of the creep behaviour of a linear viscoelastic material

Equation (15.2.3) is readily generalised to cover the case of arbitrary stress cycles $\sigma(t)$. We suppose $\sigma(t)$ to be approximated by a series of step functions of infinitesimal amplitude $\Delta\sigma(t')$ occurring at time t' (Fig. 15.2). Each step $\Delta\sigma(t')$ give rise to a strain increment at time t of

$$\Delta e(t) = \frac{\Delta\sigma(t')}{E_D} [1 + \psi(t - t')] \quad (t \geqslant t')$$

so that the total strain due to all such increments is

$$e(t) = \underset{\Delta\sigma \to 0}{\text{Lt}} \; E_D^{-1} \sum_{\Delta\sigma} \Delta\sigma(t')\,[1 + \psi(t - t')]$$

$$= E_D^{-1} \int_{-\infty}^{t} \frac{d\sigma(t')}{dt'}\,[1 + \psi(t - t')]\,dt'$$

$$= E_D^{-1} \left[\sigma(t) + \int_{-\infty}^{t} \psi(t - t')\,\frac{d\sigma(t')}{dt'}\,dt' \right]. \tag{15.2.4}$$

In obtaining this last expression it has been assumed that the material is unstressed for $t \to -\infty$. If $\sigma(t) = 0$ for $t < 0$, the integral in (15.2.4) may be modified by replacing the lower limit $t' = -\infty$ by $t' = 0$.

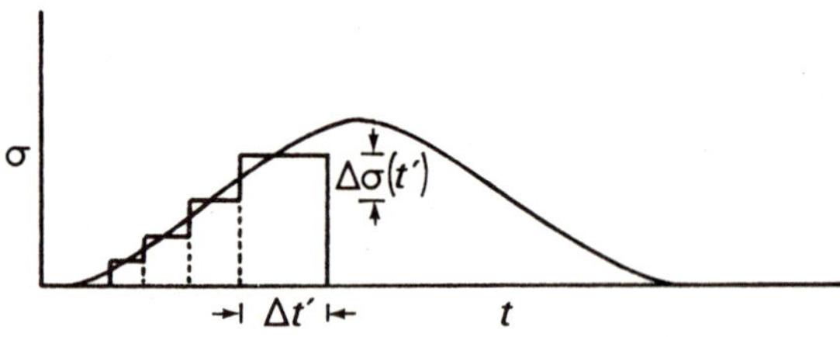

Fig. 15.2 Approximation to $\sigma(t)$ by sum of step functions.

The equation

$$e(t) = E_D^{-1} \left[\sigma(t) + \int_{-\infty}^{t} \psi(t - t')\,\frac{d\sigma(t')}{dt'}\,dt' \right] \tag{15.2.5}$$

may be taken as the basic constitutive equation of one dimensional linear visco-elasticity; there are other equivalent formulations which are discussed subsequently. The term $\sigma(t)/E_D$ represents an instantaneous elastic response while the remaining term, involving a knowledge of the past history of the stress, is due to creep. Because the current strain depends on the past history of the stress, quantities like the second term in (15.2.5) are sometimes called **hereditary integrals.**

Another formulation of the basic constitutive equation proceeds from a basic cycle for which the strain (rather than stress) assumes the step function behaviour

$$e(t) = e_0 H(t)$$

with associated stress

$$\sigma(t) = E_D e_0 \,[1 - \phi(t)] \qquad (t \geqslant 0) \tag{15.2.6}$$

Here $\phi(0) = 0$ and $\phi(t)$ is termed the **relaxation function.** [Experiments for which the strain is held constant and the stress varies with time are known as stress relaxation tests.] Experimentally $\phi(t)$ is a monotonically increasing function with monotonically decreasing slope. The value $\phi(\infty)$ is less or equal to unity;† the case where $\phi(\infty) = 1$ may be termed a completely relaxing material and $\phi(\infty) < 1$ a partially relaxing material. A completely relaxing material is one which shows long term viscous fluid behaviour in creep. A schematic diagram of the relaxation behaviour of viscoelastic materials is given in Fig. 15.3.

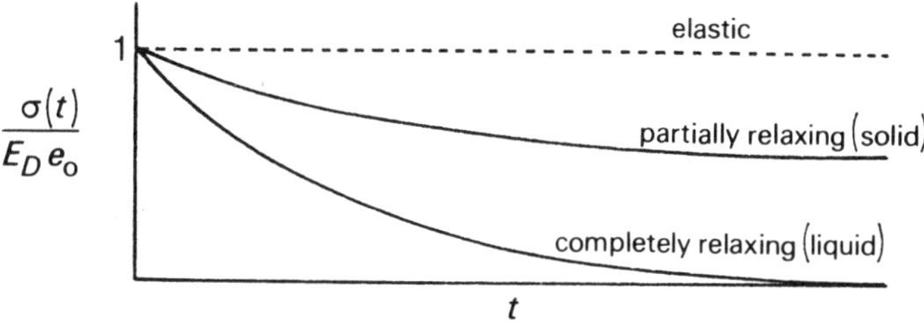

Fig. 15.3 Relaxation behaviour of viscoelastic material

For arbitrary $e(t)$, arguments identical with those used to derive (15.2.4) may be used to obtain the equation

$$\sigma(t) = E_D \left[e(t) - \int_{-\infty}^{t} \phi(t-t') \frac{de(t')}{dt'} dt \right] . \qquad (15.2.7)$$

Since equations (15.2.4) and (15.2.7) represent the same mechanical properties the functions ϕ and ψ appearing in the two equations are not independent. The relationship between ϕ and ψ may be represented by a (complicated) integral equation; however a much simpler algebraic relation exists between the Laplace transforms of the two functions. If (15.2.7) is multiplied by $\exp(-st)$ and integrated between $t = 0$ and $t = \infty$ there results

$$\overline{\sigma}(s) = E_D \left[\overline{e}(s) - \int_0^{\infty} e^{-st} dt \int_0^{t} \phi(t-t') \frac{de}{dt'} dt' \right] \qquad (15.2.8)$$

where $\overline{\sigma}(s)$ and $\overline{e}(s)$ are the Laplace transforms of $\sigma(t)$ and $e(t)$,

$$\overline{\sigma}(s) = \int_0^{\infty} \sigma(t) e^{-st} dt \; ; \qquad \overline{e}(s) = \int_0^{\infty} e(t) e^{-st} dt$$

† The result $\phi(\infty) \leqslant 1$ is an observational result which conforms with physical intuition; it seems very unlikely that an initially stress free specimen would ever support a compressive stress in stretch.

and where s is the transform parameter.† In deriving this result it has been assumed, without loss of generality, that the stress and strain cycles are initiated at $t = 0$ and that for $t < 0$ the material is stress (and strain) free. The second term on the right side of (15.2.8) may be simplified by the use of the **Faltung theorem.** This theorem is here worked through in detail. We interchange the order of integration in the double integral obtaining

$$\int_0^\infty e^{-st}dt \int_0^t \phi(t-t')\frac{de(t')}{dt'}dt' = \int_0^\infty \frac{de(t')}{dt'}dt' \int_{t'}^\infty e^{-st}\phi(t-t')dt\,. \tag{15.2.9}$$

The limits on the right side of (15.2.9) may be derived from appeal to Fig. 15.4 which shows the region of integration for the double integral to be the area between the straight line $t' = t$ and the positive t axis.

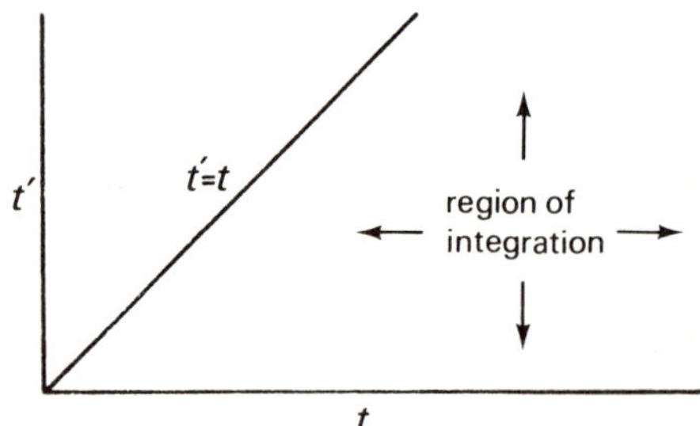

Fig. 15.4 Region of integration for the double integral (15.2.9)

The inner integral on the right side of (15.2.9) is simplified by changing the variable from t to $\tau = (t - t')$ which leads to

$$\int_{t'}^\infty e^{-st}\phi(t-t')dt = \int_0^\infty \phi(\tau)e^{-s(\tau+t')}d\tau = e^{-st'}\,\overline{\phi}(s)$$

where
$$\overline{\phi}(s) = \int_0^\infty \phi(t)e^{-st}dt\,.$$

Finally therefore

$$\int_0^\infty e^{-st}dt \int_0^t \phi(t-t')\frac{de(t')}{dt'}dt' = \overline{\phi}(s)\int_0^\infty \frac{de(t')}{dt'}e^{-st'}dt' = \overline{\phi}(s)s\,\overline{e}(s) \tag{15.2.10}$$

† There should be no confusion between the exponential function e^{-st} and the strain $e(t)$ appearing in (15.2.8), and subsequently.

where in the last step, obtained after integration by parts, we have made use of the assumption that the material is strain free at $t = 0$ and also that $\mathop{\mathrm{Lt}}_{t' \to \infty} [e^{-st'} e(t')] = 0.$

From (15.2.8), (15.2.9) and (15.2.10)

$$\bar{\sigma}(s) = E_D [1 - s\bar{\phi}(s)] \bar{e}(s) \tag{15.2.11}$$

while similarly (15.2.4) leads to

$$\bar{e}(s) = E_D^{-1} [1 + s\bar{\psi}(s)] \bar{\sigma}(s) . \tag{15.2.12}$$

Since the last two relationships must be equivalent there follows the algebraic relation

$$[1 + s\bar{\psi}(s)] \, [1 - s\bar{\phi}(s)] = 1 \tag{15.2.13}$$

connecting the Laplace transforms of $\phi(t)$ and $\psi(t)$.

The relations (15.2.11) and (15.2.12) are immensely valuable in the solution of problems. We define a **transform modulus**

$$E(s) = E_D \, [1 - s\bar{\phi}(s)] \equiv E_D [1 + s\bar{\psi}(s)]^{-1} \tag{15.2.14}$$

in terms of which the constitutive relation of one dimensional linear visco-elasticity may be written

$$\bar{\sigma}(s) = E(s)\bar{e}(s) . \tag{15.2.15}$$

The formal analogy between (15.2.15) and the elastic relation (15.2.1) provides the basis for solving wide classes of linear viscoelastic problems, discussed in Section 15.3.

Equation (15.2.13) provides a relation which determines ϕ when ψ is given, or vice-versa. However it is not possible to prescribe either completely arbitrarily. The reason is that viscoelastic properties obtain by virtue of dissipative processes occurring within the material, and this fact must be reflected in the general nature of the function ϕ (or ψ). It is possible, for example, to write down specific (unsuitable) functions for ϕ leading to materials which fail to conform even to the first law of thermodynamics i.e. materials from which it would be possible to extract mechanical energy during a stress cycle which ultimately restored the material to a stress and strain free state. Evidently no such energy source is allowable within the framework of generally accepted physical beliefs and for real viscoelastic materials only energy losses may occur. In the limiting case of an elastic material the energy losses are zero, for the entire energy stored in elastic deformation is recoverable.

For many years viscoelastic models were employed which guarantee the appropriate physical requirements. The models employ two types of element, namely a **linear spring** for which the mechanical behaviour is $\sigma = \lambda e$ and a **dashpot** with behaviour described by $\sigma = \mu\dot{e}$. Here λ and μ are constants.† The viscoelastic model is derived by coupling springs and dashpots in series and parallel. As an illustrative example consider the system displayed in Fig. 15.5 comprising a single dashpot and two springs.

† Not to be confused with the Lamé elastic constants.

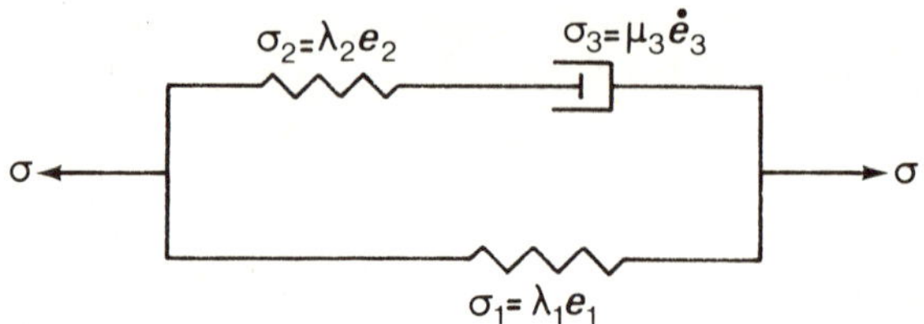

Fig. 15.5 Spring-dashpot representation of a standard linear solid.

The total stress on the system σ is compounded from the stresses in the two parallel branches. Also the stress in the spring λ_2 is the same as that in the dashpot μ_3. We have therefore the equations

$$\sigma = \sigma_1 + \sigma_2; \qquad \sigma_2 = \lambda_2 e_2 = \mu_3 \dot{e}_3 = \sigma_3 .$$

Additionally the strain e_1 occurring in λ_1 is the overall strain e in the model and also the sum strain occurring in λ_2 and μ_3 so that

$$e = e_1 = e_2 + e_3 = \sigma_1/\lambda_1$$

The quantities e_1 and e_2, σ_1 and σ_2 are now eliminated to provide a differential equation connecting $\sigma(t)$ and $e(t)$. We have

$$\dot{e} = \dot{e}_2 + \dot{e}_3 = \frac{\dot{\sigma}_2}{\lambda_2} + \frac{\sigma_3}{\mu_3} = \frac{\dot{\sigma}_2}{\lambda_2} + \frac{\sigma_2}{\mu_3}$$

$$= \frac{(\dot{\sigma} - \dot{\sigma}_1)}{\lambda_2} + \frac{(\sigma - \sigma_1)}{\mu_3} = \frac{\dot{\sigma}}{\lambda_2} + \frac{\sigma}{\mu_3} - \frac{\lambda_1 \dot{e}}{\lambda_2} - \frac{\lambda_1 e}{\mu_3}$$

On re-arranging this last equation there results

$$\dot{\sigma} + \frac{\sigma}{\tau_R} = E_D \left[\dot{e} + \frac{e}{\tau_C}\right] \qquad (15.2.16)$$

where E_D, given by $\qquad E_D = \lambda_1 + \lambda_2$

plays the role of dynamic modulus, while τ_R and τ_C which respectively play the roles of characteristic times for relaxation and creep behaviour are given by

$$\tau_R = \mu_3/\lambda_2 \qquad \tau_C = (\lambda_1 + \lambda_2)\mu_3/\lambda_1\lambda_2 . \qquad (15.2.17)$$

Equation (15.2.16) describes the so called **standard linear solid**. The equation offers a three parameter description (E, τ_C, τ_R) of a viscoelastic material and encompasses all the mechanical features of such solids. For very rapid deformations only the terms in $\dot{e}$ and $\dot{\sigma}$ are of importance and the material behaves as an elastic solid with (dynamic) Young's modulus E_D; for slow deformations only the terms in σ and e are important and again the behaviour is elastic but with (static) Young's modulus $\tau_R E_D/\tau_C$. For intermediate rates of deformation all terms in (15.2.16) are important.

We evaluate the transform modulus, creep and relaxation functions for the standard linear solid. On taking the Laplace transform of each side of (15.2.16), assuming $\sigma = e = 0$ for $t < 0$, there results

$$\bar{\sigma}(s) = E_D \left(\frac{s + \tau_C^{-1}}{s + \tau_R^{-1}} \right) \bar{e}(s)$$

with implication, from comparison with (15.2.15), that $E(s)$ is of the form

$$E(s) = E_D \left(\frac{s + \tau_C^{-1}}{s + \tau_R^{-1}} \right) . \qquad (15.2.18)$$

From this result and the first of (15.2.14) we find for $\bar{\phi}(s)$

$$\bar{\phi}(s) = (1 - \tau_R/\tau_C)\,[s^{-1} - (s + \tau_R^{-1})^{-1}]$$

with inversion

$$\phi(t) = (1 - \tau_R/\tau_C)\,[1 - e^{-t/\tau_R}] \qquad (15.2.19)$$

which is positive (since from (15.2.17) $\tau_R < \tau_C$), and monotonic increasing. From (15.2.18) and the second of (15.2.14), $\psi(t)$ is given by the similar formula

$$\psi(t) = (\tau_C/\tau_R - 1)\,[1 - e^{-t/\tau_C}] \qquad (15.2.20)$$

and the form of (15.2.19) and (15.2.20) justifies the interpretation of τ_R and τ_C as characteristic times for relaxation and creep.

Either (15.2.5) with (15.2.20) or (15.2.7) with (15.2.19) [or the differential equation (15.2.16)] may be taken as the one-dimensional constitutive equation of a standard linear solid. In fact no real material seems to be modelled adequately by the standard linear solid. More satisfactory representations of real materials are obtained by joining several standard linear solid elements (of different parameters) in series or parallel as in Fig. 15.6.

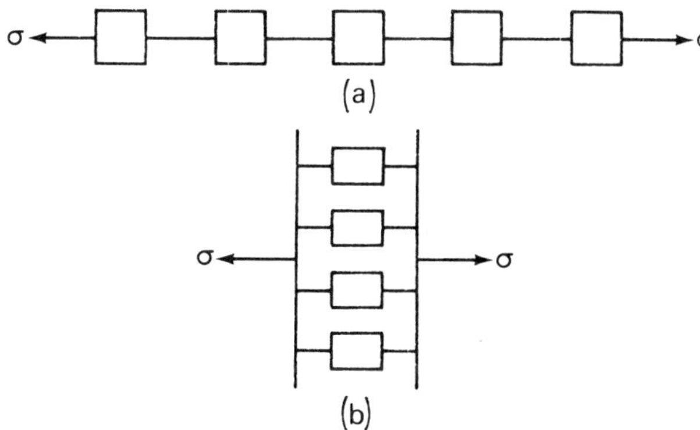

Fig. 15.6 Use of standard linear solid elements to build up more general material behaviour. Each box represents a standard linear solid element.

In the series case the stress in each element is the same while the strain is the sum of the strain in all the elements; in particular the creep function is a sum of terms of the form (15.2.20) and may be written

$$\psi = \sum_{n=1}^{N} f(\tau_n)\,[1 - e^{-t/\tau_n}] \tag{15.2.21}$$

where the $f(\tau_n)$ (dependent on the characteristics of the element) may be regarded as the contribution of a creep process with characteristic time τ_n while N is the number of elements in the chain. For the case where the elements are in parallel, the strain in each element is the same and the total stress is the sum of the stresses in all the elements; in particular the relaxation function may then be written

$$\phi = \sum_{n=1}^{N} g(\tau_n')\,(1 - e^{-t/\tau_n'})\,. \tag{15.2.22}$$

However it may be shown that any arbitrary two dimensional array of standard linear solid elements, with mixed series and parallel couplings, possess mechanically equivalent structures of each of the types of Fig. 15.6 with appropriate relations connecting τ_n, τ_n', $f(\tau_n)$ and $g(\tau_n')$. Therefore quite generally, linear viscoelastic solids possess creep and relaxation functions of the immediately above forms. In the limiting case of many τ_n and τ_n', closely packed together, the sums in the above formula may be replaced by integrals to yield for the most general linear viscoelastic material

$$\phi(t) = \int_0^\infty g(\tau)\,(1 - e^{-t/\tau})d\tau \tag{15.2.23}$$

$$\psi(t) = \int_0^\infty f(\tau)\,(1 - e^{-t/\tau})d\tau \tag{15.2.24}$$

where g and f, which are related, are called respectively the **relaxation and creep spectrum.**

The linear viscoelastic solid characterised by (15.2.23), or equivalently (15.2.24) provides a valid model for the behaviour of most polymers, rubbers and bitumens provided the strain amplitudes are not too large; larger deformations require non-linear theories which have emerged only within the last two decades, and which only now are beginning to be successfully applied to the behaviour of real materials. The subject of non-linear viscoelasticity is beyond the scope of this book.

In (15.2.23) and (15.2.24) the functions $g(\tau)$ and $f(\tau)$, which for polymers and rubbers depend on molecular processes, are of more concern to the polymer physicist than the continuum mechanist. The *form* of the functions (15.2.23) and (15.2.24) is important in proving results in the general theory of linear visco-elasticity; one such example is encountered below in the context of the complex modulus. For the continuum mechanist it is sufficient to postulate any one of $\phi(t)$, $\psi(t)$, $f(\tau)$, $g(\tau)$, $E(s)$ which provides a reasonably realistic model for real materials. The constant loss angle model, discussed subsequently, suits a variety of solids and implies a simple expression for $E(s)$ (which, curiously, of all the quantities listed above, is the one most removed from direct experimental evidence).

We turn now to the important **complex modulus** description of a linear visco-elastic solid, which, in the first instance, we suppose described by the relaxation function (15.2.22). Consider the case of a periodic oscillatory strain initiated at $t = 0$ and specified by

$$e(t) = \begin{cases} 0 & (t < 0) \\ e(\omega)\exp(i\omega t) & (t > 0) \end{cases} \tag{15.2.25}$$

where $e(\omega)$ is the amplitude of a (complex) periodic oscillation of circular frequency ω.

For the strain behaviour (15.2.25)

$$\bar{e}(s) = \int_0^\infty e(t)\exp(-st)dt = e(\omega)/[s-i\omega]$$

so that from the first of (15.2.14) and (15.2.15)

$$\bar{\sigma}(s) = E_D e(\omega)\,[1 - s\bar{\phi}(s)]\,/\,[s-i\omega]\ . \tag{15.2.26}$$

With $\phi(t)$ given by (15.2.22), $s\bar{\phi}(s)$ is readily shown to be

$$s\bar{\phi}(s) = \sum_1^N g(\tau_n')/(1+s\tau_n')$$

so that (15.2.26) becomes

$$\bar{\sigma}(s) = E(s)\,e(\omega)/(s-i\omega)$$

where

$$E(s) = E_D\left[1 - \sum_1^N g(\tau_n')/(1+s\tau_n')\right]. \tag{15.2.27}$$

Evidently $\bar{\sigma}(s)$ possesses simple poles at $s = i\omega$ and $s = -(\tau_n')^{-1}$. The Laplace inversion is readily effected to yield

$$\sigma(t) = E_D e(\omega)\left[1 - \sum_1^N g(\tau_n')/[1+i\omega\tau_n']\right]e^{i\omega t} +$$

$$+ \; E_D e(\omega) \sum_{1}^{N} g(\tau_n')/(1 + i\omega\tau_n')\, e^{-t/\tau_n'}$$

so that for very long times, after the exponential transients have decayed to yield negligible contributions to $\sigma(t)$, there results

$$\sigma(t) \;=\; E_D e(\omega)\left[1 - \sum_{1}^{N} \frac{g(\tau_n')}{1 + i\omega\tau_n'}\right] e^{i\omega t} \;\equiv\; E(i\omega)e(\omega)e^{i\omega t} \quad (15.2.28)$$

where $E(i\omega)$ is the transform modulus in which s is replaced by $i\omega$.

Equations (15.2.25) and (15.2.28) are equivalent to the statements that indefinitely persisting periodic stresses (strains) of amplitude $\sigma(\omega)$ $[e(\omega)]$ are accompanied by periodic strains (stresses) of the same frequency and of amplitude $e(\omega)$ $[\sigma(\omega)]$, where $\sigma(\omega)$ and $e(\omega)$ are related by

$$\sigma(\omega) \;=\; E(i\omega)e(\omega)\,. \quad (15.2.29)$$

Alternately equation (15.2.29) may be regarded as the Fourier transform of (15.2.5) and (15.2.7).

Equation (15.2.29) is similar in structure to the elastic relation $\sigma = Ee$ but is distinguished by the fact that $E(i\omega)$ is complex. Because of the complex nature of $E(i\omega)$ the periodic stress and strain

$$\sigma(t) \;=\; \sigma(\omega)e^{i\omega t}, \qquad e(t) \;=\; e(\omega)e^{i\omega t} \quad (15.2.30)$$

are not in phase.

Non-periodic disturbances may be synthesised out of (15.2.30) by Fourier integrals. This is particularly useful in dealing with the subject of transient visco-elastic waves (Section 15.4).

The quantity $E(i\omega)$ is universally known as the **complex modulus** and some-times written in the following form

$$E(i\omega) \;=\; E_1(\omega) \;+\; iE_2(\omega)$$

where E_1 and E_2 are real. It is readily shown that $E_1(\omega)$ is an even function and $E_2(\omega)$ is an odd function of ω, for

$$E_1(\omega) \;=\; E_D\left[1 - \sum_{1}^{N} \frac{g(\tau_n')}{(1 + \omega^2\tau_n'^2)}\right], \quad E_2(\omega) \;=\; E_D\omega \sum_{1}^{N} \frac{\tau_n' g(\tau_n')}{(1 + \omega^2\tau_n'^2)}\,.$$

$E(i\omega)$ or rather E_1 and E_2 have been the subject of much experimental investiga-tion. For real materials $E_2(\omega)$ is usually much smaller than $E_1(\omega)$ for all values of ω. However the variation of E_1 with ω may be quite startling, varying for example from 1 N/mm^2 at low frequencies to 10^3 N/mm^2 at high frequen-cies.

Fig. 15.7 sketches the typical behaviour at room temperature of a viscoelastic solid like polyethylene on a double logarithmic scale. The complex modulus is very sensitive to temperature. Commonly the effect of temperature variation preserves the shapes of the double logarithmic plots, but translates the curves along the direction of the log ω axis. Typically a 3°C increase of temperature results in a shift of one decade in frequency in the positive direction of the log ω axis.

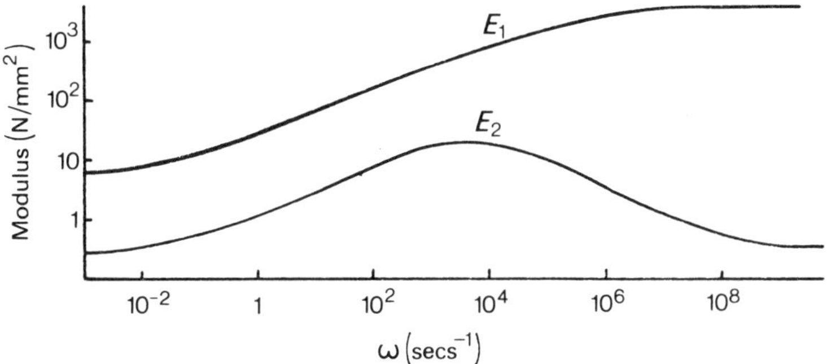

Fig. 15.7 Complex modulus of a typical viscoelastic material

Sometimes $E(i\omega)$ is written in the form

$$E(i\omega) = |E(i\omega)|\, e^{i\Delta(\omega)} \tag{15.2.31}$$

where $\Delta(\omega)$, called the **loss angle**, is defined by

$$\tan \Delta(\omega) = E_2(\omega)/E_1(\omega)\,.$$

For many viscoelastic solids a very useful specific mathematical representation of the properties is afforded by the model

$$E(s) = ks^n$$

where k and n are positive constants and $n \ll 1$. The complex modulus $E(i\omega)$ is given by

$$\log E(i\omega) = \log k + n \log(i\omega) \qquad = \log k + n \log \omega + \tfrac{1}{2}n\pi i$$

so that

$$\left.\begin{aligned} \log |E(i\omega)| &= \log k + n \log \omega \\ \Delta(\omega) &= \tfrac{1}{2}n\pi \end{aligned}\right\} \tag{15.2.32}$$

For this model Δ is independent of ω and the parameter n is given by

$$n = 2\Delta/\pi\,.$$

For $\Delta \ll 1$, characteristic of real materials,

$$|E(i\omega)| = E_1(\omega)\,[1 + O(\Delta^2)] \simeq E_1(\omega)$$

and the first of (15.2.32) may be written

$$\log E_1(\omega) = \log k + \frac{2\Delta}{\pi} \log \omega$$

or equivalently

$$\log [E_1(\omega)/E_1(\omega_0)] = (2\Delta/\pi) \log(\omega/\omega_0) \tag{15.2.33}$$

where ω_0 is an arbitrary chosen reference frequency.

The model proposed implies that a plot of $\log E_1(\omega)$ *v.* $\log \omega$ is a straight line; for the variations of E_1 with ω found experimentally as in Fig. 15.7), the model is evidently realistic for variations of frequency over one or two decades, e.g. from $\omega = 1\ \text{s}^{-1}$ to $\omega = 10^2\,\text{s}^{-1}$. This is to be contrasted with the very limited frequency range for which the standard linear solid has been found to be an adequate model. For the constant loss angle model to be valid, ω_0 must be chosen to lie within the frequency range of immediate interest.

The model fails for large values of $(2\Delta/\pi)\,|\log(\omega/\omega_0)|$; for limited ranges of ω/ω_0 where the model is valid it is possible to approximate (15.2.33) further by the simpler expression

$$E_1(\omega) = E_1(\omega_0) \left[1 + \frac{2\Delta}{\pi} \log(\omega/\omega_0)\right] . \tag{15.2.34}$$

This formula for $E_1(\omega)$, providing a valid description of many real materials, was given essentially by Kolsky in 1956 [*Phil. Mag.* 8, p 693] in an analysis of wave propagation experiments. For self evident reasons the model is known as the constant loss angle model and involves three parameters $E_1(\omega_0)$, Δ and ω_0. In application of the model to problems it is necessary to choose these parameters to suit the circumstances of the problem. This becomes clearer in the discussion of Section 15.4 where the model is used in the analysis of wave propagation problems.

15.3 Equation for three dimensional viscoelasticity – Quasi static problems

The one-dimensional constitutive equations for a linear viscoelastic solid may be taken in either the form (15.2.15) or (15.2.29). For an isotropic viscoelastic solid these equations are readily generalised from analogy with the theory of linear elasticity. Either two transform moduli $\lambda(s)$ and $\mu(s)$ are involved or alternately two complex moduli $\lambda(i\omega)$ and $\mu(i\omega)$. The quantities $\lambda(s)$, $\mu(s)$ are given by formulae of the type (15.2.27) [or their integral generalisations]

$$\lambda(s) = \lambda_D \left[1 - \sum_{1}^{N_1} g_\lambda(\tau_n^{(\lambda)})/(1 + s\tau_n^{(\lambda)})\right]$$

$$\mu(s) = \mu_D \left[1 - \sum_{1}^{N_2} g_\mu(\tau_n^{(\mu)})/(1 + s\tau_n^{(\mu)})\right] .$$

where λ_D and μ_D are dynamic Lamé constants while $\tau_n^{(\lambda)}$ and $\tau_n^{(\mu)}$ are the characteristic times associated respectively with λ and μ. In general there are no grounds for assuming that the $\tau_n^{(\lambda)}$ and $\tau_n^{(\mu)}$ are the same or that the quantities g_λ, g_μ are identical (or indeed that $N_1 = N_2$). In fact just as the theory of isotropic linear elasticity involves two elastic constants, so the corresponding viscoelastic theory entails two distinct transform moduli (or creep functions or relaxation functions or complex moduli).

Experimental evidence on the behaviour of real solids suggests that the variation of Poisson's ratio with frequency in periodic oscillations is not large, and as a matter solely of mathematical convenience it is often helpful to assume that $\lambda(s)$ is proportional to $\mu(s)$, so that the transform Poisson ratio defined by

$$\nu = \frac{\lambda(s)}{2[\lambda(s) + \mu(s)]}$$

is in fact a real constant, independent of s. In these circumstances the constitutive equations of linear viscoelasticity can be taken in the form

$$\bar{\sigma}_{ij}(s) = E(s)\left[\frac{\nu}{(1+\nu)(1-2\nu)}\,\bar{e}_{mm}(s)\delta_{ij} + \frac{1}{1+\nu}\,\bar{e}_{ij}(s)\right] \tag{15.3.1}$$

so that there is only a single creep (or relaxation) mechanism. In this case the viscoelastic aspect of 3-dimensional problems is no more complicated than in the one dimensional case.

The constitutive equations (15.3.1) need supplementing by the usual strain-displacement gradient equations

$$e_{ij} = \tfrac{1}{2}\left(\frac{\partial u_i}{\partial x_j} + \frac{\partial u_j}{\partial x_i}\right) \tag{15.3.2}$$

and the equations of motion

$$\frac{\partial \sigma_{ij}}{\partial x_j} + \rho_0 b_i = \rho_0 \frac{\partial^2 u_i}{\partial t^2}\,.$$

In what follows we neglect the body force terms, so that the equations of motion reduce to

$$\frac{\partial \sigma_{ij}}{\partial x_j} = \rho_0 \frac{\partial^2 u_i}{\partial t^2}\,. \tag{15.3.3}$$

On taking Laplace transforms of (15.3.2) we derive trivially

$$\bar{e}_{ij}(s) = \tfrac{1}{2}\left(\frac{\partial \bar{u}_i}{\partial x_j} + \frac{\partial \bar{u}_j}{\partial x_i}\right) \tag{15.3.4}$$

where account is now taken of the fact that the $\bar{e}_{ij}$ and $\bar{u}_i(x_j,s)$ are functions of position, as well as the transform variable s.

Evidently there are no real static viscoelastic problems because of the time dependent nature of the mechanical properties. However there occur problems of a

quasi-static nature for which the acceleration terms on the right side of (15.3.3) may be neglected. In such cases the equations of motion reduce to

$$\frac{\partial \sigma_{ij}}{\partial x_j} = 0$$

or on taking Laplace transforms
$$\frac{\partial \bar{\sigma}_{ij}}{\partial x_j} = 0 \,. \tag{15.3.5}$$

Again the $\bar{\sigma}_{ij}$ depend on space variables as well as the transform parameter s.

The equations of quasi-static linear viscoelasticity, for the constant ν model, are now given by (15.3.1), (15.3.4) and (15.3.5). In certain circumstances the solutions of these equations for some boundary value problems are no more complicated than for the corresponding linear elastic case. We consider one such problem below.

A thick spherical shell $a \leqslant r \leqslant b$ is subject to time dependent internal pressure $P(t)$; the external pressure is zero and it is required to determine the displacement field $\mathbf{u}$. Eliminating $\bar{\sigma}_{ij}$ and $\bar{e}_{ij}$ from (15.3.1), (15.3.4) and (15.3.5) leads to the displacement equation

$$\frac{E(s)}{2(1+\nu)(1-2\nu)} \left[\operatorname{grad} \operatorname{div} \bar{\mathbf{u}} + (1-2\nu)\nabla^2 \bar{\mathbf{u}}\right] = 0$$

from which the factor $[E/2(1+\nu)(1-2\nu)]$ may be cancelled to yield

$$\operatorname{grad}(\operatorname{div} \bar{\mathbf{u}}) + (1-2\nu)\nabla^2 \bar{\mathbf{u}} = 0 \,.$$

Since this equation for $\bar{\mathbf{u}}$ is essentially the same as equation (14.4.40) for $\mathbf{u}$ in the corresponding elastic problem, the general radially symmetric solution is given by (14.4.43)

$$\bar{u} = Ar + B/r^2 \tag{15.3.6}$$

where $\bar{u}$ is now the Laplace transform of u, the displacement in the radial (r) direction, and where $r = (x^2+y^2+z^2)^{\frac{1}{2}}$. The solution (15.3.6) differs from (14.4.43) only in that $\bar{u}$ depends on s so that we must allow $A(s)$ and $B(s)$ to depend on the parameter s. The analysis proceeds paralleling that of Section 14.4.3 except that we work entirely in terms of the Laplace transforms of quantities rather than the quantities themselves. The final form of the transformed displacement field is therefore (from 14.4.47)

$$\bar{u}(r,s) = \frac{\bar{P}(s)a^3}{E(s)(b^3-a^3)} \left[(1-2\nu)r + \tfrac{1}{2}(1+\nu)b^3/r^2\right] \tag{15.3.7}$$

while the stress transforms are given by [from (14.4.48), (14.4.49)]

$$\bar{\sigma}_{rr}(r,s) = -\frac{\bar{P}(s)a^3}{(b^3-a^3)} \left[\frac{b^3}{r^3} - 1\right] \tag{15.3.8}$$

$$\bar{\sigma}_{\theta\theta}(r,s) = \bar{\sigma}_{\phi\phi}(r,s) = \frac{\bar{P}a^3}{(b^3-a^3)} \left(1 + \frac{1}{2}\frac{b^3}{r^3}\right) . \tag{15.3.9}$$

Since in (15.3.8) and (15.3.9), s only occurs in $\overline{P}(s)$, the last two equations invert directly to yield precisely the same stress distribution as for the elastic problem. However in the present problem P may be time dependent.† On the other hand the inversion of (15.3.7) is slightly more complicated since s occurs in $\overline{P}(s)$ and $E(s)$. However it is still possible to invert (15.3.7) in general terms for an arbitrary relaxation function, for, if we write

$$E^{-1}(s) = E_D^{-1}[1 + s\overline{\psi}(s)] ,$$

there results

$$\overline{u}(r, s) = \frac{a^3}{(b^3 - a^3)E_D} [(1 - 2\nu)r + \tfrac{1}{2}(1 + \nu)b^3/r^3]\overline{P}(s)[1 + s\overline{\psi}(s)]$$

with inversion

$$u(r, t) = \frac{a^3}{(b^3 - a^3)E_D} [(1 - 2\nu)r + \tfrac{1}{2}(1 + \nu)b^3/r^3]\left[P(t) + \int_0^t \frac{dP(t')}{dt'} \psi(t - t')dt'\right] .$$

In arriving at this result it has been assumed that $P = 0$ for $t < 0$ and use has been made of the Faltung theorem discussed in Section 15.2.

The result that the stress field is the same as for the corresponding elastic problem may be generalised. Given a quasi-static viscoelastic problem for which the surface tractions are given over the entire boundary, then the viscoelastic stress field is the same as for the corresponding quasi-static elastic problem; similarly for a viscoelastic problem for which the displacements are specified everywhere on the boundary then the displacement field is the same as for the corresponding quasi-static elastic problem. These results, valid generally for the constant Poisson's ratio model are particular cases of the so-called correspondence principle [e.g. see S.C. Hunter, *Mechanics and Chemistry of Solid Propellants* p. 257–295, Pergamon (1967)].

Many problems of technological importance are solvable using the correspondence principle, which reduces the difficulties of viscoelastic stress analysis to that of the corresponding elastic problem. The only intrinsically new problems in quasi-static linear viscoelasticity are mixed boundary value problems for which the boundary conditions specify displacements over one region of the surface and stresses over the remainder, and where also the different surface regions in question are themselves time-dependent (as occurs for example in pressing a ball into a flat surface). Such problems are not discussed here.

† But not too rapidly varying since then neglect of the inertial terms in the equations of motion is not legitimate. Also of course similar quasi-static approximations are possible for the elastic problem provided P does not vary too rapidly.

15.4 One-dimensional viscoelastic waves

We conclude with a brief discussion of the propagation of longitudinal waves along rods of uniform cross section. The theory is considerably more complicated than for the corresponding elastic theory presented at the beginning of Section 14.6.4. The strain-displacement gradient equation

$$e(\equiv e_{xx}) = \partial u/\partial x \tag{15.4.1}$$

and the equation of motion

$$\frac{\partial \sigma}{\partial x} = \rho_0 \frac{\partial^2 u}{\partial t^2} \tag{15.4.2}$$

are of course identical with those of the corresponding elastic theory [equations (14.6.24) and (14.6.27)].† However the constitutive equation (14.6.25) is here replaced by (15.2.15) or its equivalent. In fact, while it is possible to proceed by taking Laplace transforms of (15.4.1) and (15.4.2) and combining the resultant equations with (15.2.15), it is much more profitable, from the viewpoint of analysing available experimental results, to work in complex modulus terms.

Accordingly we seek **steady state oscillatory** solutions for which

$$\sigma = \sigma(\omega, x)\exp(i\omega t), \qquad u = u(\omega, x)\exp(i\omega t), \qquad e = e(\omega, x)\exp(i\omega t). \tag{15.4.3}$$

The reader should not be confused between the use of the same symbols for both the physical variable and the oscillation amplitudes since the latter will be specified throughout as functions of ω and x as in (15.4.3).

Substituting from (15.4.3) into (15.4.1) and (15.4.2) leads immediately to

$$e(\omega,x) = \partial u(\omega,x)/\partial x, \ (15.4.4); \ \text{and} \ \frac{\partial \sigma(\omega,x)}{\partial x} = -\rho_0\omega^2 u(\omega,x), \ (15.4.5)$$

Finally the amplitudes $\sigma(\omega, x)$, $e(\omega, x)$ are related by the equation (15.2.29)

$$\sigma(\omega, x) = E(i\omega)\, e(\omega, x). \tag{15.4.6}$$

Of course, here, $\sigma(\omega, x)$ and $e(\omega, x)$ now depend on the space variable x.

Eliminating $\sigma(\omega, x)$ and $e(\omega, x)$ from equations (15.4.4), (15.4.5) and (15.4.6) leads to the ordinary differential equation

$$E(i\omega)\frac{d^2u(\omega, x)}{dx^2} = -\rho_0\omega^2 u(\omega, x) \tag{15.4.7}$$

with solutions

$$u(\omega, x) = A\exp[\pm i\omega(\rho_0/E(i\omega))^{\frac{1}{2}}x] \tag{15.4.8}$$

where A is a constant. The quantity $[\rho_0/E(i\omega)]^{\frac{1}{2}}$ is complex. We utilise (15.2.31) and write

$$[\rho_0/E(i\omega)]^{\frac{1}{2}} = [\rho/|E(i\omega)|]^{\frac{1}{2}}e^{-\frac{1}{2}i\Delta(\omega)} = [\rho/|E(i\omega)|]^{\frac{1}{2}}\{\cos(\tfrac{1}{2}\Delta) - i\sin(\tfrac{1}{2}\Delta)\}$$

$$= \frac{1}{c(\omega)}[1 - i\tan(\tfrac{1}{2}\Delta)]$$

† The present symbols σ and e are to be identified with σ_{xx} and e_{xx} in Section 14.6.

where $$c(\omega) \;=\; [|E(i\omega)|/\rho_0]^{\frac{1}{2}}\sec(\tfrac{1}{2}\Delta)\,. \tag{15.4.9}$$

The solutions (15.4.8) yield for the second of (15.4.3)

$$u \;=\; Ae^{i\omega(t \mp x/c(\omega)) \mp \alpha(\omega)x} \tag{15.4.10}$$

where $c(\omega)$ is given by (15.4.9) and $\alpha(\omega)$ is given by

$$\alpha(\omega) \;=\; \frac{\omega}{c(\omega)}\;\tan(\tfrac{1}{2}\Delta)\,. \tag{15.4.11}$$

Writing $A = u_0 = \text{const.}$, choosing the upper of the sign alternatives and taking the real part of (15.4.10) leads to a solution

$$u(x,\,t) \;=\; u_0\;\cos\left[\omega(t - x/c(\omega))\right]\,e^{-\alpha(\omega)x} \tag{15.4.12}$$

representing a standing wave travelling in the positive x direction of circular frequency ω with phase velocity $c(\omega)$ (dependent on ω) and with an attenuation factor $\alpha(\omega)$, so that the wave is damped. The solution (15.4.12) is realisable in an experiment in which a long filament $x \geqslant 0$ of material is subject to longitudinal displacements $u = u_0\,\cos(\omega t)$ at the accessible end $x = 0$. By measuring the resultant oscillation amplitudes and phases at various points along the filament it is possible to make experimental estimates of $\alpha(\omega)$ and $c(\omega)$. Such experiments were carried out by Kolsky and Hillier and Fig. 15.8, taken from Kolsky (*Phil. Mag.* 8, p. 693, 1956), displays results for a low density I.C.I. polyethylene at 10°C.

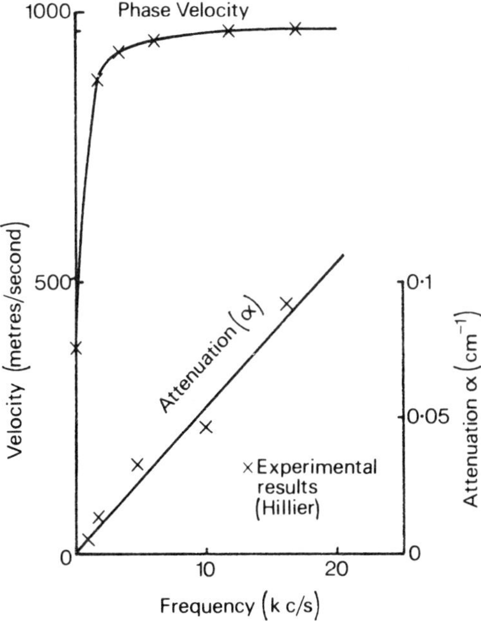

Fig. 15.8 (From Kolsky 1956). Experimental measurements of attenuation and velocity in polyethylene filaments at 10°C.

The results obtained suggest that $\alpha(\omega)$ is essentially proportional to ω while $c(\omega)$ is virtually constant; in turn these results go some way to confirming the constant loss angle model introduced in Section 15.2. For this model $c(\omega)$ given by (15.4.9) becomes [from (15.2.34) and the assumption $\Delta \ll 1$].

$$c(\omega) = c_0 \left[1 + \frac{\Delta}{\pi} \log(\omega/\omega_0) + O(\Delta^2) \right] \quad (15.4.13)$$

where

$$c_0 = [E_1(\omega_0)/\rho_0]^{\frac{1}{2}}.$$

If ω_0 is chosen somewhere in the middle of the observed frequency range (and evidently from the phase velocity results the choice is not very critical), $c(\omega)$ given by (15.4.13) is only slightly sensitive to ω and this accords with the figure except at the very lowest frequencies. Accordingly since c is almost everywhere rather insensitive to ω both experimentally and in the model, while in (15.4.11) $\alpha(\omega)$ contains the factor ω, which swamps the variation of c with ω, the theoretical attenuation is essentially proportional to ω. This is also in accord with the experimental results for $\alpha(\omega)$. The latter may now be used to estimate the (constant) value of Δ, which for the experiments in question is about 0.16.

In the discussion above the steady state solutions (15.4.10) have been compared directly with available experimental evidence. The functions (15.4.10) may also be regarded, with A as a function of ω, as a basic set of eigenfunctions for transient wave propagation. With the upper sign choice the functions represent waves travelling in the positive x direction while the lower sign represents waves travelling in the negative x direction. For a transient wave induced at $x = 0$ in a long rod $x > 0$, only the former waves are excited and the most general solution of the one dimensional viscoelastic wave problem is then given by the Fourier integral

$$u(x, t) = \int_{-\infty}^{\infty} A(\omega)\, e^{i\omega[t - x/c(\omega)] - \alpha(\omega)x}\, d\omega \quad (15.4.14)$$

where $A(\omega)$ is a function to be determined by initial conditions.

In (15.4.14) $c(\omega)$ is an even function of ω as may be verified from (15.4.9) and the result that E_1 is even while E_2 (and hence Δ) is odd (see p. 454). [It is to be noted that even for the constant loss angle model where Δ is a positive constant for $\omega > 0$, for $\omega < 0$, Δ must be chosen negative, but of the same magnitude as for $\omega > 0$. With this reservation it is evident from (15.4.11) that α is an even function of ω.]

Associated with the displacement pulse (15.4.14) are the velocity pulse

$$v(x, t) (\equiv \partial u/\partial t) = \int_{-\infty}^{\infty} i\omega A(\omega)\, e^{i\omega[t - x/c(\omega)] - \alpha(\omega)x}\, d\omega \quad (15.4.15)$$

and stress pulse

$$\sigma(x, t) = -\int_{-\infty}^{\infty} E(i\omega)\left[\frac{i\omega}{c(\omega)} + \alpha(\omega)\right] A(\omega)\, e^{i\omega[t - x/c(\omega)] - \alpha(\omega)x}\, d\omega\,.$$

With the help of previous results this latter expression may be put in the form

$$\sigma(x, t) = -\rho_0 \int_{-\infty}^{\infty} i\omega c(\omega)\,[1 - i\tan(\tfrac{1}{2}\Delta)]^{-1} A(\omega)\, e^{i\omega[t - x/c(\omega)] - \alpha(\omega)x}\, d\omega\,. \tag{15.4.16}$$

We consider the velocity pulse for an explosively induced stress pulse modelled by the initial condition

$$\sigma(0, t) = -P\delta(t)$$

where $\delta(t)$ is the Dirac delta function and P is a constant which measures the magnitude of the explosion. From (15.4.16) and the initial condition we have

$$\int_{-\infty}^{\infty} i\rho_0\omega\, c(\omega)\,[1 - i\tan(\tfrac{1}{2}\Delta)]^{-1} A(\omega)\, e^{i\omega t}\, d\omega = P\delta(t)$$

whence from Fourier's integral theorem

$$A(\omega) = [2\pi i\rho_0 c(\omega)\omega]^{-1}\,[1 - i\tan(\tfrac{1}{2}\Delta)]\,P$$

so that in particular the velocity pulse (15.4.15) becomes

$$v(x, t) = \frac{P}{2\pi\rho_0}\int_{-\infty}^{\infty} c^{-1}(\omega)\,[1 - i\tan(\tfrac{1}{2}\Delta)]\, e^{i\omega[t - x/c(\omega)] - \alpha(\omega)x}\, d\omega\,. \tag{15.4.17}$$

The integral may be evaluated analytically in the simplest approximation based on the constant loss angle model. If we write

$$c(\omega) \simeq c(\omega_0) \equiv c_0\,, \qquad \alpha = k\,|\omega|$$

where

$$k = \tan(\tfrac{1}{2}\Delta)/c_0$$

recalling that for negative ω, $\Delta(\omega) = -\Delta$ there results

$$v(x, t) = \frac{P}{\pi\rho_0 c_0}\,\frac{kx + \tan(\tfrac{1}{2}\Delta)\,[t - x/c_0]}{k^2x^2 + (t - x/c_0)^2} \tag{15.4.18}$$

or equivalently

$$v(x, t) = \frac{Pk}{\pi\rho_0}\,\frac{t}{k^2x^2 + (t - x/c_0)^2}\,. \tag{15.4.19}$$

The second expression shows the expected result $v = 0$ for $t = 0$ (except for $x = 0$); however the first expression is more illuminating. If in (15.4.18) we write $t' = t - x/c_0$ and $\tau = t'/kx$ so that τ is a dimensionless time variable which is measured from origin x/c_0 but whose scale varies inversely with x, (15.4.18) becomes

$$v = \frac{P}{\pi \rho_0 c_0 k x} \frac{1 + \tau \tan(\frac{1}{2}\Delta)}{1 + \tau^2} . \tag{15.4.20}$$

Equation (15.4.20) admits the interpretation that for fixed x, v (as a function of time) is an almost symmetric pulse, centred on origin x/c_0, and whose width broadens with increasing x.† Also because of the factor x^{-1}, the maximum amplitude decreases with x. These features are completely characteristic of experimental velocity pulses measured by Kolsky at different stations x. For the material whose properties are depicted in Fig. 15.8, a dimensionless plot of $v/v_{\max}$ as a function of τ is shown in Fig. 15.9, and compared with Kolsky's experimental profile for polyethylene for $x = 60$ cm. In making the comparison the experimental results have been plotted assuming $\tan(\frac{1}{2}\Delta) = 0.081$ and $c_0 = 950$ m/sec.†† The agreement is not entirely satisfactory for the experimental pulse is markedly more asymmetric than the theoretical one. Complete agreement with experiment is obtained by taking account of the variation of c with ω inside the exponential term (15.4.17). This was undertaken by Kolsky assuming

$$c^{-1}(\omega) = c_0 \left[1 - \frac{\Delta}{\pi} \log(\omega/\omega_0) + 0(\Delta^2)\right]$$

which is readily derived from (15.4.13). The resulting integral requires to be evaluated numerically. For the details the reader is referred to the Kolsky paper.

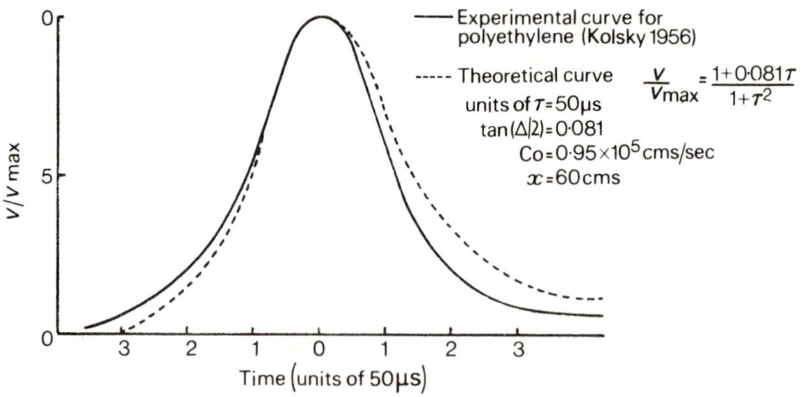

Fig. 15.9 Transient velocity pulse in polyethylene

† The pulse would be precisely symmetric if the term $\tau \tan(\frac{1}{2}\Delta)$ in the numerator was neglected.

†† $v_{\max}$ occurs at $\tau \simeq \frac{1}{2}\tan^2(\frac{1}{2}\Delta)$, i.e. effectively at $\tau = 0$.

Chapter 16

Plasticity Theory

16.1 Introduction: The experimental behaviour of metals

For sufficiently small deformations the behaviour of metals conforms to the theory of (infinitesimal) elasticity. In tensile tests, in which cylindrical specimens of uniform cross section are extended longitudinally, elastic behaviour, described by Hooke's law $\sigma = Ee$, holds provided the strain e remains always below a limiting value e^*

i.e. $$e \leqslant e^*$$

where e^*, characteristic of the metal in question, is at most of the order 5.10^{-3} (for hardened steels) and may be as small as 10^{-3} (for annealed metals). For strains in excess of e^* the typical behaviour of a metal is as shown in Fig. 16.1. Immediately beyond e^* the slope of the stress strain curve decreases slightly and this behaviour persists up to a value e_Y where e_Y is slightly larger than e^*. For strain cycles where the strain never exceeds e_Y, the mechanical behaviour of the material is defined unequivocally by the stress-strain diagram. For strain cycles where the strain exceeds e_Y, permanent deformation ensues so that even when the specimen is unloaded to the stress free state there remains a permanent strain, known as **plastic strain**.

For strain cycles where the strain exceeds e_Y the detailed stress-strain behaviour is as depicted in Fig. 16.1. We consider first the case where e increases continuously; the stress increases monotonically with strain, but usually with monotonically decreasing slope, up to the point of complete failure e_f, where the material fractures and thereby terminates the testing procedure. The quantities e^*, e_Y and e_f are usually known respectively as the limit of proportionality, the yield strain and the fracture strain. Associated with these quantities are the corresponding stresses σ^*, σ_Y – the yield stress and σ_f – the fracture stress. The solid curve of stress v strain up to $e = e_f$ is known simply as the stress-strain curve. Values of e_f vary for different metals up to a maximum of about 0.4.

In published experimental curves like that of Fig. 16.1, the strain is usually defined as $(l - l_0)/l_0$ where l is the length of the deformed specimen originally of length l_0. On occasion for experiments involving large strains (say up to $e = 0.4$) the experimental results are plotted using the measure $e = \log(l/l_0)$. For $e \ll 1$ the two measures virtually coincide, as for example in the initial elastic region.

For strain cycles where e increases up to a maximum e_1, which lies between e_Y and e_f, and then subsequently decreases until the stress free state is attained, the stress-strain path follows the stress-strain curve up to e_1; thereafter during the unloading part of the cycle, the stress-strain locus is as shown as the dotted straight line which is a line of slope E, parallel to the initial elastic slope, and passing through e_1. On complete unloading to the stress free state $\sigma = 0$ the specimen is left with a **permanent** or **residual plastic strain** e_p as indicated in Fig. 16.1.

On subsequent reloading of the specimen the path followed is the dotted line up until the stress-strain curve is reached, whereupon the stress-strain locus again joins the latter curve either to fracture or to a point similar to e_1, which is the maximum strain for the new cycle.†

A completely characteristic feature of metal stress-strain curves is that the (varying) slope in the plastic region is much smaller than E. The stress-strain curve is characteristic of the material and depends not only on chemical constitution but also on heat treatment and method of manufacture.

In uniaxial compression the behaviour of metal specimens is similar to the behaviour in tension. A complete stress-strain curve embracing tension and compression, is displayed in Fig. 16.2. For most metals the limit of proportionality and elastic limit are virtually $-\sigma^*$ and $-\sigma_Y$ and the stress-strain curve is approximately anti-symmetrical. With the logarithmic strain measure most metals display almost exactly anti-symmetric stress-strain curves.

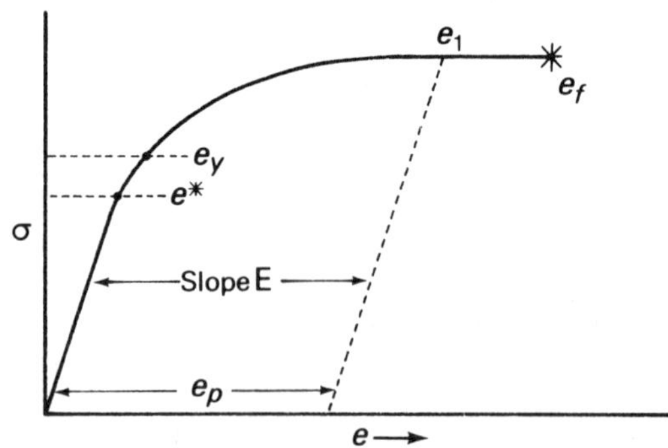

Fig. 16.1 Mechanical behaviour of metals in uniaxial tension

† The picture presented is a slight oversimplification in that for some materials the dotted curve deviates very slightly at each end from the straight line indicated in the Figure.

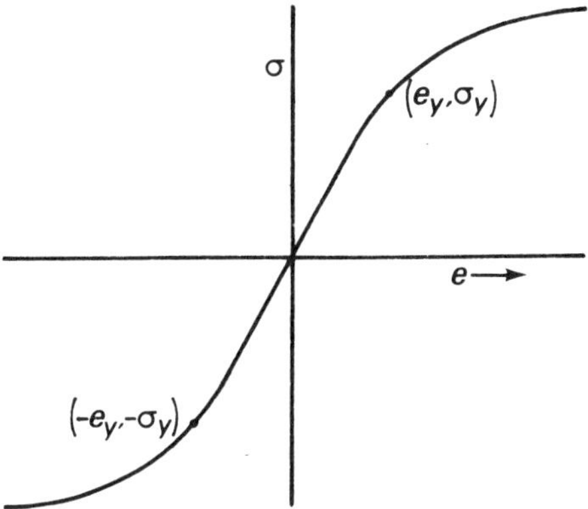

Fig. 16.2 Compressive/tensile stress strain curve

16.2 Mathematical idealisations

Theories of plasticity of varying degrees of complexity have been based on the behaviour of metals outlined above. In the discussion here we consider only a theory of elastic-perfectly plastic solids for which the curve of Fig. 16.2 is idealised by the model of Fig. 16.3. Here the (slight) distinction between σ_Y and σ^* is ignored, while during plastic deformation the stress is assumed constant with one of the values $\pm\,\sigma_Y$. The quantity σ_Y is known as the **yield stress**; commonly, and in the remainder of this chapter, the yield stress σ_Y is denoted by Y.

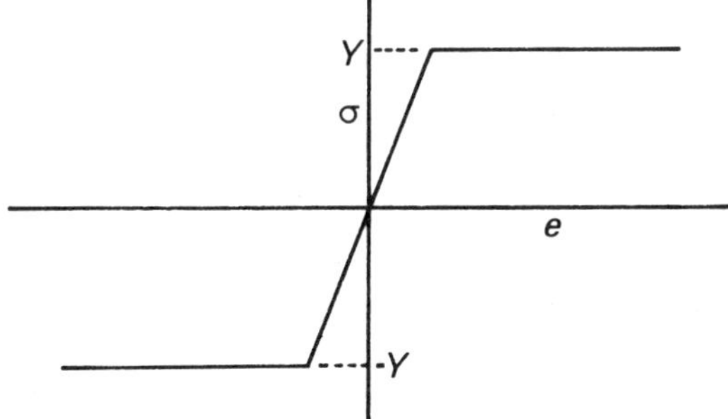

Fig. 16.3 Stress-strain curve for an elastic-perfectly plastic solid

Sometimes elastic strains are completely ignored and Fig. 16.3 is further idealised into Fig. 16.4. In this **perfectly plastic** model, for stresses of magnitude less than Y, no deformation occurs at all.

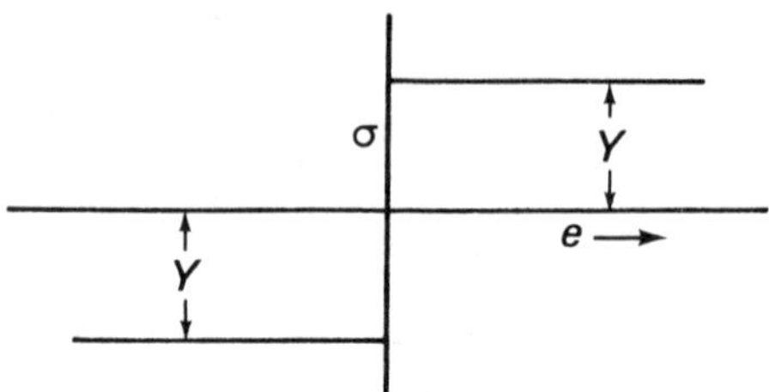

Fig. 16.4 Stress-strain curve for a perfectly plastic solid

It remains to generalise the one dimensional behaviour described above into a full three dimensional set of equations. This process is not unique and further hypotheses are required. Two problems are involved; they are the formulation of, (a) a criterion for the initiation of yield and (b) the constitutive equations for plastic deformation once yield has occurred. These problems are discussed respectively in Sections 16.3 and 16.4.

16.3 The initiation of yielding

In both of the models proposed in Fig. 16.3 and 16.4 plastic behaviour is initiated when the magnitude of the uniaxial stress σ attains the value Y. Mathematically for

$$|\sigma| < Y \tag{16.3.1}$$

the behaviour is that of an elastic (or rigid material) and for plastic deformation to occur we require

$$|\sigma| = Y. \tag{16.3.2}$$

A natural generalisation of (16.3.2) for three dimensional behaviour is the assumption that there exists a non-negative definite scalar function of the stresses (say ϕ) such that for $\phi < Y$ the behaviour is plastic while plastic deformation occurs for $\phi = Y$. The further assumption that the material is isotropic shows that ϕ is solely a function of three independent stress invariants which may be taken (for example) as the three principal stresses $\sigma(1)$, $\sigma(2)$, $\sigma(3)$ or equivalently the three quantities I_σ, II_σ, III_σ. [If the material is not isotropic the situation is more complicated.] The quantity ϕ is usually known as the **yield function.**

There exists considerable experimental evidence to suggest that initial yielding and subsequent plastic deformation is independent of hydrostatic pressure (or tension). In some classical experiments, Bridgman carried out uniaxial tensile and compressive stress-strain tests on specimens immersed in a chamber which allowed the specimens to be subject to a large external pressure P, i.e. to a stress field $\sigma_{ij} = -P\delta_{ij}$ additional to that imposed uniaxially. The resulting stress-strain behaviour was independent of P. In addition to this experimental evidence there exists today a large body of theoretical and experimental evidence on the origin of plasticity at an atomic level – all this evidence supports the Bridgman conclusions that yield is uninfluenced by pressure.† The actual mechanism of plastic deformation devolves on the passage of 'dislocations', which are lines of atomic mismatch, through the metal crystals. Dislocation motion is caused by shear stresses rather than by hydrostatic pressure (or hydrostatic tension) and the plastic deformations that result are essentially shear deformations which entail no volume changes. This last result is built into the constitutive equation theory developed in Section 16.4 for the plastic region. The reader interested further in explanations at the atomic level of the origins of metal plasticity should consult *The theory of dislocations* by A.H. Cottrell, Oxford University Press (1952).

In concurrence with the above we assume now that ϕ is independent of hydrostatic pressure P, or equivalently independent of the stress invariant $\mathrm{I}_\sigma \equiv \sigma_{mm}$. However it is not sufficient to say simply that ϕ depends on II_σ and III_σ since the latter quantities themselves vary with hydrostatic pressure, as may easily be seen by evaluating II_σ and III_σ for the two stress fields σ_{ij} and $\sigma_{ij} - P\delta_{ij}$. In the second case II_σ and III_σ depend on P and hence on superimposed hydrostatic pressure. Our objective is most easily accomplished by considering the *reduced* stress tensor defined by††

$$\sigma'_{ij} = \sigma_{ij} - \tfrac{1}{3}\sigma_{mm}\delta_{ij}\,. \tag{16.3.3}$$

It is readily verified that σ'_{ij} is independent of P (for consider replacing σ_{ij} on the right side of (16.3.3) by $\sigma_{ij} - P\delta_{ij}$). Thus, automatically, the invariants of σ'_{ij} fulfil our requirements. We have in particular

$$\mathrm{I}_{\sigma'} = \sigma'_{mm} = 0\,.$$

There remain two non-vanishing invariants $\mathrm{II}_{\sigma'}$ and $\mathrm{III}_{\sigma'}$ which are independent of each other and independent of P. Accordingly we take ϕ to be a function solely of $\mathrm{II}_{\sigma'}$ and $\mathrm{III}_{\sigma'}$ so that the yield criterion may be written

$$\phi(\mathrm{II}_{\sigma'},\ \mathrm{III}_{\sigma'}) = Y$$

where ϕ is a function whose detailed behaviour is from the present viewpoint a matter for experiment.

One restriction imposed on ϕ follows from appeal to the cases of uniaxial tension (or compression) considered in Section 16.1 and specified by just one non-vanishing (normal) stress component, say σ_{11}, which takes the value $\pm Y$. For these cases $\sigma'_{11} = \pm\frac{2}{3}Y$, $\sigma'_{22} = \sigma'_{33} = \mp\frac{1}{3}Y$ and the formulae (4.5.10), (4.5.11) yield easily $\mathrm{II}_{\sigma'} = -\frac{1}{3}Y^2$, $\mathrm{III}_{\sigma'} = \pm\frac{2}{27}Y^3$; in consequence their results the identity

$$\phi\left[-\frac{1}{3}Y^2,\ \pm\frac{2}{27}Y^3\right] = Y\,. \tag{16.3.4}$$

† Exceptionally, for very large pressures, much in excess of those encountered normally and generated only in explosive experiments, there is some evidence that the yield stress is influenced by pressure in some metals.

†† There is no connection between σ'_{ij} as defined by (16.3.3) and the σ'_{ij} appearing in (4.4.1), defining stress components under rotations of coordinate systems.

It is commonly assumed that ϕ depends only on $\mathrm{II}_{\sigma'}$ in which case ϕ may be taken to be $[-3\mathrm{II}_{\sigma'}]^{\frac{1}{2}}$. The resulting yield criterion, first proposed by von-Mises in the early twentieth century, becomes

$$(-3\mathrm{II}_{\sigma'})^{\frac{1}{2}} = Y. \qquad (16.3.5)$$

Commonly this criterion is expressed in terms of principal stresses, when it assumes the form

$$[\sigma^2(1) + \sigma^2(2) + \sigma^2(3) - \sigma(1)\sigma(2) - \sigma(2)\sigma(3) - \sigma(3)\sigma(1)]^{\frac{1}{2}} = Y.$$

or, more symmetrically,

$$[\tfrac{1}{2}\{[\sigma(1)-\sigma(2)]^2 + [\sigma(2)-\sigma(3)]^2 + [\sigma(3)-\sigma(1)]^2\}]^{\frac{1}{2}} = Y. \qquad (16.3.6)$$

Expressed in terms of all components of the stress tensor we have from the first of (4.5.10) and (16.3.5)

$$\sqrt{\frac{3}{2}(\sigma'_{ij}\sigma'_{ij})} = Y \qquad (16.3.7a)$$

which in full may be written in the form

$$\left\{\tfrac{1}{2}[(\sigma_{11}-\sigma_{22})^2 + (\sigma_{22}-\sigma_{33})^2 + (\sigma_{33}-\sigma_{11})^2] + 3(\sigma_{12}^2 + \sigma_{13}^2 + \sigma_{23}^2)\right\}^{\frac{1}{2}} = Y \qquad (16.3.7b)$$

The criterion (16.3.7) is in reasonable agreement with experiment for metals in the sense that for differently imposed stress fields initial yielding satisfies the criterion to within a few per cent.

One other criterion in general use is due originally to Coulomb (late eighteenth century) and is commonly associated with Tresca (nineteenth century). The so called **Tresca yield criterion** is most simply expressed in terms of principal stresses rather than the invariants $\mathrm{II}_{\sigma'}$, $\mathrm{III}_{\sigma'}$, and is given by

$$\mathrm{Max}\,\{|\sigma(1)-\sigma(2)|,\ |\sigma(2)-\sigma(3)|,\ |\sigma(3)-\sigma(1)|\} = Y \qquad (16.3.8)$$

i.e. the absolute maximum of the principal stress differences is equal to the yield stress. The reader will easily convince himself that the invariants $[\sigma(1)-\sigma(2)]$ etc., are also independent of any additional superimposed pressure P. Because of the complicated form of solutions of the eigenvalue cubic equation which determine $\sigma(1)$, $\sigma(2)$, $\sigma(3)$ in terms of the σ_{ij}, it is not possible to express simply the Tresca criterion in a form similar to (16.3.7).

It is shown below that the Tresca and Coulomb criteria are not too dissimilar, and in the matter of initial yield most metals conform as well to the Tresca criterion as to that of von-Mises. Perhaps it should be said that determination of experimental data in the present context is only possible within a per cent or so because of the variability of nominally similar material specimens. In these circumstances the difference between the two criteria is of only marginal significance from the experimental viewpoint. From the viewpoint of the solution of problems it is usual to use the more convenient of the two criteria. The von-Mises criterion has the advantage of being readily expressible in terms of the σ_{ij}, and is represented by a single function, while the Tresca criterion has the advantage of being linear in the principal stresses. The two criteria coalesce in the important case where two of the principal stresses are equal, as for example occurs in the case of a spherical shell subjected to internal pressure.

Formally any yield criterion may be regarded as represented by a surface in a six (or nine) dimensional stress space. For most people a more concrete picture emerges from a representation in the three dimensional principal stress space with axes $\sigma(1)$, $\sigma(2)$, $\sigma(3)$. Both the Tresca and von-Mises criterion are independent of superimposed additional stress $-P\delta_{ij}$ and hence independent of $\sigma(1) + \sigma(2) + \sigma(3)$. It follows that in principal stress space the surface (16.3.6) intersects every member of the family of planes

$$\sigma(1) + \sigma(2) + \sigma(3) = \text{const.} \tag{16.3.9}$$

in the same (closed) curves, and similarly for the surface (16.3.8). The yield surfaces are therefore cylinders whose generators lie in the normal directions of the planes (16.3.9). To determine the bounding curve of the cylinder we introduce the new set of orthogonal coordinates $\Sigma_1, \Sigma_2, \Sigma_3$ defined

$$\begin{bmatrix} \Sigma_1 \\ \Sigma_2 \\ \Sigma_3 \end{bmatrix} = \begin{bmatrix} \frac{1}{\sqrt{3}} & \frac{1}{\sqrt{3}} & \frac{1}{\sqrt{3}} \\ -\frac{1}{\sqrt{2}} & \frac{1}{\sqrt{2}} & 0 \\ -\frac{1}{\sqrt{6}} & -\frac{1}{\sqrt{6}} & \frac{2}{\sqrt{6}} \end{bmatrix} \begin{bmatrix} \sigma(1) \\ \sigma(2) \\ \sigma(3) \end{bmatrix}. \tag{16.3.10}$$

The reader will verify easily that the square matrix appearing above is proper orthogonal, so that $\Sigma_1, \Sigma_2, \Sigma_3$ are indeed a set of orthogonal coordinates in principal stress space. From (16.3.10)

$$\Sigma_1 = \frac{1}{\sqrt{3}} [\sigma(1) + \sigma(2) + \sigma(3)]$$

and from the considerations given immediately above we anticipate that the criteria (16.3.6) and (16.3.8) are independent of Σ_1. In terms of the remaining coordinates Σ_2, Σ_3

$$\Sigma_2 = \frac{1}{\sqrt{2}} [\sigma(2) - \sigma(1)] , \qquad \Sigma_3 = \frac{1}{\sqrt{6}} [2\sigma(3) - \sigma(1) - \sigma(2)]$$

the reader will verify easily that (16.3.6) may be written

$$\Sigma_2^2 + \Sigma_3^2 = \tfrac{2}{3} Y^2 \tag{16.3.11}$$

so that in the $\Sigma_2\,\Sigma_3$ plane the von-Mises criterion is represented by a circle of radius $\sqrt{\tfrac{2}{3}}\,Y$ (Fig. 16.5).

On the same diagram we have plotted the Tresca criterion (16.3.8) which in terms of Σ_2 and Σ_3 assumes the form

$$\Sigma_2 = \pm\, Y/\sqrt{2}\,, \qquad \Sigma_3 + \frac{1}{\sqrt{3}}\,\Sigma_2 = \pm\,\sqrt{\tfrac{2}{3}}\,Y\,, \qquad \Sigma_3 - \frac{1}{\sqrt{3}}\,\Sigma_2 = \pm\,\sqrt{\tfrac{2}{3}}\,Y. \tag{16.3.12}$$

The straight lines (16.3.12) form a regular hexagon, usually known as the Tresca Hexagon, which is inscribed in the von-Mises circle (16.3.11).

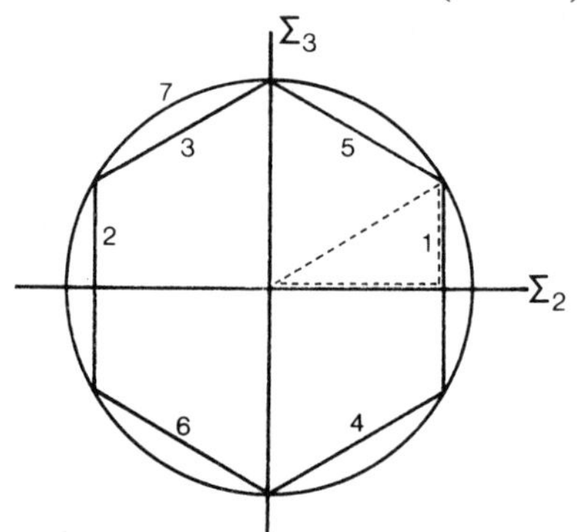

Tresca Hexagon
(1) $\Sigma_2 = Y/\sqrt{2}$ (2) $\Sigma_2 = -Y/\sqrt{2}$ (3) $\Sigma_3 - \frac{1}{\sqrt{3}}\Sigma_2 = \sqrt{\frac{2}{3}}Y$
(4) $\Sigma_3 - \frac{1}{\sqrt{3}}\Sigma_2 = -\sqrt{\frac{2}{3}}Y$ (5) $\Sigma_3 + \frac{1}{\sqrt{3}}\Sigma_2 = \sqrt{\frac{2}{3}}Y$ (6) $\Sigma_3 + \frac{1}{\sqrt{3}}\Sigma_2 = -\sqrt{\frac{2}{3}}Y$

von-Mises circle (7) $\Sigma_2^2 + \Sigma_3^2 = \frac{2}{3}Y^2$

Fig. 16.5 The von-Mises and Tresca yield criteria

In terms of distance from the origin the two yield loci agree at the six hexagon vertices and differ at most by $(\sqrt{2/3} - \sqrt{\tfrac{1}{2}})\,Y \equiv 0.109Y$ (for example along the Σ_2 axis). As Hill remarks in *Mathematical Theory of Plasticity* p.21, Oxford University Press (1960) even closer agreement, within ± 8%, between the two criteria is possible by slightly modifying the radius of the von-Mises circle. As stated earlier either criterion provides a reasonably viable description of the behaviour of real metals.

Evidently for stress points which lie inside the yield locus the behaviour is elastic, and plastic deformation ensues when the stress point lies on the bounding curve. For many engineering purposes interest in the theory of plasticity stops precisely at this point. Usually in mechanical and civil engineering design the designer's primary concern in the present context is that yielding should not occur. In these circumstances stress analysis may be based on the linear theory of elasticity and the resulting stress fields then checked everywhere against the criterion

$$\phi < Y\,.$$

In practice safety factors are introduced so that the criterion becomes

$$\phi \; < \; kY$$

where k is a positive constant, less than unity. If the design fails the criterion, then modifications are introduced as necessary so that the criterion is fulfilled.

A simple example is provided by the stress field arising in a thick spherical shell $a \leqslant r \leqslant b$ subject to internal pressure P. Here the elastic stress field is given by equations (14.4.48) and (14.4.49). Two of the principal stresses are equal $(\sigma_{\theta\theta} = \sigma_{\phi\phi})$ so that for both the von-Mises and Tresca criteria elastic behaviour is guaranteed provided

$$|\sigma_{\theta\theta} - \sigma_{rr}| \; < \; kY.$$

Since for all relevant r in the range $a \leqslant r \leqslant b$, $\sigma_{\theta\theta} > 0$ and $\sigma_{rr} < 0$ for this problem, the safety criterion becomes, from (14.4.48) and (14.4.49),

$$\frac{3}{2}\,\frac{b^3a^3P}{(b^3 - a^3)r^3} \; < \; kY.$$

For this to be true for all r in the range $a \leqslant r \leqslant b$, we must consider the largest value that is attainable by the left side. This occurs for $r = a$ and leads to

$$P \; < \; \tfrac{2}{3}kY\;[1 - a^3/b^3]$$

for the maximum allowable pressure P. If P is allowed to approach the value $\frac{2}{3}Y(1 - a^3/b^3)$ plastic deformation is initiated at the inner surface $r = a$ and with increasing P, spreads throughout the shell. In this example it may be shown that once the plastic zone reaches $r = b$ an unstable situation develops. Unless P is reduced immediately, the spherical pressure vessel expands rapidly to fail catastrophically. This occurs for the ideally plastic case: for a somewhat more realistic case, the monotonically increasing stress-strain curve of Fig. 16.1 mitigates to come extent against the catastrophy.

As stated earlier much civil and mechanical engineering design is concerned primarily with avoiding the onset of plasticity and the ensuing possibility of catastrophic failure. [However in recent years certain sophisticated structural designs have allowed for a limited amount of plastic deformation.] On the other hand in order to study metal forming processes such as the rolling of sheet metal, wire drawing, die casting etc., where the behaviour is essentially plastic, it is necessary to develop constitutive equations for the behaviour of metals undergoing plastic deformation. The equations in common use are discussed in the next section.

16.4 Constitutive equations of plasticity and the plastic potential

For either of the models of Figs. 16.3 and 16.4, plastic deformation occurs for

$$\phi(\mathrm{II}_{\sigma'},\ \mathrm{III}_{\sigma'}) \; = \; Y. \tag{16.4.1}$$

Equation (16.4.1) provides one constitutive equation relating the stresses. Five further equations are required therefore in order to specify behaviour in the plastically deforming region.

We consider first the simpler case of an ideally plastic material for which elastic strains are neglected completely so that any deformation occurring is necessarily plastic. Since it is a trivial matter to demonstrate the theoretical existence of yield states for which the stress is constant while plastic flow continues indefinitely, for example the tensile test case for which $\sigma_{11} = Y$ while all the remaining stress components vanish, evidently the constitutive equations of plasticity theory do not relate stress to deformation (or strain). A more reasonable hypothesis is one which relates (small) current increments of deformation to stress. For example such a hypothesis would be consistent with physical expectations in the situation mentioned above where $\sigma_{11} = Y$ and all other stress components vanish. Here we might expect small increments of strain to be given by $\Delta\hat{E}_{11}$, $\Delta\hat{E}_{22} = \Delta\hat{E}_{33} = -\frac{1}{2}\Delta\hat{E}_{11}$, $\Delta\hat{E}_{12} = \Delta\hat{E}_{13} = \Delta\hat{E}_{23} = 0$.† [The equality $\Delta\hat{E}_{22} = \Delta\hat{E}_{33}$ follows from symmetry while $\Delta\hat{E}_{22} = -\frac{1}{2}\Delta\hat{E}_{11}$ follows from the assumption that plastic deformation is incompressible.] If however at some instant in time we changed to a new stress state for which the only non-vanishing stress component is $\sigma_{22} = Y$ then, following the switch, we anticipate $\Delta\hat{E}_{12} = \Delta\hat{E}_{23} = \Delta\hat{E}_{13} = 0$ as before while $\Delta\hat{E}_{33} = \Delta\hat{E}_{11} = -\frac{1}{2}\Delta\hat{E}_{22}$. In this example it appears therefore that the ratios of strain increments, rather than the strain increments themselves, are determined by the stress field.

This suggests that a plausible form of constitutive relation would be one where infinitesimal increments of strain, measured from the current state, are given by equations of the form

$$\Delta\hat{E}_{ij} = f_{ij}(\sigma_{kl})\,\Delta\Lambda \tag{16.4.2}$$

where f_{ij} are functions solely of the stresses while $\Delta\Lambda$ is an (infinitesimal) proportionality factor which is not determinable in the constitutive theory.

The reason for the latter result is that with specified f_{ij} and unknown $\Delta\Lambda$ the equations (16.4.2), after eliminating $\Delta\Lambda$, provide five equations relating rate of deformation to stress. These five equations together with (16.4.1) then provide the required six constitutive equations. In fact, in application of the theory to problems, usually the quantity $\Delta\Lambda$ (or rather $d\Lambda/dt$) is determined comcomitantly with the solution for the stress and/or velocity fields. Usually in the solution $d\Lambda/dt$ is a quantity which varies from point to point. **In no sense is $d\Lambda/dt$ a material fixed constant** for if it were (16.4.2) alone would provide the complete number of required constitutive equations which, in fact, are then similar to those for a viscous fluid.

We recall from (5.4.3) that small increments of strain occurring in time Δt and measured from the current state are given by

$$\Delta\hat{E}_{ij} = d_{ij}\Delta t$$

† The notation recalls that of Section 5.4 for increments of strain measured about the current state.

where
$$d_{ij} = \tfrac{1}{2}\left(\frac{\partial v_i}{\partial x_j} + \frac{\partial v_j}{\partial x_i}\right)$$

so that equations (16.4.2) may be written in the form

$$d_{ij} = \dot{\Lambda} f_{ij}(\sigma_{kl}) \qquad (16.4.3)$$

where $\dot{\Lambda}$ is $d\Lambda/dt$.

We restrict (16.4.3) further by the assumption, discussed earlier, that plastic deformation is independent of σ_{mm} so that the f_{ij} depend on the reduced stresses σ'_{kl} rather than on σ_{kl}; i.e.

$$d_{ij} = \dot{\Lambda} f_{ij}(\sigma'_{ij})\,.$$

or in matrix notation
$$\mathbf{d} = \dot{\Lambda}\mathbf{f}(\boldsymbol{\sigma}'). \qquad (16.4.4)$$

To proceed further we make the additional hypothesis of isotropy so that if $\mathbf{d}' = \mathbf{Q}\mathbf{d}\mathbf{Q}^T$ and $\boldsymbol{\sigma}'' = \mathbf{Q}\boldsymbol{\sigma}'\mathbf{Q}^T$ where $\mathbf{Q}$ is proper orthogonal but otherwise arbitrary, then $\mathbf{d}' = \dot{\Lambda}\mathbf{f}(\boldsymbol{\sigma}'')$. This requires $\mathbf{f}(\mathbf{Q}\boldsymbol{\sigma}'\mathbf{Q}^T) = \mathbf{Q}\mathbf{f}(\boldsymbol{\sigma}')\mathbf{Q}^T$ so that $\mathbf{f}$ is an isotropic symmetric tensor function of $\boldsymbol{\sigma}'$. Invoking the representation theorem (7.2.19) shows that

$$\mathbf{d} = \dot{\Lambda}\,[\alpha\mathbf{I} + \beta\boldsymbol{\sigma}' + \gamma(\boldsymbol{\sigma}')^2]$$

or in suffix notation

$$d_{ij} = \dot{\Lambda}\,[\alpha\delta_{ij} + \beta\sigma'_{ij} + \gamma\sigma'_{ik}\sigma'_{kj}] \qquad (16.4.5)$$

where α, β and γ are functions of the invariants of σ'_{ij}.† Imposition of the incompressible condition $d_{ii} = 0$, discussed on p. 469, now leads to

$$3\alpha + \gamma\sigma'_{ik}\sigma'_{ki} = 0$$

so that (16.4.5) may be written

$$d_{ij} = \dot{\Lambda}[\beta\sigma'_{ij} + \gamma(\sigma'_{ik}\sigma'_{kj} - \tfrac{1}{3}\sigma'_{rs}\sigma'_{rs}\delta_{ij})]\,. \qquad (16.4.6)$$

A more succinct form of (16.4.6) is obtained by noting the results

$$\frac{\partial \mathrm{II}_{\sigma'}}{\partial \sigma_{ij}} = -\,\sigma'_{ij}\,, \quad (16.4.7); \qquad \frac{\partial \mathrm{III}_{\sigma'}}{\partial \sigma_{ij}} = (\sigma'_{ik}\sigma'_{kj} - \tfrac{1}{3}\sigma'_{rs}\sigma'_{rs}\delta_{ij}), \qquad (16.4.8)$$

which may be derived as follows. From the relations for the derivatives of the invariants I_C, II_C and III_C with respect to the components C_{rs} on p. 154 (and remembering in this case $\mathrm{I}_{\sigma'} = 0$) we have easily

$$\frac{\partial \mathrm{II}_{\sigma'}}{\partial \sigma'_{mn}} = -\,\sigma'_{mn}; \quad (16.4.9); \qquad \frac{\partial \mathrm{III}_{\sigma'}}{\partial \sigma'_{mn}} = \mathrm{II}_{\sigma'}\delta_{mn} + \sigma'_{mk}\sigma'_{kn}, \quad (16.4.10)$$

† The result (16.4.5) is not obtainable by appeal to material frame indifference arguments since, while Λ is a scalar, it is not clear whether or not it is a material frame indifferent scalar. To obtain (16.4.5) requires the stronger isotropy postulate, which may or may not be realised in real materials. Commonly annealed polycrystalline metals are isotropic, while single metal crystals are anisotropic.

From the chain rule for differentiation

$$\frac{\partial \mathrm{II}_{\sigma'}}{\partial \sigma_{ij}} = \frac{\partial \mathrm{II}_{\sigma'}}{\partial \sigma'_{mn}} \frac{\partial \sigma'_{mn}}{\partial \sigma_{ij}}, \qquad \frac{\partial \mathrm{III}_{\sigma'}}{\partial \sigma_{ij}} = \frac{\partial \mathrm{III}_{\sigma'}}{\partial \sigma'_{mn}} \frac{\partial \sigma'_{mn}}{\partial \sigma_{ij}}. \tag{16.4.11}$$

Also from $\sigma'_{mn} = \sigma_{mn} - \frac{1}{3}\sigma_{pp}\delta_{mn}$ we have

$$\frac{\partial \sigma'_{mn}}{\partial \sigma_{ij}} = \delta_{mi}\delta_{nj} - \tfrac{1}{3}\delta_{mn}\delta_{ij} \tag{16.4.12}$$

so that from (16.4.11), (16.4.9), (16.4.10) and (16.4.12) there follow the results

$$\frac{\partial \mathrm{II}_{\sigma'}}{\partial \sigma_{ij}} = -\sigma'_{mn}(\delta_{mi}\delta_{nj} - \tfrac{1}{3}\delta_{mn}\delta_{ij}) = -[\sigma'_{ij} - \tfrac{1}{3}\sigma'_{mm}\delta_{ij}] = -\sigma'_{ij},$$

(since $\sigma'_{mm} = 0$), and

$$\frac{\partial \mathrm{III}_{\sigma'}}{\partial \sigma_{ij}} = (\mathrm{II}_{\sigma'}\delta_{mn} + \sigma'_{mk}\sigma'_{kn})(\delta_{mi}\delta_{nj} - \tfrac{1}{3}\delta_{mn}\delta_{ij})$$

$$= \mathrm{II}_{\sigma'}\delta_{ij} - \tfrac{1}{3}\mathrm{II}_{\sigma'}\delta_{ij}\delta_{mn}\delta_{mn} + \sigma'_{ik}\sigma'_{kj} - \tfrac{1}{3}\sigma'_{mk}\sigma'_{mk}\delta_{ij} = \sigma'_{ik}\sigma'_{kj} - \tfrac{1}{3}\sigma'_{rs}\sigma'_{rs}\delta_{ij}$$

which establish the required relations (16.4.7) and (16.4.8).

Equation (16.4.6) may now be written

$$d_{ij} = \dot{\Lambda}\left[-\beta\frac{\partial \mathrm{II}_{\sigma'}}{\partial \sigma_{ij}} + \gamma\frac{\partial \mathrm{III}_{\sigma'}}{\partial \sigma_{ij}}\right] \tag{16.4.13}$$

where we recall that β and γ are solely functions of $\mathrm{II}_{\sigma'}$ and $\mathrm{III}_{\sigma'}$.

For differentiable functions β and γ it is always possible to find an 'integrating factor' $\mu(\mathrm{II}_{\sigma'}, \mathrm{III}_{\sigma'})$ so that $-\mu\beta$ and $\mu\gamma$ are expressible as the partial derivatives of another function $f(\mathrm{II}_{\sigma'}, \mathrm{III}_{\sigma'})$:—

$$-\mu\beta = \partial f/\partial \mathrm{II}_{\sigma'}, \qquad \mu\gamma = \partial f/\partial \mathrm{III}_{\sigma'}.$$

Equation (16.4.13) may therefore be written

$$d_{ij} = \dot{\lambda}\left(\frac{\partial f}{\partial \mathrm{II}_{\sigma'}}\frac{\partial \mathrm{II}_{\sigma'}}{\partial \sigma_{ij}} + \frac{\partial f}{\partial \mathrm{III}_{\sigma'}}\frac{\partial \mathrm{III}_{\sigma'}}{\partial \sigma_{ij}}\right) \tag{16.4.14}$$

$$\equiv \dot{\lambda}\frac{\partial f}{\partial \sigma_{ij}} \tag{16.4.15}$$

where $\dot{\lambda} = \dot{\Lambda}/\mu$.

The equation

$$d_{ij} = \dot{\lambda}(\partial f/\partial \sigma_{ij}) \tag{16.4.16}$$

where $f = f(\mathrm{II}_{\sigma'}, \mathrm{III}_{\sigma'})$ expresses the ratios of the d_{ij} in terms of a single scalar function f. Equations (16.4.16), usually called the **flow rule**, together with the yield criterion provide a succint form of the basic equations of ideal plasticity. The quantity f is known as the plastic potential. Evidently this latter quantity is not unique since if f is a suitable plastic potential so is $\Phi(f)$ where Φ is arbitrary. For, with the latter choice, the ratios of the d_{ij} are preserved while the as yet undetermined quantity $\dot{\lambda}$ is replaced by the original value divided by $\Phi'(f)$. We suppose without loss of generality that f is a homogenous function of the σ_{ij} of degree unity, i.e. that f satisfies

$$f(\theta\sigma_{ij}) = \theta f(\sigma_{ij}) \tag{16.4.17}$$

where θ is an arbitrary constant. With this restriction f is necessarily a function with the dimension of stress which is also the dimension of the yield function ϕ.

Even the restriction (16.4.17) does not specify f uniquely for evidently if f is a suitable yield function so is Af where A is an arbitrary constant. It is usual to choose the sign of A so that $f > 0$, and from now on we assume that this choice has been made. The precise magnitude of A is evidently irrelevant.

From (16.4.15)

$$\sigma_{ij}d_{ij} = \overset{\circ}{\lambda}\,\sigma_{ij}\frac{\partial f}{\partial \sigma_{ij}}. \tag{16.4.18}$$

However if f is a homogeneous function of degree unity then from Euler's theorem for such functions (readily established by differentiating (16.4.17) with respect to θ, followed by writing $\theta = 1$)

$$\sigma_{ij}\frac{\partial f}{\partial \sigma_{ij}} = f$$

so that

$$\dot{\lambda} = \sigma_{ij}d_{ij}/f = \operatorname{trace}(\boldsymbol{\sigma}\,\mathbf{d})/f. \tag{16.4.19}$$

In Section 4.4 it was shown that the quantity

$$\sigma_{ij}d_{ij} = \operatorname{trace}(\boldsymbol{\sigma}\,\mathbf{d}) \tag{16.4.20}$$

is the rate of supply of energy per unit volume other than that required to increase the kinetic energy of the material. For perfect fluids and elastic solids it was shown that the energy so absorbed in material deformation is interpretable as a recoverable stored energy; in contrast for the case of incompressible viscous fluids the energy so absorbed is irrecoverable. Here, in the absence of any elastic strains in the model, a similar situation to the latter case prevails, for, on unloading an element of material from yield, the element behaves as a rigid body and there is no mechanism like the release of elastic strain to enable the energy absorbed to be released. The quantity (16.4.20) is usually called the rate of plastic working per unit volume and is written as dW^p/dt, where W^p is the total plastic work absorbed per unit volume:—

$$\frac{dW^p}{dt} = \sigma_{ij}d_{ij}.$$

Physical expectations demand that $dW^p/dt \geqslant 0$ and this condition is always checked in the solution of plasticity problems. [Should it turn out that for some regions of a supposed solution the inequality is violated, then the solution must be rejected.]

The energy W_p is absorbed predominantly by internal structural changes at an atomic level. These changes entail progressive dislocation entanglement, the creation of new dislocations and the interaction of dislocations with grain boundaries, vacancies, foreign atoms etc. The entire process, which makes dislocation motion progressively more difficult and hence leads to the observed increase of yield stress with plastic strain is called work hardening.† As indicated work hardening aborbs most of the energy W_p; a small proportion (typically about 10%) is liberated as heat.

It is possible to find a 'formula' for $\dot{\lambda}$ in terms of the rate of plastic working. We have from (16.4.19) and above

$$\dot{\lambda} = \dot{W}^p/f. \qquad (16.4.21)$$

Since by hypothesis $f > 0$, the non-negative plastic work rate requirement is met by $\dot{\lambda} \geqslant 0$ and it is usually this inequality which is checked in the validation process of a supposed solution to a plasticity problem. At first sight it appears that (16.4.21) provides a formula for $\dot{\lambda}$ which could be substituted into the constitutive equations. This is illusory since (16.4.21) merely replaces the unknown $\dot{\lambda}$ by the (a priori) unknown $\dot{W}^p$. As stated earlier, in solving specific problems $\dot{\lambda}$ is determined concomitantly with the stress and velocity fields.††

In terms of f the non-negative rate of plastic working implies for $f > 0$ that

$$\sigma_{ij} \frac{\partial f}{\partial \sigma_{ij}} \geqslant 0. \qquad (16.4.22)$$

Formally the left side of (16.4.22) represents the scalar product of the two 'vectors' σ_{ij} and $\partial f/\partial \sigma_{ij}$ located in a nine dimensional stress space. The requirement that this scalar product be non-negative is equivalent to the statement that the surfaces $f = \text{const.}$ in the nine dimensional stress space are *star like* with respect to the origin. This somewhat abstract notion is realisable in concrete three dimensional terms if the surface $f = \text{const.}$ is considered in the three dimensional stress space of principal stresses. The inequality (16.4.22) may then be written

$$[\sigma(1), \sigma(2), \sigma(3)] \cdot \left[\frac{\partial f}{\partial \sigma(1)}, \frac{\partial f}{\partial \sigma(2)}, \frac{\partial f}{\partial \sigma(3)}\right] \geqslant 0$$

with the interpretation that for points on the surface $f = \text{const.}$ the scalar product of the radius vector $[\sigma(1), \sigma(2), \sigma(3)]$ with a vector whose direction is that of the outward normal to the surface

$$\left[\frac{\partial f}{\partial \sigma(1)}, \frac{\partial f}{\partial \sigma(2)}, \frac{\partial f}{\partial \sigma(3)}\right]$$

† More sophisticated mathematical plasticity models which seek to encompass the increase of yield stress with strain are called work hardening models.

†† It must be emphaised that in using the flow rule to solve problems it is not possible to write down initially a quantity λ whose time derivative is $\dot{\lambda}$. The quantity $\dot{\lambda}$, which usually varies with position, emerges in the calculation and is of significance only in respect of sign. [In this context the notation $\dot{\lambda}$ and $\dot{W}^p$ is unfortunate. If it were possible to express λ and W in terms of spatial coordinates and time, then $\dot{\lambda}$ and $\dot{W}^p$ would in fact be convective derivatives $D\lambda/Dt$ and DW/Dt. However, as should now be clear, the question of evaluating λ does not arise.]

is positive.† The inequality is characteristic of (differentiable) surfaces which are star-like with respect to the origin, i.e. surfaces for which the origin is an interior point and also for which every radius vector, centred on the origin, intersects the surface at one point only. Two dimensional realisations of star-like and non star-like surfaces are displayed in Fig. 16.6.

Two simple examples of star-like surfaces are the von-Mises circle and the Tresca hexagon of Fig. 16.5, though the latter is not differentiable at the hexagon corners. In fact both of these surfaces satisfy the stronger **convexity** condition, i.e. the surfaces are star-like with respect to *any* interior point, so that an arbitrary vector centred on an arbitrary interior point intersects the surface at one point only. [Convex surfaces are convex in the normal sense of the word convex so that, for example, neither of the surfaces of Fig. 16.6 are convex.]

Since f (like ϕ) is independent of I_σ, the surface $f = \text{const.}$ is a cylinder whose generators are parallel to those of the yield surface. The cylindrical surface $f = \text{const.}$ intersects the planes $\Sigma_1 = \text{const.}$ in curves similar to those of Fig. 16.5, i.e. the bounding curve is star-like with respect to the origin, which is located internally. Also it may be shown from consideration of interchanging the principal stress labels 1, 2, 3 that the bounding curves of $\phi = Y$ and $f = \text{const.}$ possess a six fold symmetry in the $\Sigma_2 \Sigma_3$ plane of Fig. 16.5. In fact the von-Mises circle displays entire symmetry and the Tresca Haxagon possesses twelve fold symmetry. A one twelfth segment of the Tresca Hexagon is indicated by dotted lines in the Figure. Twelve fold symmetry is a necessary consequence of

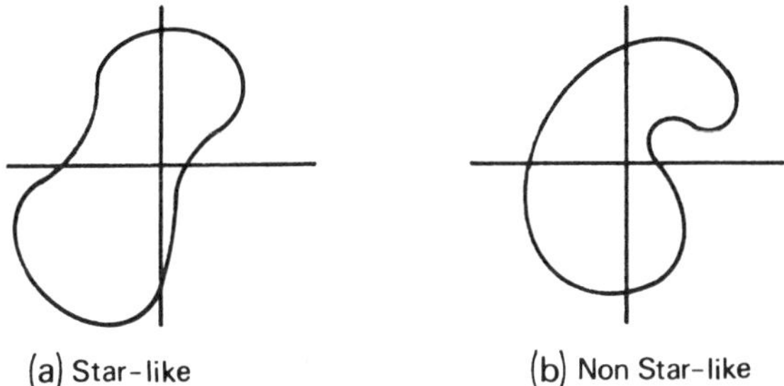

Fig. 16.6 Two dimensional realisations of (a) star-like and (b) non-star-like surfaces. In the non-star-like case some radii vectors intersect the surface more than once.

† That the gradient vector $\left[\frac{\partial f}{\partial \sigma(1)}, \frac{\partial f}{\partial \sigma(2)}, \frac{\partial f}{\partial \sigma(3)}\right]$ lies in the direction of the *outward* normal to $f = c$ may be proved as follows. From the hypothesis $f > 0$ while also $f(\theta \sigma_{ij}) = \theta f(\sigma_{ij})$ it follows that the sequence of geometrically similar surfaces $f = c_1$, $f = c_2 \ldots f = c_n, \ldots$ $(0 < c_1 < c_2 \ldots < c_n < \ldots)$ successively enclose the preceding members. Consider now the infinitesimal vector $[\delta\sigma(1), \delta\sigma(2), \delta\sigma(3)]$ along the direction of the normal which links the surfaces $f = c$ and $f = c + \delta c$. We have $\delta\sigma(1)\frac{\partial f}{\partial \sigma(1)} + \delta\sigma(2)\frac{\partial f}{\partial \sigma(2)} + \delta\sigma(3)\frac{\partial f}{\partial \sigma(3)} = \delta c > 0$. Now $[\delta\sigma(1), \delta\sigma(2), \delta\sigma(3)]$ is in the same direction or opposite direction to $\left[\frac{\partial f}{\partial \sigma(1)}, \frac{\partial f}{\partial \sigma(2)}, \frac{\partial f}{\partial \sigma(3)}\right]$. Since the scalar product is positive the gradient vector is in the same direction as $[\delta\sigma(1), \delta\sigma(2), \delta\sigma(3)]$.

any yield criterion which encompasses equal uniaxial compressive and tensile yield stresses – see R. Hill's *Mathematical Theory of Plasticity,* p.18 Oxford University Press, (1950).

On comparing the cylindrical surfaces $\phi = Y$ and $f =$ const. it is seen that in each case the bounding curve in the $\Sigma_2\Sigma_3$ plane encloses the origin with respect to which the curves are star-like, and displays six fold symmetry. While the argument is hardly mathematical it seems that if also the curves in question are reasonably smooth and do not display pathological behaviour, then the curves are likely to be similar in shape.

In fact in almost all applications of plasticity theory the plastic potential is identified with the yield function so it is assumed that

$$\phi = f .$$

There seem to be no grounds for this assumption other than those indicated above. However in order to obtain an explicit working theory the precise natures of ϕ and f require to be specified. The assumption $\phi = f$ coupled with the assumption that ϕ is given either by the von-Mises or Tresca criterion has been applied to a large number of plasticity problems.

On occasion yield functions other than those of von-Mises and Tresca have been employed; more rarely, and usually because of some mathematical simplifications, different functions ϕ and f have been used.

For the case $\phi = f$, Drucker's 'A More Fundamental Approach to Plastic Stress-strain Relations' *Proceedings of the 1st U.S. National Congress in Applied Mechanics* (1951) has adduced from thermodynamic arguments that the yield surface is necessarily convex.

In the common case where $\phi = f$ it is usually said that behaviour is governed by the yield criterion $\phi = Y$ (with a specified ϕ) together with the *associated flow rule.* For the particular case of the von-Mises criterion the associated flow rule assumes the form

$$d_{ij} = \sqrt{3/2}\,\dot{\lambda}\,\frac{\partial}{\partial \sigma_{ij}}(\sigma'_{ij}\sigma'_{ij})^{\frac{1}{2}} = \sqrt{3/2}\,\dot{\lambda}(\sigma'_{rs}\sigma'_{rs})^{-\frac{1}{2}}\sigma'_{ij} \equiv (3\dot{\lambda}/2Y)\sigma'_{ij} .$$

Commonly this result is expressed in the infinitesimal form

$$d\hat{E}_{ij} = \sigma'_{ij}d\lambda \qquad (16.4.23)$$

where the factor $(3/2Y)$ has been absorbed into the $d\lambda$. It should be emphasised, particularly in the context of large deformations, that the infinitesimal $d\hat{E}_{ij}$ on the left side of (16.4.23) is only to be interpreted as

$$d\hat{E}_{ij} = d_{ij}dt$$

i.e. as the infinitesimal strain increment occurring in time dt with reference to the current state. Exceptionally, for small displacement gradients, it is permissible to write

$$d\hat{E}_{ij} = de_{ij} \equiv \frac{\partial e_{ij}}{\partial t}\,dt$$

where e_{ij} is the total (infinitesimal) strain defined as in classical elasticity theory

$$e_{ij} = \tfrac{1}{2}\left(\frac{\partial u_i}{\partial x_j} + \frac{\partial u_j}{\partial x_i}\right)$$

and where u_i is the displacement vector.

The relations (16.4.23) are known as the Levy-von Mises equations.

It is not possible to write down simply the general flow rule for the Tresca criterion. Usually the Tresca criterion and associated flow rule are used only when it is possible to express easily the stress field of a problem everywhere in terms of principal stresses.†

In terms of principal stresses, for the side

$$\sigma(1) - \sigma(2) = Y$$

of the Tresca hexagon, the associated flow rule reads

$$d\hat{E}(1) = -\,d\hat{E}(2) = d\lambda\,, \qquad d\hat{E}(3) = 0$$

where the $d\hat{E}(n)$ $(n = 1, 2, 3)$ are principal strain increments. Similar results hold for the other five sides.

Slight complications occur at the hexagon corners. Except for these corners the flow rule is equivalent to stating that the direction of the 'vector' $(d\hat{E}(1),\ d\hat{E}(2),\ d\hat{E}(3))$ is normal to the hexagonal cylinder. At any corner two directions are defined in Fig. 16.7 normal to the adjacent sides. From consideration of a smoothed corner it seems that the possibility that $[d\hat{E}(1), d\hat{E}(2), d\hat{E}(3)]$ may lie in any direction between the two normals must be countenanced. This is indeed correct, the particular choice of direction being determined by the precise details of the problem. While the direction of the vector is not uniquely specified at a corner of the hexagon, an alternative condition, namely identity of two principal stresses, replaces the flow rule.

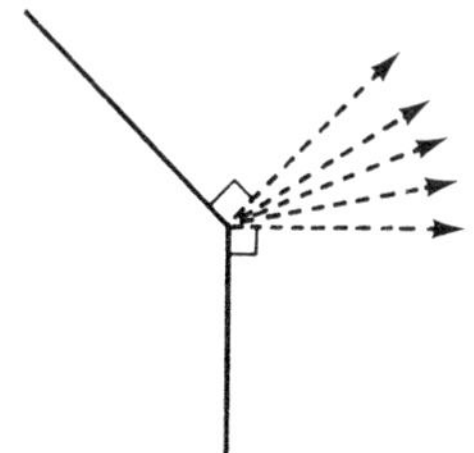

Fig. 16.7 Possible directions for the vector $(d\hat{E}(1),\ d\hat{E}(2),\ d\hat{E}(3))$ at a corner of the Tresca Hexagon

† e.g. in problems like the expansion of a cylindrical shell for which the only non-vanishing stresses σ_{rr}, $\sigma_{\theta\theta}$ and σ_{zz} constitute a set of principal stresses.

A complete statement of the constitutive equations of ideal plasticity necessitates some reference to the possibility of initial loading and subsequent unloading. This is accomplished by writing (for the case $\phi = f$)

$$\left.\begin{aligned} \phi < Y, \quad d\hat{E}_{ij} &= 0 \\ \phi = Y, \quad d\hat{E}_{ij} &= \frac{\partial \phi}{\partial \sigma_{ij}}\, d\lambda \end{aligned}\right\} \tag{16.4.24}$$

where ϕ is some specified function of the invariants $\mathrm{II}_{\sigma'}$, $\mathrm{III}_{\sigma'}$, or perhaps of the principal stress differences. Of course in the solutions of problems, equations (16.4.24) need to be augmented by the equations of motion.

While some work has been carried out in dynamic plasticity, the majority of problems solved have been of a quasi-static nature for which the acceleration terms in the equations of motion have been neglected.

Almost always in the solution of plasticity problems it is necessary to identify plastic zones where $\phi = Y$, and the complementary zones where $\phi < Y$ and for which rigid body motions only occur. An extensive and fully developed body of theory exists, in particular for the case of two dimensional problems, see R. Hill *Mathematical Theory of Plasticity* Oxford University Press (1950).

Modifications to equations (16.4.24) are necessary if it is proposed to allow for elastic deformations in loading and unloading. In the modified theory the quantities $d\hat{E}_{ij}$ are commonly supposed split into a plastic component $d\hat{E}^p_{ij}$ and an elastic component $d\hat{E}^e_{ij}$:

$$d\hat{E}_{ij} = d\hat{E}^p_{ij} + d\hat{E}^e_{ij} \cdot$$

The elastic component obeys (a differential form of) Hooke's law

$$d\hat{E}^e_{ij} = \frac{d\sigma_{ij}}{2\mu} - \frac{\nu\, d\sigma_{mm}}{2\mu(1+\nu)}\,\delta_{ij}$$

where μ is the shear modulus and ν is Poisson's ratio, while the plastic components satisfy

$$d\hat{E}^p_{ij} = \begin{cases} 0 & (\phi < Y) \\ \dfrac{\partial \phi}{\partial \sigma_{ij}}\, d\lambda & (\phi = Y)\,. \end{cases}$$

The resulting equations are not entirely satisfactory from a theoretical view because they fail to satisfy material frame indifference requirements. In recent years material frame indifferent equations for elastic-plastic materials have been formulated and in certain circumstances the equations approximate to those given above, However to date the solutions of elastic-plastic problems have been based almost entirely on the above equations.

The equations may be generalised further to encompass work hardening. The equations most commonly used in this context are those developed by Hill (loc. cit. Chapter 2). In these equations the increase of yield stress is accommodated by assuming an expanding yield locus $\phi = Y(W_p)$ where W_p is the plastic work expended; this hypothesis is evidently in general accord with the known experimental facts mentioned in Section 16.1 and with the dislocation models of the work hardening process.

Attention should be called to the important subject of soil mechanics. In some respects the behaviour of soils is similar to that of metals except that the effect of compressibility on yielding in the plastic range is often of great importance. Modified versions of plasticity theory which include the effect of compressibility on yield have been proposed and used.

16.5 Elastic-Plastic torsion of a cylindrical bar

A simple example in plasticity, which displays most of the features characteristic of problems in the theory, is provided by the elastic-plastic torsion of a cylindrical bar $0 \leqslant r \leqslant a$, $0 \leqslant z \leqslant l$. Here the calculation is limited to small strains and is based on the von-Mises yield criterion and flow rule.

The displacement field in cylindrical polar coordinates for the problem is given by

$$u_r = u_z = 0\,, \qquad u_\theta = \alpha rz/l$$

where α is the angle of twist of the section $z = l$ relative to the section $z = 0$. In Cartesian coordinates

$$u_x = -\alpha yz/l\,, \qquad u_y = \alpha xz/l\,, \qquad u_z = 0$$

and the only non-vanishing components of strain are

$$e_{xz} = -\tfrac{1}{2}\alpha y/l\,, \qquad e_{yz} = \tfrac{1}{2}\alpha x/l$$

On referring this set of components to cylindrical polar coordinates the only non-vanishing strain component is

$$e_{\theta z} = \tfrac{1}{2}\alpha r/l \tag{16.5.1}$$

with associated (elastic) stress

$$\sigma_{\theta z} = \mu\alpha\, r/l \tag{16.5.2}$$

and twisting couple

$$G = 2\pi \int_0^0 r^2\sigma_{\theta z}dr = \tfrac{1}{2}\,\pi\mu a^4\alpha/l \tag{16.5.3}$$

in agreement with the calculation of Section 14.4.6.

However for sufficiently large angles of twist (16.5.2) violates the yield criterion. From (16.5.2) $\sigma_{\theta z}$ is a maximum at $r = a$ and the yield criterion (16.3.7b) implies therefore that yield is initiated at the periphery $r = a$ for a critical angle α_0 given by

$$\mu\alpha_0\, a/l = Y/\sqrt{3} \tag{16.5.4}$$

For values of $\alpha > \alpha_0$ these last results, concerning the location of initial yield, suggest that yield spreads inwards from $r = a$ so that for $c \leqslant r \leqslant a$ the material is plastic while for $0 \leqslant r \leqslant c$ the material is elastic. Here c $(0 < c \leqslant a)$ is a parameter determined by the requirement that material located at $r = c-$ is on the point of going plastic so that $\mu\alpha c/l = Y/\sqrt{3}$ whence $c/a = \alpha_0/\alpha$.

If for the plastic region it is assumed that $e_{\theta z}$ remains the only non-vanishing strain component, then $\sigma_{\theta z}$ is constant, and, from (16.3.7b) is given by

$$\sigma_{\theta z} = Y/\sqrt{3} \qquad (c \leqslant r \leqslant a) \tag{16.5.5}$$

while for $0 \leqslant r \leqslant c$ the formula (16.5.2) still obtains. Both (16.5.2) and (16.5.5) satisfy the equilibrium equations in the absence of body forces.†

Since $\sigma_{\theta z}$ is constant in the plastic region there are no elastic strain increments and the only surviving equation from the flow rule reads

$$de_{\theta z} = de^{p}_{\theta z} = \sigma_{\theta z}\, d\lambda$$

However this equation is vacuous except to indicate that $d\lambda$ takes the sign of $d\alpha$ (on substituting for $e_{\theta z}$ from (16.5.1)), so that the proposed solution is satisfactory provided α is a monotonic increasing function of time.

The twisting couple is easily evaluated to give

$$G = \frac{2\pi Y a^3}{3\sqrt{3}}\left[1 - \frac{1}{4}\left(\frac{\alpha_0}{\alpha}\right)^3\right] \qquad (\alpha \geqslant \alpha_0)$$

so that initiation of yielding $(\alpha = \alpha_0)$ occurs for a critical couple

$$G_0 = \pi Y a^3/2\sqrt{3}$$

while the *fully plastic twisting couple* $(\alpha \gg \alpha_0)$ is $4\,G_0/3$. This latter value, which is not realised for finite α since there always remains a core of elastic material near $r = 0$, represents an upper limit to the couple sustainable by the bar.

16.6 Elastic-plastic expansion of cylindrical shell

A slightly more complicated example concerns the expansion of a cylindrical pressure vessel of initial cross-section dimensions $a_0 \leqslant r \leqslant b_0$. The tube is subject to internal pressure P at $r = a$ and zero pressure at $r = b$ so that boundary conditions for the problem are

$$\sigma_{rr} = -P \quad (r = 0)\,; \qquad \sigma_{rr} = 0 \quad (r = b)$$

† See Problem 10 of Chapter 4 for the equations of motion in cylindrical polar coordinates.

We consider the case of an incompressible medium ($\nu = \frac{1}{2}$) which results in a number of simplifications. Also we consider the case where there is no axial expansion in the z direction. The solution of the corresponding classical elastic problem is given by (14.4.59), (14.4.60) and (14.4.61), where in the latter case we have $e_{zz} = 0$.

For sufficiently small P the tube remains elastic. Here we base the plasticity calculation on the Tresca criterion which takes the simplified form

$$\text{Max}\ \{\ |\sigma_{rr} - \sigma_{zz}|,\quad |\sigma_{zz} - \sigma_{\theta\theta}|,\quad |\sigma_{\theta\theta} - \sigma_{rr}|\ \} = Y$$

Since for the elastic stress field, $\sigma_{zz} = \frac{1}{2}(\sigma_{rr} + \sigma_{\theta\theta})$, σ_{zz} is the intermediate principal stress and the Tresca criterion reduces to

$$\sigma_{\theta\theta} - \sigma_{rr} = \pm\ Y. \tag{16.6.1}$$

From (14.4.59) and (14.4.60)

$$\sigma_{\theta\theta} - \sigma_{rr} = 2a_0^2 b_0^2 P/(b_0^2 - a_0^2)\, r^2$$

so that the positive sign in (16.6.1) is to be chosen and initial yielding occurs at $r = a_0$ when the pressure attains the critical value P_0 given by

$$P_0 = \tfrac{1}{2}Y\,[1 - (a_0/b_0)^2]$$

For $P < P_0$ the tube remains elastic; for $P = P_0$ yielding is initiated at the inner boundary $r = a_0$ and it may be anticipated that with increasing P yielding spreads across the cross section to a radius c so that there is a plastic zone $a_0 \leqslant r \leqslant c$ and an elastic region $c \leqslant r \leqslant b_0$. Ultimately the plastic zone finally spreads across the entire tube ($c = b_0$) when P attains the critical value

$$P_1 = Y \log\,(b_0/a_0)\,.$$

This result is proved below. Since $b_0/a_0 > 1$ it is easily shown that $P_1 > P_0$ so that the pressure required to initiate yield is less than that required for the tube to become fully plastic.

Detailed calculations concerning the intermediate phase $a_0 \leqslant c \leqslant b_0$ are straightforward but tedious, and here we consider only the situation where the entire cross section has developed the plastic state. In the plastic state we anticipate from symmetry arguments that σ_{rr}, $\sigma_{\theta\theta}$ and σ_{zz} remain principal stresses; with the further assumption, justified a posteriori, that σ_{zz} remains the intermediate principal stress, the yield criterion remains (16.6.1) (and from continuity arguments with positive choice of sign). With a corresponding plastic potential

$$\phi = \sigma_{\theta\theta} - \sigma_{rr}$$

we deduce

$$d\hat{E}^p_{\theta\theta} = d\lambda\,,\qquad d\hat{E}^p_{rr} = -\,d\lambda\,,\qquad d\hat{E}^p_{zz} = 0 \tag{16.6.2}$$

so that in particular the last result, together with $dE_{zz} = 0$ implies $dE^e_{zz} = 0$, and hence from Hooke's law for an incompressible material

$$\sigma_{zz} = \tfrac{1}{2}\,(\sigma_{rr} + \sigma_{\theta\theta})\,,$$

so that the assumption that σ_{zz} is the intermediate principal stress is justified.

We appeal now to the equations of motion in cylindrical polar coordinates (see Problem 10 of Chapter 4). In the present problem we have $\sigma_{r\theta} = \sigma_{rz} = \sigma_{z\theta} = 0$ while the remaining stresses depend only on r. With neglect of body forces and acceleration terms, two of the equations of motion are satisfied identically and the third, in the radial direction, simplifies to

$$\frac{d\sigma_{rr}}{dr} + \frac{\sigma_{rr} - \sigma_{\theta\theta}}{r} = 0$$

Substituting from (16.6.1) for $\sigma_{\theta\theta} - \sigma_{rr}$ and integrating subject to the boundary conditions leads easily to

$$P = Y \log (b/a) \tag{16.6.3}$$

relating P to the current cross section dimensions. Initially when plasticity has just spread across the cross section $a \simeq a_0$, $b \simeq b_0$ and P takes the value P_1. However for large deformations it is important to insert current values of (b/a) in (16.6.3). Since global conservation of mass requires $(b^2 - a^2) = (b_0^2 - a_0^2)$ equation (16.6.3) may be written

$$P = \tfrac{1}{2} Y \log [1 + (b_0^2 - a_0^2)/a^2]$$

and evidently P is a decreasing function of a. Thus once the pressure has attained the value P_1, further expansion of the tube is possible with reduced pressure. This means that if the pressure is maintained at P_1, the tube must expand rapidly and in these circumstances the inertial forces (i.e. the acceleration terms in the equations of motion) need to be taken into account. The calculation is left to the reader; the final result is

$$P = [Y + \rho_0(a\ddot{a} + \dot{a}^2)] \log(b/a) - \tfrac{1}{2}\rho_0 \dot{a}^2 [1 - (a/b)^2] \tag{16.6.4}$$

where, as before, $b = (b_0^2 - a_0^2 + a^2)^{\frac{1}{2}}$.

With $P(t)$ given, (16.6.4) provides a differential equation for $a(t)$.

The theory remains valid provided $d\lambda \geqslant 0$; it may be shown easily that this is equivalent to $\dot{a} \geqslant 0$, For consider the flow rule which yields

$$d\hat{E}_{\theta\theta} - d\hat{E}_{rr} \equiv d\hat{E}^p_{\theta\theta} - d\hat{E}^p_{rr} = 2d\lambda$$

The intermediate step follows from the constancy of $\sigma_{\theta\theta} - \sigma_{rr}$. We leave the reader to show that

$$d\hat{E}_{\theta\theta} = - d\hat{E}_{rr} = \frac{a da}{r^2}$$

so that $d\lambda \geqslant 0$ provided $da \geqslant 0$, i.e. provided the cavity continues to expand.

The theory given above is valid for the von-Mises yield criterion and flow rule provided Y is replaced throughout by $2Y/\sqrt{3}$.

Problems, Chapter 16

1. In the engineering theory of the elastic bending of beams, based on the solution of Section 14.4.2, it is assumed that filaments along the beam are subject to longitudinal strain and stress

$$e_{xx} = yR^{-1} \equiv \kappa y\,, \qquad \sigma_{xx} = Ee_{xx}$$

where E is Young's modulus, $\kappa = R^{-1}$ is the curvature, and y is the distance of the filament from the neutral axis. In particular all other components of stress are assumed of negligible importance.

For completely elastic behaviour the above hypotheses lead to the bending moment-curvature relation $M = EI\kappa$ where I is the second moment of area.

In the engineering approach to plastic beam theory, the assumption $e_{xx} = \kappa y$ is retained and stresses other than σ_{xx} are ignored. Elastic theory fails therefore in this approximation when $|\sigma_{xx}|$ first attains the value Y. Evidently this occurs for the largest y values i.e. for those fibres which are furthest removed (in a perpendicular direction) from the neutral plane.

Show that for a beam of rectangular dimensions $-a \leqslant y \leqslant a$, $-b \leqslant z \leqslant b$, the beam remains elastic for

$$|M| < M_1 = \frac{4}{3}\, ba^2 Y.$$

For $|M| = M_1$ the outer fibres $y = \pm a$ become plastic; with increasing values of $|M|$ the plastic region spreads into the beam. Evidently, within the framework of the engineering approximation, plasticity does not ensue throughout the cross section, since in particular at $y = 0$ the longitudinal strain is zero and the corresponding stress remains zero and lies below the elastic limit. Therefore there exists a central core $|y| < c$ which remains elastic while for $a \geqslant y \geqslant c$ and $-c \geqslant y \geqslant -a$, the beam is plastic.† For positive M the resulting stress distribution across the cross section is shown in the diagram.

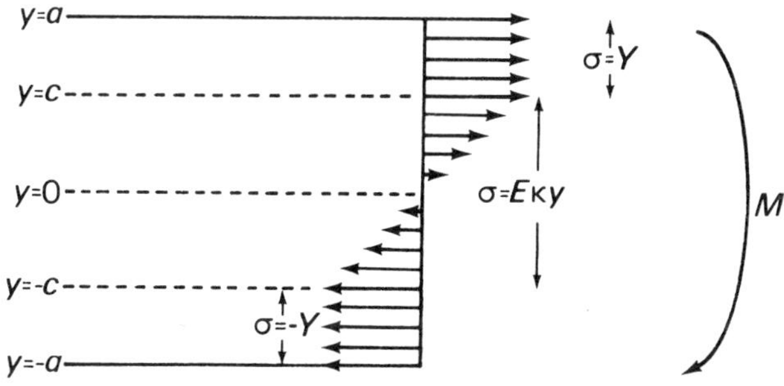

† c varies with the curvature κ.

Derive the relations

$$Ec\,|\kappa| = Y, \qquad M = \frac{4}{3}Ebc^3\kappa \pm 2b(a^2 - c^2)Y$$

where $\pm$ refer respectively to positive and negative κ. Deduce the bending moment-curvature relations

$$M = \frac{4}{3}Eba^3\kappa \qquad (|M| < M_1,\ |\kappa| < Y/Ea)$$

$$M = \pm\,(2ba^2Y - \tfrac{2}{3}bY^3/E^2\kappa^2) \qquad (M_1 < |M| < M_Y,\ |\kappa| > Y/Ea)$$

where again the $\pm$ signs refer respectively to $\kappa > 0$, $\kappa < 0$ and where

$$M_Y = 2ba^2Y$$

is the **fully plastic bending moment.**

The beam is unable to sustain bending moments of magnitude larger than M_Y. Any attempt to impose a larger moment results in rapid motion of the beam. For example for a weightless cantilever beam (Fig. 1.1) of length l the maximum supportable load is $W = M_Y/l$.

If a load of greater magnitude is imposed slowly, the maximum bending moment M_Y is maintained at the built in end. Since the beam is no longer in equilibrium, rotation occurs at the built in end $x = 0$ where a **plastic hinge** develops. (If the load is applied rapidly the situation is more complicated – initially the plastic hinge occurs at $x = l$ and then travels along the beam to $x = 0$.)

2. A cylindrical elastic tube $a \leqslant r \leqslant b$ is subjected to internal pressure P. The elastic stress analysis problem is solved in Section 14.4.4 for three cases (tube open at ends, tube closed and tube restrained from expanding axially). Assuming a Tresca yield criterion show that for all three cases yield is initiated at $r = a$ when the pressure reaches the value

$$P = \tfrac{1}{2}[1 - (a^2/b^2)]\,Y.$$

Examine also the open tube case for the von-Mises criterion.

3. For the rotating shaft problem of Section 14.4.5, principal stresses are

$$(\sigma_{rr}, \sigma_{\theta\theta}, \sigma_{zz}) = A\,[(3-2\nu)(a^2 - r^2),\ (3-2\nu)a^2 - (1+2\nu)r^2,\ 2\nu(a^2 - 2r^2)]$$

where $$A = \rho_0\omega^2/8(1-\nu)$$

Discuss the initiation of yield for both the Tresca and von-Mises criteria.

4. **(i)** Show that for the stress field

$$\sigma = \begin{bmatrix} 0 & 0 & \sigma_{zx} \\ 0 & 0 & \sigma_{zy} \\ \sigma_{zx} & \sigma_{zy} & 0 \end{bmatrix}$$

principal stresses are

$$\sigma(1) = 0, \qquad \sigma(2), \sigma(3) = \pm (\sigma_{zx}^2 + \sigma_{zy}^2)^{\frac{1}{2}}.$$

Hence deduce that the Tresca criterion becomes

$$(\sigma_{zx}^2 + \sigma_{zy}^2)^{\frac{1}{2}} = \tfrac{1}{2}Y.$$

Write down the corresponding result for the von-Mises criterion.

(ii) In the theory of torsion (Section 14.4.6) the only non-vanishing components of stress are

$$\sigma_{zx} = \mu\tau\left[-y + \frac{\partial\phi}{\partial x}\right], \qquad \sigma_{zy} = \mu\tau\left[x + \frac{\partial\phi}{\partial y}\right]$$

where ϕ is the warping function and τ the twist/unit length.

Show that in terms of χ

$$\sigma_{zx} = \mu\tau\frac{\partial\chi}{\partial x}, \qquad \sigma_{zy} = -\mu\tau\frac{\partial\chi}{\partial x}$$

where χ satisfies $\qquad \nabla^2\chi = -2.$

Deduce that the Tresca and von-Mises criteria assume the forms

$$\mu^2\tau^2\left[\left(\frac{\partial\chi}{\partial x}\right)^2 + \left(\frac{\partial\chi}{\partial y}\right)^2\right] = \begin{cases} \frac{1}{4}Y^2 & \text{(Tresca)} \\ \frac{1}{3}Y^2 & \text{(von-Mises)}. \end{cases}$$

(iii) Show that $\qquad A = \left(\frac{\partial\chi}{\partial x}\right)^2 + \left(\frac{\partial\chi}{\partial y}\right)^2$

does not possess a local maximum within the bounding contour of the cylinder.

Hint: Prove that $\nabla^2 A \geqslant 0$ and recall that conditions for a local maximum require (amongst other things)

$$\frac{\partial^2 A}{\partial x^2} < 0, \qquad \frac{\partial^2 A}{\partial y^2} < 0.$$

It follows that A takes its largest value on the boundary of the prism so that yielding is initiated on the curved surface of the prism for either criterion.

(iv) Show that for the elliptic cylinder $(x/a)^2 + (y/b)^2 = 1$ obeying the Tresca criterion, yielding is initiated either at $x = \pm a$, $y = 0$ or at $x = 0$, $y = \pm b$ when the couple attains the value

$$G = \frac{1}{4} \pi ab Y \operatorname{Max}(a, b) .$$

Answers to Problems

Answers, Chapter 2

2. The direction cosines of the vector **R** are undisturbed by the spherical motion and are given by X/R, Y/R and Z/R. The formulae for x, y and z follow immediately.

The resulting partial derivatives entering J are

$$\frac{\partial x}{\partial X} = \frac{r}{R} + \frac{X^2}{R^2}\left(\frac{\partial r}{\partial R} - \frac{r}{R}\right), \quad \frac{\partial x}{\partial Y} = \frac{XY}{R^2}\left(\frac{\partial r}{\partial R} - \frac{r}{R}\right) \text{ etc.,}$$

and

$$J = \begin{vmatrix} \frac{r}{R} + \frac{X^2\gamma}{R^2} & \frac{XY\gamma}{R^2} & \frac{XZ\gamma}{R^2} \\ \frac{XY\gamma}{R^2} & \frac{r}{R} + \frac{Y^2\gamma}{R^2} & \frac{YZ\gamma}{R^2} \\ \frac{XZ\gamma}{R^2} & \frac{ZY\gamma}{R^2} & \frac{r}{R} + \frac{Z^2\gamma}{R^2} \end{vmatrix}$$

where

$$\gamma = (\partial r/\partial R) - (r/R) .$$

The determinant is readily evaluated to yield

$$J = (r/R)^2 [r/R + \gamma] = \frac{r^2}{R^2}\frac{\partial r}{\partial R} .$$

For an incompressible material $J = 1$ and $r^2(\partial r/\partial R) = R^2$ with solution $r^3 = R^3 + h(t)$ where $h(t)$ is an arbitrary function of time.

If the material with current radius $a(t)$ was located originally at a_0, then $a^3 = a_0{}^3 + h(t)$ determining $h(t)$ in terms of $a(t)$ and yielding

$$r = (R^3 + a^3(t) - a_0{}^3)^{\frac{1}{3}} . \tag{1}$$

From this result the spatial velocity is in the radial direction and with component

$$V_r = \partial r/\partial t = a^2\dot{a}(R^3 + a^3 - a_0{}^3)^{-\frac{2}{3}}$$

so that

$$v_r = a^2\dot{a}/r^2 .$$

The formula for **f** follows directly from $\mathbf{f} = (\partial \mathbf{v}/\partial t) + (\mathbf{v}\,.\,\text{grad})\mathbf{v}$.

The result (1), above, is readily obtained by noting that for an incompressible material of uniform density the volume of material located between the radii $a(t)$ and r at time t is the same volume as that contained between the radii a_0 and R at $t = 0$.

3. Since $\lambda(0) = \Lambda(0) = 1$ and $\tau(0) = 0$, it follows that $x(0) = X$, $y(0) = Y$ and $z(0) = Z$.

Since $x^2 + y^2 = \lambda^2(X^2 + Y^2)$, circles of initial radius a, $(X^2 + Y^2 = a^2)$, become $x^2 + y^2 = \lambda^2 a^2$.

Also $\tan\theta = (y/x) = [(Y/X) + \tan(\tau Z)]/[1 - (Y/X)\tan(\tau Z)] = \tan(\Theta + \tau Z)$

To find the spatial velocity field, we have

$$V_x \equiv \partial x/\partial t = \dot{\lambda}\,[X\cos(\tau Z) - Y\sin(\tau Z)] - \lambda\dot{\tau}Z[Y\cos(\tau Z) + X\sin(\tau Z)]$$

so that in terms of lower case letters

$$v_x = (\dot{\lambda}/\lambda)x - (\dot{\tau}/\Lambda)zy \;.$$

Similarly $\quad v_y = (\dot{\lambda}/\lambda)y + (\dot{\tau}/\Lambda)zx\,, \quad v_z = (\dot{\Lambda}/\Lambda)z \;.$

From these results $\quad \operatorname{div}\mathbf{v} = 2(\dot{\lambda}/\lambda) + (\dot{\Lambda}/\Lambda)$

so that the motion occurs in an incompressible material if $2(\dot{\lambda}/\lambda) + (\dot{\Lambda}/\Lambda) = 0$ with solution $\lambda^2\Lambda = \text{const.}$ However since $\lambda(0) = \Lambda(0) = 1$, the constant is unity.

[The result $\lambda^2\Lambda = 1$ also follows from evaluating J and setting $J \equiv \rho_0/\rho = 1$.]

5. The Pfaffian form of the streamline equations is

$$\frac{dx}{x} = -\frac{dy}{y} = \frac{dz}{0}$$

with immediate integral for z, $\quad z = C \quad (C = \text{const.})$.

The remaining equation may be written $xdy + ydx = 0$, i.e. $d(xy) = 0$, with solution $xy = A$.

The choice $A = 0$ leads to the streamlines $x = 0$ and $y = 0$ which intersect on the z axis. At $x = y = 0$, the velocity field vanishes and the intersection occurs at a stagnation point.

8. (i)
$$\begin{aligned} \mathbf{f} = \frac{d\mathbf{v}}{dt} &= \ddot{\mathbf{a}} + \dot{\boldsymbol{\omega}} \wedge (\mathbf{r} - \mathbf{a}) + \boldsymbol{\omega} \wedge \frac{d\mathbf{r}}{dt} - \boldsymbol{\omega} \wedge \dot{\mathbf{a}} \\ &= \ddot{\mathbf{a}} + \dot{\boldsymbol{\omega}} \wedge (\mathbf{r} - \mathbf{a}) + \boldsymbol{\omega} \wedge \dot{\mathbf{a}} + \boldsymbol{\omega} \wedge [\boldsymbol{\omega} \wedge (\mathbf{r} - \mathbf{a})] - \boldsymbol{\omega} \\ &= \ddot{\mathbf{a}} + \dot{\boldsymbol{\omega}} \wedge (\mathbf{r} - \mathbf{a}) + \boldsymbol{\omega} \wedge [\boldsymbol{\omega} \wedge (\mathbf{r} - \mathbf{a})] \;. \end{aligned}$$

(ii) $\mathbf{f} = (\partial \mathbf{v}/\partial t) + (\mathbf{v}\,.\,\mathrm{grad})\mathbf{v}$

$$\frac{\partial \mathbf{v}}{\partial t} = \ddot{\mathbf{a}} + \dot{\boldsymbol{\omega}} \wedge (\mathbf{r} - \mathbf{a}) - \boldsymbol{\omega} \wedge \dot{\mathbf{a}}$$

$$(\mathbf{v}\,.\,\mathrm{grad})\mathbf{v} = \{[\dot{\mathbf{a}} + \boldsymbol{\omega} \wedge (\mathbf{r} - \mathbf{a})]\,.\,\mathrm{grad}\ (\boldsymbol{\omega} \wedge \mathbf{r}) = \boldsymbol{\omega} \wedge \dot{\mathbf{a}} + \boldsymbol{\omega} \wedge [\boldsymbol{\omega} \wedge (\mathbf{r} - \mathbf{a})]\ .$$

Hence $\quad \mathbf{f} = \ddot{\mathbf{a}} + \dot{\boldsymbol{\omega}} \wedge (\mathbf{r} - \mathbf{a}) + \boldsymbol{\omega} \wedge [\boldsymbol{\omega} \wedge (\mathbf{r} - \mathbf{a})]$ as above.

If $\mathbf{a} = 0$ and $\boldsymbol{\omega} = \omega(t)\mathbf{k}$

$$\frac{dx}{dt} = -\omega(t)y\,, \qquad \frac{dy}{dt} = \omega(t)x\,, \qquad \frac{dz}{dt} = 0$$

so that z is constant. Writing $\tau = \int_0^t \omega(t')dt'$ the equations for x and y become

$$\frac{dx}{d\tau} = -y\,, \qquad \frac{dy}{d\tau} = x$$

so that

$$\frac{d^2x}{d\tau^2} + x = 0\,, \qquad \frac{d^2y}{d\tau^2} + y = 0$$

with general solution $x = A\cos(\tau + \gamma)$, $y = A\sin(\tau + \gamma)$, where A and γ are constants of integration. Choosing the constants so that at $t = 0$ $(\tau = 0)$, $x = X$, $y = Y$ and $z = Z$ implies $z = Z$ and also

$$A\cos\gamma = X, \qquad A\sin\gamma = Y$$

i.e.

$$A = (X^2 + Y^2)^{\frac{1}{2}}, \qquad \gamma = \tan^{-1}(Y/X)\ .$$

Thus

$$x = R\cos(\tau + \Theta), \qquad y = R\sin(\tau + \Theta)$$

where R and Θ are polar coordinates and the motion is evidently that of a rigid body rotating about the z axis.

Answers, Chapter 3

1. **(a)** Since $\sigma_{xx} = \sigma_{yy} = \sigma_{zz} = \sigma_{xy} = 0$, the equations of equilibrium (i.e. the equations of motion with $\mathbf{f} = 0$) in the absence of body forces ($\mathbf{b} = 0$) reduce to

$$\frac{\partial \sigma_{xz}}{\partial z} = 0\,, \qquad \frac{\partial \sigma_{yz}}{\partial z} = 0\,, \qquad \frac{\partial \sigma_{zx}}{\partial x} + \frac{\partial \sigma_{zy}}{\partial y} = 0\ .$$

The first and second of these are satisfied identically (since σ_{xz} and σ_{yz} are independent of z). The last equation is satisfied provided

$$\frac{\partial}{\partial x}\left(-\frac{y}{x^2 + y^2}\right) + \frac{\partial}{\partial y}\left(\frac{x}{x^2 + y^2}\right) = 0$$

and this is readily verified.

(b) Here the equations reduce to

$$\frac{\partial\sigma_{xx}}{\partial x} + \frac{\partial\sigma_{xy}}{\partial y} = 0\,, \qquad \frac{\partial\sigma_{yx}}{\partial x} + \frac{\partial\sigma_{yy}}{\partial y} = 0$$

and these are satisfied by the given expressions.

4. In the limit $K \to \infty$

$$\rho = \rho_0\,, \qquad P = \underset{K\to\infty}{Lt}\; K[\exp(\rho_0 gz/K) - 1] = \rho_0 gz\,.$$

The latter result is also obtainable directly from $dP/dz = \rho_0 g$.

For $\rho_0 = 10^3\,\mathrm{K/m^3}$, $\quad z = 10^3\,\mathrm{m}$, $\quad K = 2.10^9\,\mathrm{N/m^2}$, $\quad \rho_0 gz/K = \frac{1}{2}.10^{-2}$.

In these circumstances, on expanding the exponential, there results

$$P = \rho_0 gz[1 + \tfrac{1}{2}(\rho_0 gz/K) + \ldots]$$

and the error afforded by the incompressibility assumption is $\frac{1}{2}(\rho_0 gz/K)$ compared to unity, i.e. approx. $\frac{1}{4}$%.

5. The equations of equilibrium are

$$\frac{\partial P}{\partial x} = \frac{\partial P}{\partial y} = 0\,, \qquad \frac{\partial P}{\partial z} = -\rho g\,.$$

Hence P is independent of x and y, and therefore

$$\frac{dP}{dz} = -\rho g = -\frac{Pmg}{RT(z)}$$

with solution $\qquad \log P = \text{const.} - \int_0^z mgR^{-1}[T(z')]^{-1}dz'\,. \qquad (1)$

Substituting $P = P_0$ at $z = 0$ leads to the stated result.

If T varies with x and y as well as z, there results from (1) a value of P which depends on x and y. This is not consistent with $\partial P/\partial x = \partial P/\partial y = 0$ so there is no solution in the absence of motion.

Answers, Chapter 4

1. **(a)** tensor of order one (vector).
(b) tensor of order zero (scalar).
(c) Not well formed.
(d) tensor of order zero (scalar)
(e) tensor of order zero (scalar)
(f) tensor equation of order two.
(g) scalar equation.
(h) tensor equation of order three.
(i) scalar equation.
(j) tensor equation of order two.
(k) Not well formed.

2. It is verified easily that $|\mathbf{A}| = 1$ and $\mathbf{A}\mathbf{A}^T = \mathbf{I}$. Also

$$\sigma' = \mathbf{A}\sigma\mathbf{A}^T = \begin{pmatrix} 1 & 0 & 0 \\ 0 & -1 & 0 \\ 0 & 0 & 2 \end{pmatrix}$$

so that the principal stresses are $1, -1, 2$.

The product of the principal stresses is -2 and it is verified easily that $|\sigma| = -2$.

3. The normal direction of the surface is

$$\mathbf{n} = \frac{1}{\sqrt{3}}(1, 1, 1)$$

so that

$$T_1 = \sigma_{11}n_1 + \sigma_{12}n_2 + \sigma_{13}n_3 = \frac{1}{2\sqrt{3}},$$

$$T_2 = \sigma_{21}n_1 + \sigma_{22}n_2 + \sigma_{23}n_3 = \frac{2}{\sqrt{3}} - \frac{1}{2},$$

$$T_3 = \sigma_{31}n_1 + \sigma_{32}n_2 + \sigma_{33}n_3 = \frac{1}{2} + \frac{1}{2\sqrt{3}}.$$

Therefore

$$|\mathbf{T}| = \left[\left(\frac{1}{2\sqrt{3}}\right)^2 + \left(\frac{2}{\sqrt{3}} - \frac{1}{2}\right)^2 + \left(\frac{1}{2} + \frac{1}{2\sqrt{3}}\right)^2\right]^{\frac{1}{2}} = (2 - \tfrac{1}{2}\sqrt{3})^{\frac{1}{2}}$$

and the unit vector in the direction of $\mathbf{T}$ is

$$\left(\frac{1}{2\sqrt{3}}, \frac{2}{\sqrt{3}} - \frac{1}{2}, \frac{1}{2} + \frac{1}{2\sqrt{3}}\right) \Big/ (2 - \tfrac{1}{2}\sqrt{3})^{\frac{1}{2}}.$$

5. The stress vector $\mathbf{T}_{(\mathbf{n})}$ may be decomposed into say a normal component of magnitude N and a shear component of magnitude S. Evidently from the Pythagoras theorem

$$S^2 + N^2 = |\mathbf{T}|^2 = T_i T_i = (\sigma_{ij} n_j)(\sigma_{ik} n_k)$$

while
$$N = \mathbf{T} \,.\, \mathbf{n} = (\sigma_{ij} n_j) n_i \,.$$

Therefore
$$S^2 = \sigma_{ij}\sigma_{ik} n_j n_k - (\sigma_{ij} n_i n_j)(\sigma_{kl} n_k n_l) \,.$$

If
$$\sigma = \text{diag}\,[\sigma(1), \sigma(2), \sigma(3)]$$

then $S^2 = \sigma^2(1) n_1^2 + \sigma^2(2) n_2^2 + \sigma^2(3) n_3^2 - [\sigma(1) n_1^2 + \sigma(2) n_2^2 + \sigma(3) n_3^2]^2$

where since $\mathbf{n}$ is a unit vector $n_1^2 + n_2^2 + n_3^2 = 1$.

Eliminating n_3 leads to

$$S^2 = [\sigma^2(1) - \sigma^2(3)] n_1^2 + [\sigma^2(2) - \sigma^2(3)] n_2^2 + \sigma^2(3) - \left\{[\sigma(1) - \sigma(3)] n_1^2 + [\sigma(2) - \sigma(3)] n_2^2 + \sigma(3)\right\}^2 .$$

Conditions for an extremum are $\partial S^2/\partial n_1 = \partial S^2/\partial n_2 = 0$ and these lead to

$$2[\sigma(1) - \sigma(3)] n_1 \{[\sigma(1) - \sigma(3)](1 - 2n_1^2) - 2[\sigma(2) - \sigma(3)] n_2^2\} = 0$$

$$2[\sigma(2) - \sigma(3)] n_2 \{[\sigma(2) - \sigma(3)](1 - 2n_2^2) - 2[\sigma(1) - \sigma(3)] n_1^2\} = 0$$

with solutions

(i) $n_1 = n_2 = 0$, **(ii)** $n_1 = 0$, $n_2^2 = \frac{1}{2}$ (and hence $n_3^2 = \frac{1}{2}$), **(iii)** $n_2 = 0$, $n_1^2 = \frac{1}{2}$ (and hence $n_3^2 = \frac{1}{2}$), **(iv)** the two quantities within { } vanish.

(i) leads to a minimum ($S^2 = 0$). **(ii)** and **(iii)** lead to the stated maxima. **(iv)** possesses no solution unless $\sigma(1) = \sigma(2)$, in which case $n_1^2 + n_2^2 = \frac{1}{2}$ when the maximum value of S is $\frac{1}{2}|\sigma(1) - \sigma(3)|$; in any event this last result is covered by (iii).

6. It is required to calculate

$$S_{\max} = \tfrac{1}{2}\,\text{Max}\,[|\sigma(1) - \sigma(2)|,\ |\sigma(2) - \sigma(3)|,\ |\sigma(3) - \sigma(1)|]$$

where
$$\sigma_{zz} \equiv \sigma(3) = \tfrac{1}{2}(\sigma_{xx} + \sigma_{yy})$$

and
$$\sigma(1),\ \sigma(2) = \tfrac{1}{2}[(\sigma_{xx} + \sigma_{yy}) \pm \{(\sigma_{xx} - \sigma_{yy})^2 + 4\sigma_{xy}^2\}^{\frac{1}{2}}] \,.$$

Evidently
$$\sigma(3) = \tfrac{1}{2}[\sigma(1) + \sigma(2)]$$

so that
$$|\sigma(3) - \sigma(1)| = |\sigma(3) - \sigma(2)| = \tfrac{1}{2}|\sigma(1) - \sigma(2)|$$

and the maximum shear stress is $\frac{1}{2}|\sigma(1) - \sigma(2)|$,

i.e.
$$S_{\max} = \tfrac{1}{2}\{(\sigma_{xx} - \sigma_{yy})^2 + 4\sigma_{xy}^2\}^{\frac{1}{2}} .$$

7. If the eigenvalues of σ_{ij} are $\sigma(1)$, $\sigma(2)$, $\sigma(3)$ then evidently the eigenvalues of the augmented matrix are $\sigma'(1) = \sigma(1) - P$, $\sigma'(2) = \sigma(2) - P$, $\sigma'(3) = \sigma(3) - P$, so that

$$\sigma'(1) - \sigma'(2) = \sigma(1) - \sigma(2) \text{ etc.,}$$

and S_{max} is independent of P.

8. The principal stresses satisfy

$$\begin{vmatrix} \frac{41}{50} - \sigma & -\frac{6}{25} & -\frac{9}{10} \\ -\frac{6}{25} & \frac{17}{25} - \sigma & -\frac{6}{5} \\ -\frac{9}{10} & -\frac{6}{5} & \frac{1}{2} - \sigma \end{vmatrix} = 0 .$$

On expanding the determinant there results the cubic equation

$$-\sigma^3 + 2\sigma^2 + \sigma - 2 = 0, \quad \text{i.e. } (\sigma - 2)(1 - \sigma^2) = 0$$

with solutions $1, -1, 2$. We write $\sigma(1) = 1$, $\sigma(2) = -1$, $\sigma(3) = 2$.

The principal stress $\sigma(1)$ acts in the direction $\mathbf{n}(1)$ where the components of $\mathbf{n}(1)$ satisfy

$$-\frac{9}{50}n_1 - \frac{6}{25}n_2 - \frac{9}{10}n_3 = 0, \quad -\frac{6}{25}n_1 - \frac{8}{25}n_2 - \frac{6}{5}n_3 = 0,$$

$$-\frac{9}{10}n_1 - \frac{6}{5}n_2 - \frac{1}{2}n_3 = 0 .$$

The first two of these equations are identical, apart from a numerical factor, and lead to $3n_1 + 4n_2 + 15n_3 = 0$ while the third equation may be written $9n_1 + 12n_2 + 5n_3 = 0$. These two equations imply $n_3 = 0$, $n_2/n_1 = -\frac{3}{4}$ so that the **unit** vector $\mathbf{n}(1)$ associated with $\sigma(1) = 1$ is

$$\mathbf{n}(1) = \pm\left(\tfrac{4}{5}, -\tfrac{3}{5}, 0\right) .$$

Similarly for the other principal stresses, unit vectors in the principal stress directions are

$$\sigma(2) = -1, \quad \mathbf{n}(2) = \pm(3, 4, 5)/5\sqrt{2}$$

$$\sigma(2) = 2, \quad \mathbf{n}(3) = \pm(3, 4, -5)/5\sqrt{2}$$

and it is readily verified that these vectors are orthogonal

i.e. $\quad \mathbf{n}(1).\mathbf{n}(2) = \mathbf{n}(2).\mathbf{n}(3) = \mathbf{n}(3).\mathbf{n}(1) = 0 .$

9. The principal stresses are evidently

$$\sigma = 25 \equiv \sigma(3) \quad \text{(say)}$$

and $\sigma(1)$, $\sigma(2)$, the roots of

$$(34 - \sigma)(41 - \sigma) - 144 = 0\,, \qquad \text{i.e. } \sigma = 25, 50\,.$$

Let us label $\sigma(2) = 25$, $\sigma(3) = 50$ so that $\sigma(1) = \sigma(2)$. From the general theory it is expected that the distinct root $\sigma(3) = 50$ is associated with a unique direction $\mathbf{n}(3)$ (apart from a choice of sign). With $\sigma = 50$ equations for the components of $\mathbf{n}(3)$ are

$$-16n_1 + 12n_2 = 0\,, \quad 12n_1 - 9n_2 = 0, \quad n_3 = 0$$

so that $n_3 = 0, n_2/n_1 = 4/3$ and the unit vector $\mathbf{n}(3)$ is

$$\mathbf{n}(3) = \pm(3, 4, 0)/5\,.$$

With $\sigma = 25$, the equations for $\mathbf{n}(1)$ and $\mathbf{n}(2)$ yield $n_2/n_1 = -3/4$ and no information about n_3. Therefore any $\mathbf{n}$ of the form $[4\alpha, -3\alpha, \pm(1 - 25\alpha^2)^{1/2}]$ is a unit vector associated with $\sigma = 25$. All these vectors are perpendicular to $\mathbf{n}(3)$.

12. Since $\mathbf{F} = \mathbf{RU}$

$$\mathbf{U}^2 = \mathbf{F}^T\mathbf{F} = \begin{bmatrix} 5 & 4 \\ 4 & 5 \end{bmatrix}.$$

The eigenvalues of $\mathbf{U}^2$ are 1 and 9. Associated with these respectively are normalised eigenvectors

$$\mathbf{n}^T = \left(\frac{1}{\sqrt{2}}, -\frac{1}{\sqrt{2}}\right), \quad \mathbf{n}^T = \left(\frac{1}{\sqrt{2}}, \frac{1}{\sqrt{2}}\right)$$

so that

$$\mathbf{U}^2 = \begin{bmatrix} \frac{1}{\sqrt{2}} & \frac{1}{\sqrt{2}} \\ -\frac{1}{\sqrt{2}} & \frac{1}{\sqrt{2}} \end{bmatrix} \begin{bmatrix} 1 & 0 \\ 0 & 9 \end{bmatrix} \begin{bmatrix} \frac{1}{\sqrt{2}} & -\frac{1}{\sqrt{2}} \\ \frac{1}{\sqrt{2}} & \frac{1}{\sqrt{2}} \end{bmatrix}$$

with square root

$$\mathbf{U} = \begin{bmatrix} \frac{1}{\sqrt{2}} & \frac{1}{\sqrt{2}} \\ -\frac{1}{\sqrt{2}} & \frac{1}{\sqrt{2}} \end{bmatrix} \begin{bmatrix} 1 & 0 \\ 0 & 3 \end{bmatrix} \begin{bmatrix} \frac{1}{\sqrt{2}} & -\frac{1}{\sqrt{2}} \\ \frac{1}{\sqrt{2}} & \frac{1}{\sqrt{2}} \end{bmatrix} = \begin{bmatrix} 2 & 1 \\ 1 & 2 \end{bmatrix}$$

The inverse of $\mathbf{U}$ is

$$\mathbf{U}^{-1} = \begin{bmatrix} \frac{2}{3} & -\frac{1}{3} \\ -\frac{1}{3} & \frac{2}{3} \end{bmatrix} \quad \text{and} \quad \mathbf{R} = \mathbf{F}\mathbf{U}^{-1} = \begin{bmatrix} \frac{3}{5} & -\frac{4}{5} \\ \frac{4}{5} & \frac{3}{5} \end{bmatrix}$$

while

$$\mathbf{V} = \mathbf{F}\mathbf{R}^T = \begin{bmatrix} \frac{26}{25} & -\frac{7}{25} \\ -\frac{7}{25} & \frac{74}{25} \end{bmatrix}$$

Answers, Chapter 5

1. Writing $\partial x_m/\partial X_i = a_{mi}$ the statement $E_{ij} = 0$ implies $\mathbf{A}^T\mathbf{A} = \mathbf{I}$ where $\mathbf{A} \equiv (a_{ij})$. It follows that $\mathbf{A}^T = \mathbf{A}^{-1}$ (since $|\mathbf{A}| \neq 0$) and since $\mathbf{A}^{-1}$ and $\mathbf{A}$ commute, then also $\mathbf{A}\mathbf{A}^T = \mathbf{I}$. Thus $\mathbf{A}$ is orthogonal.

It remains to prove that the a_{ij} are independent of the X_i. We have $a_{ij}a_{ik} = \delta_{jk}$ and on differentiating with respect to X_p

$$\frac{\partial a_{ij}}{\partial X_p}a_{ik} + \frac{\partial a_{ik}}{\partial X_p}a_{ij} = 0 .$$

Multiply by a_{qk} and use the result $a_{ik}a_{qk} = \delta_{iq}$ to obtain

$$\frac{\partial a_{qj}}{\partial X_p} + a_{ij}a_{qk}\frac{\partial a_{ik}}{\partial X_p} = 0 . \tag{1}$$

However since $\dfrac{\partial a_{qj}}{\partial X_p} = \dfrac{\partial^2 x_q}{\partial X_j \partial X_p} = \dfrac{\partial}{\partial X_j}\left(\dfrac{\partial x_q}{\partial X_p}\right) = \dfrac{\partial a_{qp}}{\partial X_j}$

the immediately preceding result may be written

$$\frac{\partial a_{qp}}{\partial X_j} + a_{ij}a_{qk}\frac{\partial a_{ik}}{\partial X_p} = 0$$

or on interchanging the roles of p and j

$$\frac{\partial a_{qj}}{\partial X_p} + a_{ip}a_{qk}\frac{\partial a_{ik}}{\partial X_k} = 0$$

i.e.

$$\frac{\partial a_{qj}}{\partial X_p} + a_{ip}a_{qk}\frac{\partial a_{ij}}{\partial X_k} = 0 . \tag{2}$$

Adding (1) and (2) leads to

$$2\frac{\partial a_{qj}}{\partial X_p} + a_{qk}\left[a_{ij}\frac{\partial a_{ik}}{\partial X_p} + a_{ip}\frac{\partial a_{ij}}{\partial X_k}\right] = 0$$

i.e.

$$2\frac{\partial a_{qj}}{\partial X_p} + a_{qk}\left[a_{ij}\frac{\partial a_{ip}}{\partial X_k} + a_{ip}\frac{\partial a_{ij}}{\partial X_k}\right] = 0$$

i.e.

$$2\frac{\partial a_{qj}}{\partial X_p} + a_{qk}\frac{\partial}{\partial X_k}(\delta_{jp}) = 0$$

i.e.

$$\partial a_{qj}/\partial X_p = 0 \; .$$

It follows that the a_{qj} are independent of the X_i so that

$$\frac{\partial x_i}{\partial X_j} = a_{ij}(t)$$

with solution

$$x_i = a_{ij}(t)X_j + b_j(t) \; .$$

The result is not surprising and is the mathematical statement of the fact that if the strains E_{ij} are everywhere zero, the motion is that of a rigid body.

2. The result

$$\mathbf{u} = \boldsymbol{\Omega} \wedge \mathbf{R} + \mathbf{b}$$

is obtained incidentally as a solution of the equations

$$\frac{\partial U_i}{\partial X_j} + \frac{\partial U_j}{\partial X_i} = 0$$

in Section 14.4.1.

The result differs from that of Problem 1 in that it is valid only for values of $|\Omega|$ satisfying $|\Omega| \ll 1$. When this inequality is satisfied the exact solution of $E_{ij} = 0$

i.e.

$$x_i = a_{ij}x_j + b_i$$

is well approximated by

$$\mathbf{u} = \boldsymbol{\Omega} \wedge \mathbf{R} + \mathbf{b}$$

(see Section 5.3).

3. From the Cartesian form of the deformation, the components of **F** are given by

$$\frac{\partial x}{\partial X} = \frac{r}{R} + \frac{X^2}{R}\left\{\frac{r'}{R} - \frac{r}{R^2}\right\}, \quad \frac{\partial x}{\partial Y} = \frac{XY}{R}\left\{\frac{r'}{R} - \frac{r}{R^2}\right\}, \quad \frac{\partial x}{\partial Z} = 0$$

$$\frac{\partial y}{\partial X} = \frac{XY}{R}\left\{\frac{r'}{R} - \frac{r}{R^2}\right\}, \quad \frac{\partial y}{\partial Y} = \frac{r}{R} + \frac{Y^2}{R}\left\{\frac{r'}{R} - \frac{r}{R^2}\right\}, \quad \frac{\partial y}{\partial Z} = 0$$

$$\frac{\partial z}{\partial X} = \frac{\partial z}{\partial Y} = 0\,, \quad \frac{\partial z}{\partial Z} = z'(Z) \; .$$

Evidently $\mathbf{F}$ is symmetric and since the polar decomposition is unique $\mathbf{F} = \mathbf{U}$ and $\mathbf{R} = \mathbf{I}$.

The proper orthogonal matrix $\mathbf{A}$ relating the axes in the R, Θ and Z directions to the Cartesian axes is

$$\mathbf{A} = \begin{bmatrix} XR^{-1} & YR^{-1} & 0 \\ -YR^{-1} & XR^{-1} & 0 \\ 0 & 0 & 1 \end{bmatrix}$$

so that referred to the new axes $\mathbf{F}$ takes the form $\mathbf{F}'$ where

$$\mathbf{F}' = \mathbf{A}\mathbf{F}\mathbf{A}^T .$$

Evaluating the matrix product leads to

$$\mathbf{F}' = \begin{bmatrix} r'(R) & 0 & 0 \\ 0 & rR^{-1} & 0 \\ 0 & 0 & z'(Z) \end{bmatrix}$$

which is evidently in principal axes form.

Since $J \equiv |\mathbf{F}| \equiv |\mathbf{F}'| > 0$, then r' and z' are either both positive or negative. In either case the principal stretches are the absolute values of the diagonal elements in $\mathbf{F}'$, that is

$$\lambda_R = |dr/dR|\,, \qquad \lambda_\theta = r/R\,, \qquad \lambda_Z = |z'(Z)|\ .$$

For an incompressible material $\lambda_R\lambda_\theta\lambda_Z = 1$ so that

$$rr'/R = [z'(Z)]^{-1} \tag{1}$$

and this is valid for either choice of sign of r' and z'. Since the left side of equation (1) above is independent of Z while the right side is independent of R, each side is constant (say a^{-1}) leading to

$$dz/dZ = a\,, \qquad rdr/dR = a^{-1}R$$

with solution $\quad z = aZ + b\,, \qquad r^2 = a^{-1}[R^2 + c]$

where a, b, c are constants (or perhaps functions of time).

4. (i) For the deformation in question

$$\mathbf{F} \equiv (\partial x_i/\partial X_j) = \begin{bmatrix} \lambda & k & 0 \\ 0 & \lambda^{-1} & 0 \\ 0 & 0 & 1 \end{bmatrix}$$

so that $|\mathbf{F}| = 1$ and there is no volume (or density) change.

Also

$$\mathbf{F}^T\mathbf{F} = \begin{bmatrix} \lambda^2 & \lambda k & 0 \\ \lambda k & \lambda^{-2} + k^2 & 0 \\ 0 & 0 & 1 \end{bmatrix} .$$

(ii) Consider an element of length dS in the fibre direction. Initially, the infinitesimal vector of magnitude dS lying in the direction N is given by

$$(dX, dY, dZ) = (\cos \Phi, \sin \Phi, 0)dS \ .$$

After deformation the vector becomes

$$(dx, dy, dz)^T = \mathbf{F}(dX, dY, dZ)^T$$

and is of length ds given by

$$ds^2 = dx^2 + dy^2 + dz^2 \equiv dx_m dx_m = (\mathbf{F}^T\mathbf{F})_{ij} dX_i dX_j \ .$$

Since the fibre is inextensible $ds = dS$, so that with the above values of (dX, dY, dZ) and $\mathbf{F}^T\mathbf{F}$

$$dS^2 = [\lambda^2 \cos^2 \Phi + (\lambda^{-2} + k^2) \sin^2 \Phi + 2\lambda k \cos \Phi \sin \Phi] \, dS^2$$

whence $\lambda^2 \cos^2 \Phi + (\lambda^{-2} + k^2) \sin^2 \Phi + 2\lambda k \cos \Phi \sin \Phi = 1$.

This last equation is a quadratic in k with solution

$$k = [-\lambda \cos \Phi + (1 - \lambda^{-2} \sin^2 \Phi)^{\frac{1}{2}}]/\sin \Phi$$

where the positive square root is implied. The second solution of the quadratic is inadmissable if λ and k are continuous functions of time for which $\lambda(0) = 1$, $k(0) = 0$.

Since λ and k are real while $\lambda > 0$, evidently it is necessary that $\lambda > |\sin \Phi|$ so that there is a limit to the possible compression of the material in the X direction.

Finally the vector $\mathbf{N} \, dS$ becomes the vector (dx, dy, dz) where

$$dx = \lambda dX + k dY, \quad dy = \lambda^{-1} dY, \quad dz = dZ$$

or since $dX = \cos \Phi \, dS, \quad dY = \sin \Phi \, dS, \quad dZ = 0$

then $dx = (\lambda \cos \Phi + k \sin \Phi) dS, \quad dy = \lambda^{-1} \sin \Phi \, dS, \quad dz = 0$.

Clearly the new orientation ϕ satisfies

$$\frac{dy}{dx} = \tan \phi = \frac{\lambda^{-1} \sin \Phi}{\lambda \cos \Phi + k \sin \Phi} = \frac{\sin \Phi}{[\lambda^2 - \sin^2 \Phi]^{\frac{1}{2}}}$$

(after substituting for k). Since ϕ is independent of position, the fibres remain parallel.

[The example is based on material from *Deformations of Fibre Reinforced Materials* by A. J. M. Spencer, Oxford University Press (1972).]

Answers, Chapter 8

2. **(i)** $\boldsymbol{\sigma} = \alpha\mathbf{I} + \beta\mathbf{B} + \gamma\mathbf{B}^2$

$$= \alpha\mathbf{I} + \beta\begin{bmatrix}\lambda_1^2 & 0 & 0\\ 0 & \lambda_2^2 & 0\\ 0 & 0 & \lambda_3^2\end{bmatrix} + \gamma\begin{bmatrix}\lambda_1^4 & 0 & 0\\ 0 & \lambda_2^4 & 0\\ 0 & 0 & \lambda_3^4\end{bmatrix}$$

where $\alpha = \alpha(\mathrm{I, II, III})$, $\beta = \beta(\mathrm{I, II, III})$, $\gamma = \gamma(\mathrm{I, II, III})$

$$\mathrm{I} = \lambda_1^2 + \lambda_2^2 + \lambda_3^2,$$

$$\mathrm{II} = \tfrac{1}{2}[(\lambda_1^2 + \lambda_2^2 + \lambda_3^2)^2 - (\lambda_1^4 + \lambda_2^4 + \lambda_3^4)] = \lambda_1^2\lambda_2^2 + \lambda_2^2\lambda_3^2 + \lambda_3^2\lambda_1^2,$$

$$\mathrm{III} = (\lambda_1\lambda_2\lambda_3)^2$$

$\sigma_{ij} = 0$ for $i \neq j$ and the diagonal components of σ_{ij} are constants so that the equilibrium equations are satisfied in the absence of body forces.

(ii) Incompressibility requires $\rho = \rho_0$, i.e. $|\mathbf{F}| = 1$, i.e. $\lambda_1\lambda_2\lambda_3 = 1$, so that $\lambda_3 = (\lambda_2\lambda_1)^{-1}$.

For an incompressible material (for which $\alpha = -P$ is arbitrary) and for which $\sigma_{33} = 0$,

$$0 = -P + \beta\lambda_3^2 + \gamma\lambda_3^4$$

determining P. Substituting for P into the formulae for σ_{11} and σ_{22} and using $\lambda_3 = (\lambda_1\lambda_2)^{-1}$ yields

$$\sigma_{11} = \beta[\lambda_1^2 - (\lambda_1\lambda_2)^{-2}] + \gamma[\lambda_1^4 - (\lambda_1\lambda_2)^{-4}]$$

$$\sigma_{22} = \beta[\lambda_2^2 - (\lambda_1\lambda_2)^{-2}] + \gamma[\lambda_2^4 - (\lambda_1\lambda_2)^{-4}]\ .$$

Writing

$$\beta = 2\rho_0\left(\frac{\partial U}{\partial \mathrm{I}} + \mathrm{I}\,\frac{\partial U}{\partial \mathrm{II}}\right), \quad \gamma = -2\rho_0\,\frac{\partial U}{\partial \mathrm{II}} \quad \text{and} \quad \mathrm{I} = \lambda_1^2 + \lambda_2^2 + (\lambda_1\lambda_2)^{-2}$$

then gives

$$\sigma_{11} = 2\rho_0\left\{[\lambda_1^2 - (\lambda_1\lambda_2)^{-2}]\,\frac{\partial U}{\partial \mathrm{I}} + [\{\lambda_1^2 - (\lambda_1\lambda_2)^{-2}\}\{\lambda_1^2 + \lambda_2^2 + (\lambda_1\lambda_2)^{-2}\} - \lambda_1^4 + (\lambda_1\lambda_2)^{-4}]\,\frac{\partial U}{\partial \mathrm{II}}\right\}$$

$$= 2\rho_0[\lambda_1^2 - (\lambda_1\lambda_2)^{-2}]\left[\frac{\partial U}{\partial \mathrm{I}} + \lambda_2^2\,\frac{\partial U}{\partial \mathrm{II}}\right].$$

Similarly $$\sigma_{22} = 2\rho_0[\lambda_2^2 - (\lambda_1\lambda_2)^{-2}]\left[\frac{\partial U}{\partial \mathrm{I}} + \lambda_1^2\,\frac{\partial U}{\partial \mathrm{II}}\right].$$

These equations may be solved for $\partial U/\partial \mathrm{I}$ and $\partial U/\partial \mathrm{II}$ in terms of experimental quantities λ_1, λ_2, σ_{11} and σ_{22}, obtained by stretching a sheet in the x_1 and x_2 directions by amounts λ_1 and λ_2. In their classical experiments described by Rivlin and Saunders *Phil. Trans. Roy. Soc.* **A243**, (1951) p. 51, using vulcanised rubber, the stretches λ_1 and λ_2 were chosen so that either $\partial U/\partial \mathrm{I}$ was determined experimentally with II constant, or $\partial U/\partial \mathrm{II}$ was determined experimentally with I constant. In this way it was shown that for $\lambda_1 > 1, \lambda_2 > 1$, for the material in question $U(\mathrm{I}, \mathrm{II})$ is of the Mooney form

$$U = A(\mathrm{I} - 3) + B(\mathrm{II} - 3)$$

where A and B are constants.

6. For the deformation in question

$$\mathbf{F} = \begin{bmatrix} 1 & a & 0 \\ b & 1 & 0 \\ 0 & 0 & 1 \end{bmatrix}, \quad \mathbf{B} = \mathbf{F}\mathbf{F}^T = \begin{bmatrix} 1+a^2 & a+b & 0 \\ a+b & 1+b^2 & 0 \\ 0 & 0 & 1 \end{bmatrix}$$

and
$$\mathbf{B}^2 = \begin{bmatrix} (1+a^2)^2 + (a+b)^2 & (2+a^2+b^2)(a+b) & 0 \\ (2+a^2+b^2)(a+b) & (1+b^2)^2 + (a+b)^2 & 0 \\ 0 & 0 & 1 \end{bmatrix}.$$

and the constitutive equations

$$\boldsymbol{\sigma} = \alpha \mathbf{I} + \beta \mathbf{B} + \gamma \mathbf{B}^2$$

imply a uniform stress field for which $\sigma_{zx} = \sigma_{zy} = 0$, so that the faces $Z = 0\ (z = 0)$ and $Z = N\ (z = N)$ are subject only to normal traction σ_{zz}.

Also from the above

$$\sigma_{xx} = \alpha + (1 + a^2)\beta + [(1 + a^2)^2 + (a + b)^2]\gamma$$
$$\sigma_{yy} = \alpha + (1 + b^2)\beta + [(1 + b^2)^2 + (a + b)^2]\gamma$$
$$\sigma_{xy} = (a + b)\beta + (2 + a^2 + b^2)(a + b)\gamma$$

and the result
$$\sigma_{xx} - \sigma_{yy} = (a - b)\sigma_{xy} \tag{1}$$
follows immediately.

Normals to the deformed slant faces $x = ay$ and $x = ay + L(1 - ab)$ are in the directions $\pm\, \mathbf{n}_1$ where $\mathbf{n}_1 = (1, -a, 0)/(1 + a^2)^{1/2}$. Similarly normals to the other pair of slant faces are in directions $\pm\, \mathbf{n}_2$ where $\mathbf{n}_2 = (b, -1, 0)/(1 + b^2)^{1/2}$. Conditions for the stress vector $\sigma_{ij}n_j$ to be normal to a surface with normal $(n_x, n_y, 0)$ are

$$\sigma_{xx}n_x + \sigma_{xy}n_y = \sigma n_x$$
$$\sigma_{xy}n_x + \sigma_{yy}n_y = \sigma n_y\,.$$

Eliminating σ (the magnitude of the normal stress) leads to

$$(\sigma_{xx} - \sigma_{yy})n_x n_y = \sigma_{xy}(n_x^2 - n_y^2).$$

From equation (1) this implies for $\mathbf{n}_1$ and $\mathbf{n}_2$

$$(a - b)n_x n_y = n_x^2 - n_y^2 \tag{2}$$

and for the surfaces in question, n_x/n_y is $-b$ or $-a^{-1}$. Substituting either result into (2) leads to $ab = 1$ and this is not allowed by the result of Problem 6 of Chapter 2.

Answers, Chapter 10

3. The solution of $$\left(\frac{d^2}{dr^2} + \frac{1}{r}\frac{d}{dr}\right) v_z = 0$$

is $$v_z = A \log r + B\ .$$

$v_z = v_0$ on $r = a$ and $v_z = 0$ on $r = b$ implies

$$A \log a + B = v_0\ , \qquad A \log b + B = 0$$

i.e. $$A = -v_0/\log(b/a)\ , \qquad B = v_0 \log b/\log(b/a)$$

and hence $$v = v_0 \log(b/r)/\log(b/a)\ .$$

The force exerted by the fluid on unit length of the inner cylinder derives from the shear stress σ_{rz} and is given by

$$F_z = (2\pi r \sigma_{rz})_{r = a}\ .$$

σ_{rz} is most easily derived from symmetry arguments, as follows

$$\sigma_{rz} = (\sigma_{xz})_{x = r,\ y = 0} = \mu(\partial v_z/\partial x)_{x = r,\ y = 0} = \mu\left(\frac{dv_z}{dr}\frac{\partial r}{\partial x}\right)_{x = r,\ y = 0}$$

$$= -\frac{\mu v_0 r^{-1}}{\log(b/a)}$$

so that the force exerted by the fluid is

$$F_z = -\frac{2\pi\mu v_0}{\log(b/a)}$$

per unit of the cylinder. To maintain the motion it is necessary to exert a counter force

$$F_z = \frac{2\pi\mu v_0}{\log(b/a)}$$

whose rate of working is $F_z v_0 = 2\pi\mu v_0^2/\log(b/a)$.

The energy dissipation rate is

$$D = 2\mu d_{ij}d_{ij} = 4\mu({d_{xz}}^2 + {d_{yz}}^2)$$

(since $d_{xx} = d_{yy} = d_{zz} = d_{xy} = 0$ and $d_{xz} = d_{zx}, d_{yz} = d_{zy}$).

Also
$$d_{xz} = \tfrac{1}{2}(\partial v_z/\partial x) = -\tfrac{1}{2}v_0 x r^{-2}/\log(b/a)\,,$$
$$d_{yz} = \tfrac{1}{2}(\partial v_z/\partial y) = -\tfrac{1}{2}v_0 y r^{-2}/\log(b/a)$$

and therefore
$$D = \mu {v_0}^2 r^{-2}/[\log(b/a)]^2\,.$$

The rate of energy dissipation per unit length is

$$2\pi\int_a^b rDdr = 2\pi\mu {v_0}^2/\log(b/a)\,.$$

8. (v)
$$d_{rr} = \frac{\partial v_r}{\partial r} = v_0\cos\theta\left[-\frac{3}{2}\frac{a}{r^2} + \frac{3}{2}\frac{a^3}{r^4}\right]$$

$$d_{r\theta} = \frac{1}{2}\left(\frac{\partial v_\theta}{\partial r} + \frac{1}{r}\frac{\partial v_r}{\partial \theta} - \frac{v_\theta}{r}\right) = \tfrac{3}{4}v_0\,(a^3/r^4)\sin\theta$$

$$P = \tfrac{3}{2}(\mu a v_0\cos\theta/r^2) + P_0\,.$$

On $r = a$, $\sigma_{rr} = -\tfrac{3}{2}(\mu v_0\cos\theta/a) - P_0$, $\quad\sigma_{r\theta} = \tfrac{3}{2}\mu v_0\sin\theta/a$

and the element of surface area is given by $dS = a^2\sin\theta\,d\theta\,d\phi$. The resisting force on the sphere is

$$F_z = \int_0^\pi d\theta\int_0^{2\pi} a^2\sin\theta\,[\sigma_{rr}\cos\theta - \sigma_{r\theta}\sin\theta\,]\,d\phi$$

$$= -3\pi a\mu v_0\int_0^\pi(\cos^2\theta + \sin^2\theta)\sin\theta\,d\theta - 2\pi a^2 P_0\int_0^\pi\cos\theta\sin\theta\,d\theta$$

$$= -6\pi a\mu v_0\,.$$

Answers, Chapter 11

1. Conservation of mass requires

$$v(s,t)\,A(s) = \dot{S}A(S)$$

where v is the velocity at s in the positive s direction and $\dot{S}$ is the velocity at S.

Thus
$$\frac{\partial v}{\partial t} = A^{-1}(s)[\ddot{S}A(S) + \dot{S}^2 A'(S)]$$

so that $$\int_0^S \frac{\partial v}{\partial t}\, ds = [\ddot{S}A(S) + \dot{S}^2 A'(S)] \int_0^S A^{-1}(s)ds \,.$$

Applying Bernoulli's equation along the streamline which terminates at one end at the fluid surface in the tank and at the other end at $s = S$ leads to

$$\tfrac{1}{2}\dot{S}^2 + g[h(s) - H] + [\ddot{S}A(S) + \dot{S}^2 A'(S)] \int_0^S A^{-1}(s)\,ds = 0 \,.$$

which is a non-linear differential equation for $S(t)$ where $A(s)$ and $h(s)$ are given. Boundary conditions required to complete the solution are $\dot{S} = 0$ at $S = s_0$ and $t = 0$.

Writing $\dot{S} = q(S)$ whence also $\ddot{S} = qq' \equiv \tfrac{1}{2}\dfrac{d}{ds}(q^2)$

the equation becomes

$$\frac{d(q^2)}{dS} + \left[\frac{2A'}{A} + \frac{1}{A\int_0^S A^{-1}(s)ds}\right] q^2 = \frac{2g(H-h)}{A\int_0^S A^{-1}(s)ds}$$

which is a first order linear equation with integrating factor $A^2 e^{\phi(S)}$

where $$\phi(S) = \int_\alpha^S A^{-1}(s)\left[\int_0^s A^{-1}(\xi)d\xi\right]^{-1} ds$$

and where α is arbitrary.

The equation may now be written

$$\frac{d}{dS}(q^2 A^2 e^{\phi}) = 2g(H-h)e^{\phi} A(S) \left[\int_0^S A^{-1}(s)ds\right]^{-1}$$

with solution

$$q^2 = 2gA^{-2}e^{-\phi} \int_{s_0}^S \frac{A(s)\,[H - h(s)]\, e^{\phi(s)}}{\int_0^s A^{-1}(\xi)\,d\xi}\, ds \,.$$

where the constant of integration has been chosen to satisfy $q = 0$ at $S = s_0$. The form of the solution implies that q is independent of the choice for α, and this is readily verifiable.

For the case $A = A_0 =$ const., $h(s) =$ const., the expression reduces to

$$q^2 = 2g(H-h)\,(S - s_0)/S$$

in agreement with calculations in the text. [Curiously the choice $\alpha = 0$ here leads to slight difficulties; the simplest choice is $\alpha = 1$.]

For the case $A(s) = A_0 e^{\gamma s}$, $h(s) = $ const., the most convenient choice for α is $\alpha = \infty$ and there results

$$q^2 = 2g(H-h)\,[e^{-\gamma S} - e^{-\gamma(2S - s_0)}]/[1 - e^{-\gamma S}]$$

For the particular case $s_0 = 0$ this reduces to

$$q^2 = 2g(H-h)e^{-\gamma S}$$

2. The second calculation, based on the fundamental equations of motion is correct. Thus the free surface is a paraboloid of revolution with a minimum, rather than maximum, at $r = 0$.

The first calculation provides a good example of misuse of Bernoulli's equation. Because the flow is *not* irrotational, as is evident physically from the fact that the fluid is rotating as a rigid body†, there is no single Bernoulli equation valid throughout the liquid. Of course different Bernoulli equations (with different constants) hold on the circular streamlines, but this information is here vacuous and not useful for determining the dependence of P on r.

Because the flow is not irrotational there arises the question as to whether such a flow is attainable. The answer is positive for if an even slightly viscous liquid is rotated in a cylindrical vessel, then ultimately the rigid body motion ensues. However, for the rigid body motion, all components of the rate of deformation tensor d_{ij} vanish, so that the liquid is now modelled adequately by the assumption of a perfect fluid.

3. Since the motion is incompressible

$$\text{div}\,\mathbf{v} = \partial v_r/\partial r + r^{-1} v_r + \partial v_z/\partial z = 0$$

so that

$$\frac{1}{r}\frac{\partial}{\partial r}(rv_r) = -\frac{\partial v_z}{\partial z}\,.$$

The left side of this equation is independent of z while the right side is independent of r. Therefore each side is at most a function of time [say $-h(t)$] whence

$$\frac{\partial v_z}{\partial z} = h(t)\,, \qquad \frac{\partial}{\partial r}(rv_r) = -\,rh(t)$$

with solutions

$$v_z = h(t)z + m(t) \qquad v_r = -\tfrac{1}{2}h(t)r + q(t)r^{-1}.$$

The motion is irrotational since $\text{curl}\,\mathbf{v} = 0$ so there exists a velocity potential ϕ such that

$$-\frac{\partial\phi}{\partial r} = -\tfrac{1}{2}hr + qr^{-1}, \qquad -\frac{\partial\phi}{\partial z} = hz + m\,.$$

† A more mathematical proof that the flow is not irrotational stems from evaluating the non-vanishing quantity $\text{curl}\,\mathbf{v} = 2\omega\mathbf{k}$.

By inspection

$$\phi = h(\tfrac{1}{4}r^2 - \tfrac{1}{2}z^2) - mz - q\log r + \gamma(t) .$$

However the quantity $\gamma(t)$ is irrelevant since it provides no contribution to $\mathbf{v}$ and enters Bernoulli's equation only as $\dot{\gamma}(t)$, which is readily absorbed into the existing arbitrary function of time appearing in the equation.

The Bernoulli equation reads

$$P/\rho_0 = f(t) - \tfrac{1}{2}\mathbf{v}^2 + \partial\phi/\partial t .$$

Substituting for $\mathbf{v}$ and ϕ now gives

$$P/\rho_0 = \chi(t) - \tfrac{1}{2}(\dot{h} + h^2)z^2 + \tfrac{1}{4}(\dot{h} - \tfrac{1}{2}h^2)r^2 - (\dot{m} + mh)z - \dot{q}\log r - \tfrac{1}{2}q^2r^{-2}$$

where $\chi = f - \tfrac{1}{2}m^2 + \tfrac{1}{2}hq.$

4. Since there is no axial motion $v_z = 0$ and $h = m = 0$, while v_r becomes

$$v_r = qr^{-1}$$

so that since for $r = a$, $v_r = \dot{a}$ then $q = a\dot{a}$ and the Bernoulli equation becomes

$$P/\rho_0 = \chi - (a\ddot{a} + \dot{a}^2)\log r - \tfrac{1}{2}a^2\dot{a}^2r^{-2} .$$

The condition $P = 0$ at $r = b$ determines χ and leads to

$$P/\rho_0 = (a\ddot{a} + \dot{a}^2)\log(b/r) + \tfrac{1}{2}a^2\dot{a}^2(b^{-2} - r^{-2})$$

so that since $P = \Pi$ at $r = a$

$$\Pi/\rho_0 = (a\ddot{a} + \dot{a}^2)\log(b/a) + \tfrac{1}{2}a^2\dot{a}^2(b^{-2} - a^{-2}) .$$

Finally eliminating b using the incompressibility condition

$$b^2 - a^2 = b_0^2 - a_0^2 ,$$

which is evident from comparison of the annular areas in the initial and deformed configuration,†

$$\Pi/\rho_0 = \tfrac{1}{2}(a\ddot{a} + \dot{a}^2)\log\left[1 + (b_0^2 - a_0^2)a^{-2}\right] - \tfrac{1}{2}\dot{a}^2(b_0^2 - a_0^2)/(a^2 + b_0^2 - a_0^2) .$$

The solution is valid provided $P \geqslant 0$ for $a < r < b$. Since the pressure is positive at $r = a$ and vanishes at $r = b$, it is sufficient to show

$$\frac{\partial P}{\partial r} \leqslant 0 \qquad a \leqslant r \leqslant b$$

† Alternately note that since $v_r = \dot{b}$ at $r = b$ then

$$q = a\dot{a} = b\dot{b}$$

so that $(b^2 - a^2)$ is independent of t and therefore $(b^2 - a^2) = (b_0^2 - a_0^2)$.

for then P decreases continuously through positive values from Π to zero. We have

$$-\rho_0^{-1}\frac{\partial P}{\partial r} = a\ddot{a}r^{-1} + \dot{a}^2 r^{-1}(1 - a^2 r^{-2})$$

which is positive if $\ddot{a} > 0$.

5. In the absence of body forces the equation of mass conservation is

$$\frac{d}{dx}(\rho v) = 0$$

with solution

$$\rho v = c \tag{1}$$

where c is constant. Also along the streamlines in the x direction the Bernoulli equation takes the form

$$\psi(\rho) + \tfrac{1}{2}v^2 = A \tag{2}$$

where A is a constant and where $\psi = \int^{\rho} (\rho')^{-1}(dP/d\rho')d\rho'$.

From (1) and (2)

$$\psi(\rho) + \tfrac{1}{2}c^2\rho^{-2} = A$$

with solution $\rho = \text{const}$. Thus from $P = P(\rho)$, P is constant while from (1) $v(x)$ is constant.

Answers, Chapter 12

1. Since

$$\bar{P} = P_1(\sigma/\sigma_0)^{\gamma} = P_1(a_0/a)^{3\gamma}$$

equation (12.2.10) yields

$$P_1 a_0^{3\gamma}\int_{a_0}^{a} a^{2-3\gamma}da = \tfrac{1}{3}P_0(a^3 - a_0^3) + \tfrac{1}{2}\rho_0 a^3\dot{a}^2$$

i.e.

$$\frac{P_1 a_0^3}{3(\gamma-1)}[1-(a/a_0)^{-3(\gamma-1)}] - \frac{P_0 a_0^3}{3}[(a/a_0)^3 - 1] = \tfrac{1}{2}\rho_0 a^3\dot{a}^2 \tag{1}$$

The cavity comes to rest when $\dot{a} = 0$ so that if at this time $a = a_f$ and we define $X = (a_f/a_0)$, equation (1) with $\dot{a} = 0$ yields

$$\frac{P_1}{(\gamma-1)}[1 - X^{-3(\gamma-1)}] = P_0[X^3 - 1]\,.$$

Using this last result to substitute for P_1 in (1) and writing $a = a_0\theta$ $(1 \leqslant \theta \leqslant X)$ gives

$$\tfrac{1}{2}\rho_0 a_0^5\theta^3\dot{\theta}^2 = \tfrac{1}{3}P_0 a_0^3\,[(X^3-1)(1-\theta^{-3(\gamma-1)})(1-X^{-3(\gamma-1)})^{-1} - (\theta^3-1)]$$

with solution

$$\left(\frac{2P_0}{3\rho_0 a_0^2}\right)^{\frac{1}{2}} t =$$

$$[1-X^{-3(\gamma-1)}]^{\frac{1}{2}} \int_1^{(a/a_0)} \frac{\theta^{3/2}\,d\theta}{[(X^3-1)(1-\theta^{-3(\gamma-1)}) - (\theta^3-1)(1-X^{-3(\gamma-1)})]^{\frac{1}{2}}}.$$

The time T for the cavity to attain the radius a_f is found by writing $X(\equiv a_f/a_0)$ in the upper limit of the integral.

2. If the tube expands radially symmetrically with a radial velocity $v(r, t)$ and there is no motion in the axial direction, the incompressible condition $(\text{div}\,\mathbf{v} = 0)$ leads to

$$\frac{\partial v_r}{\partial r} + \frac{v_r}{r} = 0$$

with solution $$v_r = g(t)/r$$

where $g(t)$ is determined from $v_r = \dot{a}$ at $r = a$. Therefore $v_r = a\dot{a}/r$ and the associated velocity potential is $\phi = -a\dot{a}\log r$. From Bernoulli's equation

$$\frac{P}{\rho_0} + \frac{1}{2}\frac{a^2\dot{a}^2}{r^2} + (a\ddot{a} + \dot{a}^2)\log r = f(t)$$

where $f(t)$ is determined by $P = P_0$ at $r = b$. With $f(t)$ so determined and using $P = \bar{P}$ at $r = a$, there results

$$\bar{P} = P_0 - \tfrac{1}{2}\rho_0\dot{a}^2[1-(a/b)^2] + \rho_0(a\ddot{a} + \dot{a}^2)\log(b/a)\,;$$

Write $$Q = (b/a)^2 = (b_0^2 - a_0^2 + a^2)/a^2$$

[since from incompressibility $(b^2 - a^2) = (b_0^2 - a_0^2)$]. Then $\log(b/a) = \tfrac{1}{2}\log Q$ and also

$$\frac{d}{dt}(\log Q) = -2Q^{-1}(b_0^2 - a_0^2)\dot{a}/a^3 \equiv -2(\dot{a}/a)[1-(a/b)^2]\,.$$

From these results the equation of motion may be written

$$\bar{P} = P_0 + \tfrac{1}{4}\rho_0 a\dot{a}\frac{d}{dt}(\log Q) + \tfrac{1}{2}\rho_0(a\ddot{a} + \dot{a}^2)\log Q\,.$$

On multiplying through by $a\dot{a}$, using $\dot{a} = 0$ at $a = a_0$ and noting

$$\frac{d}{dt}\left(\tfrac{1}{4}\rho_0 a^2\dot{a}^2\right) = \tfrac{1}{2}\rho_0[a^2\dot{a}\ddot{a} + a\dot{a}^3]\,,$$

the first integral follows immediately.

3. The velocity potential for an isolated point dipole at the origin is

$$\phi = -\boldsymbol{\mu} \cdot \text{grad}\left(\frac{1}{r}\right) = \frac{\boldsymbol{\mu} \cdot \mathbf{r}}{r^3}$$

so that if an isolated point dipole of strength $\mu[\sin\alpha, 0, \cos\alpha]$ is located at $(0, 0, h)$ the velocity potential is

$$\phi = \frac{\mu[x \sin\alpha + (z-h)\cos\alpha]}{[x^2 + y^2 + (z-h)^2]^{3/2}}.$$

Superimposing this result with the potential of the image dipole leads to

$$\phi = \frac{\mu[x\sin\alpha + (z-h)\cos\alpha]}{[x^2+y^2+(z-h)^2]^{3/2}} + \frac{\mu[x\sin\alpha - (z+h)\cos\alpha]}{[x^2+y^2+(z+h)^2]^{3/2}}.$$

It is readily verified that $(\partial\phi/\partial z) = 0$ on $z = 0$ so that on $z = 0$

$$\mathbf{v}^2 = \left(\frac{\partial\phi}{\partial x}\right)^2 + \left(\frac{\partial\phi}{\partial y}\right)^2.$$

Calculating the relevant derivatives and substituting $z = 0$ leads to

$$\left(\frac{\partial\phi}{\partial x}\right)_{z=0} = \frac{2\mu\sin\alpha}{R^3} - \frac{6\mu x\xi}{R^5}, \quad \left(\frac{\partial\phi}{\partial y}\right)_{z=0} = -\frac{6\mu y\xi}{R^5}.$$

so that

$$(\mathbf{v}^2)_{z=0} = \frac{4\mu^2\sin^2\alpha}{R^6} + \frac{36\mu^2(x^2+y^2)\xi^2}{R^{10}} - \frac{24\mu^2\xi x\sin\alpha}{R^8}$$

$$= 4\mu^2\left\{\frac{\sin^2\alpha}{R^6} + \frac{3\xi\eta}{R^8} - \frac{9h^2\xi^2}{R^{10}}\right\}.$$

At distances removed far from the dipole $\mathbf{v}^2 \to 0$ so that Bernoulli's theorem yields

$$P = P_0 - 2\rho_0\mu^2\left\{\frac{\sin^2\alpha}{R^6} + \frac{3\xi\eta}{R^8} - \frac{9h^2\xi^2}{R^{10}}\right\}.$$

4. **(a)** Kinetic energy of the fluid is

$$T = \tfrac{1}{2}\rho_0 \iiint_{\Omega} (\text{grad}\,\phi)^2 d\Omega$$

where Ω denotes the space occupied by the fluid.

From the vector identity

$$\text{div}(\phi\,\text{grad}\,\phi) = (\text{grad}\,\phi)^2 + \phi\nabla^2\phi,$$

the result $\nabla^2\phi = 0$, and Green's theorem

$$T = -\tfrac{1}{2}\rho_0 \sum_m \iint_{S_m} \phi \frac{\partial\phi}{\partial n}\, dS + \tfrac{1}{2}\rho_0 \iint_{S_R} \phi \frac{\partial\phi}{\partial n}\, dS$$

where the minus sign derives from applying Green's theorem to a region bounded internally by the solid objects.

If the surface integral over S_R vanishes as $R \to \infty$, there follows

$$T = -\tfrac{1}{2}\rho_0 \sum_m \iint_{S_m} \phi \frac{\partial\phi}{\partial n}\, dS \,.$$

(b) Without loss of generality assume the sphere velocity $\mathbf{u}$ is given by $\mathbf{u} = u\mathbf{k}$ and that the sphere is centred on the origin. The velocity potential is

$$\phi = \tfrac{1}{2}ua^3 \cos\theta / r^2$$

and the integral over S_R vanishes as $O(R^{-3})$. Therefore

$$T = -\tfrac{1}{2}\rho_0 \iint_{(r=a)} \phi \frac{\partial\phi}{\partial r}\, dS = \tfrac{1}{4}\rho_0 u^2 a \int_0^{2\pi} d\psi \int_0^{\pi} a^2 \cos^2\theta \sin\theta \, d\theta$$

$$= \tfrac{1}{3}\pi\rho_0 a^3 u^2 = \tfrac{1}{4}M'\mathbf{u}^2$$

where
$$M' = \tfrac{4}{3}\pi a^3 \rho_0 \,.$$

The rate of working of the external force $\mathbf{F}$ on the sphere required to set the fluid in motion is

$$\frac{d}{dt}\left(\tfrac{1}{4}M'\mathbf{u}^2\right) = \tfrac{1}{2}M'\dot{\mathbf{u}}\cdot\mathbf{u} \equiv \mathbf{F}\cdot\mathbf{u}\,,$$

and since this is true for arbitrary $\mathbf{u}$,

$$\mathbf{F} = \tfrac{1}{2}M'\,\dot{\mathbf{u}}\,.$$

Therefore the resistive force of the fluid on the sphere is $-\tfrac{1}{2}M'\dot{\mathbf{u}}$. [An additional force is required to overcome the inertia of the sphere.]

The present calculation, while less direct, is simpler to carry through than that of the text, and covers easily the case where $\mathbf{u}$ and $\dot{\mathbf{u}}$ are non-parallel.

5. **(a)** In the absence of the cylinder of radius a, the complex potential of the stream is $\omega = -uze^{-i\alpha}$ so that from the circle theorem, in the presence of the cylinder

$$\omega = -u[ze^{-i\alpha} + (a^2e^{i\alpha}/z)]$$

so that
$$\psi \equiv \psi_0 = \operatorname{Im}\omega = -u[r-(a^2/r)]\sin(\theta-\alpha).$$

(b) On $r = a[1+\epsilon\cos(p\theta)]$, ψ_0 takes the value

$$\begin{aligned}\psi_0 &= -2ua\epsilon\cos(p\theta)\sin(\theta-\alpha) + O(\epsilon^2)\\ &= -ua\epsilon\left[\sin[(p+1)\theta-\alpha] - \sin[(p-1)\theta+\alpha]\right] + O(\epsilon^2)\end{aligned}$$

and to make $\psi = O(\epsilon^2)$ on the cylinder boundary together with the other conditions requires a choice of ψ_1 which satisfies

$\operatorname{grad}\psi_1 \to 0 \qquad (r\to\infty)$

$\psi_1 = ua\left[\sin[(p+1)\theta-\alpha] - \sin[(p-1)\theta+\alpha]\right]$ on the boundary $r = a$

$\nabla^2\psi_1 = 0$.

These conditions lead to

$$\psi_1 = ua\left[(a^{p+1}/r^{p+1})\sin[(p+1)\theta-\alpha] - (a^{p-1}/r^{p-1})\sin[(p-1)\theta+\alpha]\right]$$

and the corresponding (total) complex potential is

$$\omega = -u[ze^{-i\alpha} + (a^2e^{i\alpha}/z)] - u\epsilon[(a^{p+2}/z^{p+1})e^{i\alpha} - (a^p/z^{p-1})e^{-i\alpha}] + O(\epsilon^2)$$

so that

$$\frac{d\omega}{dz} = -u[e^{-i\alpha} - a^2e^{i\alpha}z^{-2} - \epsilon(p+1)a^{p+2}z^{-(p+2)}e^{i\alpha} + \epsilon(p-1)a^pz^{-p}e^{-i\alpha} + O(\epsilon^2).$$

From the Blasius theorems

$$F_y + iF_x = -\tfrac{1}{2}\rho_0\oint_C\left(\frac{d\omega}{dz}\right)^2 dz, \qquad M = -\tfrac{1}{2}\rho_0\operatorname{Re}\oint_C z\left(\frac{d\omega}{dz}\right)^2 dz.$$

For $p \geqslant 2$ there are no contributions of order z^{-1} from $(d\omega/dz)^2$ so that F_x and F_y vanish. Also for $p \geqslant 3$, $z(d\omega/dz)^2$ contains a term $-2u^2a^2z^{-1}$ with a real residue, so that for $p \geqslant 3$, $M = 0$. On the other hand for $p = 2$, $z(d\omega/dz)^2$ contains the term

$$-2u^2a^2[1-\epsilon e^{-2i\alpha}]z^{-1}$$

so that
$$M = \rho_0u^2a^2\operatorname{Re}2\pi i[1-\epsilon e^{-2i\alpha}] = -2\pi\rho_0u^2a^2\epsilon\sin(2\alpha).$$

7. If r' denotes the distance from $(0, 0, \xi)$ to (x, y, z) then from geometry

$$r' = (r^2 + \xi^2 - 2r\xi\cos\theta)^{\frac{1}{2}}.$$

The potential of source of strength m located at $(0, 0, \xi)$ is m/r' and therefore for a dipole located at the same point

$$\phi = \phi_1 = \mu \frac{\partial}{\partial\xi}(r')^{-1}.$$

For $r < \xi$, $(r')^{-1}$ has the expansion

$$(r')^{-1} = \xi^{-1}[1 - 2(r/\xi)\cos\theta + (r/\xi)^2]^{-\frac{1}{2}} = \sum_{n=0}^{\infty} \frac{r^n}{\xi^{n+1}} P_n(\cos\theta)$$

so that
$$\phi_1 = -\mu \sum_{n=0}^{\infty} \frac{(n+1)r^n}{\xi^{n+2}} P_n(\cos\theta).$$

In the presence of the sphere of radius a, ϕ is modified by an additional potential which is non-singular in the field of flow and whose derivates vanish for $r \to \infty$. Write

$$\phi = \phi_1 + \sum_{n=0}^{\infty} A_n r^{-(n+1)} P_n(\cos\theta)$$

where the A_n are determined by $(\partial\phi/\partial r)_{r=a} = 0$. Since $a < \xi$ we use the above expression for ϕ_1, obtaining

$$\left(\frac{\partial\phi}{\partial r}\right)_{r=a} = -\mu \sum_{n=0}^{\infty} \frac{n(n+1)a^{n-1}}{\xi^{n+2}} P_n(\cos\theta) - \sum_{n=0}^{\infty} (n+1)A_n a^{-(n+2)} P_n(\cos\theta)$$

so that the boundary condition is satisfied by choosing

$$A_n = -na^{2n+1}/\xi^{n+2}$$

and therefore
$$\phi_2 = -\mu \sum_{0}^{\infty} \frac{na^{2n+1}}{\xi^{n+2} r^{n+1}} P_n(\cos\theta)$$

A dipole of strength $-\mu(a/\xi)^3$ located at the image point has potential

$$\phi = -\mu(a/\xi)^3 \frac{\partial}{\partial\xi'} [(\xi')^2 + r^2 - 2r\xi'\cos\theta]^{-\frac{1}{2}}$$

where $\xi' = a^2/\xi$. Since $r > \xi'$ for all relevant r

$$(\xi'^2 + r^2 - 2r\xi'\cos\theta)^{-\frac{1}{2}} = \sum_{n=0}^{\infty} \frac{(\xi')^n}{r^{n+1}} P_n(\cos\theta)$$

and

$$\phi = -\mu(a/\xi)^3 \sum_{n=0}^{\infty} \frac{n\xi'^{(n-1)}}{r^{n+1}} P_n(\cos\theta) = -\mu \sum_{n=0}^{\infty} \frac{na^{2n+1}}{\xi^{n+2} r^{n+1}} P_n(\cos\theta)$$

$$\equiv \phi_2 \,.$$

In closed form ϕ_1 is

$$\phi_1 = \mu \frac{\partial}{\partial\xi}(r')^{-1} = -\mu(\xi - r\cos\theta)\,[r^2 + \xi^2 - 2r\xi\cos\theta]^{-3/2}\,.$$

Similarly $\phi_2 = \mu(a/\xi)^3(\xi' - r\cos\theta)\,[r^2 + (\xi')^2 - 2r\xi'\cos\theta]^{-3/2}$ and the complete solution to the problem is $\phi = \phi_1 + \phi_2$.

The reader is left to show that the closed form solution yields $(\partial\phi/\partial r)_{r=a} = 0$.

The only non-vanishing component of **v** on $r = a$ is

$$v_\theta = -r^{-1}(\partial\phi/\partial\theta)\,.$$

Carrying out the differentiating and writing $r = a$ yields ultimately

$$(v_\theta)_{r=a} = \frac{-3\mu(\xi^2 - a^2)\sin\theta}{(a^2 + \xi^2 - 2a\xi\cos\theta)^{5/2}}\,.$$

From Bernoulli's theorem the pressure distribution on the sphere is

$$P = \text{const.} - \tfrac{1}{2}\rho_0 v_\theta^2\,.$$

From symmetry the only force is in the z direction and given by

$$F_z = -\int_0^{2\pi} d\psi \int_0^{\pi} P\cos\theta\, a^2 \sin\theta\, d\theta = 9\pi\rho_0 a^2\mu^2(\xi^2 - a^2)^2 I$$

where

$$I = \int_0^{\pi} \frac{\sin^3\theta\cos\theta\, d\theta}{(a^2 + \xi^2 - 2a\xi\cos\theta)^5} = \int_{-1}^{1} \frac{x(1 - x^2)dx}{(a^2 + \xi^2 - 2a\xi x)^5}\,.$$

The integral I is elementary and left to the reader; there results finally

$$F_z = 24\pi\rho_0 a^3\mu^2\xi/(\xi^2 - a^2)^4$$

so that the sphere is attracted to the dipole.

8. **(a)** the expression for ϕ_0 is given in the text in equation (12.7.24).

(b) From (12.7.24) an isolated sphere of radius b located at $z = q_2(t)$ possesses a velocity potential which is that of a dipole of strength

$$\mu(t) = \tfrac{1}{2}u_2(t)b^3$$

located at $z = q_2(t)$. With this value of μ, the boundary conditions on $R = a$ are satisfied by

$$\hat{\phi} = \phi_1 + \phi_2$$

where ϕ_1 and ϕ_2 are given in Problem 7. This leads to the stated result for $\hat{\phi}$ where both expansions are valid near $R = a$.

In constructing ϕ, use has been made of the property

$$(\partial\hat{\phi}/\partial R)_{R=a} = 0$$

so that $$T_1 = -\tfrac{1}{2}\rho_0 \iint_{S_a} \phi \frac{\partial \phi}{\partial R}\, dS = -\tfrac{1}{2}\rho_0 \iint_{S_a} (\phi_0 + \hat{\phi}) \frac{\partial \phi_0}{\partial R}\, dS\,.$$

On $R = a$ $$\phi_0 + \hat{\phi} = \tfrac{1}{2}u_1 a P_1(\cos\theta) - \tfrac{1}{2}u_2 b^3 \sum_0^\infty \frac{(2n+1)a^n}{\xi^{n+2}} P_n(\cos\theta)$$

$$\left(\frac{\partial \phi_0}{\partial R}\right)_a = -u_1 P_1(\cos\theta)$$

and

$$T_1 = \tfrac{1}{4}\rho_0 u_1^2 a^3 \int_0^{2\pi} d\psi \int_0^{\pi} P_1^2(\cos\theta)\sin\theta\, d\theta$$

$$- \tfrac{1}{4}\rho_0 u_1 u_2 a^2 b^3 \int_0^{2\pi} d\psi \sum_n \frac{(2n+1)a^n}{\xi^{n+2}} \int_0^{\pi} P_1(\cos\theta) P_n(\cos\theta)\sin\theta\, d\theta\,.$$

From the orthogonality relations for the P_n, only the term for $n = 1$ survives in the sum and there results

$$T_1 = \tfrac{1}{4}M_1' u_1^2 - \pi\rho_0 u_1 u_2 a^3 b^3 \xi^{-3}\,.$$

From symmetry a similar term is given for the surface integral over the second sphere and the total kinetic energy of the fluid becomes

$$\tfrac{1}{4}(M_1'\dot{q}_1^2 + M_2'\dot{q}_2^2) - 2\pi\rho_0 a^3 b^3 \dot{q}_1 \dot{q}_2/(q_2 - q_1)^3\,.$$

Adding the kinetic energy of the spheres leads to the result given.

(c) In the absence of potential energy Lagrange's equation for q_1 is

$$\frac{d}{dt}\left(\frac{\partial T}{\partial \dot{q}_1}\right) - \frac{\partial T}{\partial q_1} = 0$$

with a similar equation for q_2. The resulting equation of motion may be written

$$M_1\ddot{q}_1 = -\tfrac{1}{2}M_1'\ddot{q}_1 + \frac{2\pi\rho_0 a^3 b^3 \ddot{q}_2}{(q_2-q_1)^3} - \frac{6\pi\rho_0 a^3 b^3 \dot{q}_2^2}{(q_2-q_1)^4} .$$

9. (a) See Problem 7 of Chapter 10.

(b) $$\text{grad}\ \psi = \left[\frac{\partial\psi}{\partial r}, \frac{1}{r}\frac{\partial\psi}{\partial\theta}, 0\right] \quad \text{[in directions } r, \theta, \phi] .$$

It follows that $\mathbf{v}.\text{grad}\ \psi = 0$ so that $\mathbf{v}$ is perpendicular to the normal of $\psi = \text{const.}$, i.e. $\mathbf{v}$ lies in the surface $\psi = \text{const.}$ For axisymmetric flow past a stationary axisymmetric body $\mathbf{v}$ must lie parallel to the surface of the body, so that $\psi = \text{const.}$ on the surface.

(c) It is readily verified that the equation for the stream function is satisfied by the three solutions given.

For $\psi = m\cos\theta$; $v_r = m/r^2$, $v_\theta = 0$ (i.e. source of strength m at origin)

For $\psi = -\frac{1}{2}ur^2\sin^2\theta$; $v_r = u\cos\theta$, $v_\theta = -u\sin\theta$ and this is easily shown to be the same as $v_z = u$, $v_x = v_y = 0$

For $\psi = -\frac{1}{2}u\sin^2\theta\,[r^2 - (a^3/r)]$; $\mathbf{v} \to u\mathbf{k}\ (r \to \infty)$ and $v_r = u\cos\theta\,[1 - (a/r)^3]$ so that v_r vanishes on $r = a$ (or alternately $\psi = 0$ on $r = a$). Therefore the solution represents a stream flowing past a sphere.

The present solution differs from that of Problem 8 of Chapter 10 because the solution in the slow viscous flow approximation satisfies a more complicated differential equation and also take account of the adhesion condition i.e. $\mathbf{v} = 0$ on $r = a$ rather than simply $v_r = 0$ on $r = a$.

10. (i) The vortex κ_1 at $(a, 0)$ induces a velocity in the negative y direction of $\kappa_1/(a+b)$ at the position of κ_2. Similarly κ_2 induces a velocity in the positive y directions of $\kappa_2/(a+b)$ at the position of κ_1. Instantaneously the angular velocity of κ_1 about the origin is $\kappa_2/a(a+b)$ while that of κ_2 is $\kappa_1/b(a+b)$. If $\kappa_2/a = \kappa_1/b$ the two angular velocities are identical and the vortices and origin remain collinear. In these circumstances the subsequent motion is one in which both vortices continue to rotate at a common angular velocity about the origin.

(ii) The corresponding complex potential is

$$\omega = i\kappa_1 \log[z - ae^{i\Omega t}] + i\kappa_2 \log[z + be^{i\Omega t}]$$

so that at the origin $z = 0$, $\quad \partial\omega/\partial t = -(\kappa_1 + \kappa_2)\Omega = \partial\phi/\partial t$

i.e. $$(\partial\phi/\partial t)_{z=0} = -(\kappa_1+\kappa_2)\Omega = -\kappa_1\kappa_2/ab .$$

Also at $z = 0$ the fluid velocity is perpendicular to the line joining κ_1 and κ_2 and of magnitude

$$|\mathbf{v}| = |(\kappa_2/b) - (\kappa_1/a)|\,.$$

At large distances both $\partial\phi/\partial t$ and $\mathbf{v}$ tend to zero so that from Bernoulli's theorem the pressure at the origin is

$$P = P_0 - \tfrac{1}{2}\rho_0[(\kappa_2/b) - (\kappa_1/a)]^2 - \rho_0\kappa_1\kappa_2/ab = P_0 - \tfrac{1}{2}\rho_0[(\kappa_2/b)^2 + (\kappa_1/a)^2]$$

(iii) Provided $\kappa_2 \neq -\kappa_1$, κ_1 and κ_2 rotate at a common angular velocity about a common centre. The location of the centre is the centre of mass of two bodies of mass κ_1 and κ_2 placed at the corresponding vortex positions. If the distance apart of the two vortices is l the common angular velocity is $(\kappa_1 + \kappa_2)/l^2$.

Exceptionally if $\kappa_2 = -\kappa_1$, the two vortices travel at constant velocity in a direction perpendicular to their line of centres.

11. The image system consists of vortices of strengths $-\kappa$, $-\kappa$ and κ located respectively at $(x, -y)$, $(-x, y)$ and $(-x, -y)$. The first of these induces at (x, y) a velocity $(\kappa/2y, 0)$ while the second induces a velocity $(0, -\kappa/2x)$. Finally the third vortex induces a velocity $\frac{1}{2}\kappa(x^2+y^2)^{-\frac{1}{2}}(-\sin\theta, \cos\theta)$ where θ is the usual polar angle. Expressing θ in terms of x and y and adding the induced velocities leads to

$$\dot{x} = \tfrac{1}{2}\kappa x^2 y^{-1}(x^2+y^2)^{-1}, \qquad \dot{y} = -\tfrac{1}{2}\kappa y^2 x^{-1}(x^2+y^2)^{-1}.$$

From these equations $\quad x^{-3}\dot{x} + y^{-3}\dot{y} = 0, \qquad \dot{x}y - \dot{y}x = \frac{1}{2}\kappa\,.$

The first of these possesses the immediate integral

$$x^{-2} + y^{-2} = c^{-2},$$

where c^{-2} is a constant of integration. This last result yields the parametric representation

$$x = c(\cos\gamma)^{-1}, \qquad y = c(\sin\gamma)^{-1} \qquad (0 < \gamma < \tfrac{1}{2}\pi)$$

so that $\quad \dot{x} = c\dot{\gamma}\sin\gamma/\cos^2\gamma, \quad \dot{y} = -c\dot{\gamma}\cos\gamma/\sin^2\gamma\,.$

It follows that $\quad \dot{x}y - \dot{y}x \equiv c^2\dot{\gamma}(\sin\gamma\cos\gamma)^{-2} = \frac{1}{2}\kappa$

so that $\quad \dot{\gamma}\,\mathrm{cosec}^2(2\gamma) = \kappa/8c^2$

with solution $\quad \frac{1}{2}\cot 2\gamma = \frac{1}{2}A - (\kappa t/8c^2)$

or $\quad \gamma = \frac{1}{2}\tan^{-1}[A - (\kappa t/4c^2)]^{-1}$

where A is a constant of integration.

12. From Bernoulli's theorem, in the absence of body force, the pressure is given by

$$P = \rho_0 f(t) - \tfrac{1}{2}\rho_0 \mathbf{v}^2 + \rho_0(\partial\phi/\partial t) .$$

From the analysis in the text (Section 12.8.6), $\rho_0 f(t)$ yields no contribution to the force on the cylinder while $-\tfrac{1}{2}\rho_0 \mathbf{v}^2$ leads to the force (F_x, F_y) where

$$F_y + iF_x = -\tfrac{1}{2}\rho_0 \oint_C \left(\frac{\partial\omega}{\partial z}\right)^2 dz .$$

For unsteady flow there is a further force (F_x', F_y') deriving from $\partial\phi/\partial t$. We have

$$dF_x' = -\rho_0 \left(\frac{\partial\phi}{\partial t}\right) dy , \qquad dF_y' = \rho_0 \frac{\partial\phi}{\partial t} dx$$

so that

$$F_y' - iF_x' = \rho_0 \oint_C \frac{\partial\phi}{\partial t} dz .$$

However since on C, ψ is solely a function of time (since C is a streamline), then

$$\frac{\partial\phi}{\partial t} = \frac{\partial\omega}{\partial t} - i\psi'(t) .$$

Also however $\quad -i \oint_C \psi'(t) dz = -i\psi'(t) \oint_C dz = 0$

so that

$$F_y' - iF_x' = \rho_0 \oint_C \frac{\partial\omega}{\partial t} dz .$$

13. In the absence of the cylinder, the velocity potential of an isolated rectilinear vortex located at $z = f$ is $i\kappa \log(z - f)$. From the circle theorem, in the presence of the cylinder $|z| = a$,

$$\omega = i\kappa \log[z - f] + i\kappa \log z - i\kappa \log[z - a^2/\bar{f}\,]$$

where for the moment f is being regarded as complex. It is shown in the text that the image vortices lead to the real vortex rotating clockwise about the cylinder with angular velocity $\Omega = \kappa a^2/f^2(f^2 - a^2)$, where now f denotes the magnitude of the distance of the vortex from the origin. The position of the vortex at time t is $fe^{-i\Omega t}$ and the unsteady potential is

$$\omega = i\kappa \log[z - fe^{-i\Omega t}] + i\kappa \log z - i\kappa \log[z - (a^2/f)e^{-i\Omega t}] .$$

It is adequate to calculate

$$F_y + iF_x = -\tfrac{1}{2}\rho_0 \oint_C \left(\frac{\partial\omega}{\partial z}\right)^2 dz \qquad (1)$$

assuming $t = 0$ and subsequently generalising the result. The integrand $(\partial\omega/\partial z)^2$ contains poles inside C at $z = 0$ and $z = a^2/f$. The calculation is a straightforward example in the theory of residues and yields

$$F_x = 2\pi\rho_0\kappa^2 a^2/f(f^2 - a^2)\,, \qquad F_y = 0 \tag{2}$$

More generally the force on the cylinder from equation (1) is directed radially from the origin to the vortex and is of the magnitude given by F_x in (2).

Also from Problem 12 the unsteady contribution to the force is

$$F_y' - iF_x' = \rho_0 \oint_C \frac{\partial\omega}{\partial t}\, dz$$

$$= \rho_0\kappa\Omega \oint_C \left(-\frac{fe^{-i\Omega t}}{(z - fe^{-i\Omega t})} + \frac{(a^2/f)e^{-i\Omega t}}{[z - (a^2/f)e^{-i\Omega t}]} \right) dz$$

$$= 2\pi i\rho_0\kappa\Omega a^2 f^{-1} e^{-i\Omega t}\,.$$

The components F_x', F_y' imply a total force directed radially inwards from the vortex to the cylinder, of magnitude

$$2\pi\rho_0\kappa\Omega a^2/f \equiv 2\pi\rho_0\kappa^2 a^4/[f^3(f^2 - a^2)]\,.$$

The total force directed along the radius vector from the origin to the vortex is in the outward direction and of magnitude

$$\frac{2\pi\rho_0\kappa^2 a^2}{f(f^2 - a^2)}\,[1 - (a^2/f^2)] = 2\pi\rho_0\kappa^2 a^2/f^3\,.$$

14. (i) In absence of the cylinder, the velocity potential of the vortex and stream is $\omega = i\kappa\log(z - f) + iuz$. It follows from the circle theorem that

$$\omega = i\kappa\log(z - f) - i\kappa\log[(a^2/z) - f] + iu[z - (a^2/z)]$$

$$= i\kappa\log(z - f) + i\kappa\log z - i\kappa\log[z - (a^2/f)] + iu[z - (a^2/z)]$$

$+$ irrelevant constant.

Discarding the self potential of the vortex leads to

$$\omega_1(z) = i\kappa\log z - i\kappa\log[z - (a^2/f)] + iu[z - (a^2/z)]$$

and the vortex moves at velocity (V_x, V_y)

where

$$-V_x + iV_y = (d\omega_1/dz)_{z=f}\,.$$

Therefore the vortex is stationary if

$$(d\omega'/dz)_{z=f} \equiv i\kappa f^{-1} - i\kappa[f - (a^2/f)]^{-1} + iu[1 + (a^2/f^2)] = 0\,.$$

This yields $$u = \kappa a^2 f/(f^4 - a^4)\,. \tag{1}$$

(ii) If the vortex is located at $z = f + Z$ the velocity potential is

$$\omega = i\kappa \log[z - f - Z] + i\kappa \log z - i\kappa \log[z - a^2/(f + \bar{Z})\}] + iu[z - (a^2/z)]$$

and $-V_x + iV_y(\equiv -d\bar{Z}/dt)$ is given by $(d\omega_1/dz)_{z=f+Z}$. Therefore

$$-\frac{d\bar{Z}}{dt} = \frac{i\kappa}{f+Z} - \frac{i\kappa}{f + Z - \{a^2/(f+\bar{Z})\}} + iu\left[1 + \frac{a^2}{(f+Z)^2}\right]$$

Expanding the right hand side of this equation up to terms linear in Z and $\bar{Z}$ and noticing that the terms independent of Z and $\bar{Z}$ vanish (because of the equilibrium condition which obtains for $Z = 0$), there results

$$-\frac{d\bar{Z}}{dt} = -\frac{i\kappa Z}{f^2} + \frac{i\kappa f^2}{(f^2 - a^2)^2}(Z + a^2\bar{Z}/f^2 - \frac{2iua^2 Z}{f^3}$$

$$= \left(\frac{i\kappa a^2(2f^2 - a^2)}{f^2(f^2 - a^2)^2} - \frac{2i\kappa a^4}{f^2(f^4 - a^4)}\right) Z + \frac{i\kappa a^2}{(f^2 - a^2)^2}\bar{Z}\,.$$

[Here use has been made of equation (1) to eliminate u.]

Expressing this equation in terms of real and imaginary parts now leads to

$$\frac{dY}{dt} = \frac{\kappa a^2(a^4 + 3f^4)}{f^2(f^2 - a^2)(f^4 - a^4)}X\,, \qquad \frac{dX}{dt} = \frac{\kappa a^2}{f^2(f^2 + a^2)}Y$$

with solution

$$X = Ae^{\lambda t} + Be^{-\lambda t}\,, \qquad Y = \frac{f^2(f^2 + a^2)\lambda}{\kappa a^2}[Ae^{\lambda t} - Be^{-\lambda t}]$$

where λ is given by $$\lambda = \frac{\kappa a^2(a^4 + 3f^4)^{\frac{1}{2}}}{f^2(f^4 - a^4)}$$

and where A and B are constants of integration. The presence of the terms in $\exp(\lambda t)$ imply instability.

15. At $t = 0$ in the absence of the boundary, the complex potential is $\omega = i\kappa \log(z - f)$. The circle theorem yields

$$\omega = i\kappa \log[z - f] - i\kappa \log[z - (a^2/f)] + i\kappa \log z + \text{const.}\,.$$

However the term $i\kappa \log z$ is not acceptable, since it represents a real vortex in the field of flow. [The term $-i\kappa \log[z - (a^2/f)]$ represents an acceptable image vortex outside the field of flow.] However the term $i\kappa \log z$ may be removed without disturbing the boundary condition that $|z| = a$ is a streamline. There results, after discarding also the irrelevant constant,

$$\omega = i\kappa \log(z - f) - i\kappa \log[z - (a^2/f)]\,.$$

The image vortex induces a velocity at the real vortex with components

$$v_x = 0, \qquad v_y = \kappa/[(a^2/f) - f] .$$

For a vortex not located on the real axis, these components would be written more appropriately

$$v_r = 0, \qquad v_\theta = \kappa f/[a^2 - f^2]$$

so that the real vortex moves around the circle of radius f at angular velocity

$$\Omega = (v_\theta/f) \equiv \kappa/[a^2 - f^2] .$$

Therefore at time t the vortex is located at $z = fe^{i\Omega t}$ and the complex potential is

$$\omega = i\kappa \log[z - fe^{i\Omega t}] - i\kappa \log[z - (a^2/f)e^{i\Omega t}] .$$

To evaluate the thrust on the boundary we use the results of Problem 12 but with a change of sign (since the pressure is acting here on the inside of the contour). $F_y + iF_x$ is most simply evaluated at $t = 0$

$$F_y + iF_x = \tfrac{1}{2}\rho_0 \oint_c \left(\frac{\partial\omega}{\partial z}\right)^2 dz = -\tfrac{1}{2}\rho_0\kappa^2 \oint_c [(z-f)^{-2} + (z - a^2/f)^{-2}$$

$$- 2(z-f)^{-1}(z - a^2/f)^{-1}]\, dz .$$

There are no contributions from $z = a^2/f$ (singularity outside c) and none from the double pole $z = f$. There remains a contribution from the simple pole $z = f$ in the cross product term, with residue $-2[f - (a^2/f)]^{-1}$ so that $F_y = 0$ and

$$F_x = -2\pi\rho_0\kappa^2 f/(a^2 - f^2) .$$

For non-zero t, this force is of magnitude $2\kappa\rho_0\kappa^2 f/(a^2 - f^2)$ and directed radially from the current location of the vortex towards the origin.

Also

$$F'_y - iF'_x = -\rho_0 \oint_c \frac{\partial\omega}{dt}\, dz = -2\pi i\rho_0\kappa\Omega f e^{i\Omega t}$$

which yields a force of magnitude

$$2\pi\rho_0\kappa f\Omega \equiv 2\pi\rho_0\kappa^2 f/(a^2 - f^2)$$

directed in the positive radial direction from the origin to the current vortex position. The two contributions cancel and there is no net force on the surface $|z| = a$.

16. (i) The solution of the potential problem is

$$\begin{aligned}\omega &= i\kappa \log(\zeta - \zeta_0) - i\kappa \log[(a^2/\zeta) - \bar{\zeta}_0] \\ &= i\kappa \log(\zeta - \zeta_0) + i\kappa \log\zeta - i\kappa \log[\zeta - (a^2/\bar{\zeta}_0)] + g(t)\end{aligned}$$

where $g(t)$ is an irrelevant function of time, discarded subsequently.

The angular velocity calculation is given in the text (Section 12.8.5).

(ii) The transformation

$$z(\zeta) = \zeta + a^2\zeta^{-1}, \qquad \zeta(z) = \tfrac{1}{2}[z + (z^2 - 4a^2)^{\frac{1}{2}}]$$

maps the circle $\zeta = ae^{i\theta}$ on to the flat plate $|x| \leqslant 2a$, $y = 0$ and also the region $|\zeta| > a$ onto the cut z plane. The transformation is analytic except at the edges of the plate $(z = \pm 2a)$ so that a vortex located at $\zeta = \zeta_0 \equiv \zeta(z_0)$ outside the circle maps into a vortex of the same strength at $z = z_0$. Using the results of (i) therefore

$$\omega = i\kappa \log(\zeta - \zeta_0) + i\kappa \log \zeta - i\kappa \log[\zeta - (a^2/\bar{\zeta}_0)]$$

where

$$\zeta = \tfrac{1}{2}[z + (z^2 - 4a^2)^{\frac{1}{2}}].$$

Subtracting the self potential of the vortex in the z plane leads to

$$\omega_1 = i\kappa \log[(\zeta - \zeta_0)/(z - z_0)] + i\kappa \log \zeta - i\kappa \log[\zeta - (a^2/\bar{\zeta}_0)].$$

However

$$z - z_0 = (\zeta - \zeta_0)[1 - (a^2/\zeta\zeta_0)]$$

so that

$$\begin{aligned}\omega_1 &= -i\kappa \log[1 - (a^2/\zeta\zeta_0)] + i\kappa \log \zeta - i\kappa \log[\zeta - (a^2/\bar{\zeta}_0)] \\ &= -i\kappa \log[\zeta - (a^2/\zeta_0)] - i\kappa \log[\zeta - (a^2/\bar{\zeta}_0)] + 2i\kappa \log \zeta.\end{aligned}$$

The vortex velocity is given by $\quad -\dot{\bar{z}}_0 = (\partial\omega_1/\partial z)_{z=z_0}$

or since

$$\frac{\partial \omega_1}{\partial z} = \frac{\partial \omega_1}{\partial \zeta} \bigg/ \frac{\partial z}{\partial \zeta}$$

$$-\dot{\bar{z}}_0 = i\kappa \left[2\zeta_0^{-1} - [\zeta_0 - (a^2/\zeta_0)]^{-1} - [\zeta_0 - a^2/\bar{\zeta}_0]^{-1}\right] / [1 - (a^2/\zeta_0^2)].$$

Also if $\zeta_0 = \zeta(z_0)$ and conversely $z_0 = z(\zeta_0)$ then since there are no quantities involving $\sqrt{-1}$ appearing in the transformation formulae

$$\bar{z}_0 = z(\bar{\zeta}_0) \quad \text{and} \quad \frac{d\bar{z}_0}{dt} = \frac{d\bar{\zeta}_0}{dt}\frac{dz(\bar{\zeta}_0)}{d\bar{\zeta}_0} = [1 - (a^2/\bar{\zeta}_0^2)]\frac{d\bar{\zeta}_0}{dt}.$$

Combining these results and rearranging slightly leads to the differential equation

$$-\frac{d\bar{\zeta}_0}{dt} = \frac{i\kappa a^2 \bar{\zeta}_0^2 \zeta_0 [2a^2 - \zeta_0(\zeta_0 + \bar{\zeta}_0)]}{(\zeta_0^2 - a^2)^2(\bar{\zeta}_0^2 - a^2)(\zeta_0\bar{\zeta}_0 - a^2)}$$

for the image path in the ζ plane.

17. From the Blasius theorem and the form of $\omega(z)$

$$F_y + iF_x = -\tfrac{1}{2}\rho_0 \oint_C \left(\frac{i\kappa}{z} - u - \frac{m}{z - z_0} + \frac{d\omega_1}{dz}\right)^2 dz$$

$$= -\tfrac{1}{2}\rho_0 \oint_C \left\{-\frac{\kappa^2}{z^2} + u^2 + \frac{m^2}{(z - z_0)^2} + \left(\frac{d\omega_1}{dz}\right)^2 - \frac{2i\kappa u}{z} - \frac{2i\kappa m}{z(z - z_0)}\right.$$

$$\left. + \frac{2i\kappa}{z}\frac{d\omega_1}{dz} + \frac{2mu}{z - z_0} - 2u\frac{d\omega_1}{dz} - \frac{2m}{z - z_0}\frac{d\omega_1}{dz}\right.$$

where C is the bounding contour of the cylinder. Several of the contributions to the contour integral are readily evaluated to yield

$$F_y + iF_x = \rho_0\pi[-2\kappa u + 2\kappa m z_0^{-1}] - \tfrac{1}{2}\rho_0 \oint_C \left\{\left(\frac{d\omega_1}{dz}\right)^2 + \frac{2i\kappa}{z}\frac{d\omega_1}{dz} - 2u\frac{d\omega_1}{dz} - \frac{2m}{z-z_0}\frac{d\omega_1}{dz}\right\} dz\,.$$

In obtaining this last result use has been made of the fact that $z = 0$ lies inside C and $z = z_0$ is outside C.

The remaining integrand possesses only a simple pole at $z = z_0$ in the region exterior to C. Also for large $|z|$ the integrand is at most $O(z^{-2})$ so that applying the residue theorem to a region bounded by C and a circle of radius R (where ultimately $R \to \infty$), there results

$$F_y + iF_x = 2\pi\rho_0[-\kappa u + \kappa m z_0^{-1} - im(d\omega_1/dz)_{z=z_0}]\,.$$

18. (i) In the absence of the cylinder

$$\omega = -m\log(z-f)$$

so that from the circle theorem

$$\omega = -m\log(z-f) - m\log(z-a^2/f) + m\log z + \text{const.}$$

Discarding the irrelevant constant and subtracting the self potential of the source leads to

$$\omega_1 = -m\log(z-a^2/f) + m\log z$$

whence

$$\left(\frac{d\omega_1}{dz}\right)_{z=f} = m\left[\frac{1}{f} - \frac{1}{[f-(a^2/f)]}\right] = -\frac{ma^2}{f(f^2-a^2)}\,.$$

From this and the results of Problem 17

$$F_y = 0\,, \qquad F_x = 2\pi\rho_0 m^2 a^2 f^{-1}(f^2-a^2)^{-1}\,.$$

(ii) The bijective mapping

$$z(\zeta) = \zeta + a^2\zeta^{-1}, \qquad \zeta(z) = \tfrac{1}{2}[z + (z^2-4a^2)^{\frac{1}{2}}]$$

maps points exterior to $\zeta = be^{i\theta}\,(b > a)$ onto points exterior to the ellipse

$$\frac{x^2}{\gamma^2} + \frac{y^2}{\beta^2} = 1$$

where $\qquad \gamma = b + a^2b^{-1}, \qquad \beta = b - a^2b^{-1}$

so that $\gamma > \beta$ and also $a = \tfrac{1}{2}(\gamma^2-\beta^2)^{\frac{1}{2}}$.

In the ζ plane a source at ζ_0 outside $|\zeta| = b$ results in a complex potential (from the circle theorem).

$$\omega = -m\log[\zeta - \zeta_0] + m\log\zeta - m\log[\zeta - (b^2/\bar{\zeta}_0)] + \text{const.} \qquad (1)$$

and since the mappings are conformal at z_0 and ζ_0 where $\zeta_0 = \zeta(z_0)$ the problem of a line source located at z_0 outside the ellipse is solved by

$$\omega = \omega[\zeta(z)]$$

where $\omega(\zeta)$ is given by (1).

Discarding the self potential of the source, and also the irrelevant constant, leads to

$$\begin{aligned}\omega_1(z) &= \omega + m\log(z - z_0)\\ &= -m\log[(\zeta - \zeta_0)/(z - z_0)] + m\log\zeta - m\log[\zeta - (b^2/\bar{\zeta}_0)]\end{aligned}$$

However $\quad z - z_0 = (\zeta - \zeta_0)[1 - (a^2/\zeta\zeta_0)]$

so that
$$\begin{aligned}\omega_1(z) &= m\log[1 - (a^2/\zeta\zeta_0)] + m\log\zeta - m\log[\zeta - (b^2/\bar{\zeta}_0)]\\ &= m\log[\zeta - (a^2/\zeta_0)] - m\log[\zeta - (b^2/\bar{\zeta}_0)]\,.\end{aligned}$$

The quantity $(d\omega_1/dz)_{z_0}$ is most easily expressed in terms of ζ_0

$$\left(\frac{d\omega_1}{dz}\right)_{z_0} = \left(\frac{d\omega_1}{d\zeta}\right)_{\zeta_0} \Big/ \left(\frac{dz}{d\zeta}\right)_{\zeta_0}$$

$$\begin{aligned}&= m\{[\zeta_0 - (a^2/\zeta_0)]^{-1} - [\zeta_0 - (b^2/\bar{\zeta}_0)]^{-1}\}/[1 - (a^2/\zeta_0^2)]\\ &= \frac{m\zeta_0^2(a^2\bar{\zeta}_0 - b^2\zeta_0)}{(\zeta_0^2 - a^2)^2(\zeta_0\bar{\zeta}_0 - b^2)}\end{aligned}$$

Using the results of Problem 17 and substituting

$$a^2 = \tfrac{1}{4}(\gamma^2 - \beta^2), \quad b^2 = \tfrac{1}{4}(\gamma + \beta)^2$$

leads finally to

$$F_y + iF_x = \frac{\pi\rho_0 m^2 i[\gamma + \beta)^2\zeta_0 - (\gamma^2 - \beta^2)\bar{\zeta}_0]\zeta_0^2}{2[\zeta_0^2 - \frac{1}{4}(\gamma^2 - \beta^2)]^2[\zeta_0\bar{\zeta}_0 - \frac{1}{4}(\gamma + \beta)^2]}\,.$$

For the particular case of a circle $(\gamma = \beta,\ a = 0,\ \zeta_0 = z_0)$, this reduces to

$$F_y + iF_x = 2\pi\rho_0 m^2 i\gamma^2 z_0^{-1}(z_0\bar{z}_0 - \gamma^2)^{-1}$$

which is a force of magnitude

$$2\pi\rho_0\gamma^2 m^2\,|z_0|^{-1}\,[|z_0|^2 - \gamma^2]^{-1}$$

directed from the origin to the source (in agreement with the calculation in 18(i) immediately above).

19. (iv)

$$f'(z) = \left(\frac{\partial\phi}{\partial x}\right)_{x=z,\ y=0} - i\left(\frac{\partial\phi}{\partial y}\right)_{x=z,\ y=0} = 3z^2 \ .$$

Hence $f(z) = z^3 + iA$ where A is real and

$$\phi = \operatorname{Re} f(z) = x^3 - 3xy^2, \qquad \psi = \operatorname{Im} f(z) = 3yx^2 - y^3 + A\,.$$

Applying the above technique to

$$\phi = x^3 - \lambda xy^2$$

leads again to
$$f(z) = z^3 + iA$$

although evidently the only valid choice of λ is $\lambda = 3$.

The method fails for $\lambda \neq 3$ because the corresponding ϕ is then no longer a function which satisfies Laplace's equation (which is a consequence of the Cauchy-Riemann equations). In order that a given ϕ may be the real part of a function $f(z)$, it is necessary (and sufficient) that ϕ satisfies Laplace's equation. Similarly ψ satisfies Laplace's equation.

21. (ii) The vortex at $z = il$ maps into a vortex of the same strength at

$$\zeta_0 \equiv \zeta(il) = e^{\frac{1}{2}\pi i l/h}.$$

In the ζ plane the vortex is located outside the wall $\xi = 0$ so that the potential problem in the ζ plane is solved by locating an image vortex of opposite sign at the image point. The latter is located at $-\bar{\zeta}_0$. Thus in the ζ plane

$$\omega = i\kappa \log(\zeta - \zeta_0) - i\kappa \log(\zeta + \bar{\zeta}_0)$$

and to obtain ω in the z plane it is only necessary to write

$$\zeta = \zeta(z) \equiv e^{\frac{1}{2}\pi z/h}.$$

The vortex moves with the complex velocity $-V_x + iV_y$ where

$$-V_x + iV_y = (d\omega_1/dz)_{z=il}$$

and where
$$\omega_1 = \omega - i\kappa \log(z - il)\,.$$

Therefore

$$-V_x + iV_y = i\kappa\left\{\frac{d}{dz}\log\left[\frac{\zeta(z) - \zeta(il)}{z - il}\right]_{z=il} - \frac{\zeta'(il)}{[\zeta(il) + \bar{\zeta}(il)]}\right\}.$$

To evaluate the first term, write (from Taylor's theorem)

$$\frac{\zeta(z) - \zeta(il)}{z - il} = \zeta'(il) + \tfrac{1}{2}(z - il)\zeta''(il) + O(z - il)^2$$

so that

$$-V_x + iV_y = i\kappa\left[\frac{1}{2}\frac{\zeta''(il)}{\zeta'(il)} - \frac{\zeta'(il)}{\zeta(il) + \bar{\zeta}(il)}\right]$$

$$= \tfrac{1}{2}i\kappa\left\{\tfrac{1}{2}\pi h^{-1} - \tfrac{1}{2}\pi h^{-1}e^{\frac{1}{2}\pi il/h}[\cos(\tfrac{1}{2}\pi l/h)]^{-1}\right\}$$

$$= \tfrac{1}{4}\kappa\pi h^{-1}\tan(\tfrac{1}{2}\pi l/h)$$

so that $V_y = 0$ and $V_x = -\tfrac{1}{4}\kappa\pi h^{-1}\tan(\tfrac{1}{2}\pi l/h)$.

Evidently since in the z plane the origin of x may be chosen arbitrarily, the result is valid for a vortex lying at $y = l$ and an arbitrary value of x.

22. The source at $z = il$ maps into a source at $\zeta_0 = \zeta(il)$, so that following the analysis of Problem 21, but here locating a source of the *same* strength at the image point in the ζ plane, leads to

$$\omega = -m\log(\zeta - \zeta_0) - m\log(\zeta + \bar{\zeta}_0) \qquad (1)$$

where $\zeta(z) = e^{\frac{1}{2}\pi z/l}$.

However while equation (1) solves a potential problem in which a source of strength m is located at $z = il$ between two planes, the potential given by (1) includes also an unwanted stream flow parallel to the planes. This stream is annihilated by the inclusion of the sink of strength m located at $\zeta = 0$ discussed in the hint. Thus the correct solution to the problem is

$$\omega = -m\log[(\zeta - \zeta_0)(\zeta + \bar{\zeta}_0)/\zeta]$$

which after some manipulation (and discarding an irrelevant constant becomes

$$\omega = -m\log[\sinh(\tfrac{1}{2}\pi z/h) - i\sin(\tfrac{1}{2}\pi l/h)] .$$

23. On labelling the bounding streamlines ABC, $A'B'C'$ respectively by $\psi = hu$, $\psi = -hu$, the potential problem in the **v** plane is defined below.

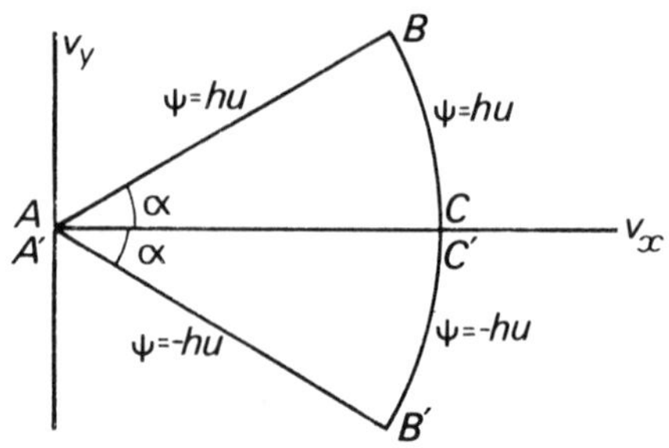

The potential problem is specified by

$$\nabla^2 \psi = 0, \qquad 0 \leqslant q < u, \qquad |\theta| < \alpha,$$

with boundary conditions

$$\begin{aligned} \psi &= -hu; & 0 \leqslant q \leqslant u, && \theta = -\alpha, \\ &= hu \quad ; & 0 \leqslant q \leqslant u && \theta = \alpha, \\ \psi &= -hu; & q = u, && -\alpha \leqslant \theta < 0, \\ &= hu \quad ; & q = u, && 0 < \theta \leqslant \alpha. \end{aligned}$$

The conditions on $\theta = \pm\alpha$ are met easily by writing

$$\psi = hu\theta/\alpha + \psi_1$$

where ψ_1 is zero on $\theta = \pm\alpha$ and

$$\begin{aligned} \psi_1 &= -hu[1 + (\theta/\alpha)] \qquad & (-\alpha < \theta < 0) \\ &= hu[1 - (\theta/\alpha)] \qquad & (0 < \theta < \alpha) \;. \end{aligned}$$

Evidently ψ_1 is antisymmetric in θ, suggesting an expansion of ψ_1 in terms of sine functions. The latter are required to vanish on $\theta = \pm\alpha$ so that the appropriate functions are

$$\sin(n\pi\theta/\alpha)\,.$$

The corresponding separation of variable solution for ψ_1 is therefore

$$\psi_1 = \sum_1^\infty a_n (q/u)^{n\pi/\alpha} \sin[n\pi\theta/\alpha]\,.$$

Here for convenience the quantity u has been incorporated into the coefficients of the Fourier series. Also terms involving negative powers of q have been rejected to ensure finite values of ψ_1 at $q = 0$.

The remaining boundary conditions on ψ_1 require

$$\sum_1^\infty a_n \sin(n\pi\theta/\alpha) = hu[1 - (\theta/\alpha)]\,, \qquad (0 < \theta < \alpha)$$

and from the theory of Fourier series

$$a_n = \frac{2hu}{\alpha} \int_0^\alpha [1 - (\theta/\alpha)] \sin(n\pi\theta/\alpha) d\theta = 2hu/n\pi\,.$$

Therefore

$$\psi = hu\left[(\theta/\alpha) + (2/\pi)\sum_{1}^{\infty} n^{-1}(q/u)^{(n\pi/\alpha)}\sin(n\pi\theta/\alpha)\right]$$

$$= hu\left[-\alpha^{-1}\,\mathrm{Im}\log(-qe^{-i\theta}) - \alpha^{-1}\log(-1)\right.$$

$$\left. - (2/\pi)\mathrm{Im}\sum_{1}^{\infty} n^{-1}(-qe^{i\pi}/u)^{n\pi/\alpha}\, e^{-in\pi\theta/\alpha}\right]$$

It follows that to within an irrelevant constant

$$\omega(v) = hu\left[-\alpha^{-1}\log v - (2/\pi)\sum_{1}^{\infty} n^{-1}(ve^{i\pi}/u)^{(n\pi/\alpha)}\right]$$

$$\equiv hu\left[-\alpha^{-1}\log v + (2/\pi)\log[1-(ve^{i\pi}/u)^{\pi/\alpha}]\right] .$$

For the particular case $\alpha = \frac{1}{4}\pi$

$$\omega(v) = (4hu/\pi)\left[-\log v + \tfrac{1}{2}\log\,[1-(v/u)^4\,]\right]$$

and

$$z = \int v^{-1}\,\frac{d\omega}{dv}\,dv$$

$$= (4hu/\pi)\,\{v^{-1} + (2iu)^{-1}\log[(v-iu)\,(v+iu)] - (2u)^{-1}\log[(u+v)/(u-v)]$$

where the constant of integration (determining only the origin of coordinates) has been taken as zero.

The points B and B' are defined respectively by

$$B'\;;\qquad v = -ue^{i\pi/4}$$

$$B\;;\qquad v = -ue^{-i\pi/4}$$

so that

$$z_B = (4h/\pi)\left\{-e^{i\pi/4} + (2i)^{-1}\log\left[\frac{-i-e^{-i\pi/4}}{i-e^{-i\pi/4}}\right] - \tfrac{1}{2}\log\left[\frac{1-e^{-i\pi/4}}{1+e^{-i\pi/4}}\right]\right\}$$

$$= (4h/\pi)\left[-(1+i)/\sqrt{2} + (2i)^{-1}\log\,[i/(\sqrt{2}+1)] - \tfrac{1}{2}\log[i\tan(\pi/8)]\right] .$$

Similarly

$$z_{B'} = (4h/\pi)\left[-(1-i)/\sqrt{2} + (2i)^{-1}\log[i/(\sqrt{2}-1)] - \tfrac{1}{2}\log[-i\tan(\pi/8)]\right]$$

In particular $x_B = x_{B'}$ (as would be expected) and

$$y_{B'} = -y_B = (4h/\pi)\left[1/\sqrt{2} - \tfrac{1}{2}\log(1+\sqrt{2}) + \tfrac{1}{4}\pi\right].$$

The contraction ratio is

$$\frac{2h}{(y_{B'} - y_B)} = \left[1 + \frac{2\sqrt{2}}{\pi} - \frac{2}{\pi}\log(1+\sqrt{2})\right]^{-1} = 0.747$$

24. Consideration of the Bernoulli equation in the absence of body forces suggests the possibility of a steady state solution for which the particle speed on all the free streamlines is u. In these circumstances consider the flow of fluid at places remote from the collision region. Here the flow is uniform so that the rate of supply of fluid from the incident jets to the collision region is $2(h_1 + h_2)u$. Similarly the rate of supply of fluid by the collision region to the emergent jets is $2(h_3 + h_4)u$. Equating these two results leads to

$$h_1 + h_2 = h_3 + h_4\,.$$

Similar considerations of incident and final momentum in the x and y directions yield the other two equations.

For a symmetrical collision $\alpha = \beta$ so that $h_3 = h_4 = \tfrac{1}{2}(h_1 + h_2)$ and

$$\cos\alpha = (h_1 - h_2)/(h_1 + h_2)\,.$$

The case $h_1 = h_2$, $\beta = \tfrac{1}{2}\pi$ has been dealt with in the text (Section 12.9.1). For non-zero values of $\cos\alpha$ the $\mathbf{v}$ plane mapping shows that on $|\mathbf{v}| = u$, $\psi(u,\theta)$ is an odd function of θ for which

$$\begin{aligned}\psi(u,\theta) &= -h_1 u \qquad (0 < \theta < \beta)\\ &= h_2 u \qquad (\beta < \theta < \pi)\end{aligned}$$

and the resultant Fourier expansion of $\psi(q,\theta)$ is

$$\psi = -\frac{2u}{\pi}(h_1 + h_2)\sum_0^\infty \frac{[1 - \cos(n\beta)]}{n}(q/u)^n \sin(n\theta)$$

$$+ \frac{4uh_2}{\pi}\sum_0^\infty \frac{(q/u)^{2n+1}}{(2n+1)}\sin(2n+1)\theta$$

which is the imaginary part of

$$\begin{aligned}\omega(v) = &-(2u/\pi)\,[h_1\log(u+v) + h_2\log(u-v)\\ &-\tfrac{1}{2}(h_1 + h_2)\log(u^2 + v^2 + 2uv\cos\beta)]\,.\end{aligned}$$

25. If the primed system moves at velocity $\mathbf{U}$ with respect to the unprimed system, observers located in the two systems observe respectively fluid velocities $\mathbf{v}$ and $\mathbf{v}'$ where $\mathbf{v}' = \mathbf{v} - \mathbf{U}$.

The additional imposed constant velocity $-\mathbf{U}$ may be derived from a velocity potential $\mathbf{U}\cdot\mathbf{r}'$ so that

$$\phi' = \phi + \mathbf{U}\cdot\mathbf{r}' .$$

Therefore $$-\operatorname{grad}_{\mathbf{r}'}\phi'(\mathbf{r}', t) = -\operatorname{grad}_{\mathbf{r}'}\phi\,[\mathbf{r}' + \mathbf{U}t, t] - \mathbf{U}$$

i.e. $$\mathbf{v}' = \mathbf{v} - \mathbf{U}$$

as expected. However also

$$\frac{\partial\phi'(\mathbf{r}', t)}{\partial t} = \frac{\partial\phi}{\partial t} + \mathbf{U}\cdot\operatorname{grad}_{\mathbf{r}'}\phi = \frac{\partial\phi}{\partial t} - \mathbf{U}\cdot\mathbf{v} .$$

In the primed system $P' = P$ so that the Bernoulli equation

$$P' + \tfrac{1}{2}\rho_0(\mathbf{v}')^2 - \rho_0(\partial\phi'/\partial t) = f^{(1)}(t)$$

becomes $$P + \tfrac{1}{2}\rho_0(\mathbf{v} - \mathbf{U})^2 - \rho_0(\partial\phi/\partial t) + \rho_0\mathbf{v}\cdot\mathbf{U} = f^{(1)}(t)$$

i.e. $$P + \tfrac{1}{2}\rho_0\mathbf{v}^2 - \rho_0(\partial\phi/\partial t) = f(t) \text{ where } f(t) = f^{(1)}(t) - \tfrac{1}{2}\rho_0\mathbf{U}^2 .$$

Answers, Chapter 13

1. For $t > 0$ the force acting on the face of the piston inside the tube is

$$A\,[P_0 + \rho_0 c v(t)]$$

where $v(t - x/c)$ is the velocity field induced in the fluid. Since the force AP_0 is counterbalanced by the external force, the equation of motion of the piston is

$$mv'(t) = -\rho_0 c A v(t)$$

with solution (satisfying $v(0) = V$)

$$v(t) = \begin{cases} Ve^{-kt}, & (t > 0) \\ 0\ , & (t < 0) \end{cases}$$

where $k = \rho_0 cA/m$. Hence behind the wave front $x = ct$

$$v_x = v(t - x/c) = Ve^{-k(t - x/c)}$$

while for $x > ct$, $v_x = 0$.

$$\text{Linear momentum of gas} = A\rho_0 \int_0^{ct} v_x dt$$

$$= (A\rho_0 Vc/k)[1 - e^{-kt}] \equiv mV[1 - e^{-kt}] .$$

Linear momentum of piston $= mVe^{-kt}$.

Total momentum $= mV$ (conserved).

NOTE. This last result, momentum conservation, appears to be confirmed exactly, as should be the case. However careful examination reveals that the exact confirmation is perhaps lucky, because a number of approximations have been insinuated into the analysis; in this context the conservation law may not have been verified exactly by an approximate analysis.

2. The first part of this problem [derivation of equation (1)] is the subject of Section 13.4.1. From equation (1) the pressure for $0 < ct < L$ is given by

$$P = P_0 + \rho_0(\partial\phi/\partial t)$$
$$= \begin{cases} P_0 + \rho_0 c v_0(t - x/c) & (t \geqslant x/c) \\ P_0 & (t \leqslant x/c) \end{cases}$$

so that the overpressure is $\rho_0 c v_0(t - x/c)$.

At $t = L/c$ the incident wave suffers a reflexion at $x = L$. For $L < ct < 2L$, the reflected wave travelling in the negative x direction at speed c occupies, together with the incident wave (1), the region $2L\text{-}ct < x \leqslant L$. Denoting the reflected wave amplitude by ψ we have therefore for $L < ct < 2L$

$$\phi = c\int_0^{(t-x/c)} v_0(t')dt' \qquad (0 \leqslant x < 2L - ct)$$

$$= c\int_0^{(t-x/c)} v_0(t')dt' + \psi(t + x/c) \qquad (2L - ct < x \leqslant L)\ .$$

The boundary condition at $x = L$ is $v(L, t) = 0$ (closed end) so that $(\partial\phi/\partial x)_{x=L} = 0$. From the above expression for ϕ this yields

$$-v_0(t - L/c) + c^{-1}\psi'(t + L/c) = 0$$

i.e.
$$\psi'(t) = cv_0(t - 2L/c)\ .$$

In particular $\quad \psi'[t + x/c] = cv_0[t - (2L - x)/c]$

and a knowledge of ψ' is sufficient to determine velocity and pressure. For $0 \leqslant x < (2L - ct)$ the velocity and pressure are as given by the incident wave

i.e.
$$\left.\begin{aligned} v(x, t) &= v_0(t - x/c) \\ P(x, t) &= P_0 + \rho_0 c v_0(t - x/c) \end{aligned}\right\} \quad (0 \leqslant x < 2L - ct)$$

while for $2L - ct < x < L$

$$\left.\begin{aligned} v(x,t) &\equiv -\partial\phi/\partial x = v_0(t-x/c) - c^{-1}\psi'(t+x/c) \\ &= v_0(t-x/c) - v_0[t-(2L-x)/c] \\ P(x,t) &= P_0 + \rho_0(\partial\phi/\partial t) \\ &= P_0 + \rho_0 c v_0(t-x/c) + \rho_0\psi'(t+x/c) \\ &= P_0 + \rho_0 c\left[v_0(t-x/c) + v_0[t-(2L-x)/c]\right] \end{aligned}\right\} (2L - ct < x < L) \qquad (2)$$

In particular for $x = L$, during the time $2L > ct > L$, $v(L, t) = 0$ (which is the imposed boundary condition) and also

$$P(L,t) = P_0 + 2\rho_0 c v_0(t - L/c)$$

so that the overpressure $2\rho_0 c v_0(t - L/c)$ is twice that associated with the incident pulse $[\rho_0 c v_0(t - L/c)]$ which would occur at $x = L$ in the absence of a reflected wave.†

At $t = 2L/c$ the reflected wave ψ meets the piston and results in a modified wave in the positive x direction. For $2L < ct < 3L$ the solution for v and P is given by equation (2) above for $L \geqslant x > 3L - ct$. For $0 < x < 3L - ct$

$$\begin{aligned} \phi &= \phi_1(t - x/c) + \phi_1[t-(2L - x)/c] + \phi_1[t-(2L + x)/c] \\ v &= v_0(t - x/c) - v_0[t - (2L - x)/c] + v_0[t - (2L + x)/c] \\ P &= P_0 + \rho_0 c\left[v_0(t-x/c) + v_0[t-(2L - x/c)] + v_0[t-(2L+x)/c]\right] . \end{aligned}$$

3. The problem is identical to that of Problem 2 for $0 \leqslant t < L/c$. For $2L/c > t > L/c$ the open end of the tube propagates a reflected wave in the negative x direction. For $L > x > 2L - ct$ write

$$\phi = c \int_0^{(t-x/c)} v_0(t')dt' + \psi(t + x/c)$$

where ψ is determined from $P = P_0$ at $x = L$. We have

$$\begin{aligned} P &= P_0 + \rho_0(\partial\phi/\partial t) \\ &= P_0 + \rho_0 c_0 v_0(t - x/c) + \rho_0\psi'(t + x/c) . \end{aligned}$$

The condition $P = P_0$ at $x = L$ leads to

$$\psi'(t + L/c) = -c_0 v_0(t - L/c)$$

and the resultant velocity and pressure fields for $L > x > 2L - ct$ are

$$\begin{aligned} v(x,t) &= v_0(t - x/c) + v_0[t - (2L - x)/c] \\ P &= P_0 + \rho_0 c\left[v_0(t - x/c) - v_0[t - (2L - x/c)]\right] . \end{aligned}$$

† Doubling of overpressure is of considerable significance in blast damage on buildings.

For the present problem the fluid velocity at $x = L$ is twice that associated with the incident wave.

The resultant velocity and pressure for $0 < x < 3L - ct$, $(2L/c < t < 3L/c)$ are

$$v(x, t) = v_0(t - x/c) + v_0[t - (2L - x)/c] - v_0[t - (2L + x)/c]$$

$$P = P_0 + \rho_0 c\left[v_0(t - x/c) - v_0[t - (2L - x)/c] + v_0[t - (2L + x)/c]\right] .$$

Evidently the pattern is clear to extend the solution to times larger than $3L/c$.

4. Define the Laplace transform of ϕ by

$$\bar{\phi}(x, s) = \int_0^\infty \phi(x, t)e^{-st}dt$$

then the Laplace transform of $v(x, t)$ is given by

$$\bar{v}(x, s) = \int_0^\infty v(x, t)e^{-st}dt \equiv -\partial\bar{\phi}/\partial x . \qquad (1)$$

Also the Laplace transform of the overpressure is

$$\rho_0 \int_0^\infty e^{-st}(\partial\phi/\partial t)dt = \rho_0 s\bar{\phi} + \rho_0[\phi(x, t)e^{-st}]_0^\infty = \rho_0 s\bar{\phi}$$

on taking account that $\phi(x, 0) = 0$ since the fluid is undisturbed at $t = 0$. Multiplying the wave equation by e^{-st} and integrating between $t = 0$ and $t = \infty$ leads to

$$\frac{d^2\bar{\phi}}{dx^2} = \frac{1}{c^2}\int_0^\infty \frac{\partial^2\phi}{\partial t^2} e^{-st}dt = \frac{s^2}{c^2}\bar{\phi} + \frac{1}{c^2}[e^{-st}\{(\partial\phi/\partial t) + s\phi\}]_0^\infty = \frac{s^2}{c^2}\bar{\phi} \qquad (2)$$

since at $t = 0$, $\partial\phi/\partial t$ and ϕ are both zero for the undisturbed fluid.

The general solution of (2) is

$$\bar{\phi} = A(s)e^{sx/c} + B(s)e^{-sx/c} \qquad (3)$$

where $A(s)$ and $B(s)$ are determined from the boundary conditions

$$v(L, t) = 0 \qquad \text{i.e.} \qquad \bar{v}(L, s) = 0$$

$$v(0, t) = v_0(t) \qquad \text{i.e.} \qquad \bar{v}(0, s) = \bar{v}_0 \equiv \int_0^\infty v_0(t)e^{-st}dt .$$

From equations (1) and (3) and the above boundary conditions there obtains

$$\overline{\phi} = (c\overline{v}_0/s)[e^{-s(2L-x)/c} + e^{-sx/c}](1 - e^{-2Ls/c}]^{-1} .$$

In particular

$$\overline{v} \equiv -\partial\overline{\phi}/\partial x = \overline{v}_0\{-e^{-s(2L-x)/c} + e^{-sx/c}\}[1 - e^{-2Ls/c}]^{-1} .$$

Expanding the term in square brackets using the binomial theorem leads to

$$\overline{v} = \overline{v}_0[e^{-sx/c} - e^{-s(2L-x)/c} + e^{-s(2L+x)/c} - e^{-s(4L-x)/c}\ldots] . \tag{4}$$

On recalling the theorem that if

$$\overline{f}(s) = \int_0^\infty f(t)e^{-st} \text{ then } \overline{f}(s)e^{-s\tau} = \int_0^\infty f(t-\tau)H(t-\tau)e^{-st}dt$$

where H is the Heaviside step function

$$H(t) = \begin{cases} 1, & (t>0) \\ 0, & (t<0) \end{cases}$$

the inversion of (4) leads to

$$\begin{aligned} v(x, t) = {} & v_0[t - x/c]H[t - x/c] - v_0[t - (2L - x)/c]H[t - (2L - x)/c] \\ & + v_0[t - (2L + x)/c]H[t - (2L + x)/c] \ldots \end{aligned} \tag{5}$$

The terms identified explicitly are those obtained above in the answer to Problem 2. Here the step functions serve to identify the instants of time $[x/c, (2L - x)/c, (2L + x)/c \ldots$ etc.) at which additional terms caused by successive wave reflections are added to the series (5).

The reader is left to undertake the parallel analysis for the overpressure.

5. The Laplace transform of the velocity potential $\phi(x, t)$ is given by

$$\overline{\phi}(x, s) = A(s)e^{sx/c} + B(s)e^{-(sx/c)} \tag{1}$$

where A and B are determined by boundary conditions. [For derivation of equation (1) see the previous answer.] The Laplace transforms of the velocity is

$$\overline{v} \equiv -\frac{\partial\overline{\phi}}{\partial x} = -(s/c)[Ae^{sx/c} - Be^{-(sx/c)}] \tag{2}$$

while that of the overpressure $P - P_0$ is

$$\overline{P - P_0} = \rho_0 s\overline{\phi} = \rho_0 s[Ae^{sx/c} + Be^{-sx/c}] .$$

Also the acceleration of M_1 is $\partial v(0, t)/\partial t$ with Laplace transform $-(s^2/c)(A - B)$.

Similarly the Laplace transform of the acceleration of M_2 is $-(s^2/c)[Ae^{sL/c} - Be^{-sL/c}]$. From Newtons laws of motion for M_1 and M_2 we have

$$\left.\begin{aligned} -(M_1 s^2/c)(A - B) &= \bar{F} - \rho_0 s\alpha(A + B) \\ -(M_2 s^2/c)[Ae^{sL/c} - Be^{-sL/c}] &= \rho_0 s\alpha[Ae^{sL/c} + Be^{-sL/c}] \end{aligned}\right\} \tag{3}$$

where the terms involving α derive from the forces due to the overpressures at $x = 0$ and $x = L$.

Equations (3) have solution

$$A = \frac{(M_2 s - \rho_0 \alpha c)c\bar{F}s^{-1}}{(M_1 s + \rho_0 \alpha c)(M_2 s + \rho_0 \alpha c)e^{2sL/c} - (M_1 s - \rho_0 \alpha c)(M_2 s - \rho_0 \alpha c)}$$

$$B = \frac{(M_2 s + \rho_0 \alpha c)c\bar{F}s^{-1}e^{2sL/c}}{(M_1 s + \rho_0 \alpha c)(M_2 s + \rho_0 \alpha c)e^{2sL/c} - (M_1 s - \rho_0 \alpha c)(M_2 s - \rho_0 \alpha c)}.$$

The velocity of M_2 is given by equation (2) with $x = L$. Substituting from the above for A and B leads to

$$\bar{v}(L, s) = \frac{2\rho_0 \alpha c\bar{F}e^{-sL/c}}{(M_1 s + \rho_0 \alpha c)(M_2 s + \rho_0 \alpha c) - (M_1 s - \rho_0 \alpha c)(M_2 s - \rho_0 \alpha c)e^{-2sL/c}}. \tag{4}$$

For $F(t) = F_0 H(t), \bar{F} = F_0 s^{-1}$.

Evidently the denominator in equation (4) may be expanded in a power series in $e^{-2sL/c}$ leading to an infinite series solution for $\bar{v}$ of the form

$$\bar{v}(L, s) = \sum_0^\infty \bar{\gamma}_n(s)e^{-(2n+1)sL/c}$$

where the $\bar{\gamma}_n(s)$ are rational functions whose inverse Laplace tranforms $\gamma_n(t)$ are readily obtained. As in Problem 4 the exponential factors modify the inversion formulae by time delays so that

$$v(L, t) = \sum_0^\infty \gamma_n[t - (2n + 1)L/c]H[t - (2n + 1)L/c] \tag{5}$$

where H is the Heaviside step function. Successive terms in the series are added as each new reflexion process is initiated at $x = L$. In particular for $0 \leqslant t < L/c$ $v = 0$, while for $L/c \leqslant t < 3L/c$

$$v(L, t) = \mathcal{L}^{-1}\left[\frac{2\rho_0 \alpha c F_0 s^{-1} e^{-sL/c}}{(M_1 s + \rho_0 \alpha c)(M_2 s + \rho_0 \alpha c)}\right]$$

where $\mathcal{L}^{-1}$ denotes Laplace inversion. The inversion is a straightforward example of partial fractions (poles at $s = 0, s = -\rho_0\alpha c/M_1, s = -\rho_0\alpha c/M_2$) and yields

$$v(L, t) = \frac{2F_0}{\rho_0\alpha c}\left[1 - (1 - \mu)^{-1}\left[e^{-(\rho_0\alpha cT/M_1)} - \mu e^{-\rho_0\alpha cT/M_2}\right]\right]H(T)$$

where $\mu = M_2/M_1$ and where $T = (t - L/c)$.

It is worth noting the economy of the Laplace transform method in problems of this type. The answer for all t is contained effectively in the closed form expression (4), whereas attempts to determine the $\Upsilon_n(t)$ of (5) from consideration of successive reflexions at both pistons in the absence of Laplace transform methods would be very time consuming.

7. Let the reflected wave in $x < 0$ be given by the velocity potential

$$\phi = g[t - (\mathbf{n}_R\,.\,\mathbf{r})/c_1]$$

and the transmitted wave in $x > 0$ by

$$\phi = h[t - (\mathbf{n}_T\,.\,\mathbf{r})/c_2]$$

where $\mathbf{n}_R = (\cos\theta_3, \sin\theta_3, 0)$ and $\mathbf{n}_T = (\cos\theta_2, \sin\theta_2, 0)$.

Then
$$\begin{aligned}\phi &= f[t - (\mathbf{n}_I\,.\,\mathbf{r})/c_1] + g[t - (\mathbf{n}_R\,.\,\mathbf{r})/c_1] \qquad (x<0)\\ &= h[t - (\mathbf{n}_T\,.\,\mathbf{r})/c_2] \qquad (x>0)\,.\end{aligned}$$

Continuity of v_x implies continuity of $\partial\phi/\partial x$ on $x = 0$. Continuity of pressure implies continuity of $\rho(\partial\phi/\partial t)$ on $x = 0$ and hence continuity of $\rho\phi$ (from integrating with respect to t and writing $\phi = 0$ for the undisturbed state).

On writing $\mathbf{n}_R = (\cos\theta_3, \sin\theta_3, 0)$ and $\mathbf{n}_T = (\cos\theta_2, \sin\theta_2, 0)$ continuity of $(\partial\phi/\partial x)$ and $\rho\phi$ at $x = 0$ lead to

$$\begin{aligned}c_1^{-1}\left\{\cos\theta_1 f'[t - (y\sin\theta_1/c_1)] + \cos\theta_3 g'[t - (y\sin\theta_3/c_1)]\right\}\\ = c_2^{-1}\cos\theta_2\, h'[t - (y\sin\theta_2/c_2)] \qquad (1)\end{aligned}$$

$$\rho_1 f[t - (y\sin\theta_1/c_1)] + \rho_1 g[t - (y\sin\theta_3/c_1] = \rho_2 h[t - (y\sin\theta_2/c_2)]\,. \qquad (2)$$

Evidently these relations are satisfied in general only if the arguments of f, g and h are identical so that

$$\frac{\sin\theta_1}{c_1} = \frac{\sin\theta_2}{c_2}, \qquad \sin\theta_3 = \sin\theta_1\,. \qquad (3)$$

If these relations obtain, equation (1) may be integrated to yield

$$c_1^{-1}[f\cos\theta_1 + g\cos\theta_3] = c_2^{-1}h\cos\theta_2$$

which in conjunction with (2) gives

$$h = \frac{\rho_1 c_2(\cos\theta_1 - \cos\theta_3)}{(\rho_1 c_1\cos\theta_2 - \rho_2 c_2\cos\theta_3)}f$$

$$g = -\frac{(\rho_1 c_1 \cos\theta_2 - \rho_2 c_2 \cos\theta_1)}{(\rho_1 c_1 \cos\theta_2 - \rho_2 c_2 \cos\theta_3)} f \,.$$

From $\sin\theta_3 = \sin\theta_1$ either $\cos\theta_3 = \cos\theta_1$ or $\cos\theta_3 = -\cos\theta_1$. The first choice implies $h = 0, g = -f$ resulting in a vacuous problem. The second (correct) choice leads to $\mathbf{n}_R = (-\cos\theta_1, \sin\theta_1, 0)$ as required and also

$$h = \frac{2\rho_1 c_2 \cos\theta_1}{(\rho_1 c_1 \cos\theta_2 + \rho_2 c_2 \cos\theta_1)} f$$

$$g = -\frac{(\rho_1 c_1 \cos\theta_2 - \rho_2 c_2 \cos\theta_1)}{(\rho_1 c_1 \cos\theta_2 + \rho_2 c_2 \cos\theta_1)} f,$$

which determine the relative amplitudes (h/f) and (g/f) of the velocity potentials. The relative amplitudes of the pressures and velocities are different. The pressure associated with the incident wave is $\rho_1 f'$ while the pressures associated with the reflected and transmitted waves are $\rho_1 g'$ and $\rho_2 h'$ respectively. Thus the ratios of the pressure amplitudes are

$$\frac{\text{Reflected wave pressure}}{\text{Incident wave pressure}} = \frac{g'}{f'} = -\frac{(\rho_1 c_1 \cos\theta_2 - \rho_2 c_2 \cos\theta_1)}{(\rho_1 c_1 \cos\theta_2 + \rho_2 c_2 \cos\theta_1)}$$

$$\frac{\text{Transmitted wave pressure}}{\text{Incident wave pressure}} = \frac{\rho_2 h'}{\rho_1 f'} = \frac{2\rho_2 c_2 \cos\theta_1}{(\rho_1 c_1 \cos\theta_2 + \rho_2 c_2 \cos\theta_1)}.$$

Similarly the magnitude of the ratios of the particle velocities are given by

$$\frac{|\text{ Reflected velocity }|}{|\text{ Incident velocity }|} = \left|\frac{c_1^{-1} g'}{c_1^{-1} f'}\right| = \left|\frac{g'}{f'}\right| = \frac{|\rho_1 c_1 \cos\theta_2 - \rho_2 c_2 \cos\theta_1|}{(\rho_1 c_1 \cos\theta_2 + \rho_2 c_2 \cos\theta_1)}$$

$$\frac{|\text{ Transmitted velocity }|}{|\text{ Incident velocity }|} = \left|\frac{c_2^{-1} h'}{c_1^{-1} f'}\right| = \left|\frac{c_1 h'}{c_2 f'}\right| = \frac{2\rho_1 c_1 \cos\theta_1}{(\rho_1 c_1 \cos\theta_2 + \rho_2 c_2 \cos\theta_1)}.$$

It is of interest that apart from the dependence on angle, the above ratios of physical quantities depend on $\rho_1 c_1$ and $\rho_2 c_2$. The quantity ρc is sometimes called the characteristic impedance and in particular for the case of liquids of identical characteristic impedances ($\rho_1 c_1 = \rho_2 c_2$) and for an incident *normal* wave ($\theta_1 = 0$), there is no reflected wave.

For the case where $(c_2 \sin\theta_1/c_1) > 1$, so that the first of (3) predicts a value of $\sin\theta_2 > 1$, no real value of θ_2 is possible. In this case it is conceivable that equations (1) and (2) admit a solution in which h is identically zero and $\sin\theta_3 = \sin\theta_1$. However apart from the trivial solution $g = -f, \theta_3 = \theta_1$ (which cancels out the incident wave) there is no solution of equations (1) and (2) for which $h = 0$. Of course there exists a genuine problem for $(c_2 \sin\theta_1/c_1) > 1$; the implications of the above are merely that it is not possible to find a solution in simple wave function terms. In fact the correct solution now entails the propagation of disturbances which are exponentially damped in the x direction. (See Problem 30 (iii) of Chapter 14 for a discussion of this situation in the context of elastic waves.)

8. Only outgoing waves are generated so ϕ is of the indicated form. From (13.4.17)

$$f(t) = \omega\epsilon c a_0^2 e^{-t/\tau} \int_0^t \cos(\omega t') e^{t'/\tau} dt' \qquad (t > 0)$$

$$= 0 \qquad (t < 0)$$

where $\tau = (a_0/c)$. On evaluating the integral there results

$$f(t) = \omega\epsilon a_0^3 [\cos(\omega t) + \omega\tau \sin(\omega t) - e^{-t/\tau}]/(1 + \omega^2\tau^2) .$$

The fluid velocity is in the radial direction and given by

$$v_r = -\frac{\partial}{\partial r}(f/r) = r^{-2} f + (cr)^{-1} f'$$

where the arguments of f and f' are $[t - (r - a_0)/c]$. For $[t - (r - a_0)/c] \gg \tau$, the contribution of the exponential term in f is negligible compared with the periodic terms. Evidently the latter imply periodic velocity potential (and hence periodic velocity) with circular frequency ω. For sufficiently large r the term $(cr)^{-1} f'$ would appear to be the dominant contribution to v_r. The precise criterion that the periodic motion due to $(cr)^{-1} f'$ is of much larger amplitude than the amplitude associated with $r^{-2} f$ is readily established to be $(\omega r/c) \gg 1$. Neglecting $r^{-2} f$ leads to

$$v_r = \frac{\omega^2 \epsilon a_0^3}{(1 + \omega^2\tau^2) cr} [-\sin(\omega T) + \omega\tau \cos(\omega T)] , \quad [T = t - (r - a_0)/c]$$

with amplitude

$$\frac{\omega^2 \epsilon a_0^3}{cr[1 + \omega^2\tau^2]^{\frac{1}{2}}} .$$

9.

$$\phi = r f(\eta), \quad \frac{\partial\phi}{\partial r} = f + \eta f', \quad \frac{\partial^2\phi}{\partial r^2} = \frac{\eta f'' + 2f'}{ct},$$

$$\frac{\partial\phi}{\partial t} = -\frac{r^2}{ct^2} f', \quad \frac{\partial^2\phi}{\partial t^2} = \frac{2r^2}{ct^3} f' + \frac{r^3}{c^2 t^4} f'' .$$

Substituting these results into the spherical wave equation leads to

$$\eta^2(1 - \eta^2) f'' + (4\eta - 2\eta^3) f' + 2f = 0 .$$

The substitution $f = q(\eta)/\eta$ leads to $\eta(1 - \eta^2) q'' + 2q' = 0$

with solution $\quad q' = A(\eta^{-2} - 1)$

which is sufficient to determine (to within the multiplicative constant A)

$$v_r \equiv -\frac{\partial\phi}{\partial r} = -(f + \eta f') = -q' = -A(\eta^{-2} - 1) .$$

The problem specified demands the boundary condition that $v_r = v_0$ at $\eta = v_0/c$, so that for $v_0 < c$,

$$v_r = \frac{v_0^3}{(c^2 - v_0^2)}\left(\frac{c^2 t^2}{r^2} - 1\right) \qquad (v_0 t \leqslant r \leqslant ct)$$
$$= 0 \qquad (r > ct)\,.$$

The requirement $v_0 \ll c$ is necessary for the acoustic approximation; in these circumstances no disturbances propagate beyond the wave front $r = ct$ so $v_r = 0$ for $r > ct$.

11. An appropriate velocity potential is

$$\phi = \cos\theta\,[-(f/r^2) - (f'/cr)]$$

where the arguments of f and f' are $[t - (r - a)/c]$. The boundary condition requires (after cancelling $\cos\theta$)

$$\left(\frac{2f}{r^3} + \frac{2f'}{cr^2} + \frac{f''}{c^2 r}\right)_{r=a} = -\dot{\xi}(t)$$

i.e.
$$\frac{\ddot{f}(t)}{c^2 a} + \frac{2\dot{f}}{ca^2} + \frac{2f}{a^3} = -\dot{\xi}(t)\,.$$

For $\xi = v_0 t$, $\dot{\xi} = v_0$ and the resulting second order linear differential equation with constant coefficients has general solution

$$f = e^{-ct/a}[A\cos(ct/a) + B\sin(ct/a)] - \tfrac{1}{2}a^3 v_0\,.$$

Imposing $f(0) = \dot{f}(0) = 0$ leads to

$$f = \tfrac{1}{2}a^3 v_0\,[e^{-ct/a}\{\cos(ct/a) + \sin(ct/a)\} - 1]\,.$$

The pressure on the sphere is

$$P = P_0 + \rho_0(\partial\phi/\partial t)_{r=a} = P_0 - \rho_0\cos\theta\left[\frac{\dot{f}(t)}{a^2} + \frac{\ddot{f}(t)}{ca}\right]$$

and the force on the sphere in the z direction is given by

$$F_z = -\iint_S P\cos\theta\,dS$$

where S is the surface $r = a$. Transforming to spherical polar coordinates r, θ, ψ, writing $dS = a^2\sin\theta\,d\theta\,d\psi$ and effecting the ψ integration leads to

$$-\iint_S P\cos\theta\,dS = 2\pi\rho_0 a^2\left\{\frac{\dot{f}}{a^2} + \frac{\ddot{f}}{ca}\right\}\int_0^{\pi}\cos^2\theta\sin\theta\,d\theta$$

$$= \tfrac{4}{3}\pi\rho_0 a^2 \left\{\frac{\dot{f}}{a^2} + \frac{\ddot{f}}{ca}\right\}$$

$$= -\tfrac{4}{3}\pi\rho_0 a^2 v_0 c e^{-ct/a} \cos(ct/a)$$

so that a counteracting force

$$\tfrac{4}{3}\pi\rho_0 a^2 v_0 c e^{-(ct/a)} \cos(ct/a)$$

in the positive z direction is required to maintain the motion.

Answers, Chapter 14

3. **(i)**
$$e_{xx} = \frac{\partial u_x}{\partial x} = A\nu x(l - z), \qquad e_{yy} = \frac{\partial u_y}{\partial y} = A\nu x(l - z)$$

$$e_{zz} = \frac{\partial u_z}{\partial z} = -Ax(l - z), \qquad e_{xy} = \frac{1}{2}\left(\frac{\partial u_x}{\partial y} + \frac{\partial u_y}{\partial x}\right) = 0$$

$$e_{yz} = \frac{1}{2}\left(\frac{\partial u_y}{\partial z} + \frac{\partial u_z}{\partial y}\right) = -\tfrac{1}{2}A(2 + \nu)xy,$$

$$e_{zx} = \frac{1}{2}\left(\frac{\partial u_z}{\partial x} + \frac{\partial u_x}{\partial z}\right) = \tfrac{1}{2}A[-\tfrac{1}{2}\nu x^2 - (1 - \tfrac{1}{2}\nu)y^2]$$

$$\Delta = e_{xx} + e_{yy} + e_{zz} = -A(1 - 2\nu)x(l - z)$$

$$\sigma_{xx} = \lambda\Delta + 2\mu e_{xx} = Ax(l - z)[-(1 - 2\nu)\lambda + 2\mu\nu] = 0 .$$

Similarly $\sigma_{yy} = 0$

$$\sigma_{zz} = \lambda\Delta + 2\mu e_{zz} = -Ax(l - z)[\lambda(1 - 2\nu) + 2\mu] = -EAx(l - z)$$

$$\sigma_{xy} = 0, \qquad \sigma_{yz} = 2\mu e_{yz} = -\frac{1}{2}\frac{EA(2 + \nu)}{(1 + \nu)}xy$$

$$\sigma_{zx} = 2\mu e_{zx} = -\frac{1}{2}\frac{EA}{(1 + \nu)}[\tfrac{1}{2}\nu x^2 + (1 - \tfrac{1}{2}\nu)y^2] .$$

(ii) $\dfrac{\partial \sigma_{xx}}{\partial x} = \dfrac{\partial \sigma_{xy}}{\partial y} = \dfrac{\partial \sigma_{xz}}{\partial z} = 0$. Therefore $\dfrac{\partial \sigma_{xx}}{\partial x} + \dfrac{\partial \sigma_{xy}}{\partial y} + \dfrac{\partial \sigma_{xz}}{\partial z} = 0$.

Similarly
$$\frac{\partial \sigma_{xy}}{\partial x} + \frac{\partial \sigma_{yy}}{\partial y} + \frac{\partial \sigma_{yz}}{\partial z} = 0$$

$$\frac{\partial \sigma_{xz}}{\partial x} = -\frac{1}{2}\frac{\nu EAx}{(1 + \nu)}, \qquad \frac{\partial \sigma_{yz}}{\partial y} = -\frac{1}{2}\frac{EA(2 + \nu)x}{(1 + \nu)}, \qquad \frac{\partial \sigma_{zz}}{\partial z} = EAx$$

$$\frac{\partial \sigma_{xz}}{\partial x} + \frac{\partial \sigma_{yz}}{\partial y} + \frac{\partial \sigma_{zz}}{\partial z} = EAx\left[1 - \frac{\tfrac{1}{2}\nu}{1 + \nu} - \frac{\tfrac{1}{2}(2 + \nu)}{1 + \nu}\right] = 0 .$$

4. (i)

$$e_{xx} = e_{yy} = e_{zz} = e_{yx} = 0$$

$$e_{xz} = \frac{1}{2}\left(\frac{\partial u_x}{\partial z} + \frac{\partial u_z}{\partial x}\right) = \frac{1}{2}\frac{\partial u_z}{\partial x} = \frac{A}{2}\frac{\partial}{\partial x}\,[\tan^{-1}(y/x)] = -\frac{\frac{1}{2}Ay}{x^2+y^2}\,.$$

Similarly

$$e_{yz} = \frac{\frac{1}{2}Ax}{x^2+y^2}\,.$$

The only non-vanishing components of stress are

$$\sigma_{xz} = -\frac{\mu Ay}{x^2+y^2}\,,\qquad \sigma_{yz} = \frac{\mu Ax}{x^2+y^2}\,.$$

(ii) Two equilibrium equations are satisfied trivially. The third equation

$$\frac{\partial \sigma_{xz}}{\partial x} + \frac{\partial \sigma_{yz}}{\partial y} + \frac{\partial \sigma_{zz}}{\partial z} = 0$$

is readily verified since $\sigma_{zz} = 0$, while

$$\frac{\partial \sigma_{xz}}{\partial x} = \frac{2\mu Axy}{(x^2+y^2)^2} = -\frac{\partial \sigma_{yz}}{\partial y}\,.$$

[The solution $u_z = A\theta$, $u_x = u_y = 0$ of the elastostatic equations gives an example of a displacement field which appears not to be single valued. Such fields are admissable however in elastic solids after the material has been partially cut or sliced. For example the present field describes the following problem. A hollow cylinder of circular cross section $a \leqslant (x^2+y^2)^{1/2} \leqslant b$ is sliced along a radius vector (say $\theta = 0$) from a to b. After slicing, the newly created pair of surfaces may be labelled $\theta = 0$ and $\theta = 2\pi$. The present solution solves the problem in which the entire surface $\theta = 2\pi$ is displaced an amount $2\pi A$ in the z direction while the surface $\theta = 0$ is undisturbed.]

5. (ii)

$$\begin{aligned}
\mathbf{u} &= 4(1-\nu)\mathbf{A} - \operatorname{grad}(\phi + \mathbf{r}\,.\,\mathbf{A})\\
\operatorname{div}\mathbf{u} &= 4(1-\nu)\operatorname{div}\mathbf{A} - \nabla^2\phi - \nabla^2(\mathbf{r}\,.\,\mathbf{A})\\
&= 4(1-\nu)\operatorname{div}\mathbf{A} - \nabla^2\phi - \mathbf{r}\,.\,\nabla^2\mathbf{A} - 2\operatorname{div}\mathbf{A}\\
&= 2(1-2\nu)\operatorname{div}\mathbf{A} \qquad (\text{since } \nabla^2\phi = \nabla^2\mathbf{A} = 0)\\
\operatorname{grad}(\operatorname{div}\mathbf{u}) &= 2(1-2\nu)\operatorname{grad}(\operatorname{div}\mathbf{A})\\
\nabla^2\mathbf{u} &= 4(1-\nu)\nabla^2\mathbf{A} - \operatorname{grad}[\nabla^2\phi + \nabla^2(\mathbf{r}\,.\,\mathbf{A})]\\
&= -\operatorname{grad}[\mathbf{r}\,.\,\nabla^2\mathbf{A} + 2\operatorname{div}\mathbf{A}] \qquad (\text{since } \nabla^2\phi = \nabla^2\mathbf{A} = 0)\\
&= -2\operatorname{grad}(\operatorname{div}\mathbf{A}) \qquad (\text{since } \nabla^2\mathbf{A} = 0)
\end{aligned}$$

Therefore

$$(\lambda+\mu)\operatorname{grad}(\operatorname{div}\mathbf{u}) + \mu\nabla^2\mathbf{u} = [2(1-2\nu)(\lambda+\mu) - 2\mu]\operatorname{grad}(\operatorname{div}\mathbf{A}) = 0$$

since

$$(1-2\nu)(\lambda+\mu) = \mu\,.$$

[The (Papkovitch-Neuber) solution

$$\mathbf{u} = 4(1 - \nu)\mathbf{A} - \text{grad}(\phi + \mathbf{r}\,.\,\mathbf{A})$$

is a completely general solution of the elastostatic equations. The solution expresses $\mathbf{u}$ in terms of four harmonic functions (ϕ, A_x, A_y, A_z) each of which satisfies Laplace's equation. By means of this general solution it is possible to reduce elastostatic problems to somewhat complicated problems in potential theory.]

9.

$$u = Ar + Br^{-2}\,.$$

At $r = b,\ u = 0$ i.e. $Ab + Bb^{-2} = 0$

At $r = a,\ \sigma_{rr} = -P$ i.e. $(3\lambda + 2\mu)A - 4\mu Ba^{-3} = -P.$

These results imply

$$A = -\frac{P}{[(3\lambda + 2\mu) + 4\mu(b/a)^3]}$$

$$B = \frac{Pb^3}{[(3\lambda + 2\mu) + 4\mu(b/a)^3]}\,.$$

It now follows that

$$u(a) = Aa + Ba^{-2} = \frac{Pa[(b/a)^3 - 1]}{[(3\lambda + 2\mu) + 4\mu(b/a)^3]}$$

$$= \frac{Pa[(b/a)^3 - 1](1 - 2\nu)}{E[1 + [2(1 - 2\nu)/(1 + \nu)](b/a)^3]}\,.$$

The other results are derived similarly.

10. The elastostatic equations for $\mathbf{u}$ including the body force are

$$(\lambda + \mu)\text{grad}(\text{div}\,\mathbf{u}) + \mu\nabla^2\mathbf{u} = (\rho_0 gr/a)\mathbf{i}_r\,.$$

For a solution of the form $\mathbf{u} = u(r)\mathbf{i}_r$, curl $\mathbf{u} = 0$ and the equations may be written

$$\text{grad}(\text{div}\,\mathbf{u}) = \frac{\rho_0 gr}{(\lambda + 2\mu)a}\,\mathbf{i}_r\,.$$

Also div $\mathbf{u} = [u'(r) + 2u(r)/r]$ only depends on r so that grad(div $\mathbf{u}$) is a vector of magnitude $d/dr[u'(r) + 2u(r)/r]$ in the direction $\mathbf{i}_r$. Hence

$$\frac{d}{dr}\left(\frac{du}{dr} + \frac{2u}{r}\right) = \frac{\rho_0 gr}{(\lambda + 2\mu)a}\,.$$

The complementary function of this equation is $Ar + Br^{-2}$, but since $u(0)$ is finite we must choose $B = 0$. It is readily verified that a particular integral is proportional to r^3. Hence

$$u(r) = Ar + \frac{\rho_0 gr^3}{10(\lambda + 2\mu)a}\,.$$

Following the method given in the text for finding σ_{rr}, $\sigma_{\theta\theta}$ and $\sigma_{\phi\phi}$ leads to

$$\sigma_{rr} = (3\lambda + 2\mu)A + \frac{(5\lambda + 6\mu)\rho_0 g r^2}{10(\lambda + 2\mu)a}$$

$$\sigma_{\theta\theta} = \sigma_{\phi\phi} = (3\lambda + 2\mu)A + \frac{(5\lambda + 2\mu)\rho_0 g r^2}{10(\lambda + 2\mu)a} .$$

The surface $r = a$ is free of traction if $(\sigma_{rr})_{r=a} = 0$ so that

$$(3\lambda + 2\mu)A = -\frac{(5\lambda + 6\mu)\rho_0 g a}{10(\lambda + 2\mu)}$$

determining A. In terms of Poisson's ratio, this leads to

$$\sigma_{rr} = -\frac{(3 - \nu)\rho_0 g}{10(1 - \nu)a}(a^2 - r^2)$$

$$\sigma_{\theta\theta} = \sigma_{\phi\phi} = -\frac{\rho_0 g}{10(1 - \nu)a}[(3 - \nu)a^2 - (1 + 3\nu)r^2] .$$

11. **(ii)** Using $u_r = Ar + Br^{-1} - \dfrac{\rho_0\omega^2 r^3}{8(\lambda + 2\mu)}$, $\dfrac{du_z}{dz} = e_{zz} = \text{const.}$

and following the analysis of the text (except here $B \neq 0$) results in

$$\sigma_{rr} = \lambda(2A + e_{zz}) + 2\mu A - 2\mu B r^{-2} - \frac{(2\lambda + 3\mu)}{4(\lambda + 2\mu)}\rho_0\omega^2 r^2$$

$$\sigma_{\theta\theta} = \lambda(2A + e_{zz}) + 2\mu A + 2\mu B r^{-2} - \frac{(2\lambda + \mu)}{4(\lambda + 2\mu)}\rho_0\omega^2 r^2$$

$$\sigma_{zz} = \lambda(2A + e_{zz}) + 2\mu e_{zz} - \frac{\lambda}{2(\lambda + 2\mu)}\rho_0\omega^2 r^2$$

$r = a$, $r = b$ are traction free if $(\sigma_{rr})_a = (\sigma_{rr})_b = 0$

i.e.
$$\lambda(2A + e_{zz}) + 2\mu A = 2\mu B a^{-2} + \frac{(2\lambda + 3\mu)}{4(\lambda + 2\mu)}\rho_0\omega^2 a^2$$

$$= 2\mu B b^{-2} + \frac{(2\lambda + 3\mu)}{4(\lambda + 2\mu)}\rho_0\omega^2 b^2$$

These last two equations lead to

$$2\mu B = \frac{(2\lambda + 3\mu)}{4(\lambda + 2\mu)}\rho_0\omega^2 a^2 b^2$$

$$\lambda(2A + e_{zz}) + 2\mu A = \frac{(2\lambda + 3\mu)}{4(\lambda + 2\mu)}\rho_0\omega^2(a^2 + b^2) \qquad (1)$$

whence

$$\sigma_{rr} = \frac{(2\lambda + 3\mu)}{4(\lambda + 2\mu)}\rho_0\omega^2[a^2 + b^2 - r^2 - (a^2b^2/r^2)] =$$

$$\frac{(3 - 2\nu)}{8(1 - \nu)}\rho_0\omega^2[a^2 + b^2 - r^2 - (a^2b^2/r^2)]$$

and similarly

$$\sigma_{\theta\theta} = \frac{\rho_0\omega^2}{8(1 - \nu)}\left[(3 - 2\nu)[a^2 + b^2 + (a^2b^2/r^2)] - (1 + 2\nu)r^2\right].$$

In order for there to be zero mean axial load we must have

$$\int_a^b \sigma_{zz} r\,dr = 0$$

i.e. $\frac{1}{2}[\lambda(2A + e_{zz}) + 2\mu e_{zz}](b^2 - a^2) = \dfrac{\lambda}{8(\lambda + 2\mu)}\rho_0\omega^2(b^4 - a^4)$

i.e. $$\lambda(2A + e_{zz}) + 2\mu e_{zz} = \frac{\lambda}{4(\lambda + 2\mu)}\rho_0\omega^2(b^2 + a^2) \qquad (2)$$

whence $$\sigma_{zz} = \frac{\lambda}{4(\lambda + 2\mu)}\rho_0\omega^2(b^2 + a^2 - 2r^2)$$

$$= \frac{\nu}{4(1 - \nu)}\rho_0\omega^2(b^2 + a^2 - 2r^2).$$

Finally eliminating A between (1) and (2) leads to the contraction

$$e_{zz} = -\tfrac{1}{2}\rho_0\omega^2\nu(a^2 + b^2)/E.$$

13. **(i)** $$\chi = A(x - a)[(x + 2a)^2 - 3y^2]$$
$$= A[x^3 - 3xy^2 + 3ax^2 + 3ay^2 - 4a^3]$$
$$\nabla^2\chi = 12aA = -2, \quad A = -(6a)^{-1}.$$

Also $\chi = 0$ when $x = a$, $x = -2a + \sqrt{3}y$, $x = -2a - \sqrt{3}y$ and the latter form the boundaries of an equilateral triangle whose sides are of length $2\sqrt{3}a$.

(ii) $\chi + \frac{1}{2}(x^2 + y^2) = -(6a)^{-1}[x^3 - 3xy^2 + 3ax^2 + 3ay^2 - 4a^3] + \frac{1}{2}(x^2 + y^2)$
$$= -(6a)^{-1}(x^3 - 3xy^2 - 4a^3)$$

which is, of course, harmonic. We therefore wish to find ϕ such that

$$\phi + i\psi = f(z) \text{ where } \psi = -(6a)^{-1}(x^3 - 3xy^2).$$

[The term $\frac{2}{3}a^2$ is an irrelevant constant in this context.] The easiest method of solution is to write

$$x^3 - 3xy^2 = r^3(\cos^3\theta - 3\cos\theta\sin^2\theta) = r^3\cos(3\theta)$$
$$= \mathrm{Im}[ir^3e^{3i\theta}] = \mathrm{Im}(iz^3), \quad (z = re^{i\theta}).$$

Therefore (to within an irrelevant constant) $f(z) = -iz^3/6a$ and

$$\phi = -\mathrm{Re}(iz^3/6a) = -\mathrm{Re}[i(x+iy)^3/6a] = y(3x^2 - y^2)/6a .$$

(iii) The couple is given by $G = 2\mu\tau \iint_A \chi \, dxdy$

where A is the area of the equilateral triangle. The couple is readily reduced to the evaluation of the double integral

$$G = -\frac{\mu\tau}{3a}\int_{-2a}^{a} dx \int_{-(x+2a)/\sqrt{3}}^{(x+2a)/\sqrt{3}} [(x-a)(x+2a)^2 - 3y^2]dy$$

$$= -\frac{\mu\tau}{3a}\int_{-2a}^{a} dx \left\{\frac{2}{\sqrt{3}}(x-a)(x+2a)^3 - \frac{2}{3\sqrt{3}}(x-a)(x+2a)^3\right\}$$

$$= -\frac{4\mu\tau}{9\sqrt{3}a}\int_{-2a}^{a}(x-a)(x+2a)^3dx = -\frac{4\mu\tau}{9\sqrt{3}a}\int_0^{3a}\xi^3(\xi - 3a)d\xi$$

$$= 9\sqrt{3}\,\mu\tau a^4/5 .$$

14. **(i)** $\chi = -\frac{1}{2}r^2 + \frac{1}{2}b^2 + ar\sin\theta - ab^2(\sin\theta/r)$.

Now $r\sin\theta$, $\sin\theta/r$ and $\frac{1}{2}b^2$ satisfy Laplace's equation so

$$\nabla^2\chi = \nabla^2(-\tfrac{1}{2}r^2) = -2 .$$

(ii) $\chi = 0$ on $x^2 + y^2 = b^2$, $x^2 + y^2 - 2ay = 0$

i.e. on $x^2 + y^2 = b^2$ and on $x^2 + (y-a)^2 = a^2$.

(iii) $\chi + \frac{1}{2}r^2 = \frac{1}{2}b^2 + ar\sin\theta - ab^2(\sin\theta/r)$

so, discarding an irrelevant constant we seek ϕ such that $\phi + i\psi = f(z)$ where

$$\psi = ar\sin\theta - ab^2(\sin\theta/r)$$
$$= \mathrm{Im}[are^{i\theta} + (ab^2/re^{i\theta})], \quad (re^{i\theta} \equiv z)$$
$$\phi = \mathrm{Re}[are^{i\theta} + (ab^2/re^{i\theta})]$$
$$= ar\cos\theta + ab^2(\cos\theta/r)$$
$$= ax[1 + (b/r)^2]$$

$|\phi| \to \infty$ as $r \to 0$ for almost all θ and since $u_z = \tau\phi$, u_z is finite everywhere only if $r = 0$ is excluded. Thus the given χ is pertinent to the problem defined by the shaded area of the figure below.

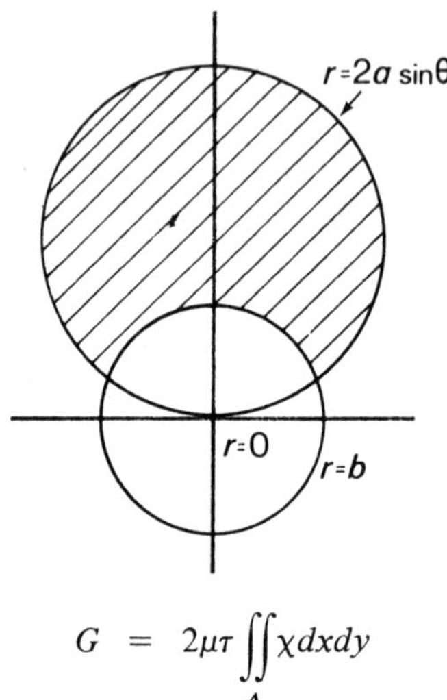

(iv)

$$G = 2\mu\tau \iint_A \chi dxdy$$

where A is the shaded area. The integral is not readily calculated unless use is made of polar coordinates. We have

$$G = 2\mu\tau \int_{\sin^{-1}(b/2a)}^{[\pi - \sin^{-1}(b/2a)]} d\theta \int_b^{2a\sin\theta} [-\tfrac{1}{2}r^2 + \tfrac{1}{2}b^2 + ar\sin\theta - ab^2(\sin\theta/r)]\,rdr$$

$$= 2\mu\tau \int_{\sin^{-1}(b/2a)}^{\frac{1}{2}\pi} d\theta \int_b^{2a\sin\theta} [-r^2 + b^2 + 2ar\sin\theta - 2ab^2(\sin\theta/r)]\,rdr$$

$$= 2\mu\tau \int_{\sin^{-1}(b/2a)}^{\frac{1}{2}\pi} d\theta(\tfrac{4}{3}a^4\sin^4\theta - 2a^2b^2\sin^2\theta + \tfrac{4}{3}ab^3\sin\theta - \tfrac{1}{4}b^4)$$

$$= \mu\tau\Big[(a^4 - 2a^2b^2 - \tfrac{1}{2}b^4)[\tfrac{1}{2}\pi - \sin^{-1}(b/2a)] + ab(\tfrac{7}{4}b^2 + \tfrac{1}{2}a^2)[1 - (b/2a)^2]^{\frac{1}{2}}\Big].$$

16. $\chi = 0$ when $-\tfrac{1}{2}r^2 + \text{Im}\,(4ib^{\frac{3}{2}}r^{\frac{1}{2}}e^{\frac{1}{2}i\theta}) = 0$

i.e. $$r = 4b[\cos(\tfrac{1}{2}\theta)]^{\frac{2}{3}}.$$

For $\theta = 0$, $r = 4b$. The curve is symmetrical about the x axis and r decreases continuously from $\theta = 0$ to $\theta = \pi$ where $r = 0$.

On calculating dy/dx from

$$\frac{dy}{dx} = \frac{\tan\theta(dr/d\theta) + r}{(dr/d\theta) - r\tan\theta}$$

there results $|dy/dx| = \infty$ for $\theta = 0$ and $dy/dx = 0$ for $\theta = \pm\pi$. Therefore there is a cusp at $\theta = \pm\pi$, $r = 0$.

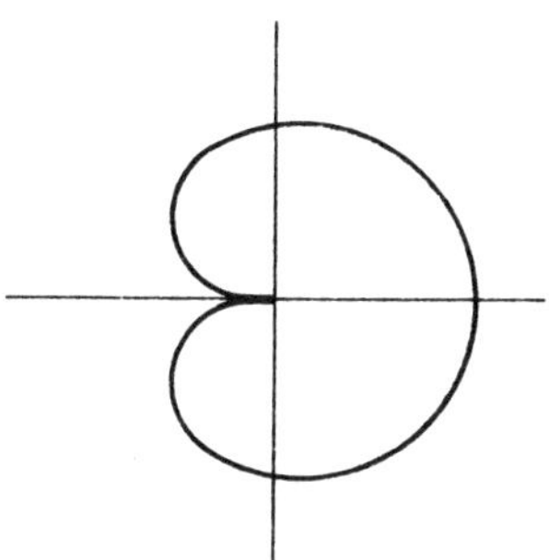

The warping function is $\quad \phi = -4b^{\frac{3}{2}}r^{\frac{1}{2}}\sin(\frac{1}{2}\theta)$

and this is single valued except conceivably for $\theta = \pm\pi$. However at these points $r = 0$ so in fact ϕ is single valued inside and on C.

Finally

$$G = 4\mu\tau\int_0^{\pi} d\theta \int_0^{4b[\cos(\frac{1}{2}\theta)]^{2/3}} [-\tfrac{1}{2}r^2 + 4b^{\frac{3}{2}}r^{\frac{1}{2}}\cos(\tfrac{1}{2}\theta)]\,r\,dr = N\mu\tau b^4$$

where

$$N = \frac{768}{5}\int_0^{\frac{1}{2}\pi}(\cos\gamma)^{\frac{8}{3}}d\gamma = \frac{384}{5}\,\frac{\Gamma(\frac{11}{6})\Gamma(\frac{1}{2})}{\Gamma(\frac{7}{3})} \simeq 107\ .$$

17. (iii) The solution of $\quad \dfrac{\partial^2\phi}{\partial z\partial\bar{z}} = 0$

has been derived previously (Section 12.8.1) in the form

$$\phi = f(z) + g(\bar{z})$$

where f and g are arbitrary. In this earlier work g was identified with $\bar{f}$ to guarantee ϕ real, but in the present context this step is delayed.† The above implies in the context of the bi-harmonic equation that

$$\frac{\partial^2\chi}{\partial z\partial\bar{z}} = \phi'(z) + \omega'(\bar{z})$$

† Also we omit the factor $\frac{1}{2}$ appearing in (12.8.5).

where, for subsequent convenience, we have written $f = \phi', g = \omega'$. A further integration now leads to

$$\frac{\partial \chi}{\partial \bar{z}} = \phi(z) + z\omega'(\bar{z}) + \bar{\psi}'(\bar{z})$$

where ψ' is arbitrary. Finally, integrating with respect to $\bar{z}$

$$\chi = \bar{z}\phi(z) + z\omega(\bar{z}) + \psi(\bar{z}) + \gamma(z) \quad \text{where } \gamma \text{ is arbitrary.}$$

There appear to be four distinct functions appearing in the solution. However since χ is real, $\chi = \bar{\chi}$, i.e.

$$\bar{z}\phi(z) + z\omega(\bar{z}) + \psi(\bar{z}) + \gamma(z) = z\bar{\phi}(\bar{z}) + \bar{z}\bar{\omega}(z) + \bar{\psi}(z) + \bar{\gamma}(\bar{z})$$

and this is met by choosing $\qquad \omega = \bar{\phi} \quad \text{and} \quad \gamma = \bar{\psi}$

so that $\qquad \chi = \bar{z}\phi(z) + z\bar{\phi}(\bar{z}) + \psi(z) + \bar{\psi}(\bar{z})$.

(iv) $$\sigma_{xx} = \frac{\partial^2 \chi}{\partial y^2} = 2\phi'(z) - \bar{z}\phi''(z) + 2\bar{\phi}'(\bar{z}) - z\bar{\phi}''(\bar{z}) - \psi''(z) - \bar{\psi}''(\bar{z})$$

$$\sigma_{yy} = \frac{\partial^2 \chi}{\partial x^2} = 2\phi'(z) + \bar{z}\phi''(z) + 2\bar{\phi}'(\bar{z}) + z\bar{\phi}''(\bar{z}) + \psi''(z) + \bar{\psi}''(\bar{z})$$

$$\sigma_{xy} = -\frac{\partial^2 \chi}{\partial x \partial y} = -i[\bar{z}\phi''(z) - z\bar{\phi}''(\bar{z}) + \psi''(z) - \bar{\psi}''(\bar{z})] \ .$$

From these results

$$\sigma_{xx} + \sigma_{yy} = 4[\phi'(z) + \bar{\phi}'(\bar{z})], \qquad \sigma_{xx} - \sigma_{yy} + 2i\sigma_{xy} = -4[z\bar{\phi}''(\bar{z}) + \bar{\psi}''(\bar{z})] \ .$$

18. (ii) The total weight of the beam is wl and by symmetry the reaction at each support is $\frac{1}{2}wl$ directed vertically upwards. To find $M(x)$ consider the beam cut at the section x; taking moments about x for the left section of the beam yields

$$M(x) - \tfrac{1}{2}wx^2 + \tfrac{1}{2}wlx = 0 \ .$$

The term $\frac{1}{2}wx^2$ stems from the weight wx of the section multiplied by the distance $\frac{1}{2}x$ to the centre of gravity of the section. Thus $M(x)$ is given by

$$M(x) = -\tfrac{1}{2}wx(l - x) \ .$$

Writing $\qquad EIY''(x) = M(x)$

yields $\qquad EIY' = -\frac{1}{4}wlx^2 + \frac{1}{6}wx^3 + A$

where A is a constant of integration.

A may be determined by noting that from symmetry arguments $Y'(\frac{1}{2}l) = 0$. This yields $A = (wl^3/24)$. Alternately A may be carried through the next integration which yields

$$EIY = -\tfrac{1}{12}wlx^3 + \tfrac{1}{24}wx^4 + Ax + B$$

where A and B are now determined by $Y(0) = Y(l) = 0$. Thus $B = 0$ and $A = wl^3/24$ as before. The final expression for Y, after substituting for I, is

$$Y = \left(\frac{w}{32a^3bE}\right) x(l - x)(l^2 + lx - x^2) .$$

(iii) The largest tensile stress occurring is at $y = -a$ and at $x = \frac{1}{2}l$ (for here the curvature is largest). The magnitude of this stress is

$$-EaY''(\tfrac{1}{2}l) \equiv aM(\tfrac{1}{2}l)/I = \tfrac{1}{8}wal^2/I = \frac{3wl^2}{32a^2b} .$$

If this is not to exceed σ_0 then the maximum length l is

$$l = [32a^2b\sigma_0/3w]^{\frac{1}{2}} .$$

20. Let R_1 and R_2 denote the (upward) reactions respectively at $x = 0$ and $x = l+L$. Then

$$R_1 + R_2 = W, \quad R_1L = R_2l$$

whence $$R_1 = Wl/(L + l), \quad R_2 = WL/(L + l) .$$

Consider the rod broken at x and for $0<x<L$ consider moments about x for the left section. There results

$$R_1x + M(x) = 0 , \quad \text{i.e. } M(x) = -Wlx/(L + l) .$$

Similarly for $L \leqslant x \leqslant l+L$, from moments about x for the right section

$$R_2(L + l - x) + M(x) = 0 , \quad \text{i.e. } M(x) = -WL(L+l-x)/(L+l) .$$

The equation for Y is

$$EIY'' = \begin{cases} -Wlx/(L + l) & (0 \leqslant x \leqslant L) \\ -WL(L + l - x)/(L + l) & (L \leqslant x \leqslant l+L) \end{cases}$$

and the boundary conditions are

$$Y(0) = Y(l + L) = 0 , \quad Y(L), Y'(L) \text{ continuous.}$$

From the differential equations, together with $Y(0) = Y(L + l) = 0$, we have

$$EIY(x) = \begin{cases} -[Wl/(l + L)](\tfrac{1}{6}x^3 + Ax) & (0 \leqslant x \leqslant L) \\ -[WL/(l + L)]\left\{\tfrac{1}{2}(L + l)[x^2 - (l+L)^2] \right. \\ \quad \left. -\tfrac{1}{6}[x^3 - (l+L)^3] + B[x - (l+L)]\right\} & (L \leqslant x \leqslant l+L) \end{cases}$$

where A and B are to be determined from continuity of $Y(L)$ and $Y'(L)$. With A and B determined there results

$$\frac{EI(L+l)Y}{W} = \begin{cases} l[(\frac{1}{6}L^2 + \frac{1}{3}lL)x - \frac{1}{6}x^3], & (0 \leqslant x \leqslant L) \\ L[-\frac{1}{6}L^2(L+l) + (\frac{2}{3}lL + \frac{1}{3}l^2 + \frac{1}{2}L^2)x \\ \quad - \frac{1}{2}(L+l)x^2 + \frac{1}{6}x^3], & (L \leqslant x \leqslant L+l) \end{cases}$$

Answers, Chapter 16

2. Open tube

Principal stresses are

$$(\sigma_{rr}, \sigma_{\theta\theta}, \sigma_{zz}) = \frac{a^2 P}{(b^2 - a^2)}\left\{-\left(\frac{b^2}{r^2} - 1\right), \left(\frac{b^2}{r^2} + 1\right), 0\right\} .$$

Evidently the maximum principal stress difference (since $r \leqslant b$) is

$$\sigma_{\theta\theta} - \sigma_{rr} = \frac{2a^2 Pb^2}{(b^2 - a^2)r^2}$$

and this attains its largest value at $r = a$. Yielding is initiated therefore at $r = a$ for $\sigma_{\theta\theta} - \sigma_{rr} = Y$, i.e. for

$$P = \tfrac{1}{2}[1 - (a^2/b^2)]Y .$$

Closed tube

Here σ_{rr} and $\sigma_{\theta\theta}$ are as above while for σ_{zz} the zero in parentheses { } above is replaced by unity. Again the largest absolute principal stress difference is $\sigma_{\theta\theta} - \sigma_{rr}$, with conclusions as above.

Zero axial expansion

Here the zero within the parentheses is replaced by 2ν and since $0 \leqslant 2\nu \leqslant 1$, again it remains true that the largest absolute principal stress difference is $\sigma_{\theta\theta} - \sigma_{rr}$, with conclusions as above.

Open tube – von-Mises criterion

Here, $\sigma_{zz} = 0$ and

$$(\sigma_{rr} - \sigma_{\theta\theta})^2 + (\sigma_{\theta\theta} - \sigma_{zz})^2 + (\sigma_{zz} - \sigma_{rr})^2 = \frac{a^4 P^2}{(b^2 - a^2)^2}\left[\frac{6b^4}{r^4} + 2\right]$$

whose maximum value is attained at $r = a$. Yielding is initiated when this maximum value is $2Y^2$ so that the critical pressure is

$$P = Y[1 - (a^2/b^2)]/[3 + (a^4/b^4)]^{\frac{1}{2}} .$$

For modest values of (b/a) near unity, this result is similar to that for the Tresca criterion while for $b \gg a$ there is a difference of about 16%.

3. Tresca criterion

The principal stress differences are

$$\begin{aligned}
\sigma_{\theta\theta} - \sigma_{rr} &= A[2(1-2\nu)r^2] \\
\sigma_{\theta\theta} - \sigma_{zz} &= A[(3-4\nu)a^2 - (1-2\nu)r^2] \\
\sigma_{rr} - \sigma_{zz} &= A[(3-4\nu)a^2 - 3(1-2\nu)r^2] .
\end{aligned}$$

For $\nu < \frac{1}{2}$ all of these differences are largest in absolute value at the extremities ($r = 0$ or $r = a$). We have

$$\begin{aligned}
\text{Max}_r|\sigma_{\theta\theta} - \sigma_{rr}| &= A[2(1-2\nu)a^2] \quad (r = a) \\
\text{Max}_r|\sigma_{\theta\theta} - \sigma_{zz}| &= A[(3-4\nu)a^2] \quad (r = 0) \\
\text{Max}_r|\sigma_{rr} - \sigma_{zz}| &= A[(3-4\nu)a^2] \quad (r = 0) .
\end{aligned}$$

In deriving these results the absolute values of the principal stress differences have been compared at $r = 0$ and $r = a$, and use made of $0 \leqslant \nu < \frac{1}{2}$.

Since $$(3-4\nu) - 2(1-2\nu) = 1 > 0$$

yielding is initiated at $r = 0$ for

$$A(3-4\nu)a^2 = Y$$

i.e. $$\omega^2 = \frac{8(1-\nu)Y}{(3-4\nu)\rho_0 a^2} .$$

Exceptionally for $\nu = \frac{1}{2}$, yielding is initiated simultaneously throughout the cross section when $\omega^2 = 4Y/\rho_0 a^2$.

von-Mises criterion

For the von-Mises criterion

$$\begin{aligned}
&(\sigma_{\theta\theta} - \sigma_{rr})^2 + (\sigma_{zz} - \sigma_{\theta\theta})^2 + (\sigma_{rr} - \sigma_{zz})^2 \\
&\quad = A^2[14(1-2\nu)^2 r^4 + 2(3-4\nu)^2 a^4 - 8(1-2\nu)(3-4\nu)a^2 r^2] .
\end{aligned}$$

For $\nu < \frac{1}{2}$, the quantity on the right side of the equation immediately above possesses a turning point at $r^2 = 2a^2(3-4\nu)/7(1-2\nu)$ which may be located in $0 < r < a$. However in any event this is a minimum so maximum values within $0 \leqslant r \leqslant a$ are taken at $r = 0$ and $r = a$. The larger of these values is for $r = 0$ and given by $2A^2(3-4\nu)^2 a^4$ so that yielding is initiated at $r = 0$ when

$$2A^2(3-4\nu)^2 a^4 = 2Y^2$$

leading to the same critical frequency as for the Tresca criterion.

Again for $\nu = \frac{1}{2}$, yielding is initiated simultaneously throughout the cross section.

4. **(i)** The principal stresses are the solutions of the eigenvalue equation

$$\begin{vmatrix} -\sigma & 0 & \sigma_{xz} \\ 0 & -\sigma & \sigma_{yz} \\ \sigma_{zx} & \sigma_{zy} & -\sigma \end{vmatrix} = 0$$

and the result follows. Evidently the maximum principal stress difference is

$$\sigma(2) - \sigma(3) = 2(\sigma_{zx}{}^2 + \sigma_{zy}{}^2)^{\frac{1}{2}}$$

and yield is initiated (for the Tresca criterion) when this difference is Y.

The von-Mises criterion is of exactly the same form except for a numerical factor, viz.:

$$(\sigma_{zx}{}^2 + \sigma_{zy}{}^2)^{\frac{1}{2}} = Y/\sqrt{3} \ .$$

(ii) Let ψ be conjugate to ϕ and use the Cauchy-Riemann equations

$$\frac{\partial\phi}{\partial x} = \frac{\partial\psi}{\partial y}, \quad \frac{\partial\phi}{\partial y} = -\frac{\partial\psi}{\partial x}$$

to obtain $\quad \sigma_{zx} = \mu\tau\left(-y + \dfrac{\partial\psi}{\partial y}\right), \quad \sigma_{zy} = \mu\tau\left(x - \dfrac{\partial\psi}{\partial x}\right) .$

However from (14.4.74) $\quad \chi = \psi - \frac{1}{2}(x^2 + y^2)$

so that $\quad \sigma_{zx} = \mu\tau\dfrac{\partial\chi}{\partial y}, \quad \sigma_{zy} = -\mu\tau\dfrac{\partial\chi}{\partial x} .$

It follows that the yield criteria become

$$\mu^2\tau^2\left[\left(\frac{\partial\chi}{\partial x}\right)^2 + \left(\frac{\partial\chi}{\partial y}\right)^2\right] \begin{array}{ll} = \frac{1}{4}Y^2, & \text{(Tresca)} \\ = \frac{1}{3}Y^2, & \text{(von Mises)} \ . \end{array}$$

(iii) $\dfrac{\partial A}{\partial x} = 2\chi_{xx}\chi_x + 2\chi_{yx}\chi_y, \quad \dfrac{\partial^2 A}{\partial x^2} = 2\chi_{xxx}\chi_x + 2\chi_{xx}{}^2 + 2\chi_{yxx}\chi_y + 2\chi_{yx}{}^2$

where suffixes denote differentiation. Similarly

$$\frac{\partial^2 A}{\partial y^2} = 2\chi_{yyy}\chi_y + 2\chi_{yy}{}^2 + 2\chi_{xyy}\chi_x + 2\chi_{xy}{}^2 \ .$$

It follows that $\nabla^2 A = 2[\chi_{xx}{}^2 + \chi_{yy}{}^2 + 2\chi_{xy}{}^2] + 2\left(\chi_y\dfrac{\partial}{\partial y} + \chi_x\dfrac{\partial}{\partial x}\right)\nabla^2\chi .$

But χ satisfies $\nabla^2\chi = -2$ so that the last term vanishes.

It follows that $\quad \dfrac{\partial^2 A}{\partial x^2} + \dfrac{\partial^2 A}{\partial y^2} \geqslant 0 \ .$

Now in order for A to have a maximum at an interior point (x, y) of C

$$\frac{\partial A}{\partial x} = \frac{\partial A}{\partial y} = 0\,, \quad \frac{\partial^2 A}{\partial x^2} < 0\,, \quad \frac{\partial^2 A}{\partial x^2}\frac{\partial^2 A}{\partial y^2} - \left(\frac{\partial^2 A}{\partial x \partial y}\right)^2 > 0\ .$$

Evidently it is necessary (but not sufficient) that $A_{xx} < 0$, $A_{yy} < 0$ which requires $\nabla^2 A < 0$ and this is not satisfied. Therefore there is no point within C where A takes a maximum. It follows that A takes its largest value on the boundary C.

(iv) For the ellipse $\chi = \dfrac{a^2 b^2}{a^2 + b^2}\left[1 - \dfrac{x^2}{a^2} - \dfrac{y^2}{b^2}\right]$

so that $$\left(\frac{\partial \chi}{\partial x}\right)^2 + \left(\frac{\partial \chi}{\partial y}\right)^2 = \frac{4a^4 b^4}{(a^2 + b^2)^2}\left[\frac{x^2}{a^4} + \frac{y^2}{b^4}\right].$$

If the ellipse is described parametrically by $x = a\cos\phi$, $y = b\sin\phi$ the quantity in square parentheses above may be written

$$\frac{x^2}{a^4} + \frac{y^2}{b^4} = \frac{1}{2}\left[\left(\frac{1}{a^2} + \frac{1}{b^2}\right) + \left(\frac{1}{a^2} - \frac{1}{b^2}\right)\cos 2\phi\right]$$

and for $b > a$ possesses a maximum when $\cos 2\phi = 1$, i.e. at $x = \pm a$, $y = 0$. The maximum value is a^{-2} leading to

$$\frac{2\mu\tau a b^2}{(a^2 + b^2)} = \tfrac{1}{2}Y\ .$$

The couple is $$G = \frac{\pi\mu\tau a^3 b^3}{(a^2 + b^2)}$$ (see p. 391)

so (eliminating τ), the couple required to induce yield is

$$G = \tfrac{1}{4}\pi a^2 b Y\ .$$

Evidently for $a > b$ yielding is initiated at $x = 0$, $y = \pm b$ for a couple $\frac{1}{4}\pi a b^2 Y$. Both results are summarised by

$$G = \tfrac{1}{4}\pi a b Y\,\mathrm{Max}(a, b)\ .$$

Appendix

Results in Vector Analysis

CARTESIANS COORDINATES

Let
$$\Phi = \phi(x, y, z) \text{ be a scalar}$$
$$\mathbf{A} = A_x(x, y, z)\mathbf{i} + A_y(x, y, z)\mathbf{j} + A_z(x, y, z)\mathbf{k}$$
$$\equiv (A_x, A_y, A_z) \text{ be a vector.}$$

Then
$$\text{grad } \Phi = (\partial\Phi/\partial x, \partial\Phi/\partial y, \partial\Phi/\partial z) \quad \text{(vector)}$$
$$\text{div } \mathbf{A} = \partial A_x/\partial x + \partial A_y/\partial y + \partial A_z/\partial z \quad \text{(scalar)}$$
$$\text{curl } \mathbf{A} = (\partial A_z/\partial y - \partial A_y/\partial z, \partial A_x/\partial z - \partial A_z/\partial x, \partial A_y/\partial x - \partial A_x/\partial y) \quad \text{(vector)}$$
$$\nabla^2 \Phi = \partial^2\Phi/\partial x^2 + \partial^2\Phi/\partial y^2 + \partial^2\Phi/\partial z^2 \quad \text{(scalar).}$$

CYLINDRICAL POLAR COORDINATES

The coordinates are (r, θ, z) where $x = r\cos\theta, y = r\sin\theta$. Unit vectors in the r, θ, z directions are denoted by $\mathbf{i}_r, \mathbf{i}_\theta, \mathbf{i}_z$ (where $\mathbf{i}_z = \mathbf{k}$).

Let
$$\Phi = \Phi(r, \theta, z) \quad \text{(scalar)}$$
$$\mathbf{A} = A_r(r, \theta, z)\mathbf{i}_r + A_\theta(r, \theta, z)\mathbf{i}_\theta + A_z(r, \theta, z)\mathbf{i}_z$$
$$\equiv (A_r, A_\theta, A_z) \quad \text{(vector).}$$

Then
$$\text{grad } \Phi = (\partial\Phi/\partial r, r^{-1}\partial\Phi/\partial\theta, \partial\Phi/\partial z)$$
$$\text{div } \mathbf{A} = \partial A_r/\partial r + r^{-1}A_r + r^{-1}(\partial A_\theta/\partial\theta) + \partial A_z/\partial z$$
$$\text{curl } \mathbf{A} = [r^{-1}(\partial A_z/\partial\theta) - \partial A_\theta/\partial z, \partial A_r/\partial z - \partial A_z/\partial r, \partial A_\theta/\partial r + r^{-1}A_\theta - r^{-1}(\partial A_r/\partial\theta)]$$
$$\nabla^2\Phi = \partial^2\Phi/\partial r^2 + r^{-1}(\partial\Phi/\partial r) + r^{-2}(\partial^2\Phi/\partial\theta^2) + \partial^2\Phi/\partial z^2$$

SPHERICAL POLAR COORDINATES

The coordinates are (r, θ, ϕ) where $x = r \sin\theta \cos\phi$, $y = r \sin\theta \sin\phi$, $z = r\cos\theta$. Unit vectors in the r, θ, ϕ directions are denoted by $\mathbf{i}_r$, $\mathbf{i}_\theta$, $\mathbf{i}_\phi$.

$$\begin{aligned}
\text{Let} \qquad \Phi &= \Phi(r, \theta, \phi) \quad \text{(scalar)} \\
\mathbf{A} &= A_r(r, \theta, \phi)\mathbf{i}_r + A_\theta(r, \theta, \phi)\mathbf{i}_\theta + A_\phi(r, \theta, \phi)\mathbf{i}_\phi \\
&\equiv (A_r, A_\theta, A_\phi) \quad \text{(vector).} \\
\text{Then grad}\,\Phi &= [\partial\Phi/\partial r, r^{-1}(\partial\Phi/\partial\theta), (r\sin\theta)^{-1}\partial\Phi/\partial\phi] \\
\text{div}\,\mathbf{A} &= \partial A_r/\partial r + 2r^{-1}A_r + r^{-1}\cot\theta A_\theta + r^{-1}(\partial A_\theta/\partial\theta) \\
&\quad + (r\sin\theta)^{-1}(\partial A_\phi/\partial\phi) \\
\text{curl}\,\mathbf{A} &= [r^{-1}(\partial A_\phi/\partial\theta) + r^{-1}\cot\theta A_\phi - (r\sin\theta)^{-1}\partial A_\theta/\partial\phi, \\
&\quad (r\sin\theta)^{-1}(\partial A_r/\partial\phi) - \partial A_\phi/\partial r - r^{-1}A_\phi, \partial A_\theta/\partial r + r^{-1}A_\theta \\
&\quad - r^{-1}(\partial A_r/\partial\theta)] \\
\nabla^2\Phi &= \partial^2\Phi/\partial r^2 + 2r^{-1}(\partial\Phi/\partial r) + r^{-2}(\partial^2\Phi/\partial\theta^2) + r^{-2}\cot\theta(\partial\Phi/\partial\theta) \\
&\quad + (r\sin\theta)^{-2}(\partial^2\Phi/\partial\phi^2)
\end{aligned}$$

USEFUL VECTOR IDENTITIES

$$\begin{aligned}
\text{div curl}\,\mathbf{A} &= 0 \\
\text{curl grad}\,\Phi &= 0 \\
\text{div grad}\,\Phi &= \nabla^2\Phi \\
\text{curl curl}\,\mathbf{A} &= \text{grad(div}\,\mathbf{A}) - \nabla^2\mathbf{A} \\
\text{div}(\Phi\mathbf{A}) &= \Phi\,\text{div}\,\mathbf{A} + \mathbf{A}\,.\,\text{grad}\,\Phi \\
\text{grad}(\mathbf{A}\,.\,\mathbf{B}) &= (\mathbf{A}\,.\,\text{grad})\mathbf{B} + (\mathbf{B}\,.\,\text{grad})\mathbf{A} + \mathbf{A} \wedge \text{curl}\,\mathbf{B} + \mathbf{B} \wedge \text{curl}\,\mathbf{A}
\end{aligned}$$

INTEGRAL THEOREMS

(*a*) *Green's theorem*

Let S be a closed surface bounding the region Ω and let $\mathbf{n}$ be the outward unit normal vector to S. (Evidently $\mathbf{n}$ varies over the surface.) Define $\mathbf{dS} = \mathbf{n}dS$ where dS is an element of area. Then for differentiable vector fields $\mathbf{A}$

$$\iint_S \mathbf{A}\,.\,\mathbf{dS} = \iiint_\Omega \text{div}\mathbf{A}\,d\Omega\ .$$

(*b*) *Stokes' Theorem*

Let C be a closed loop spanned by the (open) surface S and let **dr** be an element of the loop C. For diffentiable vector fields **A**

$$\oint_C \mathbf{A} \cdot \mathbf{dr} = \iint_S \text{curl } \mathbf{A} \cdot \mathbf{dS} \ .$$

The ambiguities concerning the direction around C, and which surface to choose for S, are removed by the following rule. On walking around C the surface with normal **n** should lie to the left hand.

Further Reference may be made to Chorlton, F., *Vector and Tensor Methods*, (Ellis Horwood, Publisher, Chichester, 1976).

Books and Articles for Further Reading

(† Denotes advanced text)

General Continuum Mechanics

1.	C. Truesdall and R.A. Toupin	"The Classical Field Theories" Vol. III/1 *Handbuch der Physik,* (Ed. S. Flugge), Springer-Verlag, Berling (1960)†
2.	C. Truesdall and W. Noll	"The Non-Linear Field Theories of Mechanics",Vol. III/3 *Handbuch der Physik,* (Ed. S. Flugge), Springer-Verlag Berlin (1963)†

Finite Elasticity

3.	A. J. M. Spencer	The Static Theory of Finite Elasticity, *Journal of the Institute of Mathematics and Its Applications,* p 164-200 Vol. 6, No. 2, June 1970
4.	A. E. Green and W. Zerna	*Theoretical Elasticity,* 2nd Edition, Clarendon Press, Oxford (1968)†
5.	A. E. Green and J. E. Adkins	*Large Elastic Deformations,* Clarendon Press, Oxford (1960)†

Classical Elasticity

6.	S. Timoshenko and N. Goodier	*The Theory of Elasticity,* 2nd Edition, McGraw Hill (1951)
7.	I. S. Sokolnikoff	*The Mathematical Theory of Elasticity,* 2nd Edition, McGraw Hill (1956)
8.	A. E. H. Love	*The Mathematical Theory of Elasticity,* 4th Edition, Cambridge University Press, (1927). Available as Dover Reprint
9.	M. I. Muskelishvili	*Some Basic Problems of the Mathematical Theory of Elasticity,* P. Noordhof, Groningen (1953)†
9.	See also Reference 4.	

Hydrodynamics

10.	L.M. Milne-Thomson	*Theoretical Hydrodynamics,* 2nd Edition, Macmillan, (1949)
11.	H. Lamb	*Hydrodynamics,* 6th Edition, Cambridge University Press, (1932). Available as Dover Reprint
12.	G. K. Batchelor	*An Introduction to Fluid Mechanics,* Cambridge University Press (1967)
13.	G. Birkhoff and E. H. Zarantenello	*Jets, Wakes and Cavities,* Academic Press (1957)

Viscoelasticity

14.	S. C. Hunter	"Viscoelastic Waves" pp. 3-56 in *Progress in Solid Mechanics Vol. I,* Editors R. Hill and I. N. Sneddon, North Holland, Amsterdam (1960)
15.	S. C. Hunter	"The Solution of Boundary Value Problems in Linear Viscoelasticity" p 257-295 in *The Mechanics and Chemistry of Solid Propellants* (Proceedings of the Fourth 1965 Symposium on Naval Structural Mechanics), Pergamon (1967)
16.	A. S. Lodge	*Elastic Liquids,* Academic Press (1964)
17.	F. J. Lockett	*Non-Linear Viscoelastic Solids,* Academic Press (1972)
18.	See also Reference 5.	

Plasticity

19.	R. Hill	*The Mathematical Theory of Plasticity,* Clarendon Press, Oxford (1950)
20.	L. M. Kachanov	*Foundations of the Theory of Plasticity,* North Holland, Amsterdam (1971)

Waves

21.	C. A. Coulson	*Waves,* Oliver and Boyd (1949)
22.	H. Kolsky	*Stress Waves in Solids,* Clarendon Press, Oxford (1953) Available as Dover Reprint
23.	J. W. S. Rayleigh	*The Theory of Sound* Volumes 1 and 2. Available as Dover Reprint (1945)
24.	H. Lamb	*The Dynamical Theory of Sound.* Available as Dover Reprint (1960)
25.	G. B. Whitham	*Linear and Non-Linear Waves,* John Wiley and Sons, New York (1974)†

Index

Index

A

acceleration of a particle in a continuum
  material/referential description 24, 39
  spatial description 28, 39
acoustic approximation 332-334
acoustics of musical instruments 346-351
acoustics of rooms 352-353
aerofoil theory 296-303
aircraft lift forces 15, 301-302
Airy stress function 58, 424
Archimedes' principle 258-259
attenuation of viscoelastic waves 461
axes
  principal 82-89
  rotation and translation of 66-68, 120-121

B

bending moment
  elastic beam 374, 393-394,
  elastic-plastic beam 487-488
bending of beam by equal terminal couples 372-373
bending of beams, technological theory of 13, 393-398
  cantilever beam with terminal load 13, 394-395
  problems in 426-428
  simply supported beam 396
  statically indeterminate problems 397
Bernoulli equations 219-220
  applications in hydraulics 221-227
  invariance of with respect to Galilean transformation 331
  role in irrotational flow 223
biharmonic equation 58, 209, 425
Blasius theorems
  steady flow 284-287, 294, 295, 301
  unsteady flow 325
body forces 46
boundary conditions
  classical hydrodynamics 238-239, 273
  linear elasticity 363-364
  Navier-Stokes equation 180

C

cantilever beam
  deflected under terminal load 13, 394-395,
  transverse vibration of 412-417, 436

Cartesian
  axes 22
  coordinates 22
  reference frames 66
  tensors 69-72
  unit vectors 22
Cauchy elasticity 144
Cauchy-Riemann equations 272, 274, 388
cavitation 89. 223-224
circle theorem 281-282, 283, 284, 298
circulation line integral 215
  around aerofoil 300, around circular cylinder 252-253, around elliptical cylinder 294, associated with rectilinear vortex 247
complex potential 274
complex variable methods; for Laplace equation 270-316, in two dimensional elastostatics 425-426
complex velocity 274
condensation 332
conformal mapping 287-289
  circle → aerofoil 298-299, circle → ellipse 293-294, circle → plate 292, Kutta-Joukowski transformation 291-292, quarter plane → half plane 289-290, strip → half plane 291, 329, wedge → half plane 290
conservation of mass 15
  equation of in material description 26, 39, 111
  equation of in spatial description 29, 30, 39, 127, 143, 179, 214
  equivalence of spatial and material descriptions 30-32
constitutive equations 16, 19, 116
  general form of 116, necessity of for complete system of equations 56, restrictions on 116-124
constitutive equations for
  barotropic fluid 89, Cauchy elasticity 145-146, 148-149, elastic materials 144-159, fluids 125-143, Green elasticity 153-156, ideal fluid 56, 89, 125, 143, incompressible inviscid fluid 129, linear elastic solid 149, 361, 363, 365-366, linear viscoelastic material 446, 447, 449, 454, 456-457, metal plasticity 473-483, Newtonian fluid 135, Reiner-Rivlin fluid 133 135, 142, Stokes' fluid 136, 142
continuity, equation of 29 *see also* conservation of mass
continuum equations of motion of 51-54, 79
  forces in 45-47
  hypothesis 21
  notion of 20
convective derivative 28
Couette flow 174-177, 193-196

couple
on aerofoil 301
in bending of beams (bending moment) 374-375, 394, 487-488
on elliptic cylinder in two dimensional flow 295
on rotating cylinders in Couette flow 196
on stationary cylinder in two dimensional flow 286-287
in torsion of an elastic prism 389-390, 403
creep function 445, 451-452
creep spectrum 452
cylinder, translatory motion of in inviscid irrotational fluid 254-258
cylindrical pressure vessel *see* pressure vessel

D

deformation 101-111
deformation gradient matric 102
density 26
dilatation in small displacement gradient approximation 111, 150, 361, 367, relation to condensation 332
diffusion equation 197
dipole (line) 246, 278, inside circular boundary 283, with stream 255
dipole (point) 237-238, 264, outside plane boundary 242, outside sphere 321
Dirac delta function 201, 203, 204, 234, 248
dispersion *see* phase dispersion
displacement of a particle in a continuum 32, 39, 364
dissipation of energy
in Couette flow of Stokes' fluid 195, in Newtonian and Stokes' fluid 142, in plastic deformation of metals 477-478, in rectilinear flow of Stokes' fluid 188
in Reiner-Rivlin fluid 141, in viscoelastic material 449-450
drag coefficient 316

E

eigenfunctions
associated with dispersive longitudinal elastic waves 439
associated with Rayleigh waves 406
associated with unsteady laminar shear flow of Stokes' fluid 200
associated with vibrating beams 416, 436
associated with wave equation 348-352
eigenvalues
of a matrix 82, 84, 90, 91, 92, 102-105, associated with eigenfunctions 200, 348, 351-352, 406, 416, 436, 462
eigenvectors of a matrix 103
elastic materials 18, 116, 144-159, 361
elastic strain energy 151-156, 158-159, 361, 437
elasticity theory
finite elasticity 144-157, linear elasticity 149-150, 158-159, 361-443
elastomer materials *see* polymer materials
energy
balance law of 15, 16, internal 140-142 (*see also* elastic strain energy), energy identity derived from equations of motion 80, 92
equation
biharmonic, consitutive, diffusion, Euler equation of motion of fluid, Laplace, conservation of mass, motion, Navier-Stokes, wave – *see under separate entries*
Euler equations of motion of an ideal fluid 126, 143, 214
Eulerian description of continuum motion 27

F

fibre reinforced material 114
flow rule (plasticity) 476
flows (of fluids)
compressible; steady state spheric[illegible] symmetric 230-232, acoustic approximation for compressible flow 332-360
incompressible; in pipes (hydraulics) 221-22[illegible] 224-228, potential flows of inviscid fluids 233-331 shear flows of Reiner-Rivlin fluid 130, 165-178, shear flows of Stokes' fluid 179-2[illegible]
fluid
barotropic 89, 127, compressible liquid 56, 127-129, ideal 18, 89, 125-130, incompressible 59, 129, 180, 233, Newtonian 18, 116, 135, perfect gas 17, 56, 59, 127-128, real 136, Reiner-Rivlin 130-135, Stokes' 18, 33, 136, 142, 179
force
on accelerating circular cylinder in fluid 257
on accelerating sphere in fluid 265-267
on aerofoil in stream with circulation 301-30[illegible]
on circular and elliptic cylinder due to to line source 327
on circular cylinder due to parallel line vortex 325
on cylinder due to line source, circulation and stream 326-327
on elliptic cylinder in stream with circulation 294-295
on plate impacted by a stream 309, 316
on sphere moving in a compressible fluid (acoustic approximation) 359
on stationary cylinders in two dimensional flows 286
forces in a continuum 45-47
Fourier integral methods 200, 356, 406, 434, 439-443, 462-464

racture criteria for metals 87
rame indifference 118-124, 131, 144, 145, 151, 75, 482
ree streamline problems 303-316, 329-330

G

eometrical dispersion 412
ravity, acceleration due to 46, 58, 183, 58-259, 267, 368-370
reen's function 204
roup velocity 407, 442

H

leaviside step function 440, 445, 446, 536
omogeneous materials 117
looke's law 16, 17, 47, 116
opkinson; dynamic fracture experiments 09-412, pressure bar 429
ydraulics 221-227
ydrodynamics – classical theory 233-331

I

nages, method of for solving potential roblems 240-242, 279-283
nternal energy of a barotropic fluid 140
nvariance form of equations with respect to ifferent coordinate frames 82, 117, 124
nvariants of a square matrix 83-84, 99-100, 33-135, 148-149, 153-156, 469-470
otropic materials 117, 132, 146, 153, 361, 56, 468, 475

J

acobian 25, 39; time differential of 30
ts (plane), 303-311, 313-315, 329-331

K

elvin; circulation theorem 215, method of ationary phase 439-443
netic energy of an incompressible inviscid uid 318
netic theory of gases 136
netic theory of rubber elasticity 156
ronecker delta; definition 67, 264, tensor ature of 71

L

Lagrangian description of continuum motion 23
Lamé constants 150, 157, 361, 365
Laplace equation 186, 197, 217, 223, 233, 270, 306, 388
    general solution for plane flows 271-272
    separation of variable solution for axisymmetry 264
    separation of variable solution in polar coordinates 242-245, 306
    simple solutions 233-234, 237-238
    uniqueness of solution 186-187, 306, 390
Laplace transform methods 355, 447-449
left stretch matrix *see* matrix
Legendre equation 261, 320
Legendre polynominals 262-263, 319-321, 358
    generating function 264, 319,
    orthogonality properties 263-264,
    recurrence relation 320
lift forces on aerofoil 15, 301-302

M

mass, conservation of *see* conservation of mass
material frame indifference *see* frame indifference
matrix (square)
    associated with rotation of axes 68, Cauchy-Green 105, Cayley Hamilton theorem 100, 149, 154, eigenvalues of 82, Hermitian 82, invariants of 83-84, left stretch matrix 105, polar decomposition theorem 89-92, proper orthogonal 68, representation of strain components 106, representation of stress components 51, representation of transformation laws of second order tensor 71, right stretch matrix 104, 146, 151
metals
    fracture criteria 87, initiation of yield 87, 468, plastic behaviour 465-468, significance of linear elasticity theory for 157
Milne-Thomson
    circle theorem 281-282, 283, 284, 298, construction of regular function from real or imaginary part 327-328
mobile derivative *see* convective derivative
modulus
    bulk 56, 127-128, 418, complex 454-455, 460-461, shear 150, 418
    Young's 13, 17, 47, 150, 365-367, 418
momentum balance laws; linear momentum 15, 16, angular momentum 15, 16
    equations resulting from balance laws 48-50, 53, 54, 55, 79, 80
motion, equations of 51-54, 57, 79, 98, 137, 167, 170, 175, 179, 214, 332, 361-363, 364, 457-458, 486

N

Navier-Stokes equations; 138, 143, 179, some solutions of 179-213
Newtonian fluid 18, 116, 135
normal stress *see* stress
normal stress effect *see* stress

P

particle paths 23, 34, 39
perfect gas 16, 17, 56, 127-128
perturbation analysis 319, 423
phase dispersion 407, 412
phase velocity 356, 404, 439-440, 461
plastic deformation of metals 87, 157, 465-490
plasticity theory 465-490
plastic potential 476
Poiseuille; flow 14, 172, 189, formula 14, 189
Poisson's ratio 150, 365-367
polar decomposition theorem 89-92, 102, 105, 145, 151
polymer materials 18, 116, 156, 157, 444
potential theory 233-331
pressure 20, 21, 56-57, 59, 89, 125-128, 129, 155-156, 166, 219-220
pressure vessel
  cylindrical elastic 14, 381-384, cylindrical elastic-plastic 484-486, initiation of yield in cylindrical vessel 485, 488, spherical elastic 377-380, spherical elastic thin walled 420, initiation of yield in spherical elastic 473, spherical viscoelastic 458-459
principal axes 86-87, 103, 133, 135
principal stresses 82-88, 379, 382
prism; suspended in earth's gravitational field 368-371, torsion of 386-393

R

rate of strain tensor *see* strain
Rayleigh equation for longitudinal elastic waves 438-439
Rayleigh surface waves 403-406
rectilinear shear flows; of Reiner-Rivlin fluid 167-174, of Stokes' fluid 180-193, 196-205
Reiner-Rivlin fluid 116, 130-135, 165-178
relaxation function 447, 451-452
relaxation spectrum 452
right stretch matrix *see* matrix
rigid body displacement 113, in small displacement gradient approximation 114, 372
rigid body motion superposed on elastic deformation 364-365
Rivlin-Saunders experiments 504
rotating shaft; 384-386, hollow 421, initiation of yielding in solid shaft 488
rotation of axes 66, 68, 120, 132
rotation of material element in deformation, 104-105; in small displacement gradient approximation 109-110
rubber 156-157, 161, 444

S

St. Venant's principle 377, 385, 393
second moment of area 374
separation of variables 198, 243, 260, 306, 328, 348, 404, 414, 436, 460
shear flows; of Reiner-Rivlin fluid 165-178, of Stokes' fluid 179-206
shear force in bending of beams 398, 413
shear stress *see* stress
similarity solution 213, 341-343, 358
slow viscous flow approximation 209-211
sound
  propagation in compressible fluid 128-130, velocity of in fluid 128, 332, 337-338
source (line) 246, 276, 278, 289
  outside cylinder with circulation 326-327, outside circular cylinder 327, outside elliptic cylinder 327, between parallel planes 329, outside plane boundary 279, inside rectangular corner 290, inside wedge 290
source (point) 234, 264
  outside plane boundary 240, inside rectangular corner 242, outside sphere 268-270
sphere
  gravitating elastic 421, motion in inviscid incompressible fluid 15, 265, two spheres moving in inviscid incompressible fluid 322-323, motion in Stokes fluid 19, 211-213, (small) motion in compressible inviscid fluid 358-359, (small) oscillations of sphere of variable radius in compressible inviscid fluid 357
spin tensor 75, 111, 112, 122, 131
stagnation point 33, 224, 312
Stokes' fluid 18, 33, 136, 142-143, 179-213
Stokes' law of force on moving sphere 19, 211-213
Stokes stream function 211, 274, 323
strain
  Cartesian components expressed in terms of displacement gradients 108, geometrical interpretation 107-108, matrix of components 106, plane strain 424-426, rate of strain tensor (rate of deformation tensor) 75, 111, 131-135, 158, small displacement gradient approximation 109, 361, tensor nature of 106
strain energy 150-151, 153, 158-159, 437
stream 233, 245, 264, 275, 290
  flowing past aerofoil with circulation 300, flowing past circular cylinder 254, 282, 286 flowing past elliptic cylinder 293 (with circulation 294-295), stream with line dipole 255, flowing past plate 293, 312-316, flowing past sphere 265
stream function 43, 44, 209, 210, 211, 272-273, 274, 305-307, 310, 311, 312, 313, 314, 323
streamline (s) 33-34, 273, 288, free streamlines 303-316

ress
Cartesian components of 50, components referred to cylindrical polar coordinates 96-98, 162-163, 173, 174, 177, 382, 383, 385, 402, 421, 484-486, 488, components referred to spherical polar coordinates 213, 380, 419-421, 430, 458, extra stress 130, stress function (Airy) 58, 424, matrix of components 51, maximum shear stress 95-96 normal component of stress 51, normal stress effect 177, plane stress 95-96, principal stress 82-83, 87, shear component of stress 51, symmetry of components 55, 57, 61, 80, tensor nature of 76-78, transformation laws under rotation of axes 76-78
ress vector 46
components referred to Cartesian axes 78, consequences of Newton's third law 47, normal component of 46, shear component of 46
retches (principal) 102-104
ıffix notation 60-61, dummy suffix 63,
ee suffix 63
ımmation convention 61-65

T

nsor (Cartesian) 65-72
addition of tensors 72, anti-symmetric second order tensors 72, contraction of indices 73, deformation gradient tensor 74, 102, 144-148, differentiation of tensors with respect to coordinates 73, differentiation of tensors with respect to time 73, frame indifferent tensors 121, invariance concepts associated with tensors and tensor equations 65, 82, 92, Kronecker delta is a tensor 71, matrix representation of transformation laws of second order tensor 71, 93, multiplication of tensors 73, tensor of order zero 70, tensor of order one 70, tensor of order two 71, tensors of order three and higher 72, quotient theorem 75, rate of strain tensor 75, 111, 122, spin tensor 75, 112, 122, strain components form a tensor 106, stress components form a tensor 76-78, symmetric second order tensor 72
ıermoelastic material 144
ırsion of an elastic prism 386-393
of circular cross section 391, of elliptic cross section 390-391, of equilateral triangle cross section 422, hollow cylinder 391-392, hollow circular cylinder 392, hollow elliptical cylinder 392-393, other cross sections 422-424, initiation of yielding in cylinder subject to torsion 489-490, dynamic torsion of circular cylinder 402-403
ırsion of an elastic-plastic circular cylinder
33-484
esca; hexagon 472, 479, 481, yield criterion ıaximum shear stress criterion) 95, 470-472,
35

U

underwater explosion problem 14, 235-237, 317
uniqueness theorem
for Laplace equation 186-187, 306, 390, in classical elastostatics 437

V

velocity of a particle in a continuum
material/referential description 24, 39, spatial description 27, 39
velocity potential 216
viscoelasticity (linear theory of) 444-464
viscosity 14, 18, coefficients of 14, 135, 136, 137
von Mises; circle 472, 479, yield criterion 470-472, 479
vortex (rectilinear) 246
motion of vortex in external velocity field 251, inside circular boundary 284, 326, outside circular boundary 282, 325, outside circular boundary with stream 325, outside flat plate 326, pair of parallel vortices 324, between parallel plates 329, near plane boundary 280, in rectangular corner 324, shedding of vortices by accelerating aerofoil 303
vortex, rectilinear with finite radius core 248-250
vorticity 112, 217
associated with rectilinear vortex 248-249, equations for 218, in two dimensional flows 250

W

wave equation 197, 334-360, 399-412
symmetric 341-343, 359, one dimensional general solution 334-335, spherically symmetric general solution 336, 357
wave reflection and transmission 345-346, 354-357, 429, 431-435
waves
cylindrical 341-343, dilatational 400, dispersive 439-443, longitudinal 408, 428, 429, 439-443, 460-464, plane 335, 338, 341, 428, 431-435, Rayleigh surface waves 403-407, 431, shear (equivoluminal) 400, 431-435, spherical 336, 343-344, 357-358, 430, torsional in circular cylinder 402-403, viscoelastic 460-464

Y

yield function 468, 479, Tresca 470, von-Mises 470
yield stress 87, 468